THERMAL ENERGY STORAGE

Systems and Applications

THERMAL ENERGY STORAGE

Systems and Applications

İbrahim Dinçer
KFUPM, Dhahran, Saudi Arabia

and

Marc A. Rosen
Ryerson Polytechnic University, Toronto, Canada

with contributions from

A. Bejan
Duke University, USA

A. J. Ghajar
Oklahoma State University, USA

K. A. R. Ismail
FEM-UNICAMP, Brazil

M. Lacroix
Université de Sherbrooke, Canada

Y. H. Zurigat
Sultan Qaboos University, Oman

JOHN WILEY & SONS, LTD

Phone (+44) 1243 779777

Email (for orders and customer service enquiries): cs-books@wiley.co.uk
Visit our Home Page on www.wiley.co.uk or www.wiley.com

Other Wiley Editorial Offices

John Wiley & Sons, Inc., 605 Third Avenue, New York, NY 10158-0012, USA

Jossey-Bass, 989 Market Street, San Francisco, CA 94103-1741, USA

Wiley-VCH Verlag GmbH, Pappelallee 3, D-69469 Weinheim, Germany

John Wiley & Sons Australia, Ltd., 33 Park Road, Milton, Queensland 4064, Australia

John Wiley & Sons (Asia) Pte Ltd., 2 Clementi Loop #02-01, Jin Xing Distripark, Singapore 129809

John Wiley & Sons Canada, Ltd., 22 Worcester Road, Etobicoke, Ontario, Canada M9W 1L1

Library of Congress Cataloging-in-Publication Data

Thermal energy storage systems and applications / edited by Ibrahim Dincer, Marc Rosen.
p. cm.
Includes bibliographcial references and index.
ISBN 0-471-49573-5
1. Heat storage. I. Dinçer, Ibrahim. II. Rosen, Marc (Marc A.)

TJ260 .T493 2001
621.402'8–dc21 2001026254

British Library Cataloguing in Publication Data

A catalogue record for this book is available from the British Library

ISBN 0-471-49573-5

Printed from PostScript files supplied by the author.
Printed and bound in Great Britain by Antony Rowe, Chippenham, Wiltshire.
This book is printed on acid-free paper responsibly manufactured from sustainable forestry in which at least two trees are planted for each one used for paper production.

Front cover image: Trigeneration Project for World Fair Expo'98 at Lisbon, Portugal with its thermal energy storage (large cylindrical tank in rear) *(Courtesy of Paragon-Litwin, France)*

✎ To my wife, Gülşen, and my children, Meliha, Miray and İbrahim Eren,
for their inspiration.
And to those who have helped and supported me in any way throughout my education and professional life.

İbrahim Dinçer

&

✎ To my wife, Margot, and my children, Allison and Cassandra, for their inspiration.
And to Frank C. Hooper and David S. Scott, two giants in the field of energy and wonderful mentors.

Marc A. Rosen

Contents

List of Contributors

Dr. Adrian Bejan
Professor
Department of Mechanical Engineering and Materials Science, Duke University
Box 90300, Durham, North Carolina, 27708-0300, USA
E-mail: abejan@duke.edu

Dr. Ibrahim Dincer
Professor
Department of Mechanical Engineering, King Fahd University of Petroleum and Minerals
Box 127, Dhahran 31261, Saudi Arabia
E-mail: idincer@kfupm.edu.sa

Dr. Afshin J. Ghajar
Regents Professor and Director of Graduate Studies
School of Mechanical and Aerospace Engineering, Oklahoma State University
218 Engineering North, Stillwater, OK 74078-5016, USA
Email: ghajar@master.ceat.okstate.edu

Dr. Kamal A. R. Ismail
Professor
Department of Thermal and Fluids Engineering, State University of Campinas
CP 6122, CEP 13083-970, Campinas (SP), Brazil
E-mail: kamal@fem.unicamp.br

Dr. Marcel Lacroix
Professor
Departement de Genie Mecanique, Universite de Sherbrooke
Sherbrooke, Quebec, Canada J1K 2R1
E-mail: marcel.lacroix@gme.usherb.ca

Dr. Marc A. Rosen.
Professor
Department of Mechanical Engineering, Ryerson Polytechnic University
350 Victoria Street, Toronto, Ontario, Canada M5B 2K3
E-mail: mrosen@acs.ryerson.ca

Dr. Yousef H. Zurigat
Associate Professor
(On leave from Jordan University, Amman, Jordan)
Department of Mechanical and Industrial Engineering, Sultan Qaboos University
Muscat - 123, Sultanate of Oman
E-mail: zurigat@squ.edu.om

Acknowledgements

We are most grateful to the contributors

- Professor Adrian Bejan
- Professor Afshin J. Ghajar
- Professor Kamal A.R. Ismail
- Professor Marcel Lacroix
- Professor Yousef H. Zurigat

who dedicated time and effort to preparing their chapters. Their thoughtful contributions span a range of topics that, taken together, permit the book to provide a comprehensive resource on thermal energy storage.

We also are especially thankful to the many companies and agencies which contributed case studies and other materials for use in the book. These valuable materials have allowed us to ensure that the book has a high degree of industrial relevance and practicality. Included among those deserving acknowledgement are the following:

- American Electric Power, USA
- Baltimore Aircoil International N.V., Belgium
- CHIYODA Corporation, Japan
- CRISTOPIA Energy Systems, France
- CRYOGEL, Ice Ball Thermal Storage, USA
- Environmental Process Systems Limited, UK
- IEA-Heat Pump Center, The Netherlands
- MARCO, Saudi Arabia
- Mitsubishi Chemical Co., Japan
- Paragon-Litwin, France
- Paul Mueller Company, USA
- WS Atkins Consultants Limited, UK

We also appreciate the support and assistance provided by our universities, King Fahd University of Petroleum and Minerals and Ryerson Polytechnic University.

In addition, we are grateful to the support provided by John Wiley & Sons, UK in the development of this book, from the initial idea through to the final product.

We also are thankful to Mr. Mohammed Mujtaba Hussain (graduate student of Dr. Dinçer) for his effort in sketching most of the figures.

Last, but not least, we thank our wives, Gülşen Dinçer and Margot Rosen, and our children Meliha, Miray and İbrahim Eren Dinçer and Allison and Cassandra Rosen. They have been a great source of support and motivation, and their patience and understanding throughout this project have been most appreciated.

İbrahim Dinçer
Marc A. Rosen

Figures 1.4; 2.7; 3.2 and 3.3 and Table 2.6 are reproduced by permission of John Wiley & Sons, Ltd. Figure 2.5 is reproduced by permission of Charles E. Bakis. Figure 3.5 is copyright Prentice Hall Inc. Figures 3.6; 3.19; 3.22 and 3.23 and Table 3.7 are reproduced by permission of IEA Heat Pump Centre. Figures 3.7; 9.7–9.9 and Tables 3.4 and 3.6 are reproduced by permission of Springer–Verlag, GmbH. Figures 3.9; 3.17; 4.1–4.3; 6.4; 6.10–6.13; 6.17; 8.14–8.23 and 8.38–8.44 and Table 4.2 are reproduced with permission from Elsevier Science. (See the individual figures and tables for reference sources.) Figure 3.16 is reproduced by permission of GSA Resources Inc. Figures 3.18; 11.19–11.22 and 11.23 are reproduced by permission of CRISTOPIA Energy Systems. Figures 3.24; 11.14 and 11.24 are reproduced by permission of Environmental Process Systems Limited. Figures 3.40; 11.11–11.13 are reproduced by permission of International District Energy Association. Courtesy of Paul Mueller Company from proceedings at the IDEA College-University Conference, New Orleans, Louisiana, February 1999. Figures 3.42–3.45 and the table on page 190 are reproduced by permission of Cryogel. Figures 3.46; 3.47; 11.7–11.10 and Tables 3.11–3.14 are reproduced by permission of Baltimore Aircoil International. Figures 5.4; 6.2; 6.16; 6.18 and 11.1 and Tables 6.1 and 11.1 are reproduced by permission of American Society of Heating, Refrigerating and Air Conditioning Engineers www.ashrae.org. Figures 6.6–6.9 6.14 and 6.15 are reproduced by permission of the American Society of Mechanical Engineers. Figures 6.19 and 8.2–8.9 and Table 4.1 are reproduced by permission of Taylor & Francis, Inc., http://www.routledge-ny.com (See the individual figures and tables for reference sources.) Figures 11.2–11.5 and Table 11.2 are reproduced by permission of WS Atkins Consultants Ltd. Figures 11.15–11.17 and Tables 11.4–11.6 are reproduced courtesy of Paragon-Litwin. Figure 11.6 is reproduced by permission of Ghaleb Abussaa'. Figures 11.25 and 11.26 are reproduced by permission of Sandia National Laboratories. Figure 11.27 is reproduced by permission of Kyushu Electric Power Co., Inc. Figure 11.28 is reproduced by permission of Rocky Mountain Research Institute. Table 2.1 is reproduced by permission of Aspes AG, Zürich. Table 2.3 is reproduced by permission of Electrosource Inc. Tables 3.2 and 3.3 are reproduced by permission of CADDET Energy Efficiency. Table 6.2 is reproduced by permission of International Solar Energy Society (ISES).

Preface

Thermal energy storage (TES) is an *advanced energy technology* that has recently attracted increasing interest for thermal applications such as space and water heating, cooling and air-conditioning. TES systems have enormous potential to facilitate more effective use of thermal equipment and large-scale energy substitutions that are economic. TES appears to be the most appropriate method for correcting the mismatch that sometimes occurs between the supply and demand of energy. It is therefore a very attractive technology for meeting society's needs and desires for more efficient and environmentally benign energy use.

This book is research-oriented, and therefore includes some practical features often not included in other, solely academic textbooks. This book is essentially intended for use by advanced undergraduate and graduate students in various disciplines ranging from mechanical to chemical engineering, and as a basic reference for practicing energy engineers. Analyses of TES systems and their applications are undertaken throughout this comprehensive book, providing new understandings, methodologies, models and applications, along with descriptions of several experimental works and case studies. Some of the material presented has been drawn from the most recent information available in the literature and elsewhere. The coverage is extensive, and the amount of information and data presented can be sufficient for several courses, if studied in detail. We strongly believe that this book will be of interest to students, engineers and energy experts, and that it provides a valuable and readable reference text for those who wish to learn about more about TES systems and applications.

The first chapter addresses general aspects of thermodynamics, fluid flow and heat transfer to furnish the reader with background information that is of relevance to the analysis of TES systems and their applications. Chapter 2 discusses the many types of energy storage technologies available. Chapter 3 deals extensively with TES methods, including cold TES. Chapter 4 addresses several environmental issues that we face today, and discusses how TES can help solve these problems. Several successful case studies are presented. Chapter 5 describes how TES is a valuable tool in energy conservation efforts that can help achieve significant energy savings. Chapter 6 delves into sensible heat storage systems and experimental and theoretical heat transfer aspects of stratified storage. Chapter 7 deals with a number of modeling aspects of latent heat storage systems, while Chapter 8 covers heat transfer with phase change in simple and complex geometries. Chapter 9 describes the thermoeconomic analysis and optimization of TES systems, and provides several illustrative examples. Chapter 10 covers energy and exergy analyses of a range of TES systems, along with various practical examples. Chapter 11 discusses many practical TES applications and case studies.

Incorporated through this book are many wide-ranging, illustrative examples which provide useful information for practical applications. Conversion factors and thermophysical properties of various materials are listed in the appendices in the International System of Units (SI). Complete references and a bibliography are included with each chapter to direct the curious and interested reader to further information.

İbrahim Dinçer
Marc A. Rosen
June 2001

1

General Introductory Aspects for Thermal Engineering

I. Dincer

1.1 Introduction

Thermal energy storage (TES) is one of the key technologies for energy conservation and therefore is of great practical importance. One of its main advantages is that it is best suited for heating and cooling thermal applications. TES is perhaps as old as civilization itself. Since recorded time people have harvested ice and stored it for later use. Large TES systems have been employed in more recent history for numerous applications, ranging from solar hot water storage to building air conditioning systems. The TES technology has only recently been developed to a point where it can have a significant impact on modern technology.

In general, a coordinated set of actions has to be taken in several sectors of the energy system for the maximum potential benefits of thermal storage to be realized. TES appears to be an important solution to correcting the mismatch between the supply and demand of energy. TES can contribute significantly to meeting society's needs for more efficient, environmentally benign energy use. TES is a key component of many successful thermal systems and a good TES should allow minimum thermal losses, leading to energy savings, while permitting the highest possible extraction efficiency of the stored thermal energy.

There are mainly two types of TES systems, i.e. sensible (e.g. water, rock) and latent (e.g. water/ice, salt hydrates). For each storage medium, there is a wide variety of choices depending on the temperature range and application. TES via latent heat has received a great deal of interest. The most obvious example of latent TES is the conversion of water to ice. Cooling systems incorporating ice storage have a distinct size advantage over equivalent capacity chilled water units because of the large amount of energy able to be stored as latent heat. TES deals with the storing of energy by cooling, heating, melting, solidifying or vaporizing a substance, and the energy becomes available as heat when the process is reversed. The selection of a TES is mainly dependent on the storage period required, i.e. diurnal or seasonal, economic viability, operating conditions, etc. In practice, many research and development activities related to energy have concentrated on efficient energy use and energy savings, leading to energy conservation. In this regard, TES appears to be one an attractive thermal application and exergy analysis as an important tool for analyzing TES performances.

We begin this chapter with a summary of fundamental definitions, physical quantities, and their units, dimensions, and interrelations. We consider introductory aspects of thermodynamics, fluid flow, heat transfer, energy and entropy.

1.2 Systems of Units

There are two main systems of units: the *International System of Units* (Le *Systéme International d'Unités*), which is normally referred to as SI units, and the *English System of Units*. SI units are used most widely throughout the world, although the English System is traditional in the United States. In this book, SI units are primarily employed. Note that the relevant unit interconversions and relationships between the International and English unit systems concerning the fundamental properties and quantities are listed in Appendix A.

1.3 Fundamental Properties and Quantities

In this section we briefly cover several general aspects of thermodynamics to provide adequate preparation for the study of TES systems and applications.

1.3.1 Mass, time, length, and force

Mass is defined as a quantity of matter forming a body of indefinite shape and size. The fundamental unit of mass is the kilogram (kg) in SI and the pound mass (lb_m) in English units. The basic unit of time for both unit systems is the second.

In thermodynamics the unit *mole* (mol) is commonly used and defined as a certain amount of a substance as follows:

$$n = \frac{m}{M} \tag{1.1}$$

where n is the number of moles, m the mass, and M the molecular weight. If m and M are given by gram and gram/mol, we get n in mol. For example, one mol of water, having a molecular weight of 18 (compared to 12 for carbon-12) has a mass of 0.018 kg.

The basic unit of length is the meter (m) in SI units and the foot (ft) in the English system.

A force is a kind of action that brings a body to rest or to change the direction of motion (e.g. a push or a pull). The fundamental unit of force is the Newton (N).

The four aspects, e.g. mass, time, length and force, are interrelated by the Newton's second law of motion, which states that the force acting on a body is proportional to the mass and the acceleration in the direction of the force, as given in Equation 1.2:

$$F = ma \tag{1.2}$$

Equation 1.2 shows the force required to accelerate a mass of one kilogram at a rate of one meter per square second as 1 N = 1 kg m/s^2.

It is important to note the value of the earth's gravitational acceleration as 9.80665 m/s^2 in the SI system and 32.174 ft/s^2 in the English system, and it indicates that a body falling freely toward the surface of the earth is subject to the action of gravity alone.

1.3.2 Pressure

When we deal with liquids and gases, pressure becomes one of the most important components. Pressure is the force exerted on a surface, per unit area, and is expressed in bar or Pascal (Pa). The related expression is

$$P = \frac{F}{A} \tag{1.3}$$

The unit for pressure in the SI is the force of one Newton acting on a square meter area (or the *Pascal*). The unit for pressure in the English system, is pounds force per square foot, lb_f/ft^2.

Here, we introduce the basic pressure definitions, and a summary of basic pressure measurement relationships is shown in Figure 1.1.

Atmospheric pressure. The atmosphere that surrounds the earth can be considered a reservoir of low-pressure air. Its weight exerts a pressure which varies with temperature, humidity, and altitude. Atmospheric pressure also varies from time to time at a single location, because of the movement of weather patterns. While these changes in barometric pressure are usually less than one-half inch of mercury, they need to be taken into account when precise measurements are essential.

Gauge pressure. The *gauge pressure* is any pressure for which the base for measurement is atmospheric pressure expressed as kPa (gauge). Atmospheric pressure serves as a reference level for other types of pressure measurements, e.g. gauge pressure. As is shown in Figure 1.1, the gauge pressure is either positive or negative, depending on its level above or below the atmospheric pressure level. At the level of atmospheric pressure, the gauge pressure becomes zero.

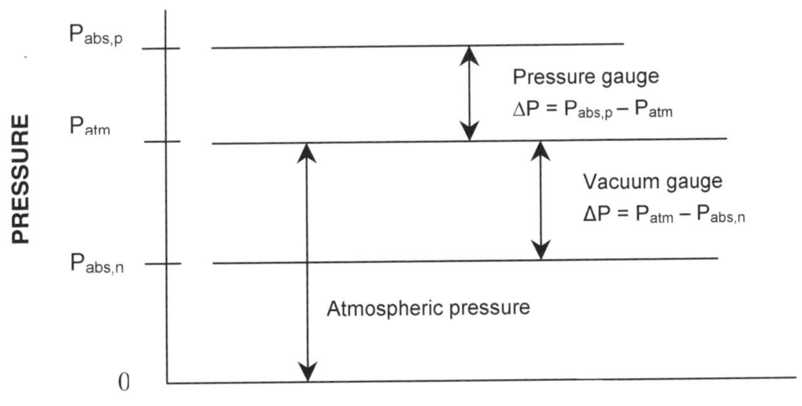

Figure 1.1 Illustration of pressures for measurement.

Absolute pressure. A different reference level is utilized to obtain a value for absolute pressure. The absolute pressure can be any pressure for which the base for measurement is full vacuum, being expressed in kPa (absolute). In fact, it is composed of the sum of the gauge pressure (positive or negative) and the atmospheric pressure as follows:

$$\textit{pressure (gauge)} + \textit{atmospheric pressure} = \textit{pressure (absolute)} \tag{1.4}$$

For example, to obtain the absolute pressure, we simply add the value of atmospheric pressure. The absolute pressure is the most common one used in thermodynamic calculations, despite having the pressure difference between the absolute pressure and the atmospheric pressure existing in the gauge being read by the most pressure gauges and indicators.

Vacuum. It is a pressure lower than the atmospheric one and occurs only in closed systems, except in outer space. It is also called the *negative gauge pressure*. In fact, vacuum is the pressure differential produced by evacuating air from the closed system. Vacuum is usually divided into four levels: (i) low vacuum representing pressures above one Torr absolute (a large number of mechanical pumps in industry are used for this purpose; flow is viscous), (ii) medium vacuum varying between 1 and 10^{-3} Torr absolute (most pumps serving this range are mechanical; fluid is in transition between viscous and molecular), (iii) high vacuum ranging between 10^{-3} and 10^{-6} Torr absolute (nonmechanical ejector or cryogenic pumps are used; flow is molecular or Newtonian), and (iv) very high vacuum representing absolute pressure below 10^{-6} Torr (primarily for laboratory applications and space simulation).

It is important to note another additional level, at which the *saturation pressure* is the pressure of a liquid or vapor at saturation conditions.

1.3.3 Temperature

This is an indication of the heat energy stored in a substance. In other words, we can identify hotness and coldness with the concept of temperature. The temperature of a substance may be expressed in either relative or absolute units. The two most common temperature scales are the Celsius (°C) and Fahrenheit (°F). In fact, the Celsius scale is used with the SI unit system and the Fahrenheit scale with the English engineering system of units. There are also two more scales, the Kelvin scale (K) and the Rankine scale (R), that are sometimes employed in thermodynamic applications.

Degree Kelvin is a unit of temperature measurement; zero Kelvin (0 K) is equal to absolute zero and equal to –273.15°C. The K and °C are equal increments of temperature. For instance, when the temperature of a product is decreased to –273°C (or 0 K), known as *absolute zero*, the substance contains no heat energy and all molecular movement stops. The saturation temperature is the temperature of a liquid or vapor at saturation conditions.

Temperature can be measured in a large number of ways by devices. In general, the following devices are common in use:

- **Thermometers**. In thermometers, the volume of the fluid expands when subjected to heat, thereby raising its temperature. In practice, thermometers work over a certain temperature range. For example, the common thermometer fluid mercury becomes solid at –38.8°C and its properties change dramatically.

- **Resistance thermometers**. A resistance thermometer (or detector) is made up resistance wire wound on a suitable former. The wire used has to be of known, repeatable, electrical characteristics so that the relationship between the temperature and resistance value can be predicted precisely. The measured value of the resistance of the detector can then be used to determine the value of an unknown temperature. Amongst metallic conductors, pure metals exhibit the greatest change of resistance with temperature. For applications requiring higher accuracy, especially where the temperature measurement is between –200°C and +800°C, the resistance thermometer comes into its own. The majority of such thermometers are made of platinum. In industry, in addition to platinum, nickel (–60°C to +180°C) and copper (–30°C to +220°C) are frequently used to manufacture resistance thermometers. Resistance thermometers can be provided with 2, 3, or 4 wire connections, and for higher accuracy at least 3 wires are required.
- **Averaging thermometers**. An averaging thermometer is designed to measure the average temperature of bulk stored liquids. The sheath contains a number of elements of different lengths, all starting from the bottom of the sheath, The longest element which is fully immersed is connected to the measuring circuit to allow a true average temperature to be obtained. There are some significant parameters namely sheath material (stainless steel for the temperature range from –50°C to +200°C or nylon for the temperature range from –50°C to +90°C), sheath length (to suit the application), termination (flying leads or terminal box), element length, element calibration (to copper or platinum curves), and operating temperature ranges. In many applications where a multi-element thermometer is not required, such as in air ducts, cooling water and gas outlets, a single element thermometer stretched across the duct or pipework will provide a true average temperature reading. Despite the working range from 0°C to 100°C, the maximum temperature may reach 200°C. To keep high accuracy these units are normally supplied with 3-wire connections. However, up to 10 elements can be mounted in the averaging bulb fittings, and they can be made of platinum, nickel or copper, and fixed at any required position.
- **Thermocouples**. A thermocouple consists of two electrical conductors, of different materials connected together at one end (the so-called *measuring junction*). The two free ends are connected to a measuring instrument, e.g. an indicator, a controller or a signal conditioner by a reference junction (the so-called *cold junction*). The thermo-electric voltage appearing at the indicator depends on the materials of which the thermocouple wires are made and on the temperature difference between the measuring junction and the reference junction. For accurate measurements, the temperature of the reference junction must be kept constant. Modern instruments usually incorporate a cold junction reference circuit and are supplied ready for operation in a protective sheath, to prevent damage to the thermocouple by any mechanical or chemical means. Table 1.1 gives several types of thermocouples along with their maximum absolute temperature ranges. As can be seen in Table 1.1, a copper-constantan thermocouple has an accuracy of ±1°C, and is often employed for control systems in refrigeration and food processing applications. The iron-constantan thermocouple with its maximum of

850°C is used in applications in the plastics industry. The chromel-alumel type thermocouples, with a maximum of about 1100°C, are suitable for combustion applications in ovens and furnaces. In addition, it is possible to reach about 1600 or 1700°C using platinum rhodium-platinum thermocouples, particularly in steel manufacture. It is worth noting that one advantage the thermocouple has over most other temperature sensors is that it has a small thermal capacity, and thus a prompt response to temperature changes. Furthermore, its small thermal capacity rarely affects the temperature of the body under examination.

- **Thermistors**. These devices are made of semi-conductors and act as thermal resistors with a high (usually negative) temperature coefficient. In use thermistors either operate self-heated or externally-heated. Self-heated units employ the heating effect of the current flowing through them to raise and control their temperature and thus their resistance. This operating mode is useful in such devices as voltage regulators, microwave power meters, gas analyzers, flow meters, and automatic volume and power level controls. Externally-heated thermistors are well suited for precision temperature measurement, temperature control and temperature compensation due to the large changes in resistance versus temperature. These are generally used for applications in the range –100°C to +300°C. Despite the early thermistors having tolerances of ±20% or ±10%, modern precision thermistors are of a higher accuracy, e.g. ±0.1°C (less than ±1%).
- **Digital display thermometers**. A wide range of digital display thermometers, e.g. hand-held battery powered displays and panel mounted mains or battery units, are available on the market. Displays can be provided for use with all standard thermocouples or BS/DIN platinum resistance thermometers with several digits and 0.1°C resolution.

It is very important to emphasize that before temperature can be controlled, it must be sensed and measured accurately. For temperature measurement devices, there are several potential sources of error, such as not only sensor properties but also contamination effects, lead lengths, immersion, heat transfer and controller interfacing. In temperature control there are many sources of error which can be minimized by careful consideration of the type of sensor, its working environment, the sheath or housing, extension leads, and the instrumentation. An awareness of potential errors is vital in the applications dealt with. Selection of temperature measurement devices is a complex task, and has been discussed only briefly here. It is extremely important to remember the following: "choosing the right tool for the right job."

1.3.4 Specific volume and density

Specific volume is the volume per unit mass of a substance, usually expressed in cubic meters per kilogram (m^3/kg) in the SI system and in cubic feet per pound (ft^3/lb) in the English system. The *density* ρ of a substance is defined as the mass per unit volume, and it is therefore the inverse of the specific volume

$$\rho = \frac{1}{v} \tag{1.5}$$

Table 1.1 Some of the most common thermocouples.

Type	Common Names	Temperature Range (°C)
T	Copper–Constantan (C/C)	–250 to 400
J	Iron–Constantan (I/C)	–200 to 850
E	Nickel Chromium–Constantan or Chromel–Constantan	–200 to 850
K	Nickel Chromium–Nickel Aluminum or Chromel–Alumel (C/A)	–180 to 1100
–	Nickel 18% Molybdenum–Nickel	0 to 1300
N	Nicrosil-Nisil	0 to 1300
S	Platinum 10% Rhodium–Platinum	0 to 1500
R	Platinum 13% Rhodium–Platinum	0 to 1500
B	Platinum 30% Rhodium–Platinum 6% Rhodium	0 to 1600

and its units are kg/m^3 in the SI system and Ib/ft^3 in the English system. That is also defined as the volume per unit mass, and density as the mass per unit volume, i.e.

$$v = \frac{V}{m} \tag{1.6}$$

$$\rho = \frac{m}{V} \tag{1.7}$$

Both specific volume and density are intensive properties and are affected by temperature and pressure.

1.3.5 Mass and volumetric flow rates

Mass flow rate is defined as the mass flowing per unit time (kg/s in the SI system and Ib/s in the English system). Volumetric flow rates are given in m^3/s in the SI system and ft^3/s in the English system. The following expressions can be written for the flow rates in terms of mass, specific volume and density:

$$\dot{m} = \dot{V}\rho = \frac{\dot{V}}{v} \tag{1.8}$$

$$\dot{V} = \dot{m}v = \frac{\dot{m}}{\rho} \tag{1.9}$$

1.4 General Aspects of Thermodynamics

In this section, we introduce briefly some general aspects of thermodynamics which are related to energy storage systems and applications.

1.4.1 *Thermodynamic systems*

A thermodynamic system is a device or combination of devices that contain a certain quantity of matter. It is important to carefully define a system under consideration and its boundaries. We can define three important types of system as follows:

- **Closed system**. This is defined as a system across the boundaries of which no material crosses. It therefore contains a fixed quantity of matter. In some books, it is also called a *control mass.*
- **Open system**. This is defined as a system in which material (mass) is allowed to cross its boundaries. The term open system is also called a *control volume*.
- **Isolated system**. This is a closed system that is not affected by the surroundings. No mass, heat or work crosses its boundary.

1.4.2 *Process*

A process is a physical or chemical change in the properties of matter or the conversion of energy from one form to another. In some processes, one property remains constant. The prefix 'iso' is employed to describe such as process, e.g. isothermal (constant temperature), an isobaric (constant pressure), and an isochoric (constant volume).

1.4.3 *Cycle*

A cycle is a series of thermodynamic processes in which the end point conditions or properties of the matter are identical to the initial conditions.

1.4.4 *Thermodynamic property*

This is a physical characteristic of a substance used to describe its state. Any two properties usually define the state or condition of the substance, from which all other properties can be derived. Some examples are temperature, pressure, enthalpy, and entropy. Thermodynamic properties are classified as intensive properties (independent of the mass, e.g. pressure, temperature, and density) and extensive properties (dependent on the mass, e.g. mass and total volume). Extensive properties per unit mass become intensive properties such as specific volume. Property diagrams of substances are generally presented in graphical form and summarize the main properties listed in the refrigerant tables.

1.4.5 *Sensible and latent heats*

It is known that all substances can hold a certain amount of heat; that property is their thermal capacity. When a liquid is heated, the temperature of the liquid rises to the boiling point. This is the highest temperature that the liquid can reach at the measured pressure. The heat absorbed by the liquid in raising the temperature to the boiling point is called *sensible heat*. The heat required to convert the liquid to vapor at the same temperature and pressure is called *latent heat*. In fact, it is the change in enthalpy during a state change (the amount of heat absorbed or rejected at constant temperature at any pressure, or the difference in enthalpies of a pure condensable fluid between its dry saturated state and its saturated liquid state at the same pressure).

1.4.6 Latent heat of fusion

Fusion is a melting of a material. For most pure substances there is a specific melting/freezing temperature, relatively independent of the pressure. For example, ice begins to melt at 0°C. The amount of heat required to melt one kilogram of ice at 0°C to one kilogram of water at 0°C is called the latent heat of fusion of water, and equals 334.92 kJ/kg. The removal of the same amount of heat from one kilogram of water at 0°C changes it back to ice.

1.4.7 Vapor

A vapor is a gas at or near equilibrium with the liquid phase—a gas under the saturation curve or only slightly beyond the saturated vapor line. *Vapor quality* is theoretically assumed; that is, when vapor leaves the surface of a liquid it is pure and saturated at the particular temperature and pressure. In actuality, tiny liquid droplets escape with the vapor. When a mixture of liquid and vapor exists, the ratio of the mass of the liquid to the total mass of the liquid and vapor mixture is called the *quality*, and is expressed as a percentage or decimal fraction. *Superheated vapor* is the saturated vapor to which additional heat has been added, raising the temperature above the boiling point. Let's consider a mass m with a quality x. The volume is the sum of both volumes of the liquid and the vapor, as defined below:

$$V = V_{liq} + V_{vap} \tag{1.10}$$

Equation 1.10 can also be written in terms of specific volumes as

$$mv = m_{liq}v_{liq} + m_{vap}v_{vap} \tag{1.11}$$

Dividing all terms by the total mass yields

$$v = (1-x)v_{liq} + xv_{vap} = v_{liq} + xv_{liq,vap} \tag{1.12}$$

where $v_{liq,vap} = v_{vap} - v_{liq}$.

1.4.8 Thermodynamic tables

The thermodynamic tables were first published in 1936 as steam tables by Keenan and Keyes, and later in 1969 and 1978 revised and republished. The use of thermodynamic tables of many substances ranging from water to several refrigerants is very common in process design calculations. In the literature they are also called either steam tables or vapor tables. In this book, we will refer to the thermodynamic tables. These tables are normally given as different distinct phases (parts), for example, four different parts for water such as saturated water, superheated vapor water, compressed liquid water, saturated solid-saturated vapor water; and two distinct parts for R-134a, such as saturated and superheated. Each table is listed according to the values of temperature and pressure, and the rest contains the values of four other thermodynamic parameters such as specific volume, internal energy, enthalpy, and entropy. When we normally have two variables, we may obtain the other data from the respective table. In learning how to use these tables, the most important point is to specify the state by any two of the parameters. In some design calculations if we do not have the exact values of the parameters, we use interpolation to find the necessary values. Beside these thermodynamic tables, recently, much attention has been paid to the computerized tables for such design calculations. Of course, despite the fact this eliminates several reading problems, the students may not understand the concepts and comprehend the subject well. That is why in thermodynamics courses it is a must for the students to know how to obtain the thermodynamic data from the respective thermodynamic tables. The *Handbook of Thermodynamic Tables* by Raznjevic (1995) is one of the most valuable sources for several solids, liquids, and gaseous substances.

1.4.9 State and change of state

The state of a system or substance is defined as the condition of the system or substance characterized by the certain observable macroscopic values of its properties such as temperature and pressure. The term *state* is often used interchangeably with the term *phase*, e.g. solid phase or gaseous phase of a substance. Each of the properties of a substance in a given state has only one definite value, and these properties of a substance in a given state have only one definite value, regardless of how the substance reached the state. For example, when sufficient heat is added or removed, most substances undergo a state change. The temperature remains constant until the state change is complete. This can be from solid to liquid, liquid to vapor, or vice versa. Figure 1.2 shows typical examples of ice melting and water boiling.

A clearer presentation of solid, liquid, and vapor phases of water is exhibited on a temperature-volume (T-v) diagram in Figure 1.3. The constant pressure line ABCD represents the states which water passes through as follows:

- **A-B**: Represents the process where water is heated from the initial temperature to the saturation temperature (liquid) at constant pressure. At point B it is fully saturated liquid water with a quality $x = 0$, with zero quantity of water vapor.
- **B-C**: Represents a constant-temperature vaporization process in which there is only phase change from saturated liquid to saturated vapor, referring to the fact that the quality varies from 0 to 100%. Within this zone, the water is a mixture of liquid water and water vapor. At point C it is completely saturated vapor and the quality is 100%.
- **C-D**: Represents the constant-pressure process in which the saturated water vapor is superheated with increasing temperature.
- **M-N-O**: Represents no constant-temperature vaporization process. The point F is called the critical point where the saturated-liquid and saturated vapor states are

identical. The thermodynamic properties at this point are called critical thermodynamic properties, e.g. critical temperature, critical pressure, and critical specific volume.

- **H-I**: Represents a constant-pressure heating process in which there is no phase change from one phase to another (only one is present), however, there is a continuous change in density.

The other process which may occur during melting of water is *sublimation*, in which the ice directly passes from the solid phase to the vapor phase. There is another important point that needs to be emphasized, which is that the solid, liquid, and vapor phases of water may be present together in equilibrium, leading to the *triple point*.

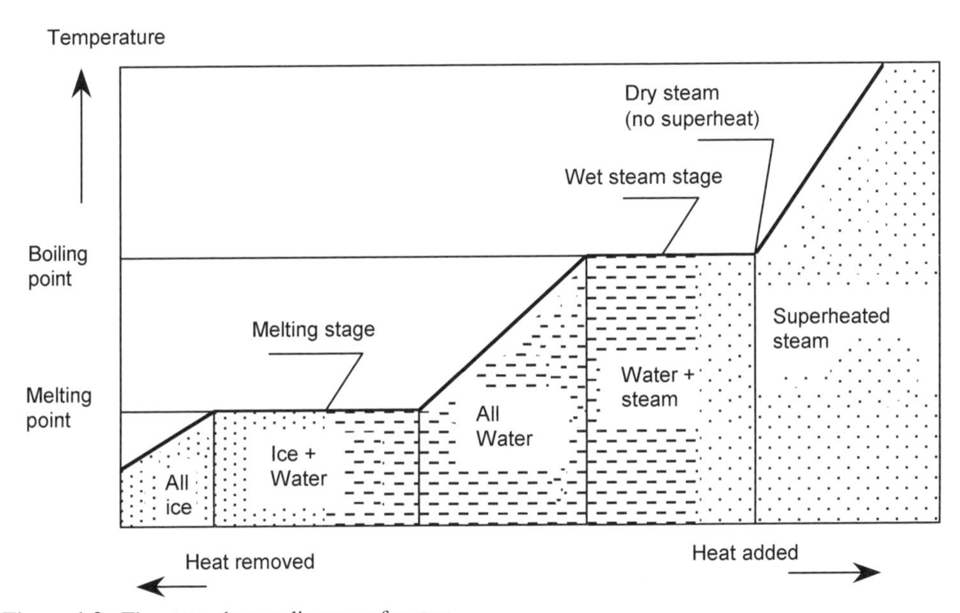

Figure 1.2 The state-change diagram of water.

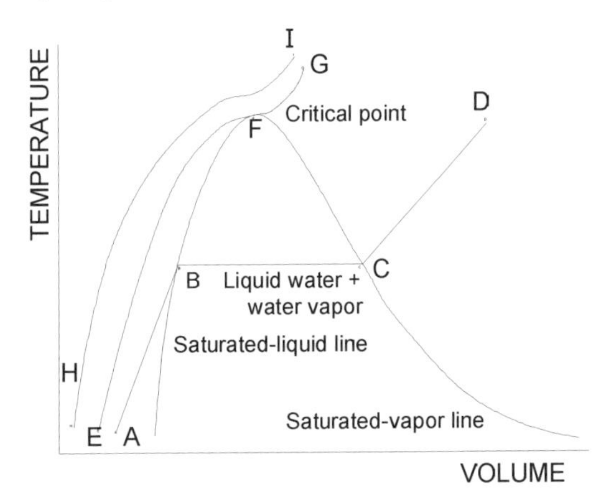

Figure 1.3 Temperature-volume diagram for the phase change of water.

1.4.10 *Specific internal energy*

Internal energy represents the molecular state type of energy. Specific internal energy is a measure per unit mass of the energy of a simple system in equilibrium as a function of $c_v dT$. In fact, for many thermodynamic processes in closed systems, the only significant energy changes are internal energy changes, and the significant work done by the system in the absence of friction is the work of pressure-volume expansion, such as the piston-cylinder mechanism. The specific internal energy of a mixture of liquid and vapor can be written in a similar form to Equation 1.12:

$$u = (1-x)u_{liq} + xu_{vap} = u_{liq} + xu_{liq,vap} \tag{1.13}$$

where $u_{liq,vap} = u_{vap} - u_{liq}$.

1.4.11 *Specific enthalpy*

Enthalpy is a measure of the energy per unit mass of a substance. Specific enthalpy, usually expressed in kJ/kg, is normally expressed as a function of $c_p dT$. Since enthalpy is a state function, it is necessary to measure it relative to some reference state. The usual practice is to determine the reference values which are called the standard enthalpy of formation (or the heat of formation), particularly in combustion thermodynamics. The specific enthalpy of a mixture of liquid and vapor components can be written as Equation 1.12:

$$h = (1-x)h_{liq} + xh_{vap} = h_{liq} + xh_{liq,vap} \tag{1.14}$$

where $h_{liq,vap} = h_{vap} - h_{liq}$.

1.4.12 *Specific entropy*

Entropy is the ratio of the heat added to a substance to the absolute temperature at which it was added, and is also a measure of the molecular disorder of a substance at a given state. The specific enthalpy of a mixture of liquid and vapor components can be written as Equation 1.12:

$$s = (1-x)s_{liq} + xs_{vap} = s_{liq} + xs_{liq,vap} \tag{1.15}$$

where $s_{liq,vap} = s_{vap} - s_{liq}$.

1.4.13 *Pure substance*

A pure substance is defined as one which has a homogeneous and invariable chemical composition. Despite having the same chemical composition, it may be in more than one phase, namely, liquid water, a mixture of liquid water and water vapor (steam), and a mixture of ice and liquid water. Each one has the same chemical composition. However, a mixture of liquid air and gaseous air can not be considered a pure substance because the composition of each phase differs from each other. A thorough understanding of the pure substance is of significance, particularly for TES applications. Thermodynamic properties of water and steam can be taken from tables and charts, in most thermodynamics books, based on experimental data or real-gas equations of state through computer calculations. It is important to note that the properties of low-pressure water are of great significance in TES systems for cooling applications, since water vapor existing in the atmosphere typically exerts a pressure less than one psi (6.9 kPa). At such low pressures, it is known that water vapor shows ideal-gas behavior.

1.4.14 Ideal gases

In many practical thermodynamic calculations, gases such as air and hydrogen can often be treated as ideal gases, particularly for temperatures much higher than their critical temperatures and for pressures much lower than their saturation pressures at given temperatures. Such an ideal gas can be described in terms of three parameters: the volume that it occupies, the pressure that it exerts, and its temperature. In fact, all gases or vapors, including water vapor, at very low pressures show the ideal gas behavior. The practical advantage of taking real gases to be ideal is that a simple equation of state with only one constant can be applied in the following form:

$$Pv = RT \tag{1.16}$$

and

$$PV = mRT \tag{1.17}$$

The ideal gas equation of state was originally established from experimental observations, and is also called the P-v-T relationship for gases. It is generally considered as a concept rather than a reality. It only requires a few data to define a particular gas over a wide range of its possible thermodynamic equilibrium states.

The gas constant (R) is different for each gas depending on its molecular weight (M):

$$R = \frac{\bar{R}}{M} \tag{1.18}$$

where $\bar{R} = 8.314$ kJ/kgK.

Equations 1.24 and 1.25 may be written in a mole-basis form as follows:

$$P\bar{v} = RT \tag{1.19}$$

and

$$PV = nRT \tag{1.20}$$

The other simplicity is that, if assumed that the constant-pressure and constant-volume specific heats are constant, the changes in the specific internal energy and the specific enthalpy can be calculated simply, without referring to thermodynamic tables and graphs, from the following expressions:

$$\Delta u = (u_2 - u_1) = c_v(T_2 - T_1) \tag{1.21}$$

$$\Delta h = (h_2 - h_1) = c_p(T_2 - T_1) \tag{1.22}$$

The following is another useful expression for ideal gases, obtained from the expression, $h = u + Pv = u + RT$:

$$c_v - c_p = R \tag{1.23}$$

For the entire range of states, the ideal gas model may be found unsatisfactory. Therefore, the compressibility factor (*Z*) is introduced to measure the deviation of a real substance from the ideal-gas equation of state, which is defined by the relation:

$$Pv = ZRT \quad \text{or} \quad Z = \frac{Pv}{RT} \tag{1.24}$$

Figure 1.4 shows a generalized compressibility chart for simple substances. In the chart, we have two important parameters: the reduced temperature ($T_r = T/T_c$) and the reduced pressure ($P_r = P/P_c$). Therefore, to calculate the compressibility factor the values of T_r and P_r should be calculated using the critical temperature and pressure values of the respective substance, which can easily be taken from thermodynamic books. As can be seen in Figure 1.4, at all temperatures $Z \rightarrow 1$ as $P_r \rightarrow 0$. This means that the behavior of the actual gas closely approaches the ideal gas behavior, as the pressure approaches zero. For the real gases, *Z* takes values between 0 and 1. If $Z = 1$, Equation 1.24 becomes Equation 1.16. In the literature, there are also several equations of state for accurately representing the P-v-T behavior of a gas over the entire superheated vapor region, namely, the Benedict-Webb-Rubin equation, van der Waals equation, and Redlich and Kwong equation. However, some of these equations of state are complicated, due to the number of empirical constants, and it requires computer software to obtain results.

There are some special cases as one of *P*, *v*, and *T* is constant. At a fixed temperature, the volume of a given quantity of ideal gas varies inversely with the pressure exerted on it (in some books it is called Boyle's law), describing compression as

$$P_1V_1 = P_2V_2 \tag{1.25}$$

where the subscripts refer to the initial and final states.

Equation 1.25 is employed by designers in a variety of situations: when selecting an air compressor, for calculating the consumption of compressed air in reciprocating air cylinders, and for determining the length of time required for storing air. Nevertheless, it may not always be practical due to temperature changes. If temperature increases with compression at a constant pressure, the volume of a gas varies directly with its absolute temperature in K as:

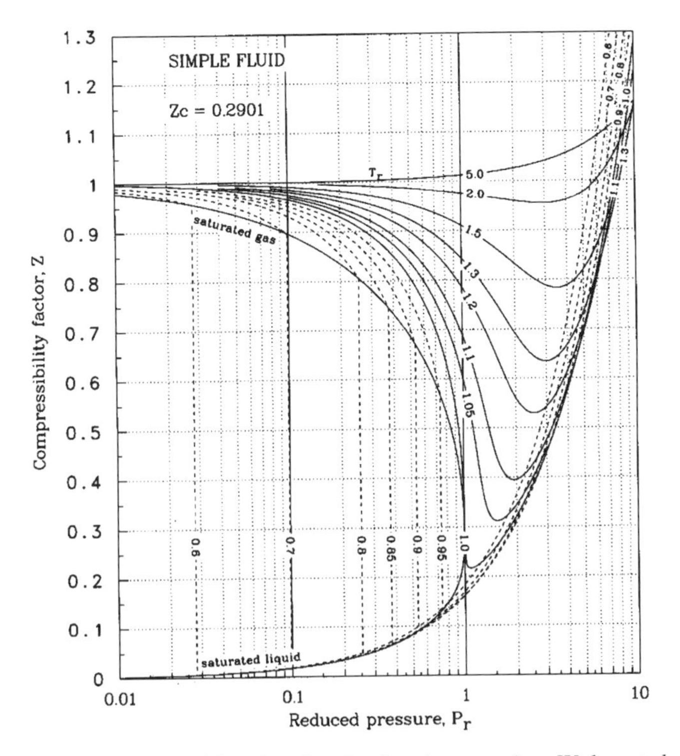

Figure 1.4 Generalized compressibility chart for simple substances (van Wylen *et al.*, 1998).

$$\frac{V_1}{T_1} = \frac{V_2}{T_2} \tag{1.26}$$

If temperature increases at a constant volume, the pressure of a gas this time varies directly with its absolute temperature in K as:

$$\frac{P_1}{T_1} = \frac{P_2}{T_2} \tag{1.27}$$

Equations 1.26 and 1.27 are known as Charles' law. If both temperature and pressure change at the same time, the combined ideal gas equation can be written as:

$$\frac{P_1 V_1}{T_1} = \frac{P_2 V_2}{T_2} \tag{1.28}$$

For a given mass, the internal energy of an ideal gas can be written as a function of temperature, since c_{v0} is constant below:

$$dU = mc_{v0}dT \tag{1.29}$$

and the specific internal energy becomes

$$du = c_{v0} dT \tag{1.30}$$

The enthalpy equation for an ideal gas, based on $h = u + Pv$, can be written as:

$$dH = mc_{p0} dT \tag{1.31}$$

and the specific enthalpy then becomes:

$$dh = c_{p0} dT \tag{1.32}$$

The entropy change of an ideal gas, based on the general entropy equation in terms of $T\,ds = du + P\,dv$ and $T\,ds = dh - v\,dP$, as well as the ideal gas equation $Pv = RT$, can be obtained in two ways by substituting Equations 1.29 and 1.30:

$$s_2 - s_1 = c_{v0} \ln\left(\frac{T_2}{T_1}\right) + R \ln\left(\frac{v_2}{v_1}\right) \tag{1.33}$$

$$s_2 - s_1 = c_{p0} \ln\left(\frac{T_2}{T_1}\right) - R \ln\left(\frac{P_2}{P_1}\right) \tag{1.34}$$

For a reversible adiabatic process the ideal gas equation in terms of the initial and final states under Pv^k = constant:

$$Pv^k = P_1 v_1^k = P_2 v_2^k \tag{1.35}$$

where k denotes the adiabatic exponent (the *specific heat ratio*) as a function of temperature:

$$k = \frac{c_{p0}}{c_{v0}} \tag{1.36}$$

Based on Equation 1.35 and the ideal gas equation, the following expressions can be obtained as:

$$\frac{P_2}{P_1} = \left(\frac{T_2}{T_1}\right)^{k/k-1} = \left(\frac{v_1}{v_2}\right)^k = \left(\frac{V_1}{V_2}\right)^k \tag{1.37}$$

Consider a closed system with ideal gas, undergoing an adiabatic reversible process with a constant specific heat. The work can be derived from the first law of thermodynamics equation as follows:

$$W_{1-2} = \frac{mR(T_2 - T_1)}{1-k} = \frac{(P_2V_2 - P_1V_1)}{1-k} \tag{1.38}$$

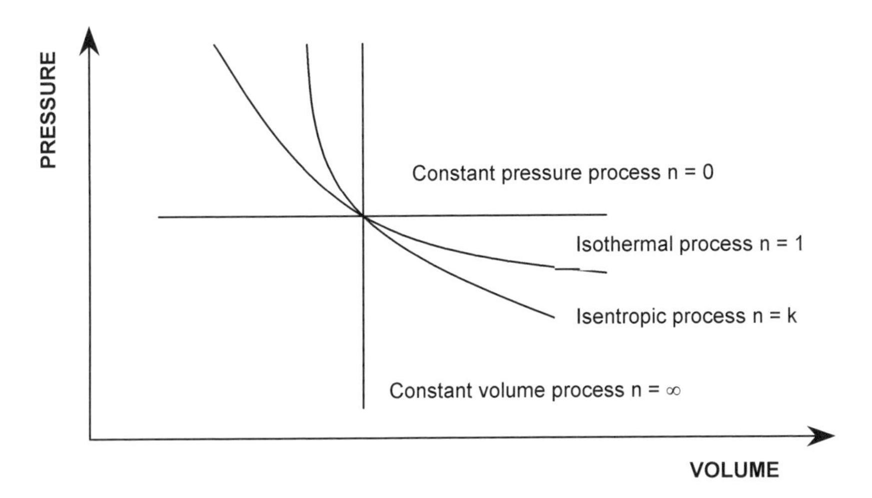

Figure 1.5 Representation of four different polytropic processes on a pressure-volume diagram.

Equation 1.38 can also be derived from the general work relation, $W = PdV$.

For a reversible polytropic process, the only difference is the polytropic exponent (n) which shows the deviation from in a $\log P$ and $\log V$ diagram, leading to the slope. Therefore, Equations 1.35, 1.37 and 1.38 can be rewritten with the polytropic exponent under Pv^n = constant as:

$$Pv^n = P_1 v_1^n = P_2 v_2^n \tag{1.39}$$

$$\frac{P_2}{P_1} = \left(\frac{T_2}{T_1}\right)^{n/n-1} = \left(\frac{v_1}{v_2}\right)^n = \left(\frac{V_1}{V_2}\right)^n \tag{1.40}$$

$$W_{1-2} = \frac{mR(T_2 - T_1)}{1-n} = \frac{(P_2V_2 - P_1V_1)}{1-n} \tag{1.41}$$

To give a clear idea it is important to show the values of n for four different types of polytropic processes for ideal gases (Figure 1.5) as follows:

- $n = 0$ for isobaric process ($P = 0$)
- $n = 1$ for iothermal process ($T = 0$)
- $n = k$ for isentropic process ($s = 0$)
- $n = \infty$ for isochoric process ($v = 0$)

As is obvious in Figure 1.5, there are two quadrants where n varies from zero to infinity and where it has a positive value. The slope of any curve drawn is an important consideration when a reciprocating engine or compressor cycle is under consideration.

In thermodynamics a number of problems involve the mixture of different pure substances (i.e. ideal gases). In this regard, it is important to understand the related aspects accordingly. Tables 1.2 and 1.3 give a summary of the relevant expressions and two ideal gas models, the Dalton model and Amagat model. In fact, in the analysis it is assumed that each gas is unaffected by the presence of other gases, and each one is treated as an ideal gas. With regard to entropy, it is important to note that an increase in entropy is dependent only upon the number of moles of ideal gases, and is independent of its chemical composition. Of course, whenever the gases in the mixture are distinguished, the entropy increases.

1.4.15 Energy transfer

Energy is the capacity for doing work. It can take a number of forms during transfer such as thermal (heat), mechanical (work), electrical, and chemical energy. Energy flows only from a higher temperature level to a lower temperature level unless energy is added to reverse the process. The rate of energy transfer per unit time is called *power*.

Table 1.2 Equations for gas and gas mixtures and relevant models.

Definition	Dalton Model and Amagat Model
Total mass of a mixture of N components	$m_{tot} = m_1 + m_2 + ... + m_N = \sum m_i$
Total number of moles of a mixture of N components	$n_{tot} = n_1 + n_2 + ... + n_N = \sum n_i$
Mass fraction for each compoent	$c_i = m_i / m_{tot}$
Mole fraction for each component	$y_i = n_i / n_{tot} = P_i / P_{tot} = V_i / V_{tot}$
Molecular weight for the mixture	$M_{mixi} = m_{tot} / n_{tot} = \sum n_i M_i / n_{tot} = \sum y_i M_i$
Internal energy for the mixture	$U_{mix} = n_1 \overline{U_1} + n_2 \overline{U_2} + ... + n_N \overline{U_N} = \sum n_i \overline{U_i}$
Enthalpy for the mixture	$H_{mix} = n_1 \overline{H_1} + n_2 \overline{H_2} + ... + n_N \overline{H_N} = \sum n_i \overline{H_i}$
Entropy for the mixture	$S_{mix} = n_1 \overline{S_1} + n_2 \overline{S_2} + ... + n_N \overline{S_N} = \sum n_i \overline{S_i}$
Entropy difference for the mixture	$S_2 - S_1 = -R(n_1 \ln y_1 + n_2 \ln y_2 + ... + n_N \ln y_N)$

Table 1.3 Comparison of Dalton and Amagat models.

Definition	Dalton Model	Amagat Model
P, V, T for the mixture	T and V are constant.	T and P are constant.
	$P_{tot} = P = P_1 + P_2 + ... + P_N$	$V_{tot} = V = V_1 + V_2 + ... + V_N$
Ideal gas equation for the mixture	$PV = n\overline{R}T$	
Ideal gas equations for the components	$P_1 V = n_1 \overline{R}T$	$PV_1 = n_1 \overline{R}T$
	$P_2 V = n_2 \overline{R}T$	$PV_2 = n_2 \overline{R}T$
	:	:
	$P_N V = n_N \overline{R}T$	$PV_N = n_N \overline{R}T$

1.4.16 Heat

The definitive experiment which showed that heat was a form of energy convertible into other forms was carried out by a Scottish physicist, James Joule. It is the thermal form of energy, and heat transfer takes place when a temperature difference exists within a medium or between different media. Heat always requires a difference in temperature for its transfer. Higher temperature differences provide higher heat transfer rates. The units for heat are Joules or kilojoules in the International system and the foot pound force in the English system. In the thermodynamic calculations, heat transfer to a system is considered *positive*, while heat transfer from a system is *negative*. If there is no heat transfer involved in a process, it is called an *adiabatic process*.

1.4.17 Work

Work is the energy that is transferred by a difference in pressure or force of any kind, and is subdivided into shaft work and flow work. Shaft work is a mechanical energy used to drive a mechanism such as a pump, compressor, or turbine. Flow work is the energy transferred into a system by fluid flowing into, or out of, the system. Both forms are usually expressed in kilojoules or a mass basis kJ/kg. Work done by a system is considered positive and work done on a system (work input) is considered negative. The unit for power is a rate of work of one Joule per second, which is a watt (W).

1.4.18 The first law of thermodynamics

It is known that thermodynamics is the science of energy and entropy, and that the basis of thermodynamics is experimental observation. In thermodynamics, such observations were formed into four basic laws of thermodynamics: the zeroth, first, second, and third laws of thermodynamics. The first and second laws of thermodynamics are the most common tools in practice, due to fact that transfers and conversions of energy are governed by these two laws, and in this chapter we will focus on these two laws.

The first law of thermodynamics can be defined as the law of conservation of energy and states that in a closed system, energy can be neither created nor destroyed. For a change of state from an initial state 1 to a final state 2, with a constant amount of matter, it can be formulated as follows:

$$Q_{1-2} = (E_2 - E_1) + W_{1-2} = (U_2 - U_1) + (KE_2 - KE_1) + (PE_2 - PE_1) + W_{1-2} \qquad (1.42)$$

where $(U_2 - U_1) = mc_v(T_2 - T_1)$, $(KE_2 - KE_1) = m(V_2^2 - V_1^2)/2$, $(PE_2 - PE_1) = mg(Z_2 - Z_1)$.

As is clear in Equation 1.42, we broadened the definition of heat energy to include kinetic and potential energies in addition to internal energy. An important consequence of the first law is that the internal energy change resulting from some process will be independent of the thermodynamic path followed by the system, and of the paths followed by the processes, e.g. heat transfer and work. In turn, the rate at which the internal energy content of the system changes is dependent only on the rates at which heat is added and work is done.

1.4.19 The second law of thermodynamics

As mentioned earlier, the first law is the energy-conservation principle. The second law of thermodynamics refers to the inefficiencies of practical thermodynamic systems, and indicates that it is impossible to have 100% efficiency in energy conversion. The classical statements as the Kelvin-Plank statement and the Clausius statement help us formulate the second law:

- **The Kelvin Plank statement**: It is impossible to construct a device, operating in a cycle (e.g. heat engine), that accomplishes only the extraction of heat energy from some source and its complete conversion to work. This statement simply shows the impossibility to have a heat engine with a thermal efficiency of 100%.
- **The Clausius statement**: It is impossible to construct a device, operating in a cycle (e.g. refrigerator and heat pump) that transfers heat from the low-temperature side (cooler) to the high-temperature side (hotter).

A very easy way to show the implication of both the first and second laws is a desktop game that consists of several pendulus (made of metal balls) one in contact with the other. When you raise the first of the balls, you give energy to the system, potential energy. Releasing this ball it gains kinetic energy at the expense of the potential energy. When this ball hits the second ball, a small elastic deformation transforms the kinetic energy, again in the form of potential energy. The energy is transferred from one ball to the other. The last ball again gains kinetic energy to go up. The cycle continues, but every time lower, until it finally stops. The first law is about why the balls keep moving, but the second law is why they do not do it forever. In this game the energy is lost in sound and heat, and it is no longer useful to keep motion.

The second law also states that the entropy in the universe is always more. As mentioned before, entropy is the degree of disorder, and every process happening in the universe is a transformation from a lower entropy to a higher entropy. Therefore, the entropy of a state of a system is proportional to (depends on) its probability, which gives us an opportunity to define the second law in a broader manner as "the entropy of a system increases in any heat transfer or conversion of energy within a closed system." That is why all energy transfers or conversions are irreversible. From the entropy perspective, the basis of the second law is the statement that the sum of the entropy changes of a system and that of its surroundings must always be positive. Recently, much effort has been spent to minimize the entropy generation (irreversibilities) in thermodynamic systems and applications.

Moran and Shapiro (1998) noted that the second law and deductions from it are useful because they provide a means for:

- predicting the direction of processes,
- establishing conditions for equilibrium,
- determining the best performance of thermodynamic systems and applications,
- evaluating quantitatively the factors that preclude the attainment of the best theoretical performance level,
- defining a temperature scale, independent of the properties of the substance, and

- developing tools for evaluating some thermodynamic properties, e.g. internal energy and enthalpy using the experimental data available.

Consequently, the second law is the linkage between entropy and the usefulness of energy. The second law analysis has found applications in a large variety disciplines, e.g. chemistry, economy, ecology, environment, sociology, etc. far removed from engineering thermodynamics applications.

1.4.20 Reversibility and irreversibility

These two concepts are highly important to thermodynamic processes and systems. The *reversibility* is defined as the statement that both the system and its surroundings can be returned to their initial states, just leading to the theoretical one. The irreversibility shows the destruction of availability, and is stated that both the system and its surroundings cannot be returned to their initial states due to the irreversibilities that have occurred, e.g. friction, heat rejection, electrical and mechanical effects, etc. For instance, as an actual system provides an amount of work that is less than the ideal reversible work, so the difference between these two values gives the irreversibility of that system. In real applications, there are always such differences, and therefore real cycles are always irreversible.

1.4.21 Exergy

Exergy is defined as the maximum amount of work (referring to availability, see Table 1.4) which can be produced by a stream of matter, heat or work as it comes to equilibrium with a reference environment. In fact, it is a measure of the potential of a stream to cause change, as a consequence of not being completely stable relative to the reference environment. For exergy analysis, the state of the reference environment, or the reference state, must be specified completely. This is commonly done by specifying the temperature, pressure and chemical composition of the reference environment. Exergy is not subject to a conservation law. Rather exergy is consumed or destroyed, due to irreversibilities in any process. Table 1.5 shows a clear comparison between energy and exergy from the thermodynamics point of view.

As pointed out by Dincer and Rosen (1999), exergy is a measure of usefulness, quality or potential of a stream to cause change and an effective measure of the potential of a substance to impact on the environment.

Exergy analysis is a method that uses the conservation of mass and conservation of energy principles together with the SLT for the design and analysis of refrigeration systems and applications. The exergy method can be suitable for furthering the goal of more efficient energy-resource use, for it enables the locations, types, and true magnitudes of wastes and losses to be determined. Therefore, exergy analysis can reveal whether or not, and by how much, it is possible to design more efficient energy systems by reducing the sources of inefficiency in existing systems. In the past exergy was called *essergy*, *availability*, and *free energy*. Table 1.4 gives the connections among essergy, availability, exergy and free energy.

Table 1.4 Connections among essergy, availability, exergy and free energy.

Name	Function	Remarks
Essergy	$E + P_0V - T_0S - \Sigma_i \mu_{i0}N_i$	Formulated for the special case in 1878 by Gibbs and in general in 1962, and later changed from available energy to exergy in 1963, and from exergy to essergy (i.e. essence of energy) in 1968 by Evans.
Availability	$E + P_0V - T_0S - (E_0 + P_0V_0 - T_0S_0)$	Formulated by Keenan in 1941 as a special case of the essergy function.
Exergy	$E + P_0V - T_0S - (E_0 + P_0V_0 - T_0S_0)$	Introduced by Darrieus in 1930 and Keenan in 1932, called the availability in steady flow by him, and exergy by Rant in 1956 as a special case of essergy.
Free Energy	Helmholtz: $E - TS$ Gibbs: $E + PV - TS$	Introduced by von Helmholts and Gibbs in 1873 as the Legendre transforms of energy to yield useful alternate criteria of equilibrium, as measures of the potential work of systems representing special cases of the essergy function.

Source: Szargut *et al.* (1988).

Table 1.5 Comparison between energy and exergy.

Energy	Exergy
• Dependent on the parameters of matter or energy flow only, and independent of the environment parameters.	• Dependent both on the parameters of matter or energy flow and on the environment parameters.
• Has the values different from zero (which is equal to mc^2 in accordance with Einstein's equation).	• Equal to zero (in dead state by equilibrium with the environment).
• Guided by the FLT for all the processes.	• Guided by the FLT for reversible processes only (in irreversible processes it is destroyed partly or completely).
• Limited by the SLT for all processes (including reversible ones).	• Not limited for reversible processes owing to the SLT.

From the energy and exergy efficiency point of view, it is important to note that if a fossil fuel based energy source was used for low-temperature thermal application like space heating or cooling, there would be a great difference between energy and exergy efficiencies such as 50–70% as energy efficiency and 5% as exergy efficiency (Dincer, 1998). One may ask why? Here we can identify the following facts:

- High quality (or high temperature) energy sources such as fossil fuels are being used for relatively low-temperature processes like water and space heating or cooling, etc.
- Exergy efficiency permits a better matching of energy sources and uses, leading to high quality energy not being used for performing high quality work.

1.5 General Aspects of Fluid Flow

For a good understanding of the operation of thermal energy storage systems and their components, as well as the behavior of fluid flows, an extensive background on fluid mechanics is essential. In addition to learning the principles of fluid flow, the student and/or engineer should develop an understanding of the properties of fluids which should enable him or her to solve practical refrigeration problems.

In practice, refrigerating engineers come into contact everyday, or at least on an occasional basis, with a large variety of fluid flow problems:

- subcooled liquid refrigerant, water, brine, and other liquids,
- mixtures of boiling liquid refrigerant and its vapor,
- mixtures of refrigerants and absorbents,
- mixtures of air and water vapor as humid air, and
- low- and high-side vaporous refrigerant and other gases.

To deal effectively with fluid flow systems it is necessary to identify flow categories, defined in predominantly mathematical terms, that will allow the appropriate analysis to be undertaken by identifying suitable and acceptable simplifications. Examples of the categories to be introduced include variation of the flow parameters with time (steady or unsteady) or variations along the flow path (uniform or non-uniform). Similarly, compressibility effects may be important in high-speed gas flows, but may be ignored in many liquid flow situations.

1.5.1 Classification of fluid flows

There are several criteria to classify the fluid flows into the following categories:

- uniform or non-uniform,
- one-, two-, or three-dimensional,
- steady- or unsteady-state,
- laminar or turbulent, and
- compressible or incompressible.

Also, the liquids flowing in open channels may be classified according to their regions, e.g. subcritical, critical, or supercritical, and the gas flows may be categorized as subsonic, transonic, supersonic, or hypersonic.

Uniform flow and non-uniform flow

If the velocity and cross-sectional area are constant in the direction of flow, the flow is uniform. Otherwise, the flow is non-uniform.

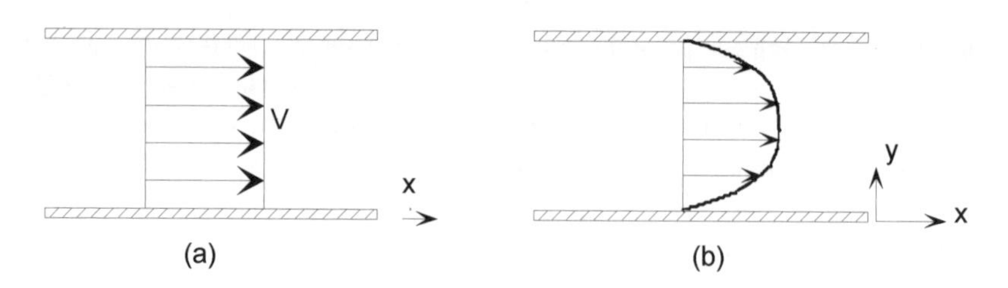

Figure 1.6 Velocity profiles for flows. (a) One-dimensional flow. (b) Two-dimensional flow.

One-, two-, and three-dimensional flow

The flow of real fluids occurs in three dimensions. However, in the analysis the conditions are simplified to either one- or two-dimensional, depending on the flow problem under consideration. If all fluid and flow parameters (e.g. velocity, pressure, elevation, temperature, density, viscosity, etc.) are considered to be uniform throughout any cross-section and vary only along the direction of flow (Figure 1.6a), the flow becomes one-dimensional. Two-dimensional flow is the flow in which the fluid and flow parameters are assumed to have spatial gradients in two directions, i.e. x and y axes (Figure 1.6b). In fact, in a three-dimensional flow the fluid and flow parameters vary in three directions, i.e. x, y and z axes, and the gradients of the parameters occur in all three directions.

Steady flow

This is defined as a flow in which the flow conditions do not change with time. However, we may have a steady-flow in which the velocity, pressure and cross-section of the flow may vary from point to point but do not change with time. That is why it requires us to distinguish this by dividing it into the *steady uniform flow* and the *steady non-uniform flow*. In the steady uniform flow, all conditions (e.g. velocity, pressure, cross-sectional area) are uniform, and do not vary with time or position. For example, uniform flow of water in a duct is considered a steady-uniform flow. If the conditions (e.g. velocity, cross-sectional area) change from point to point (e.g. from cross-section to cross-section) but not with time, it is called steady non-uniform flow. For example, a liquid flows at a constant rate through a tapering pipe running completely full.

Unsteady flow

If the conditions vary with time, the flow becomes unsteady. At a given time, the velocity at every point in the flow field is the same, but the velocity changes with time, referring to the *unsteady uniform flow*. For example, accelerating flow, a fluid through a pipe of uniform bore running full. In the unsteady uniform flow, the conditions in cross-sectional area and velocity vary with time from one point to another, e.g. a wave travelling along a channel.

Laminar flow and turbulent flow

This is one of the most important classifications of fluid flow and depends primarily upon the arbitrary disturbances, irregularities, or fluctuations in the flow field, based on the internal characteristics of the flow. In this regard, there are two significant parameters such as velocity and viscosity. If the flow occurs at a relatively low velocity and/or with a highly viscous fluid, resulting in a fluid flow in an orderly manner without fluctuations, the flow is referred to as laminar. As the flow velocity increases and the viscosity of fluid decreases, the fluctuations will take place gradually, referring to a *transition state* which is dependent on the fluid viscosity, the flow velocity, and the geometric details. In this regard, the Reynolds number is introduced to represent the characteristics of the flow conditions relative to the transition state. As the flow conditions deviate more from the transition state, a more chaotic flow field, i.e. turbulent flow, occurs. It is obvious that an increasing Reynolds number increases the chaotic nature of the turbulence. Turbulent flow is therefore defined as a characteristic representative of the irregularities in the flow field.

The differences between laminar flow and turbulent flow can be distinguished by the Reynolds number, which is expressed by:

$$\mathrm{Re} = \frac{VD}{\upsilon} = \frac{\rho VD}{\mu} \tag{1.43}$$

In fact, the Reynolds number indicates the ratio of inertia force to viscous force. One can point out that at high Reynolds numbers the inertia forces predominate, resulting in turbulent flow, while at low Reynolds numbers the viscous forces become dominant, which makes the flow laminar. In a circular duct, the flow is laminar when *Re* is less than 2100 and turbulent when *Re* is greater than 4000. In a duct with a rough surface, the flow is turbulent at *Re* values as low as 2700.

Compressible flow and incompressible flow

All actual fluids are normally compressible, leading to their density changing with pressure. However, in most cases during the analysis it is assumed that the changes in density are negligibly small. This refers to incompressible flow.

1.5.2 Viscosity

Viscosity is known as one of the most significant fluid properties, and is defined as a measure of the fluid's resistance to deformation. In gases, the viscosity increases with increasing temperature, resulting in a greater molecular activity and momentum transfer. The viscosity of an ideal gas is a function of molecular dimensions and absolute temperature only, based on the kinetic theory of gases. However, in fluids, molecular cohesion between molecules considerably affects the viscosity, and the viscosity decreases with increasing temperature due to the fact that the cohesive forces are reduced by increasing the temperature of the fluid (causing a decrease in shear stress), resulting in an increase of the rate of molecular interchange, and therefore the net result is apparently a reduction in the viscosity. The coefficient of viscosity of an ideal fluid is zero, meaning that an ideal fluid is inviscid, so that no shear stresses occur in the fluid, despite the fact that shear deformations are finite. Nevertheless, all real fluids are viscous.

As a fluid moves past a solid boundary or wall, the velocity of the fluid particles at the wall must equal the velocity of the wall; the relative velocity between the fluid and the wall at the surface of the wall is zero, which is called the *no slip* condition, and results in a varying magnitude of the flow velocity (e.g. a velocity gradient), as one moves away from the wall (see Figure 1.7).

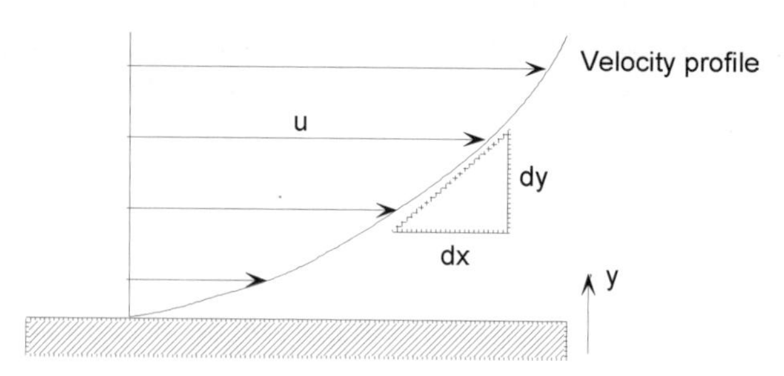

Figure 1.7 Schematic of velocity profile.

There are two types of viscosities, namely *dynamic viscosity* which is the ratio of a shear stress to a fluid strain (velocity gradient), and *kinematic viscosity* which is defined as the ratio of dynamic viscosity to density.

The dynamic viscosity, based on a two-dimensional boundary layer flow and the velocity gradient du/dy occurring in the direction normal to the flow, as shown Figure 1.7, leading to the shear stress within a fluid being proportional to the spatial rate of change of fluid strain normal to the flow, is expressed as:

$$\mu = \frac{\tau}{(du / dy)} \tag{1.44}$$

where the units of μ are Ns/m^2 or kg/ms in the SI system and $lb_f s/ft^2$ in the English system.

The kinematic viscosity then becomes

$$\upsilon = \frac{\mu}{\rho} \tag{1.45}$$

where the units of ν are m^2/s in the SI system and ft^2/s in the English system.

From the viscosity perspective, the types of fluids may be classified into the following.

Newtonian fluids
These fluids have a dynamic viscosity dependent upon temperature and pressure and independent of the magnitude of the velocity gradient. For such fluids, Equation 1.44 is applicable. Some examples are water and air.

Non-Newtonian fluids
The fluids which cannot be represented by Equation 1.44 are called *non-Newtonian fluids.* These fluids are very common in practice and have a more complex viscous behavior due to the deviation from Newtonian behavior. There are several approximate expressions to represent their viscous behavior. Some example fluids are slurries, polymer solutions, oil paints, toothpaste, sludges, etc.

1.5.3 Equations of flow

The basic equations of fluid flow may be derived from some of the fundamental principles, namely the conservation of mass, the conservation of motion (i.e. Newton's second law of motion), and the conservation of energy. Although very general statements of these laws can be written down (applicable to all substances, e.g. solids and fluids), in fluid flow these principles can be formulated as a function of some variables, namely pressure, temperature, and density. The equations of motion may be classified into two general types, namely the equations of motion for inviscid fluids (i.e. frictionless fluids) and the equations of motion for viscous fluids. In this regard, we deal with the Bernoulli equations and Navier-Stokes equations.

Continuity equation

This is based on the *conservation of mass* principle. The requirement that mass be conserved at every point in a flowing fluid imposes certain restrictions on the velocity u and density ρ. Therefore, the rate of mass change is zero, so that for a steady flow, the mass of fluid in the control volume remains constant, and therefore the mass of fluid entering per unit time is equal to the mass of fluid exiting per unit time. Let's apply this to a steady flow in a stream tube (Figure 1.8). The equation of continuity for the flow of a compressible fluid through a stream tube is

$$\rho_1 \delta A_1 u_1 = \rho_2 \delta A_2 u_2 = constant \tag{1.46}$$

where $\rho_1 \delta A_1 u_1$ is the mass entering per unit time and $\rho_2 \delta A_2 u_2$ is the mass exiting per unit time for sections 1 and 2.

In practice, for the flow of a real fluid through a pipe or a conduit, the mean velocity is used since the velocity varies from wall to wall. Thus, Equation 1.46 can be rewritten as

$$\rho_1 A_1 \bar{u}_1 = \rho_2 A_2 \bar{u}_2 = \dot{m} \tag{1.47}$$

where u_1 and u_2 are the mean velocities at sections 1 and 2.

For the fluids that are considered as incompressible, Equation 1.47 is simplified to the following, since $\rho_1 = \rho_2$:

$$A_1 \bar{u}_1 = A_2 \bar{u}_2 = \dot{V} \tag{1.48}$$

The various forms of the continuity equations for steady-state and unsteady-state cases are summarized below:

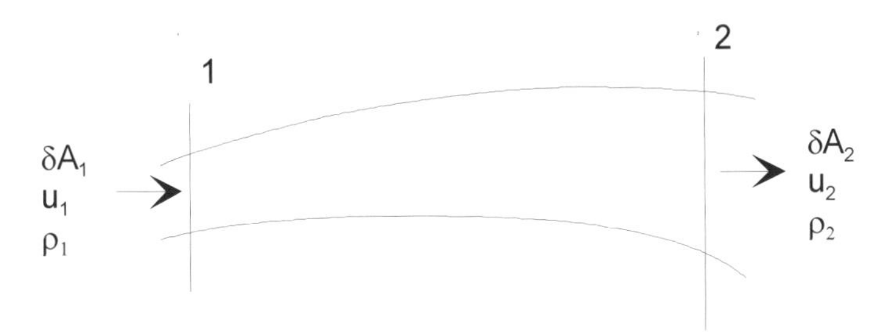

Figure 1.8 Fluid flow in a stream tube.

- The steady-state continuity equation for an incompressible fluid in a stream tube:

$$V \cdot A = \dot{V} \tag{1.49}$$

- The unsteady-state continuity equation for an incompressible fluid in a stream tube:

$$\left(\frac{dm}{dt}\right)_{sys} = \frac{d}{dt}\int_{cv} \rho dV + \int_{cs} \rho \bar{V} d\bar{A} \tag{1.50}$$

- The steady-state continuity equation for an incompressible fluid in cartesian coordinates:

$$\frac{\partial u}{\partial x} + \frac{\partial v}{\partial y} + \frac{\partial w}{\partial z} = 0 \tag{1.51}$$

- The unsteady-state continuity equation for an incompressible fluid in cartesian coordinates:

$$\frac{\partial(\rho u)}{\partial x} + \frac{\partial(\rho v)}{\partial y} + \frac{\partial(\rho w)}{\partial z} = \frac{\partial \rho}{\partial t} \tag{1.52}$$

- The steady-state continuity equation for an incompressible fluid in cylindrical coordinates:

$$\frac{\partial v_r}{\partial r} + \frac{1}{r}\frac{\partial v_\theta}{\partial \theta} + \frac{\partial v_z}{\partial z} + \frac{v_r}{r} = 0 \tag{1.53}$$

- The steady-state continuity equation for a compressible fluid in a stream tube:

$$\rho V \cdot A = \dot{m} \tag{1.54}$$

- The steady-state continuity equation for a compressible fluid in cartesian coordinates:

$$\frac{\partial(\rho u)}{\partial x} + \frac{\partial(\rho v)}{\partial y} + \frac{\partial(\rho w)}{\partial z} = 0 \tag{1.55}$$

- The steady-state continuity equation for a compressible fluid in cylindrical coordinates:

$$\frac{\partial(\rho v_r)}{\partial r} + \frac{1}{r}\frac{\partial(\rho v_\theta)}{\partial \theta} + \frac{\partial(\rho v_z)}{\partial z} + \frac{\rho v_r}{r} = 0 \tag{1.56}$$

- The unsteady-state continuity equation for a compressible fluid in a stream tube:

$$\frac{\partial(\rho A)}{\partial t} + \frac{\partial(\rho V \cdot A)}{\partial s} = 0 \tag{1.57}$$

- The unsteady-state continuity equation for a compressible fluid in cartesian coordinates:

$$\frac{\partial \rho}{\partial t}+\frac{\partial(\rho u)}{\partial x}+\frac{\partial(\rho v)}{\partial y}+\frac{\partial(\rho w)}{\partial z}=0 \tag{1.58}$$

- The unsteady-state continuity equation for a compressible fluid in cylindrical coordinates:

$$\frac{\partial \rho}{\partial t}+\frac{\partial(\rho v_r)}{\partial r}+\frac{1}{r}\frac{\partial(\rho v_\theta)}{\partial \theta}+\frac{\partial(\rho v_z)}{\partial z}+\frac{\rho v_r}{r}=0 \tag{1.59}$$

Momentum equation

The analysis of fluid flow phenomena is fundamentally dependent on the application of Newton's second law of motion, which is more general than the momentum principle, stating that when the net external force acting on a system is zero, the linear momentum of the system in the direction of the force is conserved in both magnitude and direction (the so called *conservation of linear momentum*). In fact, the momentum principle is concerned only with external forces, and provides useful results in many situations without requiring much information on the internal processes within the fluid. It may find applications in all types of flows (e.g. steady or unsteady, compressible or incompressible).

The motion of the particle must be described relative to an inertial coordinate frame. The one-dimensional momentum equation at constant velocity can be written as follows:

$$\Sigma F=\frac{d}{dt}(mV) \tag{1.60}$$

where ΣF stands for the sum of the external forces acting on the fluid, and mV stands for the kinetic momentum in that direction. Equation 1.60 states that the time rate of change of the linear momentum of the system in the direction of V equals the resultant of all forces acting on the system in the direction of V. In fact, the linear momentum equation is a vector equation and is therefore dependent on a set of coordinate directions.

The rate of change of momentum of a control mass system can be related to the rate of change of momentum of a control volume under the guidance of the continuity equation. Therefore, Equation 1.60 becomes

$$\Sigma F_t=\Sigma F_{cv}+\Sigma F_{cs}=\frac{d}{dt}\int_{cv} V\rho dV-\int_{cs} V(\rho \bar{V} d\bar{A}) \tag{1.61}$$

in which the sum of forces acting on the control volume in any one direction is equal to the rate of change of momentum of the control volume in that direction plus the net rate of momentum flux from the control volume through its control surface in the same direction.

For a steady flow, if the velocity across the control surface is constant, the momentum equation in scalar form becomes

$$\Sigma F_x=(\dot{m}V_x)_e-(\dot{m}V_x)_i \tag{1.62}$$

If $\dot{m}$ is constant, Equation 1.62 can be written as

$$\Sigma F_x = \dot{m}(V_{xe} - V_{xi}) \tag{1.63}$$

Similar expressions can be written for the other directions y and z.

Euler's equation

This is a mathematical statement of Newton's second law of motion, and finds application to an inviscid fluid continuum. It states that the product of mass and acceleration of a fluid particle can be equated vectorially with the external forces acting on the particle. Consider a stream tube, as shown in Figure 1.9, with a cross-sectional area small enough for the velocity to be considered constant along the tube.

The following is a simple form of Euler's equation for a steady flow along a stream tube, representing the relationship in the differential form between pressure p, velocity v, density ρ, and elevation z, respectively:

$$\frac{1}{\rho}\frac{dp}{ds} + v\frac{dv}{ds} + g\frac{dz}{ds} = 0 \tag{1.64}$$

For an incompressible fluid (i.e. ρ is constant) the integration of the above equation gives the following along the streamline (with respect to s) for an inviscid fluid:

$$\frac{p}{\rho} + \frac{v^2}{2} + gz = \text{constant} \tag{1.65}$$

For a compressible fluid the integration of Equation 1.64 can only be completed to provide the following:

$$\int \frac{dp}{\rho g} + \frac{v^2}{2g} + z = H \tag{1.66}$$

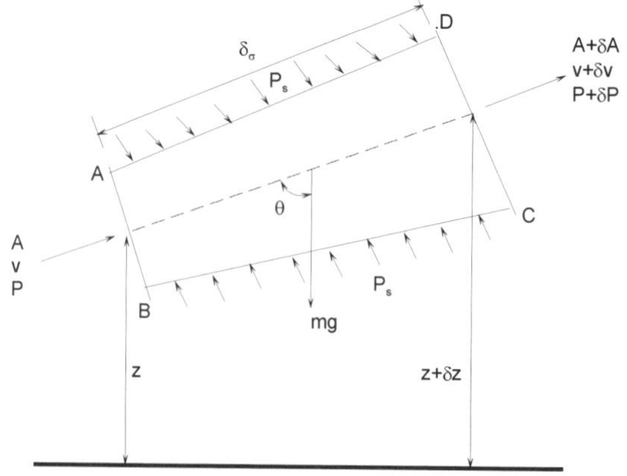

Figure 1.9 Relationship between velocity, pressure, elevation, and density for a stream tube.

It is important to note that the relationship between ρ and p should be known for the given case, and that for gases it can be in the form $p\rho^n$ = constant, varying from adiabatic to isothermal conditions, while for a liquid, $\rho(dp/d\rho) = K$, which is an adiabatic modulus.

Bernoulli's equation

This equation can be written for both incompressible and compressible flows. Under certain flow conditions, Bernoulli's equation for incompressible flow is often referred to as a mechanical-energy equation due to the fact that there is similarity to the steady-flow energy equation obtained from the first law of thermodynamics for an inviscid fluid with no heat transfer and no external work associated. It is necessary to point out that for inviscid fluids the viscous forces and surface tension forces are not taken into consideration, simply leading to the negligible viscous effects. The Bernoulli equation is commonly used in a variety of practical applications, particularly in flows in which the losses are negligibly small, e.g. in hydraulic applications. The following is the general Bernoulli equation per unit mass for inviscid fluids between any two points:

$$\frac{u_1^2}{2g} + \frac{p_1}{\rho g} + z_1 = \frac{u_2^2}{2g} + \frac{p_2}{\rho g} + z_2 = H \tag{1.67}$$

where each term has a dimension of a length or head scale. In this regard, $u^2/2g$ (kinetic energy per unit mass) is referred to as the velocity head, $p/\rho g$ (pressure energy per unit mass) as the pressure head, z (potential energy per unit mass) as the potential head (constant total head), and H (total energy per unit mass) as the total head in meters. Subscripts 1 and 2 denote where the variables are evaluated on the streamline.

The terms in Equation 1.67 represent energy per unit mass and have the unit of length. Actually, Bernoulli's equation can be obtained by dividing each term in Equation 1.65 by g. These terms, both individually and collectively, indicate the quantities which may be directly converted to produce mechanical energy.

In summary, if we compare the Equation 1.67 with the general energy equation, we see that the Bernoulli equation contains even more restrictions than might first be realized, due to the following main assumptions:

- steady flow (common assumption applicable to many flows),
- incompressible flow (acceptable if the Mach number is less than 0.3),
- frictionless flow along a single streamline (very much restrictive),
- no shaft work and heat transfer associated between 1 and 2.

Navier-Stokes equations

The Navier-Stokes equations are the differential expressions of Newton's second law of motion, which are known as constitutive equations for viscous fluids. These equations were named after C.L.M.H. Navier and Sir G.G. Stokes, who are credited with their derivation.

For viscous fluids, two force aspects, namely a body force and a pressure force on its surface, are taken into consideration. The solution of these equations is dependent upon what flow information is known. The solutions evolving now for such problems have become extremely useful. The other important point is to keep abreast of the numerical software packages which have been developed in the field of fluid flow for several engineering applications.

Exact solutions to the nonlinear Navier-Stokes equations are limited to a very few cases, particularly for steady, uniform flows (either two-dimensional or with radial symmetry) or for flows with simple geometries. However, approximate solutions may be undertaken for other one-dimensional simple flow cases which require only the momentum and continuity equations in the flow direction for the solution of the flow field. Here, we present a few cases, namely uniform flow between parallel plates, uniform free surface flow down a plate, and uniform flow in a circular tube.

Uniform flow between parallel plates Consider a two-dimensional uniform steady-flow between parallel plates which is infinite in the z direction, as shown in Figure 1.10. It is also considered that the plates are oriented at an angle of θ to the horizontal plane, which results in a body force per unit mass as the gravitational term $g\sin\theta$. Therefore, the pressure gradient is specified as $(\partial p/\partial y)$ = constant = P^*. After making necessary simplifications and integrations, we find the total flow rate per unit width and the velocity profile as follows:

$$q_f = \frac{h^3}{12\mu}(-P^* + \rho g \sin\theta) \tag{1.68}$$

$$u = \frac{1}{2\mu}(-P^* + \rho g \sin\theta)(hy - y^2) \tag{1.69}$$

If the plates are horizontally located (i.e. $\sin\theta = 0$), the above equations reduce to

$$q_f = -\frac{P^* h^3}{12\mu} \tag{1.70}$$

$$u = -\frac{P^*}{2\mu}(hy - y^2) \tag{1.71}$$

Uniform free surface flow down a plate This case is, by nature, similar to the previous case, except that the upper plate has been removed, as shown in Figure 1.11. The boundary condition at the lower boundary is the same no-slip boundary condition as before, such that the velocity is zero. However, the boundary condition at the free surface can no longer be specified as a no-slip one. The total flow rate per unit width becomes

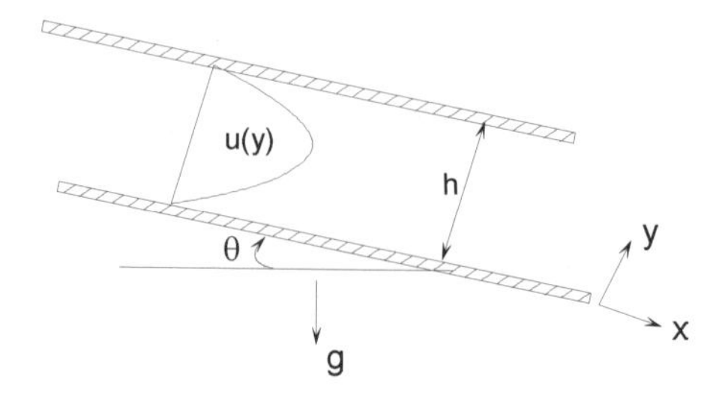

Figure 1.10 Uniform flow between two stationary parallel plates.

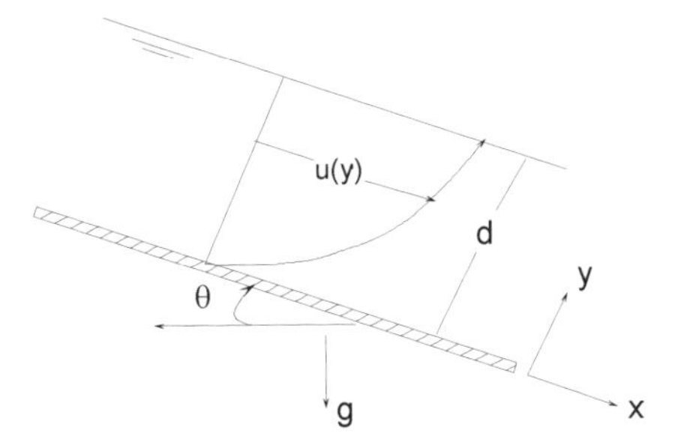

Figure 1.11 Uniform flow down a plate.

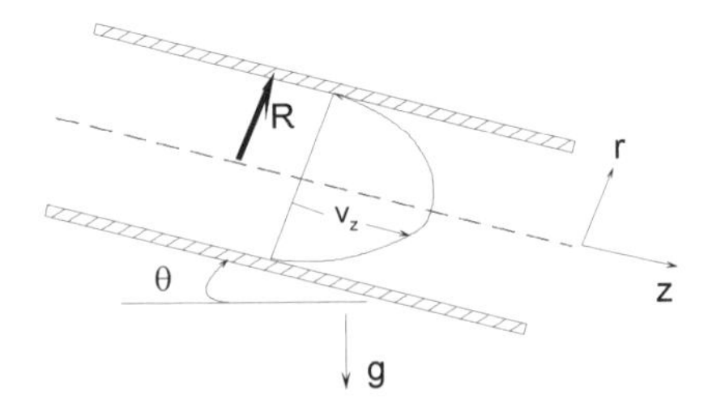

Figure 1.12 Uniform flow in a pipe.

$$q_f = \frac{\rho g d^3 \sin\theta}{3\mu} \tag{1.72}$$

while the flow velocity is

$$u = \frac{\rho g \sin\theta}{\mu}(yd - y^2/2) \tag{1.73}$$

Furthermore, the average velocity can be obtained by dividing the discharge by the flow area or depth as follows:

$$V = \frac{\rho g d^2 \sin\theta}{3\mu} \tag{1.74}$$

Uniform flow in a circular tube This case is about a uniform fluid flow in a pipe of radius R, as shown in Figure 1.12 which is the most common example in practical applications associated with pipe flows. Despite having the flow field as three-dimensional, the assumption of radial symmetry makes it two-dimensional. Therefore, the parabolic velocity distribution with the maximum velocity at the center of the pipe can be found as follows:

$$v_z = \frac{(R^2 - r^2)}{4\mu}(-P^* + \rho g \sin\theta) \tag{1.75}$$

which is known as the Hagen-Poiseuille equation. The total volumetric flow rate can be calculated, if the pressure gradient along with other flow conditions are specified, and vice-versa, as follows:

$$Q = \frac{\pi R^4}{8\mu}(-P^* + \rho g \sin\theta) \tag{1.76}$$

If the pipe is horizontally located (i.e. $sin\theta = 0$), the above equations result in

$$v_z = \frac{(R^2 - r^2)}{4\mu} \tag{1.77}$$

$$Q = \frac{\pi R^4}{8\mu} \tag{1.78}$$

1.5.6 Boundary layer

If there is an equivalence between fluid and surface velocities at the interface between a fluid and a surface, it is called a 'no slip' condition, which is entirely associated with viscous effects, as mentioned earlier. In practice, any real fluid flow shows a region retarded flow near a boundary in which the velocity relative to the boundary varies from zero at the boundary to a value that may be estimated by the potential-flow solution some distance away. This region of retarded flow is known as the *boundary layer*, which was first introduced by Prandtl in 1904. His hypothesis of a boundary layer was arrived at by experimental observations of the flow past solid surfaces.

Of course, the boundary layer can be taken as that region of the fluid close to the surface immersed in the flowing fluid, and the boundary-layer development takes place in both internal and external flows. In internal flows, it occurs until the entire fluid is encompassed, such as pipe flow and open-channel flow. In fact, the boundary-layer development is more important for external flows which exhibit a continued growth due to the absence of a confining boundary such as a flow along a flat plate. It is therefore important to assume that the velocity at some distance from the boundary is unaffected by the presence of the boundary, referring to the free-stream velocity u_s.

Let's now consider a uniform flow of incompressible fluid at a free-stream velocity approaching the plate, as shown in Figure 1.13. Since the plate is stationary with respect to the earth, when the fluid is in contact with the plate surface, it has zero velocity, 'no slip'. Later, the boundary-layer thickens in the direction of flow and a velocity gradient at a distance δ over an increasingly greater distance normal to the plate takes place between the fluid in the free stream and the plate surface. Thus, the rate of change of velocity determines the velocity gradient at the surface as well as the shear stresses. The shear stress for the laminar boundary layer becomes

$$\tau = \mu\left(\frac{du}{dy}\right)_{y=0} \tag{1.79}$$

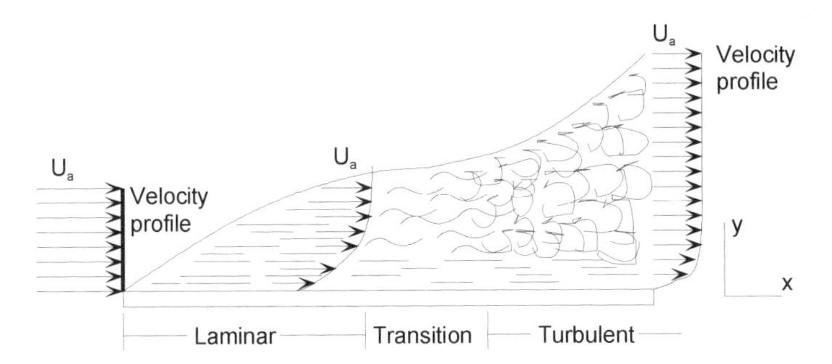

Figure 1.13 Development of boundary layer in a viscous flow along a plate.

which varies with distance along the plate by the change in velocity. Further along the plate, the shear force is gradually increased, as the laminar boundary layer thickens, due to the increasing plate surface area affected, and the fluid becomes retarded, so that a turbulent boundary layer occurs by the instabilities set in. Thus, the shear stress for the turbulent flow can be approximated as

$$\tau = (\mu + \varepsilon)\left(\frac{du}{dy}\right)_{y=0} \tag{1.80}$$

Experimental studies indicate that there are two boundary layer flow regimes; a *laminar flow regime* and a *turbulent flow regime*, which can be characterized by the Reynolds number. It should be noted that the transition from a laminar to a turbulent boundary layer is dependent mainly upon the following:

- $Re = u_s x_c / \nu$,
- the roughness of the plate, and
- the turbulence level in the free stream.

There are various boundary-layer parameters to be considered such as boundary-layer thickness, the local wall shear stress (or local friction or drag coefficient), and the average wall shear stress (or average friction or drag coefficient). The boundary-layer thickness may be expressed in several ways. The simplest approach is that the velocity u within the boundary layer approaches the free stream velocity u_s. From the experimental measurements it was observed that the boundary-layer thickness δ can be defined as the distance from the boundary to the point at which $u = 0.99u_s$.

Table 1.6 gives the values of the boundary-layer thicknesses for a laminar flow along a flat plate with respect to the different dimensionless coordinates $\eta = y(u_s/\nu x)$. These values are of practical interest in momentum analysis of fluid flow.

The momentum equations for velocity profiles can be summarized with respect to the momentum thickness, the average skin-drag coefficient and the displacement thickness in Table 1.7 for a flat plate. As can be pointed out from the table, the laminar boundary-layer thickness increases with $x^{1/2}$ from the leading edge and inversely with $u_s^{1/2}$, the local and average skin-drag coefficients change inversely with $x^{1/2}$ and $u_s^{1/2}$, and the total drag force, F

$= C_f \rho u_s^2 x/2$ per unit width, changes as the 1.5 power of the u_s and the square root of the length x. Normally, the fluid flow along a flat plate is laminar for the values of the Reynolds number up to about 300,000–500,000, depending on the plate roughness and the level of turbulence in the free stream, as mentioned earlier.

Moreover, Table 1.8 presents additional momentum equations for a boundary turbulent layer along a flat plate, including additional pipe-flow velocity profiles, and summarizes the following facts for a turbulent boundary layer on a flat plate:

- The boundary-layer thickness increases as the 4/5 power of the distance from the leading edge, as compared with $x^{1/2}$ for a laminar boundary layer.
- The local and average skin-friction coefficients vary inversely as the fifth root of both x and u_s, as compared with the square root for a laminar boundary layer.
- The total drag varies as $u_s^{9/5}$, and $x^{4/5}$, as compared with the values of corresponding parameters for a laminar boundary layer.

Initially, as the boundary layer develops, it will be laminar in form while the boundary layer will become turbulent, based on the ratio of inertial and viscous forces acting on the fluid, referring to the value of the Reynolds number. For example, in a pipe-flow, for the values of $Re < 2300$ the flow is laminar. If it goes beyond, the flow becomes turbulent. If compared to the flow along a flat plate, the only major difference in pipe flow is that there is a limit to the growth of the boundary layer thickness because of the pipe radius.

Table 1.6 The values of laminar boundary-layer thicknesses for a laminar flow over a flat plate.

η	0.0	0.6	1.2	1.8	2.4	3.0	3.6	4.2	4.8	5.4	6.0
δ	0.000	0.200	0.394	0.575	0.729	0.846	0.924	0.967	0.988	0.996	0.999

Source: Olson and Wright (1990).

Table 1.7 Momentum equations for laminar boundary layer.

Velocity Profile	δ/x	C_f	δ^*/x
$u/u_s = y/\delta$	$3.46/Re_x^{1/2}$	$1.156/Re_x^{1/2}$	$1.73/Re_x^{1/2}$
$u/u_s = 2(y/\delta) - (y/\delta)^2$	$5.48/Re_x^{1/2}$	$1.462/Re_x^{1/2}$	$1.83/Re_x^{1/2}$
$u/u_s = 1.5(y/\delta) - 0.5(y/\delta)^3$	$4.64/Re_x^{1/2}$	$1.292/Re_x^{1/2}$	$1.74/Re_x^{1/2}$
$u/u_s = \sin \pi y/2\delta$	$4.80/Re_x^{1/2}$	$1.310/Re_x^{1/2}$	$1.74/Re_x^{1/2}$
Blasius Exact solution	$4.91/Re_x^{1/2}$	$1.328/Re_x^{1/2}$	$1.73/Re_x^{1/2}$

Source: Olson and Wright (1990).

Table 1.8 Momentum equations for turbulent boundary layer on a flat plate flow and in a pipe flow.

Re_D	f	u/u_s	V/u_s	C_f	Re_x
$<10^5$	$0.316/Re_D^{1/4}$	$(y/R)^{1/7}$	49/60	$0.074/Re_x^{1/5}$	5×10^5-10^7
10^4-10^6	$0.180/Re_D^{1/5}$	$(y/R)^{1/8}$	128/153	$0.045/Re_x^{1/6}$	1.8×10^5-4.5×10^7
10^5-10^7	$0.117/Re_D^{1/6}$	$(y/R)^{1/10}$	200/231	$0.0305/Re_x^{1/7}$	2.9×10^6-5×10^8

Source: Olson and Wright (1990).

Many empirical pipe-flow equations have been developed, particularly for water. The velocity V and volumetric flow rate $\dot{V}$ equations of Hazen-Williams are the most widely used ones, as follows:

$$V = 0.850 C R_h^{0.63} S^{0.54} \quad (1.81)$$

$$\dot{V} = 0.850 C R_h^{0.63} S^{0.54} A \quad (1.82)$$

where R_h is the hydraulic radius of the pipe, P is wetted perimeter (A/P, for example, $R_h = D/4$ for a round pipe), S is the slope of the total head line, h/L, A is the pipe cross-sectional area, and C is the roughness coefficient. C takes different values for the pipes, for example, $C = 140$ for very badly corroded iron or steel pipes.

1.6 General Aspects of Heat Transfer

Thermal processes involving the transfer of heat from one point to another are often encountered in the food industry, as in other industries. The heating and cooling of liquid or solid food products, the evaporation of water vapors, and the removal of heat liberated by a chemical reaction are common examples of processes that involve heat transfer. It is of great importance for food technologists, refrigeration engineers, researchers, etc. to understand the physical phenomena and practical aspects of heat transfer, along with a knowledge of the basic laws, governing equations, and related boundary conditions.

In order to transfer heat, there must be a driving force, which is the temperature difference between the points where heat is taken and where the heat originates. For example, consider that when a long slab of food product is subjected to heating on the left side, the heat flows from the left-hand side to the right-hand side, which is colder. It is said that heat tends to flow from a point of high temperature to a point of low temperature, owing to the temperature difference driving force.

Many of the generalized relationships used in heat transfer calculations have been determined by means of dimensional analysis and empirical considerations. It has been found that certain standard dimensionless groups appear repeatedly in the final equations. It is necessary for people working in the food cooling industry to recognize the more important of these groups. Some of the most commonly used dimensionless groups that appear frequently in the heat transfer literature are given in Table 1.9.

In the utilization of these groups, care must be taken to use equivalent units so that all the dimensions cancel out. Any system of units may be used in a dimensionless group as long as the final result will permit all units to disappear by cancellation.

Basically, heat is transferred in three ways: conduction, convection, and radiation (the so-called modes of heat transfer). In many cases, heat transfer takes place by all three of these methods simultaneously. Figure 1.14 shows the different types of heat transfer processes as modes. When a temperature gradient exists in a stationary medium, which may be a solid or a fluid, the heat transfer occurring across the medium is by conduction, the heat transfer occurring between a surface and a moving fluid at different temperatures is by convection, and the heat transfer occurring between two surfaces at different temperatures, in the absence of an intervening medium, is by radiation, where all surfaces of finite temperature emit energy in the form of electromagnetic waves.

Table 1.9 Some of the most important heat transfer dimensionless parameters.

Name	Symbol	Definition	Application
Biot number	Bi	hY/k	Steady- and unsteady-state conduction
Fourier number	Fo	at/Y^2	Unsteady-state conduction
Graetz number	Gz	GY^2c_p/k	Laminar convection
Grashof number	Gr	$g\beta\Delta TY^3/v^2$	Natural convection
Rayleigh number	Ra	$Gr \times Pr$	Natural convection
Nusselt number	Nu	hY/k_f	Natural or forced convection, boiling, or condensation
Peclet number	Pe	$UY/a = Re \times Pr$	Forced convection (for small Pr)
Prandtl number	Pr	$c_p\mu/k = v/a$	Natural or forced convection, boiling, or condensation
Reynolds number	Re	UY/v	Forced convection
Stanton number	St	$h/\rho Uc_p = Nu/Re\ Pr$	Forced convection

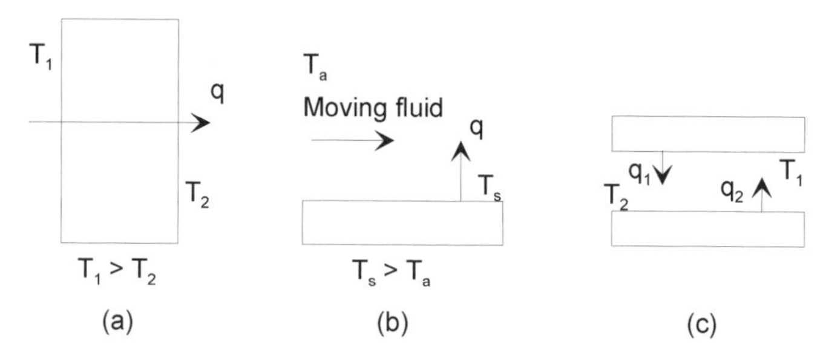

Figure 1.14 Schematic representations of heat transfer modes. (a) Conduction through a solid. (b) Convection from a surface to a moving fluid. (c) Radiation between two surfaces.

1.6.1 Conduction heat transfer

Conduction is a mode of transfer of heat from one part of a material to another part of the same material, or from one material to another in physical contact with it, without appreciable displacement of the molecules forming the substance. For example, the heat transfer in a solid object subject to cooling in a medium is by conduction. In solid objects, the conduction of heat is partly due to the impact of adjacent molecules vibrating about their mean positions and partly due to internal radiation. When the solid object is a metal, there are also large numbers of mobile electrons which can easily move through the matter, passing from one atom to another, and they contribute to the redistribution of energy in the metal object. Actually, the contribution of the mobile electrons predominates in metals, which explains the relation that is found to exist between the thermal and electrical conductivity of such materials.

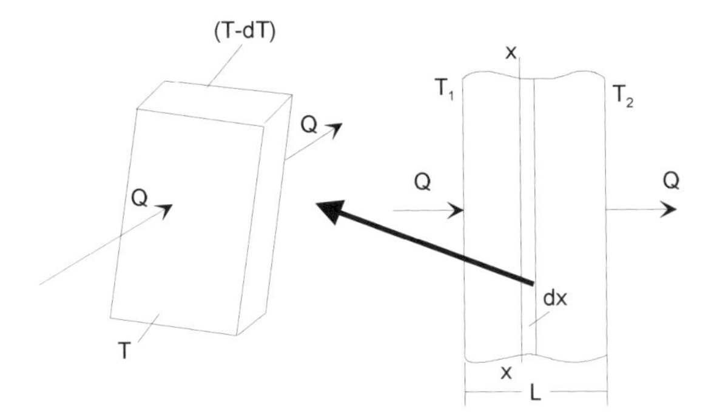

Figure 1.15 Schematic illustration of conduction in a slab object.

Fourier's law of heat conduction

Fourier's law states that the instantaneous rate of heat flow through an individual homogeneous solid object is directly proportional to the cross-sectional area A (i.e. the area at right angles to the direction of heat flow) and to the temperature difference driving force across the object with respect to the length of the path of the heat flow, dT/dx. This is an empirical law based on observation.

Figure 1.15 presents an illustration of Fourier's law of heat conduction. Here, a thin slab object of thickness dx and surface area F has one face at a temperature T and the other at a lower temperature $(T - dT)$, where heat flows from the high-temperature side to the low-temperature side, with a temperature change in the direction of the heat flow dT. Therefore, under Fourier's law the heat transfer equation results in

$$Q = -kA\frac{dT}{dx} \tag{1.83}$$

Here, we have a term *thermal conductivity, k,* of the object that can be defined as the heat flow per unit area per unit time when the temperature decreases by one degree in unit distance. Its units are usually written as W/m°C or W/mK.

Integrating Equation 1.83 from T_1 to T_2 for dT and 0 to L for dx, the solution becomes

$$Q = -k\frac{A}{L}(T_2 - T_1) = k\frac{A}{L}(T_1 - T_2) \tag{1.84}$$

Equation 1.84 can be solved when the variation of thermal conductivity with temperature is known. For most solids, thermal conductivity values are approximately constant over a broad range of temperatures, and can be taken as constants.

1.6.2 Convection heat transfer

Convection is the heat transfer mode that takes place within a fluid by mixing one portion of the fluid with another. Convection heat transfer may be classified according to the nature of the flow. When the flow is caused by some mechanical or external means such as a fan, a pump, or atmospheric wind, it is called *forced convection*. On the other hand, for *natural (free) convection* the flow is induced by buoyancy forces in the fluid that arise from density variations caused by temperature variations in the fluid. For example, when a hot object is exposed to the atmosphere, natural convection occurs, whereas in a cold place forced-convection heat transfer takes place between air flow and the object subject to this flow. The heat transfer of heat through solid objects is by conduction alone, whereas the heat transfer from a solid surface to a liquid or gas takes place partly by conduction and partly by convection. Whenever there is an appreciable movement of the gas or liquid, the heat transfer by conduction in the gas or liquid becomes negligibly small compared with the heat transfer by convection. However, there is always a thin boundary layer of liquid on a surface, and through this thin film the heat is transferred by conduction. The convection heat transfer occurring within a fluid is due to the combined effects of conduction and bulk fluid motion. Generally the heat that is transferred is the *sensible*, or internal thermal, heat of the fluid. However, there are convection processes for which there is also *latent* heat exchange, which is generally associated with a phase change between the liquid and vapor states of the fluid.

Newton's law of cooling

Newton's law of cooling states that the heat transfer from a solid surface to a fluid is proportional to the difference between the surface and fluid temperatures and the surface area. This is particular nature of the convection heat transfer mode, and is defined as

$$Q = hA(T_s - T_f) \tag{1.85}$$

where h is referred to as the *convection heat transfer coefficient* (the *heat transfer coefficient*, the *film coefficient*, or the *film conductance*). It encompasses all the effects that influence the convection mode, and depends on conditions in the boundary layer, which is affected by factors such as surface geometry, the nature of the fluid motion, and the thermal and physical properties (Figure 1.16).

In Equation 1.85, a radiation term is not included. The calculation of radiation heat transfer will be discussed later. In many heat transfer problems, the radiation effect on the total heat transfer is negligible compared with the heat transferred by conduction and convection from the surface to the fluid. When the surface temperature is high, or when the surface loses heat by natural convection, then the heat transfer due to radiation is of a similar magnitude to that lost by convection.

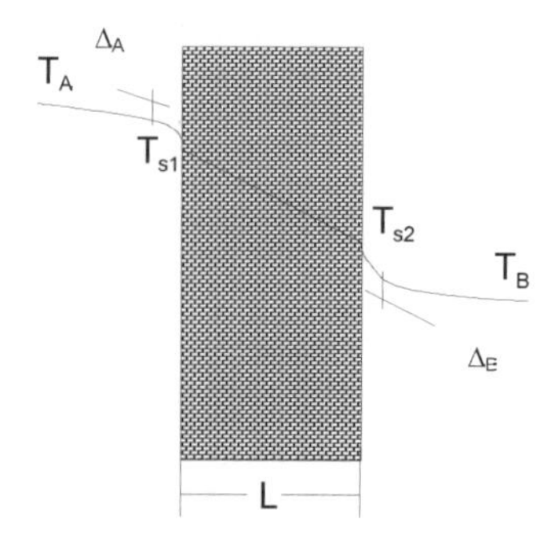

Figure 1.16 A wall subject to convection heat transfer from both sides.

To better understand Newton's law of cooling, consider the heat transfer from a high-temperature fluid A to a low-temperature fluid B through a wall of thickness x (Figure 1.16). In fluid A the temperature decreases rapidly from T_A to T_{sl} in the region of the wall, and similarly in fluid B from T_{s2} to T_B. In most cases the fluid temperature is approximately constant throughout its bulk, apart from a thin film (Δ_A or Δ_B) near the solid surface bounding the fluid. The heat transfer per unit surface area from fluid A to the wall and that from the wall to fluid B are:

$$q = h_A(T_A - T_{s1}) \tag{1.86}$$

$$q = h_B(T_{s2} - T_B) \tag{1.87}$$

Also, the heat transfer in thin films is by conduction only as follows:

$$q = \frac{k_A}{\Delta_A}(T_A - T_{s1}) \tag{1.88}$$

$$q = \frac{h_B}{\Delta_B}(T_{s2} - T_B) \tag{1.89}$$

Equating Equations 1.86–1.89, the convection heat transfer coefficients can be found to be $h_A = k_A/\Delta_A$, and $h_B = k_B/\Delta_B$. Thus, the heat transfer in the wall per unit surface area becomes

$$q = \frac{k}{L}(T_{s1} - T_{s2}) \tag{1.90}$$

For a steady-state heat transfer case, Equation 1.86 is equal to Equation 1.87, and hence to Equation 1.90:

$$q = h_A(T_A - T_{s1}) = h_B(T_{s2} - T_B) = \frac{k}{L}(T_{s1} - T_{s2}) \tag{1.91}$$

which yields

$$q = \frac{(T_A - T_B)}{(1/h_A + L/k + 1/h_B)} \tag{1.92}$$

An analogy can be made with Equation 1.85, and Equation 1.92 becomes

$$Q = HA(T_A - T_B) \tag{1.93}$$

where $1/H = [(1/h_A) + (L/k) + (1/h_B)]$. H is the overall heat transfer coefficient and consists of various heat transfer coefficients.

1.6.3 *Radiation heat transfer*

An object emits radiant energy in all directions unless its temperature is absolute zero. If this energy strikes a receiver, part of it may be absorbed and part may be reflected. Heat transfer from a hot to a cold object in this manner is known as radiation heat transfer. It is clear that the higher the temperature, the greater is the amount of energy radiated. If, therefore, two objects at different temperatures are placed so that the radiation from each object is intercepted by the other, then the body at the lower temperature will receive more energy than it radiates, and thereby its internal energy will increase; in conjunction with this the internal energy of the object at the higher temperature will decrease. Radiation heat transfer frequently occurs between solid surfaces, although radiation from gases also takes place. Certain gases emit and absorb radiation at certain wavelengths only, whereas most solids radiate over a wide range of wavelengths. The radiative properties of some gases and solids may be found in heat transfer books. Radiation striking an object can be absorbed by the object, reflected from the object, or transmitted through the object. The fractions of the radiation absorbed, reflected, and transmitted are called the *absorptivity* a, the *reflectivity* r, and the *transmittivity* t, respectively. By definition, $a + r + t = 1$. For most solids and liquids in practical applications, the transmitted radiation is negligible, and hence $a + r = 1$. A body which absorbs all radiation is called a *blackbody*. For a blackbody $a = 1$ and $r = 0$.

The Stefan-Boltzman law

This law was found experimentally by Stefan, and proved theoretically by Boltzmann. It is that the emissive power of a blackbody is directly proportional to the fourth power of its absolute temperature. The Stefan-Boltzmann law enables calculation of the amount of radiation emitted in all directions and over all wavelengths simply from knowledge of the temperature of the blackbody. This law is given as follows:

$$E_b = \sigma T_s^4 \tag{1.94}$$

where σ stands for the Stefan-Boltzmann constant, and its value is 5.669×10^{-8} W/m^{2}K^4. T_s stands for the absolute temperature of the surface.

The energy emitted by a non-blackbody becomes

$$E_{nb} = \varepsilon\sigma T_s^4 \tag{1.95}$$

Then the heat transferred from an object's surface to its surroundings per unit area is

$$q = \varepsilon\sigma(T_s^4 - T_a^4) \tag{1.96}$$

It is important to explain that if the emissivity of the object at T_s is much different from the emissivity of the object at T_a, then the gray object approximation may not be sufficiently accurate. In this case, it is a good approximation to take the absorptivity of object 1 when receiving radiation from a source at T_a as being equal to the emissivity of object 1 when emitting radiation at T_a. This results in

$$q = \varepsilon_{Ts}\sigma T_s^4 - \varepsilon_{Ta}\sigma T_a^4 \tag{1.97}$$

There are numerous applications for which it is convenient to express the net radiation heat transfer (radiation heat exchange) in the following form:

$$Q = h_r A(T_s - T_a) \tag{1.98}$$

After combining Equations 1.97 and 1.98, the radiation heat transfer coefficient can be found as follows:

$$h_r = \varepsilon\sigma(T_s + T_a)(T_s^2 + T_a^2) \tag{1.99}$$

Here, the radiation heat transfer coefficient depends strongly on temperature, whereas the temperature dependence of the convection heat transfer coefficient is generally weak.

The surface within the surroundings may also simultaneously transfer heat by convection to the surroundings. The total rate of heat transfer from the surface is the sum of the convection and radiation modes:

$$Q_t = Q_c + Q_r = h_c A(T_s - T_a) + \varepsilon\sigma A(T_s^4 - T_a^4) \tag{1.100}$$

1.6.4 *Thermal resistance*

There is a similarity between heat flow and electricity flow. At this point, electrical resistance may be associated with the conduction of electricity and thermal resistance may be associated with the conduction of heat. The temperature difference providing the heat conduction plays a role analogous to that of the potential difference or voltage in the conduction of electricity. Below we give the *thermal resistance for heat conduction*, based on Equation 1.84, and similarly the *electrical resistance for electrical conduction* according to Ohm's law:

$$R_{t,cd} \equiv \frac{(T_1 - T_2)}{Q_{cd}} = \frac{L}{kA} \tag{1.101}$$

$$R_e \equiv \frac{(E_1 - E_2)}{I} = \frac{L}{\sigma A} \tag{1.102}$$

It is also possible to write the *thermal resistance for convection*, based on Equation 1.85, as follows:

$$R_{t,c} \equiv \frac{(T_s - T_f)}{Q_c} = \frac{1}{hA} \tag{1.103}$$

In a series, total thermal resistance can be written in terms of the overall heat transfer coefficient. This is expressed for a composite wall in the following.

1.6.5 *The composite wall*

In practice, there are many cases in the form of a composite wall, for example, the wall of a cold store. Consider that we have a general form of the composite wall as shown in Figure

1.17. Such a system includes any number of series and parallel thermal resistances owing to layers of different materials. The heat transfer rate is related to the temperature difference and resistance associated with each element as follows:

$$Q = \frac{(T_A - T_1)}{(1/h_1 A)} = \frac{(T_1 - T_2)}{(L_1 / k_1 A)} = ... = \frac{(T_n - T_B)}{(1/h_n A)} \tag{1.104}$$

Therefore, the one-dimensional heat transfer rate for this system can be written as

$$Q = \frac{(T_A - T_B)}{\Sigma R_t} = \frac{\Delta T}{\Sigma R_t} \tag{1.105}$$

where $\Sigma R_t = R_{t,t} = 1/HA$. Therefore, the overall heat transfer coefficient becomes

$$H = \frac{1}{R_{t,t} A} = \frac{1}{(1/h_1 + L_1/k_1 + ... + 1/h_n)} \tag{1.106}$$

1.6.6 The cylinder

A practical common object is a hollow cylinder, and one of the most commonly met problems is the case of heat transferred through a pipe or cylinder. Consider that we have a cylinder of internal radius r_1 and external radius r_2, whose inner and outer surfaces are subjected to fluids at different temperatures (Figure 1.18). In a steady-state form with no heat generation, the governing heat conduction equation is written as

$$\frac{1}{r}\frac{d}{dr}\left(kr\frac{dT}{dr}\right) = 0 \tag{1.107}$$

Based on Fourier's law, the rate at which heat is transferred by conduction across the cylindrical surface in the solid is expressed as

$$Q = -kA\frac{dT}{dr} = -k(2\pi rL)\frac{dT}{dr} \tag{1.108}$$

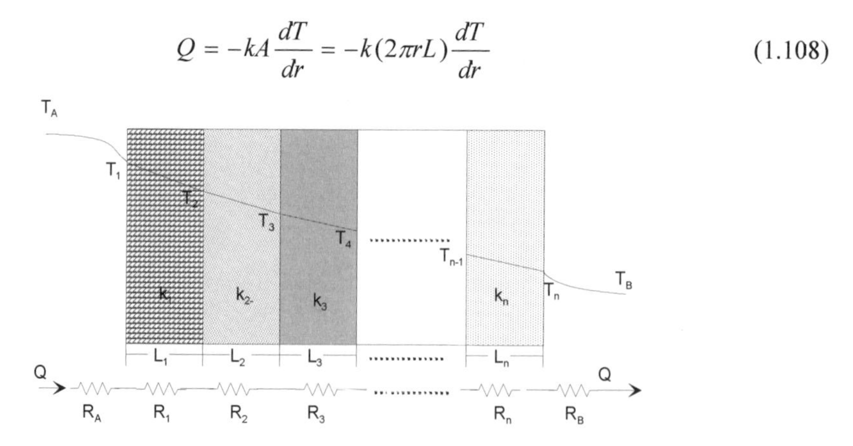

Figure 1.17 A series of a composite wall.

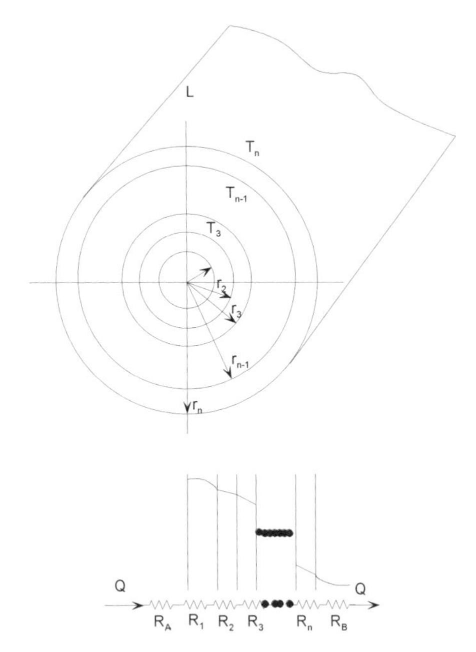

Figure 1.18 A hollow cylinder.

where $A = 2\pi rL$ is the area normal to the direction of heat transfer.

To determine the temperature distribution in the cylinder, it is necessary to solve Equation 1.107 under appropriate boundary conditions, by assuming that k is constant. By integrating Equation 1.107 twice, the following heat transfer equation is obtained:

$$Q = \frac{k(2\pi L)(T_1 - T_2)}{\ln(r_1 / r_2)} = \frac{(T_1 - T_2)}{R_t} \tag{1.109}$$

If we now consider a composite hollow cylinder, the heat transfer equation is found as follows, by neglecting the interfacial contact resistances:

$$Q = \frac{(T_1 - T_n)}{R_{t,t}} = HA(T_1 - T_n) \tag{1.110}$$

where $R_{t,t} = [(1/2\pi r_1 Lh_1) + (\ln(r_2/r_1)/2\pi k_1 L) + (\ln(r_3/r_2)/2\pi k_2 L) + \cdots + (1/2\pi r_n Lh_n)]$.

1.6.7 The sphere

The case of heat transferred through a sphere is not as common as the cylinder problem. Now consider analyzing conduction in a hollow sphere of internal radius r_1 and external radius r_2 (Figure 1.19). Also, consider the inside and outside temperatures of T_1 and T_2, and constant thermal conductivity with no heat generation. We can express the following heat conduction in the form of Fourier's law:

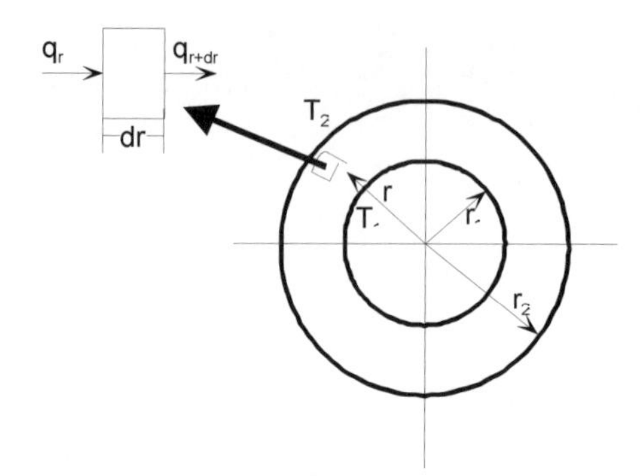

Figure 1.19 Heat conduction in a hollow sphere.

$$Q = -kA\frac{dT}{dr} = -k(4\pi r^2)\frac{dT}{dr} \tag{1.111}$$

where $A = 4\pi r^2$ is the area normal to the direction of heat transfer.

After integrating Equation 1.111, we reach the following expression:

$$Q = \frac{k(4\pi)(T_1 - T_2)}{(1/r_2 - 1/r_1)} = \frac{k(4\pi r_1 r_2)(T_1 - T_2)}{(r_2 - r_1)} = \frac{(T_1 - T_2)}{R_t} \tag{1.112}$$

If we now consider a composite hollow sphere, the heat transfer equation is found as follows, by neglecting the interfacial contact resistances:

$$Q = \frac{(T_1 - T_n)}{R_{t,t}} = HA(T_1 - T_n) \tag{1.113}$$

where $R_{t,t} = [(1/4\pi r_1^2 h_1) + (r_2 - r_1)/(4\pi r_1 r_2 k_1) + (r_3 - r_2)/(4\pi r_2 r_3 k_2) + \cdots + (1/4\pi r_2^2 h_2)]$.

1.6.8 Conduction with heat generation

The plane wall

Consider a plane wall, as shown in Figure 1.20a, in which there is uniform heat generation per unit volume. The heat conduction equation becomes

$$\frac{d^2T}{dx^2} + \frac{q_h}{k} = 0 \tag{1.114}$$

By integrating Equation 1.114 with the prescribed boundary conditions, $T(-L) = T_1$ and $T(L) = T_2$. The temperature distribution can be obtained as

$$T(x) = \left(\frac{q_h L^2}{2k}\right)\left(1 - \frac{x^2}{L^2}\right) + \left(\frac{T_2 - T_1}{2}\right)\left(\frac{x}{L}\right) + \left(\frac{T_2 + T_1}{2}\right) \tag{1.115}$$

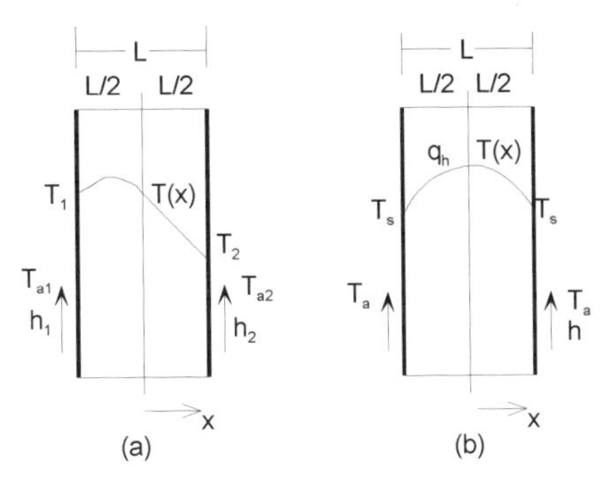

Figure 1.20 Heat conduction in a slab with uniform heat generation. (a) Asymmetrical boundary conditions. (b) Symmetrical boundary conditions.

The heat flux at any point in the wall can be found, depending on x, by using Equation 1.115 with Fourier's law.

If $T_1 = T_2 \equiv T_s$, the temperature distribution is then symmetrical about the midplane (Figure 1.20b), and results in

$$T(x) = \left(\frac{q_h L^2}{2k}\right)\left(1 - \frac{x^2}{L^2}\right) + T_s \tag{1.116}$$

At the plane of symmetry $dT/dx = 0$, and the maximum temperature at the midplane is

$$T(0) \equiv T_m = \left(\frac{q_h L^2}{2k}\right) + T_s \tag{1.117}$$

After combining Equations 1.116 and 1.117 we find the dimensionless temperature as follows:

$$\frac{(T(x) - T_m)}{(T_s - T_m)} = \left(\frac{x}{L}\right)^2 \tag{1.118}$$

The cylinder

Consider a long cylinder (i.e. Figure 1.18) with uniform heat generation. The heat conduction equation can be rewritten as

$$\frac{1}{r}\frac{d}{dr}\left(r\frac{dT}{dr}\right) + \frac{q_h}{k} = 0 \tag{1.119}$$

By integrating Equation 1.119, with the boundary conditions, $dT/dr = 0$ for the centerline ($r = 0$) and $T(r_1) = T_s$, the temperature distribution can be obtained as

$$T(r) = \left(\frac{q_h r_1^2}{4k}\right)\left(1 - \frac{r_2^2}{r_1^2}\right) + T_s \tag{1.120}$$

After making required combinations, the dimensionless temperature equation results in

$$\frac{(T(r)-T_m)}{(T_s-T_m)} = 1-\left(\frac{r_2}{r_1}\right)^2 \tag{1.121}$$

The approach mentioned previously can also be used in obtaining the temperature distributions in solid spheres and spherical shells for a wide range of boundary conditions.

1.6.9 Natural convection

As mentioned previously, heat transfer by natural (or free) convection involving motion in a fluid is due to differences in density and the action of gravity, and this causes a natural circulation and hence a heat transfer. For many problems involving a fluid across a surface, the superimposed effect of natural convection is negligibly small. It is obvious that for natural convection the heat transfer coefficients are generally much lower than for forced convection. When there is no forced velocity of the fluid, the heat is transferred entirely by natural convection (when there is negligible radiation). For some practical cases, it is necessary to consider the radiative effect on the total heat loss or gain. Radiation heat transfer may be of the same order of magnitude as natural convection, even at room temperatures, since wall temperatures in a room can affect human comfort.

It is important to mention that in many systems involving multimode heat transfer effects, natural convection provides the largest resistance to heat transfer, and therefore plays an important role in the design or performance of the system. Moreover, when it is desirable to minimize the heat transfer rates or to minimize operating costs, natural convection is often preferred to forced convection.

Natural convection is of significance in a wide variety of heating, cooling and air-conditioning equipment. Natural convection heat transfer is influenced mainly by gravitational force from thermal expansion, viscous drag, and thermal diffusion. For this reason, the gravitational acceleration, the coefficient of performance, the kinematic viscosity, and the thermal diffusivity directly affect natural convection. As shown in Table 1.8, these parameters depending on the fluid properties, the temperature difference between the surface and the fluid, and the characteristic length of the surface are involved in the Nusselt, Grashof, and Prandtl equations.

The natural convection boundary layers are not restricted to laminar flow. In many cases, there is a transition from laminar to turbulent flow. This is schematically shown in Figure 1.21 for a heated vertical plate.

Transition in a natural convection boundary layer is dependent on the relative magnitude of the buoyancy and viscous forces in the fluid. It is customary to correlate its occurrence in terms of the Rayleigh number. For example, for vertical plates the critical Rayleigh number is $Ra \approx 10^9$. As in forced convection, transition to turbulence has a strong effect on the heat transfer. Numerous natural convection heat transfer correlations for several plates, pipes, wires, cylinder, etc. along with a list of heat transfer coefficients, which were compiled from literature, are given in Table 1.10. To calculate the natural convection heat transfer coefficient, determine the Rayleigh number to find whether the boundary layer is laminar or turbulent, then apply the appropriate equation from this table.

Table 1.10 Natural convection heat transfer equations and correlations (see Dincer, 1997).

Equation or correlation

- *General equations*

$Nu = hY/k_f = cRa^n$ and $Ra = Gr\,Pr = g\beta(T_s - T_a)Y^3/\nu a$

where n is 1/4 for laminar flow and 1/3 for turbulent flow. Y is height for vertical plates or pipes, diameter for horizontal pipes and radius for spheres. $T_{fm} \equiv (T_s + T_a)/2$.

- *Correlations for vertical plates (or inclined plates, inclined up to 60°)*

$Nu = [0.825 + 0.387Ra^{1/6}/(1 + (0.492/Pr)^{9/16})^{4/9}]^2$ for an entire range of Ra

$Nu = 0.68 + 0.67Ra^{1/4}/(1 + (0.492/Pr)^{9/16})^{4/9}$ for $0 < Ra < 10^9$

- *Correlations for horizontal plates ($Y \equiv A_s/P$)*

For upper surface of heated plate or lower surface of cooled plate:

$Nu = 0.54Ra^{1/4}$ for $10^4 \le Ra \le 10^7$

$Nu = 0.15Ra^{1/3}$ for $10^7 \le Ra \le 10^{11}$

For lower surface of heated plate or upper surface of cooled plate:

$Nu = 0.27Ra^{1/4}$ for $10^5 \le Ra \le 10^{10}$

- Correlations for horizontal cylinders

$Nu = hD/k = cRa^n$

where

$c = 0.675$ and $n = 0.058$ for $10^{-10} < Ra < 10^{-2}$

$c = 1.020$ and $n = 0.148$ for $10^{-2} < Ra < 10^2$

$c = 0.850$ and $n = 0.188$ for $10^2 < Ra < 10^4$

$c = 0.480$ and $n = 0.250$ for $10^4 < Ra < 10^7$

$c = 0.125$ and $n = 0.333$ for $10^7 < Ra < 10^{12}$

$Nu = [0.60 + 0.387Ra^{1/6}/(1 + (0.559/Pr)^{9/16})^{8/27}]^2$ for an entire range of Ra

- *Correlations for spheres*

$Nu = 2 + 0.589Ra^{1/4}/(1 + (0.469/Pr)^{9/16})^{4/9}$ for $Pr \ge 0.7$ and $Ra \le 10^{11}$

- *Heat transfer correlations*

$Gr\,Pr = 1.6\times10^6 Y^3(\Delta T)$

where Y in m; ΔT in °C.

$h = 0.29\,(\Delta T/Y)^{1/4}$ for vertical small plates in laminar range,

$h = 0.19\,(\Delta T)^{1/3}$ for vertical large plates in turbulent range,

$h = 0.27\,(\Delta T/Y)^{1/4}$ for horizontal small plates in laminar range (facing upward when heated or downward when cooled),

$h = 0.22\,(\Delta T)^{1/3}$ for vertical large plates in turbulent range (facing downward when heated or upward when cooled),

$h = 0.27\,(\Delta T/Y)^{1/4}$ for small cylinders in laminar range,

$h = 0.18\,(\Delta T)^{1/3}$ for large cylinders in turbulent range.

Table 1.11 Forced convection heat transfer equations and correlations (see Dincer, 1997).

Equation or correlation
• *Correlations for flat plate in external flow*
$Nu = 0.332Re^{1/2}Pr^{1/3}$ for $Pr \geq 0.6$ for Laminar; local; T_{fm}.
$Nu = 0.664Re^{1/2}Pr^{1/3}$ for $Pr \geq 0.6$ for Laminar; average; T_{fm}.
$Nu = 0.565Re^{1/2}Pr^{1/2}$ for $Pr \leq 0.05$ for Laminar; local; T_{fm}.
$Nu = 0.0296Re^{4/5}Pr^{1/3}$ for $0.6 \leq Pr \leq 60$ for Turbulent; local; T_{fm}, $Re \leq 10^8$.
$Nu = (0.037Re^{4/5} - 871)Pr^{1/3}$ for $0.6 < Pr < 60$ for Mixed flow; average; T_{fm}, $Re \leq 10^8$.
• *Correlations for circular cylinders in cross-flow*
$Nu = cRe^nPr^{1/3}$ for $Pr \geq 0.7$ for Average; T_{fm}; $0.4 < Re < 4\times10^6$.
where
$c = 0.989$ and $n = 0.330$ for $0.4 < Re < 4$
$c = 0.911$ and $n = 0.385$ for $4 < Re < 40$
$c = 0.683$ and $n = 0.466$ for $40 < Re < 4000$
$c = 0.193$ and $n = 0.618$ for $4000 < Re < 40000$
$c = 0.027$ and $n = 0.805$ for $40000 < Re < 400000$
$Nu = cRe^nPr^s(Pr_d/Pr_s)^{1/4}$ for $0.7 < Pr < 500$ for Average; T_a; $1 < Re < 10^6$.
where
$c = 0.750$ and $n = 0.4$ for $1 < \mathrm{Re} < 40$
$c = 0.510$ and $n = 0.5$ for $40 < \mathrm{Re} < 1000$
$c = 0.260$ and $n = 0.6$ for $10^3 < \mathrm{Re} < 2\times10^5$
$c = 0.076$ and $n = 0.7$ for $2\times10^5 < \mathrm{Re} < 10^6$
$s = 0.37$ for $Pr \leq 10$
$s = 0.36$ for $Pr > 10$
$Nu = 0.3 + [(0.62Re^{1/2}Pr^{1/3})/(1 + (0.4/Pr)^{2/3})^{1/4}][1 + (Re/28200)^{5/8}]^{4/5}$ for $RePr > 0.2$ for Average; T_{fm}.
• *Correlations for spheres in cross-flow*
$Nu/Pr^{1/3} = 0.37Re^{0.6}/Pr^{1/3}$ for Average; T_{fm}; $17 < Re < 70000$.
$Nu = 2 + (0.4Re^{1/2} + 0.06Re^{2/3})Pr^{0.4}(\mu_d/\mu_s)^{1/4}$ for $0.71 < Pr < 380$ for Average; T_a; $3.5 < Re < 7.6\times10^4$; $1 < (\mu_d/\mu_s) < 3.2$.
• *Correlation for falling drop*
$Nu = 2 + 0.6Re^{1/2}Pr^{1/3}[25(x/D)^{-0.7}]$ for Average; T_a.

1.6.10 Forced convection

The study of forced convection is concerned with the heat transfer occurring between a moving fluid and a solid surface. To apply Newton's law of cooling as given in Equation

1.85, it is necessary to determine the heat transfer coefficient. For this purpose, the Nusselt-Reynolds correlations may be used. The definitions of the Nusselt and Reynolds numbers have already been given in Table 1.9. Forced air and water coolers, forced air and water evaporators and condensers, and heat exchangers are examples of the equipment involved in forced convection heat transfer.

The various kinds of forced convection, such as flow in a tube, flow across a tube, and flow across a flat plate may be solved mathematically when certain assumptions are made with regard to the boundary conditions. It is extremely difficult to obtain an exact solution to such problems, especially in the event of turbulent flow, but for approximate solutions appropriate assumptions are necessary.

The essential first step in the solution of a convection heat transfer problem is to determine whether the boundary layer is *laminar* or *turbulent*. These conditions affect the convection heat transfer coefficient and hence convection heat transfer rates.

The conditions of laminar and turbulent flows on a flat plate are shown in Figure 1.13. In the laminar boundary layer, fluid motion is highly ordered and it is possible to identify streamlines along which particles move. On the other hand, fluid motion in the turbulent boundary layer is highly irregular, and is characterized by velocity fluctuations which begin to develop in the transition region (after this the boundary layer becomes completely turbulent). These fluctuations enhance the transfer of momentum, heat, and species, and hence increase surface friction as well as convection transfer rates. In the laminar sublayer which is nearly linear, transport is dominated by diffusion and the velocity profile. There is an adjoining buffer layer in which diffusion and turbulent mixing are comparable. In the turbulent region, transport is dominated by turbulent mixing.

Here, it is useful to mention that the *critical Reynolds number* is the value of *Re* for which transition begins, and for external flow it is known to vary from 10^5 to 3×10^6, depending on the surface roughness, the turbulence level of the free stream, and the nature of the pressure variation along the surface. A representative value of *Re* is generally assumed for boundary layer calculations:

$$\mathrm{Re}_c = \frac{\rho U_a X_c}{\mu} = \frac{U_a X_c}{\upsilon} = 5\times10^5 \tag{1.122}$$

Another point is that for smooth circular tubes, when the Reynolds number is less than 2100, the flow is laminar, and when it is greater than 10 000, the flow is turbulent. The range between these values represents the transition region.

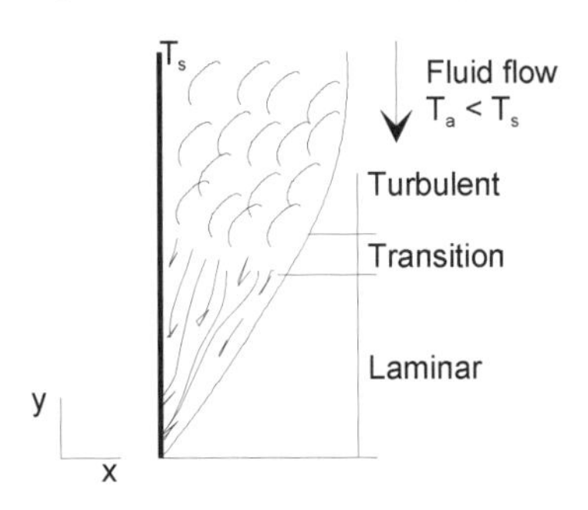

Figure 1.21 Natural convection on a vertical plate.

We give a list of various forced convection heat transfer correlations (the Nusselt-Reynolds correlations), direct convection heat transfer coefficient equations, along with the relevant parameters and remarks, which were compiled from literature, in Table 1.11. In many of these equations the *film temperature* is used, and is defined as $T_{fm} = (T_s + T_a)/2$.

1.7 Concluding Remarks

In this chapter, a summary of some general introductory aspects of thermodynamics, fluid flow and heat transfer is presented, and their fundamental definitions and physical quantities are given to provide sufficient thermal sciences background for a better understanding and energy and exergy analysis of thermal energy storage systems and applications, and their operations.

Nomenclature

a	acceleration, m/s^2; thermal diffusivity, m^2/s; absorptivity
A	cross-sectional area, m^2; surface area, m^2
Bi	Biot number
c	mass fraction; constant in Tables 1.10 and 1.11
c_p	constant-pressure specific heat, kJ/kgK
c_v	constant-volume specific heat, kJ/kgK
C_f	average skin friction coefficient
d	diameter, m; depth normal to flow, m
D	diameter, m
E	energy, W or kW; electric potential, V; constant
F	force; drag force, N
Fo	Fourier number
g	acceleration due to gravity (= 9.81 m/s^2)
G	mass flow velocity, kg/sm^2
Gz	Graetz number
Gr	Grashof number
h	specific enthalpy, kJ/kg; heat transfer coefficient, W/m^2°C; head, m
H	entalpy, kJ; overall heat transfer coefficient, W/m^2°C; head, m
I	electric current, A
k	thermal conductivity, W/m°C
K	adiabatic modules
KE	kinetic energy, W or kW
L	thickness, m
m	mass, kg; constant
$\dot{m}$	mass flow rate, kg/s
M	molecular weight, kg/kmol
n	mole number, kmol; constant exponent in Tables 1.9 and 1.10
Nu	Nusselt number
P	perimeter, m; pressure, Pa or kPa

P^*	constant-pressure gradient, Pa or kPa
Pe	Peclet number
PE	potential energy, W or kW
Pr	Prandtl number
q	heat rate per unit area, W/m^2; flow rate per unit width or depth
q_h	heat generation per unit volume, W/m^3
Q	heat transfer, W or kW
$\dot{Q}$	heat transfer rate, W or kW
r	reflectivity; radial coordinate; radial distance, m
R	gas constant, kJ/kgK; radius, m
$\bar{R}$	universal gas constant, kJ/kgK
R_t	thermal resistance, °C/W
Ra	Rayleigh number
Re	Reynolds number
s	specific entropy, kJ/kg; streamline direction; distance, m; constant exponent in Tables 1.10 and 1.11
S	entropy, kJ
St	Stanton number
t	time, s; transmittivity
T	temperature, °C or K
T_s	absolute temperature of the object surface, K
u	specific internal energy, kJ/kg; velocity in x direction, m/s; variable velocity, m/s
U	internal energy, kJ; flow velocity, m/s
x	quality, kg/kg; cartesian coordinate; variable
X	length for plate, m
v	specific volume, m^3/kg; velocity in y direction, m/s
$\bar{v}$	molal specific volume, kmol/kg
V	volume, m^3; velocity, m/s
V_x	velocity in x direction, m/s
V_r	veloxity in radial direction, m/s
V_y	velocity in y direction, m/s
V_z	velocity in z direction, m/s
V_θ	tangential velocity, m/s
$\dot{V}$	volumetric flow rate, m^3/s
w	velocity in z direction, m/s
y	mole fraction; cartesian coordinate, variable; coordinate normal to flow
Y	characteristic dimension (length), m
z	cartesian coordinate, variable
Z	compressibility factor (Equation 1.24); elevation, m

Greek letters

ϕ	temperature difference, °C or K

θ	angle
β	volumetric coefficient of thermal expansion, 1/K
δ	increment; difference
μ	dynamic viscosity, kg/ms; root of the characteristic equation
ρ	density, kg/m^3
ν	kinematic viscosity, m^2/s
Δ	thickness of the stagnant film of fluid on the surface, m
ΔT	temperature difference, K; overall temperature difference, °C or K
σ	Stefan-Boltzmann constant, W/m^2K^4; electrical conductivity, 1/ohmm
ε	surface emissivity, Eddy viscosity
τ	shear stress, N/m^2
Σ	summation
π	pi number (= 3.141)

Subscripts and superscripts

a	air; medium; surroundings
av	average
A	fluid A
b	black
B	fluid B
c	convection, critical
cd	conduction
cs	control surface
cv	control volume
D	diameter
e	electrical; end; exit
f	fluid; final; flow; force; friction
fm	film condition
h	heat generation
H	high-temperature
hs	heat storage
i	component; input
ie	internal energy
liq	liquid
l	liquid
L	low-temperature
m	midplane for plane wall; centerline for cylinder
mix	mixture
n	nth value
nb	nonblack
p	previous
r	radiation
s	surface; near surface; saturation; free stream; in direction parallel to streamline
t	total; thermal

tot total
x x direction
v vapor
vap vapor
y y direction
z z direction
0 surroundings; ambient; environment; reference
1 first value; 1st state; initial
1, 2, 3 points

References

Dincer, I. (1997). *Heat Transfer in Food Cooling Applications*, Taylor & Francis, Washington, DC.

Dincer, I. (1998). Thermodynamics, exergy and environmental Impact, *Proceedings of the ISTP-11, the Eleventh International Symposium on Transport Phenomena*, pp.121-125, 29 November-3 December, Hsinchu, Taiwan.

Dincer, I. and Rosen, M.A. (1999). Energy, environment and sustainable development, *Applied Energy* 64(1-4), 427-440.

Moran, M.J. and Shapiro, H.N. (1998). *Fundamentals of Engineering Thermodynamics*, 4th edition, Wiley, New York.

Olson, R.M. and Wright, S.J. (1991). *Essentials of Engineering Fluid Mechanics*, Harper & Row, New York.

Raznjevic, K. (1995). *Handbook of Thermodynamic Tables*, 2nd edition, Begell House, New York.

Szargut, J., Morris, D.R. and Steward, F.R. (1988). *Exergy Analysis of Thermal, Chemical, and Metallurgical Processes*, Hemisphere, New York.

van Wylen, G., Sonntag, R. and Borgnakke, C. (1998). *Fundamentals of Classical Thermodynamics*, 5th edition, Wiley, New York.

2

Energy Storage Systems

I. Dincer and M.A. Rosen

2.1 Introduction

Energy storage (ES) has only recently been developed to a point where it can have a significant impact on modern technology. In particular, ES is critically important to the success of any intermittent energy source in meeting demand. For example, the need for storage for solar energy applications is severe, especially when the solar availability is lowest, namely in winter.

ES systems can contribute significantly to meeting society's needs for more efficient, environmentally benign energy use in building heating and cooling, aerospace power, and utility applications. The use of ES systems often results in such significant benefits as

- reduced energy costs,
- reduced energy consumption,
- improved indoor air quality,
- increased flexibility of operation, and
- reduced initial and maintenance costs.

In addition, Dincer *et al.* (1997a) point out some further advantages of ES:

- reduced equipment size,
- more efficient and effective utilization of equipment,
- conservation of fossil fuels (by facilitating more efficient energy use and/or fuel substitution), and
- reduced pollutant emissions (e.g. CO_2 and CFCs).

ES systems have an enormous potential to increase the effectiveness of energy-conversion equipment use and for facilitating large-scale fuel substitutions in the world's economy. ES is complex and cannot be evaluated properly without a detailed understanding of energy supplies and end-use considerations. In general, a coordinated set of actions is needed in several sectors of the energy system for the maximum potential benefits of storage to be realized. ES performance criteria can help in determining whether prospective advanced systems have performance characteristics that make them useful and attractive and, therefore, worth pursuing through the advanced development and demonstration stages. The merits of potential ES systems need to be measured, however, in terms of the conditions that are expected to exist after research and development is completed. Care should be taken not to apply too narrow a range of forecasts to those conditions. Care also should be taken to evaluate specific storage system concepts in terms that account for their full potential impact. The versatility of some ES technologies in a number of application areas should be accounted for in such assessments.

Today's industrial civilizations are based upon abundant and reliable supplies of energy. To be useful, raw energy forms must be converted into energy currencies, commonly through heat release. For example, steam, which is widely used for heating in industrial processes, is normally obtained through converting fuel energies into heat, and transferring the heat into water. Electricity, increasingly favored as a power source, is generated predominately with steam-driven turbogenerators, fueled by fossil or nuclear energy. Power demands, in general, whether thermal or electrical, are not steady. Moreover, some thermal and electrical energy sources, such as solar energy, are not steady in supply. In cases where either supply or demand is highly variable, reliable power availability has in the past generally required energy conversion systems large enough to supply the peak-demand requirements. The results are high and partially inefficient capital investments, since the systems operate at less than capacity much of the time.

Alternatively, capital investments can sometimes be reduced if load-management techniques are employed to smooth power demands, or if energy storage systems are used to permit the use of smaller power-generating systems. The smaller systems operate at or near peak capacity, irrespective of the instantaneous demand for power, by storing the excess converted energy during reduced-demand periods for subsequent use in meeting peak-demand requirements. Although some energy generally is lost in the storage process, ES often permits fuel conservation by utilizing more plentiful but less flexible fuels such as coal and uranium in applications now requiring scarce oil and natural gas. In some cases, ES systems enable the waste heat accompanying conversion processes to be used for secondary purposes.

The opportunities for energy storage are not confined to industries and utilities. Storage at the point of energy consumption, as in residences and commercial buildings, will likely be essential to the future use of solar heating and cooling systems, and may prove important in lessening the peak-demand loads imposed by conventional electrical, space-conditioning systems. In the personal transportation sector, now dominated by gasoline-powered vehicles, adequate electrical storage systems might encourage the use of large numbers of electric vehicles, reducing the demand for petroleum.

The concept of energy storage using flywheels is not new. The ability of flywheels to smooth intermittent power impulses was recognized shortly after the invention of reciprocating engines in the eighteenth century. Special purpose locomotives have been operated with stored, externally supplied steam for about 100 years. Electric cars were early automobile competitors.

The marked increases in fuel costs in the last few years, the increasing difficulty in acquiring the large amounts of capital required for power-generation expansions, and the emergence of new storage technologies has led to a recent resurgence of interest in the possibilities for advanced energy storage systems. To the energy supplier, energy is a commodity whose value is determined by the cost of production and the marketplace demand. For the energy consumer, the value of energy is in its contribution to the production of goods and services or to personal comfort and convenience. Although discussions abound about the merits of alternative national energy production and consumption patterns in the future, it is likely that energy decisions, in general, will continue to be made based on evaluations of the costs of a alternative means to attain these needs. In particular, decisions on whether to use energy storage systems will likely be made on the basis of prospective cost savings in the production or use of energy, unless legislative or regulatory constraints are imposed. Thus, among criteria necessary for the commercialization of ES systems, potential economic viability is a major parameter.

2.2 Energy Demand

Energy demand in the commercial, industrial and utility sectors varies on a daily, weekly, and seasonal basis. Ideally, these demands are matched by various energy conversion systems that operate synergistically. Peak hours are the most difficult and expensive to supply. Peak electrical demands are generally met by conventional gas turbines or diesel generators, which are reliant on costly and relatively scarce oil or gas. ES provides an alternative method of supplying peak energy demands. Likewise, ES systems can improve the operation of cogeneration, solar, wind, and run-of-river hydro facilities. Some details on these ES applications follow:

- **Utility**. Relatively inexpensive base load electricity can be used to charge energy storage systems during evening or off-peak weekly or seasonal periods. The electricity is then used during peak periods, reducing the reliance on conventional gas and oil peaking generators.
- **Industry**. High-temperature waste heat from various industrial processes can be stored for use in preheating and other heating operations.
- **Cogeneration**. Since the closely coupled production of heat and electricity by a cogeneration system rarely matches demand exactly, excess eletricity or heat can be stored for subsequent use.
- **Wind and run-of-river hydro**. Conceivably these systems can operate around-the-clock, charging an electrical storage system during low-demand hours and later using that electricity for peaking purposes. ES increases the capacity factor for these devices, usually enhancing their economic value.
- **Solar systems**. By storing excess solar energy received on sunny days for use on cloudy days or at night, ES systems can increase the capacity factor of solar energy systems.

2.3 Energy Storage

Mechanical and hydraulic ES systems usually store energy by converting electricity into energy of compression, elevation, or rotation. Pumped storage is proven, but quite limited in its applicability by site considerations. Compressed-air ES has been tried successfully in Europe, although limited applications appear in the USA. This concept can be applied on a large scale using depleted natural gas fields for the storage reservoir. Alternatively, energy can be stored chemically as hydrogen in exhausted gas fields. Energy of rotation can be stored in flywheels, but advanced designs with high-tensile materials appear to be needed to reduce the price and volume of storage. A substantial energy penalty up to 50% is generally incurred by mechanical and hydraulic systems in a complete storage cycle because of inefficiencies.

Reversible chemical reactions can also be used to store energy. There is a growing interest in storing low-temperature heat in chemical form, but practical systems have not yet emerged. Another idea in the same category is the storage of hydrogen in metal hydrides (lanthanum, for instance). Tests of this idea are ongoing.

Electrochemical ES systems have better turnaround efficiencies but very high prices. Intensive research is now directed toward improving batteries, particularly by lowering their weight-to-storage capacity ratios, as needed in many vehicle applications. As a successor to the lead-acid battery, sodium-sulfur and lithium-sulfide alternatives, among others, are being tested. A different type of electro chemical system is the redox flow cell, so named because charging and discharging is achieved through reduction and oxidation reactions occurring in fluids stored in two separate tanks. To make the leading candidate (an iron redox system) competitive with today's batteries, its price would have to be at least halved.

Thermal energy storage systems are varied, and include designed containers, underground aquifers and lakes, bricks and ingots. Some systems using bricks are operating in Europe. In these systems, energy is stored as sensible heat. Alternatively, thermal energy can be stored in the latent heat of melting in such materials as salts or paraffin. Latent storages can reduce the volume of the storage device by as much as 100 times, but after several decades of research many of their practical problems have still not been solved. Finally, electric energy can be stored in superconducting magnetic systems, although the costs of such systems are high.

There are a number of promising areas of research in ES technology. Given the cost gap that needs to be spanned and the potential beneficial ES applications, it is clear that a sustained ES development effort is in order. For solar energy applications, advanced ES systems may not be needed for years and decades. For the near-term, many less expensive ES alternatives are available that should allow for the growth of solar energy use.

Here, some current research and development topics in the field of ES are as follows:

- Advanced energy storage and conversion systems with phase transformation, chemical and electrochemical reactions.
- Fundamental phenomena inside a single cell as well as engineering integration of whole battery packs into vehicles.
- High dielectric constant polymers.
- High K composites for capacitors.
- Polymer electrode interfaces (low and high frequency effects).
- Integrated polymer capacitors.

2.4 Energy Storage Methods

For many energy technologies, storage is a crucial aspect. If we consider the storage of fuels as the storage of the energy embedded in them, then oil is an excellent example. The massive amounts of petroleum stored worldwide are necessary for the reliable, economic availability of gasoline, fuel oil and petrochemicals.

Electric utilities also store energy using a scheme called pumped storage. Electricity generated by thermal power plants drives large electric motors to pump water uphill to elevated reservoirs during periods of low electric demand. During periods of peak demand, the water is allowed to flow back downhill to redeliver the energy through hydroelectric generation. Also, electricity can be stored in batteries. However, the present-day conventional automobile battery, to use an important and common example, is used for starting the internal combustion engine and not for locomotion.

ES includes heat storage. In thermodynamic terms, such storages hold transferred heat before it is put to useful purposes. A conventional example is hot water storage in residences and industry. Such heat storage smoothes out the delivery of hot water or steam, but it is not usually considered for periods longer than one day.

Advanced new storage devices are often an integral part of other new technologies, and these sometimes can be made more feasible by innovations in storage. Advances in storage especially benefit wind and solar energy technologies. Also, new storage technologies may facilitate the development of electric-powered automobiles.

A large variety of ES techniques are under development. We shall discuss them by categories, grouping together those techniques that store energy in the following forms: mechanical, thermal, chemical, biological and magnetic, as shown in Figure 2.1. Of course, ES devices can be classified and categorized in other ways. Here, each category considers the storage of one form of energy. Below, we examine briefly several possible storage options (Dincer, 1998).

2.4.1 Mechanical energy storage

Mechanical energy may be stored as the kinetic energy of linear or rotational motion, as the potential energy in an elevated object, as the compression or strain energy of an elastic material, or as the compression energy in a gas. It is difficult to store large quantities of energy in linear motion because one would have to chase after the storage medium continually. However, it is quite simple to store rotational kinetic energy. In fact, the potter's wheel, perhaps the first form of energy storage used by man, was developed several thousand years ago and is still being used. As seen in Figure 2.1, there are three main mechanical storage types that we will discuss in this section: hydrostorage, compressed air storage, and flywheels.

Hydrostorage (pumped storage)

Hydrostorage is a very simple ES method. At night, when energy demand is low, the pumps pump water upward from the river (Figure 2.2). The water is pumped through the pipe to the reservoir. During the day, when energy demand is high, the reservoir releases water to flow downhill. The flowing water turns the turbine to make electricity. The pump which pumps the water upward from the river can be powered by solar energy during the day. At night, when there is no solar energy, the stored water turns the turbine to make electricity. The efficiency of a pumped water storage plant is about 50%. When you pump water uphill, you lose 30% of energy. When the water flows down, you lose 20% of energy. A pumped water storage plant operates for more than 20 years. When the energy is needed the plant only needs 30 seconds to reach 100% of its power.

In this storage type, reversible pump/turbines devices pump water upward into a storage reservoir and after a period of time operate as turbines, driving generators, when the water runs back down through them. Hydrostorage has been proven itself economically viable, but its use is geographically limited to only a few percent of the total hydroelectric capacity. Pumped storage is now considered important with wind machines. The best alternative to building expensive new storage systems in the near term is often to use existing storage systems, especially those of hydroelectric installations. By holding back water that would otherwise flow from a hydroelectric dam, energy can be stored in one part of an electric network while a solar electrical energy or wind system produces energy in another part. The USA currently has 59 000 megawatts of hydroelectric capacity, and an additional 10 000 megawatts of pumped storage capacity. Pumped storage facilities consist of a pair of reservoirs; the upper one is filled with water pumped up from lower levels at times of low electricity usage.

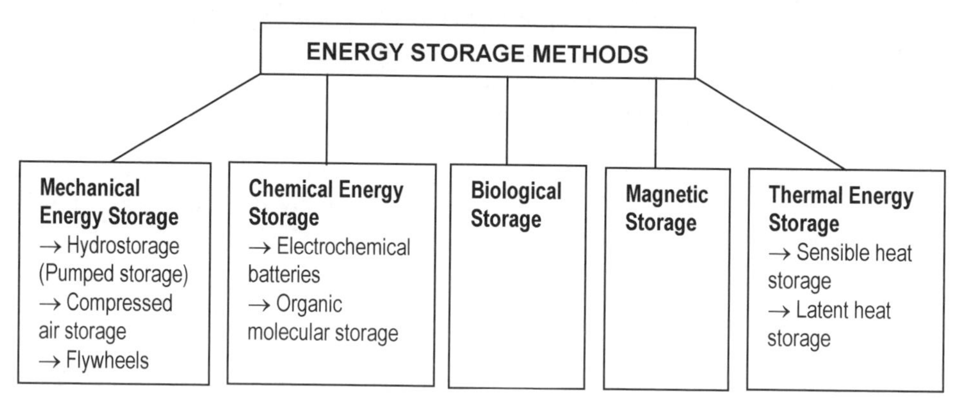

Figure 2.1 A classification of energy storage methods.

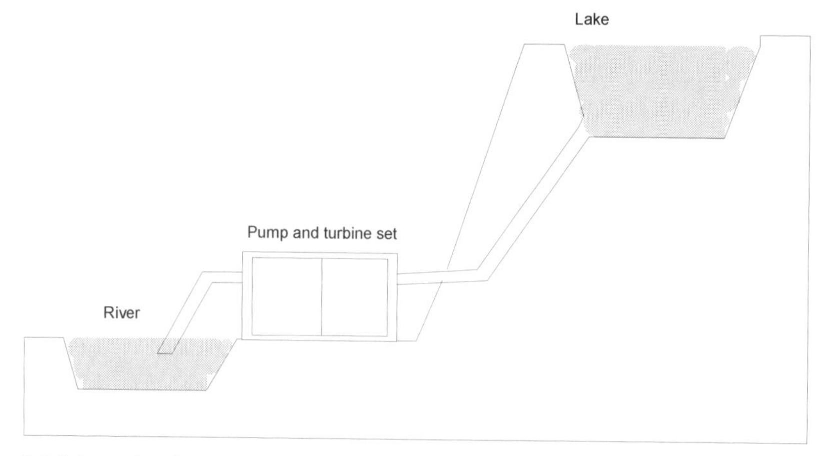

Figure 2.2 Schematic of a pumped hydrostorage plant.

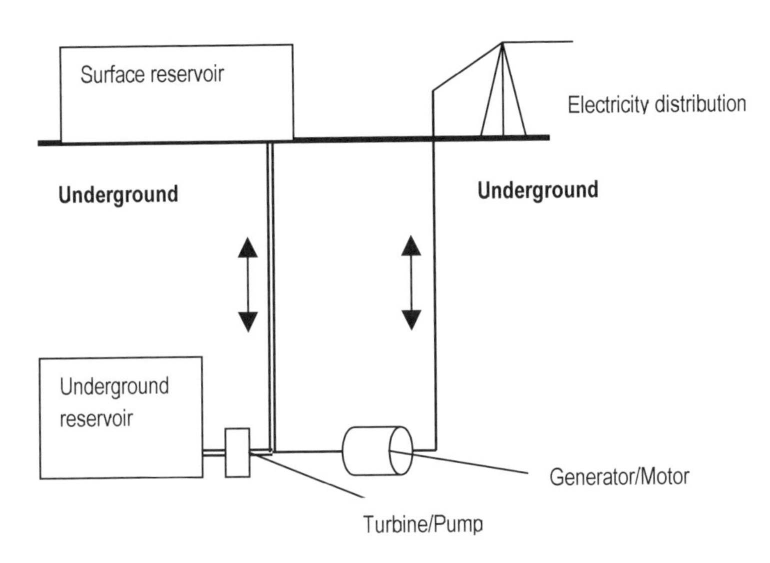

Figure 2.3 A schematic representation of underground pumped hydrostorage.

Hydrostorage is widely used in the power industry to store off-peak power for peak load periods. This technique utilizes a dam that has sufficient hydrostatic head to drive a hydroelectric power plant. Water is pumped into the reservoir in off-peak periods and drawn out during peak periods. The basic requirement is a dam with a large quantity of water at its base, or two dams with a height difference between them. Hydrostorage is relatively efficient. The energy used to pump water upward is recovered in a storage cycle with about 65–75% efficiency. Hydrostorage is often ideal for solar power storage. The solar plant produces power at the maximum rate during the day and is on standby during the night, maintaining only system temperatures so that it is ready to turn out power the next day as soon as the collector subsystem reaches operating temperature.

Pumped storage is the only well-established energy storage concept that is available on a large scale. The concept is simple. Energy is stored during evening hours by pumping water from a lower body of water to an upper reservoir behind a conventional dam. During peak demand hours the water flows down from the upper reservoir through a hydroelectric turbine back into the lower body of water. Because of the environmental concerns associated with large-scale hydroelectric facilities, however, it is doubtful that many conventional pumped hydro plants will be constructed in the future.

Underground pumped hydrostorage is a variation of this concept that has significant potential. Still in the planning stage, it would use an upper surface reservoir in conjunction with a man-made lower reservoir (Figure 2.3). The upper surface reservoir can be an existing body of water or an artificial lake formed by dikes and dams. The lower reservoir is a large cavern excavated in hard solid rock. The two reservoirs are hydraulically connected by a waterway (pen-stock) which passes through a powerhouse with a dual purpose turbine/pump and generator/motor. Except for the man-made lower reservoir, pumped hydro operates in the same manner as a conventional system. The power produced from a hydro facility is directly related to the hydraulic head (elevation difference between the reservoirs) available to the plant. Therefore, since the lower reservoir can be positioned to obtain high heads of approximately 1400 m (composed to the heads of usually less than

300 m for natural conventional hydro facilities), comparable power outputs can be achieved with significantly smaller reservoirs. Underground systems can be more acceptable since environmental impact is decreased. The area in underground pumped storage needing greatest development and having the greatest cost is the lower reservoir.

The primary criteria for location are the geological conditions for the lower reservoir. The subsurface material should be of solid hard rock, and not be located in areas that have:

- predominantly loose sedimentary rock,
- volcanic rock,
- complex geological structures with widely varying conditions over short distances,
- high seismic activity, or
- major faults.

As an example of an underground pumped storage, consider the following from Diamant (1984). Although in Great Britain nuclear power accounts for only about 3–4% of the country's total electrical generating capacity, nuclear energy supplies about 15% of the annual output of electricity. The reason for these different values is that nuclear power stations, whose main cost is the capital invested in them, are best operated at a very steady rate. Fuel costs for nuclear stations are much smaller than those for fossil fuel-fired electric generating stations. Thus, since it is difficult to vary the output from nuclear power stations significantly over short time periods, these stations are often used for generating the base-load power. The outputs of power stations fuelled by coal, gas and oil can be more easily adjusted to match changing loads, but even so, it is best to steady the load if possible. Sometimes there are very high peak load requirements which may only last a few minutes a day. To satisfy these demands it would be necessary to have costly stand-by stations which operate for only very short periods over the year. Yet not satisfying such loads may mean voltage reductions and the accompanying dangers of motors and other equipment burning out due to overheating.

If the nuclear capacity increases relative to the country's total installed electric generating capacity, these problems will be accentuated. A first attempt to address this issue is the pumped storage scheme at Dinorwic in North Wales. Two lakes existing near the town of Llanberis, the Llyn Peris lake which lies in a valley and the Marchlyn Mawr lake, which lies about 500 m higher on a mountain. Both lakes were enlarged and connected by a system of tunnels with enough capacity to drive six 313 MW turbine generators. When the turbine generators are driven in the opposite direction, powered by electricity from the national grid, they act as motor pumps which consume 281 MW of electricity per unit. When in operation, the Dinorwic station will be able to provide a constant output of 1680 MW for five hours during peak demand periods, using the potential energy in the difference in water levels between the upper and lower reservoirs to drive the generators. The pumps require six hours to transfer the water from the bottom Llyn Peris lake back to the top Marchlyn Mawr lake, an operation normally done at night using cheap off-peak electricity. The Dinorwic station can be put into operation to meet sudden surges in power demand far more rapidly than almost any alternative system. It is claimed that the station can ramp up from zero to an output of 1320 MW within ten seconds, engaged extremely quickly to correct sudden voltage drops due to power consumption momentarily outstripping production (Diamant, 1984).

Compressed air storage

In a compressed air ES system, air is compressed during off-peak hours and stored in large underground reservoirs, which may be naturally occurring caverns, salt domes, abandoned mine shafts, depleted gas and oil fields, or man-made caverns. During peak hours the air is released to drive a gas turbine generator.

The technique used by such a system to compress air to store energy is relatively straightforward. In a conventional gas turbine, high-pressure hot gas is supplied, and about two-thirds of the gross power output is used to drive the compressor. A compressed air ES system decouples the compressor and the turbine, and operates the former during off-peak hours to produce compressed air which is stored in natural caverns, old oil or gas wells, or porous rock formations. Such ES storage is advantageous when an appreciable part of the power load is carried by nuclear stations, and where suitable spent salt caverns make it easy to build the compressed gas reservoirs.

The compressed air storage technique in general takes advantage of off-peak electrical generating capacity using gas turbines in the same way that pumped storage does so using water turbines. However, with gas systems the heat generated when the air is compressed may be stored and used to preheat the expanding air, increasing efficiency.

Significant amounts of energy can be stored in the form of compressed air in underground caverns. Early studies indicate system costs to be comparable to those for hydrostorage, but the requirement of a large cavern limits the usefulness of this approach to regions where natural caverns exist, or where caverns can be easily formed, as in salt domes. The air in such a storage facility is normally compressed in a device which later serves as expander or turbine.

In practice, two general categories of compressed air ES systems are possible, depending on the storage pressure (see Figure 2.4). In the sliding pressure systems (see Figure 2.3a), pressure increases as the store is charged and decreases as the stored air is released, between maximum and minimum pressures. In the compensated pressure systems (see Figure 2.3b), an external force is used to keep the storage pressure constant throughout the operation.

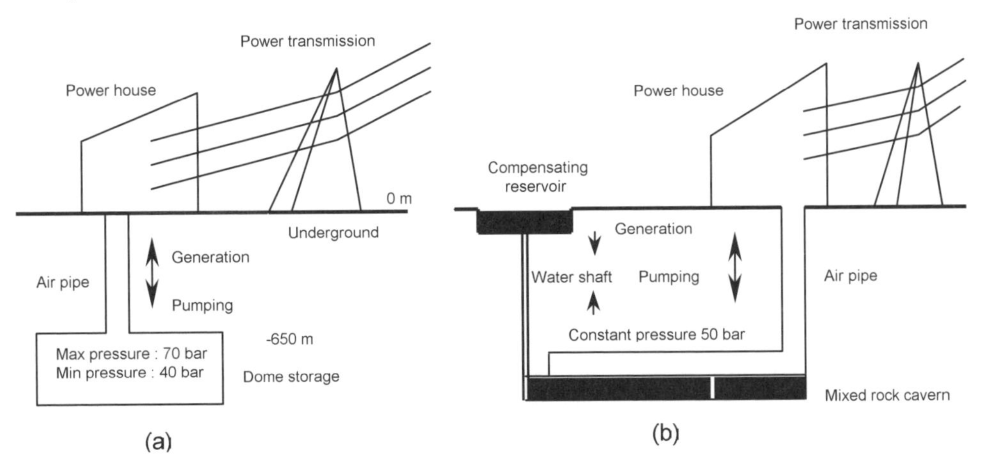

Figure 2.4 Compressed-air ES systems. (a) Sliding pressure system. (b) Compensated pressure system.

The world's first compressed air ES facility became operational in 1978 at Huntorf in Hamburg, Germany. The plant is connected to the local electric utility grid. During off-peak hours air is compressed to approximately 47,780 Pa and stored in two caverns leached out of a salt dome. The combined storage capacity is about 283,179 m^3. During peak demand periods, the air is released, heated by natural gas, and expanded through high- and low-pressure turbines. The system can generate 290 MW for up to about two hours. The Huntorf facility requires an electric energy input of 0.8 kWh for air compression and 5600 kJ of natural gas input for reheating, for each 1.0 kWh of plant output. Heat recuperation from the compressed air for subsequent addition to the expanding air can reduce fuel consumption by about 25% (Sheahan, 1981).

Conventional power stations could include a compressed air ES system. In one practical design, ambient air is compressed during off-peak operation by an axial flow compressor, intercooled and boosted to a pressure of approximately 70 bar by a high-speed centrifugal blower. Heat produced during compression is removed with conventional cooling devices. The compressed air is stored underground (ideally in leached-out salt domes or similar spaces). The pressure of the soil serves to resist the appreciable gas pressure. At peak demand periods, air passes from the underground compressed air storage cavern through a control valve, where the pressure is throttled down to 43 bar at full load. After being heated, the compressed air drives gas turbines, effectively releasing the stored off-peak electricity.

The Brown Boveri company developed the ASSET (Air Storage System of Energy Transfer) plant for this purpose. The first plant of this type was ordered by NWK (Nordwestdeutsche Kraftwerke) of Hamburg in June 1974. In 1980 the various power stations operated by this utility had a total electricity capacity of 4515 MW, broken down as follows (Diamant, 1984):

- Fossil-fuelled steam plants: 3000 MW
- Nuclear plants: 1210 MW
- Gas turbines plants: 305 MW

The installed system is the one described earlier at Huntorf. In this system, the storage capacity is sufficient for two hours of full peak load for the entire NWK system. No electric power is needed to start the unit. The compressed air is heated to 540°C in the high-pressure combustion chamber of the gas turbine plant and expanded in the gas turbine. When the gas turbine rotational speed reaches 3000 rpm, the generator connects to it via a 'synchroself-shifting clutch.' The total cold start-up time to full load is normally 11 minutes, but as short as 6 minutes in an emergency. All operations are controlled remotely. Since in this plant the heat of compression is rejected as a waste before storage, and then re-supplied using fuel before the air drives the gas turbine, the system has a relatively poor thermodynamic efficiency of approximately 46%. This ES method appears to be particularly advantageous when an appreciable part of the power load is met by nuclear stations, and where suitable compressed gas reservoirs such as spent salt caverns exist.

Flywheels

The flywheel, a wheel of relatively large mass that stores rotational kinetic energy, has long been used to smooth out the shaft power output from one- or two-cycle (stroke) engines and to adjust for uneven loads. New uses of this device, and of the other two mechanical storage techniques discussed in this section, take advantage of the reversibility of the electric motor/generator. Such a device can be designed to work both as a motor when driven by electric power and as a generator when driven by mechanical power.

In such places as islands or isolated communities, where support from a larger-area electrical grid is not possible, local electric generators are installed to meet local need. Because variation in system loading is greater when the number of customers is small, and because maintenance of system voltage requires sufficient on-line capacity to meet load, such systems are typically operated at only a fraction of nameplate capacity. For example, a large diesel M/G set is run at half capacity so that if several customers switch on load at once, the extra load can be carried at full voltage. The excess capacity is spinning reserve. The use of on-line fuel-based capacity has several disadvantages. First, when a generator is operated at partial capacity, it operates inefficiently. Excess fuel is consumed, and the price per kWh of supplied electricity rises. Second, especially in the case of such advanced modular generation technologies as fuel cells and micro-turbines, variation of load causes variations in thermal stress on the generator, shortening its useful life. Both these problems can be addressed by using flywheel energy storage as the spinning reserve, and operating the fuel-based generators at peak efficiency.

Flywheels can be used for transportation energy storage, particularly in road vehicles (e.g. buses). Flywheels can have a significant advantage in vehicles that undergo frequent start/stop operations as in urban traffic. The basic idea is that with flywheels, decelerating does not convert mechanical energy into waste heat via friction brakes. Instead, the kinetic energy is stored by setting a flywheel spinning. Then, the power surge needed for vehicle acceleration is provided from the spinning flywheel. In petrol-driven test vehicles using flywheel ES, operational economies on the order of 50% have been achieved. It is expected that similar economies would be obtainable with electric vehicles. Figure 2.5 shows a flywheel which was successfully designed and constructed at the Pennsylvania Transportation Institute, USA. More research is needed to make these devices practical.

Figure 2.5 A flywheel (*Courtesy of Prof. Charles E. Bakis, The Pennsylvania State University*).

The most important use of flywheel ES will probably be regenerative braking. A subway train, for instance, may use a flywheel to decelerate by transferring energy to it from the wheels. The transfer can be made electrically; a conducting flywheel is rotated in a magnetic field, and the generated electricity energy stored in a battery for subsequent use. Work is required to move a conductor in a magnetic field, producing this work to slow the vehicle. In subway stop-and-start operations, energy savings as high as 30% may be realized through ES (and the subway tunnels remain more comfortable in summer due to less heat dissipation by friction braking).

Flywheel systems for electrical energy storage have two properties in which they differ from present state rechargeable batteries: (i) high power mass density specified by the maximum charge (discharge) power per system mass; and (ii) high lifecycle.

In utility applications, a large flywheel located near an electric power demand could be set spinning by a motor using off-peak power and then drive the motor as a generator when additional electricity is needed. Although flywheels offer 80–90% cyclic efficiency, further research and development (e.g. on new materials and shapes) is needed to make them practical for large-scale ES.

Flywheels have also received attention elsewhere as potential energy storage devices. They have found practical application in storing the energy temporarily released from the large magnets of synchrotrons; the energy is recovered within a few seconds to re-energize the magnets. One can manage the energy system of this type simply by means of a switch. One polarity uses the dynamo as a motor, accelerating the flywheel; the other uses the dynamo as a generator, drawing energy from the kinetic energy of the flywheel.

The quantity of energy stored in a flywheel is usually small. One watt-hour of energy is equivalent to 1.8 kg of mass on a 2 m-diameter flywheel rotating at 600 rpm (Genta, 1985). New materials, such as carbon fiber composites, can withstand large centrifugal forces, and at high rotational speeds store much more energy than steel. The energy E_k stored in a flywheel (or kinetic energy of the rotor) is given by

$$E_k = \frac{1}{2}\Theta\omega^2 = \frac{1}{2}\omega^2 \int_V \rho r^2 \Delta V \tag{2.1}$$

where Θ is the moment of inertia, ω the rotational speed, ρ the density and dV the increment of the volume.

An equivalent way of viewing the energy stored in a flywheel is as the energy in the 'spring' formed by the tension created in the rim of the flywheel by the centrifugal force, which slightly expands the diameter of the flywheel. As is the case for all types of ES, the potential of the stored energy to accidentally cause damage is appreciable.

The power P of the flywheel system is determined by the size of the electrical machine, not by the flywheel. The time t ideally required for the flywheel to be charged or discharged can be expressed as

$$t = \frac{E_k}{P} \tag{2.2}$$

Using the material specific strength, the maximum circumferential speed v_{circ} of a thin rim can be determined as

Table 2.1 Some materials for flywheels and their properties.

	Density (kg/m^3)	Strength (MN/m^2)	Specific strength (MNm/kg)
Steel (AISI 4340)	7800	1800	0.22
Alloy (AIMnMg)	2700	600	0.22
Titanium ($TiAl_6Zr_5$)	4500	1200	0.27
GFRP (60 vol% E-Glass)	2000	1600	0.80
CFRP (60 vol% HT C)	1500	2400	1.60

Source: ASPES Engineering AG.

$$v_{circ} = \left(\frac{2\sigma_{ult}}{\rho} \right)^{1/2} \tag{2.3}$$

with σ_{ult} as the ultimate strength of the material.

The specific energy or energy density of a flywheel can then be expressed as

$$\frac{E_k}{m} = \frac{1}{2} v_{circ}^2 = \frac{\sigma_{ult}}{\rho} \tag{2.4}$$

For a rotating thin ring, therefore, the maximum energy density is dependent on the specific strength of the material, not the mass. The energy density of a flywheel is normally the first criterion for the selection of a material. Regarding specific strength, composite materials have significant advantages compared to metallic materials. Table 2.1 lists some flywheel materials and their properties.

The burst behavior is a deciding factor for choosing a flywheel material. With circumferential speeds up to 2000 m/s, the bursting of a metallic rotor causes large fragments to be projected. Composite-material rotors can be designed to have benign failure modes with almost no penetration of pieces into the housing (Genta, 1985).

The stored energy in flywheels has a destructive potential when released uncontrolled. Some efforts have been made to design rotors such that, in the case of a failure, many thin and long fragments are released. These fragments have little translateral energy and the rotor burst can be relatively benign. However even with careful design, a composite rotor still can fail dangerously. The safety of a flywheel ES system is not related only to the rotor. The housing, and all components and materials within it, can influence the result of a burst significantly.

Many kinds of bearing can also be running under vacuum conditions. Precision ball bearings are probably the most economic type presently. Active magnetic bearings have numerous technical benefits but are much more costly. Passive permanent magnet bearings in combination with another bearing can expand the limits of conventional bearings without the cost penalty of active magnetic bearings. Passive magnetic bearings using superconducting materials may prove useful in the future for such high-speed applications.

Choosing the most appropriate bearing system depends very much on the specific application, including such factors as lifetime, maintenance interval, rotational speed and rotor weight, vibration monitoring needs, cost, etc.

Since we deal only with high-speed flywheels running under vacuum conditions, mechanical shafts with gearboxes become beyond the scope of discussion. The flywheel is coupled with an electrical motor/generator on a common shaft. In general, flywheel ES systems use synchronous electrical machines, with permanent or dynamic excitation. Based on the rotational speed, variable frequency can be converted to the constant grid frequency with an inverter.

To facilitate mechanical energy storage by flywheel, low-loss and long-life bearings and suitable flywheel materials need to be developed. Some new materials are steel wire, vinyl-impregnated fiberglass and carbon fiber. Recently, a series of industrial flywheel ES systems using magnetic bearings have been employed. A typical example of those flywheels is capable of storing up to 3 kW of power, and consists of a series of discs, 400 mm in diameter and 200 mm deep, with a spinning mass of about 240 kg. The system is suspended on six magnetic bearings which support the unit when it is in operation, and two roller bearings which are used when accelerating the unit to working speed. Such flywheels can operate at speeds between 7500 and 15 000 rpm. A larger unit can practically be designed for a maximum power storage of 10 kW with a maximum speed of 24 000 rpm. Both the flywheel and its ball bearings should be kept under vacuum.

Feasibility studies have shown that an economic single-family flywheel ES unit capable of storing up to 30 kWh of energy could be built. Inexpensive electricity would be used during off-peak periods to set the flywheel spinning. During peak demand periods, the energy from the spinning flywheel would be discharged to generate electricity. A combined electric motor/generator is used in both operations. The entire flywheel motor system is arranged with a vertical axis in a special underground chamber such as the garage of a dwelling. A high vacuum would be maintained within the container housing the flywheel and bearings.

Here, some current research and development topics in the field of flywheel ES can be listed as follows:

- Development of high-speed, low-cost manufacturing methods for composite rotors and hubs having high specific energy densities.
- Development and experimental evaluation of novel composite rotor concepts (e.g. elastomeric matrix and elastomeric interlayer).
- Evaluation of the durability of new composite rotor materials (considering properties such as fatigue and creep).

In the future, flywheels will be used to store energy for discharge over longer periods. These applications will become a reality when flywheel power systems can be made with both acceptably low cost and acceptably low losses. Among the applications which bring a gleam to the eye of energy futurists are the all-electric vehicle run on flywheel power, the use of flywheel systems as the ES medium to couple with photovoltaics, and the use of flywheel systems in lieu of chemical batteries for backup power at telecom nodes.

2.4.2 Chemical energy storage

Energy may be stored in systems composed of one or more chemical compounds that release or absorb energy when they react to form other compounds. The most familiar chemical energy storage device is the battery. Energy stored in batteries is frequently referred to as electrochemical energy because chemical reactions in the battery are caused by electrical energy and subsequently produce electrical energy.

Some chemical storage systems are thermally charged and discharged. Many chemical reactions are endothermic and proceed forward with absorption of thermal energy. Then, when the temperature of the system falls below a certain value, the energy stored in the system during the original reaction is released as the reaction is reversed. Thus energy is stored by utilizing the heats of chemical reactions. Such chemical storage is considered for solar thermal applications, but is still at the developmental stage.

Production of a storable chemical using electricity or heat is a technical possibility. The simplest chemical to produce is conventionally considered to be hydrogen. Hydrogen can be produced electrochemically by the electrolysis of water, or thermochemically by direct chemical reactions in multistage processes. This hydrogen then can be used as a fuel to drive some energy devices. In the meantime, the hydrogen acts as an ES medium. An important use of hydrogen fuel is to generate electricity in fuel cells. Fuel cells have proven useful in manned space missions, and have demonstrated good reliability using hydrogen. Fuel cells using pipeline natural gas have not proved reliable to date. The use of a hydrogen fuel cell in conjunction with hydrogen production devices using solar energy have technical benefits, but costs appear to be prohibitive. The main reason is that hydrogen produced from solar electricity or other solar energy systems would be several times as costly as the hydrogen used today, which is reproduced by reforming hydrocarbons. Combusting the hydrogen in a fuel cell further increases the cost of the final output due to the high costs of fuel cells. So although such ES schemes are technically possible, their economics are questionable.

Another way to use hydrogen as a fuel, which involves more conventional technology, is direct burning in an engine. Burning in a thermal cycle incurs significant thermodynamic losses. For example, hydrogen can be burned in a combined cycle consisting of a Rankine cycle and a Brayton topping cycle with an overall cycle efficiency in the vicinity of 60%. This efficiency is near to that of a fuel cell, but again the capital costs of the system are high and thus creates implementation barriers.

Other chemical fuels also have the potential to act as storage media and to be produced with solar-derived (or other) electricity. One interesting possibility is aluminum. A large quantity energy can be stored in a small mass of aluminium. Aluminum in granular form can easily be stored in open piles, not needing the complex storage containers used for other chemicals, such as cryogenic tanks for liquid hydrogen or containers for other hydrogen and carbon substances. The aluminum can be readily oxidized in air in a fluidized-grate burner, like powdered coal. The combustion product, aluminum oxide, is solid and can be recovered from the stacks with high efficiency. The aluminum oxide can be stored until it is needed for reprocessing into aluminium, closing the cycle. A significant disadvantage, again, is economics.

Any reversible chemical reaction may be considered for storing energy. The driving force for the reaction is generally thermal or electrical energy. When the reaction is reversed, the driving input commodity is released.

Electrochemical batteries

Batteries chemically store energy and release it as electric energy on demand. Batteries are a stable form of storage and can provide high energy and power densities, such as those needed for transportation. The lead/sulphuric-acid battery has long been considered to have the most benefits, and has been widely applied. Recently, fuel cells have demonstrated the ability to act as large-scale chemical storages like batteries.

There are three main categories of energy applications for which batteries are potentially attractive: electric-utility load management, electric vehicles, and storage for renewable energy systems (e.g. photovoltaic and wind systems). In the first of these applications, batteries can be preferable to pumped hydro and compressed air systems when the storage capacities needed are not great, and lead to savings in transmission system costs and construction lead time. Battery facilities can be modular in construction, making it easier to match storage capacity with utility system requirements. Batteries have higher energy efficiencies than mechanical systems (70–80% versus 50–70%), but do not benefit as much from economies of scale. For large utility systems, the simultaneous use of two or more types of storage may be desirable. For example, the compressed-air option could meet the needs for electricity generation over an 8–12 hour period, while a battery facility over a 3–5 hour period. For batteries used in utility systems, lifetime costs and service life are important characteristics, while weight or mass, volume, and power density are of secondary importance.

It is difficult to use the power generated by solar, wind or hydro-power sources directly, so the electricity is usually stored in special batteries for use when it is needed. These batteries are often similar in chemistry to car batteries, but are designed differently. Car and truck batteries are designed to give short bursts of very high current to start the engine. They are not suited for use as storage batteries because they are made to be fully charged all the time, and will have a very short life if subjected to the deep discharge cycles (i.e. most of the energy stored in the battery is used before recharging) that are required with solar electric storage systems.

Solar storage batteries are often made from large individual 2-volt cells connected together, although smaller 6 and 12 volt batteries are also available. By far the most common type of battery is the lead-acid battery, and these are usually of the flooded-cell type, though sealed lead acid batteries are becoming more popular.

Nickel-cadmium (nicad) batteries are also available, and while very expensive, do have the advantages of very long life and more stable voltage during discharge.

The amount of energy that can be stored in a battery is called its capacity, and is measured in amp hours (Ah). A 100Ah battery will deliver 1 amp of current for 100 hours, 4 amps for 25 hours, and so on, although battery capacity will decrease with increasing discharge rates. Battery capacities ranging from 1–2000 Ah or more are available.

For long battery life, it is desirable to use only a small part of the total battery capacity before recharging. Each time the batteries are run down and charged up, the batteries undergo a charge/discharge cycle. If more than half the battery's stored energy is discharged before it is recharged, this is called deep cycling. For example, lead acid deep cycle batteries designed for solar storage will last anywhere from 300 to 5000 cycles (and up to 50 000 cycles for nicads), provided the discharge is limited to about 20% per cycle. Solar systems normally do one shallow cycle per day, but during 'low sun' periods may undergo much deeper discharges. For long battery life, the shallow cycle should be less than 20% of battery capacity and the deep cycle less than 80%.

It is possible to damage batteries by overcharging them. The maximum voltage that a battery should be charged to is about 2.5 volts per cell, or 15.0 volts for a 12 volt battery.

Some solar panels have an output voltage which is claimed to be low enough to stop charging above 15 volts and to be self-regulating. However, because their open circuit voltage is still 18 volts or so, they will actually continue to charge, with a much reduced current, until about 17 volts. Most conventional panels will deliver full power up to about 17 volts and so need an external regulator. It is in general difficult to store significant quantities of electrical energy practically. Conventional battery designs can store in rechargable batteries only very small amounts of the high-voltage AC power generated by either conventional or nuclear power plants. The difficulties include the following:

- The amount of energy which can be stored per unit weight or mass is small.
- The power density per unit weight or mass is low.
- All reversible batteries can withstand only a limited number of charge/discharge cycles before they are discarded. The lead-sulphuric acid cell is still the most economic option, and is the battery used in most recently designed electric cars.

The cost of battery systems is uncompetitive at the present for large-scale ES applications. For example, battery storage of electricity may cost up to 50% more than pumped hydropower. The goal of the developers of advanced batteries for utilities is to reduce costs to about half those of present heavy-duty lead-acid batteries.

Characteristics of batteries Some chemical changes are readily reversed upon the application of a voltage and from the basis of chemical batteries. Batteries have undergone development and experimentation since the late 19th century, but even then were not ready for immediate application. Articles written around 1900 suggest that cheap and long-lived batteries are just around the corner. Unfortunately that corner has not yet been turned, in terms of all desired battery characteristics, but some claim such advanced batteries are ‘almost here’. The batteries that come closest to having the lifetime characteristics and ES capacities needed for many modern applications are sodium or lithium batteries, which have recently been the subject of intensive development at a number of laboratories. These batteries must be maintained at high temperatures (several hundred degrees Celsius) to operate, but exhibit high ES capacities per unit mass. Sodium batteries can in principle store more than 200 Wh/kg, compared to the storage capacity of nickel-cadmium batteries of about 24 Wh/kg. In addition to having high storage capacity, a battery for utility functions must be able to tolerate repeated, full charge-discharge cycles. One thousand such cycles is a major challenge for today’s batteries, yet a lifetime of 10 000 cycles is desirable. A comparison of storage capacity and lifetime cycles for different types of batteries is given in Table 2.2.

Several terms are used to distinguish the characteristics of batteries in practical applications. It is significant in determining which battery is appropriate for a particular application. Here are five measures of merit for batteries for electric vehicles (Cassedy and Grossman, 1998):

- *Specific energy (or specific energy storage).* This is an important factor in determining the range of the battery, and can be defined as the amount of energy stored per unit of weight or mass of a storage technology in a particular vehicle application (often measured in Wh/kg).

Table 2.2 Some batteries and their capacities.

Type	Specific ES capacity (Wh/kg)	Lifetime cycles
Lead-acid (automotive)	33	300-1000
Lead-acid (commercial)	29	1600
NiCd	24	2000-3000
NiFe (Edison cell)	23	2000
AgZn	185-220	200
AgCd	110-165	500
NaS (250-350°C)	220	-
$LiCl_2$ (600°C)	440	-

Source: Meinel and Meinel (1977).

- *Energy density.* This refers to the amount of energy of a battery, and has a direct relation to its size, and can be defined as the total amount of energy a battery can store per liter of its volume for a specified rate of discharge (Wh/L or Wh/m^3). High-energy density batteries have a smaller size.
- *Specific power or power density.* This is an important factor for determining the acceleration of the battery, and can be defined as the rate of energy delivery or power for the storage per weight or mass of a storage technology in a particular vehicle application (often measured in W/kg). Specific power is at its highest when the battery is fully charged. As the battery is discharged the specific power decreases, and acceleration also decreases. Specific power is usually measured at 80% of the depth of discharge. The driving range of a vehicle is roughly proportional to its specific energy, whereas the speed capability is nearly proportional to its specific power.
- *Cycle life.* The total number of times a battery can be discharged and charged during its life. When the battery can no longer hold a charge over 80%, its cycle life is considered finished. The cycle of discharge and recharge for a battery can only be repeated a limited number of times due to electrode and electrolyte deterioration, which determines the life of the battery. For lead-acid batteries in automobiles, the life is around 300 cycles. If lead-acid batteries were used in a solar application with daily cycling, they clearly would last less than a year. Industrial-grade lead-acid batteries can tolerate over five times the number of life cycles, but cost more than twice as much as automobile units, and would not be economically viable for such service.
- *Battery cost.* This is expressed in dollars per kilowatt-hour ($/kWh).

Batteries are commercially available in many types and sizes. Although the most common type is lead-acid batteries, there are some other promising batteries which will be explained later. Present-day batteries generally are not suitable for most large-scale energy applications because of their weight, cost, and/or performance characteristics.

Lead-acid batteries For years the 'lead-acid' battery has dominated the field. In this battery a lead compound (lead dioxide) at one electrode interacts with the sulfuric acid solution (the electrolyte) that fills each cell. Electrons are released and flow through the circuit to the other electrode (made of lead), providing the electric power. In the process the lead compound reacts to form sulfate and the sulfuric acid is diluted with water. The battery can be recharged by forcing electric current through it (causing electrons to flow in the direction reverse). This process restores both electrodes and the sulfuric acid at an energy-conversion efficiency of about 80%.

The specific ES capacity of a typical lead-acid battery is only 126 kJ/kg, assuming 100% efficient conversion to electrical energy. The ES capacity of petrol is about 35 000 kJ/kg gross. After considering that only about 18% of this energy can be converted to mechanical energy, the net ES capacity is 6300 kJ/kg or 50 times as much as for the lead-acid battery.

The lead-acid battery operates on the principle of the galvanic cell, whose discovery goes back to the 18th century. A single-cell battery consists of two electrodes immersed in an electrolyte (Figure 2.6). Chemical reactions of the electrolyte with the substances at each of the two electrodes release electrons with the potential to do electrical work. In the lead-acid battery, one electrode is lead (Pb), the other electrode is made of lead dioxide (PbO_2), and the electrolyte is sulfuric acid (H_2SO_4). For the battery to deliver electricity, electrons are supplied to the lead dioxide electrode from the lead electrode. But as the energy is discharged, the chemical composition changes, as can be seen by writing the electrochemical equation:

$$PbO_2 + Pb + 2H_2SO_4 \rightarrow 2PbSO_2 + 2H_2O$$

The products of the discharge appear on the right side. The lead sulfate ($PbSO_2$) product is deposited on the electrodes, while water goes into the solution of the electrolyte. These products build up, of course, as the discharge progresses, and eventually the battery no longer produces electricity. However, the reaction is reversible. A storage battery can be recharged by a direct current source (generator). The charge flow in the cell reverses and so do the reactions from $PbSO_2$ back to PbO_2 on the positive electrode, to lead on the negative electrode, and to sulfuric acid in the electrolyte.

The lead-acid storage battery is generally too short-lived for many of the new applications envisaged. It is also too heavy and bulky for many uses, especially electric vehicle applications. Although proposed advanced batteries operate on the same galvanic principle as conventional batteries, major innovations have been attempted in their construction and operation. In an effort to avoid deposits and corrosion of solid electrodes, liquid electrodes were mostly designed and tested in practical applications. In the above experimental battery, the two electrodes are comprised of molten sodium and a molten sulfur-carbon mixture. The use of these liquid electrodes in the sodium-sulfur battery not only reduces corrosion, but also increases the allowable life cycles of the electrodes. But this battery has drawbacks in that it must operate at temperatures of 300°C to 350°C, which can present a safety risk in electric vehicles, and it is too heavy and costly for many practical uses.

Deep-cycle lead-acid batteries are available in three voltage sizes: 6-, 8-, and 12 volts. For range, 6-volt batteries are recommended because their specific energy is apparently higher. For performance, 12-volt batteries are recommended, and these are very popular with the newer components being used in cars with 144-volt systems. The new 8-volt battery offers a good balance between the range of the 6-volt and the acceleration capabilities of the 12-volt.

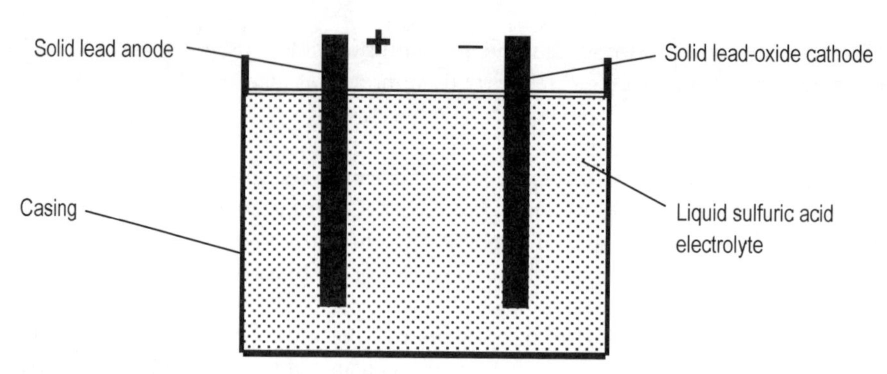

Figure 2.6 Schematic of a liquid-acid battery.

Temperature has a direct effect on the performance of a lead-acid battery. The concentration of sulfuric acid inside the battery increases and decreases with temperature. A battery being used in 0°C weather will only operate at 70% of its capacity. Likewise, a battery being used in 43.3°C weather will operate at 110% of its capacity. The most efficient temperature that battery manufacturers recommend is 25.5°C. Because the temperature factor is important in colder climates, insulated battery boxes or thermal management systems are highly recommended.

Maintaining batteries is very simple and requires little time. Each battery has three 'flooded' cells, or six, which require watering with distilled water once a month. In addition the batteries need to be cleaned once every month with a solution of distilled water and baking soda to prevent ion tracking. Ion tracking is a condition in which dirt or moisture on top of the battery forms a conductive path from one terminal to another, or to metal such as a battery rack. This can cause the ground-fault interrupter to trip on some battery chargers when the car is charging. Ion tracking is more prominent when the batteries are not stored in a protective enclosure or battery box.

Recently, Electrosource, a company in Austin, Texas developed a horizon lead-acid battery. To develop this battery, Electrosource invented a patented process to extrude lead onto fiber-glass filaments that are woven into grids in the battery's electrode plates. The results are greater power capacity, longer life cycle, deep discharge without degeneration, rapid recharge, and high specific energy. The battery is sealed and maintenance free. Each battery is 12 volts and costs about $440. Table 2.3 compares today's sealed lead-acid battery with the new horizon lead-acid battery technology.

Table 2.3 Comparison of today's sealed lead-acid battery with the new horizon lead-acid battery.*

Characteristics	Today's lead acid battery	Horizon lead-acid battery
Specific energy (Wh/kg)	33	42
Energy density (Wh/L)	92	93
Specific power (W/kg)	75	240
Recharge time (hours)	8-16	< 5
Cycle life (cycles)	400	800

* Adapted from Electrosource Inc.

Nickel-zinc (Ni-Zn), nickel-iron (Ni-Fe) and nickel-cadmium (Ni-Cd) batteries In addition to improved lead-acid batteries, two other batteries with metallic electrodes (the Ni-Zn and Ni-Fe batteries) are in advanced stages of development. The Ni-Zn battery uses dilute potassium hydroxide as an electrolyte. The present weakness of this battery is its lifetime, which is limited to 200 to 300 cycles. Researchers will try to extend this value to 300 to 500 cycles during the next few years. Such lifetimes may be sufficient for short-range urban driving where complete recharging is not often necessary. This battery's energy density of 65 Wh/kg is expected to be improved to at least 80 Wh/kg in the future, while its power density of 175 W/kg is already excellent.

The Ni-Fe battery is similar to the Ni-Zn unit, and in fact, was first invented by Thomas Edison in 1901 and only supplanted by the lead-acid battery in the 1920s. Currently, Ni-Fe batteries have storage capacities of 50–55 Wh/kg and power densities of 100 W/kg. They have excellent lifetimes (900 cycles have been demonstrated) but are somewhat bulky. A troublesome additional problem with this battery is that hydrogen gas is evolved from the electrode, which both reduces battery efficiency and creates a potential safety problem. Another difficulty with the Ni-Fe battery is that its power output drops drastically at temperatures below 10°C and it is inoperable at 0°C or below. Heaters may therefore be needed to make these batteries usable in cold climates. Nonetheless, the Ni-Fe battery will probably be an attractive power source for trucks and buses.

The Ni-Fe and Ni-Cd batteries are known as the near-term batteries. Ni-Cds are already in use in some countries, e.g. Japan and Europe. They are apparently more expensive then lead-acid batteries because nickel is costly. Ni-Cd's advantages are higher-energy density and higher cycle lifes over 1000 charges. Although they can be recharged very quickly they have a tendency to overheat; Cd is also highly toxic so recycling efforts have to be managed very carefully. Although cadmium supplies are not very high, it can be produced from sources such as copper, lead, zinc, and cadmium recycling.

Ni-Fe batteries have high energy density and are capable of over 1000 deep discharge cycles before recharging. They need to be 11% overcharged to be charged. The result of the overcharging is water loss and a build-up of hydrogen, which is a safety concern. The battery's efficiency is being improved which reduces these problems.

Lithium-iron sulfide batteries The lithium-iron sulfide battery has as a negative electrode an alloy of aluminum and lithium, and iron sulfide as a positive electrode. The electrolyte is a molten salt that must be kept well above its melting point of about 350°C. This battery is potentially one of the most suitable for electric vehicles. It is quite compact, with energy/volume projected at better than 200 Wh/liter, and has a correspondingly high energy density of 100 Wh/kg and a power density of at least 100 W/kg. Despite the use of lithium, which is a fairly unstable metal that reacts with water to release hydrogen, safety tests in which prototype cells were crushed produced no combustion. Some severe engineering problems must be solved before the battery has a more adequate life expectancy than its present value of only about 200 cycles. Commercialization of a battery with a 1000 cycle lifetime is expected.

Although the lithium-ion battery was predicted to be a long-term battery, it will soon be available for lease to fleets. The promising aspects of the battery are its low memory effect, high specific energy of 100Wh/kg, high specific power of 300W/kg, and a battery life of 1000 cycles. The battery is 28.8 volts and consists of eight metal cylindrical cells encased in a resin module. Each battery has a built-in cell controller to ensure that each cell is operating within a specified voltage range of 2.5 to 4.2 volts during charging and discharging. The cell controller communicates with the vehicle's battery controller to optimize power and energy usage. The disadvantages of the lithium-ion battery are its very high cost and the ventilation system required to keep the batteries cool. The manufacturing costs are high because the battery uses an oxidized cobalt material for the anode, a highly purified organic material for the electrolyte, and a complex cell control system.

The lithium (metal) sulfide battery is an elevated temperature battery based on a lithium alloy/molten salt/metal-sulfide electrochemical system. This system provides high specific power for better acceleration. Other advantages include its small size, low weight, and low cost per kilowatt hour. The battery is composed of iron disulfide and lithium-aluminum alloy which is completely recyclable.

The lithium-polymer battery is based on thin film technology. The battery is expected to cost 20% more then lead-acid but deliver twice the energy, with a lifespan of 50 000 miles. It has an operating temperature between 65°C and 120°C. It can be fast charged in less then 90 minutes, but can be damaged by overcharging. The major challenge confronting this technology is scaling up its size to properly power an electric car.

Sodium-sulfur (Na-S) batteries The Na-S battery (Figure 2.7), like the lithium-iron sulfide battery, requires high-temperature operation. It differs significantly from other advanced batteries in that the electrodes are liquid (molten sodium or sulfur) and the electrolyte is a solid. The two electrodes are separated by a ceramic material (beta alumina), which allows passage of (or conducts) sodium ions but not sodium atoms. The Na-S battery has adequate energy and power densities of 90 Wh/kg and 100 W/kg, respectively. Its volumetric energy density is expected to be at least 110 Wh/L. The major advantage of the liquid electrodes should be a long lifetime, because liquids are much easier to reconstitute than solids. Lifetimes of greater than 1000 cycles are expected, and the target for a sodium-sulfur battery for utility use is 2000 cycles. The Na-S battery has durability and safety problems that are yet to be solved. Sodium is a very reactive metal, and the necessary safety systems to protect a Na-S battery in case of a crash may make it too bulky for electric vehicle use. However, the battery may turn out to be of interest to utilities.

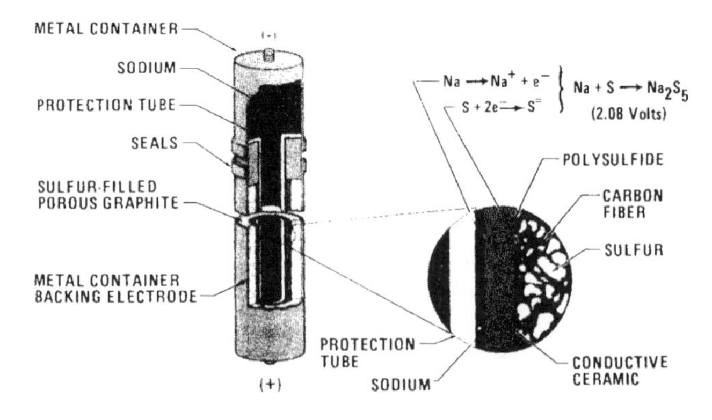

Figure 2.7 High-temperature sodium-sulfur battery (Wilbur, 1985).

The requirement that the lithium-iron sulfide and Na-S batteries be maintained at 400°C may appear to be a serious disadvantage. But because they are not 100% efficient, they do give off heat in charging. If carefully insulated, this heat is sufficient most of the time. A small auxiliary heater may be needed after a long periods of inactivity.

The NaS battery uses a ceramic beta-alumina electrolyte tube with sodium negative electrodes and molten sulfur positive electrodes within a sealed insulated container. To keep the sulfur in a molten state the battery must be kept at a temperature of 300–350°C. The batteries have been built in heaters to keep the sodium and sulfur from solidifying. Presently the battery costs seven times more then lead-acid batteries, but their price is expected to drop during high volume production because the materials to build the battery are plentiful and cheap. NaS battery research is currently being supported by various companies in the USA, UK, Canada, Germany, Sweden, etc.

The main disadvantage of the NaS battery is the high temperature which has raised safety concerns. Also the battery must be charged every 24 hours to keep the sodium and sulfur from solidifying.

Other batteries Other batteries that are even more advanced are also under investigation. One, a sodium-sulfur battery with a glass electrolyte, looks promising. There is interest in an aluminum-air battery that anticipates a high energy density. Because of its inexpensive components, this battery also offers the potential advantage of mechanical replacement of the aluminum plates instead of recharging.

One of the most promising battery technologies is the nickel-metal hydride (NiMH) battery. It is composed of nontoxic recyclable materials and is environmentally friendly. The NiMH has twice the range and cycle life of today's lead-acid batteries. It is composed of nickel hydroxide and a multicomponent, engineered alloy consisting of vanadium, titanium, nickel, and other metals. It is sealed, maintenance free, and can be charged as quickly as 15 minutes. It can withstand overcharging and over discharge abuse. Ovonic Battery of Troy, Michigan is currently manufacturing NiMH batteries.

Sodium-nickel-chloride ($NaNiCl_2$) batteries are under development by AEG Anglo batteries. The battery operates at a temperature of 300°C and is claimed by its manufacturer to be safe in accidents and operate even if one of its cells fails. The $NaNiCl_2$ battery currently meets the future goals for both energy and power density. The battery can be cooled down and reheated without damage; however, no current can be drawn from the battery if the temperature is below 270°C. Costs to produce the battery remain a problem. Various car making companies, e.g. BMW, Mercedes Benz, Opel, and VW, are testing their electric cars with $NaNiCl_2$ batteries.

The zinc-air battery developed by the Israeli firm Electric Fuel, Inc. is being used to power 40 test vans. After the battery pack or cassette is discharged it is taken out of the vehicle and replaced with another cassette. The cassette replacement is done in a matter of minutes with highly automated equipment set up at various geographical locations. The discharged cassette has its electrodes replaced with fresh ones and the cassette is used for another vehicle. The spent electrodes are recycled and used to make new electrodes. The battery has an energy density 10 times that of lead-acid.

Aluminum-air is a battery that has aluminum plates added every 200 miles to replenish the used aluminum. The aluminum plates react with oxygen in sodium hydroxide electrolyte solution to form sodium aluminate. The sodium aluminate produces a byproduct of aluminum, aluminum trihydroxide, which is removed and replaced with fresh aluminum. Aluminum-air batteries will probably be used by larger vehicles because of its size.

The nickel-hydrogen battery is currently under development by Johnson Controls. It is very expensive but has a long life and is safe to operate. It is currently being used in spacecraft and deep-water vehicles.

The nickel-zinc battery has more power then lead-acid batteries, but has a shorter discharge cycle making it only useful for short commutes.

The zinc-chloride battery has high energy but must have a complex system to recapture the chlorine released during recharging.

The zinc-bromide battery developed by Johnson Controls stores electricity by plating zinc onto a surface and then unplating it. A bromide electrolyte solution which is 80% water is pumped through the battery to cause the plating and unplating reactions. Pure bromide is extremely toxic. Safety issues were raised about the battery in a 1992 electric vehicles race when it was involved in an accident. A hose that carries the bromide electrolyte became unconnected from the battery and leaked onto the race track, releasing irritating fumes.

The nickel-zinc, zinc-chloride, and zinc-bromide batteries will probably not be used commercially.

In the near future, electric vehicles are expected to be using advanced lead-acid batteries such as the horizon sealed lead-acid battery. One of the most promising battery technologies is the nickel-metal hydride battery. Although these batteries are very expensive today, the price is expected to come down sharply. Further down the road, electric vehicles might be powered by lithium batteries such as the lithium-ion battery now under development by some Japanesse companies.

Developments of batteries We discuss below some new developments in batteries. Our understanding of batteries is about as old as our knowledge of electricity. In the not-too-distant future, 'advanced' lead-acid batteries with new design breakthroughs may achieve ES capacities of 60 Wh/kg and 1000 cycle lifetimes. However, these figures of merit, which seem to be near the limit for lead-acid batteries, are still insufficient for satisfactory vehicular range and acceleration. Hence, battery researchers have begun to look at alternative materials, especially lighter elements.

During the past decades research and development has been undertaken on batteries of quite untraditional design. The leading contenders for commercial success among these are the lithium-metal sulfide, the zinc-chloride, and the sodium-sulfur batteries. All three of these (i) use light elements in one or both electrodes to achieve higher energy densities, and (ii) operate at elevated temperatures. The zinc-chloride system must be at about 40°C for its electrolyte to work properly; the other two need to be at the 300–400°C range to operate. Development of the zinc-chloride battery was originally slowed because of the dangers of chlorine gas. A system that stores the chlorine in a frozen form (as a chlorine hydrate) seems to have solved that problem. The zinc-chloride battery will probably not be a candidate for most vehicular use due to its bulk, although it may be useful in trucks and vans if its size and shape do not cause severe design problems.

New developments in batteries are important for the stand-alone operation of solar photovoltaic and wind generators for utilities as an alternative to pumped hydropower and for electric vehicles. All of these applications require major technical improvements and cost reductions for storages. Advance battery concepts are typically variations on the conventional lead-acid battery.

It is obvious that extending the lifetimes of batteries can make them more cost competitive. Developers of advanced batteries try to attain life cycles in excess of 2000 and lifetimes of over 10 years. To put these values into context, it is noted that daily cycling would require 3650 cycles over a 10-year lifetime. For comparison, note that the lifetime of a conventional pumped storage plant can be as high as 50 years, and the number of charge-discharge cycles presents no additional limitation to that technology. This observation has important economic ramifications because the capital cost of equipment is usually amortized over periods no longer than the working life of the equipment. The relatively shorter lifetimes of present-day batteries means that the projected costs of delivered energy are more than 50% higher for batteries than for pumped hydropower. Thus, the cost to the utility ratepayer is correspondingly higher for that fraction of the electric energy delivered by batteries. In stand-alone processes utilizing batteries, such as remote wind generation, cost can be considered in a similar manner, but with different operational parameters if high availability is the objective. Regardless, achieving economic viability for large-scale wind or solar generation will require significantly greater energy storage capacities. That is, such systems must be capable of storing the equivalent of several days' worth of energy usage so that they can deliver the energy at the normal rate of daily consumption during periods of limited sunlight or wind.

To bring advanced batteries to market, a coordinated effort is now underway to improve the development of battery technology. This effort has been undertaken by the United States Advanced Battery Consortium (USABC), whose members include Ford, Chrysler, GM, utilities, the Department of Energy, national laboratories, and battery companies. The USABC has established a timeline that includes goals for battery development. The timeline is divided into mid-term and long-term goals as shown in Table 2.4.

Finally, there is a general need for the weight (or mass) and volume characteristics for batteries to be relatively competitive with those for other conventional ES technologies. Weight is often the most important technical requirement for batteries in such applications as electric vehicles where the alternative storage is a tank of gasoline.

Organic molecular storage

The intermittent availability of solar radiation, its seasonal and geographical variations, and its relatively low intensity, will limit the exploitation of that resource until it can be converted into forms of energy that can be efficiently stored and transported. However, most technologies that are presently available for the utilization of solar energy depend on the direct conversion of solar radiation into low-grade heat or electricity, both of which are difficult to store. The process of photosynthesis shows that endergonic photoreactions can convert radiative energy into storable chemical energy and, for at least a century, at least, there have been attempts to find biological systems for the conversion of solar energy. In general, these efforts have been aimed at finding cyclic processes in which an endergonic photochemical reaction is followed by exergonic regeneration. After the second reaction initial material is regenerated:

$$A + h\nu \rightarrow B$$

$$B \rightarrow A + \text{useful energy}$$

Table 2.4 The mid- and long-term goals of the USABC.

Characteristics	Mid-term goals	Long-term goals
Specific energy (Wh/kg)	80 (100 is desirable)	200
Specific power (W/kg)	150 (200 is desirable)	400
Recharge time (hours)	< 6	3 to 6
Cycle life (cycles)	600	1000
Price ($/kWh)	< 150	< 100

The general requirements for photochemical energy storage and conversion are discussed extensively elsewhere (Wilbur, 1985). Here, we restrict our discussion to selected aspects of six requirements which significantly affect the performance of organic systems for delivering heat. We point out the interdependence of these factors, and indicate areas where conflicting demands arise when trying to optimize them independently.

The efficiency of converting light into chemical energy by an endergonic photoreaction is subject to limitations that arise from the quantum nature of photochemical processes. These limitations apply to all quantum converters, including biological photosynthetic systems and photo-electric devices.

As shown in Figure 2.8, the photoreaction $A \rightarrow A^* \rightarrow B$ can be initiated only by photons whose energies E_n match or exceed the energy E^* of the excited state A^* which is involved in the photoreaction. Therefore only the fraction E^*/E_n of the absorbed radiation is available for the photoreaction. When this threshold effect is combined with the distribution of spectral intensities in the terrestrial solar spectrum, one finds that the maximum possible efficiency of a quantum process utilizing solar radiation depends on E^*, and reaches a maximum of about 44% at a value of E^* corresponding to a radiation wavelength of –1100 nm. At wavelengths that are more effective photochemically but less abundant, this efficiency decreases (e.g. to 25% at about 550 nm and 10% at about 390 nm). These efficiencies represent upper limits for any quantum converter operating with solar radiation (for details, see Sasse (1977)).

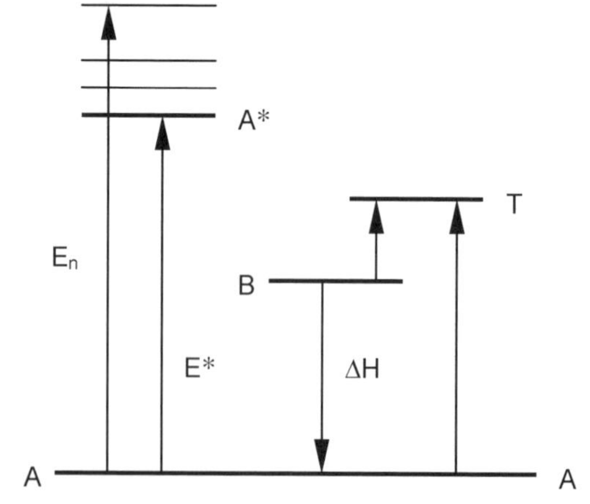

Figure 2.8 Representation of energy levels in a cyclic system converting light into heat. Only the fraction E^*/E_n of the absorbed radiation can be used in the reaction $A + h\nu \rightarrow A^* \rightarrow B$. T is the transition state of the exothermic reaction $B \rightarrow A = \Delta H$.

The potential for energy storage in an organic photo-reaction depends on the difference in the energy contents of A and B, which can generally be recognized qualitatively by inspection of the structures involved. For the reactions considered here, ES requires the formation of strained structures (i.e. photoproducts with bond lengths and angles that differ from 'normal' values) and/or an appreciable loss of 'resonance energy' (i.e. a decrease in stabilization by π-electron delocalization) in the molecules concerned. The effects of strain and loss of resonance energy on the energy contents of A and B can be qualitatively compared and quantitatively predicted.

The overall performance of a photochemical ES system delivering heat depends critically on the temperature at which the exothermic reaction occurs, which depends on the method used to induce the release of heat from ES compound B. Traditionally, heat release is caused by heating photoproduct B until the desired rate of heat evolution has been reached.

In general, the temperature ranges for heat evolution are mainly between 100°C and 200°C. Thermal damage is a problem with compounds containing olefinic double bonds at these temperatures, although aromatic dimmers function without detectable damage at higher temperatures. Unless especially stable compounds can be developed, e.g. photochemically reactive aromatic fluorocarbons, heat-release temperatures in excess of approximately 220°C probably cannot be routinely employed. However, photochemical storage systems operating between 100°C and 200°C may be sufficiently applicable in solar heating, and cooling at the upper temperature limitations may not prevent further interest from developing in photochemical ES systems. Major improvements in storage capacity are possible for all of the systems described here by improving and better understanding the photophysical and photochemical characteristics of the ES compounds. Although much quantitative work has been done on individual compounds, more research is needed into the effects of structural modifications on photochemical and photophysical characteristics.

Some recognized areas for future investigation are the design of molecules that react sequentially with photons of different energies, and the use in ES of molecules that store energy. Also, the development of heterogeneous sensitizers is likely to offer important advantages both by extending the useful range of the solar spectrum and by simplifying the thermal steps involved.

Chemical heat pump storage

The system is based on a discontinuously working solid state absorption heat pump incorporating a storage function, of which the operating principle is as follows: two chambers are connected with each other. Chamber I is the energy accumulator containing the vapor absorbing salt (Na_2S), and chamber II is the condenser/evaporator containing the working fluid (H_2O), as shown in Figure 2.9. The reversible chemical reaction in case sodiumsulfide and water are used as a working pair is:

$$Na_2S{\cdot}nH_2O + \text{heat} \Leftrightarrow Na_2S + nH_2O$$

The system is charged by supplying heat from a heat source to chamber I so the vapor is driven from the salt to chamber II, where the vapor condensates. The heat of condensation has to be removed from chamber II, then called a condenser. If the system is completely charged the discharge reaction starts on the right side of the reaction balance. When the system starts to discharge, heat is supplied to chamber II, then called an evaporator, to provide the latent heat of evaporation. At the same time, heat is released in chamber I when the vapor is absorbed by the hygroscopic salt. The result is that chamber II can be used to produce cold, while chamber I can produce heat. The discharge process stops when the fluid is completely evaporated, the reaction balance is then on the left side and the system must be charged again. The pressure-temperature curve above the sorption reaction between Na_2S and H_2O lies about 55°C from the vapor pressure curve of water. This temperature difference is defined as the equilibrium temperature difference. This is the temperature difference between the hot and the cold chamber in pressure equilibrium, i.e. when no vapor is moving from one chamber to the other and no heat is taken from or supplied to one of the chambers If the connection between the chambers is shut when the system is completely charged, the energy can be stored for an indefinite period. Both chambers may then reach equal temperatures without vapor moving from the evaporator to the accumulator; the losses are restricted to the loss of sensible heat, which is about 3–5 % of the total heat stored. (For details, see de Beijer and Klein (1992) and de Beijer, (1993).)

The system also works with other absorbers and working fluids. Although the power output level is usually in the same order of magnitude, the storage capacity depends on the physical and chemical properties of the absorber and working fluid used. For applications where heat storage is an important qualification, the use of Na_2S and H_2O leads to the highest possible energy storage density.

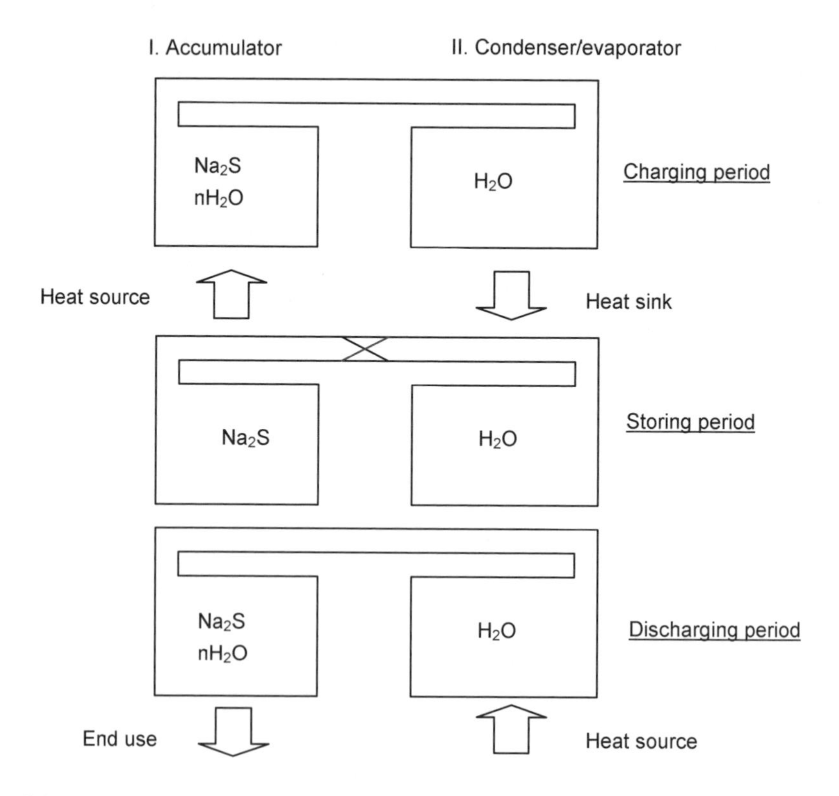

Figure 2.9 Operating principle of a chemical heat pump system (*adapted from de Beijer, 1993*).

Per kg, Na_2S 1.1 kWh of heat can be stored. During the discharge process about 1.05 kWh of heat can be released per kg of Na_2S. The cooling capacity of the system is about 0.75 kWh per kg of Na_2S, so the theoretical cooling factor is about 0.7. Cooling can be provided from ambient to 0°C; below this point the water in the evaporator will freeze. By using heat exchangers in which the vapor absorbing salt is not integrated, like those used in comparable systems in the past, maximum power outputs of about 40 W/kg salt can be reached. By integration and optimization of the heat exchanger and the crystal structure of the salt, the system has considerably improved power output per kg of salt. Power outputs of over 1.5 kW of heat and 1 kW/kg of Na_2S are reached (de Beijer, 1993).

If there is a demand for high powers only, and the internal temperature step of the heat pump or the evaporation temperature must be adapted to an application, the use of other absorbers and working fluids is recommended.

If the system is used for air conditioning in office and residential buildings, the system has the following advantages (de Beijer, 1993):

- saving of energy and costs by using residual or waste heat,
- use of difference in the electricity price (day/night rate),
- reduction of the peak power payment to the energy distribution companies,
- less capital investment in comparison to conventional systems,
- cooling of buildings using district heating, combined with heat or cold storage, and
- low maintenance cost, because the system has no moving parts.

Some of the advantages which make the system suitable for application in vehicles are:

- high power density of 150–220 W/kg,
- low-specific system weight of 4.5 kg per kWh, and
- specific system volume 6 dm^3 per kWh.

However, the basic problems which are encountered in these systems are:

- poor stability and side reactions of the working materials,
- low power output, and
- low efficiency.

In fact, the system has proved its technical applicability for heating, cooling, and air conditioning applications with low cost and high market potential. There is good opportunity to use alternative absorbents and working fluids in these chemical heat pump storage systems, and for the future such systems look promising, particularly for energy conversion systems.

2.4.3 Biological storage

Biological storage is the storage of energy in chemical form by means of biological processes and is considered an important method of storage for long periods of time. In this book we do not discuss bioconversion, as it is not presently of practical interest and very limited information is available on biological storage in this regard. Note that if the quantum efficiency of biological processes can be increased by a factor of 10 over its present efficiency of about 1%, interest in bioconversion for ES will likely increase.

2.4.4 Magnetic storage

Energy can be stored in a magnetic field (e.g. in a large electromagnet). An advanced scheme that employs superconducting materials is under development. At temperatures near absolute zero, certain metals have nearly no electrical resistance, and thus large currents can circulate in them with almost no losses. Because this scheme stores DC electricity, some losses are incurred in converting standard AC power to and from DC, and some energy is used to drive the refrigeration device to maintain the low temperatures. Overall storage efficiencies of 80–90% are anticipated for these superconducting magnetic ES systems.

Magnetic storage is considered for two main purposes. First, large superconducting magnets capable of storing 1000–10 000 MWh of electricity could be attractive as load-leveling devices for central power stations, and may be cost-effective at such capacities. Second, smaller magnets with storage capacities in the 10 kWh range may be cost-effective in smoothing out transmission line loads, to better match short-term customer demands and generating-equipment characteristics. A small superconducting magnet that can help in meeting customer peak needs at the far end of a transmission line could increase the effective load that the line can serve by as much as 25%, producing cost savings that could affect the additional costs of expanding the transmission line capability.

The potential for highly efficient electrical ES is especially attractive for utilities, particularly when energy costs increase. The storage coil in a superconducting magnetic ES would likely be helical and located below ground. In order to obtain high current densities in the coil, and thereby reduce the amount of costly superconductor needed, the proposed storage system is recommended for operation at 1.85 K. The storage unit could be charged during off-peak hours, with the electricity discharged back to the grid later to meet peaking needs. The unit would generally operate on one charging and discharging cycle per day and would be connected to a three-phase utility transmission line. As noted earlier, the storage system requires the conversion of alternating to direct current for storage in the superconducting coil. An ES capacity of 1 GWh is typical of the size that is considered commonly. As the costs are projected to be high, no large-scale superconducting magnetic ES device is expected to be built in the foreseeable future. However, a small prototype system is now being built for use on transmission lines to dampen rapid voltage variations that occur with a periodicity on the order of seconds. The system has a ES capacity of 10 kWh and is designed to respond in fractions of a second. If line variations are reliably damped, the effective total capacity of the transmission system increases. The value of this increase in capacity is expected to more than compensate for the cost of the storage system. The stabilization unit itself is similar to the large-scale system for daily energy storage, but operates at the normal boiling point of helium.

2.3.5 Thermal Energy Storage (TES)

Thermal energy may be stored by elevating or lowering the temperature of a substance (i.e. altering its sensible heat), by changing the phase of a substance (i.e. altering its latent heat), or by a combination of the two. Both TES forms are expected to see extended applications as new energy technologies are developed. TES is the temporary storage of high- or low-temperature energy for later use. Examples of TES are the storage of solar energy for overnight heating, of summer heat for winter use, of winter ice for space cooling in summer, and of the heat or cool generated electrically during off-peak hours for use during subsequent peak demand hours. Solar energy, unlike fossil fuels, is not available at all times. Even cooling loads, which nearly coincide with maximum levels of solar radiation, often are present after sunset. TES can be an important means of offseting the mismatch between thermal energy availability and demand.

Energy storage as sensible heat changes appear very promising for the high-temperature storage of large quantities of energy at fossil-fired power plants. Oil will likely be used as the storage medium for this type of system.

The ES types discussed in the preceding sections of this chapter are applicable to electrical energy. In thermal electrical generating systems, TES offers the possibility of storing energy before its conversion to electricity.

Energy quality as measured by the temperatures of the materials entering, leaving and stored within a storage is an important consideration in TES. For example, one kWh of energy can be stored by heating one metric ton of water 0.86°C, or by heating 10 kg of water 86°C. The latter case is more attractive in terms of energy quality as a wider range of tasks can be accomplished with the higher temperature medium upon discharging the storage. For comparison, it is pointed out that the quality of energy stored in mechanical ES is higher still as, for example, one kWh of energy can also be stored by lifting one metric ton of water 314 m.

Energy demands in the commercial, industrial, utility and residential sectors vary on a daily, weekly and seasonal basis. The use of TES in such varied sectors requires that the various TES systems operate synergistically, and that they are carefully matched to each specific application. The use of TES for such thermal applications as space heating, hot water heating, cooling, air-conditioning, etc. has recently received much attention. A variety of new TES techniques has been developed over the past four or five decades in industrial countries. TES systems have enormous potential for making the use of thermal equipment more effective and for facilitating large-scale substitutions of energy resources economically. In general, a coordinated set of actions is needed in several sectors of the energy system for the maximum potential benefits of storage to be realized.

Sensible heat changes in a material are dependent on its the specific heat capacity and the temperature change. Latent heat changes are the heat interactions associated with a phase change of a material, and occur at a constant temperature. Sensible heat storage systems commonly use rocks or water as the storage medium. Latent heat storage systems can utilize a variety of phase change materials, and usually store heat as the material changes phase from a solid to a liquid. We discuss each TES type in detail separately in the next chapter.

2.5 Hydrogen for Energy Storage

Energy can be stored in chemical form as hydrogen. The potential versatility of this low-density gas for storing and transmitting energy has made it the subject of extensive research over the last few decades, and brought it increasingly into the energy news. The variety of end uses for which hydrogen is well adapted includes being a fuel for electricity and/or heat production in fuel cells or combustion engines, and for powering transportation devices, in addition to being a chemical commodity.

2.5.1 *Storage characteristics of hydrogen*

Hydrogen is not a source of energy. It does not exist in pure form to an appreciable extent on this planet. It will be useful as a heat intermediate form (an 'energy carrier'), much as electric energy is used today. It has many advantages over electricity in this intermediate energy competition. It can be stored more easily, transported more cheaply, and can deliver a much greater variety of end use forms (including electricity). It only suffers in the comparison when it comes to production costs.

Hydrogen has advantages and disadvantages as a medium for storing energy. Its energy density on a mass basis is high (116 300 kJ/kg) as compared with, for instance, about 46520kJ/kg for jet aviation fuel and liquid methane. Hydrogen has a very low volumetric energy density, however, and thus requires a large storage volume usually. For example, the volumetric energy density of liquid hydrogen is only 20.9×10^6 kJ/m^3, compared to 34.84×10^6 kJ/m^3 for gasolin.

2.5.2 *Hydrogen storage technologies*

It is more difficult to store hydrogen than most conventional chemical fuels. Whereas gasoline can be stored in a relatively inexpensive tank, for example, hydrogen requires either an expensive high-pressure tank in which the steel is 100 times heavier than the hydrogen stored, or a refrigerated, vacuum-insulated system which is both expensive and energy consuming. An alternative hydrogen storage system absorbs hydrogen in metallic powders, forming hydrogen metallic compounds (metal hydrides). These can be decomposed by the application of heat, releasing hydrogen. An additional favorable property of metal hybrids is that they release hydrogen steadily at approximately a constant pressure. In a given volume, metal hydrides can hold an amount of hydrogen about the same as can be stored as a liquid. However, metal hybrids are dense, so that most of the weight in a storage system is due to the metal, not the hydrogen. A good candidate for metal hybride storage is iron-titanium.

2.5.3 *Hydrogen production*

Most commercially available hydrogen is produced at present from such hydrocarbons as methane or other gases, oil or similar petroleum products. The raw material from which most hydrogen will likely be produced in the future is water, which is abundantly available. By the application of the required amount and form of energy, water can be separated into hydrogen and oxygen. In fact, a significant amount of hydrogen is already produced by the electrolysis of water. Water electrolysis is a relatively simple process in which positive and negative electrodes are immersed in water and a voltage is applied. Gas bubbles then appear, hydrogen at the negative electrode and oxygen at the positive electrode. The gases are prevented from diffusing to the other electrode by diaphragms that allow electric current to pass, but not gas. In a perfectly efficient electrolytic cell, 94 kWh of energy is needed to produce 28.3 m^3 of hydrogen (at atmospheric pressure). Only 79 kWh of that energy has to be electrical; the other 15 kWh of energy can be heat. This latter requirement can be exploited by power engineers, as a large electric power plant can provide an electrolysis plant with both electrical and thermal energy, the latter which is otherwise unused. At present, most electrolysis cells operate at energy efficiencies of about 60–75%, although it should be noted that they use a high-quality form of energy (electricity) to make a slightly lower-quality form (chemical energy as hydrogen). The future prospects for large-scale electrolysis do not offer much hope as a future process. A process that generates inexpensive electricity will likely make electrolytic production of hydrogen commercially attractive.

Water can be decomposed directly by thermal energy, but the temperature required for a reasonable yield, 2500°C, is not readily available from nuclear reactors. It is possible, though not likely, that fusion reactors or even high-temperature gas-cooled fission reactors could provide such temperatures in the future.

A more practical approach for using heat to dissociate water is through a thermochemical process. In such a process a series of chemical reactions takes place where the required energy is supplied as heat. Most of the chemical products of the reactions are consumed in the other reactions, and the net reaction is decomposition of water into hydrogen and oxygen. However, practical thermochemical reactions (e.g. those that produce at least half as much hydrogen per unit of primary energy as electrolysis) appear to need temperatures of 900°C or higher.

Direct thermal dissociation or these more complicated thermochemical water decompositions are attractive because they upgrade heat to hydrogen. By bypassing the less efficient production of electricity needed for electrolysis, they have an advantage Unfortunately, however, this advantage depends on temperature, and the routine availability of such temperatures from suitable energy source may necessitate the successful development of advanced nuclear reactors or large-scale solar installations using lenses or mirrors to concentrate energy.

Another technique for hydrogen production that is undergoing laboratory development and which holds some future promise is photo-electrolysis. In this process the electrodes are semiconductors which are immersed in an aqueous solution, and which produce an electric current and voltage when exposed to sunlight. This current then drives an electrolysis process. The process works well at laboratory scale, but some difficulties need to be resolved. For instance, a physical separation process needs to be developed for the hydrogen and oxygen generated, as these gases are presently mixed in the electrolyte, posing an explosion hazard.

Other hydrogen production processes that are only at the early research stages of development or of uncertain practicality are being considered. In another process, biophotolysis, enzymes from algae help split water.

In general, for hydrogen to compete with electricity as an intermediate energy source, it appears that highly efficient water electrolysis systems using very inexpensive baseload electricity will have to become available, or technological breakthroughs in other hydrogen production processes will have to occur.

Table 2.5 A technical comparison of ES technologies.

ES technology	ESD (kJ/m^3)	MPD (kW/m^3)	SE (kJ/kg)	MSP (kW/kg)	CSL	TDT (s)	UL	AA
Mechanical ES								
- Flywheels	5×10^4	5×10^3	6	0.6	Days	1-10^3	20 yrs	RB/SA
- Compressed air[a]	10^4	1	—	—	Weeks	10^4-10^5	20 yrs	EU
- Pumped hydro storage[b]	10^3	0.1	1	1×10^{-4}	Weeks	10^4-10^5	20 yrs	EU
Thermal ES								
- Sensible heat								
High temp. oil	10^5	10	100	0.01	Days	10^4-10^5	20 yrs	EU
Low temp. rocks	3×10^4	3	10	0.001	Days	10^4-10^5	20 yrs	RH
- Latent heat								
Various salts	3×10^5	30	100	0.01	Weeks	10^4-10^5	10^2-10^3 cycs	RH
Electromagnetic ES								
- Capacitors[c]	100	10^8	0.3	3×10^5	Days	10^{-6}-10^{-3}	10^4 cycs	SA
- Magnets/coils[d]	10^5	10	1000	0.1	Days	10^{-3}-10^5	20 yrs	EU/SA
Chemical ES								
- Batteries[e]								
Pb acid	5×10^4	5	100	10	Weeks	10-10^4	10^3 cycs	ES/SLI/AP
Others[f]	10	10	500	—	Weeks	10-10^5	10^3-10^5 cycs	EU/SA/AP
- Hydrogen[g]	9×10^6	—	10^5	—	Months	10^4-10^5	20 yrs	EU/SA/AP
- Methanol	2×10^7	—	3×10^4	—	Months	10^4-10^5	20 yrs	EU/SA/AP

(ESD: Energy storage density; MPD: Maximum power density (based on minimum possible discharge times, except for those technologies where the device is for the electric utility diurnal storage application); SE: Specific energy; MSP: Maximum specific power (based on minimum possible discharge times, except for those technologies where the device is for the electric utility diurnal storage application); CSL: Charged shelf life; TDT: Typical discharge time (minimum discharge times are usually for partial discharges. For specific technologies, smaller units can be discharged faster than larger units.); UL: Useful life in cycles and years (for economic considerations, a lifetime of twenty years is equivalent to an infinite life. There is no indication that the various devices given a 20-year life expectancy in the table will wear out in 20 years.); AA: Application areas; EU: Electric utility diurnal storage; ES: Emergency service for electric power system failure, and so on; SLI: Automotive-starting, lighting, and ignition; RB: Regenerative braking; RH: residential heating; SA: special applications; AP: Automotive propulsion)

[a] Data for compressed gas are based on the plant installed at Huntorf, Germany.

[b] Data for pumped-hydro storage are based on the facility at Ludington, Michigan.

[c] Data are for oil-impregnated-paper capacitive ES systems.

[d] Most data are for a 104 MWh superconducting coil constructed underground and used for diurnal storage on an electric power system. Energy density is based on excavated volume.

[e] Most data are based on diurnal ES application. Automotive propulsion batteries will have higher power rating.

[f] Several advanced battery systems were proposed for automotive propulsion and electric utility applications.

[g] Storage efficiency depends on the type of hydrogen storage, and storage cost depends on the medium used. Liquid hydrogen storage is less efficient but cheaper than metal hydride storage. Power and storage capacities are almost unrelated for hydrogen storage.

Source: Hassenzahl (1981).

Table 2.6 Capital costs and efficiencies for ES devices.

Technology	State-of-the-art technology		Future technology goal	
	Capital cost ($/kW)	Efficiency (%)	Capital cost ($/kW)	Efficiency (%)
Pumped hydro				
- Aboveground	300-500	71-74	—	—
- Underground	—	—	500[a]	75
Compressed air	770	50-52	600[b]	65-70
Superconducting magnetic	—	—	2600[c]	80-90
Batteries	600	75	35-55[d]	50-60
Flywheels	—	—	1200	70
Thermal				
- Residential low temperature	200	100	150	100
- Industrial high temperature[b]	400	70	200	85-95
- Commercial cooling	400	85-100	100	85-95
- Seasonal storage	—	—	200	60-70
Hydrogen				
- Daily storage	40-50	68-84	45	68
- Electrolytic production[e]	400-500	60-65	200-300	80-85

[a] Single high-head pump turbine.
[b] Cost basis is system with 8-hour storage capacity (with no fuel consumption).
[c] Cost basis is system with 11-hour storage capacity.
[d] Cost basis is vehicle battery with 2-hour storage capacity
[e] Efficiencies for electrolysis plant only.
Source: (Wilbur, 1985).

2.6 Comparison of ES Technologies

Many attempts have been made in the past to compare ES technologies based on such factors as efficiency, cost, application and numerous other technical characteristics. Two such comparisons are presented in Tables 2.5 and 2.6. These tables consider the main ES technologies discussed in this chapter, as well as others which, for several reasons, were not mentioned. The uncertainties about many ES technologies are evident in the tables, regarding present capabilities and future prospects and expectations.

2.7 Concluding Remarks

The different ES systems discussed in this chapter have their own advantages and disadvantages, and are suitable for different energy forms. Where thermal energy is available and thermal demands exist, but these are not necessarily coincident, thermal energy storage is the most direct means of ES as it avoids the need to convert energy from one form to another and the ensuing conversion losses. For the case of solar energy in particular, thermal energy storage appears to be the most efficient and effective means of storage for solar thermal applications.

References

Cassedy, E.S. and Grossman, P.Z. (1998). *Introduction to Energy, Resources, Technology, and Society*, Cambridge University Press, Cambridge.

de Beijer, H.A. (1993). The economic analysis of the salt water energy accumulation and transformation (SWEAT) system, *HPC Workshop Proceedings on Heat Pumps and Thermal Storage*, 24–25 May, Fukuoka, Japan, pp.111–118.

de Beijer, H.A. and Klein, J.W. (1992). SWEAT-a chemical heat pump for heat or cold storage, *IEA Heat Pump Center Newsletter* 10(1), 11–12.

Diamant, R.M.E. (1984). *Energy Conservation Equipment*, The Architectural Press, London.

Dincer, I. (1999). Evaluation and selection of energy storage systems for solar thermal applications, *International Journal of Energy Research* 23, 1017–1028.

Genta, G. (1985). *Kinetic Energy Storage, Theory and Practice of Advanced Flywheel Systems*, Butterworths Co., London.

Hassenzahl, W.V. (1981). *Mechanical, Thermal, and Chemical Storage of Energy*, Hutchinson Ros. Publ. Co., Stroudsburg, Pennsylvania.

Wilbur, L.C. (1985). *Handbook of Energy Systems Engineering*, John Wiley & Sons, New York.

Sasse, W.H.F. (1977). Organic molecular energy storage reactions. In: *Solar Power and Fuels* (Ed. J.R. Bolton), Academic Press, New York, pp.227–245.

Sheahan, R.T. (1981). *Alternative Energy Sources*, Aspen Publication Corp., Rockville.

3

Thermal Energy Storage (TES) Methods

I. Dincer and M.A. Rosen

3.1 Introduction

Energy demands in the commercial, industrial and utility sectors vary on daily, weekly and seasonal bases. These demands can be matched with help of TES systems that operate synergistically. The use of TES for thermal applications such as space and water heating, cooling, air-conditioning, etc. has recently received much attention. A variety of TES techniques have developed over the past four or five decades as industrial countries have become highly electrified. Such TES systems have an enormous potential to make the use of thermal energy equipment more effective and for facilitating large-scale energy substitutions from an economic perspective. In general, a coordinated set of actions in several sectors of the energy system is needed if the potential benefits of storage are to be fully realized.

Many types of energy storage play an important role in energy conservation. In processes which yield waste energy that can be recovered, energy storage can result in savings of premium fuels.

Energy may be stored in many ways, as pointed out in Chapter 2, but since in much of the economy in many countries, energy is produced and transferred as heat, the potential for thermal energy storage warrants study in detail.

Chemically-charged batteries became common in the mid-nineteenth century to provide power for telegraphs, signal lighting, and other electrical devices. By the 1890s, central stations were providing both heating and lighting, and many did both. Electric systems were almost all direct current (DC), so incorporating batteries was relatively easy. In 1896, Toledo inventor Homer T. Yaryan installed a thermal storage tank at one of his low-temperature hot-water district heating plants in that city to permit the capture of excess heat when electric demand was high. Other plants used steam storage tanks, which were generally not as successful. Other forms of TES were used to power street cars in the 1890s, including compressed air and high-temperature water that was flashed into steam to drive a steam engine. Electric cars and trucks were quite common prior to World War I, after which gasoline-powered internal combustion engines became prevelant.

TES deals with the storing of energy by cooling, heating, melting, solidifying or vaporizing a material, the thermal energy becoming available when the process is reversed.

Storage by causing a material to rise or lower in temperature is called sensible heat storage; its effectiveness depends on the specific heat of the storage material and, if volume is important, on its density. Storage by phase change (the transition from solid to liquid or from liquid to vapour with no change in temperature) is a mode of TES known as latent heat storage. Sensible storage systems commonly use rocks or water as the storage medium, the thermal energy stored by increasing the storage-medium temperature. Latent heat storage systems store energy in phase change materials, with the thermal energy stored when the material changes phase, usually from a solid to a liquid. The specific heat of solidification/fusion or vaporization and the temperature at which the phase change occurs are of design importance. Both sensible and latent TES also may occur in the same storage material. In this chapter, TES is considered to include the storage of heat through the reversible scission or reforming of chemical bonds.

Phase change materials (PCMs) are either packaged in specialised containers such as tubes, shallow panels, plastic bags, etc., or contained in conventional building elements (e.g. wall board, ceiling) or encapsulated as self contained elements.

The oldest form of TES probably involves harvesting ice from lakes and rivers and storing it in well insulated warehouses for use throughout the year for almost all tasks that mechanical refrigeration satisfies today, including preserving food, cooling drinks and air conditioning. The Hungarian Parliament Building in Budapest is still air conditioned with ice harvested from Lake Balaton in the winter.

TES has always been closely associated with solar installations, including both solar heating and photovoltaic applications. Today, compressed-air storage, batteries, chilled and hot water storage, ice storage, and flywheels are used, all designed to meet one or more of the purposes listed above. Many utilities provide direct incentives for energy storage applications, while time-of-day rates and high demand charges indirectly entice customers to consider these opportunities.

TES generally involves a temporary storage of high- or low-temperature thermal energy for later use. Examples of TES are storage of solar energy for overnight heating, of summer heat for winter use, of winter ice for space cooling in summer, and of heat or coolness generated electrically during off-peak hours for use during subsequent peak demand hours. Solar energy, unlike energy from fossil fuels, is not available all the time. Cooling loads, which nearly coincide with maximum levels of solar radiation, are often present after sunset. TES can help offset this mismatch of availability and demand.

Energy plays a major role in the economic prosperity and the technological competitiveness of nation. Because predicting future availability, demand and price of energy forms is at best approximate and often imprecise, it is important to have a broad array of technologies available to meet the energy needs of the future. Furthermore, the technologies developed should be those that ensure energy security, efficiency and environmental quality for a nation. TES is one such technology that is being promoted because it can substantially reduce total energy consumption, thereby conserving indigenous fossil fuels and reducing costly oil imports. Once technical and economic problems and risks have been reduced through proven performance, TES is expected to be accepted as an attractive option in the industrial and commercial sectors that will lead toward, among other benefits, increased energy efficiency and environmental benefits. TES has been identified as a method for substantially reducing peak electrical demands, thereby helping to ameliorate predicted peak-power shortages in the future. TES provides a potentially economic means of using waste heat and climatic energy resources to meet a significant portion of our growing needs for heating and cooling, especially for industrial facilities and commercial buildings. Environmental benefits also accompany the use of TES in many applications.

TES technology has been used in various forms and specific applications. Some of the more common applications have been the use of sensible heat (oils, molten salts) or latent heat (ice, phase-change material) storage for refrigeration and/or space heating and cooling needs. Research activities on TES are continuing at various national laboratories, universities and research centers throughout the world.

3.2 Thermal Energy

Thermal energy quantities differ in temperature. As the temperature of a substance increases, the energy content also increases. The energy required E to heat a volume V of a substance from a temperature T_1 to a temperature T_2 is given by

$$E = C(T_2 - T_1)V$$

where C is the specific heat per unit volume of the substance. A given amount of energy may heat the same weight or volume of other substances, and increase the temperature to a value greater or lower than T_2. The value of C may vary from about 1 kcal/kg°C for water to 0.0001 kcal/kg°C for some materials at very low temperatures. Further information on such materials is available in section 3.6.

The energy released by a material as its temperature is reduced, or absorbed by a material as its temperature is increased, is called the sensible heat.

Latent heat is associated with the changes of state or phase change of a material. For example, energy is required to convert ice into water, to change water into steam, and to melt paraffin wax. The energy required to cause these changes is called the heat of fusion at the melting point and the heat of vaporization at the boiling point. To illustrate, let us consider water, and suppose that we wish to evaporate 1 kg of ice by converting it to liquid and then heating it until it boils away. In this case, 80 kcal is required to melt the ice at 0°C to water at 0°C; then about 100 kcal is needed to raise the temperature of the water to 100°C; finally, 540 kcal is needed to boil the water, giving a total energy need of 720 kcal. The sensible heat for a given temperature change varies from one material to another. The latent heat also varies a great deal between different substances for a given type of phase change.

It is relatively straightforward to determine the value of the sensible heat for solids and liquids, but the situation is more complicated for gases. If a gas restricted to a certain volume is heated, both the temperature and the pressure increase. The specific heat observed in this case is called the specific heat at constant volume, C_v. If instead the volume is allowed to vary and the pressure is fixed, the specific heat at constant pressure, C_p, is obtained. The ratio C_p/C_v and the fraction of the heat produced during compression that can be saved significantly affect the storage efficiency.

3.3 Thermal Energy Storage

As an *advanced energy technology*, TES has attracted increasing interest for thermal applications such as space heating, hot water, cooling and air conditioning. TES systems have the potential for increasing the effective use of thermal energy equipment and for facilitating large-scale switching. Of most significance, TES is useful for correcting the mismatch between the supply and demand of energy.

There are mainly two types of TES systems, sensible (e.g. water, rock) and latent (e.g. water/ice, salt hydrates). The selection of a TES system mainly depends on the storage period required, e.g. diurnal or seasonal, economic viability, operating conditions, etc. Many research and development activities on energy have concentrated on efficient energy use and energy conservation, and TES appears to be one of the more attractive thermal technologies that has developed.

TES is basically the temporary 'holding' of energy for later use. The temperature at which the energy is held in part determines the potential application. Examples of TESs are storage of solar energy for night and weekend use, of summer heat for winter space heating, and of ice from winter for space cooling in summer. In addition, the heat or cool generated electrically during off-peak hours can be used during subsequent peak demand hours. Solar energy, unlike energy from fossil, nuclear and some other fuels, is not available at all times. Even cooling loads, which coincide somewhat with maximum levels of solar radiation but lag by a time period, are often present after sunset. TES can provide an important mechanism to offset this mismatch between times of energy availability and demand.

Increasing societal energy demands, shortages of fossil fuels, and concerns over environmental impact are providing impetus to the development of renewable energy sources such as solar, biomass and wind energies. Because of their intermittent natures, effective utilization of these and other energy sources is in part dependent on the availability of efficient and effective energy storage systems.

TES involves the storing of energy by heating (or cooling), melting or vaporizing (or solidifying or liquefying) a material, or through thermochemical reactions. The energy is recovered as heat or cool when the process is reversed. Storage by causing a temperature rise is known as sensible heating, by causing a phase change as latent TES. Thermochemical thermal storage in which a reversible chemical reaction absorbs energy, is described in detail in Chapter 2.

TES has a wide variety of applications, the majority of which relate to heating and cooling applications. TES provides a link and buffer between a heat source and a heat user. A common example of a TES is the solar hot water storage system. The energy source is solar radiation and the heat user the person demanding hot water. In this situation storage is required because the energy supply rate is small compared with the instantaneous demand, and because solar radiation is not always available when hot water is demanded.

As an example of the cost savings and increased efficiency achievable through the use of TES, consider the following case. In some climates it is necessary to provide heating in winter and cooling in summer. Typically these services are provided by using energy to drive heaters and air conditioners. With TES it is possible to store heat from the warm summer months for use in winter, while the cold ambient temperatures of winter charge cool store and subsequently provide cooling in summer. This is an example of seasonal storage and its use to help meet the energy needs caused by seasonal fluctuations in temperature. Obviously, such a scheme requires a great deal of storage capacity because of the large storage time scales. The same principle can be applied on a smaller scale to smooth out daily temperature variations. For instance, solar energy can be used to heat tiles on a floor during the day. At night, as the ambient temperature falls, the tiles release their stored heat to slow the temperature drop in the room. Another example of a TES application is the use of thermal storage to take advantage of off-peak electricity tariffs. Chiller units can be run at night when the cost of electricity is relatively low. These units are used to cool down a thermal storage, which then provides cooling for air conditioning throughout the day. Not only are electricity costs reduced, but the efficiency of the chiller is increased because of the lower night-time ambient temperatures, and the peak electricity demand is reduced for electrical-supply utilities.

3.3.1 Basic principle of TES

The basic principle is the same in all TES applications. Energy is supplied to a storage system for removal and use at a later time. What mainly varies is the scale of the storage and the storage method used. Seasonal storage requires immense storage capacity. One seasonal TES method involves storing heat in underground aquifers. Another suggested method is circulating warmed air into underground caverns packed with solids to store sensible heat. The domestic version of this concept is storing heat in hot rocks in a cellar. At the opposite end of the storage-duration spectrum is the storage of heat on an hourly or daily basis. The previously mentioned use of tiles to store solar radiation is a typical example, which is often applied in passive solar design.

TES processes

A complete storage process involves at least three steps: charging, storing and discharging. A simple storage cycle can be illustrated as in Figure 3.1, in which the three steps are shown as distinct. In practical systems, some of the steps may occur simultaneously (e.g. charging and storing), and each step may occur more than one in each storage cycle.

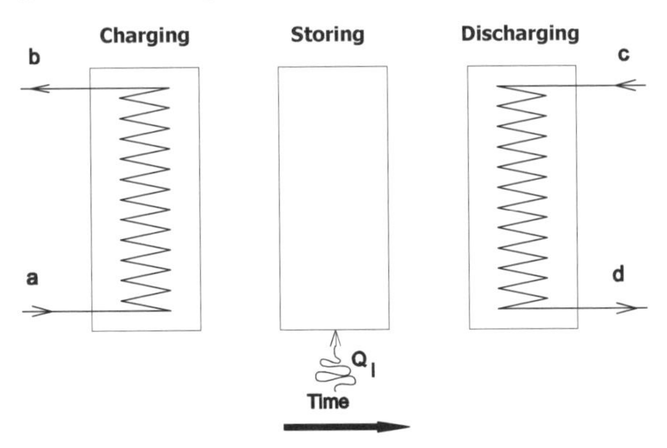

Figure 3.1 The three processes in a general TES system: charging (left), storing (middle) and discharging (right). The heat leakage into the system Q_l is illustrated for the storing process, but can occur in all three processes.

In terms of storage media, a wide variety of choices exists depending on the temperature range and application. For sensible heat storage, water is a common choice because, among its other positive attributes, it has one of the highest specific heats of any liquid at ambient temperatures. While the specific heat of water is not as high as that for many solids, it has the advantage of being a liquid which can easily be pumped to transport thermal energy. Being a liquid, water also allows good heat transfer rates. Solids have the advantage of higher specific heat capacities, which allow for more compact storage units. When higher temperatures are involved, such as for preheating furnace air supplies, solids become the preferred sensible heat store. Usually refractories are then used as the storage material. If the storage medium needs to be pumped, liquid metals are often used.

TES using latent heat changes has received a great deal of interest. The most common example of latent heat storage is the conversion of water to ice. Cooling systems incorporating ice storage have a distinct size advantage over equivalent-capacity chilled-water units because of the relatively large amount of energy that is stored through the phase change. Size is the major advantage of latent heat thermal storage. Recently, NASA has been considering using lithium fluoride salts to store heat in the zero-gravity environment of the space shuttle. Another interesting development is the use of phase change materials in wall paneling. These panels incorporate compounds that undergo solid-to-solid structural phase changes. With the right choice of material, the phase change occurs at ambient temperature. Then, these materials, when incorporated into the panels, act as high-density heat sinks/sources that resist changes in ambient room temperature.

The other category of storing heat is through the use of reversible endothermic chemical reactions, and in some literature it is considered with TES. However, we include it in Chapter 2 separately as chemical heat storage technology. In this method, the reactions involve the breaking and forming of chemical bonds, so a great deal of energy can be stored per unit mass of storage material. Although not currently viable, a variety of reactions are being explored. These include catalytic reactions such as the steam reforming reaction with methane and the decomposition of sulfur trioxide; and thermal dissociation reactions involving metal oxides and metal hydrides. These reactions are expected to be useful in high-temperature nuclear cycles and solar-energy systems, and as topping cycles for industrial boilers. At present, lower temperature reactions (<300°C) have not proven promising. TES can be an effective way of reducing costs and increasing efficiency. While effective thermochemical storage is still some way off, latent and sensible heat storage are already well established. In these cases, TES has the potential to produce significant benefits, particularly for low-temperature heating and cooling applications. These benefits should cause TES gain wider acceptance.

Topics of investigation

TES systems combined with heating, cooling and air conditioning applications have attracted much interest in recent years. Many related studies have been carried out in a variety of countries, particularly in the USA, Europe and Japan. These studies address technical issues arising from new TES concepts and the improvements required in the performance of existing TES systems. Studies have also investigated the design of compact TES systems and the use of TES in solar applications. TES research and development has been broad based and productive, and directed towards both the resolution of specific TES issues and the potential for new TES systems and storage materials. The following discussions summarize many investigations and indicate the scope of TES studies.

During the past two decades, many articles have appeared in the literature reporting investigations of TES systems and their applications (especially with solar energy), field performance characteristics and evaluations, design fundamentals, transient behavior and thermal analyses, and system and process optimization. In addition, theoretical, experimental and numerical studies have been undertaken on the thermophysical properties of new TES materials, TES selection criteria, the integration of TES systems into solar power plants, and the economics and environmental effects of TES. Some details on these studies are given below:

- For sensible heat storage, the performance and thermal characteristics of the packed-bed storage systems, the use of different storage materials, and uses for aquifer TES, water TES, and solar ponds have been investigated, as have operating conditions, effectivenesses, economics, etc., for these systems.
- Thermal analyses of PCMs and their uses for energy conservation in buildings have been carried out. Experimental and theoretical investigations and performance evaluations of the PCMs in latent heat storage applications have also been undertaken.
- Aspects of TES systems and materials during operation have been studied, including heat and mass transfer, and transient behavior, and second-law optimization and performance.
- Many practical applications of solar heating using the TES have also been reported.
- Numerous investigations have considered specific TES systems and applications, as well as the general objectives of TES and the energy conservation and related benefits of different TES methods.

There is a growing interest in the use of diurnal, or daily, thermal energy storage for electrical load management in both new and existing buildings. TES technologies allow electricity consumption costs to be reduced by shifting electrical heating and cooling demands to periods when electricity prices are lower, usually during the night. Load shifting can also reduce demand charges, which can represent a significant proportion of total electricity costs for commercial buildings.

Many TES studies aim to inform professionals concerned with heating and cooling systems in buildings about the characteristics of TES, and to examine the development of relevant technologies and assess their application in the field using data from case studies in different countries. Other studies also consider factors influencing technology adoption.

Space heating using electric TES has been used extensively in Europe and North America. The storage media can include ceramic brick, crushed rock, water and building mass, and systems can be either room or centrally based. Many improvements have been introduced in such systems in the past few years, including the development of new phase-change materials for latent heat storage which have recently become available commercially.

Cool storage using ice, water or eutectic salts as the storage media are widely used in the USA, where summertime cooling requirements are high. It is also used in Europe, often in combination with heat recovery and hot water storage, and in Australia, Canada, Korea, Japan, Taiwan and South Africa.

TES systems can be installed in both residential and commercial buildings, and can be cost-effective. Results from many of the monitored projects demonstrate payback periods of less than three years. If time-of-use tariffs exist, electricity costs to the consumer can be reduced by shifting the main electrical loads to periods when electricity prices are lower. If demand charges are implemented, a shifting or spreading of the load can reduce these significantly. To be effective, each storage system must be sized and controlled to minimize electricity costs and other system costs.

District heating and cooling systems often also incorporate TES and can benefit from its careful integration into the overall system.

Benefits from TES also accrue to the electricity utilities. The shifting of loads to off-peak periods not only spreads the demand over the generating period, but may enable electricity output from the more expensive or at least generating stations to be reduced. Worldwide, electrical utility incentive programs promoting the use of storage technologies exist, many of them within demand-side management programs. Such programs can greatly influence the economic feasibility of installing thermal storage by offering financial rebates, information, or special electricity rates for consumers.

Research and development programs for TES are needed in a number of areas, including the following related to utility-based TES applications:

- Establishing quantitative models for ascertaining the effects and benefits of changes in storage-capacity levels on service reliabilities, optimum generation mixes and reserve margins. These models need to incorporate detailed demand projections for the future that reflect probable changes in utility load characteristics.
- Establishing the benefits of dispersed storage as a function of the geographic land use, and demand characteristics of utilities. The works should be carried out on a regional basis where possible, and on an average national basis where necessary.
- Establishing the possible interactions between storage and load control for areas of differing load characteristics.
- Relating dispersed storage to dispersed generation, particularly in the context of total energy systems.

In parallel with such studies, there is a need for research and development support of promising storage concepts in order to ensure their timely availability and that they include a wide range of operating characteristics.

3.3.2 Benefits of TES

Although TES is used in a wide variety of applications, all are designed to operate on a cyclical basis (usually daily, occasionally seasonally). The systems achieve benefits by fulfilling one or more of the following purposes:

- **Increase generation capacity**: demand for heating, cooling, or power is seldom constant over time, and the excess generation available during low-demand periods can be used to charge a TES in order to increase the effective generation capacity during high-demand periods. This process allows a smaller production unit to be installed (or to add capacity without purchasing additional units), and results in a higher load factor for the units.

- **Enable better operation of cogeneration plants**: combined heat and power, or cogeneration, plants are generally operated to meet the demands of the connected thermal load, which often results in excess electric generation during periods of low electric use. By incorporating TES, the plant need not be operated to follow a load. Rather it can be dispatched in more advantageous ways (within some constraints).
- **Shift energy purchases to low cost periods**: this is the demand-side application of the first purpose listed, and allows energy consumers subject to time-of-day pricing to shift energy purchases from high-to low-cost periods.
- **Increase system reliability**: any form of energy storage, from the uninterruptable power supply of a small personal computer to a large pumped storage project, normally increases system reliability.
- **Integration with other functions**: in applications where on-site water storage is needed for fire protection, it may be feasible to incorporate thermal storage into a common storage tank. Likewise, apparatus designed to solve power-quality problems may be adaptable to energy-storage purposes as well.

One may ask what is the most significant benefit of a TES system. A common answer is reducing electric bills by using off-peak electricity to produce and store energy for daytime cooling. Indeed, TES is successfully operating in offices, hospitals, schools, universities, airports etc. in many countries, shifting energy consumption from periods of peak electricity rates to periods of lower rates. That benefit is accompanied by the additional benefit of lower demand charges.

3.3.3 Criteria for TES evaluation

There are numerous criteria to evaluate TES systems and applications such as technical, environmental, economic, energetic, sizing, feasibility, integration and storage duration. Each of these items should be considered carefully for a successful TES implementation.

Technical criteria for TES

Independent technical criteria for storage systems are difficult to establish since they are usually case specific and are closely related to and generally affected by the economics of the resultant systems. Nevertheless, certain technical criteria are desirable, although appropriate trade-offs must be made with such other criteria as

- storage capacity,
- lifetime,
- size,
- cost,
- efficiency,
- safety,
- installation, and
- environmental standards.

Before proceeding with a project, a TES designer should possess or obtain technical information on TES such as the types of storage available, the amount of storage required, the effect of storage on system performance, reliability and cost, and the storage systems or designs available.

TES is difficult to employ at sites that have severe space restrictions. Also, TES tanks often have significant first capital costs. Financial analysis for TES-based projects can be complex, although most consulting energy engineers are now capable of performing financial calculations and evaluating TES benefits.

Environmental criteria for TES

The basic design, materials, and operational practices that are used for TES should preferably not overly impair public health or the natural ecology and environment. Materials should not be used that are toxic or dangerous if released, or that could adversely affect the environment during the manufacture, distribution, installation or operation of the storage system.

Economic criteria for TES

The economic justification for storage systems normally requires that the annualized capital and operating costs for TES be less than those required for primary generating equipment supplying the same service loads and periods. In general, TES systems accrue fuel cost savings relative to primary generating equipment, but often at the expense of higher initial capital costs.

The key performance characteristics involved in an evaluation of the cost-effectiveness of TES include:

- hourly thermal loads for the peak day,
- the electrical load profile of the base-case system against which TES is being compared, and
- the size of the storage system and the control methods used.

Economic information that is needed includes:

- electricity demand charges and time-of-use costs,
- the costs of the storage, and
- financial incentives available.

Economic evaluation and comparison parameters often determined include the simple payback period. Other methods are also used to compare the annualized investment cost of a TES with annual electricity cost savings.

Energy savings criteria for TES

The past few years have seen a radical shift in understanding about the economics and environmental impact of TES. Today, well-designed TES systems can actually reduce first costs on some projects. Moreover, they can reduce the amount of electricity a facility uses, not just the amount it pays. Many project engineers now find that TES technology can significantly reduce energy use and demand, which translates into lower operating costs.

In addition, TES provides environmental advantage by using electricity produced at night, when utilities are generally operating their most efficient plants. As a result, significant savings of 'source energy' (coal, natural gas, oil or nuclear fuel) accrue to electric utilities and reduce pollutant emissions.

Air conditioning typically accounts for a large portion of a building's energy use. One or more chillers are usually used to match the cooling load, which rises during the day and peaks in mid-afternoon. With TES, peak electric use can be reduced, so smaller chillers can handle the load. In addition, air-cooled equipment can take advantage of lower night-time ambient temperatures to increase operating efficiency. Also, since water from TES systems may be colder than conventional chilled water, smaller pipes, pumps and air handlers may be integrated into the building design.

Stored cooling capacity can be used alone to meet the entire cooling load (so chillers remain off during the day) or to supplement chillers by satisfying part of the load.

Perhaps the greatest shift in thinking about the potential uses of TES comes from the growing use of cold-air distribution. Cold air distribution typically requires 30% less cold air than conventional cooling systems for a similar load, resulting in smaller fans, ducts and risers. The related cost savings often offset the first-cost premium of TES.

The ability of cool storage to shift large amounts of peak electric demand to off-peak periods usually leads to the greatest interest in TES. When combined with cold-air distribution, TES systems can yield significant cost savings and improve a building's energy efficiency and comfort level. Yet building owners often hesitate in using TES, erroneously believing that installing it is expensive. TES paired with cold-air distribution offers a more efficient, and often more cost-effective, cooling option which is competitive with conventional technology and which offers promising payback periods. By integrating TES with cold-air distribution during the design and construction of office and other buildings, building owners add floor space for useful purposes or rentals. This space becomes available because such systems require smaller air-handling units and often do not need a mechanical room.

Careful designing of a TES often yields first-cost savings by permitting the use of

- smaller air-handling units,
- smaller ducts, and
- smaller VAV boxes.

Cold-air distribution can increase the capacity of existing air conditioning systems or be used to replace older chillers. Generally, existing fans and ductwork can be used, though in some cases more duct insulation and upgraded diffusers or mixing boxes may be needed.

Fan energy usage in cold air distribution is typically reduced 30–40%. Actual long-term energy-cost savings often can be even greater, because compressor energy consumption is shifted to less expensive, off-peak hours. Research by the Electric Power Research Institute (EPRI) in the USA indicates that overall HVAC operating costs can be lowered by 20–60% by combining TES with cold-air distribution. In such combined systems, conditioned spaces are maintained at 35–45% relative humidity, as compared to the 50–60% humidity often found with conventional systems. At the lower humidity, occupants perceive improved air quality. In humid climates, the cooling energy requirement may increase slightly, but this is often offset by the inherent efficiencies of cold-air distribution.

Sizing criteria for TES

A need exists for improved TES-sizing techniques as analyses of projects reveal both undersized and oversized systems. Undersizing can result in poor levels of indoor comfort, while oversizing results not only in higher than necessary initial costs, but also in the potential wasting of electricity if more energy is stored than is required. Another requirement for successful TES that affects sizing is proper installation and control. Using state-of-the-art equipment, properly designed and controlled storage systems often do not use more energy than conventional heating and cooling equipment.

Performance data describing the use of TES for heating and cooling by shifting peak loads to off-peak periods are limited, although the potential for such technologies is substantial. The initial costs of such systems can be lower than those for other systems. To yield the benefits, new construction techniques are required together with the use of more sophisticated thermal-design calculations that are, as yet, unfamiliar to many designers.

The costs of TES systems for heating range from USD 20 to USD 60/kWh in Canada. A major development in recent years has been improved controls. Modern storage systems can shift nearly all of the space-heating energy use to off-peak hours whereas, with conventional systems, only about 50% of the energy for heating is consumed during off-peak periods. Off-peak periods typically last for more than seven hours (CADDET, 1997).

Energy use for conventional heating systems tends to be less than for storage heaters for rooms but greater than for central brick- and water-storage systems. Base-case heating systems vary greatly and include direct electric-resistance baseboard heating, heat pumps and electric central furnaces. Detailed comparisons of energy use for TES and base-case systems are often complicated by variations in other parameters such as the age and thermal integrity of buildings and homes.

The use of off-peak storage for heating in commercial buildings is growing as manufacturers produce more sophisticated storage equipment that is suited to larger buildings. Sizing of systems can, however, be difficult because of the more complex energy considerations and variable occupancy patterns for such buildings. Developments in energy control systems are enabling building operators to control HVAC and lighting systems more effectively. These improvements, in turn, allow better integration of heat storage with other energy systems. The payback periods for such systems often range from one to ten years, depending on the capacity and application.

TES systems for cooling capacity have been most successful in larger buildings, although research is underway on the development of smaller units. Unlike heat storage, part of the cost of cold storage can be paid for through savings derived from the installation of a smaller chiller than would be required for a conventional cooling system. Costs for cold TES often range from about USD 15 to USD 50/kWh, and many designers claim that initial costs for cold storage systems are below those for conventional cooling (CADDET, 1997). Cold TES systems for most buildings use ice or chilled-water on floor mass as the storage medium. The systems may include full, partial and demand-side storage. Payback periods for some systems, based on measured data and/or estimates, vary from less than one year to 15 years.

Feasibility criteria for TES

A variety of factors are known to influence dramatically the selection, implementation and operation of a TES system. Therefore, it is necessary to carry out a comprehensive feasibility which takes into consideration all parameters which impact on the cost and other benefits of TES systems considered. However, it is not always possible to conduct all the steps in a feasibility study for an application and, in such instances, as many items should be considered and studied accordingly. In such TES feasibility studies, a checklist can be helpful in ensuring that significant issues related to the project are addressed and details are assessed correctly regarding the evaluation, selection, implementation and operation of the system. Figure 3.2 provides a checklist that can be beneficial to the TES industry and analysts involved in TES projects. A checklist completed in the preliminary stages of the project guides technical staff.

CHECKLIST FOR EVALUATING A GENERAL TES PROJECT

Please tick (✔) items which are available or known.

( ) 1. Management objectives
( ) 2. Economical objectives
( ) 3. Financial parameters of the project
( ) 4. Available utility incentives
( ) 5. Status of TES system — a) New ( ) b) Existing ( )
( ) 6. Net heating or cooling storage capacity
( ) 7. Utility rates and associated energy charges
( ) 8. Loading type of TES system — a) Full ( ) b) Partial ( )
( ) 9. Best possible TES system options
( ) 10. Anticipated operating strategies for each TES system option
( ) 11. Space availability for TES system (e.g. tank)
( ) 12. Type of TES system — a) Open ( ) b) Closed ( )
( ) 13. General implementation logistics of TES system under consideration
    ( ) 13.1. Status of TES system
    ( ) 13.2. TES system location
    ( ) 13.3. Structural impact
        ( ) 13.4. Heat exchanger requirements
        ( ) 13.5. Piping arrangement
        ( ) 13.6. Automatic control requirements
        ( ) 13.7. New electrical service requirements
        ( ) 13.8. Others

Signature:

Project Leader:

Date:

Project Title and Number:

Figure 3.2 Checklist for evaluating a TES project (Dincer, 1999).

CHECKLIST FOR INTEGRATING TES INTO AN EXISTING THERMAL FACILITY

Please tick (✔) items which are available or known.

DATABASE

( ) 1. Utility's maximum incentive

( ) 2. Facility occupancy hours

( ) 3. Facility operating requirements

( ) 4. Existing physical constraints

( ) 5. Facility peak-day load and monthly average requirements

( ) 6. Historic energy consumption rates

ANALYSIS

( ) 1. Best possible TES system options

( ) 2. General implementation logistics of each TES system under consideration

( ) 3. Plant's yearly energy consumption with or without a TES system

( ) 4. Size utilization factor for TES

( ) 5. Projected operating cost reduction for each TES system under consideration

CONCLUSION

( ) 1. Financial analysis

( ) 2. System implementation recommendation

( ) 3. Others

Signature:

Project Leader:

Date:

Project Title and Number:

Figure 3.3 Checklist for integrating TES into an existing thermal facility (Dincer, 1999).

Integration criteria for TES

When considering the integration of TES into an existing thermal facility, an additional checklist (Figure 3.3) can assist in conducting the feasibility study for the TES system and its incorporation into the facility. For a facility under design, the information required is generally the same as in Figure 3.3, except that the database must be developed or estimated due to the lack of actual operating data.

Storage durations criteria for TES

In practice, it is useful to characterize different types of TES depending on the storage duration: short-, medium- or long-term.

Short-term storage is used to address peak power loads lasting a few hours to a day in order to reduce the sizing of systems and/or to take advantage of energy-tariff daily structures. Short-term is often called *diurnal storage*.

Table 3.1 Storage durations for small- and large-scale applications.

	Small-scale/decentralized storage		Large-scale/centralized storage	
	New	Existing	New	Existing
Diurnal LTES (e.g. salt hydrates..)	++	++		
Diurnal STES	++	+		++
Seasonal LTES				
Seasonal STES (e.g. aquifers, rocks)	+++		+++	+

+++ High probability, ++ Medium probability, + Low probability.

Medium- or long-term storage is recommended when waste heat or seasonal energy loads can be transferred, with a delay of a few weeks to several months. Long-term storages which take advantage of seasonal climatic variations are often referred to as annual or seasonal storage.

TES can be separated into high- and low-temperature systems, where low-temperature TES is storage where heat enters and leaves at temperatures below 120°C. The storage of cold or cooling capacity is also considered within this category. Low-temperature storage often permits efficient utilization of heat that otherwise would have been partially or entirely wasted. Low-temperature TES also permits the storage of heat obtained from solar radiation from day to night or from summer to winter, and permits the storage of heat from central power plants, from times of low-demand to hours of high-demand on both diurnal and seasonal bases.

TES also permits the storage of cold for air conditioning purposes from night to day, and from winter to summer. On a diurnal basis, the storage energy efficiency can exceed 90%, while on a seasonal basis it will usually not be much above 70%. Low-temperature TES has wide applicability in domestic hot-water systems.

Another kind of diurnal TES is the use of electric heaters that produce heat at night using low-cost electricity and store the thermal energy in the mass of bricks.

Table 3.1 gives information on diurnal and seasonal TES techniques for small- and large-scale applications. Diurnal storage equipment often reduces HVAC costs or avoids the need for back-up heating or cooling equipment. Diurnal storage is increasingly used as a demand side management technology (especially with cold storage in air conditioned buildings). Wider use of air conditioning in buildings combined with a need to reduce peak energy demands are motivating TES development.

Table 3.2 Comparison of potential TES implementation in France and the USA.

Country or region	Market area	Maximum potential deployment by 2010	Units in which deployment measured
France	Seasonal storage	25–30	Number of sites
	Diurnal storage	500 000	Number of dwellings
USA	Diurnal storage	50	GW

Source: (Anon, 1990; Piette, 1990).

Non-energy regulations are rather neutral towards TES, although large-scale seasonal storage is subject to a large number of codes and standards that may conflict with plans of building managers. For short-term storage, safety (e.g. related to PCM handling) or hygiene regulations may to a certain extent hinder the development of TES markets. Insurance companies are often reluctant to support the incorporation of products or equipment which have not received appropriate certification, as is often the case with innovative devices.

3.3.4 TES market considerations

Since TES is an energy efficiency option that complements other energy conservation strategies, government support could enhance the mutual development of such technologies. At present, TES products or equipment do not contribute significantly to international trade. A good level of knowledge of TES exists in developed countries. The main area where trade could grow is ice storage, which is already widely marketed in the USA.

Widespread implementation of large seasonal storage appears unlikely in the near future. In France and other southern European countries, large-scale TES applications will likely remain limited in part because there are few district heating (or cooling) networks. TES can be a valuable complement to such networks. Table 3.2 compares TES application potential in France and the USA. In practice, large-scale TES will face implementation challenges similar to those for district heating, due to their complementary natures. Another challenge for TES technology is the absence of an industry support group which can extol the benefits of TES, promote its use and provide information. Such support groups are common and important in other sectors.

Barriers to TES adoption

Although TES technology is proven and economically viable, it is sometimes not readily accepted. Barriers to TES adoption can be categorized into three types:

- lack of proper information,
- high initial investment, and
- infrastructure constraints.

The first of these barriers can be overcome with appropriate information dissemination and related activities (e.g. on-site visits, publication of independent monitoring results, etc.). High initial costs can only be overcome as a barrier by establishing reimbursable funds dedicated to large-project investments. In the near term, TES developments are not likely to significantly reduce TES initial costs. For example, the European Community could arrange with banks a third-party financing scheme for TES investment. Infrastructure constraints, such as the types of buildings and energy infrastructures largely in place, seem to be the most difficult to overcome because they are linked to wider policy considerations, especially in the case of large-scale TES. For short-term TES, support of renewable energy technologies in the building sector and of passive solar design can help overcome this barrier.

Other barriers that hinder greater use of TES can be summarized as follows:

- decision-makers rarely think about using TES, with the possible exception of ice storage in air conditioned buildings,
- long-term (seasonal) TES is perceived to be high risk,
- demonstration of economical long-term TES systems of local interest are lacking (e.g. in the immediate future emphasis is likely to be on aquifer and bore hole storage),
- short-term TES using PCMs is not fully developed, and safety and hygiene concerns exist for PCMs, and
- insurance companies are not enthusiastic about systems using innovative products like TES that have not yet received certification, and obtaining certification is expensive.

To overcome these barriers, priority should be given to the following:

- developing appropriate information packages on commercially developed TES systems aimed at those with no previous knowledge of storage systems,
- carrying out technical development to improve performance and reduce costs,
- focusing in the short term on technologies for which local experience has been gained and keeping systems as simple as possible,
- providing training on these systems for designers, operators and builders,
- providing risk capital for construction of systems close to commercialization, and
- carrying out R&D on financially promising TES systems, including PCMs suitable for use in the building industry and innovative systems suitable for solar applications.

Maturity of TES

Much TES technology has reached a level of maturity and begun to establish markets. PCMs are still undergoing research and development, but they nevertheless have been demonstrated and are commercially available. Thus, TES is mature and its lack of disseminatination mostly relates to barriers other than technical (e.g. financial). For large-scale seasonal storage, cost reductions can likely be obtained (e.g. by improving excavating techniques and large-store construction methods). Demonstrations are needed of the use of PCMs in diurnal applications, especially by integrating them into building components (ceiling, tiles, wallboards, shutters, etc.). A comparison of TES system characteristics for diurnal applications is given in Table 3.3.

Market position of TES

TES applications are at various states of development, depending on the country considered. Diurnal heat storage has taken on a large market share in any country, with the exception of cold storage in air conditioned buildings for demand-side management purposes (mostly in the USA, Canada and Japan, and in some commercial premises in Europe and pilot-project 'zero energy' houses in Finland). Diurnal cool storage in the USA has allowed about 15 GW of cooling power to be shifted to off-peak periods. Seasonal storage has been investigated and tested in the 1980s, and main sites are located in northern European countries (especially Sweden) and associated with district-heating networks. Seasonal storage in aquifers is well developed in China.

Large-seasonal TES facilities have been installed either to complement solar collection systems in large-scale building projects or in association with district heating plants. Long-term seasonal TES development is thus linked to the market status of these technologies.

Table 3.3 Thermal and technical data selected TES techniques.

Thermal and technical data	Concrete or steel tank	Basin with total insulation	Basin with top insulation	Rock cavern	Aquifer	Earth bed	Vertical tubes in clay	Drilled wells
Specific thermal capacity (kWh/m^{3}K)	1.16	1.16	1.16	1.16	0.75	0.70	0.80	0.63
Reference ΔT (°C)	55	55	55	55	55	55	15	55
Typical storage efficiency	0.90	0.85	0.70	0.80	0.75	0.60	0.70	0.70
Conversion factor (kWh/m^3)	57	54	45	51	31	23	8	24
Size range (m^3)	0–100 000	0–75 000	0–50 000	50 000–300 000	50 000–500 000	0–100 000	50 000–300 000	50 000–400 000
Invesment cost (ECU[a]/m^3)	150–250	120–220	40–60	80–120	20	50–100	5–8	30–40
Cost of energy supplied by TES (ECU/kWh)	0.2–0.4	0.15–0.25	0.05–0.1	0.12–0.20	<0.05	0.16–0.40	0.05	0.09–0.12

[a] 1 ECU was approximately 0.85 US$ in 1990.
Source: (Anon, 1990; Piette, 1990).

Individual houses or commercial (including office) buildings can use diurnal TES either for heat or cold applications. Demonstration projects and information dissemination are nevertheless needed to increase building managers' awareness of these options. The needs of passive, hybrid and other low-energy cooling techniques for some type of short-term cold TES should foster the development of this kind of TES.

Market demands for TES are linked to general energy policy. For example, increased support for and use of renewable energy would lead to large efforts to inform designers of active solar technologies and others of the possibilities offered by TES.

Relatively high risks are perceived as existing for seasonal TES (especially when used with an aquifer). Possible technical difficulties with pumps and exhaust water systems (clogged pump strainers, choked filters) can lead to performance and safety problems, which in turn can induce higher initial costs.

Ice storage is an exception to these concerns as many systems have been implemented and demonstrated a high cost effectiveness. When economic evaluations of TES are based on current energy-price structures, risks are associated with modified price structures. Widely disseminated information on successful operations can help reduce the perceived risks. Transfer and exchange of knowledge internationally through detailed specifications can limit the risks of projects that have been encountered in the past, and thus foster the development of TES.

Environmental legislation that directly controls the development of TES does not exist, but growing concern for the water quality in aquifers and the use of various PCMs has caused some TES technologies to be scrutinized and hindered decision makers from adopt them.

Financial factors affect TES marketability. For example, access to funds or the ability to raise funds at reasonable interest rates may hinder potential investors. Access to funds does not constitute a major barrier for TES technology, but the projected pay-back periods are often not appealing to many decision makers in the building sector. Sometimes, a financial consortium is needed for large TES facilities. Such consortia may include local authorities, which may or may not have access to funds, and energy utilities, which can usually obtain necessary funding.

Long-term TESs with large storage capacities (>1000 MWh) have been in use since the 1980s. The storage medium is generally water or rocks, and the 'container' is often aquifers, rock caverns and bore holes. The main markets for long-term (seasonal) TES have been located in northern European countries, particularly Sweden, where it is used in association with large-scale district heating schemes.

Short-term cool storage for air conditioning has been found to be cost-effective in applications in the USA, Canada, Japan and Europe have also used these systems. PCMs are not yet fully developed for the building sector, although they have been used for certain industrial applications.

In the future, TES appears to be able to help reduce the amount of heating and cooling systems by 50–70%, which in turn can yield energy savings due to better efficiency of systems working near to reference values. But TES losses can curtail some of this savings potential, lowering the energy conservation potential of TES to approximately 5–10%.

3.3.5 TES heating and cooling applications

The use of TES systems for thermal applications such as heating and cooling has recently received much attention. In various energy sectors the potential benefits of TES for heating and cooling applications have been fully realized. In the following two subsections we describe heating TES and cooling TES in detail.

Heating TES

Electricity, natural gas, propane or fuel oils can produce space heating. Heating TES using electricity operates resistance heaters at night when electricity rates are low, to produce heating capacity for use during the day. Electric-resistance efficiencies (near 100% on an energy basis) combined with lower off-peak electric rates can produce heating at a fraction of the cost of conventional systems. Heat produced by electrical resistance heaters at night is stored in such storage media as earth materials or ceramic bricks in insulated containers. The production of heat at night takes advantage of electric off-peak rates, which are generally 33–75% less expensive than peak rates. When heat demand rises (e.g. for space heating), heat is recovered from the storage unit and transferred into the room.

The use of earth as a TES medium is usually restricted to new construction, since the application requires that electric resistance grids be placed 0.5 to 1 m in the ground, beneath a structure. The need to place the grids under a building makes retrofit work extremely difficult for any facility without a basement or crawlspace. For new construction applications, approximately 2 m of earth directly below the structure is used for storage of heat produced by the grid. A rigid and waterproof insulating material is placed vertically around the perimeter of the building and extends approximately four feet below the earth grade level. The insulation ensures that heat stored in the ground is radiated mainly into the structure and not into the surrounding earth. The electric resistance grid is placed 0.5 to 1 m below the earth's surface and covered with about two inches of sand and earth materials.

Ceramic bricks provide an excellent heat storage medium for retrofit as well as new construction applications due to their modular sizes, ease of installation and high heat-retention abilities. These units are normally manufactured in various sizes and transported to building sites. The construction of this type of TES normally consists of an insulated box, about the size of a conventional radiant hot water or steam heating unit, filled with ceramic bricks. The number of bricks in a module depends on the heat storage requirement. The unit also includes a small fan. The ceramic bricks contain electric resistance strip heaters in their holes. During charging, the strip heating units produce heat which is absorbed by the ceramic bricks. The insulation surrounding the bricks restricts heat losses from them. During the day, a conventional thermostat is used to control the fractional horsepower fan which circulates air from the room across the ceramic bricks to recover the stored heat and transports it into the room. The thermostat controller shuts the circulating fan when the room temperature is acceptable.

Heating TES systems can be justified economically for most facilities which have significant space heating needs and are billed under time-of-use electric rate schedules that have large differentials between peak and off-peak electric consumption.

Cooling TES

Cooling TES can reduce cooling energy costs while maintaining a comfortable environment. Summer air conditioning bills have two components: an electric demand charge and an electric usage charge. The usage and demand charges are often further divided into peak and off-peak periods. The peak operating period of electric air conditioning systems normally occurs during the high-cost demand and usage periods (i.e. the summer afternoon). TES systems are designed to shift the peak operating period of electric air conditioning systems to the less expensive night periods.

Air conditioning systems cool by removing heat via a chilled-water network or directly from an air stream. Most air conditioning systems produce a cooling effect precisely when cooling is needed in a building or room. Cool TES-based air conditioning systems operate similarly, but remove heat from an intermediate substance when the building does not need cooling, producing a cool reservoir that is stored until there is a need for cooling. The intermediate substance is normally water, ice or eutectic salt solutions.

The most popular thermal storage medium is ice. The conversion of one kg of water at 0°C requires the removal of 152 kJ of heat. Similarly, adding 152 kJ of heat to the ice causes water at 0°C to be formed. Ice TES operates in this fashion. At night, heat is removed from water to produce ice (i.e. charging of the storage occurs). During the day when the building requires cooling, heat is removed from the building and added to the ice (i.e. discharge occurs). The melted ice is reused during the next charging period. The advantage of this cooling scheme is that the main electrically driven device in cooling systems, namely the compressor motor, is operated during low-electrical cost periods.

The previous example illustrates one design possibility. In another common design, the compression system operates 24 hours per day to provide stored cooling at night and to partially meet the cooling load during the day. This design usually requires the least investment.

Many cooling TES systems use chilled-water systems to transfer the cooling capacity from the storage to the building air-distribution system. Although chilled-water distribution systems are usually confined to large buildings with conventional air conditioning, chilled-water distribution systems with TES are now being designed for smaller buildings.

Cooling TES systems are generally advantageous for a new facility which will have large daytime cooling loads and little or no cooling load at night. For retrofit situations, cooling TES is usually difficult to justify unless the cooling system is being replaced because of old age or inadequate capacity.

TES can be a beneficial component of the refrigeration-based cooling technologies. TES can significantly reduce the size of a refrigeration system and its electrical use during peak demand periods. TES is typically employed at power plants which mainly meet peak loads or which have significantly differing revenue structures for off-peak vs. peak power.

For example, if a peaking gas turbine operates only four hours per day, it usually does not make much sense to build a refrigeration plant that also operates at full load for only four hours per day. This is because cooling capacity, unlike electricity, can be easily stored. It is usually more sensible to size the refrigeration system to meet about 20% of the peak cooling load, and slowly to make and store cooling capacity during other periods so that it can be discharged during peak conditions.

TES and gas cooling

An alternative to cold TES is the use of gas engine-driven chillers or gas-fired absorption chillers for peak cooling loads. The higher first cost of these refrigeration systems is comparable to the high capital costs of TES tanks and infrastructure. The high-electrical load of a standard vapor-compression refrigeration system is displaced by either of these gas cooling alternatives. A gas-engine chiller is often advantageous for a site that has limited operating hours per day per year. A gas-fired absorption chiller can make sense for sites that have longer operating hours.

An additional benefit of TES and gas-cooling technologies is that they permit changes to a site's electrical infrastructure to be minimized. One challenge to back-fitting electric vapor-compression systems to most sites is the placement and connection of a large transformer for the new electrical loads. With TES and gas cooling technologies, the electrical load of the new equipment is significantly reduced, and there may be sufficient capacity in the plant's existing control centers for the smaller loads.

During the past decade there has been increasing interest in gas cooling with TES in power generation industries, and cold TES and gas-cooling technologies are both expected to take more prominent roles in advanced cooling projects in the future.

A combustion-turbine inlet-air cooling system, with and without storage, is shown in Figure 3.4. The main components are the chiller, the cooling tower, the air coil, and interconnecting piping. Cold fluid from the chiller is pumped through the air coil, where the coolant is heated and returned to the chiller, while the inlet air is cooled prior to entering the compressor. The cooling tower provides cooling water to the chiller condenser. Alternatively, an evaporative condenser can be used with some types of chillers. Including storage and its associated piping loop increases the number of system components, but allows the chiller and cooling tower components to be downsized, assuming that cooling is not conducted 24 hours per day. Storage also significantly reduces peak power consumption for electrically driven chiller systems.

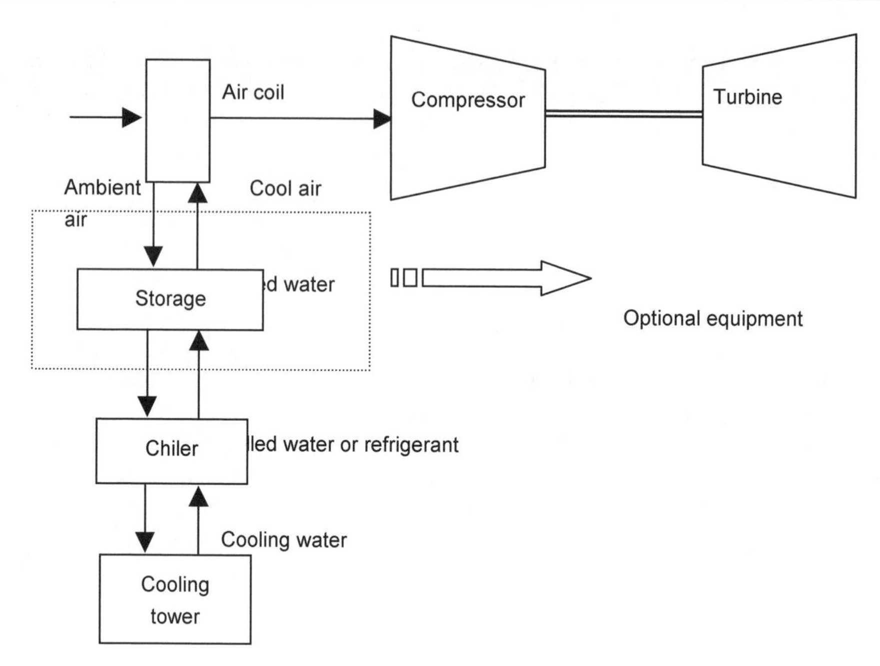

Figure 3.4 Generic inlet-air cooling system. (The cold storage inside dotted line is optional.) (*Adopted from Brown et al., 1996*).

The fundamental benefit of inlet-air cooling is that it can increase the efficiency of the gas turbine system (see Figure 3.4). However, the technique entails additional costs. The main cost is for the purchase and periodic maintenance of the inlet-air cooling system hardware. The energy used to drive the chiller also results in a significant expense, although the cost varies significantly depending on whether the chiller is thermally or electrically driven and the source of thermal or electric energy. Inclusion of an air cooling coil within the inlet duct to the compressor causes an additional pressure loss, with negative consequences to gas-turbine power output and efficiency, but the impacts are generally less than 0.5% (Brown *et al.*, 1996).

A common alternative to the inlet-air cooling system in Figure 3.4 is evaporative cooling. Direct-contact evaporative cooling, accomplished by passing the inlet air through a wet media, can be particularly effective in drier climates. Analyses of inlet-air cooling normally consider evaporative cooling as an option. Note that evaporative and refrigeration-based approaches should be considered independently. Direct evaporative cooling followed by refrigerative cooling does not reduce the refrigeration cooling load. Rather, it substitutes latent load for sensible load.

Incorporating storage into an inlet-air cooling system is desirable for downsizing the chiller and heat-rejection components and significantly reducing peak electricity consumption for electrically-driven systems. The chiller is usually the most expensive system component of such systems, so reducing its size and cost at the expense of adding storage and related piping can be cost-effective. Reducing peak electricity consumption is important because increasing peak power output is usually the primary objective of combustion-turbine inlet-air cooling. Chilled water and ice are the preferred storage media for inlet-air cooling systems. Both are applicable to diurnal storage, and ice storage is also applicable to weekly storage cycles. Seasonal storage of ice via engineered ice or snow ponds or of chilled water in naturally occurring aquifers also is possible for inlet-air cooling, but these concepts suffer from site-specific limitations, and have had only limited successful applications to date. Eutectic salts are another storage medium possibility, but the salts are more expensive than water, suffer availability losses on charge and discharge, and also suffer from limited application experience. Steel or concrete cylindrical tanks can be used for water or ice storage. External insulation is usually sufficiently thick to avoid condensation. Chilled water is normally added and removed from the bottom of water storage tanks, while warm water is added or removed from the top, so as to form a thermally stratified tank. The preferred ice-making method uses a harvesting approach that periodically passes hot refrigerant from the compressor through the evaporator to release ice from the evaporator surface. The ice falls from the evaporator and makes a pile within the tank. Several evaporators are used to aid in distributing the ice. An alternative approach is to build up logs of ice around evaporator coils that run back and forth throughout the tank. Although its defrost cycle increases the effective cooling load by about 15%, the ice harvester is less costly to build because it requires much less evaporator surface and refrigerant inventory.

Selection of the storage media for inlet-air cooling depends partly on the chiller type. Lithium-bromide absorption chillers can only use water storage. Either water or ice storage is possible for vapor-compression chillers. The principal advantage of ice is its greater cold storage density, and the advantage of water is the mechanical simplicity of the storage system. Ice storage generally allows the inlet air to be cooled to a lower temperature than water storage, but ice generation requires a lower chiller evaporator temperature, which results in poorer chiller efficiency and higher chiller cost.

3.3.6 *TES operating characteristics*

A major factor in determining the feasibility of TES is the shape of the daily electrical load curve for a utility and its impact on the availability of energy for charging a TES. For example, an urban utility with a summer peak demand may have a peak of moderate duration during the day and a much reduced load for a short duration at night. Thus, the storage device may be required to supply energy for 10 h during the day while significant charging capacity is only available for 6 h at night. Such a storage device would either have to be charged at a faster rate than it is discharged, or operate on a weekly cycle with part of the charge taking place over the weekend.

The diverse operating conditions for different utilities cause substantial variations in the duty cycles of TES systems. Table 3.4 summarizes the typical range of cycle characteristics for different applications. The values assume 75% storage efficiency.

Diurnal vs. seasonal TES

The primary characteristic of a seasonal storage system is the very large capacity that is required (in the order of a hundred times the capacity of a daily storage). Thermal losses become very important for such long-term storage. More care is therefore taken to prevent thermal losses in a seasonal system than in a system for daily storage. While diurnal systems can generally be installed within a building, seasonal storage requires such large storage volumes that separate or additional locations may frequently be required.

Table 3.4 Sample TES cycle characteristics.

Duty-cycle characteristics	Type of operation			
	Peaking duty		Regular duty	
	Daily cycle	Weekly cycle	Daily cycle	Weekly cycle
Discharge time (h/day)	2–8	2–8	9–14	9–14
Charge time (h/day)				
Weekday	5–9	5–9	5–9	5–9
Weekend	—	14–34	—	14–34
Charge/discharge ratio	0.8–2.1	0.1–2.1	1.3–3.7	0.8–2.4
Storage period (h)	2–8	4–26	9–14	17–47
Annual operating time (h) (discharge time)	350–1600	350–1600	2300–3600	2300–3600

Source: Dinter *et al.* (1991).

The capital costs associated with the size and insulation necessary for seasonal storage systems have generally kept them from becoming economical. New technologies, increased energy costs, and the general desire to conserve scarce energy sources may warrant a review of seasonal storage systems. However, diurnal and other short-term storage will probably find broader application, and will have greater impact in a country at any given time.

Individual vs. aggregate TES systems

A size relationship similar to that for diurnal/seasonal applications results from consideration of individual versus aggregate use of TES systems. When individual units are aggregated into a system large enough for buildings, the total storage volume becomes the sum of the volumes of the individual units. With a larger volume, the lower surface-volume ratio of the aggregate unit reduces the thermal losses for identical storage periods.

Lower unit-storage costs also generally result as size is increased by the aggregation of individual systems. Depending on building and load density, however, thermal transmission costs and losses can eliminate any savings arising from aggregation. Control can be more difficult when a single aggregate storage system is used for several buildings. If individual accounts are maintained, cost billings can be difficult as allocation of energy costs in an aggregated situation requires a fair and inexpensive way of prorating thermal energy consumption.

Preferences for individual or aggregate systems depend on the circumstances of each particular case. For applications characterized by small loads, building ownership and a wide diversity in consumption patterns, such as single family residences, aggregate systems are unlikely to be favored.

3.3.7 ASHRAE TES standards

The American Society of Heating, Refrigerating and Air Conditioning Engineers (ASHRAE) provides testing standards for TES systems. ASHRAE TES standard 943 is called the 'Method of Testing Active Sensible TES Devices Based on Thermal Performance.' This sensible TES standard (ASHRAE, 2000) is the second half of a revision of ASHRAE's first TES standard, 94-77, 'Methods of Testing Thermal Storage Devices Based on Thermal Performance.' The first half was issued as ANSI/ASHRAE 94.1-1985, 'Method of Testing Active Latent Heat Storage Devices Based on Thermal Performance.' Sensible TES usually applies to heat storage in water or rocks or cool storage in chilled water. This standard was prepared by a group of volunteers representing users, design engineers, manufacturers, scientists, and the US Federal Government. Research sponsored by the US Department of Energy and the Electric Power Research Institute in solar and off-peak energy programs at government laboratories and universities has assisted TES standards development. The major change in the new standards is to provide wider freedom to the supplier of the test device in the choice of cycling time, capacity, maximum flow rate and temperature range. Another important revision is in making the stand-by heat loss test more practical and accurate.

Since storage is usually part of a system, total performance depends on more than just the performance of the storage component. However, the results of this test procedure are intended to provide stand-alone specification of the storage-components' performance. The Standards Project Committee that developed these TES standards hopes they will be used by manufacturers and others.

3.4 Solar Energy and TES

Solar energy is an important alternative energy source that likely will be more utilized in the future. One main factor which limits the application of solar energy is that it is a cyclic time-dependent energy resource. Therefore, solar systems require energy storage to provide energy during the night and overcast periods. Although the need for TES also exists for many other thermal applications, it is particularly notable for solar applications.

3.4.1 TES challenges for solar applications

TES is important to the success of any intermittent energy source in meeting demand. This problem is especially severe for solar energy because it is usually needed most when solar availability is lowest, namely in winter. TES complicates solar energy systems in two main ways. First, a TES subsystem must be large enough to permit the system to operate over periods of inadequate sunshine. The alternative is to have a back-up energy supply, which adds a capital cost and provides a unit that remains idle. In the short run, solar energy can and probably needs be integrated into systems that also use conventional energy sources, such as fossil fuels. In the long run, however, stand-alone solar energy systems may be desired.

The second major complication imposed by TES is that the primary collecting system must be sufficiently large to build the supply of stored energy during periods of adequate insolation. Thus additional collecting area (and its additional capital cost) is needed. Examinations of typical sunshine records show that even in the desert the periods of cloudy and clear weather are about equally spaced, a few days of one followed by a few days of the other. Partly cloudy days can greatly affect performance and make the difference between practical and impractical energy storage. If the total energy of a partly cloudy day can be collected, then the periods requiring energy storage are greatly reduced.

Concentrating solar systems must cope with the intermittent nature of direct sunlight on a cloudy day. Consequently, absorbers and boilers must be designed with care to avoid burn-out problems when the sun suddenly returns with full brilliance. Non-concentrating systems face the fundamental problem of trying to provide sufficiently high efficiency at medium temperatures to yield energy output at reasonable cost. Thus, TES costs must be reasonable.

3.4.2 TES types and solar energy systems

In solar energy applications, TES can provide savings in systems involving either simultaneous heating and cooling or heating and cooling at different times of the year. Most TES applications involve a diurnal storage cycle, however, weekly and seasonal storage is also used. Solar energy applications require storage of thermal energy for periods ranging from very short durations (e.g. buffer storage of minutes for solar thermal power plants) to annual cycle time scales. Most solar systems use diurnal storage, where energy is stored for at most a day or two. Diurnal storage offers a number of advantages:

- capital investments for storage and energy loss are usually low,
- devices are smaller and can easily be manufactured off-site, and
- sizing of daily storage for an application is not as critical as sizing for larger annual storages.

Seasonal storages do, nevertheless, have some advantages. Larger storages have lower heat losses because of their lower surface-to-volume ratios. The need for back-up systems can be eliminated, since periods of adverse weather have little effect on the long-term thermal energy availability. Collector areas can consequently be reduced. Also, annual TES systems complement well-designed energy management systems in which excess heat or coolness from the environment or adjacent structures is saved for later use.

A TES designed primarily for the storage of solar energy is not necessarily restricted to that. It may be used to store surplus energy from the power plants, usually in the form of waste water, waste energy from air conditioners or industrial processes, and so on. Such storage use may not be applicable for small houses, but could be useful for large-scale central heating systems.

A variety of active and passive systems for storage have been developed for the effective utilization of solar energy. Passive systems, which do not need pumps, are often suitable for small-scale domestic applications, and are widely used throughout Europe and the USA. Such passive systems are of five main types:

- direct heat gain,
- heat collection and storage,
- sun space,
- roof-top heat storage, and
- thermosyphon.

Effective use of solar energy relies to such an extent on TES that solar systems without TES facilities are probably only utilizable in the most rudimentary applications. Some examples of solar thermal applications which do not need storage include solar grain driers, solar distillers and solar kilns. In these systems the solar heat is used immediately as it becomes available. However, in solar space-heating applications the situation is different because the solar system normally provides more heat than demanded by the building during the collection period. Storage is required to make such solar energy systems viable and attractive, in the long- or short-term.

3.4.3 Storage durations and solar applications

Annual solar energy TES systems are designed to collect solar energy during the summer months and retain the heat in storage for use during the following winter. Although the technology exists to construct annual storage systems, and some have been demonstrated, it is difficult to make these systems cost effective. The main impediment is the lack of a cost-effective means to contain heat for long periods (e.g. three months). Economic breakthroughs in TES are of course possible, possibly via annual storage on a community-wide scale, which could reduce costs and dramatically improve the reliability of solar heating.

Short-term solar energy TES systems are designed to store heat for up to a few days. Although solar energy systems utilizing annual storage can contribute close to 100% of building heating needs, short-term TES systems rarely contribute in excess of 60%. Nevertheless, short-term solar energy TES systems can operate on a competitive basis with conventional fuels (Dincer *et al.*, 1997a,b).

Any solar energy system has some degree of TES, either deliberately provided as a device in which to store energy, or through the thermal inertia of the extensive system of collectors and heat transfer fluid. TES is considered cost-effective only for short periods (hours to days) which is generally not enough to carry a system through much winter weather. Making TES capacity large enough normally is economically prohibitive. One exception is the special case of saline solar ponds, which act both as solar collectors and TESs having weeks of storage capacity.

TES is often thought of in conjunction with solar energy because the latter is often associated only with technologies that translate sunlight directly into thermal energy or electricity. Solar technology today includes a broad spectrum of concepts that differ greatly in their requirements for storage. Plant matter, or biomass, is ideally storable for long periods. Wind power is another form of indirectly derived solar energy which, even though it is intermittent, is available more continuously through the day and night than direct sunlight in many regions. TES is often essential for solar water heating, heating of buildings and industrial heating processes.

A factor that may facilitate the storage problem is the movement of the energy economy toward greater coordination of energy supplies and demands. Although many people will not readily give up energy choices, the flexibility of use that occurred in eras of inexpensive energy may be changing. Time-of-day electric pricing, already introduced in some areas, is one example. The use of computer systems that automatically manage the energy load in large office buildings is another. Such changes occur gradually and tend to create a social climate in which solar energy is more acceptable.

Solar energy may also develop without storage in the future as there exist some configurations in which solar power sources can be integrated with energy systems, particularly electric systems, without bulk storage. Nevertheless, even with optimistic rates of growth, solar energy's contribution will remain relatively small in the near future. Thus, traditional energy systems can be used to compensate for the fluctuations in solar energy supply. Solar energy can also be used to match fluctuating energy demand, particularly for electricity, without endangering the stability of electrical networks. Since large quantities of electric power are routinely transmitted long distances, the electric network is particularly well-suited for smoothing and balancing solar-source power fluctuations. If solar-derived energy grows to too large a fraction of the total, the overall stability of an electric network might be adversely affected, but many studies indicate that this limitation is unlikely to be a problem until the solar power penetration reaches 15–20%.

3.4.4 Building applications of TES and solar energy

The ability to store thermal energy is important for effective use of solar energy in buildings. Today, much interest is focussed on passive systems for space heating and active systems for water heating. For building heating, conventional passive TES materials include water, rocks, masonry and concrete. To perform well, these storage materials must be massive because their allowable temperature swings are limited by comfort conditions that must be maintained inside the building.

With lightweight-building construction practices commonplace in the USA, a lightweight latent TES system which is easily installed in a building could be beneficial. One problem is the effective and economic containment of a PCM in its liquid phase. Tubes, rods, bottles and canisters containing PCMs that melt in the room-temperature range have been studied with varying degrees of success; most have proved uneconomical. A more interesting approach is a wallboard containing a PCM. With the wallboard providing PCM containment as well as serving an architectural function, the economics are improved. Further, the large heat transfer area of the wallboard supports large heat fluxes driven by small temperature differences.

Seasonal storage has been pilot-tested and used in a number of countries to store solar energy for providing winter space heating, in conjunction with district heating systems. Sweden has implemented many such systems. A TES system examined by the University of Massachusetts at Amherst (Tomlinson and Kannberg, 1990) started using a long-term seasonal thermal storage of solar heat in a subsurface clay formation for heating a local athletic center. Use of seasonal storage can substantially reduce the cost of providing solar energy systems that can supply 100% of energy needs because of the reduced collector area required. In high latitudes, storage is virtually essential to provide a large percentage of heating from solar energy. The economics of scale favors relatively large systems.

Commercial buildings are now becoming more complex. Not only are architectural features such as atriums and skylights common, but sophisticated controls on HVAC equipment are being used in an attempt to provide superior comfort and lower energy costs. These situations make it difficult to determine the economic feasibility of TES systems and require computer simulation programs to model the complex systems, controls, and economic parameters of commercial buildings.

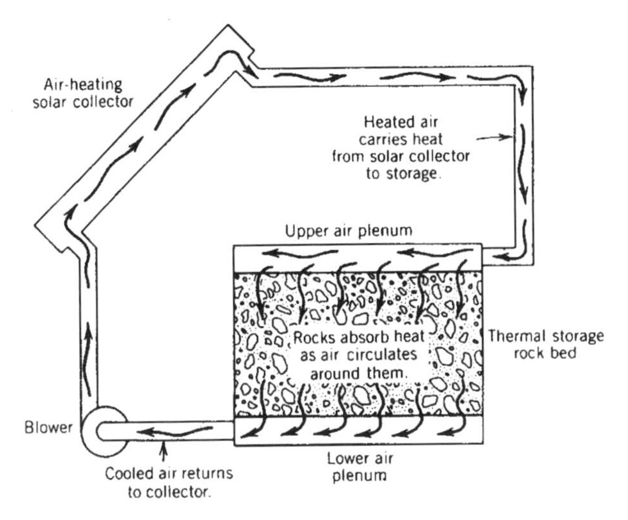

Figure 3.5 A solar rock-bed TES system (Harris *et al.*, 1985).

The storage system is the heart of a solar heating and cooling system. Storage evens out the extremes of temperature caused by the daily cycle of solar availability, permitting indoor temperatures to remain comfortable during the day while providing heat at night. The storage component of a solar system significantly affects design and construction and therefore cost. The storage medium and the 'reliability' required of a system determine the size of the storage and, to some degree, its location. Most storage systems are not sized to provide 100% of heating needs when sunshine is not available. Since solar unavailability and availability are difficult to predict in most cases, 100% storage systems would be very large, but most of the capacity would remain unused most of the time. Being willing to tolerate some daily temperature swings can reduce the size of such TES.

All substances have a thermal capacity and ability to hold a certain amount of heat. Water can store large amounts of heat. Rock has about one-fifth the thermal capacity of water, but like brick and concrete, is more dense. Rock, when used as a storage medium, is usually placed in an insulated bed under or attached to a building. Figure 3.5 illustrates a solar rock-bed TES system with air-heating collectors. As can be seen from the figure, the air from the collectors carries the absorbed solar heat to the rocks. As the heated air flows around the rocks, they absorb the heat, and the cooled air returns to the solar collectors to be heated again.

3.4.5 Design considerations for solar energy-based TES

The energy from solar collectors tends to be at low temperatures and requires a large storage mass when stored as sensible heat. Although the energy efficiency of solar collectors increases (and the collector cost probably decreases) as the temperature of the collector output reduces towards the space-conditioning comfort range of 20–25°C, the overall mass and volume of the storage device increases further. For most locations, however, a space-conditioning system using solar energy needs to be supplemented by an electric- or fuel-powered auxiliary energy source. Thus, optimal and synergistic use of both energy sources via TES is important in designing building structures and TES systems.

A major question in the design of solar TES systems involves the quantity of solar energy to be stored. The storage system must be adequate to supply heat not only during the night, but also for several consecutive cloudy days if complete independence from external energy sources is desired. In many regions, winter periods without sunshine are so long that complete independence is not feasible. Furthermore, the solar-collector system, to achieve independence, must be sufficiently large to heat the structure even while storing more heat for the next sunless period. Days can be less than eight hours during northern winters, and under such conditions, the costs of the combined collector and storage system may effectively limit the extent to which the solar system can economically supply the needed energy. If an alternate heating system with another energy source is needed as a backup for extended sunless periods, then solar energy use may be even further limited. However, if a storage system is economically justified, perhaps because it takes advantage of low off-peak electricity prices, then storage of solar energy in addition may enhance that benefit.

The energy from solar collectors need not be used directly in space-conditioning systems. One alternative and promising application is to use it as a heat source for a heat pump in a solar-augmented heat-pump system. For such uses, the solar-collector outlet temperatures can be lower than with direct heating, thereby increasing energy efficiency and probably reducing cost. Additionally, having higher source temperatures available for the heat pump increases its coefficient of performance (COP) and reduces its electricity consumption. How the solar-augmented heat-pump system operates depends, in part, on the energy storage provisions. With minimum or no storage, the solar collector only improves the heat-pump efficiency during hours of sunlight. With greater storage, the solar input also provides a reservoir of higher source temperatures for heat-pump operations during sunless periods. If the overall system is designed to limit the heat-pump operation to off-peak hours, then a dual storage system is necessary (one system for the solar input storage and a second to store the heat-pump output energy for round-the-clock use in space conditioning).

3.5 TES Methods

TES can aid in the efficient use and provision of thermal energy whenever there is a mismatch between energy generation and use. Various subsets of TES processes have been investigated and developed for building heating and cooling, industrial applications, and utility and space power systems. The period of storage is an important factor. Diurnal storage systems have certain advantages: capital investment and energy losses are usually low, and units are smaller and can easily be manufactured off-site. The sizing of daily storage for each application is not nearly as critical as it is for larger annual storage. Annual storage, however, may become economical only in multi-dwelling or industrial park designs, and often requires expensive energy distribution systems and novel institutional arrangements related to ownership and financing. In solar TES applications, the optimum energy storage duration is usually the one which offers the final delivered energy at minimum cost when integrated with the collector field and back-up into a final application.

Some of the media available for sensible and the latent TESs are classified in Table 3.5.

Table 3.5 Available media for sensible and latent TESs.

Sensible		Latent
Short term	Long term (annual)	Short term
Rock beds	Rock beds	Inorganic materials
Earth beds	Earth beds	Organic materials
Water tanks	Large water tanks	Fatty acids
	Aquifers	Aromatics
	Solar ponds	

Table 3.6 Thermal capacities at 20°C of some common TES materials.

Material	Density (kg/m^3)	Specific heat (J/kgK)	Volumetric thermal capacity (10^6 J/m^3K)
Clay	1458	879	1.28
Brick	1800	837	1.51
Sandstone	2200	712	1.57
Wood	700	2390	1.67
Concrete	2000	880	1.76
Glass	2710	837	2.27
Aluminum	2710	896	2.43
Iron	7900	452	3.57
Steel	7840	465	3.68
Gravelly earth	2050	1840	3.77
Magnetite	5177	752	3.89
Water	988	4182	4.17

Source: Norton (1992).

3.6 Sensible TES

In sensible TES, energy is stored by changing the temperature of a storage medium such as water, air, oil, rock beds, bricks, sand or soil. The amount of energy input to TES by a sensible heat device is proportional to the difference between the storage final and initial temperatures, the mass of the storage medium and its heat capacity. Each medium has its own advantages and disadvantages. For example, water has approximately twice the specific heat of rock and soil. The high heat capacity of water ((~4.2 kJ/kg°C) often makes water tanks a logical choice for TES systems that operate in a temperature range needed for building heating or cooling. The relatively low heat capacity of rocks and ceramics (~0.84 kJ/kg°C) is somewhat offset by the large temperature changes possible with these materials, and their relatively high-density (Tomlinson and Kannberg, 1990).

Sensible TES consists of a storage medium, a container, and input/output devices. Containers must both retain the storage material and prevent losses of thermal energy.

Thermal stratification, the existence of a thermal gradient across storage, is desirable. Maintaining stratification is much simpler in solid storage media than in fluids.

Sensible TES materials undergo no change in phase over the temperature range encountered in the storage process. The amount of heat stored in a mass of material can be expressed as

$$Q = mc_p\Delta T \qquad \text{or} \qquad Q = \rho c_p V \Delta T$$

where c_p is the specific heat of the storage material, ΔT is the temperature change, V is the volume of storage material, and ρ is the density of the material.

The ability to store sensible heat for a given material depends strongly on the value of the quantity ρc_p. Water has a high value and is inexpensive but, being liquid, must be contained in a better-quality container than a solid.

Some common TES materials and their properties are presented in Table 3.6. To be useful in TES application, the material normally must be inexpensive and have a good thermal capacity. Another important parameter in sensible TES is the rate at which heat can be released and extracted. This characteristics is a function of thermal diffusivity. For this reason, iron shot is an excellent thermal storage medium, having both high heat capacity and high thermal conductance.

For high temperature sensible TES (i.e. up to several 100°C) iron and iron oxide have thermal properties that are comparable to those of water per unit volume of storage. The cost is moderate for either pellets of the oxide or metal balls. Since iron and its oxide have similar thermal characteristics, the slow oxidization of the metal in a high-temperature liquid or air system would not degrade its performance.

Rock is a good sensible TES material from the standpoint of cost, but its volumetric thermal capacity is only half that of water. Past studies have shown that rock storage bins are practical, their main advantage being that they can easily be used for heat storage at above 100°C (Dincer, 1997a).

3.6.1 Thermally-stratified TES tanks

TES tanks for use in heating, air conditioning and other applications have in general received increasing attention in recent years. Thermally-stratified storage tanks, which have gradually seen more widespread use recently, are now described.

Figure 3.6 shows, for a thermally-stratified storage tank, the positions of the inlet and outlet for well and poorly designed cases. Also shown is the thermally-effective quantity of water that results from these positions. Since the tank stores thermal energy for periods of at least several hours, heat loss/gain occurs from the tank. The thermal-retaining performance of a tank is an important factor in its design.

Types and features of various stratified TES tanks

The TES system most commonly employed at present is sensible TES utilizing water as the storage medium. The term 'thermal' storage is used instead of 'heat' storage because the former implies storage of heat or cold and the latter just heat. An effective TES tank utilizing water as the storage medium satisfies the following three general requirements:

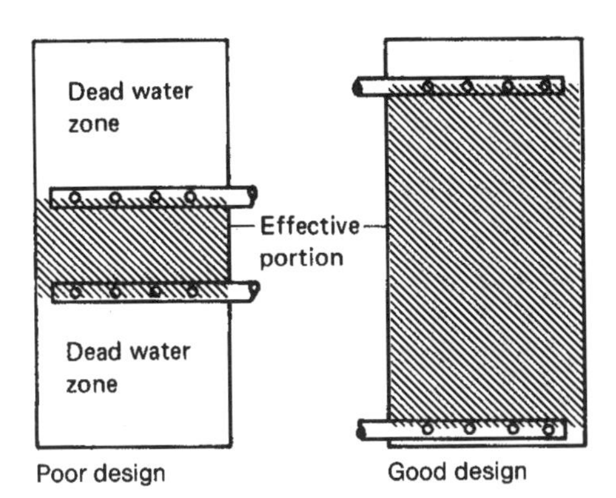

Figure 3.6 Position of inlet and outlet, and effective quantity of water (hatched regions), for a thermally stratified TES (*Shimizu and Fujita, 1985*).

- The tank should be stratified, i.e. hold separate volumes of water at different temperatures. Mixing of the volumes should be minimal, even during charging and discharging periods.
- The effective storage capacity should minimize the amount of dead water volume in the tank (see Figure 3.6).
- The heat loss/gain from the tank should be minimized.

Many types of TES tanks have been developed to satisfy these requirements, the principal of which are listed in Table 3.7.

A thermally naturally stratified storage tank has no inside partitions and has the following principle of operation. Warm water has low density and floats to the top of the tank, while cooler water with higher density sinks to the bottom. The storage volume with this type of system is reduced relative to other systems, because the dead water volume is relatively low and the energy efficiency relatively high.

Design considerations for stratified TES tanks

When designing a thermally-stratified thermal storage tank, the following criteria can guide the design process:

- **Geometrical considerations**: A deep water-storage container is desirable to improve thermal stratification. The water inlet and outlet should be installed in a manner that produces a uniform flow of water to avoid mixing. To minimize dead water volume, the outlet and inlet connections should be located as close as possible to the top and bottom of the storage volume. The surface area in contact with the storage water should be minimized.
- **Operating considerations**: the temperature difference between the upper and lower parts of the tank should be large, at least 5–10°C. Controls can be used to maintain fixed water temperatures in the upper and lower parts of the tank if desired. The velocity of the water flowing into and out of the tank should be low.
- **Other considerations**: the insulating and water-proofing characteristics of the tank should be designed to meet appropriate specifications.

Table 3.7 Types and features of various stratified TES tanks.

Type	Schematic representation of cross section	Efficiency	Remarks
Continuous multi-tank type	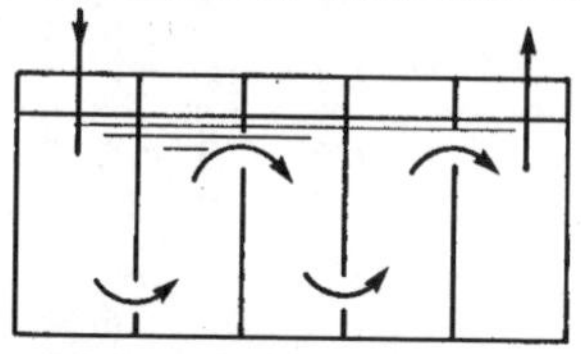	Medium	Underground beam space can be used effectively. Insulation is difficult to install.
Improved dipped weir type	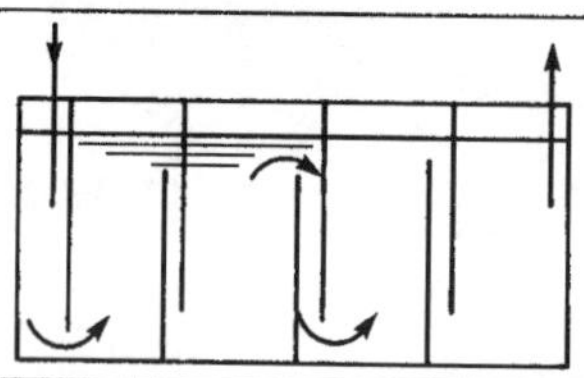	Medium ~ High	Construction is difficult.
Thermally stratified type	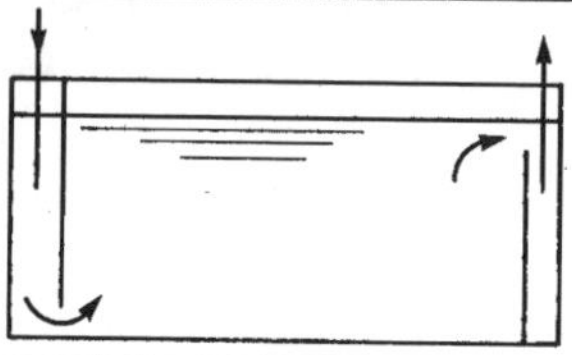	High	Best suited for large-size tank built aboveground.
Movable diaphragm type	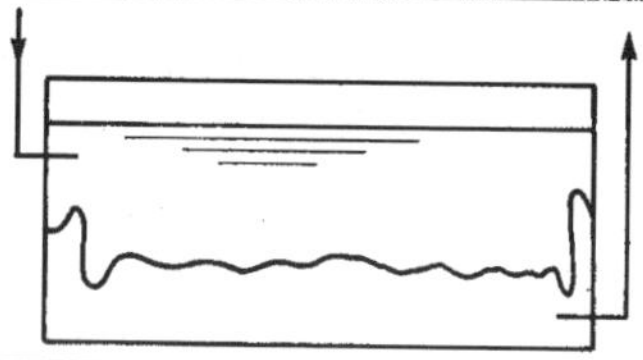	High	Diaphragm material is problematical. Not easily adapted to tanks with internal pillars and beams.
Multi-tank water renewing type	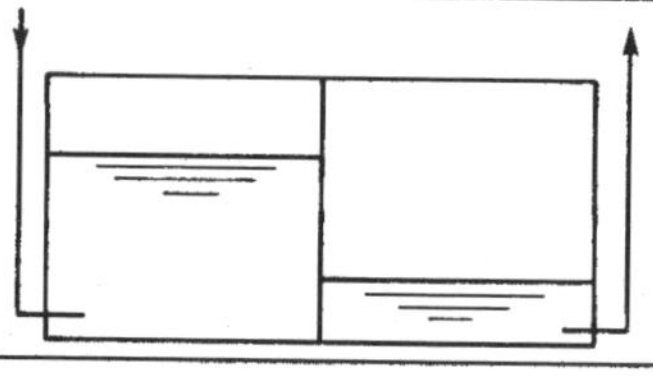	High	Underground beam space can be utilized to some extent. Heat loss is large.

Source: Shimizu and Fujita (1985).

Stratified TES tank configurations

Several possible stratified storage concepts are depicted in Figure 3.7. The advantage of the single-medium storage systems, in which the heat transfer fluid is also the storage medium, is that no internal heat exchange between transfer fluid and storage medium is necessary, thus avoiding the consequential temperature losses. If the liquid has low thermal conductivity and permits good thermal stratification (e.g. for water or thermal oil), the one tank thermocline concept requires the least tank volume, since the hot and cold media are contained in a single vessel. When the storage-medium thermal conductivity is higher, as in molten salts or sodium, rapid equilibration of hot and cold temperature regions occurs, making separate hot and cold tanks necessary. Since in that case, twice as much tank volume as fluid content is required, a three-tank system in which there is only 1.5 times as much tank volume as fluid content is often recommended. However, such systems are difficult to control, involve extensive piping, and are subject to increased heat losses from higher surface-to-volume ratios. Dual-medium concepts employ different transfer and storage media, often because the storage medium (usually solid) is less expensive than the transfer fluid. The transfer medium exchanges heat through direct or indirect contact with the storage medium, forming a thermocline. Apart from the temperature drop between charging and discharging due to the intermediate heat exchange, the dual-medium concept has another operational drawback relative to single-medium hot/cold tank systems. While the latter keep constant outlet temperatures at charging and discharging until the tank is empty, in a dual medium storage system, the outlet temperature of the heat-transfer medium increases the more it is charged and decreases the more it is discharged, leading to unusable storage capacities (Dinter *et al.*, 1991).

3.6.2 *Concrete TES*

Concrete is sometimes chosen because of its low cost, availability throughout the world and easy processing. Inexpensive aggregates to the concrete are widely available. Concrete has the following characteristics as a storage medium:

- high specific heat,
- good mechanical properties (e.g. compressive strength),
- a thermal expansion coefficient near that of steel (pipe material), and
- high mechanical resistance to cyclic thermal loading.

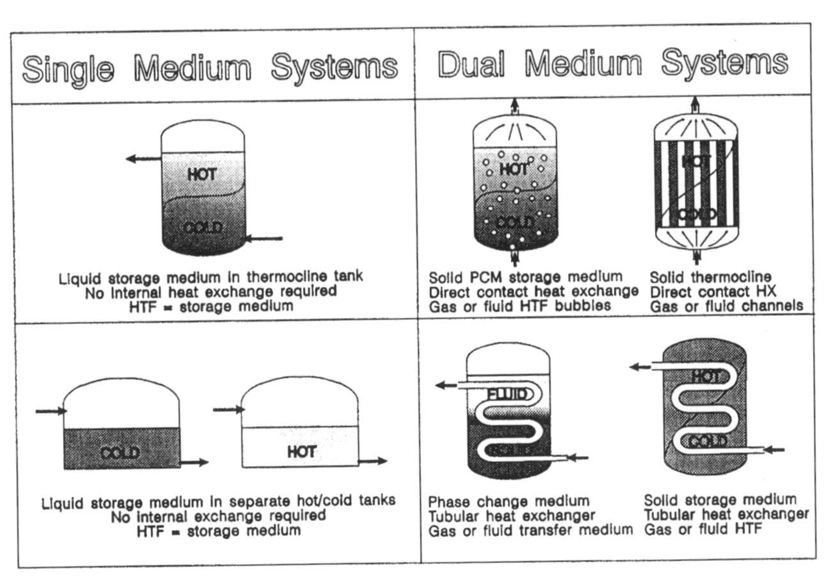

Figure 3.7 Schematics of some different tank configurations (HX: heat exchanger; HF: heat-transfer fluid) (*Dinter et al., 1991*).

When concrete is heated, a number of transformations and reactions take place which influence its strength and other physical properties. When concrete is heated to about 100°C, water is expelled (up to 130 kg of water per m^3 of concrete). Remaining water (50 to 60 kg of water per m^3 concrete), either physically bound in smaller pores or held by chemisorption, is expelled as temperatures rise from 120 to 600°C. Most dehydration occurs between 30 and 300°C. This water loss reduces the weight of concrete by 2–4%. The specific heat decreases in the temperature range between 20 and 120°C, and the thermal conductivity decreases between 20 and 280°C. The mechanical properties are also influenced slightly by the loss of water; compressive strength decreases about 20% at 400°C compared to that at ambient temperature. Resistance to thermal cycling depends on the thermal expansion coefficients of the materials used in the concrete. To minimize such problems, a basalt concrete is sometimes used. Steel needles and reinforcement are sometimes added to the concrete to impede cracking. By doing so, the thermal conductivity is increased about 15% at 100C and 10% at 250°C (Dinter *et al.*, 1991).

Such concrete storage can be supplied as prefabricated plates. Alternatively, the concrete may be poured on-site into large blocks, leading to easier and more economical construction. Whether prefabricated plates or on-site pouring is advantageous depends on local conditions.

3.6.3 Rock and water/rock TES

Rock is an inexpensive TES material from the standpoint of cost, but its volumetric thermal capacity is much less than that of water. Rock storage bins in home air-heating systems are practical (Dincer, 1999). The advantage of rock over water is that it can easily be used for TES above 100°C. Rock-bed and water storage types can both be utilized in many ways. For example, they may be used in conjunction with heat pumps to improve the efficiency of heat recovery, or with more elaborate heat exchangers, or with each other.

Eldighidy (1991) conducted extensive theoretical and experimental investigations on solar storage tanks with TES and indicated that water, rock beds and PCMs are the most suitable storage media. Three different configurations of storage tanks (Figure 3.8) for TES applications were examined. The most common TES system has a water-filled container in direct contact with both the solar collector and the house heating system (Figure 3.8a). Cool water from the bottom of the tank is circulated to the collector for solar heating and then returned to the top of the tank. Warm water from the top of the tank is circulated directly through baseboard radiators or radiant heating panels inside the rooms. Figure 3.8b shows another system which consists of a copper coiled-finned tube immersed in the tank of solar heated water. Rocks are the most widely used storage medium for air collectors.

One attractive storage method that uses both water and rocks as storage media is known as Harry Thomason's method (Figure 3.8c). In this system, heated water from the solar collector enters a water tank at the top, sinks as it cools and finally leaves at the tank bottom as it is recirculated to the collectors. The water tank is surrounded by river rock through which air is circulated to carry the heat into the house; the entire rock and water tank assembly is contained within insulated walls. The advantage of this system is the high heat capacity of the water, and the extensive area of the rock's container leads to efficient transfer of heat to the air.

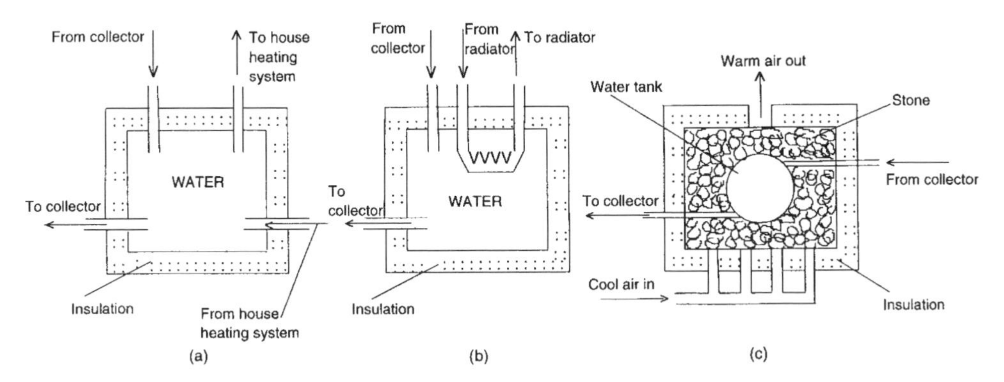

Figure 3.8 Solar storage tanks: (a) heat storage tank tied directly to both the collector and the house heating system, (b) sensible TES system using a heat exchanger to extract solar heat from a storage tank, and (c) Harry Thomason's technique using both water and stone as storage media.

Rocks are sometimes the preferred storage medium over water for solar systems situated in northern latitudes. However, air/rock solar systems can, and in some instances do, make provision for partial heat storage in water for domestic hot water use. Storage of heat in rocks requires approximately three times as much space as an equivalent amount of heat in water. This disadvantage is often overcome by cost advantages. Water containment in a TES or a swimming pool is expensive. Combined with the higher capital and maintenance costs of a liquid collector, the economics quickly favor the use of air collectors with rock storage for domestic heating applications. Liquid solar systems were chosen during the early stages of solar energy development because of the superior heat transfer characteristics of liquid over air. Initial solar energy collection systems were able to capture slightly more than 50% of the incident solar radiation with liquid collectors but barely 30% with collectors of the air type. Present air collectors, which incorporate developments in selective coatings are beginning to match the energy efficiency of liquid collectors while retaining simplicity, minimal maintenance and low costs. The volume occupied by a rock store is equal to approximately 1.6 m^3 per 1m^2 of collector. The average collector array for a typical residence occupies an area of 30.4 m^2. A rock store for this size collector occupies a volume equal to 4.6 m^3 (e.g. rectangle of dimensions 1.5m×1.5m×2 m). Such a size usually does not present a significant design problem with respect to interior living spaces.

Design considerations for TES in rocks

The rule of thumb for determining the size of a storage rock bed specifies that a collector area of 1 m^2 is needed for an amount rock of 1.6 m^3. This rule assumes a home insulated to high standards and situated in regions experiencing average levels of solar radiation. Regions experiencing significantly higher levels of radiation may benefit from increasing store size if there is no corresponding decrease in ambient temperature. An increase in store size in proportion to an increase in solar radiation enables the system to store more heat for use during non-solar periods of little or no solar insolation. However, if the increase in solar radiation is accompanied by colder weather, such as is experienced in mountainous terrain, there is probably no advantage in increasing the size of the store. This is because the additional heat gain as a result of the increased radiation is dissipated faster due to the accelerated heat loss caused by the lower ambient temperatures.

Storage configurations that are nearly cubic or slightly rectangular generally perform well. This type of geometry allows for a high ratio of volume to surface area of containment. In heat stores, an increase in surface area produces a corresponding increase in heat loss. Thus, heat retention can be enhanced by keeping the surface area of the store to a minimum. A way of further reducing surface area is to utilize cylindrical containment.

Washed gravel is preferable to crushed stone as a storage medium because the resistance to air flow of crushed stone is up to three times greater than for washed gravel. The major constraint associated with rock beds is that the rock should be uniform in size, which requires some sort of screening. Thus, the recommended particle diameter is 2.5 cm.

The air flow through a vertical heat store should be from top to bottom while charging and from bottom to top while recuperating heat. By reversing the flow in this manner, we are able to stratify the store. Stratification can provide a warmer air supply for the home and a cooler air supply to a solar collector or other heat source. With this arrangement, efficiency is increased.

The pressure drop through a rock store should be significantly higher (minimum 5 times) than in the plenums leading to the store, so as to assure a good flow distribution through the bed. One rock bed installation, for example, is designed for a 21.2 m^2 single glazed, selective coated, back-pass collector array. The air stream is designed for two stages with a maximum air flow rate of 4.65 m^3/s. A store for this configuration with 2.5 cm diameter would have a pressure drop of 5.84 mm of water. Additional facts pertaining to pressure drop worth noting follow (Dincer and Dost, 1996):

- Pressure drop through a rock bed is directly proportional to air speed.
- Pressure drop through a rock bed is inversely proportional to area of cross section.
- A 0.028 m^3 store containing 2.5 cm diameter rocks has approximately 0.82 m^2 of surface area.
- A 0.093 m^2 store containing 7.6 cm diameter rocks has about 1.3 m^2 of surface area.

Pressure drop and heat transfer coefficients can be altered by adjusting these parameters.

Water-rock beds

A recent study (Moschatos, 1993) discussed in detail a new type of TES, which is a combination of rock-bed and water storage. The configuration consists of a water storage, surrounded by a rock-bed storage instead of insulation, as shown in Figure 3.9. The basic concept is that, for a house heated by solar collectors, heat losses could be offset by the thermal energy of a water storage or by auxiliary energy. Part of the ventilation thermal losses can be recovered by passing fresh air through the bed storage.

The general idea of combined storage was developed for a number of houses by Thomason. The differences between the storage combination of Thomason (see Figure 3.8c) and that in Figure 3.9 are:

- Thomason utilizes both water and rock storage to hold solar energy, with the thermal load of the house mainly met by rock storage. The main solar TES is the water storage, which is covered by a rock bed instead of insulation. In this way, water-storage thermal losses preheat the cold fresh air for ventilation.
- Thomason's bed storage has no specific geometrical shape.

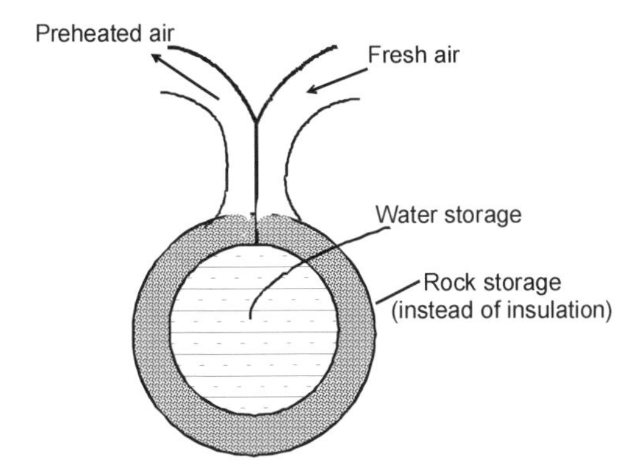

Figure 3.9 A schematic of a cylindrical combined water-rock storage (Moschatos, 1993).

With the combined storage (a) fresh air is preheated, (b) low-temperature thermal energy contained in the bed storage is utilized, and (c) water storage thermal losses are partially recovered.

Combining water with air/rock storage systems is becoming common for solar energy applications. Such systems aim to provide a portion of the energy needs for domestic hot water without a significant reduction in the solar energy supply for space heating needs. Conventional air-rock solar systems for space heating see service during the heating season only. Consequently, the amortization period is lengthened because the solar system is inoperative for approximately six months of the year. Adding domestic hot water capability to the basic air/rock system increases the capital costs by a small fraction. So, for a relatively small additional investment, the solar system can provide 100% of the domestic hot water during summer and less during other seasons. The configuration for heating domestic hot water with an air/rock solar system can be as simple as placing a steel tank in the rock store. For such systems, tanks with glass linings, which are readily available at low prices since they are mass produced for conventional hot water heating, are often used. The glass-lined water storage tank is preferred for its long life expectancy, often greater than 20 years in solar applications.

There is relatively little thermal shock on storages associated with solar use. In conventional use, the sudden application of high-grade heat to a water store can harm the tank over time, depending on the mineral content of the water and other related factors. Compared to oil, gas or electric heated water, which is at a very high temperature for a short period of time, solar heated water is at a low temperature for a long period of time. The actual firing or heat supply temperatures are approximately 60°C for solar energy and over 800°C for conventional systems.

It is advisable to provide access for visual monitoring and replacement of storage. The latter may never be needed, but if it is, it is easier to gain access through a panel than to have to destroy the storage.

In certain instances, solar collection efficiency can be improved by placing a second water reservoir in the cold end of the store. This option is sometimes considered when the water supply to the home is at a fairly low temperature (5°C). The effect is to preheat the water entering the tank situated in the hot end of the store, and to keep the cold end of the store below the average temperature. This increases the solar heat gain and the efficiency of the collector. An additional benefit of the second water tank is an increase of thermosyphoning (natural convection) within the store, which transfers heat from the rocks to the water reservoir during periods when the solar house fan is inoperative.

There are two potential disadvantages to using this type of system. First, it may prove difficult to provide access to a water tank located under a pile of rocks. Secondly, the air from the house returning to the store during the heating season may nullify all or part of the pre-heating effect. Consequently, the addition of the secondary water reservoir should be considered for select applications only.

Storage bins may be constructed utilizing containers of water as a replacement for rocks. The containers have to be small, one gallon or less, in order to have effective heat transfer between the air stream and the contained water. Also, the container must be of a type that withstands thermal degradation.

The most important advantage of a hybrid heating system with a water-to-air heat exchanger and storage unit is its multiple applications, including water heating, drying, space heating, space cooling (by letting cold water flow through the system) during different times of the year. A rock-bed heat exchanger and storage unit has been used successfully for heat transfer (from water to air) and heat storage in the Thomason residential heating systems. Such systems can operate for several years, but few design investigations and parametric studies have been conducted. In a recent work, Choudhury and Garg (1995) conducted a performance evaluation of the system with optimized design parameters, both with and without a domestic hot water load on the system. The load is the standard hot water demand of a four person (two adults and two children) residential building in India. The system has been evaluated under both continuous and intermittent (i.e. air flow rate = 0 when the room temperature is exceeds 27°C, which is taken to the upper limit of the comfortable range for residential buildings) air circulation conditions. No auxiliary energy is used to satisfy the space heating and the hot water demand.

3.6.4 Aquifer Thermal Energy Storage (ATES)

An aquifer is a ground water reservoir. The word aquifer derives from the Latin words 'aqua' meaning water and 'ferre' meaning to carry. The material in an aquifer is highly permeable to water, and the boundary layer consists of more impermeable materials such as clay or rock. Aquifers are found throughout the world. For example, two types of aquifers are found in Sweden. The most common type consists of sand and gravel deposits left by retreating ice from the Ice Age. The second type consists of sandstone or limestone, and can be found mainly in the southern part of Sweden. Water from precipitation continuously seeps down to an aquifer and flows slowly through it until finally reaching a lake or the sea. Aquifers are often used as fresh water sources.

Aquifers often have large volumes, often exceeding millions of cubic meters, and as they consist of about 25% water they have a high TES capacity. When heat extraction and charging performances are good, high heating and cooling powers can be achieved. The amount of energy that can be stored in an aquifer depends on local conditions, such as

- allowable temperature change,
- thermal conductivity, and
- natural ground water flows.

ATES is a concept that has been known for several decades. Aquifer systems have received worldwide attention because of their potential for large-scale and long-term TES. In its most common form, the technique involves storing excess heat in an aquifer and recovering it later during periods of heat demand. With growing concerns about global warming, the concept is receiving renewed attention as a viable means of conserving energy and reducing fossil fuel use. This increasing interest is reflected in accelerated research activities in northern European countries and Canada. In developing ATES systems, the physical processes governing the behavior of thermal energy transport in groundwater must be well understood. Numerical simulation models serve a key role in contributing to this understanding, and are thus indispensable in the design of efficient ATES facilities.

The injection of heated water into an aquifer may also be necessary for reasons other than energy conservation. For example, where water rights issues or concerns about land or water quality arise, water extracted for cooling purposes may be reinjected into the aquifer. In these situations, it is important to understand the effect of heat on the groundwater system, and to be able to predict consequences such as the accelerated precipitation of dissolved substances, or changes in the biological regime.

Research into ATES has been ongoing for about two decades in North America and in Europe. Work in Denmark, for example, shows a high level of development. In the United States, pioneering research has been done at Auburn University and at the General Electric Center for Advanced Studies at Santa Barbara, California. Research into ATES is also being conducted in many other European countries, including Sweden, Germany, Denmark and France.

Figure 3.10 illustrates the operation principle of simple ATES systems. Both heating and cooling cycles for a building are considered. Wells have been drilled to transport water to underground aquifers (underground areas of water-bearing rock, sand or gravel). Well spacing, depth and size are functions of the aquifer.

An aquifer storage system can be used for storage periods ranging from long to short, including daily, weekly, seasonal or mixed cycles. To avoid undesired permanent changes of the temperature level in the aquifer, the input and output of heat must be of the same magnitude at least after a number of cycles. The system should be designed to be adjustable in case the long-term energy flows do not balance. The TES capacity should be appropriate to the heating/cooling loads.

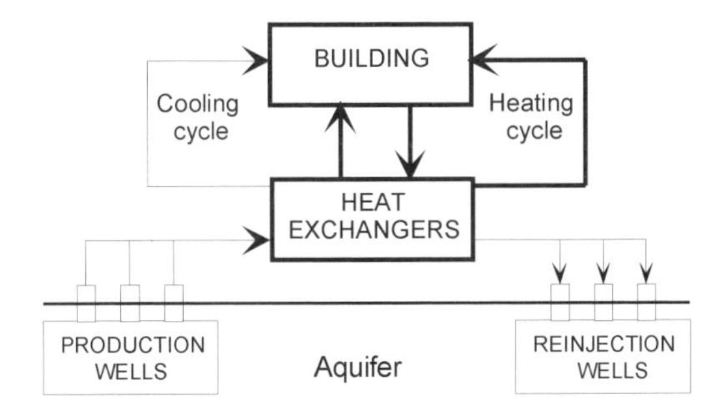

Figure 3.10 Schematic of how an ATES system works.

Extensive investigations and test runs are usually needed to predict the performance of an ATES before the design of the energy system. Such preparatory work can in some cases be relatively costly.

Aquifers with groundwater, or the ground itself (soil, rock), can be used as a storage medium in ATES systems. Aquifer stores are most suited to high capacity systems. The existing capacities range in size from less than 50 to over 10,000 kW.

Utilization of ATES

ATES or groundwater energy systems utilize groundwater, and do not necessarily deplete non-renewable resources. In some systems, external thermal energy is stored in an ATES. In others, the natural groundwater temperatures are used. In the latter case, the system requires production wells to supply groundwater to a series of heat exchangers. In winter, heat from underground is used to pre-heat outside building air. In summer, direct cooling is achieved by transferring heat from buildings to the groundwater using the same principle in reverse. In both cases, the groundwater is returned to the subsurface aquifer through reinjection wells without having been exposed to any form of contamination. Over time, a thermal balance is maintained in the aquifer.

These groundwater systems can be applied to new and existing heating and cooling systems in such facilities as government buildings, business parks, residential complexes, educational institutions, hospitals and industrial complexes. In addition, ground water systems can be integrated into industrial process plants such as paper and textile mills, and pharmaceutical manufacturing, mining and food processing facilities.

Groundwater energy systems provide an efficient and reliable supply of low cost energy that can complement conventional heating and cooling systems. Furthermore, the groundwater systems are environmentally benign technology and can reduce stack emissions (e.g. carbon dioxide) and the use of CFCs, and can reduce electrical usage during peak demand periods.

ATES has been used successfully around the world for the seasonal storage of heat and cold energy for the purpose of heating and cooling buildings. Domestic and large-scale ATES systems are used quite widely in some countries, e.g. Canada, Germany.

When heat storage is needed to match thermal demand and supply, the ground has proven to be a useful storage medium, especially for large quantities of heat and long storage durations, like seasons. Many plants use ATES to store summer solar heat for use in winter heating, and ATES for waste heat is now emerging. The efficiency of heat storage depends upon the temperature level achieved and upon the quantity of thermal losses. While heat storage in ATES in the range of 10–40°C was demonstrated successfully, storage at higher temperatures (e.g. up to ca. 150°C) was shown to cause many problems in experimental and pilot plants in the 1980s.

Deep confined aquifers

Traditionally, deep confined aquifers have been preferred for ATES facilities. Some suggest that such aquifers are advantageous because (i) regional groundwater flow usually being negligible, the injected hot water is not displaced, (ii) the thickness of overburden is such that the storage is not disturbed by seasonal surface temperature variations, and (iii) the initial temperature is higher (due to the natural geothermal gradient), thus significantly reducing heat losses to confining layers. Nonetheless, unconfined aquifers are being used for ATES systems in Sweden, France and the USA. Shallow, unconfined aquifers are easier to delineate. They are more common in many regions, and can be inexpensive when it comes to well installation and monitoring. Therefore, the use of unconfined aquifers for ATES can be advantageous, provided the energy recovery efficiency is adequate.

The aquifers used in ATES are permeable, water-bearing rock formations. Where aquifers are unavailable, a network of plastic tubing inserted into boreholes drilled into the earth can be used as underground storage, usually with water. With an aquifer system, two well fields are often tapped: one for cold storage and the other for heat. These wells, which are usually around 200 m deep, are capable of maintaining storage temperatures between 4°C and 90°C.

During the summer months, cool groundwater is extracted from one aquifer and circulated through building systems to lower the air temperature. The water, which is heated during the process, is then returned to the other aquifer for eventual use in heating in winter—a cyclic process that can be repeated indefinitely. An ATES system typically reduces cooling costs by 80% and heating costs by 40% or more, while significantly reducing emissions of greenhouse gases and ozone-depleting substances.

Environment Canada (1999) in the Atlantic region of Canada has developed a variety of tools and procedures for implementing this energy-efficient technology—including tests to determine the thermal properties of boreholes, water-treatment technologies for high-temperature applications, and environmental screening techniques—and plays an active role in the transfer of ATES technologies, both nationally and internationally.

In the mid-1990s, Environment Canada teamed up with the New Brunswick Department of the Environment, New Brunswick Power, the Canadian Electricity Association, the Panel of Energy Research and Development, and the local community to launch an aquifer-based ATES demonstration project at the Sussex Hospital in New Brunswick. The first hospital in Canada and the second in North America to adopt such a system, it has saved nearly $50,000 a year in energy consumption costs since the project began, and reduced carbon dioxide emissions by 720,000 kg annually. These savings were accomplished after the hospital had already cut its energy consumption in half by installing an energy management system. During an emergency, only small auxiliary generators are required to operate the water pumps used in heating and cooling using the ATES system.

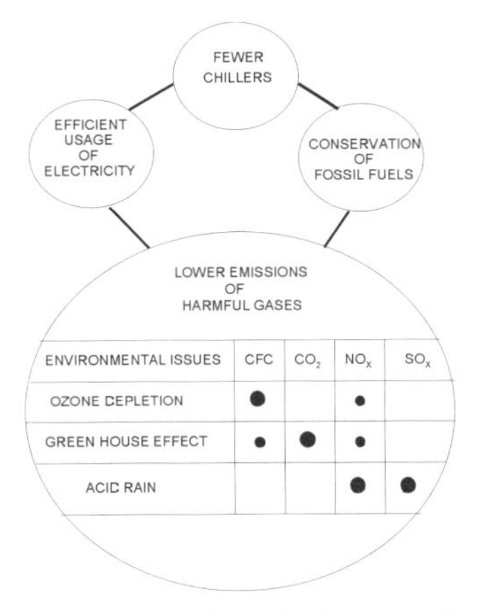

Figure 3.11 Some environmental benefits of ATES (Environment Canada, 1999).

ATES is commonly used in China and parts of western Europe—particularly the Netherlands and Sweden, where its use is growing by 25% a year. Although office buildings remain the primary market for the technology, it is gaining ground in industrial and agricultural applications, and a new market is developing in the de-icing and cooling of roads and bridges.

The environmental benefits of ATES are illustrated in Figure 3.11, considering the important global problems. The sizes of the circles represent the relative impact of the use of ATES technology on the emissions indicated.

The success of the Sussex Hospital ATES and other ATES projects in Canada, such as those at Saskatoon Airport and Carleton University in Ottawa, has spurred the development of several new initiatives in Ontario, Nova Scotia and British Columbia. Other situations exist where a system of this kind can yield considerable long-term environmental benefits and reduced usage of fossil fuels for heating and electricity.

Annual ATES systems use aquifers which are water-bearing rocks found near the Earth's surface. These are large geological formations that can be tapped by wells. Although some technical problems (e.g. biofouling and clogging of pipes and wells) remain to be solved, aquifers potentially offer low-cost storage for large-scale systems. The main cost is associated with access wells. However, relatively low cyclic ATES efficiencies (~70%) may limit their use to the storage of such low-cost energy supplies as industrial waste heat.

Performance of ATES

The viability and cost effectiveness of an ATES system depend on the mechanical design and thermodynamic efficiency of the above-ground installation, as well as on the physical, chemical and biological processes within the aquifer. The spatial and temporal variations of temperature, as well as total energy, have been found to be critical factors in determining ATES performance.

Physical processes affecting heat transport within a storage aquifer include advection, dispersion and diffusion. The diffusion of heat depends on the thermal conductivity and heat capacity of the aquifer. In a composite medium such as an aquifer, the properties of both the fluid phase and the solid phase play important roles in heat transport, and ultimately control the recovery efficiency of stored energy. In addition, heat transfer from the aquifer system to adjacent aquitards or through the unsaturated zone to the atmosphere can be a significant process for removing heat from the ATES system.

Chemical reactions resulting from changes in temperature and from mixing of the injection water with the resident groundwater can change the porosity and permeability of the aquifer, decrease well efficiency, and increase the cost of operation and maintenance of the heat exchangers. Temperature changes can also affect the activity of subsurface micro-organisms, which can lead to a decrease in permeability and increased maintenance costs.

Expansion of ATES applications

Possible heat sources and heat users for ATES applications are listed in Table 3.8. The heat sources for many promising systems can be divided into two main groups:

- **Renewable energy sources**: solar heat can be used with ATES to supply heat to district heating networks, along with back-up auxiliary heating systems. Solar heat can also be used with heat pumps, avoiding the need for auxiliary heating. Another ATES use may be for geothermal heating, allowing storage of excess production in summertime and to cover peaks in winter, or for storing waste heat from geothermal power plants.

- **Waste or excess heat**: storage of waste heat from co-generation or industrial processes (Table 3.8) may be needed on a seasonal or other cycle. ATES can also be applied as a back-up for processes that use industrial waste heat, to cover heat load during periods when the industrial process is stopped (for production breaks, repairs, etc.). Similarly, ATES can be used for load levelling in a district heating system, where the store is always charged at times of low heat demand and unloaded during peak heating periods.

Table 3.8 Some possible heat sources and heat users for ATES.

Possible heat sources	Possible heat users
Renewable energy • Solar thermal (solar collectors, but also road surfaces, etc.) • Geothermal (hydrogeothermal, but also waste heat from geothermal power plants, e.g. hot dry rock) • Others (e.g. biomass combustion) Waste heat • Heat and power cogeneration (likely only with high electrical efficiency) • Industrial process heat (e.g. from paper mills, steel works, etc.) • Waste incineration • Others Load levelling • in district heating systems (for short- to medium-term periods)	Space heating • District heating • Large buildings (houses, offices, hospitals, hotels, airports, etc.) Industrial heat • Batch or seasonal processes such as in sugar refineries • Drying in the food industry • Miscellaneous heat needs in many industries Agriculture • Greenhouse heating • Food drying applications • Aquaculture De-icing and snow-melting • On roads, sport centers, airports, runways, etc.

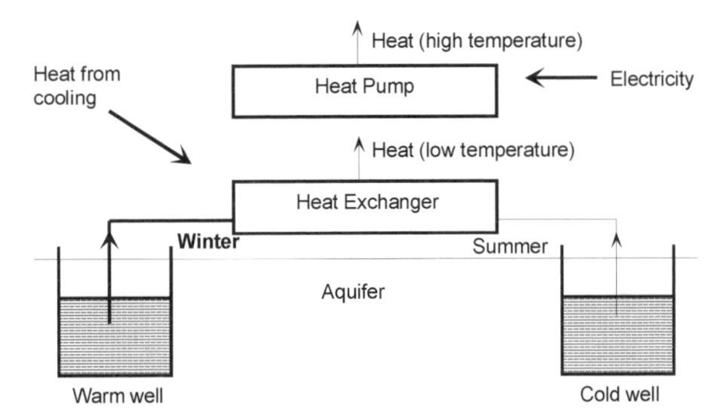

Figure 3.12 An ATES system combined with a heat pump.

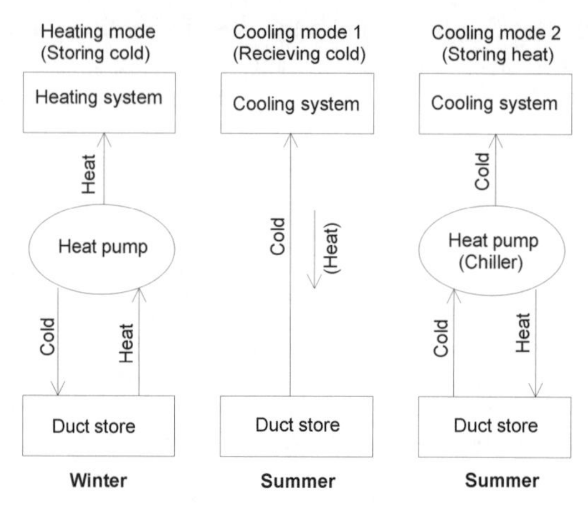

Figure 3.13 Operational principles for ATES systems using heat pumps.

The IEA study mentioned in the previous subsection shows that technical problems related to higher temperatures in ATES system may be overcome (IEA, 1990). A remaining issue is the changes in water chemistry with the drastically changing temperatures in ATES systems, resulting in clogging, scaling, corrosion and leaching. It is possible to design and build reliable ATES plants today, but caution is necessary when working with groundwater. In the future, a range of suitable methods for various hydrogeological/ hydrochemical situations and system requirements is desirable. Many opportunities exist, for promising ATES system concepts yield energy savings and reduced emissions.

ATES using heat pumps

The concept of TES in an aquifer combined with a heat pump is relatively simple (see Figure 3.12). During the summer cold water is extracted from a cold well and warmed by cooling a building, and then returned to a warm well in the aquifer. A heat pump can be used to chill the cold water further, if necessary. The warmed water diffuses out from the warm well, gradually raising the temperature of the aquifer. During the winter the process is reversed, with heat drawn from the warm well and the temperature boosted with the heat pump if necessary.

The basic operational scheme of a ground-coupled heat pump with seasonal cold storage is shown in Figure 3.13 (Sanner and Chant, 1992). During heating, the ground or ground-water is cooled, while heat is supplied to the building. At the end of the heating season, enough cold is stored to run a cooling system directly (mode 1) with cold ground-water from the injection well or cold brine from earth heat exchangers. For peak cooling loads, a back-up system using the heat pump in reversed mode can be operated (mode 2). After continuously running the heat pump for space cooling for more than a few hours, temperatures in the ground may be too high for mode 1 cooling. The system should then be operated only as a conventional heat pump plant, storing heat in the ground until the beginning of the next heating season. The most cost-effective and efficient energy performance can usually be obtained by running the system in heating mode and cooling mode 1 only.

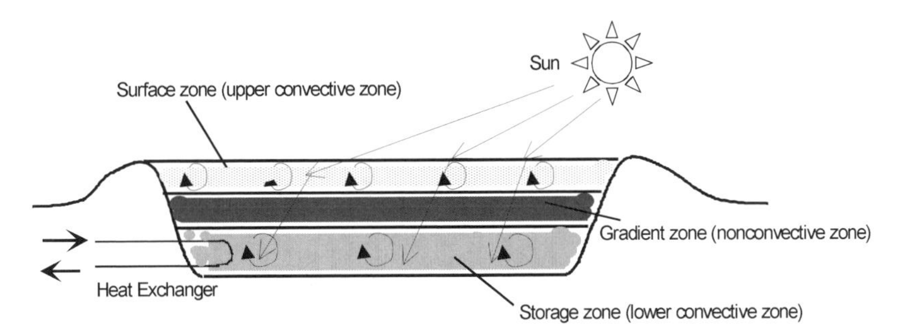

Figure 3.14 A cross-section representation of a typical salinity-gradity solar pond.

The cost-effectiveness of seasonal storage depends significantly on the thermal and seasonal characteristics of the load. Storage for cooling is more cost-effective for cooling systems with relatively high operating temperatures (up to 15°C) than with more conventional temperatures (6–8°C). By operating at higher temperatures, storage for cooling experiences reduced thermal losses, increased loading availability in winter, and reduced climatic influences on performance.

3.6.5 Solar ponds

A salinity gradient solar pond is an integral collection and storage device of solar energy. By virtue of having built-in TES, it can be used irrespective of time and season. In an ordinary pond or lake, when the sun's rays heat up the water this heated water, being lighter, rises to the surface and loses its heat to the atmosphere. The net result is that the pond water remains at nearly atmospheric temperature. The solar pond technology inhibits this phenomenon by dissolving salt into the bottom layer of this pond, making it too heavy to rise to the surface, even when hot. The salt concentration increases with depth, thereby forming a salinity gradient. The sunlight which reaches the bottom of the pond remains entrapped there. The useful thermal energy is then withdrawn from the solar pond in the form of hot brine. The pre-requisites for establishing solar ponds are: a large tract of land (it could be barren), a lot of sunshine, and cheaply available salt (e.g. NaCl) or bittern.

Salt-gradient solar ponds may be economically attractive in climates with little snow and in areas where land is readily available. In addition, sensible cooling storage can be added to existing facilities by creating a small pond or lake on the site. In some installations this can be done as part of property landscaping. Cooling takes place by surface evaporation and the rate of cooling can be increased with a water spray or fountain. Ponds can be used as an outside TES system or as a means of rejecting surplus heat from refrigeration or process equipment.

Being large, deep bodies of water, solar ponds are usually sized to provide community heating. Solar ponds differ in several ways from natural ponds. Solar ponds are filled with clear water to ensure maximum penetration of sunlight. The bottom is darkened to absorb more solar radiation. Salt is added to make the water more dense at the bottom and to inhibit natural convection. The cooler water on top acts as insulation and prevents evaporation. Salt water can be heated to high temperatures, even above the boiling point of fresh water.

Figure 3.14 represents a cross-section view of a typical salinity-gradity solar pond which has three regions. The top region is called the surface zone, or upper convective zone. The middle region is called the gradient zone, or nonconvective zone. The lower region is called the storage zone or lower convective zone. The lower zone is a homogeneous, concentrated salt solution that can be either convecting or temperature stratified. Above it the nonconvective gradient zone constitutes a thermally insulating layer that contains a salinity gradient. This means that the water closer to the surface is always less concentrated than the water below it. The surface zone is a homogeneous layer of low-salinity brine or fresh water. If the salinity gradient is large enough, there is no convection in the gradient zone even when heat is absorbed in the lower zone, because the hotter, saltier water at the bottom of the gradient remains denser than the colder, less salty water above it. Because water is transparent to visible light but opaque to infrared radiation, the energy in the form of sunlight that reaches the lower zone and is absorbed there can escape only via conduction. The thermal conductivity of water is moderately low, and if the gradient zone has substantial thickness, heat escapes upward from the lower zone very slowly. This makes the solar pond both a thermal collector and a long-term storage device.

Solar ponds were pioneered in Israel in the early 1960s, and are simple in principle and operation. They are long-lived and require little maintenance. Heat collection and storage are accomplished in the same unit, as in passive solar structures, and the pumps and piping used to maintain the salt gradient are relatively simple. The ponds need cleaning, like a swimming pool, to keep the water transparent to light. A major advantage of solar ponds is the independence of the system. No backup is needed because the pond's high heat capacity and enormous thermal mass can usually buffer a drop in solar supply that would force a single-dwelling unit to resort to backup heat.

3.6.6 Evacuated solar collector TES

Selection of a storage medium is often based on availability, durability and cost, and tends to favor solar TES using water or rocks. One of the most promising applications of sensible TES is for solar-heated residential and commercial buildings.

As mentioned earlier, in some cases rocks are preferred over water for solar energy systems. However, air/rock solar systems can allow for partial heat storage in water for domestic hot water use. Heat stored in rocks, compared with an equivalent amount of heat stored in water, occupies approximately three times as much volume. However, liquid containment systems have higher initial and maintenance costs. Economics usually favor the use of air collectors with rock storage for domestic heating applications. Water and air/rock systems are often combined for solar energy applications (Dincer *et al.*, 1997a), so as to provide a portion of energy needs for domestic hot water and space heating. Conventional air/rock solar systems for space heating are used only during the heating season, which usually lasts no more than half of the year. Hybrid hot water and air/rock systems have slightly increased the capital costs, but operate year round by providing domestic hot water during summer.

Active solar systems with standard flat-plate solar heat collectors for domestic and commercial applications have some technical and operational problems. Evacuated-tube solar collectors have some advantages, and since 1980, several million tubes have been installed in over thirty countries. In 1991, the Thermomax Mazdon evacuated solar collector (Figure 3.15a) was developed. Its manifold configuration is particularly suited to North American conditions and plumbing practices. These collectors avoid many of the thermal and operational problems that occur in the flat-plate collectors and have better efficiencies (up to 80%), mainly because of its thermophysical properties (e.g. its heat transfer rate is thousands of times greater than the best solid heat conductors of the same size). Some of the features of these solar collectors are as follows:

- high efficiency,
- low heat capacity and high heat transfer rate,
- control of maximum temperature,
- high durability,
- availability for a variety of thermal applications ranging from hot water to industrial process heat,
- freedom from corrosion problems, maintenance, and cold weather/frost problems, and
- easy installation.

Figure 3.15b shows an integrated tank system, consisting of 15 evacuated solar collector tubes and a 170 L hot water storage tank. The tank is made from a high-grade stainless steel, having at least 2.5% molybdenum content, sealed in a weather resistant metal cover. There is a thick polyurethane foam insulation jacket that lies between the tank and the steel cover. This solar energy system produces high water temperatures, eliminating the need for large and cumbersome water storage tanks. A large volume of warm water can be obtained by mixing cold water with the hot water from the unit. During the operation, the maximum chloride concentration of the water in the tank should not exceed 40 ppm in order to avoid corrosive effects.

3.7 Latent TES

Effective utilization of time-dependent energy resources requires appropriate TES methods to reduce the time and rate mismatch between energy supply and demand. TES provides a high degree of flexibility since it can be integrated with a variety of energy technologies, e.g. solar collectors, biofuel combustors, heat pumps and off-peak electricity generators.

The heat transfer which occurs when a substance changes from one phase to another is called the latent heat. The latent heat change is usually much higher than the sensible heat change for a given medium, which is related to its specific heat. When water turns to steam, the latent heat change is of the order of 2 MJ/kg.

Most practical systems using phase change energy storage involve solutions of salts in water. Several problems are associated with such systems, including the following:

- Supercooling of the PCM may take place, rather than crystallization with heat release. This problem can be avoided partially by adding small crystals as nucleating agents.
- It is difficult to build a heat exchanger capable of dealing with the agglomeration of varying sizes of crystals which float in the liquid.
- The system operation cannot be completely reversed.

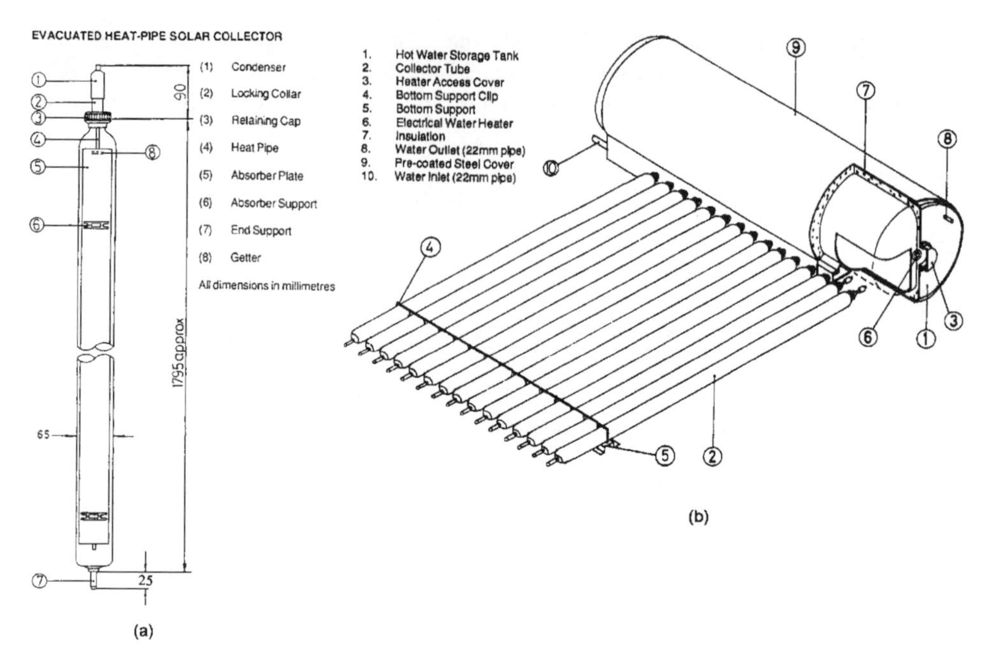

Figure 3.15 Evacuated solar collector TES system. (a) Evacuated solar collector, (b) An integrated tank system.

Any latent heat TES system must possess at least the following three components (Abhat, 1983):

- A heat storage substance that undergoes a phase transition within the desired operating temperature range, and wherein the bulk of the heat added is stored as latent heat.
- A containment for the storage substance.
- A heat-exchange surface for transferring heat from the heat source to the storage substance and from the latter to the heat sink, e.g. from a solar collector to the latent TES substance to the load loop.

Some systems use as their storage media either $Na_2SO_4 \cdot 10\ H_2O$ or $CaCl_2 \cdot 6H_2O$ crystals, and employ a heat-exchange oil. The oil is pumped in at the bottom of the storage and rises in globules through the fluid without mixing. Other promising latent TES reactions are those of inter-crystalline changes. Many of these take place at relatively high temperatures.

Solar energy applications require large TES capacities in order to cover a minimum of 1–2 days of thermal demand. This capacity is commonly achieved by sensible heat storage in large water tanks. An alternative is offered by latent heat storage systems, where thermal energy is stored as latent heat in substances undergoing a phase transition, e.g. the heat of fusion in the solid-liquid transition. The main advantages of latent TES systems are high TES capacities per unit mass compared to sensible heat systems, and a small temperature range of operation since the heat interaction occurs at constant temperature. There is no gradual decline in temperature as heat is removed from the PCM.

Salt compounds that absorb a large amount of heat during melting are useful for energy storage. Eutectic salts and salt hydrates are widely used. Glauber's salt (sodium sulfate decahydrate) is a leading PCM because it has a high heat-storage capacity (280 kJ/kg) and a phase-change temperature that is compatible with solar systems (31.5°C).

PCM can be contained in rods or plastic envelopes to facilitate the freeze-thaw cycle. These small modules, and the small number of modules required for storage, make phase-change storage especially suitable for use in conventional design and for retrofitting.

In the Mattapoisett house in Massachusetts, which is 60% heated by passive solar gain, 8.5 m^3 of Glauber's salt is located in the ceiling. The phase-change storage has been demonstrated to lower the peak indoor temperatures. Table salt (NaCl) can be added to the PCM, lowering the melting point to about 22.7°C. Because heat is then absorbed at a lower temperature, the house only reaches about 26.6°C on a clear winter's day instead of the 29.4°C or so often seen in passive solar structures (Lane, 1988).

The three most favored storage media for solar systems, water, rock, and Glauber's salt, vary considerably in price and required volume. For 9 m^2 of collector area (assuming a 20°C temperature swing), the storage cost for water is $54 and for Glauber's salt it is $146. However, only 0.18 m^3 of Glauber's salt is required, which is one-quarter of the necessary 0.72 m^3 of water. Rock storage for the same collector would cost $217 at $8/ton for the required 2.46 m^3. This option would seem the least appealing, but it is nonetheless established and widely used. For example, the Lof home in Denver has been heated by a rock storage system for more than 20 years (Tomlinson and Kannberg, 1990).

Latent TES is a promising storage technique since it provides a high energy storage density, second only to chemical energy storage, and can store and release heat at a constant temperature, corresponding to the phase transition temperature of the heat-storage medium.

Another important material category capable of storing energy through phase change is paraffin waxes. These have the advantage of very high-stability over repeated cycles of latent TES operation without degradation. Several types of heat exchanger have been used to retrieve the energy from this storage medium: a cylindrical pipe, single radial-finned pipe and multiple radial-finned pipe. Experimental tests have been performed for each of the different configurations, and the relative merits of each discussed in terms of heat-exchange properties and total energy exchanged.

3.7.1 Operational aspects of latent TES

At present, problems concerning the inability to reverse completely the storing process limit the practical applications of chemical storage media. On the other hand, storage media that undergo physical processes can usually have the storage process totally reversed, but involve less TES per unit weight of device than chemical storage. Latent TES systems thus have the advantage of compactness in comparison with sensible TES devices (as well as the operational advantage of a nearly constant storage-cycle temperature).

Among the thermodynamic phase changes at a constant temperature with the absorption or release of latent heat, the most suitable ones for TES are the solid-liquid and solid-solid transitions. Solid-gas transitions, even though they often involve the largest heat interactions per unit weight, present the disadvantage of very large volume changes.

The most important criteria to be met by the storage material for a latent TES in which the material undergoes a solid-liquid or a solid-solid phase transition are as follows:

- high transition enthalpy per unit mass,
- ability to fully reverse the transition,
- adequate transition temperature,
- chemical stability and compatibility with the container (if present),
- limited volume change with the transition,
- non-toxicity, and
- low cost, in relation to the foreseen application.

3.7.2 Phase Change Materials (PCMs)

When a material melts or vaporizes, it absorbs heat; when it changes to a solid (crystallizes) or to a liquid (condenses), it releases this heat. This phase change is used for storing heat in PCMs. Typical PCMs are water/ice, salt hydrates, and certain polymers. Since energy densities for latent TES exceed those for sensible TES, smaller and lighter storage devices and lower storage losses normally result.

Like ice and water, eutectic salts have been used as a storage media for many decades. Perhaps the oldest application of a PCM for TES was the use of seat warmers for British railroad cars in the late 1800s. During cold winter days, a PCM sodium thiosulfate pentahydrate which melts and freezes at 44.4°C was used. The PCM was filled into metal or rubber bags.

Other early applications of PCMs included 'eutectic plates' used for cold storage in trucking and railroad transportation applications. Another important application of PCMs was in association with space technology, with NASA sponsoring a project on PCM applications for thermal control of electronic packages.

The first experimental application of PCMs for cool storage occurred in the early 1970s at the University of Delaware. The University's Institute of Energy Conversion undertook the design and construction of a solar energy laboratory that became known as Solar One, located on the Newark, Delaware, campus. Dr. Maria Telkes, a leader in the research and development of eutectic salts, was selected to direct the institute's Thermal Energy Applications program. The design of the Solar One building incorporated several hydrate-based storage systems. One material was used in association with solar collectors for heat storage. In the same sheet metal bin were placed containers of a Glauber's salt mixture that melts and freezes at 12.7°C.

During the late 1970s and 1980s, several organizations offered phase change products for solar heat storage. For instance, Dow Chemical provided a technically successful product that melts and freezes at 27.2°C, whose market presence declined with the solar industry in general.

In 1982, Transphase Systems Inc. installed the first eutectic salt storage system for cool storage to serve a commercial or industrial building (Ames, 1989).

Like chilled water storage systems, the 8.3°C eutectic phase change temperature requires conventional chilled-water temperatures (5.5°C) to charge the TES system. These temperatures allow new or existing centrifugal, screw or reciprocating chillers to be used to charge the TES system, and make eutectics particularly appropriate for retrofit applications. The 5.5°C charging temperatures also enable the chiller to operate at high suction temperatures and compressor efficiencies as low as 0.55 kW/ton.

Advantageous use of a PCM's latent heat of fusion allows a TES to be moderate in size (about 0.155 m^3 per ton for the entire TES system, which includes piping headers and water in the tank). The storage capacity is based upon the amount of PCM frozen, not a temperature difference across the cooling coils.

The PCM is often contained in rugged, self-stacking, water impermeable containers made of a high-density polyethylene. In Ames (1989), the containers measure 0.6m×0.2m×0.045m, and are hermetically and redundantly sealed. The containers are designed with a surface-to-volume ratio of 24 m^{-1} for maximum heat transfer; and provide 0.00635 m of space for water to pass between the containers in a meandering flow pattern. The eutectic salt does not expand or contract when it freezes and melts, so there is no fatigue on the plastic container. The eutectic salt-filled containers are placed in a tank, typically in a below-grade concrete or gunite structure. The containers occupy about two-thirds of the tank's volume, so that one-third of the tank is occupied by the water used as the heat-transfer medium. No glycol or other water-freezing-point depressant is used. The eutectic salt density is about 1.5 times that of water, so that the containers do not float or expand when the PCM freezes and the heat transfer/container spacing arrangement is maintained throughout the melt/freeze cycle. The top of the tank is designed for heavy truck traffic, and used as a parking lot. Thus, the tank does not use or alter above-grade space and may be placed away from the chiller plant.

Longevity, or the repeated use of a containerized PCM over 20 or more years and thousands of freeze/thaw cycles is one of eutectic salt's greatest strengths as a thermal storage medium. Because it is a passive system with no mechanical parts, the storage system is maintenance-free (apart from water treatment).

The PCM with the 8.3°C phase change temperature, the eutectic has been cycled in the field for 6.5 years with complete retention of thermal storage capacity Also, the PCM has been subjected to accelerated life cycling equivalent to 12 years of performance with no loss of capacity. With the physical equilibrium of the PCM established after the first few cycles, the phase change appears to be stable and the TES capacity constant indefinitely, or at least as long as the life of chiller equipment used to freeze the PCM.

The normal paraffins

A large number of organic compounds suitable to be storage media for solar heating have been investigated in recent years. The most promising candidates seem to be the normal paraffins. Their solid-liquid transitions (fusions) satisfactorily meet seven important criteria for PCMs. For example, the heat of fusion has a mean value of 35–40 kcal/kg, there are no problems in the reversing the phase change, and the transition points vary considerably with the number of carbon atoms in the chains. Also, the normal paraffins are chemically inert, non-toxic and available at reasonably low cost. The change of volume with the transition, which is in the order of 10%, could represent a minor problem.

The zeolites

The zeolites are naturally occurring minerals. Their high heat of adsorption and ability to hydrate and dehydrate while maintaining structural stability have been found to be useful in various thermal storage and solar refrigeration systems. This hygroscopic property, coupled with the rapid exothermic reaction that occurs when zeolites are taken from a dehydrated to a hydrated form (when the heat of adsorption is released) makes natural zeolites an effective storage material for solar and waste heat energy.

Low energy density and time of availability have been key problems in the use of solar energy and waste heat. Commercial TES systems have been developed incorporating GSA zeolites to overcome these problems. These systems are capable of utilizing solar heat, industrial waste heat and heat from other sources, thereby converting under-utilized resources into useful energy. A typical means of using zeolites as a heat storage medium is illustrated in Figure 3.16. The capacity of natural zeolites to store thermal energy and adsorb the water vapor used in that energy interaction comes from their honeycomb structure and resultant high internal surface area.

When charged, GSA zeolites can store energy as latent heat indefinitely if maintained in a controlled environment and not exposed to water vapor. This stored energy can be liberated as needed simply through the addition of controlled amounts of water vapor, which initiates the exothermic reaction. Most other storage media do not possess this property. TES units using natural zeolites can reduce the dependence on secondary/backup heating systems, and allow for efficient and safe use of waste heat.

Requirements of PCMs

Latent TES in the temperature range 0–120°C is of interest for a variety of low-temperature applications, such as space heating, domestic hot water production, heat-pump assisted space heating, greenhouse heating, solar cooling, etc. The development of dependable TES systems for these and other applications requires a good understanding of heat-of-fusion storage materials and heat exchangers.

Knowledge of the melting and freezing characteristics of PCMs, their ability to undergo thermal cycling and their compatibility with construction materials is essential for assessing the short- and long-term performance of a latent TES. Using two different measurement techniques (e.g. differential scanning calorimetry and thermal analysis) the melting and freezing behaviour of PCMs can be determined. Commercial paraffins are characterized by two phase transitions (solid-liquid and solid-solid) which occur over a large temperature range, depending on the paraffin concerned. n-paraffins are usually preferred in comparison to their iso-counterparts, as the desired solid-to-liquid phase transition is generally restricted to a narrow temperature range. Fatty acids are organic materials with excellent melting and freezing characteristics and may have a good future potential, if their costs can be reduced. Inorganic salt hydrates, on the other hand, must be carefully examined for congruent, 'semi-congruent', or incongruent melting substances with the aid of phase diagrams. Incongruent melting in a salt hydrate may be 'modified' to overcome decomposition by adding suspension media, or extra water, or other substances that shift the peritectic point. The use of salt hydrates in hermetically sealed containers is normally recommended. Also, the employment of metallic surfaces to promote heterogeneous nucleation in a salt hydrate is seen to reduce the supercooling of most salt hydrates to a considerable extent. Thermal cycling and corrosion behavior are also of importance in the appropriate choice of materials as they affect the life of a latent heat store

Characterization of PCMs

Many characteristics are desired of a PCM. Since no material can satisfy all of the desires, the choice of a PCM for a given application requires careful examination of the properties of the various candidates, weighing of their relative merits and shortcomings, and, in some cases, a certain degree of compromise. The properties of many PCMs were investigated and reported earlier (Lane, 1988). It should be noted, however, that properties of industrial-grade products, which are used in practical applications, may deviate broadly from reported values because of the presence of impurities, composition variations (mixtures, distillation cuts), and chain-length distribution (in the case of polymers). Dilution by additives, such as the stabilizing agents required by salt hydrates, also modify thermal properties and, in particular, TES capacity. Selections should be based on assayed values of fully formulated products, whenever feasible.

Difficulties with PCMs

Although significant advances have been made, some major hurdles still remain in the development of reliable and practical storage systems utilizing salt hydrates and similar inorganic substances:

- Difficulties in obtaining an optimal match between transition zone and operating range, because of the small number of materials available in the temperature range of interest.
- Uncertainties concerning the long-term thermal behavior, despite testing over a number of cycles generally much below the number of cycles that can be expected during the useful life of a storage system.
- Increased costs and reduced effective storage capacities because of the diluting effect of stabilizing additives.
- The potential for slow loss from water for encapsulated hydrates, and a resulting drift in thermal behavior.

During the last decades, attention has focused on a new class of materials: low-volatility, anhydrous organic substances such as polyethylene glycol, fatty acids and their derivatives, and paraffins. Those materials were not viewed as high-potential candidates in early studies because they are more costly than common salt hydrates, they have somewhat lower heat storage capacities per unit volume, and, possibly, because of a bias against petroleum derivatives. Recent studies have shown that some of these materials have advantages that outweigh these shortcomings. The advantages include physical and chemical stability, good thermal behavior and an adjustable transition zone. Paraffins appear particularly well suited for applications related to energy conservation in buildings and solar energy.

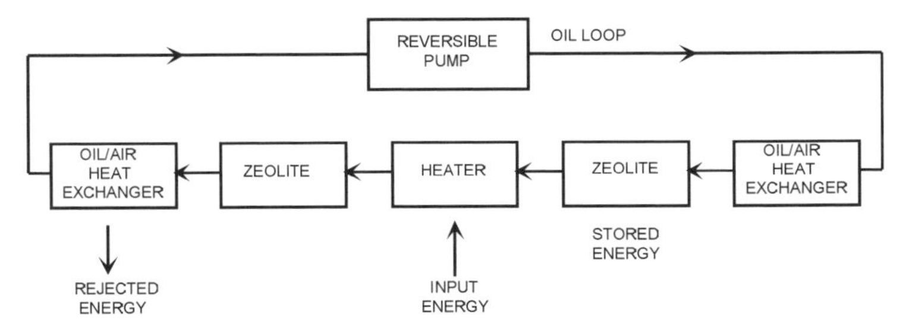

Figure 3.16 A process for using GSA zeolites as heat storage media (GSA, 2000).

Expectations of PCMs

Numerous organic and inorganic PCMs melt with a high heat of fusion in the temperature range 0–120°C. However, for their employment as heat storage materials in TES systems, PCMs must also possess certain desirable thermodynamic, kinetic, chemical, technical and economic characteristics. Some of the criteria considered in evaluating PCMs follow (Abhat, 1989, Dincer *et al.*, 1997b):

Thermodynamic criteria:

- a melting point at the desired operating temperature,
- a high latent heat of fusion per unit mass, so that less amount of material stores a given amount of energy,
- a high density, so that less volume is occupied by the material,
- a high specific heat, so that significant sensible TES can also occur,
- a high thermal conductivity, so that small temperature differences are needed for charging and discharging the storage,
- congruent melting, i.e. the material should melt completely, so that the liquid and solid phases are homogeneous (this avoids the difference in densities between solid and liquid that causes segregation, resulting in changes in the chemical composition of the material), and
- small volume changes during phase transition, so that a simple containment and heat exchanger can be used.

Kinetic criterion:

- little or no supercooling during freezing, i.e. the melt should crystallize at its freezing point. This criterion can be achieved through a high rate of nucleation and growth rate of the crystals. Supercooling may be suppressed by introducing a nucleating agent or a cold trigger into the storage material.

Chemical criteria:

- chemical stability,
- no susceptibility to chemical decomposition, so that a long operation life is possible,
- non-corrosive behavior to construction materials, and
- non-toxic, non-flammable and non-explosive characteristics.

Technical criteria:

- simplicity,
- applicability,
- effectiveness,
- compactness,
- compatibility, and
- reliability.

Economic criteria:

- commercial availability, and
- low cost.

Applications of PCMs

There are many different methods of using PCM storage systems for heating or cooling buildings, as demonstrated by the following concepts (Lane, 1988):

- A PCM melting at 5–15°C could be used for cool TES. The PCM is frozen by operating a chiller at night, when electricity demand and prices are low, and melted during the day to cool the building.
- A PCM melting near room temperature, e.g. $CaCl_2 \cdot 6H_2O$ which has a melting point (m.p.) of 27°C, could be incorporated into a building structure to temper diurnal swings in ambient temperature.
- A building could be heated and cooled using heat pumps that are connected to circulating water tempered by a PCM melting at 20–35°C, e.g. $CaCl_2 \cdot 6H_2O$ (m.p.=27°C).
- A solar hot-air heating system could use a PCM melting at 25–30°C, e.g. $CaCl_2 \cdot 6H_2O$ to provide night-time heating and as a preheat for daytime heating.
- A solar hot-air heating system could use a PCM melting at 40–60°C, e.g. $Mg(NO_3)_2 \cdot 6H_2O$-$MgCl_2 \cdot 6H_2O$ eutectic (m.p.=58°C), for day and night heating.
- Domestic hot water could be preheated in a tank filled with an encapsulated PCM melting at 55–70°C, e.g. $Mg(NO_3)_2 \cdot 6H_2O$-$MgCl_2 \cdot 6H_2O$ eutectic.
- A solar hot-water baseboard system could employ a PCM melting at 60–95°C, e.g. $Mg(NO_3)_2 \cdot 6H_2O$ (m.p.=89°C).
- Off-peak electricity could be used to melt a PCM (with a m.p. above 25°C) to heat a building during later periods.
- Concentrated solar energy could be used with a PCM melting at 100–175°C, e.g. $Mg(NO_3)_2 \cdot 6H_2O$ (m.p.=117°C), to drive an absorption air conditioner.

Evaluation of PCMs

In the evaluation and selection of a PCM, a good understanding of various aspects, including freezing or solification, supercooling, nucleation, thermal cycling, encapsulation, and compatibility is needed.

- **Freezing or solification.** Many materials are not suitable as PCMs because of their freezing or solidification behavior. Some have incongruent freezing behavior. Others crystallize exceedingly slowly, form viscous mixtures, or are not stable over the range of temperatures. Some materials have near-congruent freezing behavior and so may be suitable as PCMs. To overcome the solidification problem, the nucleation rate can be increased by two methods: homogeneous nucleation and heterogeneous nucleation.
- **Supercooling.** A major problem associated with salt hydrates as PCMs is the fact that they tend to supercool considerably. This behavior is of major importance, and has not always received sufficient attention from workers in the field. The reason for the high degree of supercooling is the fact that either the rate of nucleation (of crystals from the melt) or the rate of growth of these nuclei (or both) is very slow. As the melt is cooled, therefore, it does not solidify at the thermodynamic melting point. Thus, the advantage of the material for heat storage is reduced. The reason for the strong tendency to supercool is well understood. Some empirical evidence connects the tendency to supercool with the viscosity of the melt at the melting point. Materials with high

viscosity in the liquid state have low diffusion coefficients for their constituent atoms (or ions), and these may be unable to rearrange themselves to form a solid; instead the liquid supercools. Supercooling occurs when, in attempting to freeze the material, the temperature drops well below the melting point before freezing initiates. Once the freezing process begins, the temperature rises to the melting point where it remains until the material is entirely frozen. Supercooling behavior is undesirable in heat storage systems. If excessive, it can prevent the withdrawal of stored heat from the PCM.

- **Nucleation.** Supercooling can often be mitigated by adding nucleating materials. Some success has been attained by using additives with a crystal structure similar to that of the PCM. Often this fails. Usually nucleating additives are discovered by trial and error. Using these approaches, effective nucleating additives have been discovered for zinc nitrate hexahydrate, calcium chloride hexahydrate, magnesium chloride hexahydrate, magnesium nitrate hexahydrate and its eutectics with magnesium chloride hexahydrate or ammonium nitrate. In homogeneous nucleation, the rate of nucleation of crystals from the melt is increased without adding any foreign materials. One way in which this can be done is to use ultrasonic waves in which we stir the liquid, and thus increase the diffusion of ions in the melt; they also create cavities which act as nucleation centers and, finally, they break up the forming crystals and distribute them through the melt, creating new nucleation centers. Many crystals, such as sodium thiosulfate and potassium alums, have been nucleated from more dilute solutions. Also, other classes of materials, such as metals and organic crystals, have been nucleated by ultrasonic waves. In heterogeneous nucleation, the walls of the container, or some impurity present within the melt, act as a catalyst for nucleation by providing a substrate on which the nuclei can form. In order for an impurity to be an effective nucleating agent, it should satisfy the following criteria:
  - have a melting point higher than the highest temperature reached by the energy storage material in the storage cycle,
  - be insoluble in water at all temperatures,
  - not form solid solutions with the salt hydrate,
  - have a crystal structure similar to that of the salt hydrate,
  - have unit cell dimensions that do not vary by more than 10% from those of the salt hydrate, and
  - is not chemically react with the hydrate.
- **Thermal cycling.** Practical heat storage PCMs must be able to undergo repetitive cycles of freezing and thawing. Some unsuitable materials exhibit changes within a relatively short time. Some experimental results demonstrate this. After 21 cycles from 95°C to 20°C, palmitic acid, normally white, takes on a yellowish colour, and the melting point diminishes by 2°C. The eutectic of 25.1% propionamide and 74.9% palmitic acid progressively discoloures from white to yellow, orange, and then black, though little change occurs in the freezing curve (Lane, 1988).
- **Encapsulation.** Successful utilisation of a PCM requires a means of containment. For active solar systems with a liquid heat-transfer medium, tanks with coil-type heat exchangers are appropriate. For passive or air-cooled active solar systems, much effort has centered on the packaging of a mass of PCM in a sealed container, which itself

serves as the heat exchange surface. Potential containers include steel cans, plastic bottles, polyethylene and polypropylene bottles, high-density polyethylene pipe, flexible plastic film packages, and plastic tubes. The choice of the construction material for container of a PCM is important. Appropriate tests that are realistic and representative of usage conditions are needed in any product development. The container material should be an effective barrier that prevents loss of material or water or, when the PCM is hygroscopic, gain of water. Oxygen penetration and subsequent oxidization may also be detrimental. The encapsulating material should also be a good heat conductor so it facilitates effective heat transfer, and be mechanically resistant to damage from handling, processing and transport. Systems based on salt hydrates may sometime have encapsulation problems, particularly in early designs due to corrosion and fatigue for metals, or water loss through plastics.

- **Compatibility.** To determine the suitability of encapsulant materials, PCMs are subject to compatibility tests with packaging materials. For example, a plastic-aluminium foil laminate is not suitable for use with organic PCMs because the heat-sealed seam is attacked by the organic material, and it is unsuitable for temperatures above about 70°C. Paraffins are compatible with most metals and alloys, but they can impregnate and soften some plastics.

Characteristics and thermophysical properties of PCMs

The precise determination of the transition zone at fusion and at solidification is essential for the optimal design of a latent TES unit, since the transition zone of the PCM must match, sometimes exactly, the intended operating temperature span of the latent TES unit. Pure substances have an exact melting and freezing point, but they are affected by impurities even in small amounts. Polymers and mixtures have a more complex behavior, with a true, and sometimes very broad, transition zone over which fusion or solidification are progressive. Salt hydrates, when pure, behave as pure substances and afford little design flexibility. Design options based on salt hydrates are rather limited in the range 15–65°C, because their fusion-solidification behavior cannot be easily modified. Thermophysical properties of paraffins generally change monotonously as a function of chain length (e.g. melting temperature). An important advantage of paraffins is that neighboring homologues are totally compatible and that stable, homogeneous mixtures can be readily prepared. It is thus possible to design paraffin-based PCMs meeting a wide variety of transition-zone specifications (Paris *et al.*, 1993). Formulating a compound PCM for a given transition range is a key task in which composition, thermophysical properties and most other characteristics are fixed. However, properties of mixtures cannot be calculated from simple interpolation of properties of pure components, and therefore experimental characterization is required.

The heat storage capacity of a latent TES is primarily a function of the heat of fusion of the PCM. The heat of fusion of paraffins increases with chain length until 24 and is generally higher than that of salt hydrates. The heat of fusion of salt hydrates increases with the degree of hydration, but higher hydration often comes with incongruent or semicongruent fusion (Lane, 1988).

The specific heats of the solid and liquid PCMs are not critical factors to the performance of a system unless the operating span of the storage unit far exceeds the transition zone of the PCM. Then the mode of operation tends toward sensible heat storage and the relative contribution of latent heat to its performance diminishes. Since a PCM changes temperature during operation, heat transfer over a charge (or discharge) cycle includes sensible heat. Paraffins generally have higher specific heats in both the solid and liquid states than salt hydrates.

Table 3.9 Measured thermophysical data of some PCMs.

Compound	Melting temp. (°C)	Heat of fusion (kJ/kg)	Thermal conductivity (W/mK)	Density (kg/m^3)
Inorganics				
$MgCl_2·6H_2O$	117	168.6	0.570 (liquid, 120°C)	1450 (liquid, 120°C)
			0.694 (solid, 90°C)	1569 (solid, 20°C)
$Mg(NO_3)_2·6H_2O$	89	162.8	0.490 (liquid, 95°C)	1550 (liquid, 94°C)
			0.611 (solid, 37°C)	1636 (solid, 25°C)
$Ba(OH)_2·8H_2O$	78	265.7	0.653 (liquid, 85.7°C)	1937 (liquid, 84°C)
			1.255 (solid, 23°C)	2070 (solid, 24°C)
$Zn(NO_3)_2·6H_2O$	36	146.9	0.464 (liquid, 39.9°C)	1828 (liquid, 36°C)
			—	1937 (solid, 24°C)
$CaBr_2·6H_2O$	34	115.5	—	1956 (liquid, 35°C)
			—	2194 (solid, 24°C)
$CaCl_2·6H_2O$	29	190.8	0.540 (liquid, 38.7°C)	1562 (liquid, 32°C)
			1.088 (solid, 23°C)	1802 (solid, 24°C)
Organics				
Paraffin wax	64	173.6	0.167 (liquid, 63.5°C)	790 (liquid, 65°C)
			0.346 (solid, 33.6°C)	916 (solid, 24°C)
Polyglycol E400	8	99.6	0.187 (liquid, 38.6°C)	1125 (liquid, 25°C)
			—	1228 (solid, 3°C)
Polyglycol E600	22	127.2	0.189 (liquid, 38.6°C)	1126 (liquid, 25°C)
			—	1232 (solid, 4°C)
Polyglycol E6000	66	190.0	—	1085 (liquid, 70°C)
			—	1212 (solid, 25°C)
Fatty acids				
Stearic acid	69	202.5	—	848 (liquid, 70°C)
			—	965 (solid, 24°C)
Palmitic acid	64	185.4	0.162 (liquid, 68.4°C)	850 (liquid, 65°C)
			—	989 (solid, 24°C)
Capric acid	32	152.7	0.153 (liquid, 38.5°C)	878 (liquid, 45°C)
			—	1004 (solid, 24°C)
Caprylic acid	16	148.5	0.149 (liquid, 38.6°C)	901 (liquid, 30°C)
			—	981 (solid, 13°C)
Aromatics				
Biphenyl	71	119.2	—	991 (liquid, 73°C)
			—	991 (liquid, 73°C)
Naphthalene	80	147.7	0.132 (liquid, 83.8°C)	976 (liquid, 84°C)
			0.341 (solid, 49.9°C)	1145 (solid, 20°C)

Source: Lane (1980).

Heat transfer to and from a storage unit depends strongly on the thermal conductivities of the solid and liquid PCM. The higher the conductivities, the more efficient is the heat transfer for a given design. However, the heat-transfer phenomena during fusion or solidification of a PCM are very complex because of the moving solid-liquid interface, the density and conductivity differences between the two phases, and the induced movements in the liquid phase. Salt hydrates have higher thermal conductivities than paraffins.

The density of a PCM is important because it affects its storage effectiveness per unit volume. Salt hydrates are generally more dense than paraffins, but are slightly more effective on a per volume basis, despite a slightly lower heat of fusion. The rate of crystallization of a salt hydrate can be low, and can become the limiting factor in the rate of heat storage and restitution. Crystallization is generally more rapid for paraffins, and heat-transfer mechanisms are then the limiting factors. In addition, paraffins exhibit little or no supercooling, which is frequent and often significant in magnitude with salt hydrates.

Paraffins have very low vapor pressures, which leads to low long-term loss of material and flammability. Salt hydrates have significantly higher vapor pressures, which induce water loss and progressive changes of thermal behavior. The vapor pressure of salt hydrates increases with the degree of hydration, and salt hydrates exhibit variable chemical stability and can be subject to long-term degradation by oxidization, hydrolysis, thermal decomposition and other reactions. Some salt hydrates are very corrosive in the presence of water. Paraffins are very stable and unreactive, but slow oxidization may occur when they are exposed to air at elevated temperatures over extended periods. Paraffins are not corrosive.

Salt hydrates are neither toxic nor flammable. They can be irritants, and contact with skin and eyes should be avoided. Paraffins are innocuous, being neither toxic nor irritants. Although paraffins are flammable in the presence of oxygen and at elevated temperature, because of their low vapor pressure they are considered low fire hazards Compliance with safety codes for products containing paraffin-based PCMs should not present particular difficulties if this point is properly addressed in the design stage.

Table 3.9 presents the experimental data on melting temperature, heat of fusion, thermal conductivity and density data for several organic and inorganic compounds, aromatics and fatty acids.

Latent TES using PCMs provides an effective way to store thermal energy from a range of sources, high storage capacity and heat recovery at almost constant temperatures. The relatively constant temperature of storage can maximize solar collector efficiency, where relevant.

Performance of latent TES with PCMs

To improve the performance of latent TES with PCMs, nucleating agents and thickeners have been used to prevent supercooling and phase separation. Also, extended heat transfer surfaces can be used to enhance the heat transfer from PCM to the heat transfer tubes. While many studies on the latent TES systems have been performed at relatively low temperatures (below 100°C) for TES in home heating and cooling units, few studies have been undertaken for higher temperature heat (above 200°C), as is applicable for some solar energy systems and intermediate-temperature latent TES. Magnesium chloride hexahydrate ($MgCl_2 \cdot 6H_2O$), with a melting temperature of 116.7°C, is an attractive high-temperature PCM in terms of cost, material compatibility and thermophysical properties (specific latent heat=168.8 kJ/kg, specific heat=2.25 kJ/kgK in the solid state and 2.61 kJ/kgK in the liquid state, thermal conductivity=0.704 W/mK in the solid state and 0.570 W/mK in the liquid state, and density=1570 kg/m^3 in the solid state and 1450 kg/m^3 in the liquid state) (Choi and Kim, 1995).

Sharma *et al.* (1992) showed that the acetamide-sodium bromide eutectic is a promising latent TES material which could find use in such applications as commercial and laundry water heating, process heating, domestic water and air heating, crop drying and food warming. Vaccarino *et al.* (1987) conducted an experimental study on the low-temperature latent TES and found that two mixtures containing $Ca(NO_3)_2 \cdot 4H_2O$ and KNO_3 or $Mg(NO_3)_2 \cdot 6H_2O$ are suitable PCMs for passive heating of such facilities as buildings and greenhouses in the temperature range 15–35°C, and for domestic hot water production and active space heating in the range 25–55°C.

Although PCMs must usually be placed in capsules or other vessels to prevent them from mixing with heat carriers after melting, concepts have recently been developed that permit PCMs to come into direct contact with heat carriers, thereby providing good heat exchange without the usual heat exchangers or capsules. In such a system, an immiscible fluid is bubbled into the storage bottom of the fused PCM and heat is transferred as droplets rise. Here the immiscible fluid agitates the PCM so that the disadvantage of supercooling is minimized (Sokolov and Keizman, 1991).

Over 20,000 compounds and/or mixtures have been considered in PCM, including single-component systems, congruent mixtures, eutectics and peritectics. Criteria being investigated include melting point, phase-diagram characteristics, toxicity, stability, corrosivity, flammability, safety, availability and cost. Over 200 compositions, organic and inorganic compounds, eutectics and other mixtures have been considered as promising (Lane, 1988).

Sokolov and Keizman (1991) developed an attractive PCM application for hot water heating. The system contains a solar pipe consisting of two concentric pipes with the space between them filled with PCM (Figure 3.17). Solar radiation is absorbed directly on the outer surface and then transmitted to the PCM, where it is stored as sensible and latent heat. During energy release, heat is exchanged between a water flow through the inner tube and the PCM storage, and hot water is delivered at the discharge of the solar pipe.

In this system, direct solar radiation absorption onto the PCM container and direct heating of water eliminate the need for energy transport media. Some advantages of the system include:

- simple construction,
- efficient and compact latent heat storage,
- elimination of expensive components such as a water tank, pump and control devices,
- suitability of the system for modular construction and installation, and
- protection against freezing.

Solar TES as latent heat has an enormous potential. With latent heat, the temperature of the phase change medium stays fixed. The most common PCM for solar applications is Glauber's salt ($Na_2SO_4 \cdot 10H_2O$).

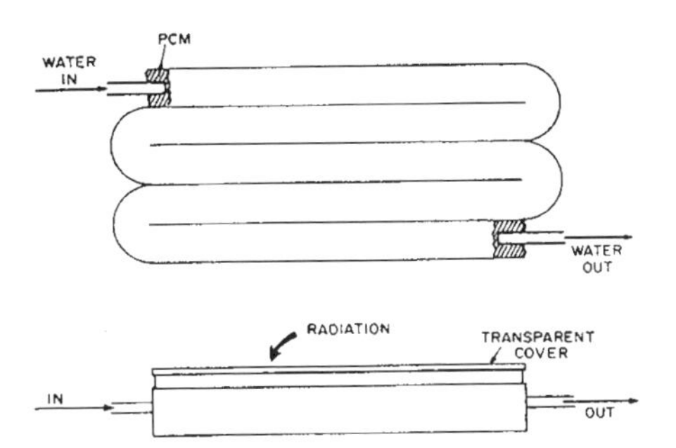

Figure 3.17 A solar TES pipe using a PCM (Sokolov and Keizman, 1991).

Two approaches are under development to improve the performance of phase change storage. One approach is to increase the temperature at which PCMs such as salts change phase, and the other is to employ the use of a heat pump. The heat pump raises the temperature of the heat extracted from the phase change storage to a temperature high enough to satisfy the thermal needs.

A further advantage of using a PCM is its capability to store up to nine times more energy for the same volume of containment occupied by rocks. This characteristic is particularly advantageous for retrofits (e.g. adding a solar system to an existing dwelling). One of the prime difficulties in solar retrofits is locating and charging the store. The volume and mass of sensible TES in rocks or water limits applications. On the other hand, the relatively low mass and volume of PCM often makes it a good choice for retrofits. These characteristics permit latent TES to be located in attic spaces, closets or even sandwiched between floors and ceilings.

Selection of PCMs for latent TES

No material has all the optimal characteristics for a PCM, and the selection of a PCM for a given application requires careful consideration of the properties of various substances. Among PCMs, sodium acetate trihydrate deserves special attention for its large latent heat of fusion-crystallization (264–289 kJ/kg) and its melting temperature of 58–58.4°C. However, this substance exhibits significant subcooling, preventing most practical large-scale applications, even though attempts have been made to find ways of suppressing or reducing this phenomenon.

Energy balance simulations of a PCM wall as a TES in a passive direct gain solar house suggest that the PCM melting temperature should be adjusted from the climate-specific optimum temperature to achieve maximum performance of the storage. A non-optimal melting temperature significantly reduces the latent heat storage capacity, e.g. a 3°C non-optimality temperature causes a 50% loss (Dincer and Dost, 1996).

Of practical importance in the selection of PCMs for solar and other thermal applications are thermophysical properties such as:

- heat of fusion,
- heat capacity of solid and liquid,
- thermal conductivity of solid and liquid, and
- density of solid and liquid.

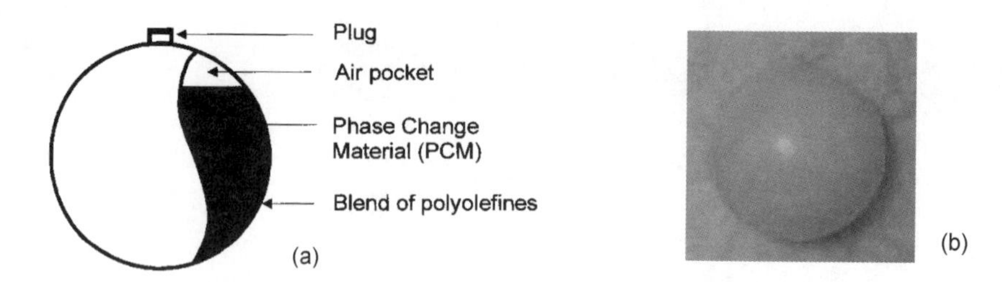

Figure 3.18 (a) The nodule for STL system (*Courtesy of Cristopia Energy Systems*), and (b) a capsule for the new STL latent TES (*Courtesy of Mitsubishi Chemical Co.*).

Another important factor in PCM selection is supercooling and nucleation. Several PCMs exhibit supercooling, that is, on attempting to freeze the material the temperature drops well below the melting point before freezing initiates. Although the temperature rises to the melting point once freezing begins, supercooling is nevertheless undesirable in latent TES because it can prevent the withdrawal of stored heat. For this reason, nucleative materials are sometimes used.

The STL system

Peak cooling loads for buildings are often at a level two or more times higher than the average daily load. Some industrial processes also have load peaks that are much greater than the average load. Since many electric utilities impose demand charges based on a customer's highest power demand during on-peak hours and/or during the entire billing cycle, TES can be beneficial. The STL latent TES can be an efficient solution for these applications. The STL is composed of a tank filled with spherical nodules. The tank has upper manholes to allow the filling with nodules. A lower manhole allows emptying. Inside the tank two diffusers (inlet and outlet) spread the heat transfer fluid along the tank. The pressure drop through the tank is 2.5 mWG. The inlet in charge mode must be via the lower diffuser in order to ensure the natural stratification.

The tanks are manufactured in black steel (test pressure between 4.5 to 10 bar), are delivered empty and positioned on site or, if access to the site is impossible, constructed on site. The nodules are spherical with a diameter of 77 mm, 78 mm or 98 mm (depending on the nodule type).

The nodules (Figure 3.18a) contain the PCM. The mechanical and chemical characteristics of the nodule shell (manufactured in polyolefin) are well adapted to the conditions encountered in air conditioning or refrigeration systems. Once filled with PCM the nodule plugs are sealed by ultrasonics to ensure perfect watertightness. The nodules are delivered in 22 kg bags. Tanks are filled on site. The filling is regular and homogeneous. Filling procedures are described elsewhere (for details, see Cristopia, 2000).

The tank shape is usually cylindrical in order to withstand service pressure higher than 3 bar. The test pressure varies between 4.5 to 10 bar. The spherical shape also allows an easy filling. The nodule diameter has been calculated to meet economical and technical requirements. The size allows high exchanges until the end of the cycle.

The use of modern technologies permits quality control. The materials used are completely neutral to the PCMs and heat transfer fluid. This product development work has led to a very high reliability of the STL. The temperature range offered is –33°C to + 27°C.

It is important to mention that in the STL system the quantity of energy stored for each type of nodule is proportional to the storage volume. The number of nodules in a system determines the heat exchange rate between the nodules and the heat transfer fluid.

In the terminology, an STL is determined by the phase change temperature and the volume (i.e. the storage capacity and the heat exchange rate). There are three types of nodules characterized by their diameter: AC (98 mm), IC and IN (78 mm) (Cristopia, 2000), such as:

- **STL-AC.OO-15**: where AC: 98 mm diameter nodules, 00: phase change temperature in OC, and 15: tank volume in m^3.
- **STL-IN.15-50**: where IN: 78 mm diameter nodules for negative temperature, 15: the phase change temperature (melting) is −15°C, and 50: tank volume in m^3.
- **STL-IC.27-100**: where IC: 78 mm diameter nodules, 27: the phase change temperature (melting) is +27°C, and 100: tank volume in m^3.

The features and charateristics of the nodules can be summarized as follows (Cristopia, 2000):

- material: blend of polyolefins,
- chemically neutral towards eutectics and heat transfer fluid,
- thickness 1.0 mm: no migration of the heat transfer fluid,
- sphere obtained by blow moulding: no leakage,
- sealing of the cap by ultrasonic welding,
- exterior diameter: (98 mm for air conditioning and 78 or 77 mm for industrial cooling or back-up),
- exchange surface: (diameter 78 or 77 mm for 1.0 m^2/kWh stored and diameter 98 mm for 0.6 m^2/kWh stored),
- air pocket for expansion: no stress on the nodule shell, and
- useful number of nodules per m^3: (diameter 77 mm for 2548 nodules per m^3, diameter 78 mm for 2444 nodules per m^3 and diameter 98 mm for 1222 nodules per m^3.

The charateristics of the STL tanks can be summarized as follows (Cristopia, 2000):

- black steel,
- horizontal or vertical,
- outside, inside, buried, built on site,
- rustproof exterior paint,
- insulation on site,
- efficient diffuser system,
- high service pressure,
- pressure drop of 2.5 mWG, and
- made to measure according to site requirements.

The Energy Research Group in Japan in collaboration with Mitsubishi Chemical Co. is investigating the performance of the STL system in combination with storage tanks developed for solar energy utilization. The STL system consists of salts and hydrates contained in plastic capsules (Figure 3.18b), where the thermal energy is stored during hydration of the thermally dehydrated salt. Different phase change materials which operate at a wide range of storage temperatures are suitable for various applications such as domestic hot water, space heating and cooling.

Traditional refrigeration systems are designed to satisfy the peak cooling demand, which occurs only a few hours per year, and thus spend most of their operational lives working at reduced capacity and low efficiency. The STL system, which is suitable for any air conditioning system or refrigeration plant, allows installed chiller capacity (and the size of other components) to be significantly reduced–typically between 40% and 60%. The STL system provides the shortfall of the energy when demand exceeds chiller capacity. Thus chiller operation is continuous and at maximum efficiency. The result is reduced operating costs for refrigeration and building air conditioning by taking advantage of lower cost off-peak electricity and reducing demand charges. The STL system allows real management of the cooling energy according to the demand. In addition, the STL allows the consumption of night-time electricity produced with a higher efficiency (2300 kcal/kWh compared to 3500–4000 kcal/kWh at peak hours), resulting in a reduction in CO_2 emissions and energy consumption. The reduction in the chiller size also reduces the quantity of refrigerant used, which is important with increasingly strict laws on refrigerants.

The heat pump latent TES

A heat pump was integrated with latent TES to enable quick room temperature increases and defrosting. This inverter-aided room air-conditioner/heat pump was put on the market by Daikin Industries Ltd. in Japan in 1989. The latent TES consists of the PCM, polyethylene glycol, which surrounds a rotary compressor, as shown in Figure 3.19. Heat released from the compressor is transferred to the TES through a finned-tube heat exchanger. The TES is used during start-up and during defrosting. During start-up, the TES reduces the time required to reach a 45°C discharge air temperature by 50%. During defrosting, a heating capacity of 3.5 kW is made available by the TES, which avoids a drop in room temperature. Integrating the system improved heat capacity by about 10% and COP by 5%. The sound characteristics of the system were also improved with TES. The installation space required is the same as for a conventional heat pump/air conditioner.

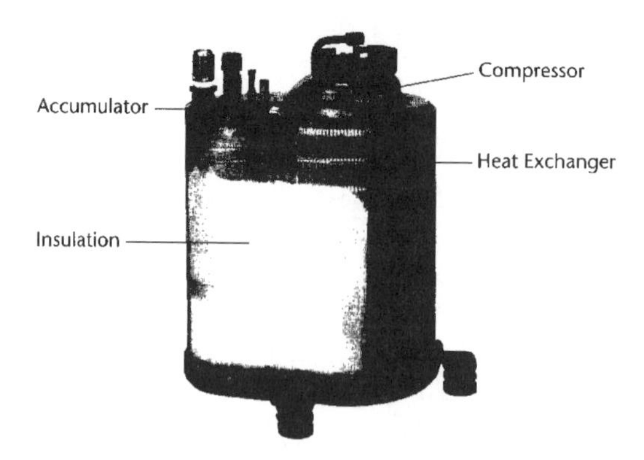

Figure 3.19 A latent TES integrated with a heat pump (IEA, 1990).

3.8 Cold TES (CTES)

Cooling capacity can be stored either by chilling or freezing water (or such other materials as glycol and eutectic salts). Water is the storage material of choice for a variety of practical and thermodynamic reasons, including its ready availability, relative harmlessness, and its compatibility with a wide availability of equipment for its storage and handling. The choices between whether the water should be used in sensible or latent types, which equipment should be used, if eutectic salts should be applied to raise freezing temperatures, etc., are often not simple. Options are numerous, and answers are not clear cut. Ultimately, the CTES method selected must meet the particular needs and constraints of the specific facility in which it is installed.

CTES is an innovative way of storing night-time off-peak energy for daytime peak use. In many locations, demand for electrical power peaks during summer. Air conditioning is the main reason, in some areas accounting for as much as half of the power demand during the hot mid-day hours when electricity is most expensive. Since, at night, utilities have electrical generating capacity spare, electricity generated during this 'off-peak' is much cheaper.

In essence, one can air-condition during the day using electricity produced at night. CTES has become one of the primary means of addressing the electrical power imbalance between high daytime demand and high night-time abundance. If properly designed, installed, operated, and maintained, CTES systems can be used to shift peak cooling loads to off-peak periods, thereby evenly distributing the demand for electricity and avoiding shortages usually encountered during peak periods.

Although the phrase 'cool TES' may appear to be contradictory, it is not. The phrase TES is widely used to describe storage of both heating and cooling energy. TES for heating capacity usually involves using heat at above environment temperatures from a variety of sources to heat a storage medium for later use. In contrast, cold TES uses off-peak power to provide cooling capacity by extracting heat from a storage medium, such as ice, chilled water or PCMs. Typically, a CTES system uses refrigeration equipment at night to create a reservoir of cold material which is tapped during the day to provide cooling capacity. In this book, the abbreviation CTES represents both cool and cold TES.

CTES has many advantages. Lower night-time temperatures allow refrigeration equipment to operate more efficiently than during the day, reducing energy consumption. Lower chiller capacity is required, which leads to lower equipment costs. Also, using off-peak electricity to store energy for use during peak demand hours, daytime peaks of power demand are reduced, sometimes deferring the need to build new power plants.

CTES systems are presently operational in many commercial and industrial buildings in various countries. Many are not achieving expected design performance, often because TES design engineers and construction companies in the past lacked experience. Now, package-type TES systems are available and are commonly used, which involve more straightforward design and installation for conventional air conditioning systems. Therefore, the economic and other benefits of TES operating performance at design conditions for system owners is more likely.

3.8.1 Working principle

CTES systems, which have the potential to provide substantial operating cost savings, are most likely to be cost-effective in situations where:

- a facility's maximum cooling load is much greater than the average load,
- the utility rate structure has higher demand charges for peak demand periods,
- an existing cooling system is being expanded,
- an existing tank is available,
- limited on-site electric power is available,
- backup cooling capacity is desirable, and
- cold-air distribution is desirable or advantageous.

It is difficult to generalize about when cool storage systems will be cost-effective, but if one or more of the above criteria are satisfied, a detailed analysis may prove worthwhile.

Some CTES systems generate ice during off-peak hours and store it for use in daytime cooling. Until recently, decreasing electricity costs and an abundance of reliable cooling equipment had slowed the development of this technology, which has been around for more than half a century. Today, increases in maximum power demands, major changes in electric rate structures and the emergence of utility-sponsored incentive programs have inspired a renewed interest in CTES. For instance, utility companies often experience peak electrical demands for four to six hours on hot summer afternoons, when air conditioning loads also peak, and apply time-of-use rates to discourage energy consumption during these peak demand periods. One objective is for the air chiller to be shut down during peak-load hours and to have a TES system provide cooling for the facility at those times.

An ice-ball system uses chillers to build ice at night. The ice balls float in a glycol solution that runs through chillers in the evening. These chillers, which are set at –7.5 to –6.5°C, freeze the ice balls in the storage tanks, and the glycol circulates around the ice balls. Chilled glycol is pumped into the bottom of the tank to freeze the ice balls, and warms as it rises and extracts heat from the ice balls. Later, the cycle is reversed and the glycol is pumped into the top of the tank and past the ice balls. The cold glycol solution then passes through heat exchangers and connected to the building's chilled water system, which interfaces with the air handler. Cooling is thus obtained while only operating the fans on the air handlers, but cooling is the same if chillers are operated with the air handling units.

3.8.2 Operational loading of CTES

Several strategies are available for charging and discharging a storage to meet cooling demand during peak hours. The strategies are full storage and partial storage.

Full-storage CTES

A full-storage strategy shifts the entire peak cooling load to off-peak hours (Figure 3.20a). The system is typically designed to operate on the hottest anticipated days at full capacity during all non-peak hours in order to charge storage. This strategy is most attractive when peak demand charges are high or the peak period is short.

Full storage (load-shifting) designs are those that use storage to fully decouple the operation of the heating or cooling generating equipment from the peak heating or cooling load. The peak heating or cooling load is met through the use (i.e. discharging) of storage while the heating or cooling generating equipment is idle. Full storage systems are likely to be economically advantageous only under one or more of the following conditions:

- spikes in the peak load curve are of short duration,
- time-of-use energy rates are based on short-duration peak periods,
- there are short overlaps between peak loads and peak energy periods,
- large cash incentives are offered for using TES,
- high peak-demand charges apply.

For example, a school or business whose electrical demand drops dramatically after 5 p.m. in an electric utility territory where peak energy and demand charges apply between 1 p.m. and 9 p.m. can usually economically apply a full CTES. Cooling during the four-hour period between 1 p.m. and 5 p.m. can be full shifted, i.e. can be met with a relatively small and cost-effective CTES system and without oversizing the chiller equipment.

Partial-storage CTES

In a partial-storage method, the chiller operates to meet part of the peak-period cooling load, and the rest is met by drawing from storage. The chiller is sized at a smaller capacity than the design load. Partial-storage systems may operate as load-leveling or demand-limiting operations. In a load-leveling system (Figure 3.20b), the chiller is sized to run at its full capacity for 24 hours on the hottest days. The strategy is most effective where the peak cooling load is much higher than the average load. In a demand-limiting system, the chiller runs at reduced capacity during peak hours, and is often controlled to limit the facility's peak demand charge (Figure 3.20c). Demand savings and equipment costs are higher than they would be for a load-leveling system, and lower than for a full-storage system.

Partial storage is more often the most economic option, and therefore represents the majority of thermal storage installations. Although partial storage does not shift as much load (on a design day) as a full-storage system, partial storage systems can have lower initial costs, particularly if the design incorporates smaller equipment by using low-temperature water and cold-air distribution systems.

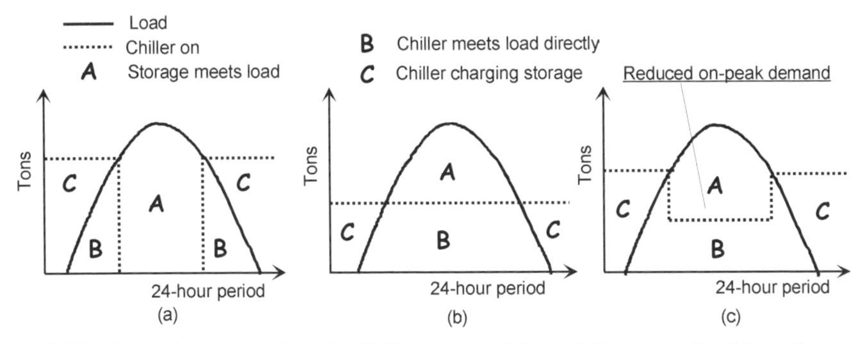

Figure 3.20 Operating strategies. (a) Full-storage, (b) partial-storage load-leveling, and (c) partial-storage demand-limiting.

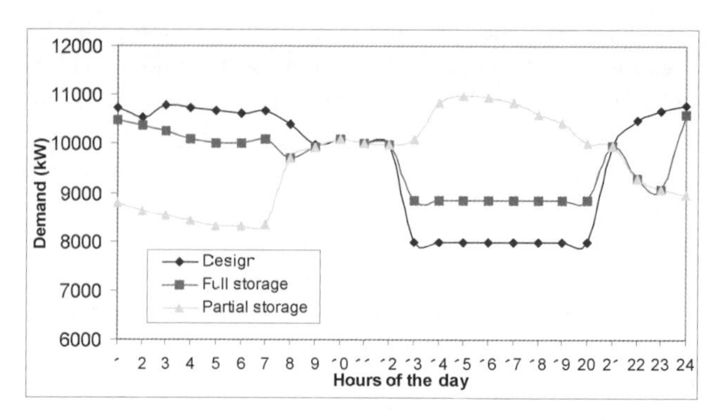

Figure 3.21 Example profiles for the design, full storage and partial storages.

For many applications, a form of partial storage known as load-leveling can be used with minimum capital cost. A load-leveling system is designed with the heating or cooling equipment sized to operate continuously at or near its full capacity to meet design-day loads. Thus, equipment having minimum capacity (and cost) can be used. During operation at less than peak design loads, partial storage designs can function as full-storage systems. For example, a system designed as a load-leveling partial storage for space heating at winter design temperatures may function as a full storage (with a full demand shift) on mild spring or autumn days.

3.8.3 Design considerations

CTES can take many forms to suit a variety of applications. This section addresses several groups of CTES applications: off-peak air conditioning, industrial/process cooling, off-peak heating, and other applications.

Selecting a storage and its characteristics usually requires a detailed feasibility study. The analysis is involved, and best accomplished following an established procedure. Data needed for feasibility analysis can include (i) an hour-by-hour 24-hour building-load profile for the design day, and (ii) a description of a baseline non-storage system, including chiller capacity, operating conditions and efficiency. The description of a CTES often stipulates the following:

- the sizing basis (full storage, load-leveling, or demand limiting),
- the sizing calculations showing chiller capacity and storage capacity, and considering required supply temperature,
- the design operating profile, showing load, chiller output, and amount of heat added to or taken from storage for each hour of the design day,
- the chiller operating conditions while charging the storage, and, if applicable, when meeting the load directly,
- the chiller efficiency under each operating condition,
- a description of the system control strategy, for the design-day and part-load operation.

An operating-cost analysis includes

- an evaluation of demand savings,
- a determination of changes in energy consumption and cost, and
- a description and justification of the assumptions used for annual energy demand and use estimates.

Storage equipment manufacturers often provide simulations of storage performance for a given load profile and chiller temperature, to assist design efforts.

Although applications and technologies vary significantly, certain characteristics and design options are common to all TES systems. Whether for heat or cool storage, and whether for storing sensible or latent heat, storage designs follow one of two control strategies: full storage or partial storage (Figure 3.21).

The following steps should be considered for all ice TES refrigeration systems:

- **Design for part-load operation.** Refrigerant flow rates, pressure drops, and velocities are reduced during part-load operation. Components and piping must be designed so that, at all load conditions, control of the system can be maintained and the working fluid can be returned to the compressors.
- **Design for pull-down load.** Because ice-making equipment is designed to operate at water temperatures approaching 0°C, a higher load is imposed on the refrigeration system during the initial start-up, when the inlet water is warmest. The components must be sized to handle this higher load.
- **Plan for chilling versus ice making.** Most ice-making equipment has a much higher instantaneous chilling capacity than ice-making capacity. This higher chilling capacity can be used advantageously if the refrigeration equipment and interconnecting piping are properly sized to accommodate alternative modes of operation.
- **Protect compressors from liquid slugging.** Ice producers and ice harvesters tend to contain more refrigerant than chillers of similar capacity used for non-storage systems. The opportunity therefore exists for compressor liquid slugging. Care should be taken to oversize suction accumulators and equip them with high-level compressor cutouts and suction heat exchangers to evaporate any remaining liquid.
- **Oversize the receivers.** The system should be made easy to maintain and service. Maintenance flexibility may be provided in a liquid overfeed system by oversizing the low-pressure receiver and in a direct-expansion system by oversizing the high-pressure receiver.
- **Prevent oil trapping.** Refrigerant lines should be arranged to prevent the trapping of large amounts of oil anywhere in the system, and to ensure its return to compressors under all operating conditions, especially during periods of low compressor loads. All suction lines should slope toward the suction line accumulators, and all discharge lines should pitch toward the oil separators. Oil tends to collect in the evaporator because that is the location at the lowest temperature and pressure. Because refrigerant accumulators trap oil as well as liquid refrigerant, the larger accumulators needed for ice storage systems require special provisions to ensure adequate oil return to the compressors.

It is important to fully analyze design trade-offs; comparisons of CTES to other available cost- and energy-saving opportunities should be fair. For example, if energy-efficient motors are assumed for a CTES design, they should be assumed for the non-storage system being compared. Current designs should be considered in comparisons, whether based on HCFCs, HFCs, or ammonia refrigerants, or on any of the proven refrigeration techniques (e.g. direct expansion, flooded, or liquid overfeed).

Means are often sought to reduce electrical costs. One way to do this is to lower electrical consumption at peak times of the day. Conventional air-conditioning systems often contribute a large percentage to this peak load because they typically run during peak hours. Shifting air conditioning loads to off-peak times when demand costs are lower cuts demand costs significantly. CTES can help shift air conditioning loads in this way.

CTES is a proven and workable technology. Over 10,000 CTES systems currently operate in the USA. TES relies on an inexpensive storage medium using high specific or latent heat to store cooling. The most common types of storage units use chilled water or ice. Due to the difference in energy density of storage, the ice storage units are smaller.

3.8.4 CTES sizing strategies

The decision of which sizing strategy to use for a CTES is generally an economic rather than technical one. There are three basic storage-sizing strategies:

- full storage,
- load-leveling partial storage, and
- demand-limiting partial storage.

The full-storage strategy supplies all of a facility's peak cooling needs using a storage unit by shifting all of the electrical demand caused by cooling to off-peak hours. Calculating the design-day cooling requirement (tons per hour) during peak times and dividing that by the tank's efficiency factor determines the size of storage tank needed. Initial costs are usually expected to be high.

The load-leveling partial storage strategy supplies only part of a building's cooling load during peak hours. This method levels the building's electrical demand caused by cooling over the design day. Compared to the other two strategies, this method minimizes the size of storage and refrigeration equipment needed to cool a building, resulting in lower equipment costs. However, this strategy does not create such large operating savings as the others.

Demand-limited partial storage requires less storage capacity than full storage, but more than load-leveling strategies, and lowers a building's peak electrical demand to a predetermined level. This level is normally equal to the peak demand imposed by non-cooling loads. To effectively keep the total electric demand below the predetermined level, real-time controllers are required to monitor the building's non-cooling loads and control the ratio of storage- and chiller-supplied cooling.

Reduced operating costs are the primary benefit derived from using CTES. Energy cost can be reduced for cooling a facility by as much as 70%. Typical payback periods using CTES usually range from two to six years. Predicting cost savings from using CTES requires the following building-specific information:

- hour-by-hour power usage,
- the performance of the proposed cool storage system, and
- the local electrical utility's rate structure.

Particularly in the USA, utility companies sometimes offer incentive programs reduced rates or free feasibility studies related to saving energy. Incentive payments are cash payments or rebates to the customer for installing a functional TES unit. Retrofits are usually eligible for higher incentives than new construction. Feasibility studies are an indirect inducement that may be offered free or at a reduced rate by the utility.

3.8.5 Load control and monitoring in CTES

Two key control issues unique to CTES systems are monitoring and regulating the water or ice temperature and controlling tank-water level (Maust, 1993). Temperature sensors inside chilled-water storage tanks provide data which the control system uses to evaluate the tank's cooling capacity. Temperature monitoring also gives the operator information to make load-management and chiller-operation decisions. For instance, encapsulated ice systems evaluate cooling capacity by monitoring tank liquid level and the amount of ice that remains. By trend-logging the cooling demand and the rate at which stored cold is used, an operator can run a chiller plant and the storage system to achieve the desired objectives.

It is also important to monitor the tank water level because it can indicate faulty operation. Large volumes of water are often involved, and a leak or break in the treated-water system could lead to large and expensive losses. Sensors monitoring tank level, performance of the make-up water system and pressure regulating components provide important diagnostic information.

Small ice CTES systems have built-in control systems. For large ice storage systems, such as the encapsulated systems, control systems are integrated into the building automation system that controls chillers, pumps and air-handling units. These control systems are normally custom designed and constructed on site. Systems typically consist of a set of microprocessors connected to a high-speed, local-area network and an operator's workstation. Network control systems usually require minimal operator training. Software continuously monitors building load, storage capacity and outdoor conditions and compares them with historical data. The program determines the best use of chillers and stored cool water. A control system also can provide troubleshooting capabilities.

Monitoring and control of CTES systems are generally straightforward. For example, consider two basic modes of operation such as near-full storage and partial storage. In near-full storage, chillers are permitted to operate as little as possible during peak hours. In partial storage, chillers are allowed to operate up to a pre-determined load limit during peak hours, with storage making up the remainder of a daily load.

3.8.6 CTES storage media selection and characteristics

The storage medium determines how large the storage tank will be and the size and configuration of the HVAC system and components. The main options include chilled water, ice and eutectic salts. Ice systems offer the densest storage capacity, but have the most complex charge and discharge equipment. Water systems offer the lowest storage density, and are the least complex. Eutectic salts have intermediate characteristics. Some details on each storage medium follow:

- **Chilled water**. Chilled water systems require the largest storage tanks, but can easily interface with existing chiller systems. Chilled water CTES use the sensible heat capacity of water to store cooling capacity. They operate at temperature ranges (3.3–5.5°C) compatible with standard chiller systems and are most economical for systems greater than 2000 ton-hours in capacity.
- **Ice**. Ice systems use smaller tanks and offer the potential for the use of low-temperature air systems, but require more complex chiller systems. Ice CTES systems use the latent heat of fusion of water (335 kJ/kg) to store cooling capacity. To store energy at the temperature of ice requires refrigeration equipment that provides charging fluids at temperatures below the normal operating range of conventional air conditioning equipment. Special ice-making equipment or standard chillers modified for low temperature service are used. The low chilled-water-supply temperatures available from ice storage allow the use of cool-air distribution, the benefits of which include the ability to use smaller fans and ducts and the introduction of less humid air into occupied spaces. With ice as the storage medium, there are several technologies available for charging and discharging the storage: Ice harvesting systems feature an evaporator surface on which ice is formed and periodically released into a storage tank that is partially filled with water. External melt ice-on-coil systems use submerged pipes through which a refrigerant or secondary coolant is circulated. Ice accumulates on the outside of the pipes. Storage is discharged by circulating the warm return water over the pipes, melting the ice from the outside. Internal melt ice-on-coil systems also feature submerged pipes on which ice is formed. Storage is discharged by circulating warm coolant through the pipes, melting the ice from the inside. The cold coolant is then pumped through the building cooling system or used to cool a secondary coolant that circulates through the building's cooling system. Encapsulated ice systems use water inside submerged plastic containers that freeze and thaw as cold or warm coolant is circulated through the storage tank holding the containers. Ice slurry systems store water or water/glycol solutions in a slurry state (a partially frozen mixture of liquid and ice crystals that looks like slush). To meet a cooling demand, the slurry may be pumped directly to the load or to a heat exchanger cooling a secondary fluid that circulates through the building's chilled water system. Internal melt ice-on-coil systems are the most commonly used type of ice storage technology in commercial applications. External melt and ice-harvesting systems are more common in industrial applications, although they can also be applied in commercial buildings and district cooling systems. Encapsulated ice systems are also suitable for many commercial applications. Ice slurry systems have not been widely used in commercial applications.
- **Eutectic salts**. Eutectic salts can use existing chillers but usually operate at warmer temperatures than ice or chilled-water systems. Eutectic salts use a combination of inorganic salts, water, and other elements to create a mixture that freezes at a desired temperature. The material is encapsulated in plastic containers that are stacked in a

storage tank through which water is circulated. The most commonly used mixture for thermal storage freezes at 8.3°C, which allows the use of standard chilling equipment to charge storage, but leads to higher discharge temperatures. These temperatures, in turn, limit the operating strategies that may be applied. For example, eutectic salts may only be used in full-storage operation if dehumidification requirements are low.

Water vs. ice CTES

To increase compactness, CTES systems have been developed that utilize the latent heat of in phase changes (usually solid-liquid) of substances. Presently in practice, latent storage utilizing water and ice are employed most widely.

Figure 3.22 shows that the cooling capacity of an ice CTES system under total freezing is 18 times as high as that of a water CTES operating between 12 and 7°C. Consequently, the thermal storage volume can be substantially reduced. However, because of practical difficulties in melting ice, it is often not advantageous to turn all the water into ice. Figure 3.23 shows how the Ice Packing Factor (IPF) affects the tank volume for ice CTES in comparison to water CTES. As shown, if 10% of the water is converted to ice, the tank volume is 32% of that for a water storage tank.

Currently, buildings in Japan have a cooling load to heating load ratio of approximately 2:1 to 3:1 due to the heating load of office equipment. If a TES tank is used as a hot water tank in winter, the same tank can be used with an IPF of about 10% for balanced service in winter. As a result of the reduced tank volume, it is possible to install it on the roof. A further advantage of the reduced volume of ice storage is lower surface area, which leads to heat losses relative to a water CTES.

Conventional water CTES systems utilize sensible heat and thus need large tanks. These are often located under a building. In many cases, however, it is difficult to secure such a large tank space, and the use of conventional systems is thus restricted. The design and development of more compact thermal storage tanks has therefore become important.

A major drawback of ice thermal storage is that, since ice must be produced, the chiller evaporator temperature must be lower than that for a water CTES, so chiller capacity and COP are decreased. For water CTES, the evaporation temperature of an ordinary chiller is in the vicinity of 0°C. For ice CTES, however, evaporator temperatures are often below −10°C so that the capacity and COP are reduced to about 56% and 71%, respectively, of those of a water CTES system. A capacity drop leads to an increase in the size and cost of the chiller, while a COP drop leads to an increase in energy costs. To control both these costs, it is desirable to store the minimum of ice required by the available space.

Two basic types of commercial ice storage exist: static (ice building) and dynamic (ice shucking) systems. In the static system, ice is formed on the cooling coils within the storage tank itself (ice-on-coil). Such systems are normally manufactured in packaged units and connected to the building's chilled water system. Dynamic systems, which are becoming popular, make ice in chunks or crushed (so-called *ice harvesting*) form and deliver it for storage in large pits similar to those used in chilled water systems. Dynamic systems may also include formation of an ice glycol slurry which can be stored in a tank. Static ice storage systems are designed to form ice on the surface of evaporator rubes, and to store it until chilled water is needed for cooling. The ice is melted by the warm return water, thereby re-cooling the water before it is pumped back to the coils in the building. Other static systems use a brine which circulates through tubes in an ice block or around containers filled with frozen water. These systems have the advantage of being closed and not open to the atmosphere. An examination of the relative costs of ice CTES systems versus comparable water storage systems sometimes indicates that the ice CTES system costs less, mainly due to the considerably reduced storage volume requirements. Ice CTES has the disadvantage of lower evaporator temperatures and higher electrical energy requirements to achieve the necessary freezing of the water.

Based on years of experience, many companies believe that chilled water and encapsulated ice are the most practical storage methods. Chilled CTES requires a large vertical tank with a capacity of approximately 0.283 m^3 per ton hour of storage. Encapsulated ice requires a much smaller storage capacity, approximately 0.071 m^3 per ton hour stored.

Both systems have advantages which are site-specific. Generally, chilled water storage becomes practical in capacities exceeding 25,000 ton hours. In addition, companies note the following:

- Encapsulated ice CTES systems produce much colder water, so pumps, piping, heat exchangers and cooling coils can be smaller. Chiller compressors must be of the positive displacement type, such as rotary screws or reciprocating. Centrifugal compressors are not practical for this application.
- Chilled CTES systems may use any type of system as long as there is a 20°C difference, during the discharge cycle, between supply and return water. Each cooling coil should be equipped with a dedicated pump and DDC control system.

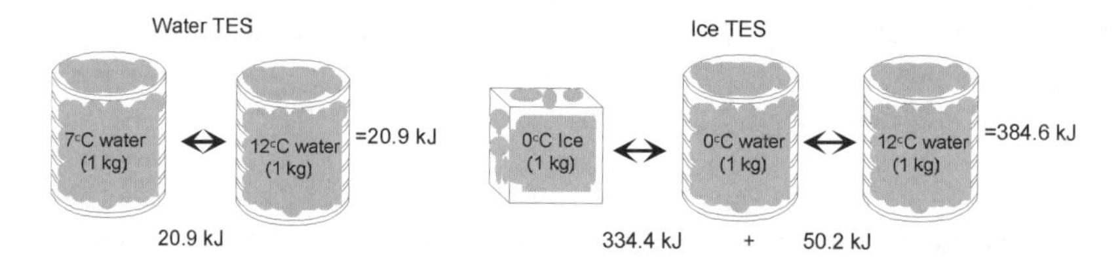

Figure 3.22 The capacity of an ice CTES is 18 times as high as with water CTES (Kuroda, 1993).

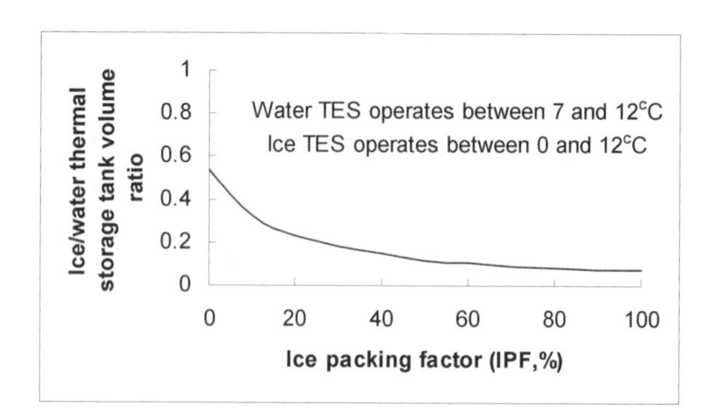

Figure 3.23 Variation in tank volume (expressed as the ratio between ice and water thermal storage tank volume) with ice packing factor (Kuroda, 1993).

PCMs (eutectic salts) for CTES

PCMs have recently received significant attention for CTES, although they have been considered for heating since the early 1980s. PCMs suitable for CTES are eutectic salts that undergo liquid/solid phase changes at temperatures as high as 8.3°C, and absorb and release large amounts of energy during the phase change. Stored in hermetically sealed plastic containers, PCMs change to solids as they release heat to water or another fluid that flows around them. At these temperatures (up to 8.3°C), chillers can operate more efficiently than at the low temperatures required by ice CTES systems. PCMs also have about three times the storage capacity a typical chilled-water CTES system. The choice of ice, chilled water, or PCMs depends on the services needed. If low temperatures are needed, ice is likely preferable. If more conventional temperatures are needed and space is available, chilled water CTES can be installed. If space is limited, but low temperatures are not required and a passive, easy-to-maintain system is desired, PCM may be a good choice.

Ice CTES is one example of using the latent heat characteristic of a storage medium. The significance of latent heat is that far more energy may be added or removed from a storage material as it undergoes a change of phase than can be added or removed from a material that remains in a single liquid phase, such as water, or a single solid phase, such as rock or brick.

Two commercially available materials that enhance the phase change process that ordinarily occurs between water and ice are eutectic salts and gas hydrates. Eutectic salts are mixtures of inorganic salts, water and additives. Gas hydrates are produced by mixing gas with water. Both of these materials work by raising the temperature at which water freezes. These materials have the advantage of a freeze point of 8.3°C or 8.8°C, which reduces energy requirements for freezing. PCMs can therefore provide a highly desirable TES storage medium for cooling purposes. PCMs also provide most of the storage space advantages associated with ice storage systems. By freezing and melting at 8.3°C or 8.8°C, the phase change materials can be easily used in conventional chilled water systems with centrifugal or reciprocating chillers. The storage tank can be placed above or below grade. In addition, chiller power requirements are reduced when phase change materials are used for TES because evaporative temperatures remain fairly constant.

Gas hydrates, still in the development stage for large, commercial installations, have some advantages over eutectic salts. Gas hydrates have high latent heat values, which lead to size and weight advantages. Gas hydrates require only one-half to one-third of the space, and are approximately one-half the weight of an equivalent eutectic salt system.

Like ice and chilled water storage systems, hydrated salts have been in use for many decades. PCMs with various phase-change points have been developed. To date, the hydrated salt most commonly used for CTES applications changes phase at 8.3°C, and is often encapsulated in plastic containers. The material is a mixture of inorganic salts, water, and nucleating and stabilizing agents. It has a latent heat of fusion of 95.36 kJ/kg and a density of 1489.6 kg/m^3. A CTES using this PCM latent heat of fusion requires a capacity of about 0.155 m^3/ton-hour for the entire tank assembly, including piping headers.

PlusICE™ PCMs The disadvantages of conventional ice CTES (low temperature chillers to build ice) and water CTES (large volume of water storage) can be overcome by utilizing the latent heat capacity of various eutectic salts, otherwise known as PCMs. Table 3.10 lists several types of commercially available PlusICE™ substances and their thermophysical data. For different phase change temperatures, some other types are also available. PlusICE™, mixtures of non-toxic eutectic salts, have freezing and melting points higher than those of water, and the temperature range offered by this concept provides the following benefits:

- Space efficient coolness and heat recovery CTES.
- Utilization of existing chiller and refrigeration equipment for new and retrofit CTESs.
- Elimination of low temperature glycol chillers.
- Improvement of system efficiency due to higher evaporation temperatures and possible charging by means of free cooling (i.e. without operating the chiller).

PlusICE™ beams (Figure 3.24a) provide a static CTES system, and the self-stacking modular tube concept offers flexibility for the size and location. They can be manufactured either as a single or multi-pass arrangement for the most economical capacity and duty balancing for any given new and retrofit application. Site assembled modular self-stacking design offers flexibility and simple installation. Future capacity increases can be easily accommodated by simply adding more beams into an existing module with minimal modification work. In the operation of the system, water or refrigerant is circulated within the inner tube and the excess capacity from this fluid is stored in the form of latent heat by the eutectic salts during the charging mode. This operation is reversed during the discharge mode to supplement the system load. Beams can be applied either as a totally centralized CTES, similar to a conventional storage tank, or totally localized, i.e. spread over the system as part of the pipe runs, or a combination of the two (Figure 3.24b). This enables a reduction or even elimination of the central storage module.

Table 3.10 Commercially available PlusICE™ substances.

PlusICE Type	Phase change temperature (°C)	Density (kg/m^3)	Heat of fusion (kJ/kg)	Latent heat (MJ/m^3)
A4	4	766	227	174
E7	7	1542	120	185
E8	8	1469	140	206
A8	8	773	220	170
E10	10	1519	140	213
E13	13	1780	140	245
E21	21	1480	150	222
A22	22	775	220	171
A28	28	789	245	193
E30	30	1304	201	262
E32	32	1460	186	272
E48	48	1670	201	336
E58	58	1280	226	289
E89	89	1550	163	253
E117	117	1450	169	245

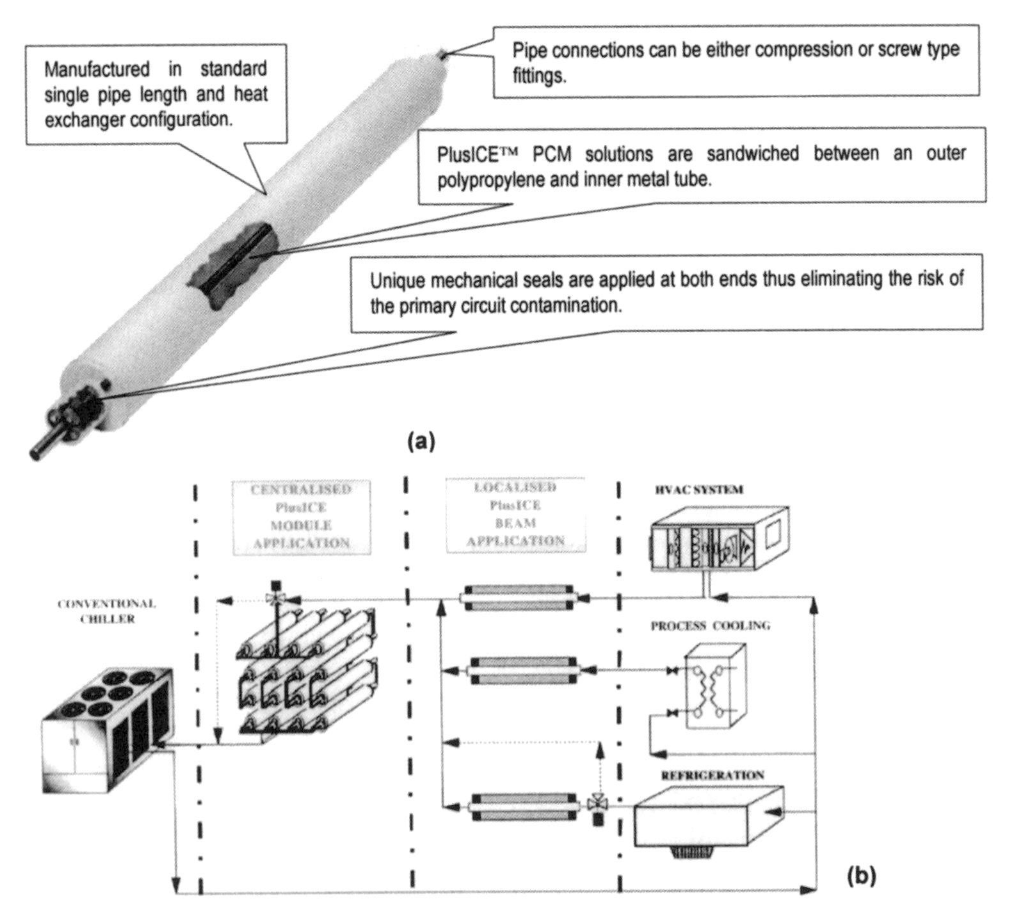

Figure 3.24 (a) PlusICE™ beam for CTES applications. (b) Centralized and localized PlusICE™ applications (*Courtesy of Environmental Process Systems Limited*).

3.8.7 *Storage tank types for CTES*

Storage tanks must have the strength to withstand the pressure of the storage medium, and be watertight and corrosion resistant. Above-ground outdoor tanks must be weather resistant. Buried tanks must withstand the weight of the soil covering and any other loads that might occur above the tank, such as the parking of cars. Tanks may also be insulated to minimize thermal losses (which are typically 1–5% per day).

The options for tank materials include the following:

- **Steel**. Large steel tanks, with capacities up to several million cubic meters, are typically cylindrical in shape and field-erected of welded plate steel. Corrosion protection, such as an epoxy coating, is usually required to protect the tank interior. Small tanks, with capacities of less than 100 m^3, are often rectangular in shape and typically made of galvanized sheet steel. Cylindrical pressurized tanks are generally used to hold between 10 and 200 m^3.

- **Concrete**. Concrete tanks may be precast or cast-in-place. Precast tanks are most economical, in sizes of one million gallons or more. Cast-in-place tanks can often be integrated with building foundations to reduce costs. However, cast-in-place tanks are more sensitive to thermal shock. Large tanks are usually cylindrical in shape, while smaller tanks may be rectangular or cylindrical.
- **Plastic**. Plastic tanks are typically delivered as prefabricated modular units. UV stabilizers or an opaque covering are required for plastic tanks used outdoors to provide protection against the ultraviolet radiation in sunlight.

Steel and concrete are the most commonly used types of tanks for chilled water storage. Most ice harvesting systems and encapsulated ice systems use site-built concrete, while external melt systems usually use concrete or steel tanks, internal melt systems usually use plastic or steel, and concrete tanks with polyurethane liners are common for eutectic salts.

CTES tanks often use insulation. Because of the low temperatures associated with ice storage, insulation is a high priority. Ice storage tanks located above ground are normally insulated to limit standby losses. For external melt ice-on-coil systems and some internal melt ice-on-coil systems, the insulation and vapor barrier are part of the factory-supplied containers; most other storage tanks require field-applied insulation and vapor barriers. Below-ground tanks used with ice harvesters sometimes do not need insulation beyond a below one meter grade. Because the tank temperature does not drop below 0°C at any time, there is no danger of freezing and thawing groundwater. All below-ground tanks using fluids below 0°C during the charge cycle should have insulation and a vapor barrier system, generally on the exterior. Interior insulation is susceptible to damage from the ice and should be avoided.

Exposed tank surfaces should be insulated to help maintain the temperature in the tank. Insulation is especially important for smaller storage tanks, because the ratio of surface area to stored volume is relatively high. Heat transfer between the stored water and the tank contact surfaces (including divider walls) is a primary source of loss. Not only does the stored fluid lose heat to (or gain heat from) the ambient by conduction through the floor and wall, but heat also flows vertically along the tank walls from the warmer to the cooler region. Exterior insulation of the tank walls does not inhibit this heat transfer.

The cost of chemicals for water treatment may be significant, especially if the tank is filled more than once during its life. A filter system helps keep the stored water clean. Exposure of the stored water to the atmosphere may require the occasional addition of biocides. While tanks should be designed to prohibit leakage, the designer should account for the potential impact of leakage on the selection of chemical water treatment. Storage circulating pumps should be installed below the minimum operating water level to ensure flooded suction. The required net positive suction pressure must be maintained to avoid sub-atmospheric pressure conditions at the pumps.

3.8.8 Chilled-water CTES

A chilled-water CTES system uses the sensible heat in a body of water to store energy. Given its specific heat of 4.187 kJ/kg°C, about 0.2 m^3 of water are needed to absorb 12,000 kJ and provide 1 ton-hour of cooling if the coil raises the water temperature by 20°C.

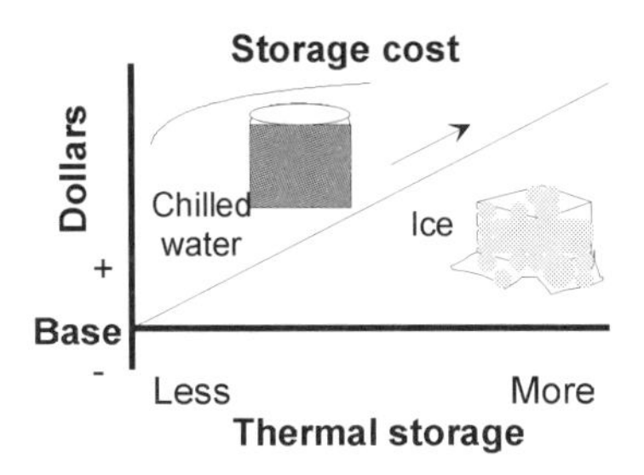

Figure 3.25 CTES cost relationship (AEP, 2001).

By contrast, the same ton-hour of cooling can be provided with just 0.042 m^3 of ice, since each kg of ice absorbs 152 kJ as it melts. Therefore, a CTES system that uses chilled water rather than ice will require 6–7 times more installed storage volume. Figure 3.25 plots the cost of CTES components as a function of the ton-hours of cooling stored. The sizable cost penalty imposed by the significantly larger storage tank volume required for chilled water is readily apparent compared with ice storage. Note, however, that the cost of the water storage tank is mainly a function of its surface area, while the capacity of the tank is mainly a function of its volume. Therefore, as the size required of a chilled-water storage tank increases, the per-ton-hour cost of the storage tank actually decreases. Consequently, it appears that chilled water may be competitive with ice in applications that require more than 10,000 ton-hours of thermal storage.

A chilled-water storage system can be viewed as a simple variation of a decoupled chiller system. Since the same fluid water is used to both store and transfer heat, few accessories are needed and the system is simple. As shown in Figure 3.26, a decoupled system separates the production and distribution of chilled water. The balance of flow between the constant-volume production of chilled water and its variable-volume distribution is handled with a bypass pipe commonly called a 'decoupler.' The decoupler re-directs surplus chilled water to storage when production exceeds distribution and withdraws storage water when distribution exceeds supply.

Chilled-water CTES systems offer a number of benefits, especially for larger systems. In addition, the large storage volume of water can be incorporated into fire safety systems, and in fact, some sprinkler systems use storage water in their design. The disadvantages of chilled-water storage which relate to the tank design, weight, location and space requirements can pose challenges.

The relative installed cost of a chilled-water CTES (see Figure 3.27) shows the significance of storage tank expense. While somewhat prohibitive for most applications under l0,000 ton-hours, the decreasing unit cost of chilled-water CTES systems can be very attractive for large central plants and industrial installations.

Series storage tanks for chilled-water CTES

The simplest form of chilled-water storage places one or more tanks in series. This concept is illustrated in Figure 3.28 as a single baffled tank. When the chillers produce more chilled water than the system requires, the excess is diverted to the series tank where it displaces the warmer water there. Likewise, when chilled water demand exceeds the quantity produced, chilled water is drawn from the tank by displacing it with warm return water.

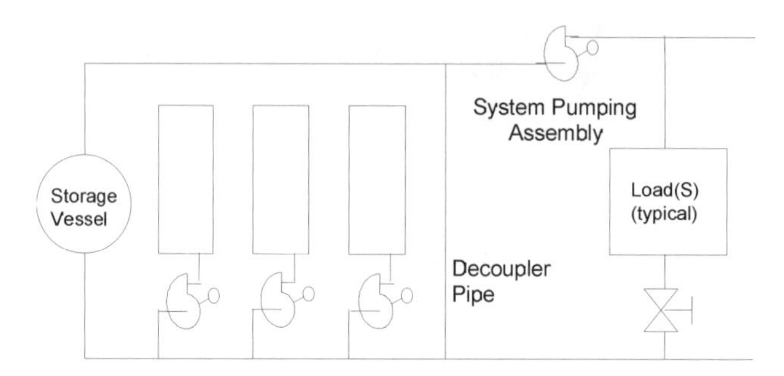

Figure 3.26 Schematic of a chilled-water production and distribution system (AEP, 2001).

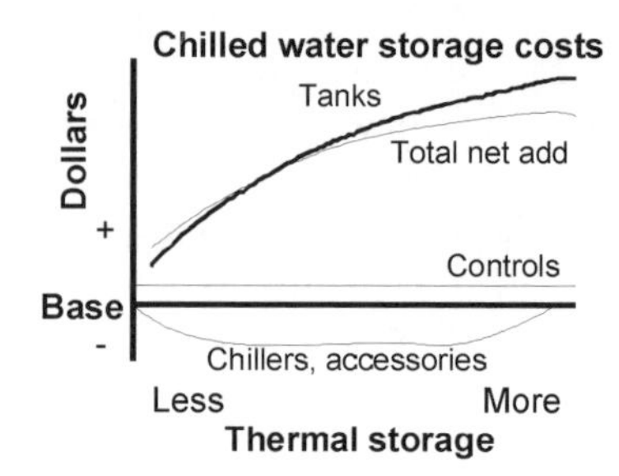

Figure 3.27 Pros and cons tank configuration (AEP, 2001).

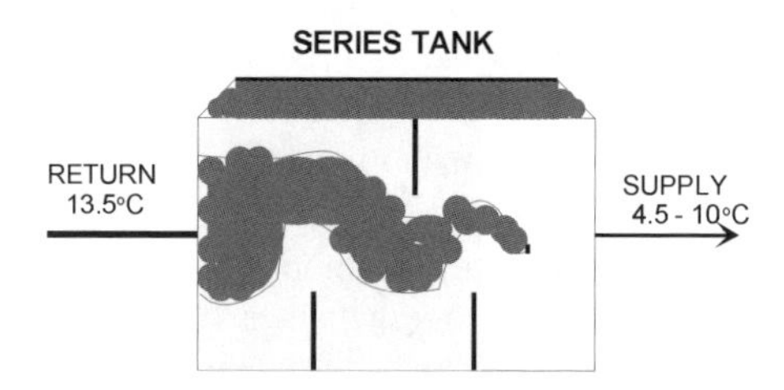

Figure 3.28 Series tank configuration for chiller-water CTES (AEP, 2001).

A number of chilled-water CTES systems with designs similar to that in Figure 3.28 have been installed and proven to be effective in reducing peak electrical demand. However, series-tank designs can cause the water to stratify or become stagnant. Stagnation is the tendency of some water to shortcut its way through the tank, and renders large volumes of tank water ineffective for storage. Also, inter-compartmental mixing can raise the tank-water exit temperature, reducing the tank effectiveness during its final hours of discharge.

Parallel storage tanks for chilled-water CTES

The problems of mixing and stratification can be minimized with a multiple-tank parallel design (see Figure 3.29). This arrangement replaces the bypass pipe or decoupler with a number of separate tanks piped in parallel between the 14.5°C return water from the cooling coils and the 4.5 to 5.5°C supply water from the chiller(s). Each of these tanks has individually controlled drain and fill valves.

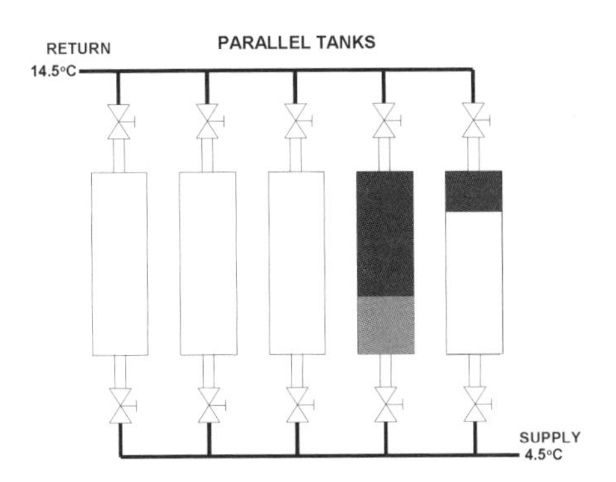

Figure 3.29 Parallel tank configuration for chilled-water CTES (AEP, 2001).

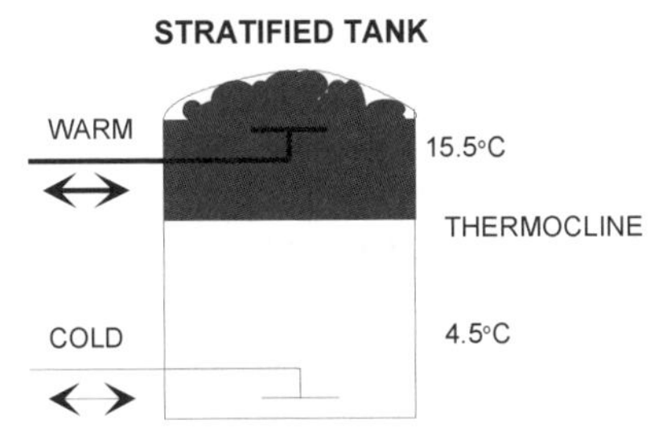

Figure 3.30 Stratified tank configuration (AEP, 2001).

In practice, one of the parallel-piped tanks is empty when the storage is changed, and that tank's drain and fill valves are closed. When the discharge cycle starts (i.e. when the system starts to use chilled water), the empty tank's fill valve opens to allow it to receive warm return water. The drain valve on any one of the tanks filled with previously chilled water opens so that, as warm return water fills the empty tank, an equal flow of cold water is drawn from the tank with the open drain valve.

Proper valve sequencing is especially important when the receiving tank is nearly full and the draining tank is almost depleted. In this valve control sequence:

1. The drain valve on a new tank previously filled with chilled water must open.
2. The drain valve on the just-emptied tank must close as its fill valve opens, allowing the tank to receive warm return water.
3. The fill valve on the once-empty tank that is now full of warm return water must close. (This tank is now ready for off-peak recharging.)

A building automation system and an accurate method of measuring tank volume are required to facilitate this control task. Although the multiple-parallel-tank scheme eliminates many of the problems associated with mixing and tank stratification, its complexity can add to the cost of a chilled-water CTES system.

Stratified storage tanks for chilled-water CTES

Most chilled-water storage systems installed today are based on designs that exploit the tendency of warm and cold water to stratify. Then, cold water is added to or drawn from the bottom of the tank, while warm water is returned to or drawn from the top. A boundary layer, or thermocline, often 0.2 to 0.4 m in height, is established between the resultant warm and cold zones (Figure 3.30). Specially engineered diffusers or any array of nozzles can ensure laminar flow within the tank. This laminar flow is necessary to avoid mixing and promote stratification, since the respective densities of the 15.5°C return water and 4.5 to 5.5°C supply water are almost identical.

Advantages of chilled-water CTES

The most important advantage of water thermal storage is the fact that water is a well understood and familiar medium. No danger is associated with handling water as a TES material. Basically, heating and cooling applications require only a moderate temperature range (5°C to 45°C). Water can be used both as a storage medium for cooling and heating, allowing the same TES material to be used throughout the year. The relatively high thermal capacity of water is also attractive for TES applications. These advantages give water CTES economic advantages over other air conditioning systems, including those using ice CTES system.

For example, an economic comparison between water CTES and ice CTES is based on a model building of 10,000 m^2, situated in Tokyo (Narita, 1993), and indicates the following:

- The initial cost of water thermal storage is approximately 20% less than that of ice CTES. The main reason is the cost of the ice making unit in the TES tank.
- In the ice making mode, the COP is approximately 20% lower than for the water cooling mode because of the lower evaporating temperature. Consequently, more electricity is consumed in ice CTES.
- The ice CTES system has little thermal storage capacity for potential use in the heating season.
- The operating cost of water thermal storage is approximately 20% less than that of ice CTES.

A major disadvantage of water CTES is volume. For the same amount of thermal capacity, a water CTES system requires approximately three times as much tank volume as ice CTES. To somewhat circumvent this problem, alternative design configurations are sometimes considered.

Heat pumps and chilled-water CTES

The heat pump is a key technology for energy conservation. Combining heat pumps and water CTES has many advantages, including the economical operation of the heat pump as well as a load-leveling effect on electricity demand. Such combined systems can overcome some of the disadvantages of water CTES, and have been introduced in actual plants.

In conventional air conditioning systems using heat pumps, the heat pump must be operated during the day when cooling demand exist. This operation contributes to electricity demand during the same period. Cooling demand is often responsible for approximately one-third of electricity demand at the peak period. In a typical water CTES system, half of the daily cooling load can covered by night operation of heat pumps.

The combination of a heat pump and a water CTES can give the following benefits:

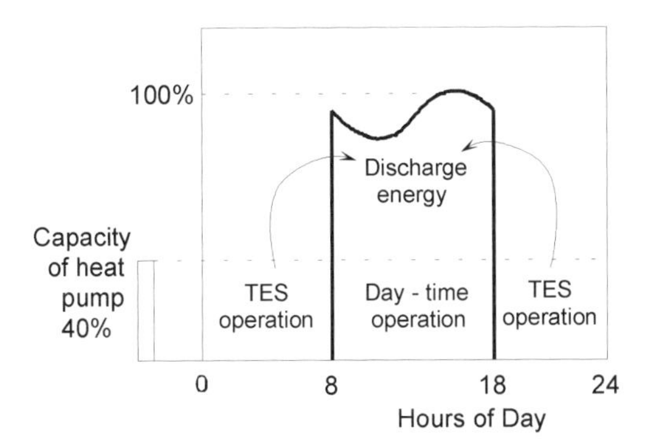

Figure 3.31 Operation of heat pump CTES.

- **Load leveling of electricity demand for air conditioning.** If 50% of the air conditioning load is shifted to the night on the peak day, the annual dependence on night-time electricity can be as high as 70% for cooling and 90% for heating. Thus, the CTES system achieves peak shifting on the peak day and improves the annual load factor of electricity-supply facilities. At the same time, customers accrue economic advantages with a CTES system when electric power companies provide discount rates for night-time electricity.
- **Efficient operation of the heat pump unit.** Heat pumps and other such devices as chillers and boilers have a maximum efficiency point of operation. In commercial and residential applications of heat pumps, heat pump operation cannot be maintained at the most efficient point because the cooling and heating load varies temporally. This variation reduces the seasonal efficiency of heat pumps. CTES can help to avoid this problem by allowing operation of heat pumps at the most efficient point, because heat pumps can operate independently of the cooling and heating load of buildings.
- **Reduction of heat pump capacity.** As Figure 3.31 shows, for the same amount of air conditioning load, the longer operating hours allow for smaller size of heat pump, which leads to a reduction in demand for electricity. Thus both initial and operating costs decrease.

3.8.9 Ice CTES

Ice CTES systems can be economically advantageous and require less space than water CTES systems. This space advantage often allows heating and cooling capacity to be enlarged within the often restricted area of existing machine rooms.

An ice CTES system with heat pump is composed of a heat pump, an ice-making system, a storage tank, and an air conditioning system which can be a conventional central system.

Ice CTES systems are often classified as static or dynamic, according to the way ice is delivered to the storage tank. Each of these ice CTES types are discussed below.

- **Static systems.** In static systems, an ice-making pipe is installed in the storage tank where ice is formed and later melted. Ice-in-tube systems produce ice inside the ice making coil. Ice-on-coil systems produce ice outside the coil. The ice may be melted using either an external melting system which melts ice from the outer side, or an internal melting system (for ice-on-coil only). Internal melting may be achieved by feeding the warmed brine returning from the air conditioning equipment into the coil. Alternatively, a refrigerant sub-cooling system can pass high-temperature and high-pressure refrigerant from the condenser through the coil. Another type of static system is the ice-tube float system which uses a polyurethane tube filled with water.

- **Dynamic systems.** In the dynamic type of ice CTES system, ice is produced outside the storage tank and removed from the ice-making surface continuously or intermittently by various means. The ice harvest system feeds refrigerant which has passed through the expansion valve into an ice-making plate where ice is peeled off and dropped into a storage tank. The ice chip system continuously produces ice which is scraped off as ice chips. In the liquid-ice system, fine ice crystals are formed in an ethylene glycol solution (ice-slurry) by cooling with refrigerant. Alternative systems use PCMs. The sherbet-ice system produces very fine (sherbet-like) ice by a water or brine sub-cooling system.

Ice-on-pipe CTES

Ice-on-pipe CTES designs produce ice by pumping very cold liquid refrigerant (usually HCFC-22 in commercial applications or ammonia in industrial refrigeration) through an array of pipes immersed in a tank of water. A system configuration, including the chiller components, is shown in Figure 3.32. Technically, ice-on-pipe CTES resembles process refrigeration. System components, i.e. the compressor and condenser(s), high- and low-pressure receivers, refrigerant pumps, evaporators and ice tanks, are individually selected for the application and integrated to provide a reliable refrigeration system.

Unlike direct-expansion systems, which rely on additional heat transfer surface area to separate refrigerant vapor from liquid refrigerant, ice-on-pipe systems use a low-pressure receiver and a method called liquid-overfeed to accomplish this.

A liquid-overfeed system works as follows. Chilled water and/or ice is produced by pumping cold liquid refrigerant to a chiller evaporator or an ice tank at a rate 1.3–1.6 times faster than required to evaporate it. A 'two-phase' solution of refrigerant liquid and vapor results, and is returned to the low-pressure receiver. This 30–60% higher refrigerant flow rate is why the system is referred to as liquid-overfeed.

The refrigerant returned to the low-pressure receiver is nearly saturated (see Figure 3.33). Refrigerant liquid that does not 'boil off' in the evaporator is sent back for a second pass.

Note that the open or 'atmospheric' design of an ice-on-pipe system dictates the use of a heat exchanger to separate ice water from the building cooling water loop. The cooling loop is normally a closed system.

On the high-pressure side of the system (Figure 3.34), the cold refrigerant vapor that collects at the top of the low-pressure receiver is drawn off by the compressor. After compression, the pressurized (and now hot) vapor passes to the condenser, where cooling tower water circulating through the shell causes the refrigerant to condense. The liquid refrigerant, still at high pressure, exits the condenser and enters a high-pressure receiver where it is stored for later use. Refrigerant flow from the high-pressure receiver is regulated by a refrigerant metering device to ensure that a minimum liquid level is maintained in the low-pressure receiver.

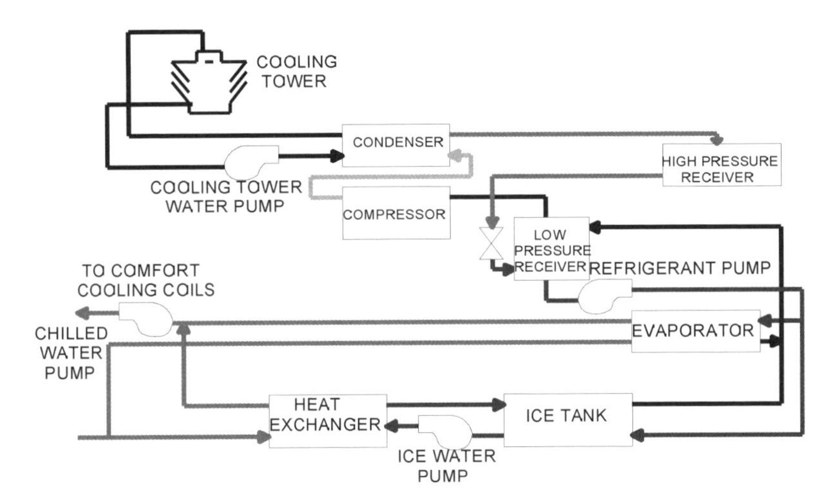

Figure 3.32 Ice-on-pipe CTES for process refrigeration (AEP, 2001).

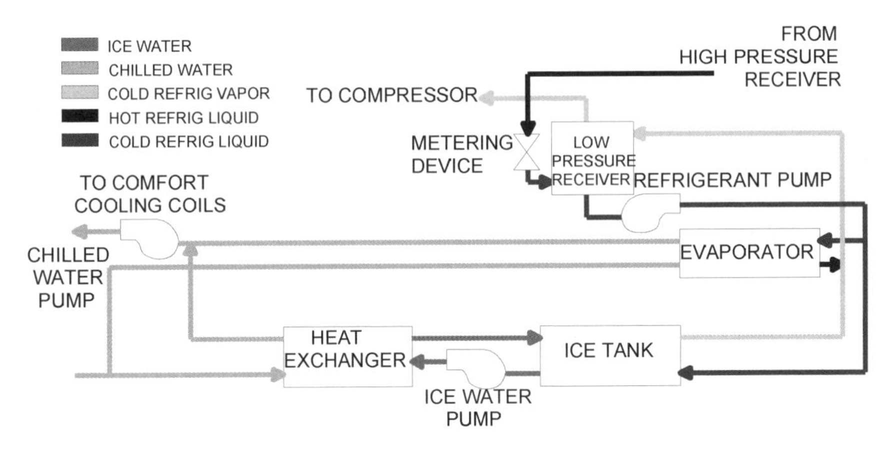

Figure 3.33 Low-pressure portion of an ice-on-pipe CTES for process refrigeration (AEP, 2001).

The ice produced by an ice-on-pipe system forms on the exterior surfaces of an ‘ice coil’. This coil is actually a series of steel pipes immersed in a tank of water. Cold refrigerant (usually HCFC-22) is then pumped through these pipes to freeze the water that surrounds them. Bubbles flow around the steel pipes to agitate the water in the tank, sometimes by injecting air at the bottom. The rising air bubbles promote dense, even ice formation during the freezing cycle and uniform melting when the tank is discharged.

The low-pressure receiver plays a critical role in liquid-overfeed ice-on-pipe systems: it separates the two-phase refrigerant solution returning from the ice coil (or chiller evaporator) into liquid and vapor. Gravity induces this separation, causing the liquid refrigerant and oil to settle to the bottom of the receiver while pure refrigerant vapor collects at the top. As the compressor draws this vapor from the receiver, the liquid level falls. To ensure that there is always sufficient liquid in this vessel, a liquid level control adds refrigerant from the high-pressure receiver as needed.

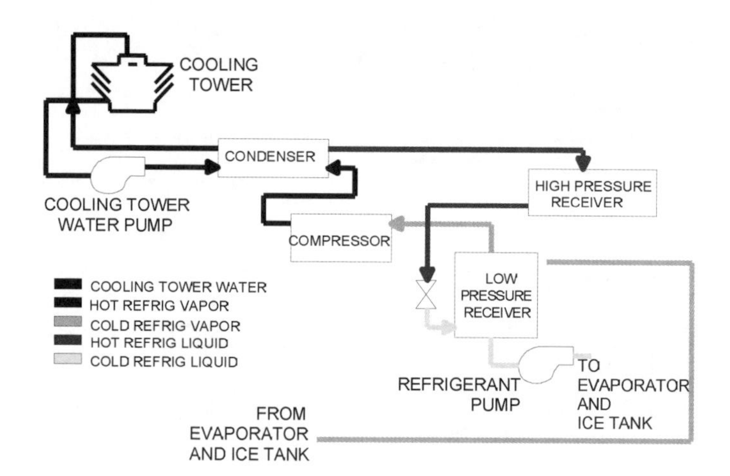

Figure 3.34 High-pressure portion of an ice-on-pipe CTES (AEP, 2001).

Liquid-overfeed systems require a separate oil return/recovery system. This is because the preferred compressor type (helical rotary/screw) expels significant amounts of oil into the discharge line. Entrained in the refrigerant, the oil makes its way through the condenser and high pressure receiver, eventually ending up in the low-pressure receiver. There, the oil collects at the bottom of the tank (along with the liquid refrigerant) and cannot return to the compressor through the suction line. A separate oil-recovery system is needed to capture, distill and return the oil to the compressor. This must be carefully addressed in the system design. The complexity of the liquid overfeed ice-on-pipe system translates into significant fixed costs that are independent of the quantity of ice produced and stored. The refrigerant and oil inventory control systems, refrigerant pumps and other system accessories, plus the field labor required to install them, constitute a sizable investment.

Depending on the system size, the tank can be either pre-manufactured to include both the ice coil and tank, or field-assembled by installing the ice coil in a field-erected concrete tank. While the latter option makes the per-ton-hour cost of the tank attractive, it only partially offsets the combined cost of field labor and accessories, even when the lower compressor cost is considered. Liquid overfeed ice-on-pipe systems are expensive because they require not only large inventories of oil and refrigerant, but refrigerant containment equipment as well. The high costs of engineering and installing liquid-overfeed ice-on-pipe systems typically limit their use to larger applications.

Ice-on-pipe systems offer a number of benefits over chilled-water storage. The storage volume required is considerably less, since each ton-hour of cooling stored can occupy as little as 0.085 m^3. Also, unlike their chilled water counterparts, ice storage systems can operate at any return-water temperature.

Ice harvesters

Ice harvesters circumvent the problems associated with liquid-overfeed ice-on-pipe systems by combining all of the components and accessories required for ice production in a single manufactured package. This device, called an ice harvester (see Figure 3.35), is installed above an open tank that stores a combination of water and flakes of ice.

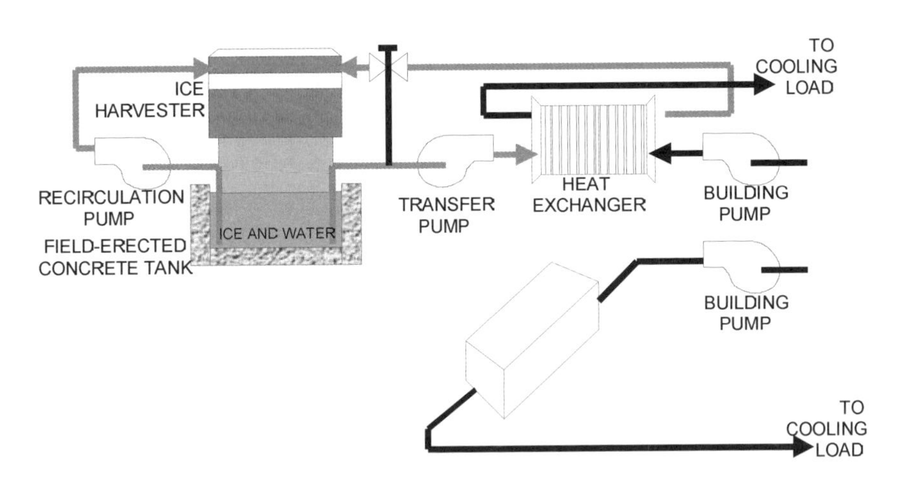

Figure 3.35 Ice harvester system (AEP, 2001).

To produce ice, 0°C water is drawn from the storage tank and delivered to the ice harvester by a recirculation pump at a flow rate of 1.75 to 2.75 m^3/h per ton of ice-producing capacity. Once inside the ice harvester, the recirculated water flows into a drain pan positioned over a series of refrigerated plates. Each of these plates is constructed of two stainless steel sheets welded together at their circumference. A refrigeration system integral to the ice harvester maintains the plates at a temperature of –9.5 to –6.5°C.

As the water leaves the drain pan, it flows freely over both sides of the refrigerated plates where it freezes to a thickness of 0.3 to 1.0 cm. On reaching a given thickness (or at the initiation of a time), the ice is dislodged from the plates by a hot-gas defrost cycle, and falls into the tank below. When cooling is required, a transfer pump draws iced water from the storage tank and delivers it to a building heat exchanger.

It is possible to use the ice harvester as a water chiller by raising the suction temperature of the refrigeration system and pumping warm water from the building heat exchanger over the refrigerated plates. In fact, operating at this higher suction temperature improves the ice harvester's efficiency. Unfortunately, the ice harvester cannot produce chilled water without melting ice stored in the storage tank.

This inability to operate as a true water chiller in a chilled-water system poses a significant efficiency penalty on ice storage systems with ice harvesters. To address this inefficiency, ice harvesters are commonly used in tandem with conventional water chillers. Ice harvesting systems separate ice formation from ice storage. Ice is typically formed on both sides of a hollow, flat plate or on the outside or inside (or both) of a cylindrical evaporator surface. The evaporators are arranged in vertical banks located above the storage tank. Ice is formed to thicknesses between 0.5 and 1 cm. This ice is then harvested, often through the introduction of hot refrigerant gas into the evaporator, the gas warms the evaporator, which breaks the bond between the ice and the evaporator surface and allows the ice to drop into the storage tank below. Other types of ice harvesters use a mechanical means of separating the ice from the evaporator surface.

Ice is generated by circulating 0°C water from the storage tank over the evaporators for a 600–1800 s build cycle. The defrost time is a function of the amount of energy required to warm the system and break the bond between the ice and the evaporator surface. Depending on the control methods, the evaporator configuration, and the discharge conditions of the compressor, defrost can usually be accomplished in 20–90 s. Typically, the evaporators are grouped in sections that are defrosted individually so that the rejected heat from the active sections provides the energy for defrost.

In load-leveling applications, ice is generated and the storage tank charged when there is no building load. When a building load is present, the return chilled water flows directly over the evaporator surface, and the ice generator functions either as a chiller or as both an ice generator and a chiller. Cooling capacity as a chiller is a function of the water velocity on the evaporator surface and the entering water temperature. The defrost cycle must be energized any time the exit water from the evaporator is within a few degrees of freezing. In chiller operation, maximum performance is obtained with minimum system-water flows and highest entering-water temperatures. In load-shifting applications, the compressors are turned off during the electric utility on-peak period.

Positive displacement compressors are usually used with ice harvesters, and saturated suction temperatures are usually between –7.5 and –5.5°C. Condensing temperatures should be kept as low as possible to reduce energy consumption. The minimum allowable condensing temperature depends on the type of refrigeration system used and the defrost characteristics of the system. Several systems operating with evaporatively cooled condensers have operated with a compressor specific power consumption of 0.9 to 1.0 kW/ton.

Ice harvesting systems can melt stored ice quickly. Individual ice fragments are characteristically less than 15.5×15.5×0.65 cm, and provide a minimum of 1.5 m^2 of surface area per ton-hour of ice stored. When properly wetted, a 24-hour charge of ice can be melted in less than 30 minutes for emergency cooling demands.

During the ice-generation mode of operation, the system is energized if the ice is below the high ice level. A partial-storage system is energized only when the entering water is at or above a temperature that will permit chilling during the discharge mode; otherwise, the system does not operate, and the ice tank is discharged during the peak period to meet the load. The high-ice level sensor can be mechanical, optical or electronic. The entering-water temperature thermostat is usually electronic.

When ice is floating in a tank, the water level remains nearly constant, so it is very difficult to measure ice inventory by measuring water level. The following general methods are used to determine ice inventory:

- **Water conductivity method**. As water freezes, dissolved solids are forced out of the ice into the liquid, thus increasing their concentration in the water. Accurate ice-inventory information can thus be maintained by measuring conductivity and evaluating the ice level.
- **Heat balance method**. The cooling effect of a system may be determined by measuring the mechanical power input to, and heat of rejection from, the compressor. The cooling load on the system is determined by measuring coolant flow and temperature. The ice inventory is then determined by integrating cooling input minus load and evaluating the ice level. A variant of the heat balance method involves a heat balance on the compressor only, using performance data from the compressor manufacturer and measured load data.

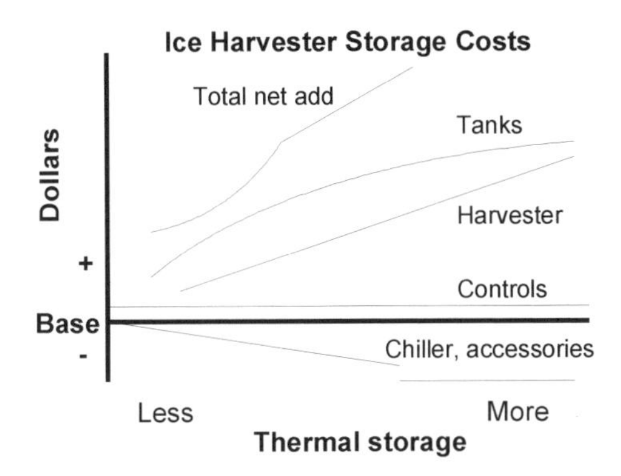

Figure 3.36 Cost relationship for ice harvesters (AEP, 2001).

Optimal performance of ice harvesting systems may be achieved by recharging the ice storage tank over a maximum amount of time with minimum compressor capacity. Ice inventory measurement and the known recharge time are used to determine such specifications. Efficiency can also be increased with proper selection of multiple compressors and unloading controls.

Design of the storage tank is important to the operation of the system. The amount of ice stored in a storage tank depends on the shape of the storage, the location of the ice entrance to the tank, the angle of repose of the ice (which normally is between 15° and 30°, depending on the shape of the ice fragments), and the water level in the tank. If the water level is high, voids occur below the water surface due to the buoyancy of the ice. Ice-water slurries have been reported to have a porosity of 0.50 and typical specific storage volumes of 0.08 m^3 per ton-hour.

Cost A cost analysis of ice harvester systems indicates the increasing costs for both the tank and the harvester as the quantity of ice stored increases (see Figure 3.36). Given their high dollar-per-ton cost, ice harvester systems are usually used to provide additional capacity in retrofit applications, or in large installations.

Advantages Ice harvesters present the system designer with a number of benefits. As with other ice storage systems, the space requirement and cost of the volume stored are less than for chilled-water systems. In addition, the ice harvester is a packaged device, leading to simplified installation, reduced installed costs, and in most cases, factory-tested performance.

Disadvantages Ice harvesters have some limitations, particularly since the harvester and tank are open to the atmosphere. For example, the plates and chassis of the ice harvester are normally constructed of stainless steel, a material that adds significantly to the ice harvester's already high cost. Water treatment is also necessary because of the open nature of the tank and drain pan. The complexities of evenly distributing the ice in the bin and the prevention of piling and bridging add to the cost and operation of this system. Finally, the ice harvester's inability to produce chilled water without depleting the ice in the storage tank may be an economic deterrent.

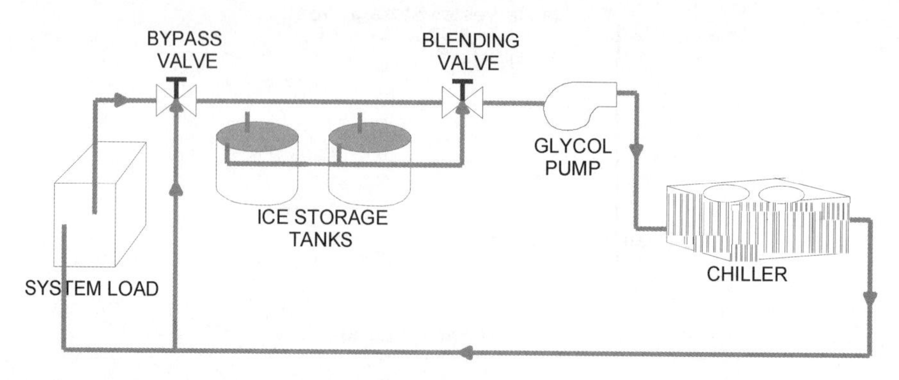

Figure 3.37 Glycol ice CTES system (AEP, 2001).

CTES glycol systems

Glycol systems freeze water by circulating ethylene or propylene glycol through storage tanks (Figure 3.37). The glycol ice storage system is simple. Few accessories are needed, and conventional water chillers are used. Instead of water, a glycol solution (in this case, 25% ethylene glycol) is pumped through the chiller coils and ice storage tanks in the chilled-water loop. The −5.5 to −4.5°C ethylene glycol produced by the packaged chiller freezes the water contained inside the ice storage tanks.

Glycol ice storage systems are available from all major chiller manufacturers. Though similar in concept, they may be packaged differently. Glycol ice storage systems can generally be divided into two major categories: modular and encapsulated ice storage.

Cost Glycol ice storage systems have low installed costs since the same packaged chiller that provides space cooling also doubles as the ice maker. The storage tanks themselves are the only significant additional cost in these systems. In fact, glycol ice storage systems sometimes reduce chiller costs.

Advantages Glycol ice storage systems present numerous benefits, including the ability to use a standard packaged chiller. They also offer an opportunity to reduce pump work, and require few ancillary devices. The choice of modular storage tanks or encapsulated ice systems offers application flexibility, but costs choices and reliable performance as well. Simple control schemes can be used and, as for all ice storage systems, volume and space requirements per ton-hour of storage are considerably lower than those for chilled water storage.

Disadvantages Glycol ice storage systems have some problems, the most significant of which is the need for a heat transfer system that uses ethylene (or propylene) glycol rather than water. Glycol ice-storage systems presently enjoy a great deal of market popularity because of their simplicity and low installed cost. They can operate a wide variety of CTES technologies and equipment, and these are now discussed

Modular ice storage for glycol systems Modular ice storage tanks can be constructed in many sizes or shapes. Two designs are common: a cylindrical polyethylene tank with circular polyethylene heat exchangers; and a rectangular metal tank with polypropylene heat exchangers. Some modular ice storage tanks are illustrated in Figure 3.38, along with some details of their storage capacities and key physical characteristics. In both modular ice storage designs, the heat exchanger separates the glycol solution from the water in the tank. The water is frozen by circulating –6.5 to –4.5°C glycol through the heat exchanger. The differences in tank geometry and heat exchanger design pose different problems for the design engineer. For example, the shape of circular ice storage tanks allows heat exchangers with fewer circuits of longer length, and permits freezing or melting at lower flow rates and higher temperature differences. Low flowrate freeze cycles enable the designer to better match the capacities of the storage tanks and chiller. Rectangular tank designs, on the other hand, incorporate high flowrate, low pressure-drop heat exchangers that operate with a lower temperature difference during freezing. These characteristics not only place additional design constraints on chiller selection, but require individual flow balancing for each storage tank.

Both modular ice storage tank designs share the advantage of pre-engineering and factory manufacture. Factory design and testing increase the likehood of reliable performance. Piping two or more modular tanks in parallel can increase capacity as required.

Encapsulated ice storage for glycol system The other class of glycol-system storage, encapsulated ice, offers a wide degree of latitude in the design of the ice containment vessel. Various construction materials and geometries can be exploited and designed to conform to the space available and building architecture.

Encapsulated ice designs store the water to be frozen in a number of plastic containers. These containers may be thin and rectangular, spherical or annular. Figure 3.39 illustrates some encapsulated ice storage containers, and provides information on their storage capacities and design-related characteristics. The number of containers or units required for an application depends on their individual storage capacities. For example, as seen in Figure 3.38, one ton-hour of storage can be provided with approximately 20 ice 'trays' or by 70 of the 10 cm diameter spheres, called 'ice balls.' Some commercially available rectangular containers are approximately 3.5×30.5×76.5 cm. Other designs are also available.

Perhaps the greatest advantage of this type of glycol system is the degree of application flexibility it affords the system designer. By selecting or designing a specifically adapted containment vessel, the storage system can be customized to the application. If desirable, it can even be placed below ground.

Encapsulated ice units consist of plastic containers filled with ionized water and an ice nucleating agent. These primary containers are placed in storage tanks, which may be either steel pressure vessels, open concrete tanks, or fiberglass or polyethylene tanks. In tanks with spherical containers, water usually flows vertically through the tank, and in tanks with rectangular containers, water flows horizontally. The type, size and shape of the storage tank is limited only by the need to achieve an even flow of heat transfer fluid between the containers.

A fluid coolant (e.g. 25% ethylene glycol or propylene glycol and 75% water) is cooled to –4.5 to –3.5°C by a liquid chiller, and circulates through the tank and over the outside surface of the plastic containers, causing ice to form inside the containers. The chiller must be capable of operating at this reduced temperature. The plastic containers must be flexible to allow for change of shape during ice formation; the spherical type has preformed dimples in the surface, and the rectangular type is designed for direct flexure of the walls. During discharge, coolant flows either directly to the system load or to a heat exchanger, thereby removing heat from the load and melting the ice within the plastic containers. As the ice melts, the plastic containers return to their original shape.

Ice inventory is measure and controlled by an inventory/expansion tank normally located at the high point in the system and connected directly to the main storage tank. As ice forms, the flexing plastic containers force the surrounding secondary coolant into the inventory tank. The liquid level in the inventory tank may be monitored to account for the ice available at any point during the charge or discharge cycle.

Ice-slurry CTES systems

Ice-slurry is a very versatile cooling medium. The handling characteristics, as well as the cooling capacities, can be matched to suit any application by means of simply adjusting the percentage of ice concentration. At 20–25% ice concentration, ice-slurry flows like conventional chilled water while providing five times the cooling capacity. At 40–50% ice concentration, it demonstrates thick slurry characteristics and at 65–75% ratio, slurry-ice has the consistency of soft ice cream. When slurry-ice is produced in dry form (i.e. 100% ice), it takes the form of non-stick pouring ice crystals which can be directly used whatever the product and process.

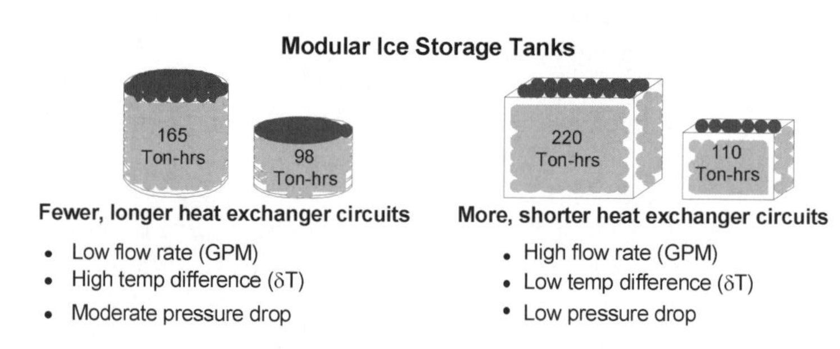

Figure 3.38 Modular ice storage tanks (AEP, 2001).

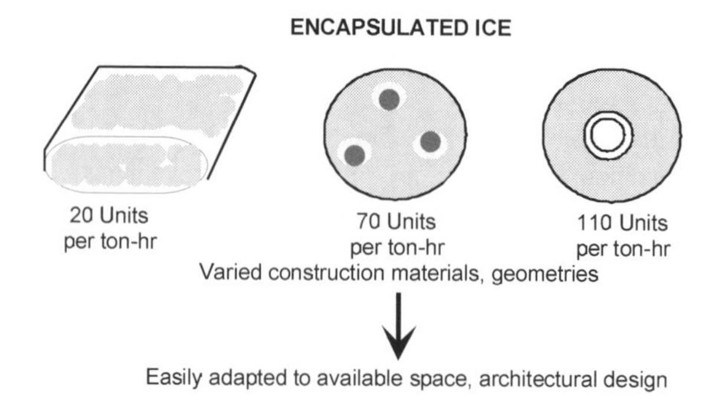

Figure 3.39 Encapsulated ice storage tanks (AEP, 2001).

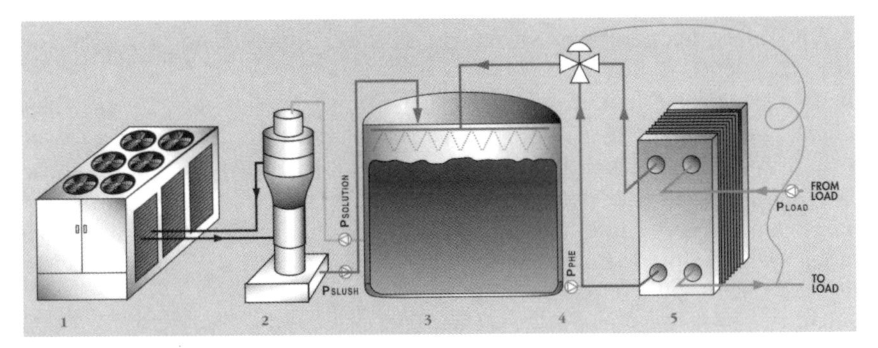

Figure 3.40 A MaximICE ice-slurry system (*Courtesy of Paul Mueller Company*).

Ice-slurry is a crystallized water-based ice solution which can be pumped, and offers a secondary cooling medium for CTES while remaining fluid enough to pump. It flows like conventional chilled water whilst providing 5–6 times higher cooling capacity. The ice-slurry system is a dynamic type ice CTES system which offers the pumpable characteristic advantage over any other type of dynamic systems. Compact equipment design and the pumpable characteristics offer tremendous flexibility for the location of the storage tank(s) and the most economical capacity and duty balancing for any given application. The storage tank can be placed under, beside, inside, or on top of a building, and can be in any shape and size to match the building, and architectural requirements. Multiple small storage tanks can be used instead of a single large static-type ice storage tank.

Figure 3.40 shows a MaximICE ice-slurry CTES system and its components. In the operation of such a system, a compressor/condenser (1) supplies refrigerant to the evaporator. A MaximICE orbital rod evaporator (2) uses a freeze-depressant solution to produce a pumpable ice slurry. Low temperature slurry makes MaximICE ideal for use with low-temperature air systems. An insulated ice storage tank (3) separates ice manufacturing from ice usage. The tank contains a freeze-depressant solution which is converted to an ice slurry in the MaximICE evaporator. The slurry melts as the stored ice absorbs the heat of the cooling load. A load control pump and valve (4) control the supply temperature to the load. A plate heat exchanger (5) separates the storage tank from the cooling load and prevents cross-contamination between the ice-melting loop and the cooling load. It is important to note that the solution can be supplied from the ice storage tank to the load at various temperatures to satisfy specific application needs.

Ice-slurry CTES systems offer a large number of advantages:

- **Higher energy efficiency.** These systems provide higher energy efficiency unlike static ice systems where ice adheres to the heat transfer surface, ice slurry produced by the ice-slurry generator does not adhere to any heat transfer surface, and unlike ice harvesters, defrost is not required to harvest the ice for storage in tanks.
- **Cost-effective tank design.** Ice slurry can be pumped into storage tanks, reducing the need for extra structural support, as required for ice harvesters located above the storage tanks.
- **Compact equipment.** Small evaporator footprint offers space savings in the refrigeration equipment room.

- **Lower supply temperature.** System offers lower supply water temperature versus other ice systems.
- **Flexible ice storage tank design.** Ice slurry can be stored in tanks of any shape. For example, the height of an ice storage tank can be increased, resulting in a reduction of the tank footprint which leads to valuable floor space savings. This is difficult to achieve in static and other dynamic ice storage systems.
- **Maintenance-free ice tank design.** Unlike ice-on-coil systems, ice slurry systems do not require miles of pipe or tubing in the storage tanks. This eliminates the need for repairing leaking pipes or tubing inside the tank.
- **Satisfies large loads.** Large loads for short durations can be met by ice slurry systems due to the quick melting of ice which is achieved by a large area of contact between the warm return solution and stored ice.
- **Ease of modification.** System can be easily adapted to changing needs. For example, facility expansion may occur and facility use and/or utility rates may change.

CYFLIP as a new CTES system

Among the various systems to facilitate night-time power usage, the CHIYODA Corp. has developed the CYFLIP CTES system which reduces 40% of the refrigerator duty of a normal air conditioning system. CYFLIP is a static CTES system which makes ice in a polyethylene tube filled with water. These tubes are placed in a TES vessel where low temperature brine is circulated. CYFLIP has a simple design and low equipment costs. The total investment can be less than for a conventional air conditioning system, since the capacity of the refrigerator becomes smaller. The CYFLIP system is regarded as reliable, because of its ease of operation and durability. The CYFLIP system is illustrated in Figure 3.41, where several operating parameter values are given.

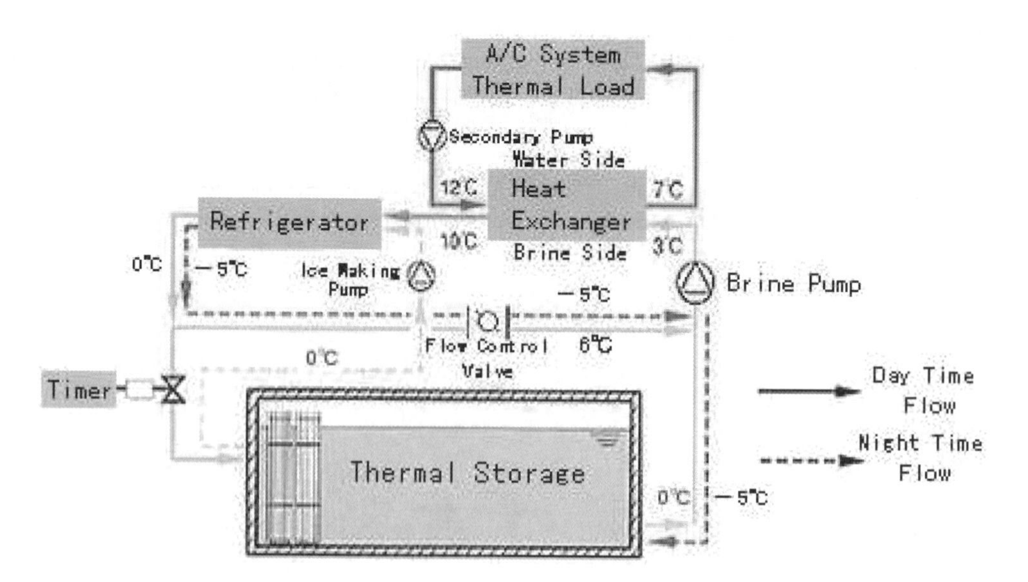

Figure 3.41 The CYFLIP ice CTES (*Courtesy of CHIYODA Corp.*).

Practical example I: Performance curves for an ice TES

In this section, a practical design application for a school is given to point out the design and technical aspects.

In this project, the chiller capacity of 225 tons is the capacity at ice making conditions. At the lower temperatures required to make ice, the chiller is de-rated to 225 tons.

The 288 tons listed in the example is chiller output as supplemental chiller capacity. This means that the chiller is running to supply cooling to the load directly, and not making ice. In this mode, the chiller would be running at standard conditions, at the higher normal air conditioning temperatures. A chiller that is capable of producing 225 tons at ice making conditions would be capable of approximately 320 tons at standard conditions.

Because this is a partial storage example, the chiller supplies cooling directly to the load between the hours of 6:00 am and 1:00 pm (supplemental chiller operation) The thermal storage system then takes over and supplies cooling during the peak electrical period from 1:00 to 8:00 pm (storage discharge-ice melting). Finally, during the off-peak hours, the chiller will run at the reduced capacity of 225 tons to replace the ice burned off during the day. The final hour of the charge is lower as the chiller unloads at the end of the cycle.

During the ice making mode, we have changed the table to provide for a chiller running at 225 tons for 7 hours and at 97 tons for 1 hour. This provides a total ice build of 1672 ton hours. This is the same amount of energy removed from the storage system during storage discharge. This is also known as the rated capacity.

The nominal capacity of the system is 1800 ton hours. Therefore, following the discharge of 1672 ton hours from storage, there is a remaining and unused capacity of 128 ton hours still remaining. This difference between nominal and rated capacity, and the unused ice left in storage, is normal for all thermal storage systems. Indeed, it is impossible to discharge 100% of the ice in storage at useful rates because the discharge rates fall off rapidly as we approach 100% discharge. The facts of life for thermal storage dictate that it is never possible to burn 100% of the ice at rates that are high enough to satisfy a typical load. Unfortunately, there are still manufacturers and designers who insist that thermal storage is 100% efficient. This myth leads them to install smaller and cheaper thermal storage systems wherein the nominal and rated capacities are the same. Such systems run out of ice before the end of the discharge period and the discharge temperatures from storage climb prematurely (Ott, 2001).

Figures 3.42 and 3.43 illustrate performance curves for charging (freezing) and discharging (melting) for the Cryogel ice balls. The details of a CTES using Cryogel ice balls for a typical application (a sample input/output design report), along with a load profile table:

Project: Typical Elementary School

Storage rating:

	Rated capacity:	1,672 Ton-hours
	Nominal capacity:	1,800 Ton-hours

Charge-cycle specifications:

	Entering-fluid temperature:	−5.55°C
	Leaving-fluid temperature:	−1.66°C
	Tank temperature difference:	−13.88°C
	Tank LMTD:	−14.55°C

Chiller capacity:	225 Tons
Charging period:	8 hours
Chiller shut off:	–6.66°C

Discharge-cycle specifications:

Entering-fluid temperature:	12.22°C
Leaving-fluid temperature:	3.33°C
Tank temperature difference:	–8.8°C
Tank LMTD:	–10.94°C
Peak discharge rate:	288 Tons

Load profile:

Hour	Building Load	Supplemental Chiller	Storage Discharge	Storage Ton-hours	Ice Making Charge	Net Storage Inventory
Mid to 1	0	0	0	0	225	1,028.00
1 to 2	0	0	0	0	225	1,253.00
2 to 3	0	0	0	0	225	1,478.00
3 to 4	0	0	0	0	225	1,800.00
4 to 5	0	0	0	0	97	1,800.00
5 to 6	0	0	0	0	0	1,800.00
6 to 7	118.00	118.00	0	0	0	1,800.00
7 to 8	205.00	205.00	0	0	0	1,800.00
8 to 9	257.00	257.00	0	0	0	1,800.00
9 to 10	252.00	252.00	0	0	0	1,800.00
10 to 11	255.00	255.00	0	0	0	1,800.00
11 to 12	279.00	279.00	0	0	0	1,800.00
12 to 1	288.00	288.00	0	0	0	1,800.00
1 to 2	285.00	0	285.00	285.00	0	1,515.00
2 to 3	287.00	0	287.00	572.00	0	1,228.00
3 to 4	262.00	0	262.00	834.00	0	966.00
4 to 5	216.00	0	216.00	1,050.00	0	750.00
5 to 6	208.00	0	208.00	1,258.00	0	542.00
6 to 7	207.00	0	207.00	1,465.00	0	335.00
7 to 8	207.00	0	207.00	1,672.00	0	128.00
8 to 9	0	0	-	1,672.00	0	128.00
9 to 10	0	0	-	1,672.00	245	353.00
10 to 11	0	0	-	1,672.00	245	578.00
11 to Mid	0	0	-	1,672.00	245	803.00
TOTAL	**3,326.00**	**1,654.00**	**1,672.00**		**1,921.00**	

Note: Chiller must run fully loaded at storage entering fluid temperature as specified above. Estimated sizing may change with actual load profile. All units are tons, except the totals across the bottom and the data under the columns 'storage-ton hours' and 'net storage inventory' which are in ton-hours (*Courtesy of CRYOGEL Ice Ball Thermal Storage*).

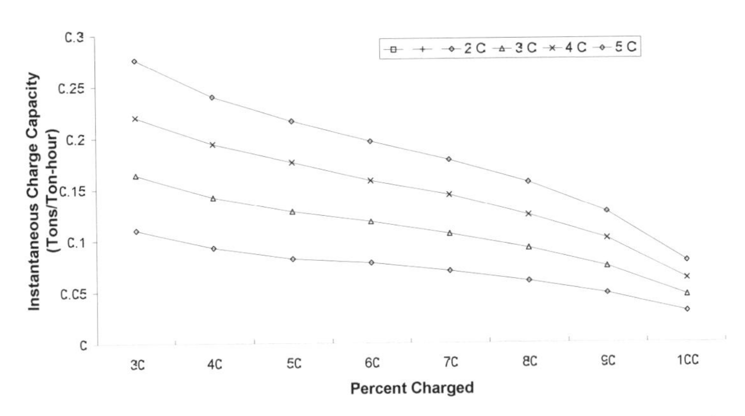

Figure 3.42 Instantaneous charge capacity of Cryogel ice balls (in tons per ton-hour of storage capacity), as a function of storage inlet and outlet LMTDs (*Courtesy of CRYOGEL Ice Ball Thermal Storage*).

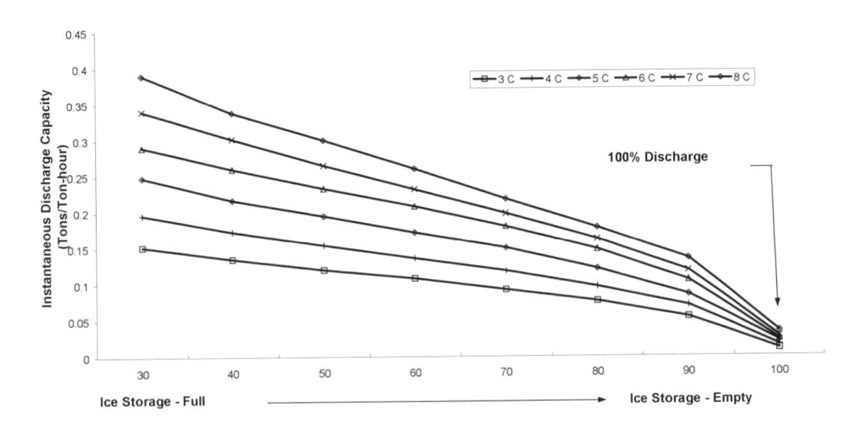

Figure 3.43 Instantaneous discharge capacity of Cryogel ice balls (in tons per ton-hour of storage capacity), as a function of storage inlet and outlet LMTDs (*Courtesy of CRYOGEL Ice Ball Thermal Storage*).

The curves in Figures 3.42 and 3.43 are based on typical heat exchanger sizing methods using Log Mean Temperature Differentials (LMTDs). Indeed, ice CTES systems are like large heat exchangers but the surface area of the thermal storage media changes as ice is frozen or melted. Therefore, a degradation is observed in Figures 3.42 and 3.43 in instantaneous performance as ice melts during discharge. Also observed is a reduction in instantaneous performance as ice gets thicker and impedes the heat transfer process during charge. This behavior is typical of all ice-based CTES systems.

The LMTD is calculated using the storage tank inlet (T_i) the storage outlet (T_o) and the ice temperature (T_c) as follows:

$$LMTD = \frac{[(T_i - T_c) - (T_o - T_c)]}{\ln[(T_i - T_c)/(T_o - T_c)]}$$

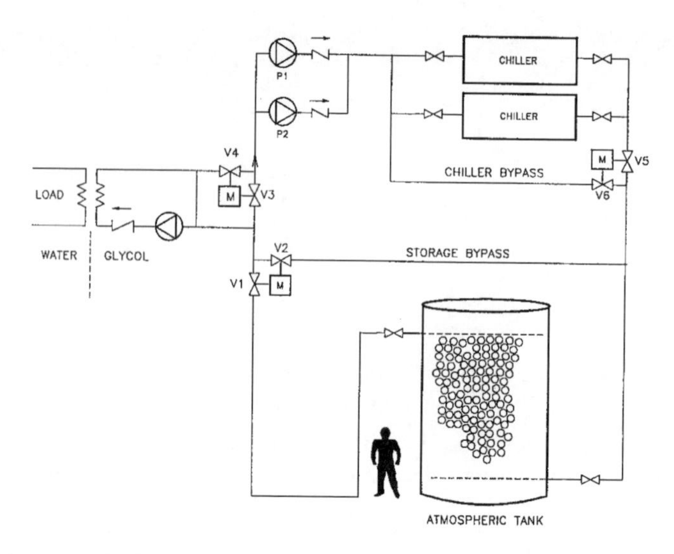

Figure 3.44 An atmospheric ice ball, single tank (series) TES system (*Courtesy of CRYOGEL Ice Ball Thermal Storage*).

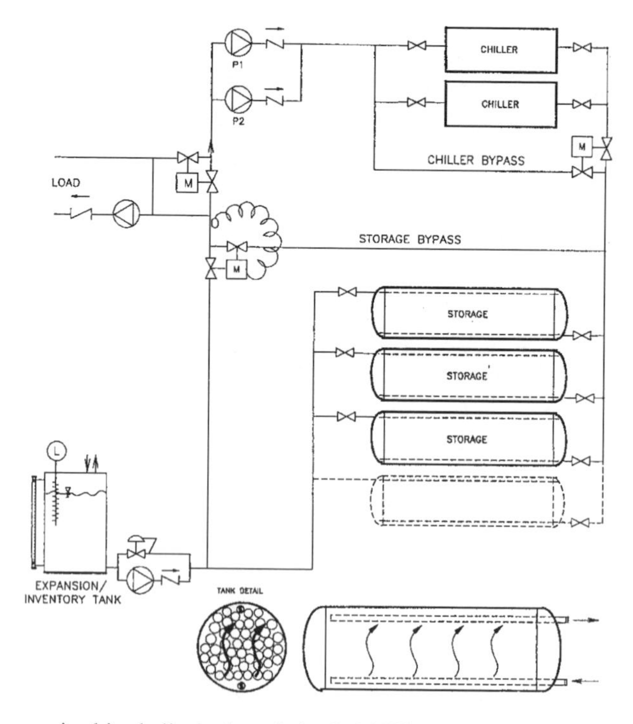

Figure 3.45 A pressurized ice ball, single tank (series) TES system (*Courtesy of CRYOGEL Ice Ball Thermal Storage*).

Using LMTD together with the performance curves enclosed allows one to predict instantaneous capacity at any point during the discharge of the storage system. This assessment is critical to be certain that a system has enough ice to meet the most demanding periods of each daily cycle. In a similar manner, although less critical, one can determine proper chiller sizing to match the ability of the ice balls to release heat during the charge process. Although the CTES market is generally not interested in this level of detail, and focuses on the practical output of the sizing methodology, this information helps understand some of the design details for ice CTES systems (e.g. Figures 3.44 and 3.45).

Practical example II: Design and operational loads

The following design example was taken from Baltimore Aircoil Company (BAC, 1995). This is done using ice chiller Thermal Storage Unit Selection (TSU-M) software. The selected TES, substance, and chiller are:

- BAC thermal storage: 1 TSU-476M 1 TSU-761M
- Glycol charge: 4776 liters of 25% ethylene glycol
- Specified chiller: Model A-chiller 12 at approximately 514 kW (for details, see Table 3.11).

Before going into detail, it is better here to introduce some definitions which are crucial in design calculations, including:

- **LBT:** chiller Leaving Brine Temperature.
- **EDB:** dry-Bulb Air Temperature (for air-cooled chillers used in ice building).
- **CWT:** Condenser Water Temperature (for water-cooled chillers used in ice building).
- **Parasitic load(s):** the heat input to the thermal storage device from other devices in the TES system. Internal melt ice-on-coil systems do not have such parasitic loads. However, for external melt ice-on-coil systems, the parasitic auxiliary device is the air pump used for agitation of the storage media. In this case, the parasitic heat gain must include the heat input to the thermal storage device as the air is cooled from its source temperature to the temperature of the storage media.
- **Operational modes:** status of the thermal devices in the cooling system regarding meeting the required load. Modes include: cooling with chillers (compressor) only, cooling with TES device + chiller (in partial storage systems), building ice without cooling, building ice with cooling, cooling with ice only (full storage systems), and finally, no chiller + no TES (in case of cooling load = 0).
- **Chiller load:** cooling load supplied by the chiller compressor.
- **Base load:** load to be handled completely by the chillers. Selection of base load chillers should be so that the chillers are running at maximum efficiency.
- **Total load:** total thermal load of the building / process to be delivered by the cooling system (chillers and ice thermal energy storage system).
- **Heat rejection:** heat dissipated by through the condenser of the chiller to achieve the required chiller cooling required from the chiller.
- **Charge and discharge rates:**
  Charge Rate. The rate (typically expressed in tons kW) at which energy (heat) is removed from the storage device during the charge period.

Discharge Rate. The rate (typically expressed in tons kW) at which energy (heat) is added to the storage device during the discharge period.

- **Ambient loss:** sum of conduction, convection and radiation heat quantity escaping from the thermal energy storage tank to the surroundings.
- **Net storage:** actual quantity of ice present in the ice TES tank at a particular time.
- **Flow rate:** brine flow rate within the ice TES system.
- **Inlet temperature:**
  Inlet Temperature for the Thermal Storage Device. Temperature of the brine entering the ice thermal energy storage tank.
  Inlet Temperature for the Thermal Storage System. Temperature of the brine entering the ice thermal energy storage system (chiller and tank).
- **Exit temperature:**
  Exit Temperature for the Thermal Storage Device. Temperature of the brine leaving the ice thermal energy storage tank.
  Exit Temperature for the Thermal Storage System. Temperature of the brine leaving the ice thermal energy storage system (chiller and tank).
- **Pressure drop:** pressure loss due to brine flow through the coil of the ice thermal energy storage system.
- **Chiller peak load:** maximum load delivered by the chiller at certain ambient and brine temperatures.
- **Chiller average load:** average load delivered by the chiller at certain ambient and brine temperatures.
- **COP:** Coefficient of Performance.

Based on the load profile and operating strategy, the selection will deliver a peak of 1000 kW at design conditions of 38 L/s of 25% by weight ethylene glycol from 10.2°C to 3.4°C. The maximum presure drop across the storage unit is 79.1 kPa and 75.3 kPa in the ice build and melt modes, respectively. Tables 3.12–3.14 presents summary sheets for the loads, flow, temperature and pressure parameters, and energy demand and use.

3.8.10 Ice forming

The ice making equipment is expected to form the ice and deliver it to the storage tank accordingly. Three main systems are available for forming ice:

- direct expansion, which forms ice by directly feeding refrigerant into the ice-making heat exchanger,
- the brine system, which exchanges heat between refrigerant and brine in the heat pump evaporator to produce low-temperature liquid brine which is fed to the ice forming coil, and
- the heat pipe system, which exchanges heat between the refrigerant evaporator and water, and forms ice on the heat pipe surface.

In the USA, direct expansion is used in the majority of large-scale systems, while the brine system is mainly used for medium- and small-scale equipment. In Japan, the brine system is favored regardless of building size.

Table 3.11 Details for chiller A (38 L/s).

LBT (°C)	EDB (°C)	Capacity (kW)	Power (kW)	COP
–5.6	23.9	374	154	2.43
–5.6	35.0	331	166	2.00
6.7	23.9	568	180	3.15
6.7	35.0	513	199	2.57

(*Courtesy of Baltimore Aircoil International N.V.*).

Table 3.12 Design and operational loads and TES system parameters.

Hours	Mode	EDB (°C)	Chiller Load (kW)	Base load (kW)	Total load (kW)	Chiller load (kW)	Heat rejection (kW)	Charge rate (kW)	Discharge rate (kW)	Ambient loss (kW)	Net storage (kW-h)
0:00	B	38	0	0	0	347	520	347	—	1.8	2081
1:00	B	38	0	0	0	345	518	345	—	1.8	2426
2:00	B	37	0	0	0	346	517	346	—	1.8	2769
3:00	B	36	0	0	0	346	517	346	—	1.8	3113
4:00	B	36	0	0	0	345	516	345	—	1.8	3457
5:00	B	36	0	0	0	344	514	344	—	1.8	3801
6:00	B	36	0	0	0	215	375	215	—	1.8	4143
6:37	S	36	0	0	0	—	—	—	—	1.8	4357
7:00	S	37	0	0	0	—	—	—	—	1.8	4357
8:00	I	38	800	0	800	464	661	—	336	1.8	4354
9:00	I	39	840	0	840	460	661	—	380	1.8	4016
10:00	I	41	860	0	860	466	673	—	394	1.8	3635
11:00	I	43	890	0	890	465	675	—	425	1.8	3239
12:00	I	45	950	0	950	463	679	—	487	1.8	2812
13:00	I	47	980	0	980	459	679	—	521	1.8	2323
14:00	I	47	1000	0	1000	458	679	—	542	1.8	1800
15:00	I	48	990	0	990	455	677	—	535	1.8	1256
16:00	I	47	880	0	880	444	662	—	436	1.8	719
17:00	I	47	660	0	660	414	623	—	246	1.8	281
18:00	B	45	0	0	0	339	526	339	—	1.8	33
19:00	B	44	0	0	0	340	524	340	—	1.8	371
20:00	B	42	0	0	0	342	523	342	—	1.8	709
21:00	B	41	0	0	0	345	523	345	—	1.8	1049
22:00	B	39	0	0	0	346	522	346	—	1.8	1392
23:00	B	38	0	0	0	347	522	347	—	1.8	1736

Since storage device(s) is passive, there are no parasitic loads.
Mode: Mode of operation. B: Ice build. S: Standby. I: Cooling (ice with compressor). C: Cooling (compressor only) (*Courtesy of Baltimore Aircoil International N.V.*).

Table 3.13 Flow, temperature and pressure parameters.

Hours	Mode	Charge rate (kW)	Discharge rate (kW)	Flow rate (L/s)	Inlet temp. (°C)	Exit temp. (°C)	Pressure drop (kPa)	Total load (kW)	Flow rate (L/s)	Inlet temp. (°C)	Exit temp. (°C)	Net storage (kW-h)
0:00	B	347	—	38.0	-3.9	-1.5	78.6	—	0	—	—	2081
1:00	B	345	—	38.0	-4.0	-1.7	78.6	—	0	—	—	2426
2:00	B	346	—	38.0	-4.2	-1.8	78.6	—	0	—	—	2769
3:00	B	346	—	38.0	-4.3	-2.0	78.6	—	0	—	—	3113
4:00	B	345	—	38.0	-4.4	-2.1	78.6	—	0	—	—	3457
5:00	B	344	—	38.0	-4.5	-2.2	78.6	—	0	—	—	3801
6:00	B	215	—	38.0	-4.6	-2.2	78.6	—	0	—	—	4143
6:37	S	—	—	—	—	—	—	—	—	—	—	4357
7:00	S	—	—	—	—	—	—	—	—	—	—	4357
8:00	I	—	336	38.0	4.3	2.0	75.2	800	38.0	7.4	2.0	4354
9:00	I	—	380	38.0	4.6	2.0	75.1	840	38.0	7.7	2.0	4016
10:00	I	—	394	38.0	5.5	2.8	74.8	860	38.0	8.6	2.8	3635
11:00	I	—	425	38.0	5.9	3.0	74.7	890	38.0	9.1	3.0	3239
12:00	I	—	487	38.0	6.5	3.2	74.5	950	38.0	9.7	3.2	2812
13:00	I	—	521	38.0	6.8	3.3	74.4	980	38.0	9.9	3.3	2323
14:00	I	—	542	38.0	7.1	3.4	74.4	1000	38.0	10.2	3.4	1800
15:00	I	—	535	38.0	7.1	3.4	74.4	990	38.0	10.1	3.4	1256
16:00	I	—	436	38.0	5.9	3.0	74.7	880	38.0	8.9	3.0	719
17:00	I	—	246	38.0	3.7	2.0	75.3	660	38.0	6.5	2.0	281
18:00	B	339	—	38.0	-2.3	0.0	77.6	—	0	—	—	33
19:00	B	340	—	38.0	-2.6	-0.3	77.8	—	0	—	—	371
20:00	B	342	—	38.0	-3.0	-0.6	78.0	—	0	—	—	709
21:00	B	345	—	38.0	-3.2	-0.9	78.2	—	0	—	—	1049
22:00	B	346	—	38.0	-3.5	-1.2	78.3	—	0	—	—	1392
23:00	B	347	—	38.0	-3.7	-1.4	78.5	—	0	—	—	1736

Since storage device(s) is passive, there are no parasitic loads.
Mode: Mode of operation. B: Ice build. S: Standby. I: Cooling (ice with compressor). C: Cooling (compressor only) (*Courtesy of Baltimore Aircoil International N.V.*).

Compared to the other types, brine systems require more space and special materials, and have lower COPs because of the additional temperature difference between refrigerant and brine, and the power consumption by the brine pump. However, they do have the advantage that the tank and heat pump can be located separately. Direct expansion systems require a much larger refrigerant charge, are more expensive, and are difficult to maintain.

3.8.11 Ice thickness controls

The oldest type of ice storage is the refrigerant-fed ice builder, which consists of refrigerant coils inside a storage tank filled with water. The tank water freezes on the outside of the chiller evaporator coils to a thickness of up to 0.065 m. Ice is melted from the outside of the formation (hence the term *external melt*) by circulating water through the tank, causing it to become chilled. Air bubbled through the tank agitates the water to promote uniform ice buildup and melting.

Table 3.14 Energy demand and usage

Hours	Mode	Chiller used	Loading (%)	Chiller peak load (kW)	Chiller average load (kW)	Electric rate
0:00	B	A	100	173	173	1
1:00	B	A	100	173	173	1
2:00	B	A	100	172	172	1
3:00	B	A	100	171	171	1
4:00	B	A	100	170	170	1
5:00	B	A	100	170	170	1
6:00	B	A	100	170	104	1
6:37	S	—	—	—	—	—
7:00	S	—	—	—	—	—
8:00	I	A	100	197	197	1
9:00	I	A	100	201	201	1
10:00	I	A	100	206	206	1
11:00	I	A	100	210	210	1
12:00	I	A	100	216	216	1
13:00	I	A	100	220	220	1
14:00	I	A	100	221	221	1
15:00	I	A	100	222	222	1
16:00	I	A	100	218	218	1
17:00	I	A	100	209	209	1
18:00	B	A	100	186	186	1
19:00	B	A	100	184	184	1
20:00	B	A	100	181	181	1
21:00	B	A	100	178	178	1
22:00	B	A	100	176	176	1
23:00	B	A	100	174	174	1

Electric rate structure: 1: Peak demand period, 2: Shoulder peak demand, 3: Off peak period. Mode: Mode of operation (*Courtesy of Baltimore Aircoil International N.V.*).

Instead of refrigerant, a secondary coolant (e.g. 25% ethylene glycol and 75% water) can be pumped through the coils inside the storage tank. The coolant has the advantage of greatly decreasing the refrigerant inventory. However, a refrigerant-to-coolant heat exchanger between the refrigerant and the storage tank is required in such instances.

Some major concerns specific to the control of ice-on-coil systems are:

- limiting ice thickness (and thus excess compressor energy) during the build cycle, and
- minimizing the bridging of ice between individual tubes in the ice bank.

Bridging is avoided because it restricts the free circulation of water during the discharge cycle. While not physically damaging to the tank, this blockage reduces performance, raising water exit temperature due to reduced heat transfer surface.

Regardless of the refrigeration method (direct-expansion or secondary coolant), the compressor is controlled by (i) a timer or controller which restricts operation to periods dictated by the utility rate structure, and (ii) an ice-thickness over-ride control, which stops the compressor(s) at a predetermined ice thickness. At least one ice-thickness controller should be installed per ice bank. If there are multiple refrigeration circuits per ice bank, one ice-thickness control per circuit should be installed. Placement of the ice-thickness controller(s) should be determined by the ice-bank manufacturer, based on circuit geometry and flow pressure drop, to minimize bridging.

Ice-thickness controls are either mechanically or electrically operated. Mechanical controls typically consist of a fluid-filled probe positioned at a desired distance from the coil. As ice builds, it encapsulates the probe, causing the fluid to freeze and apply pressure. The pressure signal controls the refrigeration system via a pneumatic-electric switch. Electric controls operate by sensing difference between electrical conductivities of ice and water. Multiple probes are installed at the desired thickness, and the change in current flow between probes provides a control signal. Consistent water treatment is essential to maintaining constant conductivity and thus accurate control.

An Ice Logic Ice Quantity (ILIQ) controller (Figure 3.46) is commonly used, allowing accurate setting of the minimum and maximum ice quantity as a function of the expected cooling load. This enables maximum control, system efficiency and operating flexibility of the cooling equipment. The setting can easily be done manually on the ILIQ controller, or automatic (remote) control can be used. The different ice quantities are measured and displayed: 0, 20, 40, 60, 80, 100% of nominal storage capacity. The ILIQ controller has the necessary output contacts which can be connected to a building management control system to allow automatic control of the cooling system. As an option, a 4–20 mA output signal is possible. The ILIQ controller contains the necessary logic to prevent unwanted chiller cycling after the desired ice build is complete. Between the control box and the sensors, the intermediate cabling has steel reinforcement and a PVC cover. The cables possess a screening against interference. Ice TES units can be designed for two different TES system concepts: external or internal melt (the ILIQ controller is the same for both except that for each concept a specific sensor (Figure 3.47) was designed).

- **Ice thickness sensor for external melt systems**. A series of accurately positioned electrodes detect the ice thickness on the coil tube. The measurement is based on difference in electrical conductivity between ice and water. Combined with this sensor the ILIQ controller limits the maximum ice thickness to the desired level.

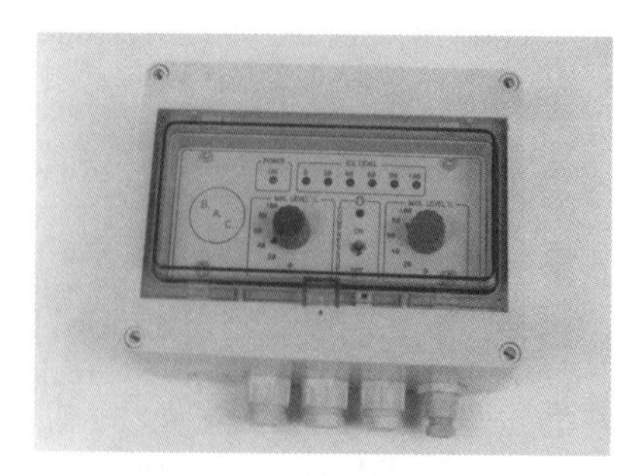

Figure 3.46 ILIQ controller for manual control (*Courtesy of Baltimore Aircoil International N.V.*).

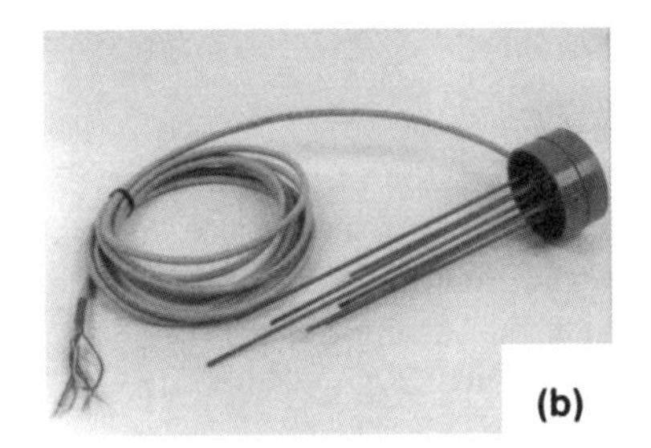

Figure 3.47 (a) Ice thickness sensor for external melt systems, and (b) water level sensor for internal melt systems (*Courtesy of Baltimore Aircoil International N.V.*).

- **Water level sensor for internal melt systems**. The water level in the sensor is proportional to the ice quantity in the ice chiller tank. A series of specially designed probes will measure the water level in the water level sensor. The measurement is based on the difference of conductivity between air and water. Combined with the ILIQ controller, output signals to control the operation of the ice chiller unit are also available.

Because energy use is related to ice thickness on the coil, a partial-load ice inventory management system is often considered. This system keeps the ice inventory at the minimum level needed to supply immediate future cooling needs, rather than topping off the inventory after each discharge cycle. It also helps prevent bridging by ensuring that the tank is completely discharged at regular intervals, thereby allowing ice to build evenly.

The most common method of measuring ice inventory is based on the fact that ice has a greater volume than water. Thus, a sensed change in water level indicates a change in the amount of stored ice. Because water increases 9% in volume when it changes to ice, the water level varies directly with the amount of ice in the tank as long as all of the ice remains submerged. This water displaced by the ice must not be frozen, or it freezes above the original water level. Therefore, no heat exchange surface area can be above the original water level. The change in water level in the tank due to freezing and thawing is typically 0.153 m. This change can be measured with either a pressure gage or a standard electrical transducer. For projects with multiple tanks, a reverse-return piping system ensures uniform flow through all the tanks, so measuring the level in one tank is sufficient to determine the overall proportion of ice remaining.

3.8.12 Technical and design aspects of CTES

While conventional cooling systems operate chillers to satisfy predictable, instantaneous cooling loads, CTES systems operate chillers to produce cold during off-peak hours. Conventional chillers typically operate 8–12 hours a day to meet cooling requirements, representing a large portion of a building's daytime electrical demand. Shifting electrical use through thermal energy storage reduces high peak demand charges. Utilities benefit from lower peak demand and gain off-peak load, often allowing a shift from less efficient to more efficient equipment. Thus utilities increase plant efficiencies and delay need for future expansion.

Deciding on a specific type of CTES system is dependent on many factors. The first issue is usually whether CTES is appropriate for a facility. If so, the next question usually considers the most appropriate type. System options often include ice or water storage, but the most appropriate choice depends on project characteristics. Some other key factors that affect system choices include building type, utility rate structure and energy-conservation grants and incentives.

Building type, not building size, normally determines whether CTES is cost effective. A building used 24 hours a day with a fairly even load demand is not a good candidate. Office buildings or schools that close at night, or buildings with short, intense power demands, such as convention centers, sports facilities and religious buildings are often good candidates. In general, the higher the peak load, the more likely a candidate the facility is to benefit from CTES. Local energy rates also play a key role in determining CTES cost effectiveness. Most utilities charge for peak demand and have significantly higher rates for electricity used during peak, daytime hours. Therefore, careful analysis of rates and load distribution are essential when considering CTES.

Client attitudes can also affect technology choices. Although CTES systems are not entirely new, some clients may not be familiar and therefore comfortable with this technology, or may not have staff willing to take on different operating and maintenance tasks. In some applications, system redundancy is required (at least for an initial period) to overcome apprehension about CTES.

Considering and selecting a CTES system for a client or facility, therefore, an engineer should have or obtain a thorough understanding of the following design issues (Maust, 1993):

- Is CTES economically and physically feasible?
- Is there room on site for a chilled-water or ice or other CTES system?
- Does the local utility have peak-demand charges, low off-peak rates or rebates for avoiding peak demand?
- Is the building-load profile flat or is there a strong peak?
- Will the client be comfortable maintaining and operating a CTES?

CTES is often best considered relatively later in the design process. While CTES must be integrated with many other aspects of building design (e.g. heating or cooling loads, mass, layout and HVAC equipment) other efficiency measurements may level cooling loads and/or heating needs sufficiently that otherwise attractive storage schemes no longer make economic sense.

Most efficiency and many load management options cost less than CTES in retrofits, and should therefore be implemented first where appropriate. In new buildings, however, installation of an ice storage or a water tank can be synergistic with other design choices that reduce both energy consumption and peak load. Such benefits are especially achievable using fully integrated and computer-aided design.

For CTES, full storage is sometimes preferable to partial storage. Similarly, ice storage sometimes has advantages over chilled water storage (when equally well designed). Skill and experience are needed to ensure appropriate implementation.

Significant recent progress has been made in improving the cost-effectiveness and reliability of CTES equipment, especially for ice harvesters and stratified storages.

Ice storage uses more energy because of the lower temperatures needed compared to chilled-water systems and higher heat leaks through the tank insulation. But ice storage uses less energy for several other reasons: the chiller always runs in its design efficiency or not at all (rather than varying continually in response to changing building loads); the ambient dry and wet bulb temperatures in contact with the condenser are lower at night; and the very cold melt water from the ice greatly reduces the energy needed for both chilled-water pumping and air handling. The net effect is usually a net decrease in electrical energy consumption with ice CTES as low as a few percent to a few tens of percent.

The economics of CTES in buildings of all types and sizes frequently hinges on two non-energy factors: (i) the expected political and economic stability of utilities time-of-use rates; and (ii) demand charges, and the physical space available and structural load tolerance for the storage tank.

3.8.13 Selection aspects of CTES

Selecting a CTES for a particular application is often difficult. International standards have not yet been fully developed to provide a basis for the rating and performance testing of these systems. In general, however, it is clear that a CTES system must cool a building in the same way as the chiller it replaces or supplements.

The selection of a CTES system should account for the load, i.e. the discharge rate, the energy discharged at all times during the discharge cycle and required storage inlet and outlet temperatures. Information on temperatures, energy stored, and energy discharged vs. discharge cycle time describes the performance of a CTES system well. With this information, a performance data schedule should be established to fully describe the system needs (Bishop, 1992). Clearly, rating a CTES system on only energy storage capacity does not accurately describe the system needs. The system must be capable of providing the same performance as the conventional chiller that it supplements or replaces. In particular, CTES must provide stored energy when the building requires it. Performance specifications that recognize this complexity can help in achieving successful CTES projects.

The selection among available system options depends upon site-specific factors. Some of the factors which can enter into this decision are as follows (Wylie, 1990):

- *Space availability*: ice storage systems require approximately one-third to a quarter of the space of chilled-water systems. PCMs also have similar storage advantages over chilled-water systems.
- *Efficiency*: water chillers use less energy than ice-makers due to higher evaporating temperatures. PCMs require slightly less energy than chilled-water systems.
- *Chilled water temperatures*: ice storage systems can supply lower water temperatures, and systems can be designed for larger supply/return temperature differences, resulting in lower flow rates and pumping costs and smaller central plants.
- *Refrigeration compressor size*: reciprocating compressors associated with ice storage are size limited. The use of multiple units in large plants is usually uneconomical, requiring the use of screw-type compressors instead.
- *Maintenance costs*: generally, chilled-water storage tanks have lower maintenance costs. Maintenance personnel are usually more familiar with chilled-water systems.

- *Experience of contractor and operator*: contractors and building operation engineers are usually unfamiliar with the large, built-up, field-erected direct-expansion or flooded coil refrigeration systems required for ice storage.

3.8.14 Cold air distribution in CTES

Reducing the temperature of the distribution air in an air conditioning system is attractive because smaller air-handling units, ducts, pumps, and piping can be used, resulting in lower initial costs. In addition, the reduced ceiling space required for ductwork can significantly reduce building height, particularly in high-rise construction. These cost reductions can make thermal storage systems competitive with non-storage systems on an initial cost basis.

The optimum supply air temperature is usually determined through an analysis of initial and operating costs for various design options. Depending on the load, the additional latent energy removed at the lower discharge-air temperature may be offset by the reduction in fan energy associated with the lower air flow rate.

The minimum achievable supply air temperature is determined by the chilled-water temperature and the temperature rise between the cooling plant and the terminal units. With some ice storage systems, the fluid temperature may rise during discharge; therefore the supply temperatures normally achievable with various types of ice storage plants should be carefully investigated with the equipment supplier.

A heat exchanger is sometimes required with storage tanks that operate at atmospheric pressure or between a secondary coolant and a chilled-water system. The rise in chilled water temperature between the cooling-plant discharge and the chilled-water coil depends on the length of piping and the amount of insulation. A smaller temperature difference can be achieved with more cooling-coil rows or a larger surface area on the cooling coil, but extra heat transfer surface is often uneconomical.

A blow-through configuration provides the lowest supply air temperature and the minimum supply air volume. The lowest temperature rise achievable with a draw-through configuration is 2–3°C because heat from the fan is added to the air. A draw-through configuration should be used if space for flow straightening between the fan and coil is limited. A blow-through unit should not be used with a lined duct because air with high relative humidity enters the duct.

Face velocity determines the size of the coil for a given supply air volume, and the coil size determines the size of the air-handling unit. A lower face velocity generates a lower supply air temperature, whereas a higher face velocity results in smaller equipment and lower initial costs. The face velocity is limited by moisture carry-over from the coil. The face velocity for cold air distribution systems is usually 1.5–2.5 m/s. Cold primary air can be tempered with room air or plenum return air by using fan-powered mixing boxes or induction boxes. The primary air should be tempered before it is supplied to the space. The energy use of fan-powered mixing boxes is significant, and negates the savings from downsizing central supply fans.

Diffusers designed for cold air distribution can provide supply air directly to the space without causing drafts, thereby eliminating the need for fan-powered boxes. If the supply air flow rate to occupied spaces is expected to be below 0.02 m^3/s per m^2, fan-powered or induction boxes should be used to boost the air circulation rate. At supply air rates of 0.02

to 0.03 m^3/s per m^2, a diffuser with a high ratio of induced room air to supply air should be used to ensure adequate dispersion of ventilation air throughout the space. A diffuser that relies on turbulent mixing rather than induction to temper the primary air may not be effective at these flow rates.

Cold air distribution systems normally maintain space humidity between 30% and 45% as opposed to the 50–60% generally maintained by other systems. At this lower humidity level, equivalent comfort conditions are provided at higher dry-bulb temperatures. The increased dry-bulb set point generally results in decreased energy consumption.

The surfaces of any equipment that may be cooled below the ambient dew point, including air-handling units, ducts and terminal boxes, need to be insulated. Any vapor-barrier penetrations should be sealed to prevent migration of moisture into the insulation. Prefabricated, insulated round ducts are normally insulated externally at joints where internal insulation is not continuous. If ducts are internally insulated, access doors also need to be insulated.

Duct leakage is undesirable because it represents cooling capacity that is not delivered to the conditioned space. In coldair distribution systems, leaking air can cool nearby surfaces to the point at which condensation forms. Designers should specify acceptable methods of sealing ducts and air-handling units, and establish allowable leakage rates and test procedures. These specifications must be checked through on-site supervision and inspection during construction.

A thorough commissioning process is important for the optimal operation of any large space conditioning system, particularly thermal storage and cold air distribution systems. Reductions in initial and operating costs are major features for cold-air distribution, but successful performance can be compromised by reducing initial costs by avoiding commissioning. In fact, comprehensive commissioning can decrease costs by reducing future system malfunctions and troubleshooting expenses, and provide increased value by ensuring optimal system operation.

Advantages of cold-air distribution and TES

- **Demand reduction.** Cooling load is often the largest contributor to the peak electrical demand in a building. Since many electric utilities experience their peak electrical demands in the summer months, attributable mainly to building cooling loads, electric utilities often promote the use of CTES systems to shift the electrical demand in buildings from peak to off-peak periods. Continually increasing peak demands from customers require the utilities to either build additional generating capacity or rely on less efficient peaking plants. With higher demand charges and seemingly high 'ratchet clauses' associated with electric utility rates, building owners can find CTES options attractive in retrofit situations. This advantage is especially significant when incentive or rebate programs are offered towards CTES.
- **Lower capital costs.** Thermal storage and cold air distribution systems require considerably less volume to meet a cooling load relative to conventional systems that use 12.7°C supply air. Thus, the mechanical system including chillers, air handling units, pumps, fans, fan motors and ductwork can be downsized and still satisfy the same cooling load as a conventional system. Reduced equipment size results in significant capital cost savings.

- **Lower operating costs.** The reduced size of the mechanical equipment associated with a thermal storage and cold air distribution system leads to significant savings in the operating costs associated with running the system. Because the chillers, fans and pumps used in the system consume less electricity than their counterparts in conventional systems, the thermal storage and cold air distribution systems consume less energy over the operating life of the equipment, and thus provide additional operating cost savings to building owners.
- **Improved comfort.** Thermal storage and cold air distribution systems typically provide 5.5–7.5°C supply air into the space for comfort cooling. This low temperature supply air improves comfort levels by lowering the relative humidity in the occupied areas. Typically, the dry bulb temperature is 5.5°C off the coil, and results in a relative humidity of about 36% in the occupied areas, compared to about 50% for a conventional system. This reduction in relative humidity along with room conditions of 25.5°C (typical for low temperature primary air systems) result in a cool feeling in the occupied zone. Also, the smaller air handling equipment and insulated ductwork reduce transmission of noise to the occupied zones, and thus increase the comfort.
- **Increased usable space.** Typically, in a retrofit project, the architectural impact resulting from a modification to a conventional HVAC system is significant. The increased cooling load associated with increased occupancy loads or computer and equipment loads require additional chiller capacity, air-handling capacity, and ductwork (or a combination of these). A low temperature air distribution system allows for small equipment and use of existing vertical shafts to carry low temperature air to occupied zones, thereby reducing major architectural impact due to modifications associated with the HVAC system. The smaller mechanical equipment frees up some existing space in the mechanical room for other systems or purposes.
- **Increased leasability and marketability.** Reductions in the demand charges and energy costs associated with operating a thermal storage and cold air distribution system may attract tenants because these savings can be passed on to them through leasing-cost reductions or lower utility bills. Similarly, when the owner of a building sells the property, the CTES increases the marketability of the building.
- **Minimum disruption.** Because thermal storage and cold air distribution systems can utilize existing ductwork in the building, this option minimizes work disruption, inconvenience to building occupants and, sometimes, the costs associated with occupant relocations.

Disadvantages of cold air distribution and TES

- **Condensation.** The main difficulty in using thermal storage and cold air distribution is the condensation that may occur because of the low temperature air passing through the ductwork, causing ceiling damage and mold. Properly insulated and sealed ductwork, sealed plenum spaces, and special collars at the duct connection to the Variable Air Volume (VAV) boxes (if used) or diffusers can prevent condensation, and ensure proper operation of the cold air distribution system.
- **Insufficient air.** The reduction in volume of air supplied to the occupied space may cause feelings of stuffiness and stagnation in the room. At the present time, building owners/designers have two alternatives to address this problem. First, fan-powered

VAV boxes can be used in conjunction with the low-temperature air to mix proper amounts of ceiling plenum (return) air with low temperature supply air to produce 12.7°C and to maintain minimum air circulation in the room. Secondly, low temperature diffusers that are offered by some leading manufacturers can be used. These low temperature diffusers have unique thermal characteristics that can prevent the condensation problems mentioned above, and provide high induction rates to maintain room air circulation. The noise levels produced by such high-induction unit diffusers are within acceptable limits.

- **Dumping of air.** Because of the decrease in supply air temperature and the low buoyancy of cold air jet, air from the outlets may be dumped into the occupied zone causing occupant discomfort due to concentrations of cold air. Dumping is simply the discharge of an air jet with little velocity. This problem can be addressed by using the two methods discussed above (fan-powered VAV boxes or high-induction diffusers).

3.8.15 Potential benefits of CTES

During the past two decades, CTES technology has matured and is now accepted by many as a proven energy-conservation technology. However, the predicted payback period of a potential cool storage installation is often not sufficiently attractive to give it priority over other technology options. This determination often is reached because full advantage is not made of the many potential benefits of CTES, or because the CTES sizing is not optimized.

Several steps can be taken to optimize the payback period of cold TES systems follow. In general, cool storage should be integrated closely with the overall building and its energy systems to take full advantage of its potential benefits. These benefits depend on whether the application is for a new or existing facility.

For new facilities, the potential benefits include:

- use of low-temperature chilled-water distribution to reduce pipe and pump sizes and operating costs,
- use of low temperature air distribution to reduce duct and fan sizes and operating costs,
- use of smaller chillers and electrical systems to reduce initial costs, and
- a gain in usable building space due to less space being required for the mechanical system components.

For existing facilities, the potential benefits which should be evaluated include:

- the possible advantages of modifying the existing chillers to make ice versus the purchase of a new machine,
- the use of spare chiller capacity in a chilled-water CTES system,
- the use of cold storage to gain more cooling capacity in situations where the chiller and electrical service capacity are fully utilized, and
- the possibility of using low temperature air to advantage, where practical.

For both new and existing facilities, the sizing of the cool storage system should be optimized, as opposed to the typical process of considering full storage and one or two levels of partial storage versus a conventional system. A practical method to assist in determining the optimum system size should be developed. Also, the value should be accounted for of the gain in usable building space due to less space being required for mechanical system components when cool TES is used.

Most CTES systems reduce electricity demand during peak times by reducing (or eliminating) the need to run the air conditioning compressor during the day when electricity costs are highest. Thus, buildings can be cooled effectively during the day, while taking advantage of lower off-peak electricity rates. The chilled medium remains in storage until needed, typically during the day when the building is occupied. Then the water or ice is used to cool the building. Cold air distribution can be used with CTES to allow the user to take advantage of the colder temperatures supplied by some CTES systems. With cold air distribution, the size of ductwork can be reduced, lowering construction costs.

Consequently, a CTES system can be of benefit to users in three main ways:

- **Lower electricity rates**. With CTES, chillers can operate at night to meet the daytime cooling needs, taking advantage of lower off-peak electricity consumption rates.
- **Lower demand charges**. Many commercial customers pay a monthly electrical demand charge based on the largest amount of electricity used during any 30-minute period of the month. CTES reduces peak demands by shifting some of those demands to off-peak periods. Furthermore, some utilities provide a rebate for shifting electrical demand to night or other off-peak periods.
- **Lower air conditioning system and compressor costs**. Without CTES, large compressors capable of meeting peak cooling demands are needed, whereas smaller and less expensive units are sufficient when CTES is used. Also, since water from a CTES may be colder than conventional chilled water, smaller pipes, pumps and air handlers may be integrated into the building design to reduce costs further.

In some situations, one can reduce air conditioning costs by up to one-half with CTES.

3.8.16 Electric utilities and CTES

Electric utilities are placing increased emphasis on peak-load pricing strategies, such as time-of-day rates and marginal cost-based rates. These rate strategies present consumers with significant opportunities for cost control through load management in building design and operation. Consumers can take advantage of peak-load time-of-use rates, while reducing cooling and heating costs and maintaining comfort by installing a CTES system for space conditioning.

CTES systems can reduce HVAC equipment sizes, and lower air supply temperature (to as low as 8.3°C, with 30% relative humidity). This low humidity can provide adequate comfort with a 26.6°C ambient thermostat setting. By comparison, the supply air for conventional systems is at 12.7°C and 50% relative humidity, which permits a 23.3°C thermostat setting. CTES systems can also lower the first cost of equipment for energy distribution (e.g. pumps, pipes, fans, air ducts) and the operating costs for fans and pumps. The chilled-water storage media in some systems can also provide back-up for fire control systems. The 30% relative humidity supply air can in some instances be passed through heat exchangers to cool incoming outdoor air from 35°C to as low as 20°C using direct evaporative cooling. Chiller condenser water (especially, during ice making) can also be used in winter to reduce heating requirements.

Actual CTES systems are designed, evaluated and approved in situations where the concerns and preferences of the electric utility are of considerable importance, since utilities often provide financial incentives to promote the inclusion of CTES in new and retrofitted air conditioning systems. Special off-peak rate schedules are also sometimes available to larger customers for which separate metering of air conditioning loads is practical. Rebates are normally predicated on electrical demand (in kW) shifted to off-peak periods. Substantial payments are common. For example, Southern California Edison's present CTES-rebate limit is $300,000, based on $200 per kW shifted, while at Wisconsin Electric, rebates are variable, typically $350 per kW shifted and $0.08 per kWh shifted during the May to September cooling period. The combined effect of such rebates and the reduced operating costs achieved with a properly designed CTES system lead typically to simple payback periods of 2–5 years based on the CTES implementation cost.

3.9 Seasonal TES

Many heating systems, notably those using heat pumps, and including virtually all active solar space heating systems, incorporate TESs capable of storing heat from times when excess amounts are available to times when it is unavailable or expensive. The period of storage can vary from a few hours for diurnal storage cycles, to many months for seasonal (annual) cycles. Seasonal TES in solar heating systems is particularly favored in high-latitude locations, where

- solar energy is much more available during the long summer days, when it is not needed, than during the short winter days, when heating demands must be met; and
- cold ambient conditions, often below 0°C, are available during the winter with its short days, when cooling is not needed, than during the long summer days, when space cooling demands must be met.

The potential exists for seasonal storage of heating capacity from summer until winter in the former case, while the potential exists for seasonal storage of cooling capacity from winter until summer in the latter case.

3.9.1 Seasonal TES for heating capacity

Water is often favored as the storage medium in seasonal TES systems because it can function both as the heat transport and the heat storage medium, eliminating the cost and thermodynamic losses of one heat exchange operation, and also meeting the engineering preferences for a low cost, non-toxic, non-flammable, non-corrosive, chemically stable, non-viscous, high specific heat fluid with known characteristics.

Many storage containers are used for seasonal TES, including tanks, caverns and aquifers. Seasonal storage requirements for heating capacity are often met by large, insulated-tank, hot-water systems. They are typically of substantial size, usually above 500,000 litres, because the decreasing surface-to-volume ratio with increasing size reduces both the cost and the heat loss, on a unit of storage capacity basis.

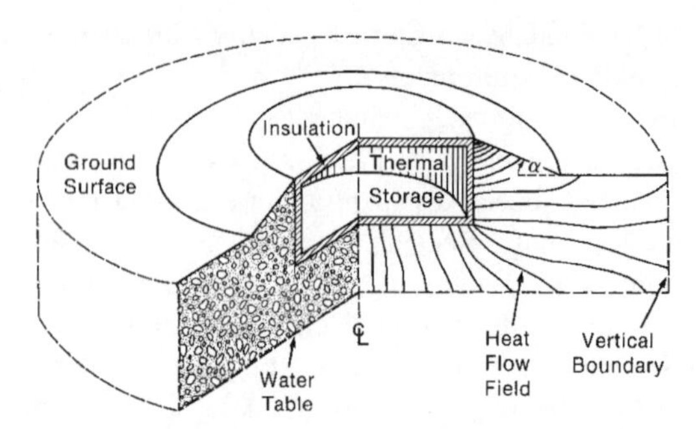

Figure 3.48 Illustration of partially buried, bermed heat storage tank. Approximate lines of heat flow outside the tank are shown.

The optimal form for such large tanks appears to be the right circular cylinder with vertical axis. Tanks of this form have low surface-to-volume ratios. They can be built to rest on the ground surface, or to be partially or fully buried. Often the tops of the tanks in buried configurations are modified for other purposes (e.g. paved for use as a parking lot or landscaped for use as a park). All heat losses or infiltrations from buried tanks flow through the soil and ultimately reach one of two heat sinks: the ground-air interface and the water table. It is noted for this cylindrical geometry that the condition where height and radius are approximately equal tends to minimize the overall tank heat loss, since it gives the minimum surface-to-volume ratio. The inside of the tank is often covered with a layer of insulation.

In some cases, tanks that are partially buried can have some or all of the excavated material bermed against the side walls (see Figure 3.48), to provide both physical support and a degree of insulation for the upper part of the tank wall. This configuration also provides for good surface water drainage, often has hydraulic advantages associated with the lift requirements of the system pumps, and avoids the need to haul and dispose of excavated soil from the site.

Methodologies for the analysis of the heat loss characteristics of several long-term storages have been studied (Hooper *et al.*, 1980; Rosen, 1990, 1998a). In particular, the fully buried tank, with its top flush with the surface of the ground, and the on-ground tank, have already received attention, and design methodologies for these cases have been developed. The thermal properties of the soil surrounding the tank are sometimes dependent on position, time and temperature. For example, changes in soil moisture content can occur during rains or melting of snow and ice, and can significantly alter the soil's thermal properties. Some additional energy interactions that are present include the latent heat of fusion of soil moisture during freezing and thawing, and the latent heat of vaporization during surface drying. The structural materials of the tank walls, often re-inforced concrete, sometimes have similar thermal properties as the soil regime.

3.9.2 Seasonal TES for cooling capacity

Several systems for seasonal storing of cold have been proposed and tested. Most of these are based on ice storage, and are applicable in climates where temperatures are below 0°C for much of the winter.

In one system, an ice store is built up throughout the winter by spraying water into a tank that is exposed to the ambient atmosphere when conditions are suitable for freezing water. During other times, the tank, which is insulated, is closed. In the summer, when space cooling is needed, the ice is allowed to melt and the cold water is circulated as required for cooling. The tank is designed to permit enough ice to be created in the winter to meet most or all of the summer cooling needs. Such a system has been tested in Ottawa, Canada.

3.9.3 Illustration

A seasonal thermal storage for heating capacity was installed at the Aylmer Senior Citizens residence in Ontario, Canada. That storage is a cylindrical tank, with its axis vertical and with a height of 6.7 m and a radius of 7.5 m. The tank is uniformly insulated with 0.15 m of foamed polyurethane having a conductivity of 0.0346 W/mK, and is buried. The storage fluid, water, ranges in temperature from about 80°C at the end of summer when the storage is almost fully charged, to about 30°C at the end of winter when the storage has been for the most part discharged.

The surrounding conditions of the Aylmer area are typical of south-western Ontario. The seasonal mean temperatures of the ambient air are about 21°C in summer, –7°C in winter and 7°C in both spring and autumn. The water table is relatively constant in temperature at about 7°C, and the thermal conductivity of the soil is about 1.73 W/m K.

In a preliminary economic analysis of cylindrical tanks such as the one considered here (Rosen, 1998b), tanks that are buried and have soil berms applied were shown, in most instances, to be superior to other tank configurations. This analysis determined whether the initial cost savings derived from using a bermed tank, instead of an in-ground tank, are greater or less than the additional costs associated with the greater heat losses for the bermed tank over the life of the tank. The factors considered in the analysis included the increased excavation cost associated with an in-ground tank, the increased wall thickness required for an in-ground tank, the haulage of excavated soil for an in-ground tank compared to the cost of placing the soil into a berm, and the increased heat loss associated with a bermed tank.

3.10 Concluding Remarks

Although energy may be stored in many ways (e.g. as mechanical, kinetic or chemical), since much of the economy involves thermal energy, the storage of thermal energy warrants attention. TES deals with the storing of energy by cooling, heating, melting, solidifying or vaporizing a material, the energy becoming available as heat when the process is reversed. TES is a temporary storage of high- or low-temperature energy for later use. There are mainly two types of TES systems: sensible (e.g. water, rock), and latent (e.g. salt hydrates). Storage by causing a material to increase or decrease in temperature is called sensible heat storage. Its effectiveness depends on the specific heat of the material and, if volume is important, the density of the storage material. Storage by phase change, the transition from solid to liquid or from liquid to vapor with no change in temperature, is known as latent heat storage.

Short-term storage (diurnal storage) is used to manage peak power loads of a few hours to a day, in order to reduce the sizing of systems and or to take advantage of energy tariffs. Medium- or long-term storage is more common when waste heat or seasonal energy loads can be transferred, with a delay of a few weeks to several months, to cover seasonal needs. This type of TES is called seasonal storage.

The selection of TES systems mainly depends on the storage period required (e.g. diurnal or seasonal), economic viability, operating conditions, etc. Some specific parameters that influence the viability of a TES include facility thermal loads, thermal and electrical load profiles, availability of waste or excess thermal energy, electrical costs and rate structures, type of thermal generating equipment, and building type and occupancy. The economic justification for TES systems usually requires that annual capital and operating costs are less than the costs for primary generating equipment supplying the same service loads and periods. Substantial energy savings can be realized by taking advantage of TES to facilitate using waste energy and surplus heat, reducing electrical demand charges, and avoiding heating, cooling or air conditioning equipment purchases.

Today TES is considered an advanced energy technology. The use of TES systems has been attracting increasing interest in several thermal applications, e.g. active and passive solar heating, water heating, cooling and air conditioning. TES is often the most economical storage technology for HVAC applications.

For solar thermal applications, the use of TES systems is essential because of fluctuations in the solar energy input. Several classes of storage may be required for a single installation, depending on the type and scale of the solar power plant, and the nature of its integration with conventional utility systems.

TES can help correct the mismatch between supply and demand of energy, and can contribute significantly to meeting society's needs for more efficient, environmentally benign energy use. TES plays an important role in energy conservation, and can yield great saving of premium fuels.

TES exhibits an enormous potential to make the use of thermal energy equipment more effective, and for facilitating large-scale energy substitutions economically. A coordinated set of actions is needed in several sectors of energy systems in order to realize the maximum benefits of storage.

References

Abhat, A. (1983). Low temperature latent heat thermal energy storage: heat storage materials, *Solar Energy* 30(4), 313–332.

AEP (2001). *HVAC Central System-Thermal Storage*, American Electric Power, http: www.aep.com/EnergyInfo/ces_html

Ames, D. (1989). The past, present and future of eutectic salt storage systems, *ASHRAE Journal*, May 1989, 26–27.

Anon. (1990). *Guide to Seasonal Storage*, Swiss Federal Energy Office, July 1990.

ASHRAE (2000). *Standard 943*, American Society of Heating, Refrigerating and Air-Conditioning Engineers, Inc., Atlanta, Georgia.

BAC (1985). *Ice Chiller Thermal Storage Unit Selection*, Baltimore Aircoil International N.V., Belgium.

Brown, D.R., Katipamula, S. and Koynenbelt, J.H. (1996). *A Comparative Assessment of Alternative Combustion Turbine Cooling Systems*, Pacific Northwest National Laboratory, Project Report: PNNL-10966/UC-202, Richland, WA.

Bishop, D. (1992). How to select thermal storage, *Heating/Piping/Air-conditioning*, January 1992, 87–88.

CADDET (1997). Learning from Experiences with Thermal Energy Storage: Managing Electrical Loads in Buildings, Center for the Analysis and Dissemination of Demonstrated Energy Technologies (CADDET), Analyses Series 4, Paris.

Choudhury, C. and Garg, H.P. (1995). Integrated rock bed heat exchanger-cum-storage unit for residential-cum-water heating, *Energy Conversion and Management* 36(10), 999–1006.

Choi, J.C. and Kim, S.D. (1995). Heat transfer in latent heat-storage system using $MgCl_2$•$6H_2O$ at the melting point, *Energy* 29, 13–25.

Cristopia (2000). *Thermal Energy Storage Technical Manual*, Cristopia Energy Systems, Vence.

Dincer, I. (1999). Evaluation and selection of thermal energy storage systems for solar thermal applications, *International Journal of Energy Research* 23(12), 1017–1028.

Dincer, I. and Dost, S. (1996). A perspective on thermal energy storage systems for solar energy applications, *International Journal of Energy Research* 20(6), 547–557.

Dincer, I., Dost, S. and Li, X. (1997a). Performance analyses of sensible heat storage systems for thermal applications, *International Journal of Energy Research* 21(10), 1157–1171.

Dincer, I., Dost, S. and Li, X. (1997b). Thermal energy storage applications from an energy saving perspective, *International Journal of Global Energy Issues* 9(4-6), 351–364.

Dinter, F., Ger, M. and Tamme, R. (1991). *Thermal Energy Storage for Commercial Applications*, Springer-Verlag, Berlin.

Eldighidy, S.M. (1991). Insulated solar storage tanks, *Solar Energy* 47, 253–268.

Environment Canada (1999). The Earth for storing energy, *Science and the Environment Bulletin*, No. 14.

GSA (2000). *Heat Storage*, GSA Resources Inc., http://www.gsaresources.com/heat.htm.

Harris, N.C., Miller, C.E. and Thomas, I.E. (1985). *Solar Energy Systems Design*, John Wiley, New York.

Hsieh, J.S. (1986). *Solar Energy Engineering*, Prentice-Hall, New Jersey.

Hooper, F.C., McClenahan, J.D. and Williams, G.T. (1980). *Solar Space Heating Systems using Annual Heat Storage*. Final Report No. DOE-CS-32939-12 prepared for U.S. Department of Energy, by Department of Mechanical Engineering, University of Toronto.

IEA (1990). Thermal storage, *IEA Heat Pump Center Newsletter* 8(2), 10.

Kuroda, H. (1993). Ice thermal storage systems, *IEA Heat Pump Center Newsletter* 11(4), 22–24.

Lane, G.A. (1980). Low temperature heat storage with phase change materials, International Journal of Energy Research 5, 155–160.

Lane, G.A. (1988). *Solar Heat Storage: Latent Heat Materials, Vol. 1: Background and Scientific Principles*, CRC Press, Florida.

Maust, D. (1993). A tale of two systems: ice versus cold water thermal storage, *Heating/Piping/ Air-conditioning*, September 1993, 30–36.

Moschatos, A.E. (1993). Combined thermal storage, *Solar Energy* 51(5), 391–399.

Narita, K. (1993). Combined system of heat pump and water thermal storage, *Proceedings of the Heat Pumps and Thermal Storage*, pp. 69–75, 24–25 May 1993, Fukuoka, Japan,

Norton, B. (1992). *Solar Energy Thermal Technology*, Springer-Verlag, London.

Ott, V.J. (2001). Personal communication, President of CRYOGEL Ice Ball Thermal Storage, San Diego.

Paris, J., Falardeau, M. and Villeneuve, C. (1993). Thermal storage by latent heat: a viable option for energy conservation in buildings, *Energy Sources* 15, 85–93.

Piette, M.A. (1990). *Learning from Experience with Diurnal Thermal Energy Storage Managing Electric Loads in Buildings*, CADDET Analysis Support Unit, June 1990.

Rosen, M.A. (1990). Evaluation of the heat loss from partially buried, bermed heat storage tanks, *International Journal of Solar Energy* 9(3), 147–162.

Rosen, M.A. (1998a). A semi-empirical model for assessing the effects of berms on the heat loss from partially buried heat storage tanks, *International Journal of Solar Energy* 20, 57–77.

Rosen, M.A. (1998b). The use of berms in thermal energy storage systems: energy-economic analysis, *Solar Energy* 63(2), 69–78.

Sanner, B. and Chant, V.G. (1992). Seasonal cold storage in the ground using heat pumps, *IEA Heat Pump Center Newsletter* 10(1), 4–7.

Sharma, S.K., Sethi, P.B.S. and Chopra, S. (1992). Kinetic and thermopyhsical studies of acetamide sodium bromide eutectic for low temperature storage applications, *Energy Conversion and Management* 33(2), 145–150.

Shimizu, M. and Fujita, K. (1985). Actual efficiencies of thermally-stratified thermal storage tanks, *IEA Heat Pump Center Newsletter* 3(1/2), 20–25.

Sokolov, M. and Keizman, Y. (1991). Performance indicators for solar pipes with phase change storage, *Solar Energy* 47(5), 339–346.

Thomason, H.E. (1960). Solar space heating and air conditioning in the Thomason house, *Solar Energy* 4, 11–19.

Tomlinson, J.J. and Kannberg, L.D.(1990). Thermal energy storage, *Mechanical Engineering* 112, 68–72.

Wylie, D. (1990). Evaluating and selecting thermal energy storage, *Energy Engineering* 87(6), 6–17.

4

Thermal Energy Storage and Environmental Impact

I. Dincer and M.A. Rosen

4.1 Introduction

Energy supply and use are related not only to problems such as global warming, but also to such environmental concerns as air pollution, ozone depletion, forest destruction, and emissions of radioactive substances. These and other environmental issues must be taken into consideration if humanity is to develop in the future while maintaining a healthy and clean environment. Much evidence suggests that the future will be negatively impacted if people and societies continue to degrade the environment.

There is an intimate connection between energy, the environment and sustainable development. A society seeking sustainable development ideally must utilize only energy resources which cause no environmental impact (e.g. which release no emissions or only harmless emissions to the environment). However, since all energy resources lead to some environmental impact, it is reasonable to suggest that some (but not all) of the concerns regarding the limitations imposed on sustainable development by environmental emissions and their negative impacts can be overcome through increased energy efficiency. A strong correlation clearly exists between energy efficiency and environmental impact since, for the same services or products, less resource utilization and pollution is normally associated with higher efficiency processes.

TES systems can contribute significantly to meeting society's needs and desires for more efficient, environmentally benign energy use in such applications as building heating and cooling and electric power generation and distribution. The use of TES systems often results in significant benefits, such as the following (Dincer *et al.*, 1997; Rosen *et al.*, 2000):

- reduced energy costs,
- reduced energy consumption,
- improved indoor air quality,
- increased flexibility of operation,
- decreased initial and maintenance costs,

- reduced equipment size,
- more efficient and effective utilization of equipment,
- conservation of fossil fuels (by facilitating more efficient energy use), and
- reduced pollutant emissions (e.g. CO_2 and CFCs).

In this chapter, anticipated patterns of future energy use and consequent environmental impacts (focusing on acid precipitation, stratospheric ozone depletion and the greenhouse effect) are comprehensively discussed. Also, some solutions to current environmental issues are identified, including energy conservation, increased efficiency and use of renewable energy technologies, and some theoretical and practical limitations to increased efficiency are explained. The relations between energy and sustainable development and between the environment and sustainable development are described, and several illustrative examples are presented. Throughout the chapter several issues relating to energy, environment and sustainable development are examined from both current and future perspectives. Finally, several conclusions are drawn and recommendations are made which are useful to scientists, researchers, engineers and policy makers in the field of TES.

4.2 Energy and the Environment

Achieving solutions to the environmental problems that we face today requires long-term planning and actions, particularly if we are to approach sustainable development. In this regard, renewable energy resources appear to represent one of the most advantageous solutions. Hence, a strong connection is often reported between renewable energy and sustainable development.

Energy is in many ways the convertible currency of technology. Without energy resources, our societies cannot function and would crumble. The effect of only a 24-hour cut in electricity supply to a city shows how totally dependent we are on that particularly useful form of energy. Computers and elevators cease to function, hospitals reduce to a basic care and maintenance level, and lights go out.

The average annual growth rate in world population is approximately 2%, and many countries exceed that level. As populations grow, the need for more and more energy resources is exacerbated. Enhanced lifestyles and energy demand usually rise together. Presently, there is a great disparity between the populations of countries and their wealth and energy use. The wealthy industrialized economies of the world which contain 25% of the world's population, consume about 75% of the world's energy supply (Dincer, 1998).

World population is expected to double by the middle of the 21st century, and economic development will almost certainly continue to grow. Global demand for energy services is expected to increase by as much as an order of magnitude by 2050, while primary-energy demands are expected to increase by 1.5–3 times. Simultaneously, concern will likely increase regarding energy-related environmental concerns such as acid precipitation, stratospheric ozone depletion and global climate change (Dincer, 2000).

Environmental considerations have been given increasing attention in recent decades by energy industries and the public. The concept that consumers share responsibility for pollution and its impact and cost has been increasingly accepted. In some jurisdictions, the prices of many energy resources have increased over the last 20 years, in part to account for environmental costs.

An often-proposed solution to possible energy-resource shortage is increased use of TES technologies, and the application of TES technologies can be quite advantageous in many instances. To achieve the benefits, engineers and designers must carefully consider the practicality, reliability, applicability and economy. Scarcity of energy supply and public acceptability should also be considered accordingly.

More broadly, it is pointed out that energy is one of the main factors that must be considered in discussions of sustainable development. Several definitions of sustainable development have been put forth, including the following common one: "*development that meets the needs of the present without compromising the ability of future generations to meet their own needs*" (Anon., 1987). Thus, to achieve sustainable development, a supply of energy resources is required that is fully sustainable (i.e. that, in the long term, is readily and sustainably available at reasonable cost and can be utilized for all required tasks without causing negative societal impacts) (Norton, 1991; MacRae, 1992; Rosen, 1996, Dincer and Rosen, 1998). Although a secure supply of energy resources is generally agreed to be a necessary but not sufficient requirement for development within a society, sustainable development further demands a sustainable supply of energy resources and effective and efficient utilization of energy resources. In this regard, the advantages of TES utilization are clear.

4.3 Major Environmental Problems

During the past two decades the risks and reality of environmental degradation have become more apparent. The environmental impact of human activities has grown dramatically due to a combination of factors such as increasing world population, energy consumption, industrial activity, etc. Throughout the 1970s most environmental analysis and control instruments concentrated on conventional pollutants such as SO_2, NOx, particulates and CO. Recently environmental concern has extended to the control of micro- or hazardous air pollutants, which are usually toxic chemical substances that are harmful in small doses, as well as globally significant pollutants such as CO_2. Despite advances in environmental science, developments in industrial processes and structures have led to new environmental problems such an increase in the effects of NOx and Volatile Organic Compound (VOC) emissions. Details on these gaseous and particulate pollutants and their impacts on the environment and human bodies may be found in Dincer (1998).

Environmental problems span a continuously growing range of pollutants and hazards, and include ecosystem degradation over ever wider areas. The major areas of environmental concern may be classified as follows (Dincer, 1998):

- acid rain,
- stratospheric ozone depletion,
- global climate change (greenhouse effect),
- hazardous air pollutants,
- ambient air quality,
- water and maritime pollution,
- land use and siting impact,
- radiation and radioactivity,

- solid waste disposal, and
- major environmental accidents.

Among these environmental issues, the most internationally known and most significant ones are usually considered to be acid precipitation, stratospheric ozone depletion and global climate change. Consequently, we will focus on these in this section.

Some gaseous pollutants are listed in Table 4.1, along with their environmental impacts.

4.3.1 Acid rain

Acid rain (acid precipitation) refers to the end result of certain pollutants which are produced by the combustion of fossil fuels, particularly from both stationary devices such as smelters for nonferrous ores and industrial boilers, and transportation vehicles, and which are transported over great distances through the atmosphere and deposited via precipitation on the earth. In the atmosphere, these substances react to form acids, which are often deposited on ecosystems that are exceedingly vulnerable to damage from excessive acidity. This acid rain is mainly attributable to emissions of SO_2 and NOx (Dincer, 1998). These gases react with water and oxygen in the atmopshere to form such substances as sulfuric and nitric acids (Figure 4.1). The control of acid precipitation requires appropriate control of mainly SO_2 and NOx emissions.

These pollutants have caused only local concerns related to health in the past. However, awareness of their contributions to the regional and trans-boundary problem of acid precipitation has recently grown, and attention has also began focusing on other contributing substances such as VOCs, chlorides, ozone and trace metals. These substances may participate in the complex set of chemical transformations in the atmosphere that result in acid precipitation and the formation of other regional air pollutants. The types of damage caused by acid precipitation are as follows (Dincer and Rosen, 1998):

Table 4.1 Main gaseous pollutants and their impacts on the environment.

Gaseous pollutant	Greenhouse effect	Stratospheric ozone depletion	Acid precipitation	Smog
Carbon monoxide (CO)	•	•	•	•
Carbon dioxide (CO_2)	+	+/–	•	•
Methane (CH_4)	+	+/–	•	•
Nitric oxide (NO) and				
Nitrogen dioxide (NO_2)	•	+/–	+	+
Nitrous oxide (N_2O)	+	+/–	•	•
Sulfur dioxide (SO_2)	–	+	•	•
Chlorofluorocarbons (CFCs)	+	+	•	•
Ozone (O_3)	+	•	•	+

Note: + denotes a contributing effect, – denotes that the substance exhibits an impact which varies with conditions and chemistry and may not be a general contributor, and • denotes no impact.
Source: Speight (1996).

- acidification of lakes, streams and ground waters,
- damage to forests, agricultural crops, and plants due to the toxicity of excessive acid concentration,
- damage to fish and aquatic life,
- deterioration of materials, e.g. buildings, metal structures and fabrics, and
- alterations of the physical and optical properties of clouds due to the influence of sulfate aerosols.

Some energy-related activities are major sources of acid precipitation. For example, electric power generation, residential heating and industrial energy use account for 80% of SO_2 emissions. Another source of acid precipitation is sour gas treatment, which releases H_2S that reacts to form SO_2 when exposed to air. Most of the NOx emissions are due to fossil fuel combustion in stationary sources and road transport. VOCs, which are generated by a variety of sources and comprise a large number of diverse compounds, also contribute to acid precipitation. Countries in which these energy-related activities occur widely are significant contributors to acid precipitation. The largest contributors are the USA, the countries from the former Soviet Union, and China.

The problem of acid precipitation is very complex because the acid precipitation produced by some countries' emissions often happens to fall on other countries, where it exhibits its damaging effects on the ecology of water systems and forests, and on historical and cultural artifacts.

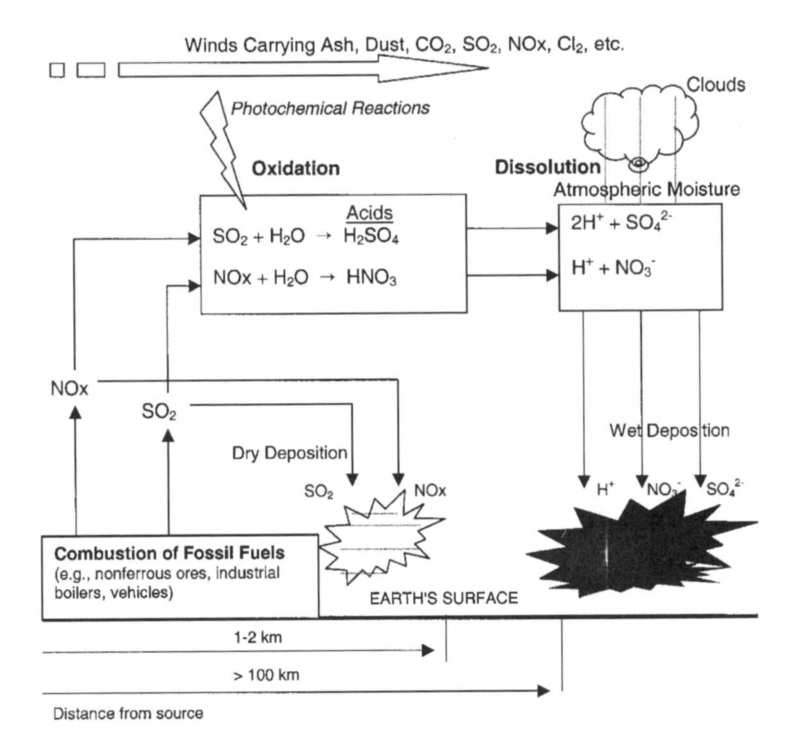

Figure 4.1 Illustration of the formation, transport, and impact of acid precipitation (Dincer, 2000).

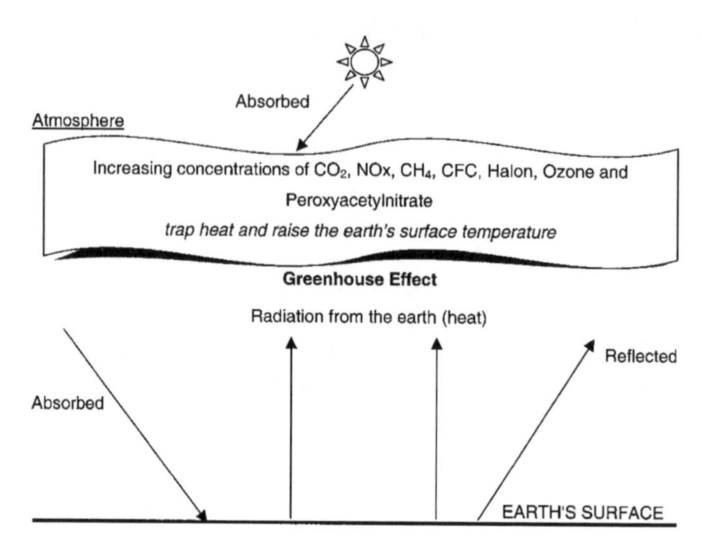

Figure 4.2 Illustration of greenhouse effect (Dincer, 2000).

4.3.2 Greenhouse effect (global climate change)

Although the term *greenhouse effect*, also known as global climate change and global warming, has generally been used in the past to refer for the role of the atmosphere (mainly water vapor and clouds) in keeping the surface of the earth warm, it is now associated with the warming contribution of CO_2. Currently, it is estimated that CO_2 emissions are responsible for about 50% of the anthropogenic greenhouse effect. Other gases such as CH_4, CFCs, halons, N_2O, ozone and peroxyacetylnitrate, produced by industrial and domestic activities, also contribute to the greenhouse effect, which results in a rise in the earth's temperature (Figure 4.2). CO_2 and these other gases are often referred to as *greenhouse gases*.

The greenhouse effect is potentially the most important environmental problem relating to energy utilization. The greenhouse effect is associated with increasing atmospheric concentrations of greenhouse gases, which increase the manner in which the atmosphere traps heat radiated from the earth's surface, thereby raising the surface temperature of the earth. The earth's surface temperature has increased about 0.6°C over the last century, and as a consequence the sea level is estimated to have risen by perhaps 20 cm (Colonbo, 1992). Such changes can have wide-ranging effects on human activities all over the world. Current knowledge of the role of various greenhouse gases is summarized in Dincer and Rosen (1998).

Humankind is contributing through many of its economic and other activities to the increase in atmospheric concentrations of various greenhouse gases. For example, CO_2 releases from fossil fuel combustion, methane emissions from increased human activity, CFC releases and deforestation all contribute to the greenhouse effect. Most scientists agree that there is a cause-effect relationship between the observed emissions of greenhouse gases and global warming. Furthermore, many scientists predict that if atmospheric concentrations of greenhouse gases continue to increase, as present trends in fossil fuel consumption suggest, the earth's temperature may increase in the next century by another 2°C, and perhaps by up to 4°C. If this prediction is realized, the sea level could rise between 30 and 60 cm before the end of the 21st century (Colonbo, 1992). The impact of such a phenomenon could be dramatic, including flooding of coastal settlements, displacement of fertile zones for agriculture and food production toward higher latitudes, and decreasing availability of fresh water for irrigation and other essential uses. Such consequences could jeopardize the survival of entire populations.

Discussions on averting global climate change must consider the costs of reducing carbon emissions. From a developing-country perspective, discussions of costs and benefits should account for the need for policies promoting rapid economic growth. Achieving a balance between economic development and emissions abatement requires domestic policies aimed at improving the efficiency of energy use and facilitating fuel switching, and international policies enabling easier access to advanced technologies and external resources.

Arguments about the magnitude of the greenhouse effect at present and the possible levels in the future have gone on for some time. Some believe that the earth is doomed to a rise in temperature, while others believe that we can go polluting the atmosphere without consequence. However, most agree that greenhouse gas emissions are harmful to the environment (Bradley *et al.*, 1991), although some contradictory reports have been published, making the issue complicated to study. Many state that the environment should be considered to be an extremely limited resource, and the discharge of chemicals into it should be subject to severe constraints.

In Table 4.2, the role of various greenhouse gases in the processes involved in the greenhouse effect is summarized.

4.3.3 Stratospheric ozone depletion

It is known that the ozone present in the stratosphere, roughly between altitudes of 12 and 25 km, plays a natural, equilibrium-maintaining role for the earth, through absorption of ultraviolet (UV) radiation of wavelengths (240–320 nm) and absorption of infrared radiation (Dincer, 1998). A global environmental problem is the distortion and regional depletion of the stratospheric ozone layer which has been shown to be caused by emissions of CFCs, halons (chlorinated and brominated organic compounds) and nitrogen oxides (NOx) (Figure 4.3). Ozone depletion in the stratosphere can lead to increased levels of damaging ultraviolet radiation reaching the ground, causing increased rates of skin cancer, eye damage and other harm to many biological species.

Energy- and non-energy related activities are only partially responsible (directly or indirectly) for the emissions which lead to stratospheric ozone depletion. CFCs, which are used in air conditioning and refrigerating equipment as refrigerants and in foam insulation as blowing agents, and NOx emissions which are produced from fossil fuel and biomass combustion processes, natural denitrification, nitrogen fertilizers and aircraft, play the most significant role in ozone depletion. Though scientific debate on ozone depletion has occurred for over a decade, only in 1987 was an international landmark protocol signed in Montreal to reduce the production of CFCs and halons. Much scientific evidence confirming the destruction of stratospheric ozone by CFCs and halons has recently been gathered, and commitments for more drastic reductions in their production were undertaken at many international conferences (e.g. the 1990 London Conference). Many researchers and scientists have undertaken comprehensive studies on the current status of the stratospheric ozone layer. Topics that have been investigated include the history of the problem, chemical and physical phenomena associated with ozone depletion, mapping of ozone losses in the stratosphere, and hypotheses for the causes of the problem and its impacts.

Replacement equipment and technologies that do not use CFCs are gradually coming to the fore, and may ultimately allow for a total ban of CFCs. An important consideration in such a CFC ban is the need to distribute fairly the economic burdens deriving from the ban, particularly with respect to developing countries, some of which have invested heavily in CFC-related technologies. In order to eliminate or minimize the impacts of NOx emissions, the solutions mentioned in the previous section can be implemented accordingly.

Table 4.2 Roles of some substances in the greenhouse effect.

Substance	ARIRR[a]	Atmospheric concentration (ppm)		AGR[b] (%)	SGEHA[c] (%)	SGEIHA[d] (%)
		Pre-industrial	In 1990s			
CO_2	1	275	346	0.4	71	50±5
CH_4	25	0.75	1.65	1	8	15±5
N_2O	250	0.25	0.35	0.2	18	9±2
R-11	17,500	0	0.00023	5	1	13±3
R-12	20,000	0	0.00040	5	2	13±3

[a] Ability to retain infrared radiation relative to CO_2.
[b] Annual growth rate.
[c] Share in the greenhouse effect due to human activities.
[d] Share in the greenhouse effect increase due to human activities.
Source: Dincer and Rosen (1999).

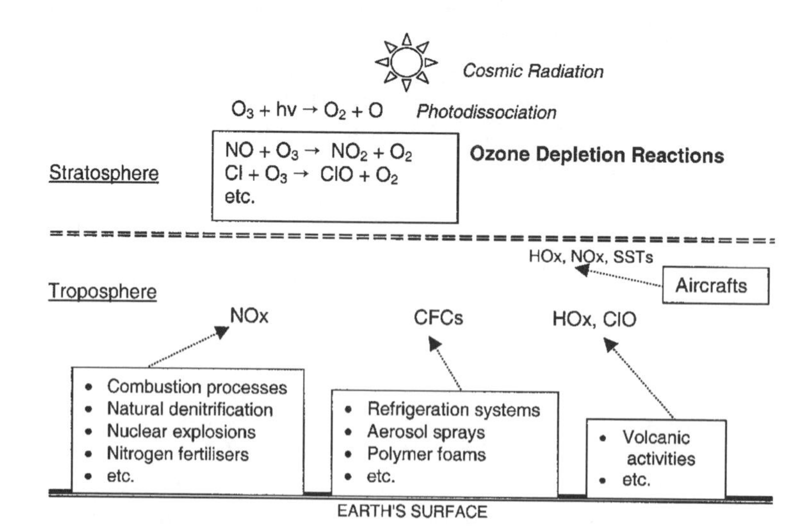

Figure 4.3 Illustration of sources of natural and anthropogenic ozone depleters (Dincer, 2000).

To conduct a successful environmental study, a clear outline of activities is needed, including the following significant steps:

- definition of the main goals, both short- and long-term,
- measurement or estimation of the data needed with as much accuracy as desired and possible,
- evaluation and assessment of the data,
- processing of the data to generate new and reliable information and knowledge, as well as conclusions and recommendations, and
- reporting of the results without exaggeration or bias.

4.4 Environmental Impact and TES Systems and Applications

TES systems can contribute significantly to meeting society's desire for more efficient, environmentally benign energy use, particularly in the areas of building heating and cooling and electric power generation. By reducing energy consumption, the utilization of TES systems results in two significant environmental benefits:

- the conservation of fossil fuels through efficiency increases and/or fuel substitution, and
- reductions in emissions of such pollutants as CO_2, SO_2, NOx and CFCs.

TES can impact air emissions at building sites by reducing (i) the amount of ozone-depleting CFC and HCFC refrigerants in chillers, and (ii) the amount of combustion emissions from fuel-fired heating and cooling equipment. Each of these impacts is considered. TES helps reduce CFC use in two main ways. First, since cooling systems with TES require less chiller capacity than conventional systems, they use fewer or smaller chillers with less refrigerant. Second, using TES can offset the lost cooling capacity that can sometimes occur when existing chillers are converted to more benign refrigerants, making building operators more willing to switch refrigerants.

The potential aggregate air-emission reductions at power plants due to TES have been shown to be significant. For example, TES systems have been shown to reduce CO_2 emissions in the United Kingdom by 14% to 46% by shifting electric load to off-peak periods (Beggs, 1994), while an EPRI co-sponsored analysis found that TES could reduce CO_2 emissions by 7% compared to conventional electric cooling technologies (Reindtl, 1994). Also, using California Energy Commission data indicating that existing gas plants produce about 0.06 kg of NOx and 15 kg of CO_2 per 293,100 kWh of fuel burned, and assuming that TES installations save an average of 6% of the total cooling electricity needs, TES could possibly eliminate annual emissions of about 560 tons of NOx and 260,000 tons of CO_2 statewide (CEC, 1996).

4.5 Potential Solutions to Environmental Problems

In this section, we list a number of general potential solutions to environmental issues as whole, and in the following, TES related solutions and their consequences are explained in detail.

4.5.1 General solutions

Potential solutions to the current environmental problems associated with the harmful pollutant emissions have recently evolved, including:

- Use of TES technologies
- Use of renewable energy technologies
- Energy conservation and increasing the efficiency of energy utilization
- Cogeneration and district heating and cooling
- Use of alternative energy forms and sources for transport
- Energy-source switching from fossil fuels to environmentally benign energy forms
- Use of coal cleaning technologies
- Optimum monitoring and evaluation of energy indicators
- Policy integration
- Recycling
- Process change and sectoral modification
- Acceleration of forestation
- Application of carbon and/or fuel taxes
- Materials substitution
- Promoting public transport
- Changing lifestyles
- Increasing public awareness of energy-related environmental problems
- Increased education and training.

4.5.2 TES-related solutions

Among the potential solutions listed above, the most important to the present discussions is the use of TES technologies. TES and its role in reducing environmental impact is discussed here.

An important step in moving toward the implementation of TES technologies is to identify and remove barriers. Several barriers have in the past been identified to the development and introduction of cleaner energy processes, devices and products. These barriers must be removed for TES and associated technologies to be more widely used. The barriers can also affect the financing of efforts to augment the supply of TES technologies. Although there are a number of barriers in practice, each TES project or application has its own difficulties and obstacles to tackle.

Some of the barriers faced by many TES technologies include (Dincer, 1999):

- technical constraints,
- financial constraints,
- limited information and knowledge of options,
- lack of necessary infrastructure for recycling, recovery and re-use of materials and products,
- lack of facilities,
- lack of expertise within industry and research organizations, and/or lack of coordinated expertise,

- poorly coordinated and/or ambiguous national aims related to energy and the environment,
- uncertainties in government regulations and standards,
- lack of adequate organizational structures,
- lack of different electrical rates to encourage off-peak electricity use,
- mismanagement of human resources,
- lack of societal acceptability of new TES technologies,
- absence of, or limited consumer demand for, TES products and processes.

Establishing concrete methods for promoting TES technologies requires analysis and clarification about how to combine environmental objectives, social and economic systems, and technical development to attain effective solutions. It is important to create and employ tools that encourage technological development and diffusion, and to align government policies in such areas as energy and environmental standards and government purchasing.

4.6 Sustainable Development

Energy resources are needed for societal development, and sustainable development requires a supply of energy resources that is sustainably available at reasonable cost and which can cause no negative societal impacts. Thus, energy resources such as fossil fuels are finite with a lack of sustainability, while others such as renewable energy sources with TES are sustainable over the relatively long term.

Environmental concerns also factor into sustainable development, as activities which degrade the environment are not sustainable. As much environmental impact is associated with energy, sustainable development requires the use of energy resources which cause as little environmental impact as possible. Clearly, limitations on sustainable development due to environmental emissions can in part be overcome through increased efficiency, as it leads to less environmental impact for the same services or products.

Consequently, there is such a diversity of choices that the use of TES technologies plays a key role in the context of sustainable development.

4.6.1 Conceptual issues

Sustainability has become a fashionable word in the 1990s, a legacy of concerns about the environment expressed during the 1970s and 1980s. Also common is *sustainable development*—another term made fashionable. The media often refer to sustainable architecture, sustainable diets, sustainable fisheries, sustainable food production, sustainable futures, sustainable communities, sustainable economic development, sustainable economic growth, sustainable policies, and even sustainable debt.

Sustainable has come to mean good. Part of the problem with *sustainable* is that it is taken to mean good, at least in political discourse. One synonym of *sustainable* is *supportable*. But the originators of the term *sustainable development* had another meaning in mind: *capable of being continued*. Thus *sustainable development* is *politically supportable development*, on the one hand, or *development that is capable of being continued* on the other hand.

Further uncertainty in meaning arises from consideration of what is *unsustainable*, i.e. cannot be continued. If transportation trends are stated to be unsustainable (to anticipate later parts of this chapter), does this mean that the transportation trends cannot be continued or that society as we know it cannot continue if the transportation trends continue?

Development is also a somewhat ambigious word (in English). It can mean both *progress* and *happening*. The former use has the connotation of movement to a better state; the latter use is less laden with value. Consequently, a minimal, value-free meaning of *sustainable development* might be "activity that is capable of being continued" (OECD, 1996).

4.6.2 The Brundtland Commission's definition

The term *sustainable development* was introduced in 1980, popularized in the 1987 report of the World Commission on Environment and Development (the Brundtland Commission), and given the status of a global mission by the United Nations Conference on Environment and Development (UNCED) in Rio de Janeiro in 1992.

The Brundtland Commission defined *sustainable development* as "development that meets the needs of the present without compromising the ability of future generations to meet their own needs." The Commission noted that its definition contains two key concepts: *needs,* meaning "in particular the essential needs of the world's poor," and *limitations*, meaning "limitations imposed by the state of technology and social organisation on the environment's ability to meet present and future needs" (OECD, 1996).

The Brundtland Commission's definition was thus not only about sustainability and its various aspects, but also about equity, equity among present inhabitants of the planet and equity among generations. *Sustainable development* for the Brundtland Commission includes environmental, social and economic aspects, but considers remediation of current social and economic aspects an initial priority. The chief tools cited for remediation were "more rapid economic growth in both industrial and developing countries, freer market access for the products of developing countries, lower interest rates, greater technology transfer, and significantly larger capital flows, both concessional and commercial." Such growth was said to be compatible with recognized environmental constraints, but the extent of the compatibility was not explored.

4.6.3 Environmental limits

The report of the Brundtland Commission stimulated debate about the environmental impacts of industrialization and about the legacy of present activities for coming generations. The report also increased interest in what might be the physical or ecological limits to economic growth.

Further definitions of sustainability proposed give priority to such limits. One business writer, Paul Hawken, suggested the following (OECD, 1996): "The word sustainability can be defined in terms of carrying capacity of the ecosystem, and described with input-output models of energy and resource consumption. Sustainability is an economic state where the demands placed on the environment by people and commerce can be met without reducing the capacity of the environment to provide for future generations. It can also be expressed in the simple terms of an economic golden rule for the restorative economy: Leave the world better than you found it, take no more than you need, try not to harm life or the environment, make amends if you do."

The author of the above quotation and others have drawn on the work of Herman Daly, formerly of the World Bank, in considering how environmental limits might be characterized. Daly suggests that the limits on society's material and energy throughputs might be set as follows (OECD, 1996):

- The rates of use of renewable resources should not exceed their rates of regeneration.
- The rates of use of non-renewable resources should not exceed the rates at which renewable substitutes are developed.
- The rates of pollution emissions should not exceed the assimilative capacity of the environment.

4.6.4 Global, regional and local sustainability

Sustainability—or unsustainability—must also be considered in terms of its geographic scope. Activity may be globally unsustainable. For example, it may result in climate change or depletion of the stratospheric ozone layer, and so affect several geographic regions, if not the whole world. Activity may be regionally unsustainable, perhaps by producing and dispersing tropospheric ozone or acidifying gases that can kill vegetation and cause famine in one region but not in other parts of the world. Activity may be locally unsustainable, perhaps because it results in hazardous ambient levels of carbon monoxide locally, or because the noise it produces makes habitation impossible. In the long term, sustainability appears to be more a global than a regional or local concern. If an environmental impact exceeds the carrying capacity of the planet, then life as we know it is threatened. If it is beyond the carrying capacity of one area, then that area may become uninhabitable, but life can most likely continue elsewhere.

4.6.5 Environmental, social and economic components of sustainability

The focus of this discussion on physical limits does not deny the social and economic aspects to sustainability. A way of life may not be worth sustaining under circumstances of extreme oppression or deprivation. Moreover, oppression or deprivation can interfere with efforts to make human activity environmentally benign. Nonetheless, if ecosystems are irreparably altered by human activity, then subsequent human existence may become not merely unpalatable, but impossible. Thus the environmental component of sustainability is essential.

The heterogeneity of the environmental, social, and economic aspects of sustainability should also be recognized. Environmental and social considerations often refer to *ends*, the former having perhaps more to do with the welfare of future generations and the latter with the welfare of present people. Economic considerations, often taken to refer to ends, can perhaps more helpfully be seen as a *means* to the various ends implied by the environmental and social considerations.

The environmental aspects of sustainability, particularly those of a global nature, are often not focused on by spokespeople for the world's poorer countries. They argue that a preoccupation with the environment ignores their real needs to eliminate poverty and may be associated with plans to prevent development in general and industrialization in particular. The reality is that the poorer half of the world's human population has contributed relatively little to the degradation of global and regional environments. Of the remaining two to three billion people whose activities do degrade global and regional environments, the several hundred millions who live in OECD Member countries contribute a disproportionately large share. They are also the people who can more afford the luxury of thinking about the future. Thus a focus within the OECD on the environmental aspects of sustainability may be viewed as appropriate (OECD, 1996).

4.6.6 Energy and sustainable development

A secure supply of energy resources is generally agreed to be a necessary but not sufficient requirement for development within a society. Furthermore, sustainable development demands a sustainable supply of energy resources. The implications of these statements are numerous, and depend on how sustainable is defined.

One important implication of these statements is that sustainable development within a society requires a supply of energy resources that, in the long term, is readily and sustainably available at reasonable cost and can be utilized for all required tasks without causing negative societal impacts. Supplies of such energy resources as fossil fuels (coal, oil, and natural gas) and uranium are generally acknowledged to be finite; other energy sources such as sunlight, wind and falling water are generally considered renewable and therefore sustainable over the relatively long term. Wastes (convertible to useful energy forms through, for example, waste-to-energy incineration facilities) and biomass fuels are also usually viewed as sustainable energy sources.

A second implication of the initial statements in this section is that sustainable development requires that energy resources be used as efficiently as possible (MacRae, 1992; Rosen, 1996). In this way, society maximizes the benefits it derives from utilizing its energy resources, while minimizing the negative impacts (such as environmental damage) associated with their use. This implication acknowledges that all energy resources are to some degree finite, so that greater efficiency in utilization allows such resources to contribute to development over a longer period of time, i.e. to make development more sustainable. Even for energy sources that may eventually become inexpensive and widely available, increases in energy efficiency will likely remain sought to reduce the resource requirements (energy, material, etc.) to create and maintain systems and devices to harvest the energy, and to reduce the associated environmental impacts.

The first implication, clearly being essential to sustainable development, has been and continues to be widely discussed. The second implication, which relates to the importance and role of energy efficiency in achieving sustainable development, is somewhat less discussed and understood.

4.6.7 Environment and sustainable development

Environmental concerns are an important factor in sustainable development. For a variety of reasons, activities which continually degrade the environment are not sustainable over time, e.g. the cumulative impact on the environment of such activities often leads over time to a variety of health, ecological and other problems.

A large portion of the environmental impact in a society is associated with its utilization of energy resources. Ideally, a society seeking sustainable development utilizes only energy resources which cause no environmental impact (e.g. which release no emissions to the environment). However, since all energy resources lead to some environmental impact, it is reasonable to suggest that some (not all) of the concerns regarding the limitations imposed on sustainable development by environmental emissions and their negative impacts can be in part overcome through increased efficiency. Clearly, a strong relation exists between efficiency and environmental impact since, for the same services or products, less resource utilization and pollution is normally associated with increased efficiency.

Improved energy efficiency leads to reduced energy losses. Most efficiency improvements produce direct environmental benefits in two ways. First, operating energy input requirements are reduced per unit output, and pollutants generated are correspondingly reduced. Second, consideration of the entire life cycle for energy resources and technologies suggests that improved efficiency reduces environmental impact during most stages of the life cycle.

4.6.8 Achieving sustainable development in larger countries

Can sustainable development he pursued by larger countries with varied political and administrative systems? Wealth and advanced technology may make it easier for the industrialized countries to strive for sustainable development, but as the reversal in the trend towards declining carbon emissions after the oil-price decline in 1986 illustrates, the basic motivations, dreams and desires which require higher energy use (and yield corresponding increases in emissions) have not changed. Transforming these behavioral and decision-making patterns requires a recognition that current development paths are not sustainable. History suggests that such a recognition occurs only when short-term consequences are obvious, as in the case of an 'oil-price shock' or a drought. In order to successfully mobilize the resources necessary to reduce the risks posed by a changing global climate, the public must perceive the potential long-term consequences associated with present behavior patterns. Translating the future threats associated with continual increases in energy use and carbon emissions into immediate priorities is and will be one of the most difficult challenge faced by policy analysts.

4.6.9 Essential factors for sustainable development

The main concept of sustainability, which often inspires local and national authorities to incorporate environmental considerations into their energy program, and which has different meanings in different contexts, embodies a long-term perspective. Future energy systems will largely be shaped by broad and powerful trends that have their roots in basic human needs. Combined with increasing world population, the need will become more apparent for successful implementation of sustainable development.

Various parameters are essential to achieving sustainable development in a society, some which are as follows (Dincer, 1999):

- *Public awareness*: improving public awareness of need is an initial and crucial step in making a sustainable energy program successful. This step should be carried out through the media and by public and/or professional organizations.
- *Information*: necessary information on energy utilization, environmental impact, renewable energy resources, etc. should be provided to the public through public and government channels.
- *Environmental education and training*: this activity complements the provision of information. Any approach which does not include as integral education and training is likely to fail, so this activity can be considered as crucial to a sustainable energy program. For this reason, a wide scope of specialized agencies and training facilities should be made available to the public.
- *Innovative energy strategies*: such strategies should be included where appropriate in an effective sustainable energy program. In parallel, efficient dissemination of information is required of the new methods through public relations, training and counseling.
- *Renewable energy resources and cleaner technologies*: in developing an environmentally benign sustainable energy program, renewable energy sources and cleaner technologies (including TES) should be promoted at every stage. Such activities form a basis for short- and long-term policies.
- *Financing*: financing is an important tool for achieving the main goal of sustainable energy development in a country and accelerating the implementation of environment friendly energy technologies.
- *Monitoring and evaluation tools*: in order to assess how successfully a program has been implemented, it is of great importance to monitor each step and evaluate the data and findings obtained. In this regard, appropriate monitoring and evaluation tools should be used.

4.7 Illustrative Examples and Case Studies

Five examples are considered of energy systems that incorporate TES, and by so doing, mitigate some environmental impacts. The examples are based on actual cases.

4.7.1 The South Coast Air Quality Management District (SCAQMD) (California, USA)

The South Coast Air Quality Management District (SCAQMD) identifies TES as one way to reduce site emissions, and TES can also significantly reduce combustion air emissions from power plants. Indeed, in California where natural gas is usually the fuel used for marginal power generation, the reductions in power plant emissions are comparable to emission reductions due to the energy savings from TES. Assuming 20% market penetration by 2005, TES could avoid 260,000 tons of CO_2 emissions annually, statewide. Just as importantly, TES could eliminate about 1.6 tons of NOx emissions per day in the SCAQMD. These NOx emission reductions are equivalent to the reductions achieved by using almost 100,000 electric vehicles (CEC, 1996). At the building site, TES can help reduce chlorofluorocarbons (CFCs) and combustion emissions. TES can also help in the transition to non-CFC air-conditioning refrigerants. For example, when existing chillers, are converted to a non-CFC refrigerant, the chillers' effective cooling capacity may be reduced. Some key facility managers see TES as making up the cooling-capacity difference. In addition, use of partial-storage TES often can reduce the required chiller capacity by half, which means half as much refrigerant is necessary.

4.7.2 Anova Verzekering Co. Building (Amersfoort, The Netherlands)

This application shows how an aquifer TES system provides energy savings and reductions in environmental pollutants. This seasonal system was introduced in the Netherlands by an insurance company, Anova Verzekering, in its newly renovated office in Amersfoort (IEA, 1994). The system is considered an innovative space-conditioning system. A schematic diagram of the system for both heating and cooling modes of operation is shown in Figure 4.4. In winter, an electric heat pump supplies a hydronic heating and cooling distribution system using a warm ground water aquifer as a heat source. In the heating mode, the ground water is cooled to 8°C and stored in a second water layer location. At the end of the heating season, sufficient cool water has been stored to meet the office cooling needs in summer without requiring operation of the heat pump or any other active cooling system. When cooling is needed, the cooled ground water is pumped up and used to extract heat from the hydronic system. The ground water then returns to the warm well at temperatures of between 17 and 20°C, thus providing a useful source of stored heat. Very efficient heat pump operation is expected, since the system uses a warm heat source coupled with a heat-distribution temperature of 35–40°C.

A key issue in the success of this design is the utilization of an advanced ceiling-mounted hydronic heating and cooling distribution system. This system offers significant energy savings (as detailed in section 5.7.3) and environmental benefits, as shown in Table 4.3. With a subsidy of US$212,000 from the Dutch government, representing over 20% of the total installation cost, the savings in energy costs will pay back the additional investment costs within 6.5 years. This assessment is relative to the conventional alternative technology of gas heating and electric air conditioning.

4.7.3 The Trane Company's Technology Center (La Crosse, Wisconsin, USA)

In 1994, the US Environmental Protection Agency (EPA) launched the Energy Star Showcase Buildings Program to encourage and highlight improved building energy efficiency and reduce the environmental pollution.

The Trane Company's Technology Center in La Crosse, Wisconsin, was selected as one of 24 charter participants in the program. Under the terms of an agreement signed at EPA headquarters in July 1994, Trane committed to making major improvements in its 18,580 m^2 Technology Center. Built in 1954, this building contains laboratories, product design facilities and offices. The strategy was to first make all the appropriate improvements in the building shell, lighting systems and HVAC distribution system; then to make necessary changes in the chillers and chilled water system.

HVAC improvements under the Energy Star Program consisted of improving the air distribution system in the building by evaluating and tuning up fans, dampers and actuators to make sure they were working properly. It turned out that only minimal structural changes to the air distribution system were necessary. After these changes were made, the chilled water system was assessed.

The existing cooling system for the building used two chillers with 17 modular ice storage containers. The chillers were a two-stage 300-ton centrifugal (rated at 1.0 kW/ton) and a 1984 three-stage 250-ton centrifugal (0.70 kW/ton). The newer chiller had been used as the lead chiller and the older machine in a backup capacity. In the icemaking mode, the 250-ton machine was rated at 150 tons (0.85 kW/ton). The new chiller configuration featured a 230-nominal-ton chiller used for icemaking, with a duty capacity of 152 tons at –4.5°C leaving water, and a second 400-nominal-ton machine with a rated efficiency of 0.497 kW/ton at 100% load. The second chiller was equipped with an adaptive frequency drive to further reduce demand under part-load conditions.

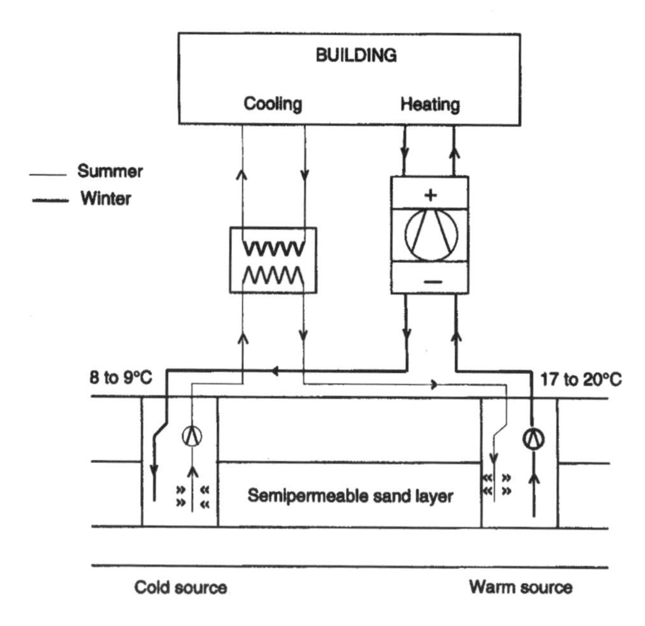

Figure 4.4 An aquifer TES space air-conditioning system (*Adapted from IEA-HPC, 1994*).

Table 4.3 Environmental benefits of the aquifer TES system*.

Emissions	Conventional system	TES-based system	Reduction for TES-based system
CO_2 (kg)	608,000	346,000	262,000 (43%)
NOx and SO_2	—	—	— (40%)

* Adapted from IEA-HPC (1994), where further details are available.

The total cost of the building improvement project was $350,000. Of this, $62,000 was rebated by the local utility, Northern States Power Company. The forecast savings from the improvements were $63,125 per year. In 1995, during only a part of which the improvements were complete, the actual savings were $36,835. The summer of 1995 was exceptionally warm, with cooling degree days over 50% higher than the previous summer. Yet actual energy performance of the building met expected levels, and a projected simple payback of 4.5 years was confirmed. This represents an internal rate of return on investment of 21.4% for the Trane Company.

From an environmental perspective, the reduced energy requirements by building improvements over 30 years represented a reduction in utility emissions as follows:

- 24.5 million kg of CO_2,
- 313,000 kg of SO_2, and
- 115,000 kg of NOx.

Additionally the project represented an electric demand reduction of over 300 kW, one-third of a megawatt of generating capacity that would not need to be built or purchased over the lifetime of the building (for details, see Trane, 1997).

4.7.4 The Ministry of Finance building (Bercy, France)

In 1987, the French government decided to build the new Ministry of Finance on the site of Bercy. The new building (260,000 m^2) is occupied by 5000 persons. It includes restaurants, banks, a post office, a nursery, etc. It is highly equipped with high technologies (security systems, automatic document carrier, etc.) The building is air conditioned by five chillers of 1100 kW each (25,000 kWh stored daily in eight tanks of 56 m^3 each). The STL is used to shave the peak demand in summer. Technical data can be summarized as follows:

- Daily cooling energy consumption: 120,000 kWh
- Maximum cooling demand: 9000 kW
- Cooling energy stored: 25,000 kWh
- Store type: STL - 00 - 448
- Number of tanks: 8

The STL is used to shave the peak demand between 8 h and 19 h. The STL is comprised of eight tanks in parallel. The STL is charged during the night by three chillers (2500 kW at −5.5/−1°C). The STL allows the Ministry of Finance to save on the operating costs and to have a backup system. The technical advantages of the TES system include:

- chiller size reduced by 40%,
- higher electrical generation plant efficiency,
- smaller heat rejection plant,
- increased lifetime of equipment,
- reduced electrical installation,
- backup at disposal, and
- reduced maintenance.

The financial advantages of the TES system can be listed as follows:

- Lower operating cost.
- Use of low cost electricity.
- Saving on demand charge.
- Saving on maintenance.

Consequently, the environmental advantages of the TES system can be highlighted as follows:

- reduced use of refrigerant by 40%, leading to ozone layer protection, and
- reduced emission of CO_2, SO_2 and N_2O (greenhouse effect).

Further information on the application may be found elsewhere (Cristopia, 1999).

4.7.5 The city of Saarbrucken (Saarbrucken, Germany)

This case study considers the city of Saarbrucken, Germany, which implemented a new energy and environment strategy in the 1980s to reduce energy consumption and CO_2 emissions. The strategy mainly involved district heating, seasonal energy storage and the use of renewable energies. The city's achievements were classified into energy savings (as detailed in section 5.7.4) and CO_2 emission reductions as follows: a 15% reduction in CO_2 emissions for the heating and electrical requirements of the municipality. As a result of this successful implementation, the city Saarbrucken received a Local Government Honor at the *United Nations Conference on Environment and Development* in Rio de Janeiro in June 1992. For further details on this case study, see OECD (1995).

4.8 Concluding Remarks

Several concluding remarks can be drawn from this chapter.

There are a number of environmental problems that we face today. These problems span a continuously growing range of pollutants, hazards and ecosystem degradation over ever wider areas. The most significant problems are acid precipitation, stratospheric ozone depletion, and global climate change. The latter is potentially the most important environmental problem relating to energy utilization. Increasing atmospheric concentrations of greenhouse gases are increasing the manner in which these gases trap heat radiated from the earth's surface, thereby raising the surface temperature of the earth and, as a consequence, sea levels.

Recently, a variety of potential solutions to the current environmental problems associated with the harmful pollutant emissions has evolved. TES appears to be one of the most effective solutions, and plays a significant role in environment policies.

Sustainable development demands a sustainable supply of energy resources that, in the long term, is readily and sustainably available at reasonable cost and can be utilized for all required tasks without causing negative societal impacts. TES systems can contribute significantly to meeting society's desire for more efficient, environmentally benign energy use, and for sustainable development of the society, particularly in the areas of building heating and cooling and electric power generation. By reducing energy consumption, the utilization of TES systems results in two significant environmental benefits: (i) the conservation of fossil fuels through efficiency increases and/or fuel substitution; and (ii) reductions in emissions of such pollutants as CO_2, SO_2, NOx and CFCs.

References

Anon. (1987). *Our Common Future*, World Commission on Environment and Development, Oxford University Press, Oxford.

Beggs, C.B. (1994). Ice thermal storage: impact on United Kingdom carbon dioxide emissions, *Building Services Engineering Research and Technology* 15(1), 756–763.

Bradley, R.A., Watts, E.C. and Williams, E.R. (1991). *Limiting Net Greenhouse Gas Emissions in the United States*, US Department of Energy, Washington, DC.

CEC. (1996). *Source Energy and Environmental Impacts of Thermal Energy Storage*, California Energy Commission, Technical Report No. P500-95-005, California.

Colonbo, U. (1992). Development and the global environment. In: *The Energy-Environment Connection* (ed. J.M. Hollander), Island Press, Washington, 3–14.

Cristopia. (1999). An STL Application: French Ministry of Finance, *Cristopia Energy Systems Catalog*, Vence, France.

Dincer, I. (1998) Energy and environmental impacts: present and future perspectives, *Energy Sources* 20(4/5), 427–453.

Dincer, I. (1999). Environmental impacts of energy, *Energy Policy* 27(14), 845–854.

Dincer, I. (2000). Renewable energy and sustainable development: a crucial review, *Renewable and Sustainable Energy Reviews* 4(2), 157–175.

Dincer, I., Dost, S. and Li, X. (1997). Performance analyses of sensible heat storage systems for thermal applications, *International Journal of Energy Research* 21(10), 1157–1171.

Dincer, I. and Rosen, M.A. (1998). A worldwide perspective on energy, environment and sustainable development, *International Journal of Energy Research* 22(15), 1305–1321.

Dincer, I. and Rosen, M.A. (1999). Energy, environment and sustainable development, *Applied Energy* 64(1–4), 427–440.

IEA-HPC. (1994). Energy Storage, International Energy Agency, *IEA-Heat Pump Center Newsletter* 12(4), p. 8.

MacRae, K.M. (1992). *Realizing the Benefits of Community Integrated Energy Systems*, Canadian Energy Research Institute, Alberta.

Norton, R. (1991). *An Overview of a Sustainable City Strategy*, Report Prepared for the Global Energy Assessment Planning for Cities and Municipalities, Montreal, Quebec.

OECD. (1995). *Urban Energy Handbook*, Organization for Economic Co-Operation and Development, Paris.

OECD. (1996). *Pollution Prevention and Control: Environmental Criteria for Sustainable Transport*, Organization for Economic Co-Operation and Development (OECD), Report: OECD/GD(96)136, Paris.

Reindl, D.T. (1994). *Characterizing the Marginal Basis Source Energy Emissions Associated with Comfort Cooling Systems*, Thermal Storage Applications Research Center, Report No. TSARC 94–1, USA.

Rosen, M.A. (1996). The role of energy efficiency in sustainable development, *Technology and Society* 15(4), 21–26.

Rosen, M.A., Dincer, I. and Pedinelli, N. (2000). Thermodynamic performance of ice thermal energy storage systems, *ASME-Journal of Energy Resources Technology* 122(4), 205–211.

Speight, J.G. (1996). *Environmental Technology Handbook*, Taylor & Francis, Washington, DC.

Trane. (1997). Energy star building performance showcased at Trane Technology Center, *Trane Commercial/Industrial Case Studies Catalog*, USA.

5

Thermal Energy Storage and Energy Savings

I. Dincer and M. Rosen

5.1 Introduction

Thermal Energy Storage (TES) is a key component of many successful thermal systems. TES should allow for the minimum reasonable thermal energy losses and the corresponding energy savings, while permitting the highest appropriate extraction efficiency of the stored thermal energy. This chapter deals with the methods for describing and assessing TES systems, and practical energy-saving applications provided by using TES systems.

Consequently, the TES systems and their practical applications and design and selection criteria are examined in this chapter. Further, energy-saving techniques and applications are discussed and highlighted with illustrative examples.

TES is considered by many to be an *advanced energy technology*, and there has been increasing interest in using this often essential technology for thermal applications such as hot water, space heating, cooling, air conditioning, etc. TES systems have enormous potential for permitting more effective use of thermal energy equipment and for facilitating large-scale energy substitutions. The resulting benefits of such actions are especially significant from an economic perspective. In general, a coordinated set of actions has to be taken in several sectors of an energy system for the maximum potential benefits of TES to be realized. TES appears to be the best means of correcting the mismatch that often occurs between the supply and demand of thermal energy. More broadly, TES can contribute significantly to meeting society's needs for more efficient, environmentally benign energy use. Each of the two main types of TES systems, sensible (e.g. water, rock) and latent (e.g. water/ice, salt hydrates), offers economic and other advantages, depending on the application. The selection of a TES system mainly depends on the storage period required, i.e. diurnal or seasonal, and such other factors as economic viability, operating conditions, etc. In practice, many research and development activities have concentrated and continue to concentrate on efficient energy use and energy savings, leading to a broad array of energy-conservation measures. In this regard, TES appears to have a major role to play as it often is an attractive thermal technology.

TES generally involves the temporary storage of high- or low-temperature thermal energy for later use. Examples of TES applications include the storage of solar energy during the day for overnight heating, of summer heat for winter use, of winter ice for space cooling in summer, and of heat or cool generated electrically during off-peak hours for use during subsequent peak demand hours. Solar energy, unlike energy from fossil fuels, is not available at all times. Even cooling loads, which nearly coincide with maximum levels of solar radiation, are often present after sunset. TES provides an important mechanism to offset the mismatch between thermal-energy availability and demand in this application.

TES can also aid in the efficient use and provision of thermal energy in other situations where there is a mismatch between energy generation and use. Various TES processes have been investigated and developed for building heating and cooling, industrial energy-efficiency improvement, and utility power systems. The period of storage is clearly an important factor. Diurnal storage systems have certain advantages: capital investment and energy losses are usually low, units are smaller and can easily be manufactured off-site, and the sizing of daily storage for each application is not nearly as critical as it is for larger annual storage systems. Annual storage systems are likely to be economical only in multi-dwelling or industrial park designs. Such systems often require expensive energy distribution networks and novel institutional arrangements related to ownership and financing. In solar TES applications, the optimum energy storage duration is usually the one which offers the final delivered thermal energy at minimum cost, when integrated with the collection system and back-up in the final application.

The economic justification for TES systems usually requires that the annualized capital and operating costs be less than the annualized costs of primary generating equipment supplying the same service loads and periods. TES is usually installed for two major reasons: (i) to lower initial costs, and (ii) to lower operating costs. Lower initial costs are usually possible when the thermal load is of short duration and there is a long time gap before the load returns because a small storage is adequate in such instances. Secondary capital costs may also be lower for systems incorporating TES. For example, the electrical service capacity can sometimes be reduced because energy demand is lower.

In order to perform a comprehensive economic analysis of TES, the initial costs must be determined. Equipment costs can be obtained from relevant manufacturers and estimates of installation costs made. The cost savings and the net capital costs can be analyzed using the life cycle cost method or other suitable methods to determine which system is most suistable for a given application.

Other items to be considered in TES economic analyses are space requirements and system reliability, and the interface to the delivery system for the application. An optimal energy storage application achieves a balance between maximizing the savings accrued in utility charges and minimizing the initial cost of the installation needed to achieve the savings. Consequently, the decision to install a storage system must be based on anticipated system loads, load characteristics, and generating capacity mix for an extended period. Uncertainty about the future economic outlook, life style changes, and the availability of low-cost energy charging the storage system may lead to differing investment decisions if alternative technical solutions are feasible. These uncertainties may vary temporally and spatially. The technical characteristics of alternative technologies for situations in which TES systems are potentially attractive may also affect decisions.

In this chapter, TES systems and their applications are examined from an energy savings perspective, and possible energy saving technologies are discussed in detail and highlighted with illustrative case studies of actual systems.

5.2 TES and Energy Savings

TES systems are an important element of many energy saving programs in a variety of sectors, residential, commercial, industrial and utility, as well as in the transportation sector.

TES can be employed to reduce energy consumption or to transfer an energy load from one period to another. The consumption reduction can be achieved by storing excess thermal energy that would normally be released as waste, such as heat produced by equipment and appliances, by lighting, and even by occupants. Energy-load transfer can be achieved by storing energy at a given time for later use, and can be applied to TES for either heating or cooling capacity.

The main objective of most TES systems, which is to alter energy-use patterns so that financial savings occur, can be achieved in several ways (Dincer *et al.*, 1997a):

- The consumption of purchased energy can be reduced by storing waste or surplus thermal energy available at certain times for use at other times. For example, solar energy can be stored during the day for heating at night.
- The demand of purchased electrical energy can be reduced by storing electrically produced thermal energy during off-peak periods to meet the thermal loads that occur during high-demand periods. There has been an increasing interest in the reduction of peak demand or transfer of energy loads from high- to low-consumption periods. For example, an electric chiller can be used to charge a chilled water storage system at night to reduce the electrical demand peaks usually experienced during the day.
- The use of TES can defer the need to purchase additional equipment for heating, cooling or air conditioning applications and reduce equipment sizing in new facilities. The relevant equipment is operated when thermal loads are low to charge the TES, and energy is withdrawn from storage to help meet the thermal loads that exceed equipment capacity.

Each of these points is discussed separately in the following three subsections.

5.2.1 Utilization of waste or surplus energy

If a TES system is installed and charged using waste heat otherwise released to the environment, and if the energy is held and later used in place of added primary energy, overall energy consumption is reduced. To be economically feasible the cost of the replaced primary energy should exceed the capitalization, maintenance and operating costs of the TES system. The stored energy can in a sense be considered free, since it would otherwise be lost.

Useful waste or surplus thermal energy is available from many sources. Some examples are (i) hot or cold water drained to a sewer, (ii) hot flue gases, (iii) exhaust air streams, (iv) hot or cold gases or waste gases, (v) heat collected from solar panels, (vi) ground source thermal energy, (vii) heat rejected from the condenser of refrigeration and air conditioning equipment, and (viii) the cooling effect from the evaporator of a heat pump.

Many of the TES applications in this category are designed for load leveling rather than waste energy recovery. The thermal energy stored is then in a higher-grade rather than waste condition, being drawn from the conversion equipment during periods of low end-use demand for thermal energy. Such TES systems do not reduce energy use, and may actually cause it to increase due to TES inefficiencies. For example, the overall energy consumption for a task supplied using a storage having an overall energy efficiency of 75%, will be one-third (i.e. 1/0.75 – 1) greater than the energy consumption using a direct primary energy supply. The objectives of such systems are clearly not to reduce energy consumption, but rather to either reduce costs or allow the displacement of scarce fuels by more abundant fuels in an energy process.

Tomlinson and Kannberg (1990) point out that industrial production uses about a third of the total energy consumed in the USA, much of it as hydrocarbon fuels. Therefore, energy efficiency improvements in the industrial sector can have a substantial impact on national energy consumption levels. TES represents an important option for improving industrial energy efficiency. By storing and then using thermal energy that would otherwise be discharged in flue gases to the environment, less purchased fuel is used, plant thermal emissions are reduced, and product costs associated with fuel use are decreased. The following six industries, which account for approximately 80% of total US industrial energy use, have the highest potential for energy savings through implementation of TES: aluminum, brick and ceramic, cement, food processing, iron and steel, and paper and pulp. Most existing TES systems in industry are found in iron and steel plants where they are used as regenerators to preheat air to about 600°C. Opportunities exist for the reclamation of waste heat from stack gases in other industries as well. Estimates have shown that TES can result in potential energy savings in US industry of as much as 3 EJ per year.

In general, TES can reduce the time or rate mismatch between energy supply and energy demand, thereby playing a vital role in improved energy management. TES use can lead to savings of premium fuels and make a system more cost-effective by reducing waste energy. TES can improve the performance of thermal systems by smoothing loads, and increasing reliability. Therefore, TES systems are becoming increasingly important in many utility systems.

5.2.2 Reduction of demand charges

A major application of TES is to lower electrical demand and thus reduce electrical demand charges. Reduction in demand charges is accomplished by eliminating or limiting electrical input to electrically operated heating or cooling devices during the peak electrical demand periods for a facility. The devices are operated before the peak occurs (e.g. overnight) to charge TES systems. During the peak demand period the heating or cooling equipment does not operate or operates at reduced levels, and the thermal loads are met with the heating or cooling capacity of the storage.

The electrical source which powers the heating or cooling equipment can be shut off, or have power limiters installed, to reduce electrical demand during the peak periods. A number of devices can be energized and de-energized in accordance with TES operating strategies. Some examples are equipment for building heating, cooling and air conditioning, domestic water heating, process heating and cooling, refrigeration, snow melting, drying, ice-making, etc.

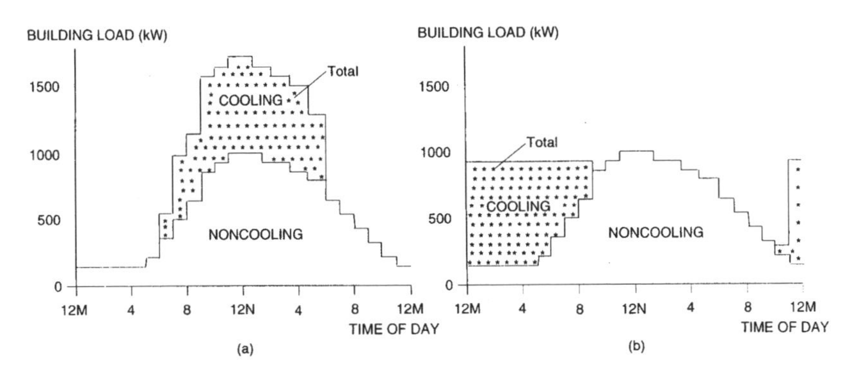

Figure 5.1 Daily load profiles of a building energy load: (a) no storage and (b) full storage (Dincer *et al.*, 1997b).

The fundamental purpose of cool storage is to provide a buffer between the chiller and the building cooling load, thereby decoupling the chiller capacity and operating schedule from the building load profile, leading to energy consumption and demand savings and economic benefits through electrical load management. This application of TES can be beneficial in several ways, regardless of the chiller energy source. In many practical applications, the intention is to maximize the utilization of efficient base load generating plants and avoid the need for additional capacity. The benefits often justify offering rate structures that favor load shifting and peak shaving, and sometimes, financial incentives to reduce the cost of storage.

Figure 5.1 shows example daily load profiles for a building, with and without cool storage. Figure 5.1a represents the case with no storage, and Figure 5.1b shows a full-storage case. In the latter case, the TES provides enough storage capacity to meet the peak (i.e. 9:00 am to 9:00 pm) cooling load, shifting the entire electrical demand for cooling to off-peak hours when there is little cooling load. This particular application achieves maximal benefits in terms of demand charge reduction and use of favorable off-peak rates, both of which to lower operating costs.

5.2.3 Deferring equipment purchases

The capacity of heating and cooling equipment is normally selected to match the part of the design-day load when the requirements for heating and cooling are close to maximum. These peak design loads occur for only short periods of time, resulting in excess capacity on average days. TES systems take advantage of the difference between the peak and average thermal loads to provide an opportunity to defer equipment purchases in a retrofit application, or reduce the equipment size in a new installation.

For example, consider a building with an average cooling load of 500 kW and a peak cooling load of 650 kW. The capacity of the existing chiller is 750 kW. A proposed expansion will increase the average cooling load to 700 kW and the peak to 850 kW. The new average load could be satisfied with the existing equipment, but not the peak. An additional chiller with a capacity 100 kW is required, based on the conventional method. As an alternative to providing a new chiller, a TES system could be incorporated to satisfy the peak cooling load. During off-peak hours, when the thermal load is less than the capacity of the existing chiller, the chiller would operate to maintain the desired building or process conditions, and excess capacity would be used to charge a chilled water TES system (or other cool TES). When the cooling load exceeds the chiller capacity, chilled water would be drawn from storage. The benefits of the TES option can include capital savings and reduced operating costs. The reduced operating costs result from limiting the peak electrical load (and corresponding demand charge) to that required to provide only 750 kW of cooling instead of 850 kW, and from having less chilling equipment to maintain. Note that in either case the annual electrical consumption for cooling increases roughly in proportion to the new cooling loads (Anon, 1985).

The technique illustrated in the above example can also be used in new facilities. Then, TES permits the capacity of the thermal equipment to be selected closer to the average rather than the peak condition.

5.3 Additional Energy Savings Considerations for TES

The complete assessment of a TES application in a given facility requires an appreciation of several other criteria: energy requirements for heating, refrigeration and heat pump equipment, storage size limitations, thermal load profiles, and optimization of conventional systems (Anon., 1985; Dincer *et al.*, 1997b). These topics are discussed in the following subsections

5.3.1 Energy for heating, refrigeration and heat pump equipment

Electricity can be converted to thermal energy by electric resistance elements or by mechanical means. In resistance heating systems, each unit of electricity (1 kWh) is converted to the heat (1 kWh), and the conversion efficiency is at or very near to 100%. Typical examples are electric baseboard heaters, electric water heaters and slab heating systems.

Where refrigeration and/or heat pump systems are used to produce heating and/or cooling, the conversion efficiency, which is referred to as a coefficient of performance (COP), is normally greater than 100%. Heat pump systems require a heat source, which can be the outdoor environment or a waste heat stream. In many typical systems the COP is approximately 3.5, where each unit of electrical input (e.g. 1 kWh) to the equipment produces about 3.5 kWh of heating or cooling. For example, a refrigeration system with a COP of 3.5 producing cooling at a rate of 350 kW has an electric power requirement of 100 kW (350 kWh/3.5). This behavior is important when TES systems, which aim at demand reduction, are being considered. When existing systems are being considered for conversion to TES arrangements, actual equipment COPs should be obtained from the manufacturers. Note that high-efficiency heat pump and refrigeration systems can have COP values well in excess of 3.5.

Refrigeration system energy consumption per unit of cooling capacity increases as the evaporator temperature is reduced and as the condenser temperature increases. Producing ice at 0°C, therefore, requires more energy than producing chilled water at 4°C. Conversely, a heat pump requires more electrical power to produce hot water at 50°C than at 35°C. When the temperature variations are minor, the variations in electrical consumption can be approximated as negligible in many practical applications.

5.3.2 *Storage size limitations*

TES systems find application possibilities in a range of capacities, from only a few hours to seasonal storage. An example of seasonal storage is the collection of solar energy available during the summer for use in winter heating. Practical limitations such as space requirement and capital cost often restrict TES schemes to storage durations of a few hours to a few days.

As an example of these difficulties, consider a building or process requiring 1000 kW of chilling capacity for 900 h of full-load operation. For a TES, the annual cooling energy that must be stored (ignoring losses) is 900 h × 1000 kW = 900,000 kWh. If ice is used as the storage medium, and hence the quantity of ice required is 9,720,000 kg or 9720 m^3, since about 3 kg of ice is required to store 1 MJ of cooling capacity. Assuming a perfect tank with no standby heat losses and no oversize margin, a tank 2 m deep would cover an area the size of a football field. A water storage system would be at least ten times larger. Costs for such large systems are unacceptable. For reasons illustrated in this example, greater attention has been paid to short term or diurnal TES systems, especially latent heat TES systems using PCMs.

5.3.3 *Thermal load profiles*

When thermal loads fluctuate, the potential exists for storing thermal energy to meet later energy requirements. Many buildings and units have load profiles conducive to TES.

Office buildings with low cooling requirements overnight and during the morning, and high cooling demands in the late afternoon, exhibit the optimum profile for TES. Often, the air-conditioning systems of these buildings are shut off at night. The combination of daytime part loads and night-time shutdown provides the daily equivalent of 15–20 h when a chilled water storage could be charged to meet the demand period.

Busy facilities, e.g. hotels, hospitals and industrial plants, are less likely candidates for full-storage systems as their load profiles are flatter. Less time is therefore available for charging the storage between the long cooling periods. However, these facilities are often suited to partial TES and peak shaving systems. For example, the base thermal load may be met by a chiller, and load peaks reduced using a combination of chiller and storage operation. In this case, the chiller contributes to the peak electrical demand, but by a lower amount than without TES.

The return on capital investment for a TES system can be maximized by careful sizing equipment. Often, compressor waste heat can be used for preheating domestic or process hot water. This by-product utilization can further improve the economics of a system.

Weather also affects the thermal load profile of a building, and is consequently a major factor in determining the feasibility of TES. Cooling storage is often advantageous in facilities where the summer weather profile includes a limited number of peak demand days and large temperature variations during a given 24 h period. Where TES for heating is considered, a minimum of 2200 degree days below 18°C are usually required to make the project viable (Dincer *et al.*, 1997b).

5.3.4 Optimization of conventional systems

Existing heating and cooling systems should be upgraded where possible and properly maintained to reduce energy inefficiencies before implementing an active TES system. When reviewing conventional systems, the possibility should be considered of heat recovery from exhaust streams, e.g. boiler flue gases. The possibility should also be examined of changing from batch to continuous industrial processes so that direct heat recovery, without intermediate TES, can be used. This mode of operation often results in more effective heat recovery, no standby thermal losses, and reduced capital expenditures.

Control flexibility is an important tool for optimizing building TES systems because exact operating modes and schedules vary continuously and are difficult to predict accurately. Systems designed with manual control or state-of-the-art computerized controls can be equally effective, provided system operation can be readily adjusted to meet actual site conditions.

Monitoring of TES system operation is required to track system performance and to identify operating problems and potential areas for future improvements. Small systems can be monitored with standard meters, gauges and manual entry logs. Electronic instrumentation and control systems with automatic data logging, trend analysis and other features are generally used in larger systems.

5.4 Energy Conservation with TES: Planning and Implementation

TES plays an important role in many energy conservation initiatives. In processes with large energy wastes, energy storage can result in a savings of premium fuels.

Energy may be stored in many ways, e.g. mechanical and chemical energy storage, but since in many economies energy is produced and transferred as heat, the potential benefits of TES in energy conservation warrants detailed study. Thermal energy can be stored by cooling, heating, melting, solidifying or vaporizing a material, the thermal energy becoming available when the process is reversed. Thermal storage by causing a rise or drop in material temperature is called sensible heat storage. Its effectiveness depends on the specific heat of the material and, if volume is important, the density of the storage material. Storage by phase change (solid to liquid or liquid to vapor) with no change in temperature is known as latent heat storage. Short-term storage is often used to manage peak power loads of a few hours to a day long in order to reduce the sizing of systems and/or to take advantage of the daily structure of energy tariffs. Long-term storage is possible when seasonal energy loads can be transferred over periods of weeks to several months, to cover seasonal needs. This type of storage is also called seasonal storage.

TES has a significant role to play in energy conservation efforts, and the following are the main steps in implementing an energy conservation strategy involving TES:

1. **Defining the main direct goals:** in a systematic way, identify clearly the goals of the project. This step should use an organized framework which facilitates deciding priorities and identifying the resources needed to achieve the goals.

2. **Identifying community goals:** community priorities and issues involving energy use, energy conservation, the environment and other local issue should be identified. Also, the institutional structures and barriers, and financial instruments should be identified.
3. **Scanning the environment:** the main objective in this step is to develop a clear picture of the community and to identify energy- and resource-related problems facing the community and its electrical and gas utilities, the existing organizational structures and base data for evaluating the future progress of the program. Communication with local and international financial institutions, project developers and bilateral aid agencies can help capture new initiatives and explain lessons learned and viewpoints on problems and potential solutions.
4. **Increasing public awareness:** governments can increase potential customers' awareness and acceptance of energy conservation programs by entering into performance contracts for government activities and publicizing the results. Also, international workshops to share experiences help to overcome the initial barrier of unfamiliarity in countries.
5. **Building community support:** obtaining the participation and support of local industries and public communities for an initiative requires understanding the nature of conflicts and barriers between given goals and local actors; improving information flows; promoting education and advice activities; identifying institutional barriers; and involving a broad spectrum of citizen and government agencies.
6. **Analyzing information:** this step includes defining available options and comparing possible options in terms of factors, e.g. program implementation costs, funding availability, utility capital deferral, potential for energy efficiency, compatibility with community goals, environmental benefits, and so on.
7. **Adopting policies and strategies:** high-priority projects need to be identified through approaches which are the best for the community. The decision process should evaluate options in terms of savings in energy costs, generation of businesses and tax revenues and the number of jobs created, as well as their contribution to energy sustainability and other community and environmental goals.
8. **Developing the plan:** a specific plan of measures and activities should be developed. Once the draft plan has been adopted, it is important for the community to review it and comment upon it. The public consultation process may vary, but a high level of approval should be sought.
9. **Implementing future programs:** this step involves deciding which future programs to concenrate on, with long-term aims being preferred over short-term aims. The options that have the greatest impact should be focused on, and all details defined. Potential financial resources need to be identified to implement the programs.
10. **Evaluating success:** the final stage involves evaluating and assessing how well the strategy performs, and helps to detect its strengths and weaknesses and to determine who is benefiting from it.

5.5 Some Limitations on Increased Efficiency

In terms of increased energy efficiency, there are a number of theoretical and practical limitations that apply to TES as well as other processes.

5.5.1 Practical and theoretical limitations

The contributions that increased energy efficiency can make towards sustainable development are theoretically limited, because there exists a limit on the maximum efficiency attainable for any process. Such limitations are a consequence of the laws of thermodynamics (Moran, 1989). This concept, when applied to TES, implies that an ideal storage is one in which all of the input energy is restored after storage with no degradation of quality (i.e. temperature) and with complete recovery of energy used to drive the process (e.g. electricity to pumps).

In conventional engineering, the goal when selecting energy sources and utilization processes is not to achieve maximum efficiency. Rather, the goal is to achieve an optimal trade-off between efficiency and such factors as economics, sustainability, environmental impact, safety, and societal and political acceptability. Consideration of these factors leads to practical limitations on increased energy efficiency. For energy efficiency to have an increased contribution towards sustainable development, the position of the optimum among these factors will have to shift towards increased energy utilization efficiency (while recognizing the theoretical limitations on increased energy efficiency).

To assess the potential of increased energy efficiency as a measure for promoting sustainable development, the limits imposed by the existence of maximum theoretical energy efficiencies must be clearly understood. Lack of clarity on this issue has in the past often led to confusion and misunderstanding. Part of the reason for this problem is that conventional energy analysis often does not evaluate efficiencies as a measure of how nearly the performance of a process approaches the ideal, or maximum possible. The difficulties inherent in energy analysis are in part attributable to the fact that such an analysis methodology considers only energy quantities, and ignores energy quality and the fact that energy quality is continually degraded during processes. Here, higher quality energy forms are taken to be those that can be used for a wider range of tasks, e.g. high-temperature steam is more useful than lower temperature steam as the hotter steam can satisfy all the heating uses of the lower temperature steam and more.

5.5.2 Efficiency limitations and exergy

One way to deal with energy forms of different qualities is to consider exergy. The exergy of a quantity of energy or a substance is a measure of its usefulness or quality, or a measure of its potential to cause change. Exergy analysis has recently been proposed by many scientists and engineers as a technique for thermodynamic assessment that overcomes most, if not all, of the problems associated with energy analysis (e.g. Moran, 1989; Rosen and Dincer, 1997a; 1997b). In practice, the authors feel that a thorough understanding of exergy and how exergy analysis can provide insights into the efficiency and performance of energy systems are required for the engineer or scientist working in the area of energy systems and the environment.

As environmental concerns such as pollution, ozone depletion and global climate change became major issues in the 1980s, environmental concerns come to represent another factor related to efficiency limit. Consequently, interest developed in the link between energy utilization and the environment. Since then, there has been increasing attention to this linkage. Many scientists and engineers suggest that the impact of energy-resource utilization on the environment is best addressed by considering exergy. Exergy appears to be an effective measure of the potential of a substance to impact the environment. Although many studies exist concerning the close relationship between energy and the environment, there have been limited works on the link between exergy and environment concepts (Rosen and Dincer, 1996; 1997a).

Another use of exergy in TES work is in comparisons. For many years TES systems have been investigated and applied. These experiences have shown that although many technically and economically successful TES systems exist, no broadly valid basis for comparing the achieved performance of one storage with that of another operating under different conditions has found general acceptance. The development of such a basis for comparison has been receiving increasing attention, especially using exergy methods. Exergy analysis, which is identified as one of the most powerful ways of evaluating the thermal performance of TES systems, is based primarily on the second law of thermodynamics, as compared to energy analysis which is based on the first law, and takes into account the quality of the energy transferred.

5.6 Energy Savings for Cold TES

In a TES for cooling capacity, 'cold' is stored in a thermal storage mass. As shown in Figure 5.2, the storage can be incorporated in an air conditioning or cooling system in a building. In most conventional cooling systems, there are two major components (Dincer and Rosen, 2001):

- a chiller—to cool a fluid such as water, and
- a distribution system—to transport the cold fluid from the chiller to where it cools air for the building occupants.

In conventional systems, the chiller operates only when the building occupants require cold air. In a cooling system incorporating TES, the chiller also operates at times other than when the cooling is needed.

During the past two decades, TES technology, especially cold storage, has matured and is now accepted by many as a proven energy conservation technology. However, the predicted payback period of a potential cold storage installation is often not sufficiently attractive to give it priority over other energy efficient technologies. This determination is often made because full advantage is not made of the many potential benefits of cold storage or because the cold storage sizing is not optimized. Some recommendations for optimizing the payback period of cold TES systems follow.

For new facilities, cold storage should be integrated carefully into the overall building and its energy systems so that full advantage is taken of the potential benefits of cold TES, including:

- reduced pipe and pump sizes for chilled water distribution,
- reduced duct and fan sizes for low temperature air distribution,
- reduced operating and maintenance costs,
- reduced electricity consumption and therefore energy costs, and
- increased flexibility of operation.

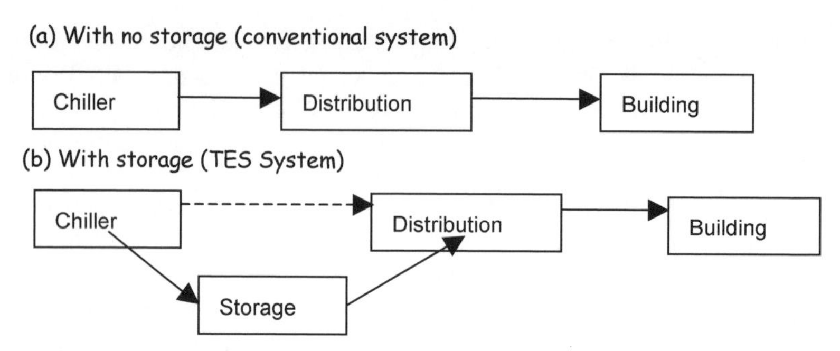

Figure 5.2 Schematic representation of two building cooling TES systems: (a) with no storage and (b) with storage.

Smaller chiller and electrical systems lead to initial cost advantages. The sizing of the cold storage system should be optimized, as opposed to the typical process of considering full storage and one or two levels of partial storage versus a conventional system. A practical method to assist in determining the optimum system size should be developed. Also, the value should be accounted for of the gain in usable building space due to less space being required for mechanical system components when cold TES is used.

For existing facilities, potential advantages of cold TES which should be evaluated include:

- modifying the existing chillers to make ice versus the purchase of a new machine,
- using spare chiller capacity by adding a cold TES system,
- using cold storage to increase cooling capacity in situations where chiller and electrical service capacity are fully utilized,
- sizing the cold storage system optimally as opposed to taking the best of only a few options, and
- using available low temperature air and water to advantage through 'free cooling', where practical.

In summary, a cold TES system can benefit users in three ways:

- **Lower electricity rates**. With cold TES, chillers can operate at night to meet the daytime cooling needs, taking advantage of lower off-peak electricity consumption rates.
- **Lower demand charges**. Many commercial customers pay a monthly electrical demand charge based on the largest amount of electricity used during any 30-minute period of the month. Cold TES reduces peak demands by shifting some of those demands to off-peak periods. Furthermore, some utilities provide a rebate for shifting electrical demand to night-time or other off-peak periods.
- **Lower air conditioning system and compressor costs**. Without cold TES, large compressors capable of meeting peak cooling demands are needed, whereas smaller and less expensive units are sufficient when cold TES is used. Also, since water from a cold TES may be colder than conventional chilled water, smaller pipes, pumps and air handlers may be integrated into the building design to reduce costs further.

5.6.1 Economic aspects of TES systems for cooling capacity

TES-based systems are usually economically justifiable when the annualized capital and operating costs are less than those costs for primary generating equipment supplying the same service loads and periods. TES is mainly installed to lower initial costs of the other plant components and operating costs. Lower initial equipment costs are usually obtained when large durations occur between periods of energy demand. Secondary capital costs may also be lower for TES-based systems. For example, the electrical service equipment size can sometimes be reduced when energy demand is lowered.

In complete economic analyses of systems including and not including TES, the initial equipment and installation costs must be determined, usually from manufacturers, or estimated. Operating cost savings and the net overall costs should be assessed using life cycle costing or other suitable methods to determine which system is the most beneficial.

Utilizing TES can enhance the economic competitiveness of both energy suppliers and building owners. For example, one study for California indicates that, assuming 20% statewide market penetration of TES, the following financial benefits can be achieved in the state (CEC, 1996):

- For energy suppliers, TES leads to lower generating equipment costs (30% to 50% lower to serve air conditioning loads), reduced financing requirements (US$1–2 billion), and improved customer retention.
- For building owners statewide, TES leads to lower energy costs (over one half billion US dollars annually), increased property values (US$5 billion), increased financing capability (US$3–4 billion), and increased revenues.

5.6.2 Energy savings by cold TES

Cold TES has been shown to be able to reduce building cooling costs, which can be significant. Stored cooling capacity can be used either to meet the total air conditioning load so that chillers remain off during the day, or to supplement the chiller so it only has to satisfy part of the load.

Numerous cities throughout the world, including many in the United States, are faced with increasingly high energy costs. Often, these costs are in large part due to electrical demand charges in addition to energy consumption costs. Many electrical utilities experience difficulties in maintaining sufficient capacity to meet the peak customer demand while at the same time supplying reasonably priced electricity. One way to defer or avoid the construction of new power plants is to level local electrical loads over time. Such leveling can be achieved in part by shifting the electrical loads in buildings due to heating, ventilating, and air conditioning equipment to periods of lower overall electrical usage. This load shifting can be accomplished by applying TES technologies, and utility companies and governments in many countries offer incentives to encourage such uses of TES.

Strong interest in TES systems for commercial buildings led to the Air-Conditioning and Refrigeration Institute (ARI) in the USA to establish in May 1997 a new product section, Thermal Storage Equipment, to promote the attributes of TES and to develop a standard for rating the efficiency of TES equipment. Members of the product section have identified many TES case studies illustrating the technical impacts and financial benefits of TES use. The ARI is the national trade association representing manufacturers of more than 90% of US produced central air conditioning and commercial refrigeration equipment.

Energy saving strategies for cold TES

Three basic strategies are typically employed for reducing peak electricity use with TES, as shown in Figure 5.3: full, near full, and partial storage. With full storage, the chiller and storage tank are sized so that the chiller does not run during the peak hours even on the hottest days, while with partial storage that equipment is downsized and the smaller chiller runs continuously on hot days. Thus full storage allows electricity costs to be lowered significantly, while partial storage reduces TES system capital costs. To achieve some of the benefits of both modes of operation, one can also utilize a near-full storage strategy, in which the chiller runs at a reduced level during peak hours.

With these strategies, five major types of TES system are usually used, as shown in Table 5.1. The first type, which uses chilled water as the storage medium, has the advantage of being compatible with existing chillers, and is usually more efficient than the other types. However, this TES type requires larger storage tanks than the other types, which use different storage media. The second type of TES, which uses a 'eutectic salt' water solution as the storage medium, stores cooling capacity by freezing the storage solution at a temperature typically near 8.3°C. The main advantages of this TES type are that (i) by storing cold through a phase change (freezing) smaller tanks are required than for chilled water, and (ii) by freezing at 8.3°C, standard chillers producing 5°C chilled water in commercial facilities can be used. The main disadvantage is that the tank typically cools the water for the distribution system to only 8.5–10°C, which accomplishes less building dehumidification and requires more pumping energy.

The last three types in Table 5.1 have ice as the storage medium, and differ in how the 'cold' from the ice is distributed throughout the building. Before considering the differences in the distribution systems, consider the features of their common components (ice storage and chiller).

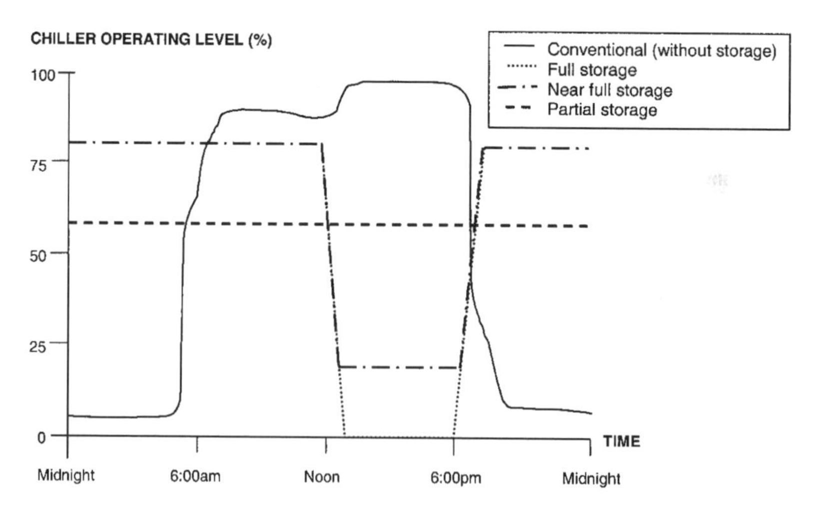

Figure 5.3 Comparison of conventional and cold TES systems for electricity use.

Table 5.1 A list of major TES cooling system types.

Chiller	Storage	Distribution
Conventional	Chilled water	Conventional water
Conventional	Eutectic salt/water solution	Conventional water
Ice-making	Ice	Conventional water
Ice-making	Ice	Cold air
Ice-making	Ice	Unitary (Rooftop)

Source: CEC (1996).

The main advantage of ice storage is compactness, which can be a significant benefit where space is a premium, as ice tanks often are 10% to 20% of the size of comparable chilled water tanks, and 30% to 50% the size of eutectic salt tanks. Additional benefits of ice storage systems, when used with cold air or rooftop distribution systems, are increased dehumidification and fan energy savings. The major disadvantage of ice systems is that they are not compatible with most conventional chillers that use cold water, and so ice-making chillers must be used, which use more electricity than conventional water chillers because of the lower temperatures required to freeze water.

Although ice storage systems can be used with conventional chilled-water distribution systems, they are particularly beneficial when the distribution system (fans and ducts) is designed to take advantage of the lower temperatures available to produce cold air, and correspondingly downsized. The benefits of such downsizing include lower distribution-system initial costs, lower distribution system energy use for fans and pumps (by 40% or more), and smaller duct passages, which can mean lower floor-to-floor heights in buildings, allowing architects to design additional floors without increasing building height, and lower net costs per unit area of floor space.

The first four TES types listed in Table 5.1 are used mostly with typical chilled-water distribution systems in larger buildings. The last type is used with unitary systems, including those used in typical single-family residences having an outdoor condensing unit and indoor coil as well as gas or electric heating, or having heat pumps and related air handlers. Unitary systems also include single-package systems that are roof mounted on low-rise commercial buildings and, in certain geographical locations, some residences. These unitary systems use a 'direct expansion' process in which the refrigerant, not chilled water, cools the air that is delivered directly to the structure. Unitary systems are typically small, air cooled and not as efficient as most of the water-cooled chilled water systems used in larger buildings. Because of their lower efficiencies, air-cooled unitary systems may undergo significant improvement efforts in the near future.

Analysis of TES savings

A major focus of this study is determining the increase or decrease in energy use and demand due to TES. TES generally reduces the fuel or energy required and changes the time at which electricity is used. In order to quantify the source energy impact of TES and to calculate the source energy savings, the incremental energy method (or marginal plant method) can be used (CEC, 1996). The incremental energy method is consistent with evaluation methods for several TES programs. With the method, decisions about which resource to use are based on how the costs of providing power change for a marginal or incremental change in electricity use from current levels. Many believe that marginal costs should be used in the design of electric rates so that they lead energy users to utilize energy resources wisely.

Following these concepts, a Standard Practice Methodology (SPM) is commonly used for evaluating the cost-effectiveness of both new supply resources and Demand-Side Management (DSM) measures (CEC, 1996). DSM measures include programs for energy efficiency (often aimed at reducing electrical energy use) and load management (primarily aimed at reducing electrical peak demand), are often considered as a special type of supply-side resources in resource planning decisions. The SPM evaluates DSM programs by comparing electrical energy and demand savings to the marginal costs for providing those quantities. This methodology has gained international acceptance as a rational way to evaluate DSM programs—including TES. With the SPM approach, DSM program saving can be expressed as follows:

$$D = \sum_{i=1}^{n}[(\text{kWh savings})_i(\text{marginal cost of kWh})_i] + \sum_{i=1}^{n}(\text{kWh savings}) \tag{5.1}$$

where n denotes the number of time periods in the program. That is, DSM program cost savings are evaluated by determining the energy consumption and demand savings in each of n time periods, multiplying each of those savings by the marginal costs for that time period, and then summing the cost savings over all time periods. Using the SPM, the year can be divided into n time periods, each with different marginal costs (for both kW and kWh). Note that utility companies often define the summer peak period as working weekdays from noon to 6:00 p.m., and that the winter peak period is much less dominant than the summer peak in determining new (marginal) capacity decisions.

The 'marginal electrical energy cost' (in \$/kWh), referred to in Equation 5.1 as 'the marginal cost of a kWh', for a time period equals the cost of fuel (in \$/kWh) multiplied by the average heat rate (or the incremental energy rate R). Thus, R can be expressed (in kWh fuel/kWh electricity):

$$R = \text{marginal electrical energy cost (\$/kWh)/marginal cost of plant's fuel (\$/kwh)} \tag{5.2}$$

Equation 5.2 can be substituted into the energy terms in Equation 5.1, after dividing all marginal energy cost terms in Equation 5.1 by the marginal fuel cost, to develop an expression for source energy savings:

$$DS = \sum_{i=1}^{n}[(\text{kWh savings})_i R_i] \tag{5.3}$$

When evaluating source energy savings with Equation 5.3 or other benefits of TES, care should be taken to account appropriately for the difference between utility source energy use (i.e. fuel used at the power plant to generate electricity), the electricity generated at the power plant, and the electrical energy provided to the user site. To transfer electricity over power lines from the power plant to the user, energy is lost due to resistance in the power lines (line losses). Line losses are often neglected even though they are sometimes significant (e.g. 10%). Of particular significance in TES assessments are the facts that line losses vary temporally, being greater when the lines are more fully loaded and when the ambient temperature is higher. Thus, line losses are often higher during summer peak hours, so TES can reduce energy use by shifting electricity use to times of lower line losses.

Assessments do not always account for line losses as utilities are concerned about marginal costs sometimes at the power plant (or generation) level, and at other times at the user site (or distribution) level. When evaluating the marginal energy cost (M) at the distribution level, the generation-level marginal costs are increased to reflect the line losses to the distribution level, as follows:

$$M(\$/\text{kWh at site}) = M \times R \times LLF \tag{5.4}$$

where MFP is the fuel marginal price at the power plant (in \$/kWh of fuel) and LLF is the line loss factor, evaluated as the ratio of kWh electric exiting the power plant to the kWh electric delivered to the site.

Finally, the TES-derived source energy savings (T) can be determined considering all assessment periods and accounting for line losses as

$$T = \sum_{i=1}^{n}[(\text{kWh electric savings})_i \times R_i \times LLF_i \tag{5.5}$$

An alternate method that air conditioning engineers can use to characterize this information involves evaluating the fractional source energy savings due to TES. This fraction can be calculated as

$$F_{TES} = F_{ES} \times F_{EST} \tag{5.6}$$

where F_{TES} is fractional source energy savings due to TES for the annual cooling load, F_{ES} is fractional source energy savings per kWh electric shifted, and F_{EST} is fraction of the annual kWh electric shifted by TES.

In Equation 5.6, the second term on the right-hand side varies with the TES system, typically ranging from about 0.40 for hospitals with partial storage systems to about 0.65 for office buildings with full storage systems (CEC, 1996).

5.6.3 *Case studies for TES energy savings*

Thousands of cold TES systems have been operating in the world, particularly in developed countries, for years in hospitals, public and private schools, universities, airports, government facilities and private office buildings, and in industrial process cooling applications. Described below are several case studies reported by the IEA-HPC (1994), OECD (1995), CEC (1996), ARI (1997), Mathaudhu (1999) and Dincer and Rosen (2001) that demonstrate how TES systems provide energy savings and reduce the environmental impact, and that illustrate some clever applications of TES equipment in new buildings to reduce initial costs. In this section, several examples are given to illustrate the energy savings achievable through TES.

TES energy conservation project (California, USA)

This case study considers a project incorporating TES and other energy conservation features into systems for using electricity and water and into the design of the building envelope. The TES project was applied to a Center in California, USA. The center's major areas include a central operations control center and computer room (which operate 24 hours per day), administrative and engineering offices, and clerical and conference rooms (all of which have 10 hour/day operation). To comply with California State energy efficiency standards, packaged rooftop heat pump units were selected for the base mechanical system. A 20-year life-cycle analysis was performed to review alternate mechanical systems using a Variable-Air-Volume (VAV) system with electric reheat; with shut-off VAV boxes; and with fan-powered VAV boxes. This analysis determined that fan-powered VAV boxes with electric reheat using an air-cooled chiller were the most cost effective, with a 5.5-year payback period.

At the design stage, the client requested a study to consider incorporating a TES system. The selected system consists of a fan-powered, variable-volume system with electric reheat, low-temperature supply air (6°C), a chilled water plant, air handlers with variable speed drives and ice storage tanks. This system resulted in 30% greater energy savings than mandated by the California State Energy Efficiency Standards and ANSI/ASHRAE/IESNA Standard 90.1.

Designing the HVAC system for the building (Figure 5.4) efficiently was a challenge due to the constantly fluctuating number of people present (50 to 370). Some of the areas have 24-hour occupancy, and many employees are dispatched to the field after coming to work in the morning and then meet at the end of the day for reporting.

In this application, seven thermal ice storage tanks with a total capacity of 1330 ton-hours, provide 1095 ton-hours of full load off-peak cooling. Other energy-conservation measures used to increase efficiency include the following (Mathaudhu, 1999):

- Low temperature supply air at 6°C used instead of conventional supply air at 13°C to help reduce the supply fan size from 20,760 L/s to 14,745 L/s.
- Variable speed drives are provided for supply, return/relief and outside air fans. All air handling unit fans and pumps have high-efficiency motors. The outside air fan speed is controlled by CO_2 sensors located in the main return-air duct.
- Fan-powered VAV boxes mix filtered plenum air with primary air supplied at 6°C. Additional heating requirements are provided by three-stage electric heaters in VAV boxes. The secondary fan in the VAV box has a variable-speed controller to fine-tune the plenum air quantity for recirculation.

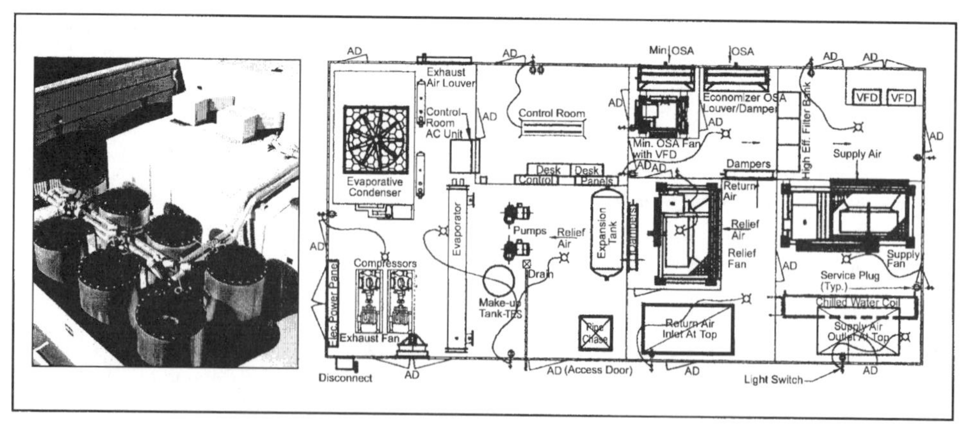

Figure 5.4 Custom built unit and sensible ice storage tanks (left) and custom built unit layout (right) (Reprinted from Mathaudhu (1999) by permission of ASHRAE).

- A direct digital control energy management system helps optimize the system operation.
- A high-efficiency heat pump unit maintains comfort level after normal operating hours in the evenings, on weekends and holidays, and for the continuously occupied central control room.
- An evaporative condenser is used rather than a cooling tower to increase system efficiency.

Further, energy-conservation features include R-19 wall insulation, R-30 roof insulation, low-emissivity glass windows, high-efficiency T-8 lamps, daylight sensors, occupancy light sensors, skylights, and low-water consumption plumbing fixtures and metering faucets.

Locating the air-handling unit, chiller, pumps and cooling tower became a challenge, because the architect had counted on the HVAC system being packaged rooftop heat pump units, and had not planned for the additional space. The installed cost for all of the central plant equipment with an architectural enclosure, but excluding the energy management controls, was over budget by at least $130,000. An alternate plan including a custom unit to house all the equipment was determined to be more cost effective. The unit would be factory-piped and wired with a complete energy management control system. The custom unit consists of a supply fan (plug type), a return/relief fan (plug type), minimum outside air fan (FC fan), pre-filters and high-efficiency filters, two multiple-stage reciprocating compressors, evaporative condenser cooler, variable speed drives for supply, return/relief and outside air fans, water/glycol circulating pumps, and a control room with complete direct digital control panels and a computer. The control room within the custom unit is designed to be accessible at all times so that clients can observe the system operation. This control room is cooled by a through-the-wall air conditioner that is factory-mounted on the common wall to the evaporative condenser area and the control room.

Although the application of packaged roof-mounted heat pumps would have been the least costly option for installation, the client preferred using a chiller system with a VAV system. Therefore, this system was used as the base case. For the TES system, the time-of-use load leveling strategy provided a payback period of 3.5 years. The time-of-use peak-hour shift option provided a payback period of 7.8 years, and the time-of-use mid- and peak-hour shift provided a payback period of 17.7 years, compared to the conventional base case of off-peak cooling. A 20-year life-cycle analysis was used to project these payback periods. The client chose to use the time-of-use peak-hour shift strategy. The supply air temperature of 6°C helped reduce the sizes and related costs for the supply fan, return fan, air-distribution duct, duct insulation, fan motors and variable frequency drives, resulting in about $56,000 savings per year. The custom air-handler/chiller system helped reduce mechanical central plant costs by about $150,000, to $550,000 from $700,000.

The seven thermal ice storage tanks (Figure 5.4) provide the flexibility to isolate a tank due to any failure and still provide over 85% off-peak cooling. The system installed in the facility has demonstrated overall a high level of occupant comfort, while proving itself as an energy efficient and cost-efficient choice. In fact, the building has been operating for almost three years (April 1997 to March 1998), and its performance is slightly better than their projected (the study predicted annual electrical consumption of 623,400 kWh, while the actual value observed was 566,900 kWh).

Table 5.2 Annual energy savings and emission reductions for the case study*.

Commodity	Conventional system	TES-based system	Reduction for TES-based system
Consumptions			
Natural gas (m^3)	215,800	95,500	120,300 (56%)
Electricity (kWh)	395,550	511,500	−84,000 (−21%)
Primary energy (m^3)[a]	322,000	179,000	143,000 (44%)

[a] Primary energy is calculated as the equivalent amount of natural gas based on the assumption that 0.25 m^3 gas is used in the generation of 1 kWh of electricity.

* Adapted from IEA-HPC (1994), where further details are available.

California Energy Commission's TES program (California, USA)

In this section, we consider the California Energy Commission's 'Opportunity Technology Commercialization (OTCOM)' program to increase the market penetration of energy technologies such as TES that offer, among other factors, compelling energy benefits. OTCOM's TES Systems Collaborative Program Commission requested an analysis of the source energy (power plant fuel) savings of electric TES systems in California and of other TES impacts. Besides environmental and economic development benefits, the study identified significant potential for energy savings. In many California TES installations, 40–80% of the annual electricity used for air conditioning can be shifted from day to night, yielding source energy savings per kWh shifted ranging from 12–43%, depending on the estimation method employed. The results indicate that if TES achieves 20% market penetration by the year 2005, enough source energy would be saved from load shifting (ignoring energy impacts) to supply the energy needs of over a fifth of all new air conditioning growth projected by the Commission during the next decade. When the site energy savings are combined with the TES source energy savings from shifting load noted above, TES can achieve even greater energy savings. Again, assuming 20% market penetration by 2005, TES could in total save enough energy to supply over a third of the new air conditioning load projected by the Commission. Of course, the source energy savings for a particular TES system in a particular building depend on a number of factors to do with the system and its environment (CEC, 1996).

Anova Verzekering Co. Building (Amersfoort, The Netherlands)

This recent TES provided energy savings and reductions in pollutant emissions (Table 5.2). In the application, a groundwater aquifer TES system was installed as part of a space conditioning unit in a newly renovated office building of Anova Verzekering Co. An electric heat pump is used to supply hydronic heating and cooling. Accounting for the subsidy of US$212,000 received from the Dutch government, which is equivalent to 20% of the total initial system costs, the reduced energy costs due to TES are expected to lead to a payback period of 6.5 years for the additional investment costs due to TES. In the case study, primary energy consumption decreases due to TES by over 40%, even though electricity use increases.

City of Saarbrucken (Saarbrucken, Germany)

The city of Saarbrucken, Germany implemented a strategy in the 1980s to reduce energy consumption through seasonal TES, district heating, and by increasing the use of renewable energy (e.g. solar thermal, solar photovoltaic energy, and small hydropower). Between 1980 and 1990, the city achieved significant energy use savings, including a 15% reduction in overall heating demand and a 45% reduction in heating consumption for the municipal buildings. Some of the energy savings resulting from this successful energy program are attributable to TES (OECD, 1995).

Kraft General Foods Headquarters Building (Northfield, Illinois, USA)
All daytime air conditioning loads are presently being met at this facility by melting ice that is made and stored overnight. It is anticipated that additional loads from future expansion will be met by operating some of the chillers during the day as well as at night. The building was designed to pump 2.2°C water to the air handling units, which in turn provide 7.2°C air to the building. These temperatures, which are lower than for non-TES based systems, permit the use of smaller pipes, pumps, air handling units and ductwork, resulting in lower initial capital costs for the system. Annual electric bills for this building are nearly US$200,000 lower than for an almost identical building just three miles away, which does not use a TES system.

Chrysler Motors Technology Development Center (Auburn Hills, Michigan, USA)
Since opening in 1990, the Chrysler Motors Corporation's new technology development center has achieved both equipment and operating cost savings by using a 68,000 ton-hour chilled-water TES system. The TES capacity allowed the center's chiller plant to be downsized from 17,710 tons, which would have been needed to meet peak cooling loads, to 11,385 tons. Chilled water is stored in the TES system at night and supplements chiller operation during peak cooling conditions the following day. Reduced chiller costs more than offset the cost of the TES installation, resulting in initial savings of US$3.6 million. In addition, the TES system shifts over 5000 kW of peak electrical demand to off-peak periods, saving over US$1 million per year.

San Francisco Marriott Hotel (San Francisco, California, USA)
Using a TES system in tandem with a real-time pricing strategy from the local utility, this hotel is expected to save US$135,000 in annual cooling costs. Only 1800 ton-hours of ice storage are needed, enough to satisfy the 450 ton cooling load during the daily peak-rate time period, which lasts only two to three hours under the real-time pricing schedule. Over one-third of the installed cost of the TES system will be covered by a rebate from the utility; the rest is expected to be recouped in less than two years of operation.

Texas A&M University (Corpus Christi, Texas, USA)
In August 1997 the Central Power & Light (CPL) Co. presented Texas A&M University-Corpus Christi with a US$431,800 incentive award for the university's participation in a TES program that is resulting in substantial energy savings. The university system uses the bulk of its electricity during off-peak evening hours, allowing CPL to shift some of its electric load from peak usage times and to share the annual cost savings of up to US$150,000 with the university. Texas A&M-Corpus Christi invested approximately US$900,000 to purchase and install the TES equipment, which became operational in January 1995, and expects to recover that cost through energy savings within five years. The US$431,800 incentive includes US$20,400 for installing high-efficiency equipment for cooling and heating the campus. The remaining incentive was provided through CPL's

energy efficiency program, wherein the university was offered US$200 for every kW of electricity load shift from CPL's peak daytime load to off-peak evening hours. The university reduced electric peak demand by approximately 2057 kW, earning a US$411,400 incentive. CPL worked with the university to install an 11,800 ton-hour thermal storage system with a water storage capacity of 5300 m^3. The university also installed a 500 ton industrial heat pump for heating, which CPL estimates, will save the university approximately US$90,000 a year in energy costs. The heat pump captures waste heat from the university's 3000 ton chiller plant, and recirculates it into areas of the campus needing heating. The TES tank stores chilled water, which is produced by the conventional air conditioning system during the night, and is then used to cool buildings during the day, when the highest demand is placed on the air conditioning system.

Gillette Capital Corporation (Gaithersburg, Maryland, USA)
The addition of 1050 ton-hours of latent ice storage to this facility in August 1994 saved the building owners both initial system costs and subsequent operating expenses. The project cost just over US$121,000, with 57% of this expense paid for by utility incentives, including a US$350 per kilowatt demand-reduction rebate. An air-cooled reciprocating chiller was originally used to cool the 5760 m^2 building, with the chiller operating 14–15 hours per day during the hottest summer months. With the full storage system, the cooling load from 8:00 a.m. to 5:30 p.m. is now supplied by ice alone, with the chiller normally running only five hours during the night to replenish the ice supply. The peak electrical load during the four-month summer peak period is reduced by 198 kW, avoiding the US$12.95 per kW demand charge, and resulting in over US$2500 in monthly operating savings. Accounting for initial financial incentives as well as operating savings, a simple payback period of 3.5 years is expected for the TES system.

Miller Electric Company (Appleton, Wisconsin, USA)
In 1990, this industrial manufacturer of welding equipment converted its cooling system from once-through well-water coolers and conventional rooftop direct-expansion units to ice thermal storage. Conversion eliminated the owner's concerns with the high sewer costs associated with the well-water units and the pending phase-out of the chlorofluorocarbons (CFCs) used in the direct expansion units. The 46,450 m^2 of conditioned space has an air-conditioning peak design capacity of 3000 tons. The cooling load is handled by 1380 tons of ice harvesting equipment using ammonia as the refrigerant and operating on a weekly load-shift strategy in which ice is made only during the off-peak weekend and weeknight hours. The owner received a US$905,000 rebate from the local electric utility, and realizes a 65% (US$140,000) reduction in annual air conditioning costs.

Kirk Produce Company (Placentia, California, USA)
In operation since 1987, two TES units are charged with 465 tons of ice cooling capacity during weekends and an additional 360 tons during off-peak weekday times. The ice is stored with 605.6 m^3 of water in a 1161.2 m^3 storage tank. The water at 0°C is pumped from the tank through spray nozzles to cool air to 1°C, for use in cooling fresh-picked strawberries prior to storage and shipping. Beyond controlling humidity, which is vital to the preservation of strawberries, this process cooling system annually saves $153,000, or more than half of the refrigeration plant operating costs.

These case studies demonstrate that TES technology offers to owners and regions compelling energy savings as well as environmental, diversity, and economic benefits. As TES now seems poised for wider commercialization, institutional policies, such as those previously identified, should be considered for implementation to increase the market penetration of TES beneficially.

5.7 Concluding Remarks

The main concluding remarks of this chapter can be summarized as follows:

- Substantial energy savings can be realized by taking advantage of TES when implementing the techniques such as using waste energy and surplus heat, reducing electrical demand charges, and avoiding heating, cooling or air conditioning equipment purchases. These savings in energy can be realized despite the fact that the storage energy efficiency, the ratio of thermal energy withdrawn from storage to the amount input, is less than 100%. Storage energy losses are often small, e.g. energy efficiencies up to 90% can be achieved in well-stratified water tanks that are fully charged and discharged on a daily cycle.
- TES plays a significant role in meeting society's needs for more efficient use in various sectors as it permits mismatches between supply and demand of energy to be addressed.
- With TES, peak-period demand for electrical energy can be reduced by storing electrically produced thermal energy during off-peak periods and using it to meet the thermal loads that occur during high demand periods. For example, a chiller can charge TES at night to reduce the peak electrical demands experienced during the day.
- TES exhibits enormous potential for more effective use of TES equipment, and for facilitating large-scale energy substitutions economically. The economic justification for TES systems normally requires the annual income needed to cover capital and operating costs to be less than that required for primary generating equipment supplying the same service loads and periods. A coordinated set of actions is required in several energy sectors to realize the maximum benefits of storage.

References

Anon. (1985). *Thermal Storage*, Energy Management Series No.19, Energy, Mines and Resources Canada.

ARI. (1997). *Thermal Energy Storage: A Solution for Our Energy, Environmental and Economic Challenges*, The Air-Conditioning and Refrigeration Institute, Arlington, VA.

CEC. (1996). *Source Energy and Environmental Impacts of Thermal Energy Storage*, California Energy Commission, Technical Report No. P500-95-005, California.

Dincer, I., Dost, S. and Li, X. (1997a). Performance analyses of sensible heat storage systems for thermal applications, *International Journal of Energy Research* 21, 1157–1171.

Dincer, I., Dost, S. and Li, X. (1997b). Thermal energy storage applications from an energy saving perspective, *International Journal of Global Energy Issues* 9, 351–364.

Dincer, I. and Rosen, M.A. (2001). Energetic, environmental and economic aspects of thermal energy storage systems for cooling capacity, *Applied Thermal Engineering* 21, 1105–1117.

IEA-HPC. (1994). Energy Storage, International Energy Agency, *Heat Pump Center Newsletter* 12(4), p. 8.

Mathaudhu, S.S. (1999). Energy conservation showcase, *ASHRAE Journal*, April 1999, 44–46.

Moran, M.J. (1989). *Availability Analysis: A Guide to Efficient Energy Use*, revised edition, American Society of Mechanical Engineers, New York.

OECD. (1995). *Urban Energy Handbook*, Organization for Economic Co-Operation and Development, Paris.

Rosen, M.A. and Dincer, I. (1996). Linkages between energy and environment concepts, *Proceedings of TIEES-96, the First Trabzon International Energy and Environment Symposium*, (Eds. T. Ayhan, I. Dincer, H. Olgun, S. Dost and B. Cuhadaroglu), Vol. 3, pp. 1051–1057, 29–31 July, Karadeniz Technical University, Trabzon, Turkey.

Rosen, M.A. and Dincer, I. (1997a). On Exergy and environmental impact, *International Journal of Energy Research* 21(7), 643–654.

Rosen, M.A. and Dincer, I. (1997b). Sectoral energy and exergy modelling of Turkey, *ASME Journal of Energy Resources Technology* 119(3), 200–204.

Tomlinson, J.J. and Kannberg, L.D. (1990). Thermal energy storage, *Mechanical Engineering* 112, 68–72.

6

Heat Transfer and Stratification in Sensible Heat Storage Systems

Y.H. Zurigat and A.J. Ghajar

6.1 Introduction

The importance of Thermal Energy Storage (TES) as an energy conservation and management tool has been discussed in previous chapters. In this chapter we focus on the problem of sensible heat storage in liquids for low-to-medium temperature ranges and the associated developments to date. The choice of the type of liquid for sensible TES depends on its specific heat, mass density, toxicity, corrosion resistivity, and cost, and on the operating temperature range. The volumetric heat capacity, ρC_p, defined as the heat storage per unit volume and unit temperature difference, determines the volume of the storage device, while the working pressure of the storage system determines the operating temperature range. High storage temperatures require low-vapor-pressure liquids or pressurized tanks. Both options are often costly to implement (Wyman *et al.*, 1980). Water, due to its abundance, low cost, high specific heat and benign characteristics, is the most widely used storage medium in the low-to-medium thermal-storage temperature range. This temperature range covers chilled water storage at about 4°C and hot water storage below 100°C. Also, since water is the working fluid in many energy systems, its choice as a thermal storage medium is natural. Hot and chilled water storage can be easily integrated with existing building heating and cooling systems. This way, the use of heat exchangers is eliminated, thereby avoiding the extra cost of heat exchangers and the thermal losses associated with their use. The practical problems associated with using water are the possibility of freezing in cold weather and the corrosion of steel storage vessels and piping. Circumventing these problems is relatively straightforward, i.e. insulating the storage device or locating it indoors or underground at a safe depth and using corrosion inhibitors and other corrosion protection measures. Therefore, in this chapter we focus on sensible heat storage in water and the associated heat transfer phenomena. This involves both hot and cold storages in practical applications, e.g. air conditioning, solar energy technologies, heat pumps, gas turbines, and other energy systems.

In the majority of solar energy collection systems water is heated during the day and stored for use during the night, thus extending the use of solar energy over a larger part of the day. Chilled water, on the other hand, is used for cooling in the air conditioning systems in buildings and in gas turbine power plants to cool the inlet air. Chilled water storage can shift part of the cooling requirements to off-peak hours, resulting in improved utility load factors, in addition to allowing the chillers to operate during the cooler night temperatures, resulting in improved coefficients of performance. The economical impact extends beyond this to the owners who not only avoid the extra demand charges during peak demand periods, but also take advantage of discounted night-time rates. Moreover, since the equipment does not have to handle the peak load its size is minimized, resulting in savings in capital investment. That is, instead of installing two chillers to operate at full load during the peak hours and at partial load during the rest of the demand period, one may install a smaller single chiller operating 24 hours per day to charge the storage tank during off-demand periods. The tank in turn assists the chiller during the demand period. Some gas-turbine power plants utilize chilled water to cool inlet air in order to boost the power output during the hot season. That is, at night when the demand is low and the air is cool, water is chilled and stored for use during the daytime peak demand period. This way the gas turbine operates longer at its high efficiency level, and the need for additional equipment to handle the peak demand is minimized.

Hot or chilled water is stored in tanks, which vary in design as dictated by different factors, like thermal performance, and architectural, retrofit and economical constraints. However, all existing thermal storage tank systems share the same objective of maintaining the thermodynamic availability of stored energy so that it can be extracted at the same temperature at which it was stored.

The separation of any fluids at different temperatures in storage tanks is the key factor in achieving this objective. The *two-tank* system (also called the *empty tank* design) is one obvious way of achieving the separation. In this system two identical tanks are used: the first tank is in the charge mode to store the heated or chilled water; and the second tank is used to store the discharged water as it exits the load. Once the stored water is fully discharged the first storage tank becomes empty and ready to be charged with water from the second tank via the heat source or the chiller. Although this design ensures separation of hot and cold water, it is not the best choice with regard to simplicity, economic feasibility, and space utilization. Other schemes have been designed and implemented. These include a single tank with a *flexible diaphragm* mounted either horizontally or vertically, *labyrinth tanks* in which the water is forced to flow through a maze, and the single *stratified tank* in which use is made of the natural process of stratification that permits the hot water to float on top of the cold water. The single tank with a flexible diaphragm has been used in several installations. Although the diaphragm prevents blending of hot and cold water, concerns about membrane maintenance, durability and cost have been raised (Tamblyn, 1980). The *labyrinth tanks* concept was developed in Japan, where many buildings have earthquake protection structures under the basement floor in the form of intersecting high-tie beams.

With little modification the resulting space compartments can be used to store chilled or hot water, and they are connected in such a way to force the water in a plug flow with minimal mixing between the hot and cold regions. A model tank employing this concept was tested by Tamblyn (1980), and on a single-pass test it proved to be efficient in separating the hot and cold liquid water, but its performance was found to deteriorate in cyclic tests. Of course, unless the suitable structure is available for no extra cost as in Japan, the economics of this concept is unlikely to compete with equally efficient anti-blending systems.

Because of the modest temperature ranges involved, the storage of significant amounts of thermal energy involves relatively large tanks, which must therefore be simple and cheap in construction in order to be economically viable. Also, to apply this technology to residential use, the operation of the tanks must be simple, reliable, and low in maintenance; it cannot involve elaborate monitoring, valving, and control systems. The last concept in the list, the single *stratified tank* (see Figure 6.1), satisfies these requirements, and thus is the most attractive choice in low-to-medium temperature thermal-storage applications due to its simplicity and low cost. Furthermore, the research and development efforts have led to performance comparable with the other storage types employing physical barriers (Tran *et al.*, 1989). Stratified tanks as large as 4 million gallons (15,140 m^3) have been installed in the US for chilled water storage. In the US chilled water storage constitutes about 34% of the cooling capacity of all cool storage systems. 60% of chilled water storage systems utilize stratified tanks (Musser and Bahnfleth, 1998). Clearly, stratified thermal storage has become an essential element in load management and energy conservation technology.

In this chapter the experimental and modeling efforts and the resulting advances in the technology of stratified thermal storage in water are presented. The next section introduces the flow and heat transfer phenomena followed by the performance measures and experimental and theoretical foundations. One- and two-dimensional models of flow and heat transfer are then discussed. We then conclude with a summary on design recommendations.

6.2 Fluid Flow and Heat Transfer Aspects

The principle of operation of stratified thermal storage tanks is based on the natural process of stratification, and hence the fluid flow within these tanks involves both forced and natural convection. In heat storage applications the cold fluid withdrawn from the bottom of the tank is heated at the heat source, i.e. solar collector, heat pump, or gas-fired or electric resistance heaters, and is returned to the top of the tank at relatively higher temperature (see Figure 6.1). Assume for now that the temperature of the incoming stream remains constant at its elevated value. The incoming flow possessing momentum will tend to mix with the fluid in the tank. However, being at a higher temperature, the resulting buoyancy tends to lift the stream restricting its motion to the surface region. This way mixing is restricted to a limited region at the surface near the inlet. As more fluid is introduced, the fluid in the mixing region is pushed down, leaving behind a region of uniform temperature equal to the inlet temperature. The region of intermediate temperatures separating this uniform temperature region from that initially in the tank is termed a *thermocline.* It is defined as the region of steepest temperature gradient separating the hot and cold fluid regions in the tank. The buoyancy arising from the stable density gradient across the thermocline region inhibits mixing between the hot and cold fluid regions on either side. Thus, the thermocline acts as a physical barrier. The thickness of the thermocline region is an important indicator of how well the stratified tank is designed.

This thickness is a function of several variables: the geometry of the tank and the inlet(s), and the hydrodynamic and thermal characteristics of the flow in the tank. The way the flow is introduced and the balance between buoyancy and inertia forces are detrimental to the formation of a thin thermocline. Once the thermocline is formed it travels down as the charging continues until it exits the tank, indicating full charge.

In the discharge mode the load flow loop is activated and the process described above is reversed. That is, the hot water is withdrawn from the top and is replaced by cold water introduced from the bottom. This could be the discharged water cooled at the thermal load or the makeup water in the case of hot water consumption. The flow momentum tends to blend the incoming fluid with the fluid in the tank while the buoyancy now acting downward tends to make the incoming stream flow in a gravity-current form below the relatively warmer fluid. A thermocline is formed and it travels up the tank separating the cold and hot fluid regions. A portion of the thermocline may exit the tank depending on the allowable temperature at the load. Frequently, a load flow loop may operate simultaneously alongside the charging loop. In this case, the tank may experience a net charge or discharge depending on the relative magnitudes of the flow rates of the corresponding loops.

The same phenomenon occurs in chilled water storage tanks, but the charge and discharge flow directions are reversed. One major difference, however, exists. That is, the operating temperature is relatively low, and consequently the density differences are very small (see Figure 6.2) and the stratification is weak, leading to a tendency for chilled water to mix excessively with warmer water in the tank, if disturbed by uncontrolled inlet flows.

In the foregoing discussion it is assumed that the inlet temperature remains constant. In solar collector systems this condition is never satisfied unless some measures are used to control the flow rate through the collector. In reality, water heated by solar panels varies continuously in temperature, resulting in buoyant flows which seek equilibrium with the fluid in the tank. This enhances mixing in the tank, and the thermocline region defined earlier is no longer clearly visible. That is why the term thermocline is reserved for the constant inlet temperature condition. To avoid excessive mixing the flow must be inserted into the stratified thermal storage tank at the proper level with minimum mixing on the way. This was the object of several designs that remove the momentum of the inlet stream while allowing it to distribute itself to the proper stratification level (see section 6.4).

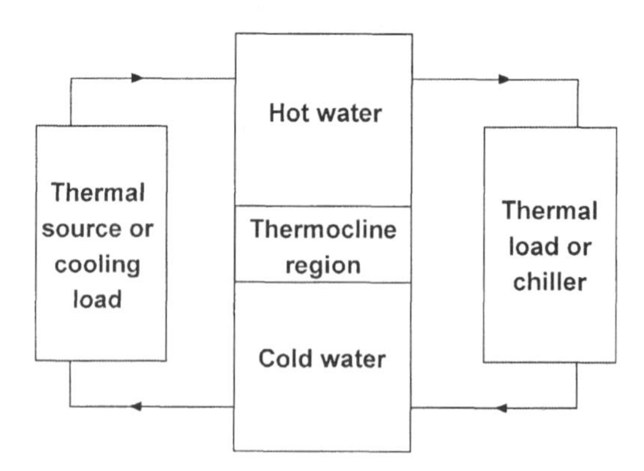

Figure 6.1 Single stratified thermal storage tank integrated with heating or cooling systems.

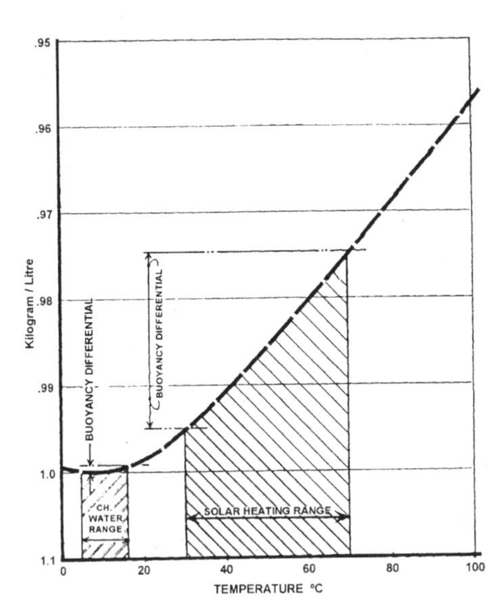

Figure 6.2 Buoyancy differentials typical of solar heating and chilled water applications (Tamblyn, 1980).

Maintaining stratification in storage tanks is essential for better performance of energy systems with which these tanks are integrated. Solar collectors operate at higher efficiency as the collector inlet temperature is decreased (Duffie and Beckman, 1980). Stratification improves the overall performance of solar collector systems by reducing the average absorber plate operating temperature. Performance improvements reported in the literature are 10% (Davis and Bartera, 1975), 5–15% (Sharp and Loehrke, 1979), and 5–20% (Cole and Bellinger, 1982). Simulations with ideally stratified and fully mixed storage tank models show improvements in annual collector system performance ranging from 11.5–18.5% when using the ideally stratified model (Wuestling *et al.*, 1985). Improvements as high as 37% were also reported by Hollands and Lightstone (1989). If a thin thermocline is maintained in chilled water storage tanks the water delivered to the cooling system is at the lower temperature for most of the discharge period. This way, smaller flow rates and pumping power are needed to satisfy the cooling requirements as opposed to the case of a chilled water tank with a high degree of mixing. As a result, maintaining stratification was the object of many research works, both experimental and analytical. These will be discussed in the following sections.

The loss of stratification in liquid thermal storage tanks is associated with several factors that manifest themselves in two ways: the mixing introduced by the inlet streams during charge or discharge; and the heat transfer that may take different paths. In thermocline TES tanks (constant inlet temperature), the mixing during the formation of thermocline at the inlet for the charge and the discharge is the major contributor to the loss of thermodynamic availability of stored energy. This mixing remains difficult to evaluate as it depends on the type of inlet and the flow conditions. A discussion of the modeling efforts of this process is introduced later. In static and dynamic modes of operation of stratified thermal storage tanks, three non-mixing heat transfer paths are present. Nonetheless, they may lead to convective currents and subsequent mixing. These are the heat transfer to the ambient surroundings through the tank envelope and insulation, heat diffusion in the water body through the thermocline, and heat leakage from the high temperature to the low temperature regions by means of axial conduction through the wall. These heat transfer processes may seriously affect thermal stratification and lead to its degradation. The governing parameters in these are the temperature difference between the hot and cold regions of stored fluid, the wall thermal conductivity, the wall thickness, the insulation type, the size of the tank, and the temperature of the ambient surroundings. It should be noted that some experimental works were successful in isolating the effect of a particular path, while others studied the combined effect of the three heat transfer paths.

In the analyses of stratified tank problems, several dimensionless numbers arise. Noting that the flow in the thermal storage tank is of the mixed convection type, the relative magnitudes of the buoyancy and inertia forces play a major role in the flow development. This was expressed in what is known as the dimensionless Richardson (also called Archimedian) number as $Ri = Ar = Gr / Re^2 = g\beta\Delta T\ell_r / u_r^2$, where the subscript r denotes a reference quantity, Gr is the Grashof number arising in free convection flows and Re is the Reynolds number. Also, the Peclet number, Pe, is used to characterize the relative magnitudes of the thermal energy transported by fluid motion to that transported by molecular diffusion. In terms of other numbers, $Pe = RePr$, where Pr is the Prandtl number. These numbers are often written with a subscript showing the reference length scale. For example, Re_D and Pe_H mean that the Reynolds number is based on the diameter and the Peclet number is based on the height. Frequently, the Froude number, Fr, is used, which is equal to the square root of the ratio of the inertia and gravity forces, $Fr = u_r / \sqrt{g\ell_r}$. A modified Froude number, Fr_m, is also used which is based on the buoyant force per unit mass, $g\beta\Delta T$, instead of the gravitational force per unit mass, g. That is, $Fr_m = u_r / \sqrt{g\beta\Delta T\ell_r}$. The square of the modified Froude number may be expressed in terms of the numbers already defined, i.e. $Fr_m^2 = 1/Ri = u_r^2 / g\beta\Delta T\ell_r$. Sometimes, the temperature difference in Ri and Fr_m is expressed in terms of the density difference as $\beta\Delta T = (\rho_r - \rho)/\rho_r$, where ρ_r is the reference density. Using this substitution in the modified Froude or Richardson numbers, we then talk of the densimetric modified Froude, Fr_{dm}, and densimetric Richardson, Ri_d, numbers. The reference quantities used in these numbers need to be ascertained when interpreting the results in the literature. This is because different investigators use different reference quantities. For example, the Reynolds and Richardson numbers used by Cabelli (1977) were based on the inlet port velocity and the height of the tank. Lavan and Thompson (1977) based the Grashof number on the diameter of the tank and the Reynolds number on the inlet port diameter. As a result, for the same test conditions a number quoted by one investigator may be quoted by another as being of several times the order of magnitude.

6.3 Performance Measures

Typically, the experimental and computational results of stratified thermal energy storage presented in the literature consist of transient temperature profiles under different thermal, hydrodynamic, and geometric conditions. To quantify the effects of these conditions on thermal stratification, different performance measures have been devised, depending on the conditions under consideration. For example, in thermocline TES tanks of the same geometry the effects of different flow conditions and inlet configurations may be judged based on the *thermocline thickness* they produce. Although a thicker thermocline is associated with a larger degradation of stored energy, the thermocline thickness does not give a quantitative measure of how large this degradation is. Also, this measure cannot be used in the variable inlet temperature condition in which a well-defined thermocline is not present. Furthermore, the thermocline thickness cannot be used for judging competing designs of different geometries. Thus, some other measures have to be used. In this section we look at different measures of performance used by different investigators. This material should help in the proper interpretation of the results cited in this chapter.

In many instances the effect of different parameters on stratification is judged in reference to the temperature profiles predicted by theoretical models such as the ideal model of plug flow, the fully stratified model, or the fully mixed flow model. These are discussed in section 6.5. The relative performance of different designs is then gauged by the departure of their corresponding test results from those predicted by these models. This is a very common approach used by many investigators because it gives quick visual comparisons. Abu-Hamdan *et al.* (1992) used this technique to compare the performance of three different inlet configurations under variable inlet temperature conditions. They calculated the instantaneous thermal efficiency of a simulated solar collector fed by water from the bottom of storage tank at three different outlet temperature profiles. These were the measured profile and those calculated for the same tank from the fully mixed and the fully stratified models. Also, the *mix number* of Davidson *et al.* (1994), discussed later in this section, is based solely on the above-mentioned technique.

One of the other measures was the *degree of stratification* used by Sliwinski *et al.* (1978), and it is characterized by the temperature gradient $\Delta T/\Delta X$ in the thermocline region. Once the thermocline is formed this gradient was calculated at any time by locating points on the temperature profile such that the gradient was less than the maximum gradient for that profile by 10%. For each experiment a mean gradient was calculated by averaging the gradients so calculated. Then the mean gradient is non-dimensionalized by the initial overall temperature gradient defined by $\Delta T_o/H$, where $\Delta T_o = (T_h - T_l)$, where T_h and T_l are the initial high and low temperatures, respectively, and H is the distance between the inlet and the outlet. This ratio was later called the *stratification number*, and was used by other investigators (e.g. Al-Najem, 1993) to quantify the rate of decay of thermal stratification in the static mode of operation.

The *extraction efficiency* is another performance indicator used in the dynamic mode of operation. It was first defined for the discharge of hot water storage tanks as

$$\eta = \frac{\dot{\forall} t_d}{\forall_T} \tag{6.1}$$

where $\dot{\forall}$ and $\forall_T$ are the volumetric flow rate and the internal volume of the tank, respectively, and t_d is the discharge time required for the outlet-to-inlet temperature difference to drop to a pre-assigned percentage of its value at the start of the discharge. Lavan and Thompson (1977) used a 10% drop, i.e. $(T_e(t) - T_{in})/(T_o - T_{in}) = 0.9$. The extraction efficiency so defined represents the useful fraction of the initially stored volume. Therefore, it does not quantify the recovered fraction of stored heating or cooling capacities. The extraction efficiency may also be interpreted as the dimensionless discharge time, $t^* = \dot{\forall}\, t_d / \forall_T$, that has a value of unity for tanks with the ideal plug flow and less than unity for actual tanks. That is, in the plug flow case with no thermal losses of any kind the thermocline thickness is zero, and all the tank volume will be extracted during the discharge. The extraction efficiency in terms of the tank height, H, and the mean vertical velocity in the tank, V, is $\eta = t^* = V\, t_d / H$. In addition to the above expression for the extraction efficiency, Ismail *et al.* (1997) used the ratio of the integrated heat discharged over that which results from a plug flow with zero thermocline thickness. That is,

$$\eta = \frac{1}{\Delta T_o} \int_0^1 \left(T_e(t) - T_\ell\right) dt^* \tag{6.2}$$

The integration time limit is that at which one tank volume is discharged. The extraction efficiency defined by Equations 6.1 and 6.2 were both used by Ismail *et al.* (1997), and negligible difference in the results was observed. This is, of course, typical of thermocline thermal storage in the discharge mode, where the temperature profile experiences a sharp drop towards the low-temperature value resulting in the small difference found.

To evaluate the effects of stratification degradation mechanisms in a static hot water storage tank, Abdoly and Rapp (1982) used the *fraction of recoverable heat, F(t)*, as a measure of the heating capacity of an initially charged tank at temperature T_h. $F(t)$ was defined as the ratio of the heating capacity available at any time, $Q(t)$, to that initially stored, Q_o. Thus,

$$F(t) = Q(t)/Q_o \tag{6.3}$$

The recoverable heat $Q(t)$ is calculated based on an arbitrarily set criterion. That is, the tank is subdivided into a number of small uniform temperature regions, and the heat stored in any fluid region is considered useless, i.e. $Q(t) = 0$, if its temperature drops below a specified useful temperature dictated by the load requirements. Abdoly and Rapp (1982) considered the heat of any fluid element, J, useful if its temperature, T_J, did not drop below its high initial value, T_h, by more than 20% of the initial high-to-low temperature difference, i.e. $T_J \geq T_\ell + 0.8(T_h - T_\ell)$. Thus, the heat recovered from a fluid region J of mass m_J and temperature T_J is:

$$Q_J(t) = \begin{cases} 0 & \text{if } (T_J - T_\ell)/(T_h - T_\ell) < 0.8 \\ m_J C_p (T_J - T_\ell) & \text{if } (T_J - T_\ell)/(T_h - T_\ell) \geq 0.8 \end{cases} \tag{6.4}$$

The total heat recoverable, $Q(t)$ from the tank with total water mass, M_T, is found by integrating over all fluid elements, i.e.

$$Q(t) = \Sigma Q(t)_J \tag{6.5}$$

and

$$Q_o = M_T\, C_p\, (T_h - T_\ell) \tag{6.6}$$

Murthy *et al.* (1992) used the above method for assessing the effect of wall thermal conductance on stratification in model storage tanks. The advantage of this method is its simplicity. Also, it is an accurate measure of comparative performance. Nelson *et al.* (1999) applied the same method for evaluating the performance of chilled water storage. They defined Percent Cold Recoverable (PCR), instead. The useful temperature considered by Nelson *et al.* (1999) was that which does not rise above the low (cold) water charging temperature, T_ℓ, by more than 20% of the initial temperature difference. That is, $T_J \leq T_\ell + 0.2(T_h - T_\ell)$. The PCR is calculated from Equation 6.3 with Equations 6.5 and 6.6 as before, and Equation 6.4 is rewritten as

$$Q(t)_J = \begin{cases} 0 & \text{if } (T_J - T_\ell)/(T_h - T_\ell) > 0.2 \\ m_J C_p (T_h - T_J) & \text{if } (T_J - T_\ell)/(T_h - T_\ell) \leq 0.2 \end{cases} \tag{6.7}$$

Obviously, in the static mode this measure is a function of time. The rate of stratification degradation increases with the increase in the rate of decrease of *F(t)*. This measure may also be applied in the dynamic mode of operation. For example, during the discharge of an initially stored hot water at T_h the outlet temperature is not allowed to drop below the useful temperature mentioned above. Likewise, for chilled water storage the outlet temperature is not allowed to rise above the useful temperature defined above. The transient outlet temperature profile is then used to calculate the cumulative discharge heating or cooling capacity, Q_d, as:

$$Q_d = \int_0^{t_d} \dot{m} C_p \left| T_e(t) - T_{in} \right| dt \tag{6.8}$$

The discharge time limit is now determined by the useful temperature as defined above. Also, the absolute value of the temperature difference is used to enable the equations to be valid for heat or cold storages. In the discharge mode the percent heat or cold recoverable as defined by Equation 6.3, now without the time dependence, is termed the *discharge efficiency*, η_d:

$$\eta_d = Q_d / Q_o \tag{6.9}$$

Clearly, η_d is dependent on the useful temperature considered, and therefore the useful temperature used should be quoted alongside. In the charging mode the charging efficiency is used.

The *charging efficiency* (also called the *storage efficiency*) is defined as the ratio of the net stored energy at the end of charging to the maximum energy that may be stored in a perfectly stratified tank, i.e. plug flow with no heat transfer between the hot and cold fluids:

$$\eta_c = Q_c / Q_o \tag{6.10}$$

where

$$Q_c = \int_0^{t_c} \dot{m} C_p \left| (T_e(t) - T_{in}) \right| dt \tag{6.11}$$

Again, the charging time limit, t_c, may be determined by the allowable outlet temperature. Wildin and Truman (1985a) defined the thermal efficiency of a chilled water storage tank for a full cycle of discharge and charge as the ratio of the integrated discharging to the integrated charging capacities, i.e.

$$\eta_{th} = Q_d / Q_c \tag{6.12}$$

where Q_c and Q_d are determined as before. For stratified chilled water storage in full-scale and scale-model tanks, cycle efficiencies ranging from 80–90% were obtained.

For the charge-discharge cycle thermal efficiency as defined by Equation 6.12 to be a valid performance indicator, it was pointed out by Wildin and Truman (1985a) that the temperature distributions in the tank at the end of discharge and at the start of charge should be identical. Failure to achieve this would result in a net heat addition or withdrawal, leading to inaccurate values of cycle efficiency. Also, the cycle efficiency does not account for internal heat transfer and mixing. The integrated cooling capacity of an initially charged tank would not recognize the effect of temperature blending. For this reason, and because it is always difficult in practice to duplicate the same experimental conditions, another measure of performance less sensitive to the initial temperature distributions was devised. This was called the *Figure Of Merit (FOM).* This index, first used by Tran *et al.* (1989) to evaluate the thermal performance of chilled water storage, was subsequently used by other investigators (Wildin, 1989; Wildin and Truman, 1989). It is defined as the ratio of the cooling capacity during the discharge (or equivalently, the energy added to the tank by the load) to the maximum cooling capacity theoretically available in the fully charged tank. This latter quantity is calculated based on the temperature difference between the average inlet temperature during discharging and the average inlet temperature during charging. Thus, the *FOM* is calculated by

$$FOM = \left[\sum \dot{m} C_p \left(T_{in} - T_e\right) \Delta t\right]_{discharge} / MC_p \left(T_{in,d} - T_{in,c}\right)_{average} \tag{6.13}$$

where M is the total mass of the water in the charged tank.

In the variable inlet temperature condition a measure termed the *mix number* (*MIX*) was devised by Davidson *et al.* (1994). *MIX* is based on the energy distribution level in the tank, and is determined by what is termed the *first moment of energy*, defined by analogy with the first moment of mass. That is, $M_E = \int_0^H y dE$ by analogy with $M_m = \int_0^L x dm$. M_E may be approximated by

$$M_E = \sum y_i E_i \tag{6.14}$$

where $1 < i \leq n$, and n is the number of uniform-temperature segments the tank is divided into, y_i is the distance from the center of the ith segment of volume $\forall_i$, to the bottom of the tank and $E_i = \rho \forall_i C_p T_i$. According to this definition, perfectly-stratified tanks have the largest M_E, while fully-mixed tanks have the smallest M_E and actual stratified tanks have a value in between. Thus, M_E may be used to characterize the stratification level in thermal storage tanks. The *mix number* is defined by the following dimensionless expression:

$$MIX = (M_{E,stratified} - M_{E,actual}) / (M_{E,stratified} - M_{E,fully-mixed}) \tag{6.15}$$

In view of Equation 6.14, the *mix number*, *MIX*, considers both the energy level and the temperature distribution. Clearly, the mix number is zero when an actual tank's performance approaches that of the perfectly stratified tank, i.e. $M_{E,actual} = M_{E,stratified}$ and it is unity when actual tanks are fully mixed. Also, because of the transient nature of the temperature profile the mix number varies with time. But this does not affect its usefulness when one is concerned with the relative performance of different tank designs under the same flow and thermal conditions. The three quantities in the above expression for *MIX* are determined using Equation 6.14 as follows: $M_{E,actual}$ is calculated from the measured temperature profile, $M_{E,stratified}$ and $M_{E,fully\text{-}mixed}$ are calculated based on the temperature profiles calculated from the perfectly-stratified and fully-mixed tank models (see section 6.4) respectively for the same actual test conditions. Using the mix number, Davidson *et al.* (1994) compared the performance of two inlet configurations under variable inlet temperature conditions. These inlets were a conventional inlet (drop-tube inlet located at the top of the tank) and a rigid porous manifold. The results for three different test conditions show lower values of *MIX* for the manifold inlet compared with the drop-tube inlet, indicating the consistency of *MIX* in measuring the stratification level achieved with different inlets. In the variable inlet temperature case, Philips and Dave (1982) used a coefficient which characterizes the effect of mixing in a stratified tank on the daily thermal performance of the solar energy collection system. A review of this and other measures may be found in Davidson *et al.* (1994).

Lately, exergy and exergy efficiency have been used to evaluate the performance of TES systems. Exergy is a measure of usefulness or quality of energy. It is defined as the maximum work potential of any particular form of energy in relation to its environment where this work is obtained by reversible processes. From the first and second laws of thermodynamics, the exergy of a material flow has the following form:

$$Ex = (h - h_a) - T_a\,(s - s_a) \tag{6.16}$$

The exergy of a heat flow is given by:

$$Ex = Q(T - T_a)/T \tag{6.17}$$

Here, h and s are the enthalpy and entropy, respectively, evaluated at the system temperature T and pressure p, while h_a and s_a are the enthalpy and entropy evaluated at the ambient temperature T_a and pressure p_a and Q is the heat transfer.

Rosen (1991) used energy and exergy efficiencies to examine the effect of temperature levels on the performance of a simple sensible TES system. A closed system was considered, involving only heat interactions: heat charge, heat discharge and heat losses to the environment. He defined the exergy efficiency as $\eta = Ex_d / Ex_c$, where Ex_d and Ex_c are obtained from Equation 6.17. Noting that the energy efficiency is $\eta = \eta_{th} = Q_d / Q_c$, an expression for the ratio ψ / η was derived as:

$$\psi / \eta = (T_d - T_a)\,T_c \;/\; (T_c - T_a)\,T_d \tag{6.18}$$

Here, T_c, T_d, and T_a are the charge, discharge, and ambient temperatures, respectively. He calculated the ratio ψ / η for different values of charging and discharging temperatures, the results are given in Table 6.1. It is seen that the energy and the exergy efficiencies are equal only when the charging and discharging temperatures are equal, which is unachievable in practice. The energy and exergy efficiencies differ as T_d is decreased below T_c, and the difference becomes more pronounced as the difference between T_d and T_c increases. This shows the importance of temperature in performance evaluation of TES systems via exergy efficiency, which is sensitive to the temperatures at which heat is charged and discharged.

Rosen and Hooper (1991) evaluated the exergy contents of a stratified tank having water as the storage fluid, and a fixed energy content. The temperatures at top and bottom of the TES were held constant. The TES was subjected to three types of temperature distribution between these locations; linear, stepped and continuous-linear. From their evaluation, while the energy content was fixed, the values of exergy were different for the three types of temperature distribution. The system with more stratification had the highest value of exergy, and the least stratified had the lowest value of exergy. So, exergy gives a quantitative measure of stratification as it changes with the stratification level in the tank.

6.4 Experimental and Theoretical Foundations

The single stratified thermal storage tank has been the subject of numerous experimental and theoretical studies that were motivated in the early 1970s by solar energy storage applications. Brumleve (1974) confirmed the feasibility of using a natural thermocline to achieve separation of hot and cold water inside a single container. Many experimental and theoretical works have appeared since then. Lavan and Thompson (1977) experimentally studied the effect of several geometric and dynamic parameters, i.e. inlet port location and geometry, mass flow rate, tank height to diameter ratio, and the difference in temperature between the inflow and the liquid in the tank. Stratification was found to improve with increasing tank aspect ratio (height to diameter ratio), temperature difference and inlet port diameter. The increase in flow rate had an adverse effect on stratification. Best results were obtained when the inlet and outlet ports were located near the top and bottom walls and the flow was directed towards these walls.

For the same storage volume tall tanks maintain better stratification than short ones. Also, for the same thickness of the thermocline region less fluid in taller tanks is wasted to this intermediate temperature region. However, a limit exists as taller tanks tend to have a larger surface area per unit volume, thereby increasing the heat loss to the surroundings and the insulation cost.

Table 6.1 Values of the ratio ψ / η for a range of practical values of T_d and T_c (Rosen, 1991).

Discharging temperature,	Charging temperature, T_c (°C)			
T_d(°C)	40	70	100	130
40	1.00	0.55	0.40	0.32
70	-	1.00	0.72	0.59
100	-	-	1.00	0.81
130	-	-	-	1.00

The reference environment temperature is set at T_a = 10°C = 283 K.

To provide a high level of stratification without excessive thermal loss an aspect ratio of 4 was recommended by Cole and Bellinger (1982). A ratio of 10 was recommended by Abdoly and Rapp (1982). However, this value would result in a high surface area-to-volume ratio and subsequently increase the heat loss and the insulation cost. In their experiments with tanks with an aspect ratio of between 2.0 to 3.5, Nelson *et al.* (1999) found that improvements in thermal performance were negligible beyond an aspect ratio of 3.0. Analytical studies have shown that little improvement in stratification was achieved for an aspect ratio greater than 3.3 (Al-Najem *et al.*, 1993), and 4.0 (Ismail, *et al.*, 1997; Hahne and Chen, 1998). It was pointed out earlier by Lavan and Thompson (1977) that an aspect ratio between 3 and 4 constitutes a reasonable compromise between performance and cost.

Sliwinski *et al.* (1978) found that the position and sharpness of the thermocline are functions of the Richardson and Peclet numbers. A critical value of the Richardson, Ri, number of 0.244 was found to be the limit below which stratification does not occur. Ri was based on the tank height and the inlet flow velocity. The extent to which mixing occurs in stratified tanks as well as the design improvements to minimize it were investigated by Baines *et al.* (1982). Based on their experiments with fresh-saline and hot-cold water systems it was determined that there are two factors which limit the approach to ideal stratification: the critical layer thickness which defines the volume of fluid that must be introduced before mixing across the thermocline ceases, and the thermocline thickness. Both factors were found to be controlled by the design of the inlet system. To enhance thermal stratification several inlet designs were tested. Cole and Bellinger (1982) tested five different inlet designs. They concluded that the dual radial diffusers are the best. Also, they defined the best inlet design for the thermocline TES as that which introduces the flow horizontally at the top or bottom with minimum velocity. In order to maintain a high degree of stratification in solar collector systems the tank inlet temperature should remain constant. Accordingly, Cole and Bellinger (1982) recommended a new collection strategy that ensures a high degree of stratification. This strategy is based on limiting the flow through the tank to one tank turnover per day and controlling the flow rate to maintain a constant inlet temperature.

Because of the low thermal conductivity of water, conduction across the thermocline was found to be a minor factor in degradation of stratification as compared to other factors, i.e. mixing during the initial stages of charge and discharge, heat loss to the surroundings, and vertical conduction through the walls. One of the earliest studies of the effect of conducting wall on the decay of thermal stratification was that of Miller (1977). Experiments were conducted on two laboratory cylindrical tanks of slightly different sizes and with different wall materials: aluminum and glass. After establishing a thermocline by filling the lower half of the tank with cold water and introducing hot water at the top, temperature measurements along the centerline and at the side wall were taken as a function of time. The results indicate that the degradation of thermocline in the metal tank is six times greater than that in the glass tank. To explain this, Miller (1977) compared the measured temperature profiles with those calculated numerically for the case of heat diffusion across the thermocline. The discrepancy was too large, indicating heat transfer other than diffusion across the thermocline is responsible. This was the conduction along the wall cooling the hot liquid region close to the wall while warming that in the cold region. Consequently, a horizontal non-uniformity in temperature results, leading to buoyancy induced convective currents that enhance mixing and broaden the thermocline.

Noting that the thermal conductivity of water and glass are about the same, it was concluded that the tank wall must be made of a material of thermal conductivity not much greater than that of the stored liquid. Of course, smaller thermal conductivity is preferable. Later, Hess and Miller (1982) studied the effect of wall axial conduction on stratification decay by measuring the velocity field close to the wall using an LDV system. The heat diffusion through the thermocline and the heat loss to the ambient surroundings were isolated in these experiments. That is, the experiments were conducted on an initially uniform low-temperature water tank subjected to axial heat conduction that was obtained by imposing a constant temperature on the cross section of the tank vertical wall. The velocity measurements have shown clearly that for the conditions tested, convective currents are responsible for the degradation of thermocline.

The thermal decay of an initially stratified fluid in two plexiglass insulated rectangular tanks was investigated by Jaluria and Gupta (1982). Several initially stratified temperature distributions were established and the thermal decay was monitored in terms of the temperature distribution along the tank's vertical axis as a function of time and height for several values of the heat loss parameter, $S = UPH^2/A_c k$. Here U is the overall heat transfer coefficient, k is the fluid thermal conductivity, and P, H, and A_c are the tank perimeter, height, and cross-sectional area, respectively. The measured temperature distribution in static tests was found to exhibit horizontal uniformity. Although there was some variation close to the wall due to heat loss to the ambient surroundings, the buoyancy induced motion acts to re-establish the uniformity quickly. This was the basis used for justifying the one-dimensional heat conduction model developed by the authors. Also, the temperature distribution in the top region maintains vertical uniformity as it cools with time. This uniformity is attributed to the heat loss from the top, which cools the uppermost layer causing it to sink and mix with the fluid layers below. This process is repeated causing the average temperature of the initially high-temperature top region to decrease with time. The initially cold bottom region experiences an increase in temperature as a result of cooling the upper region close to the wall. This locally cooled fluid sinks down causing a mixing which leads to a fully mixed condition at the bottom region. The one-dimensional nature of the temperature distribution in a stratified TES tank was recognized from early studies (Close, 1967; Brumleve, 1974). The radial measurements of temperature distribution in a stratified tank has also confirmed this (Gross, 1982). Therefore, many of the one-dimensional modeling efforts find justification on this basis (see section 6.5).

It is clear that the degradation of the thermocline in static mode under partial charge or discharge is faster if the thermal conductivity of storage wall material is higher than that of the stored fluid. In such tanks it is recommended that an insulation layer be applied at the interior surface of the tank. In storage tanks made of concrete (Wildin and Truman, 1985a), stratification was maintained over a wide range of operating strategies of charge, discharge and recycling. The fluid-to- wall thermal conductivity ratio is the governing parameter. The higher this ratio the better the stratification.

The previous findings regarding the effect of the conducting wall were based on static tests with partially charged tanks. In the dynamic mode of operation the heat transfer through the tank wall was found to have a negligible effect on stratification (Lightstone *et al.*, 1989). Murthy *et al.* (1992) conducted experiments with three cylindrical scaled model tanks of the same internal diameter. Two of the tanks were made of mild steel of different thickness (1.0 and 2.4 mm) and the third one was made of aluminum of 1.0 mm thickness.

The three tanks were insulated with the same thickness of glass wool mats on all sides. Experiments covered both static and dynamic tests, with the latter tests covering both the discharge (only the load loop was on) and the simultaneous discharge and charge cycles (both the load and the heat source loops were on). Temperature measurements at ten locations along the axes of the tanks were taken at different time intervals. Their results demonstrated that thermocline decay is faster for the aluminum tank. At an inlet-to-outlet temperature difference of 48°C the extraction efficiency was about 5% less for the aluminum tank. This decrease is less at lower temperature differences. These results are in accord with the results of Lightstone *et al.* (1989).

The mixing introduced by the inflow during the charge and discharge was recognized as a major contributor to the degradation of the stored energy. Therefore, different inlet configurations were designed and tested. These designs may be classified into two categories: the constant-inlet temperature flow diffusers and the variable-inlet temperature flow distributors. These are shown in Figures 6.3 and 6.4 for the former and in Figure 6.5 for the latter. As stated earlier, the former condition is more common in chilled water storage and in solar energy collection systems with controlled output temperature. In this case inlets that introduce the flow with minimum velocity and in a gravity or surface current forms are the most effective in producing a thin thermocline. Extensive research on diffusers for chilled water storage was conducted by a group of researchers (see Wildin and Truman, 1985a; Wildin, 1990, 1991, 1996). Flow visualisation and temperature measurements in a laboratory-scale model chilled-water storage tank were conducted by Yoo *et al.* (1986). The tank was equipped with a linear-slot diffuser spanning one side of the tank bottom. Also, the slot height was variable to obtain different values of inlet flow Froude number. It was concluded that for better stratification the inlet densimetric Froude number, Fr, should not exceed 2, where Fr is based on the slot height and the average velocity of the flow from the slot. A value of unity or less was recommended by Wildin and Truman (1985b). By adding a settling mesh to single- and multi-tube diffusers, Al-Marafie *et al.* (1991) obtained better stratification with mesh than without it. Clearly, the design of the inlet in a thermocline TES tank is critical for better performance.

The effect of the inlet geometry on stratification in thermocline TES tanks was the subject of several research works carried out at Oklahoma State University. Using a fresh-saline water system, three radial diffusers of different geometry were tested by Zurigat *et al.* (1988). These were a solid disk radial diffuser, a perforated disk, and a perforated disk with a solid center. The tank, filled initially with fresh water, is charged with saline water from the bottom resembling the charge of a chilled water storage TES tank. Using a conductivity probe installed at the exit from the tank, the transient salt concentration profiles were obtained. The advantage of the experiments with a fresh-saline water system is the isolation of two stratification degradation mechanisms: the heat loss to the surroundings and the conduction along the tank wall. The results indicated that the perforated radial diffuser with a solid center gives the thinnest thermocline. Further experiments using a hot-cold water system were conducted by Zurigat *et al.* (1991) for the charging of a hot-water thermocline TES tank. Three inlet configurations were tested. These are the side inlet, the impingement inlet, and the perforated inlet (see Figure 6.4). Characterization of the mixing introduced by these inlets was carried out using an inlet mixing parameter introduced in a one-dimensional flow model (see section 6.5). The effect of the inlet geometry was also studied numerically by Zurigat (1988) (see also Ghajar and Zurigat, 1991) using a two-dimensional flow and heat-transfer model (see section 6.6).

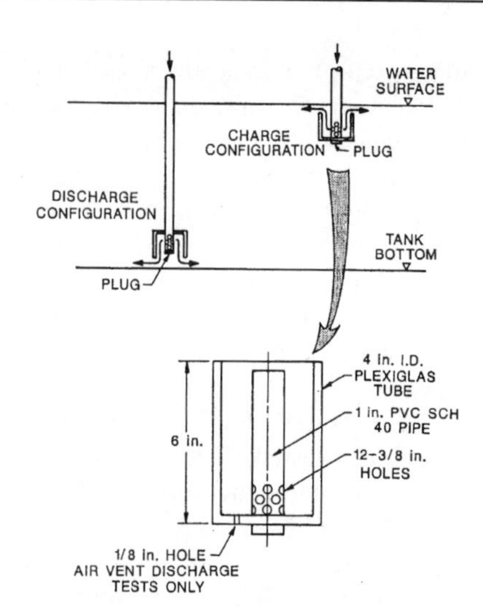

Figure 6.3 Schematics of the cup diffuser (Loehrke *et al.*, 1979).

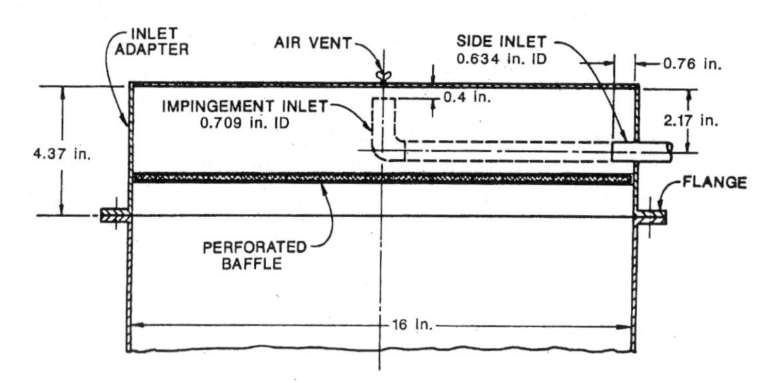

Figure 6.4 Schematics of the inlet configurations tested by Zurigat *et al.* (1991).

Most of the experimental work reported in the literature was conducted under constant inlet temperature, i.e. thermocline TES. The case of variable inlet temperature (stratified TES) has been studied in relation to solar energy collection systems. Inlets designed for the variable inlet-temperature case are termed *inlet distributors*. The common feature of these distributors is the use of perforated manifolds to house the incoming stream while allowing it to reach its temperature level where it exits the manifold through side perforations (see Figure 6.5). Notable design attempts are those which consist of rigid or flexible distribution manifolds (Sharp and Loehrke, 1979; Gari *et al.*, 1979; Davidson *et al.*, 1994; Davisdon and Adams, 1994), light flexible hose (Van Koppen *et al.*, 1979), rigid solid deflector baffles (Davis and Bartera, 1975), and rigid perforated distributor baffle (Abu-Hamdan *et al.*, 1992). Under variable inlet temperature the tests conducted with rigid porous manifolds have shown better performance than the conventional fixed inlets (Gari *et al.*, 1979). This was also shown by Davidson *et al.* (1994) by comparing the performance of a rigid porous manifold with a conventional drop-tube inlet. Under different test conditions the performance was monitored using the *mix number, MIX,* introduced in section 6.3. Despite their better performance the rigid porous manifolds are not in general use because of their inflexibility in adapting to flow conditions other than those for which they are designed. A remedy was suggested by Davidson and Adams (1994), who designed a fabric manifold which they showed to be 4% more effective in maintaining stratification than rigid porous manifolds and 48% more effective than the conventional drop-tube inlet.

The inlet distributor of Abu-Hamdan *et al.* (1992) consisted of a perforated circular baffle with a solid portion located about its mid-height. The baffle was fitted inside the test tank resulting in an annulus space of 2.5 cm. The incoming flow enters the tank through 32 inlets located around the circumference of the test tank at mid-height. The flow then impinges on the solid portion of the baffle which in turn deflects it either up or down depending on the flow temperature. The deflected stream then flows through the perforated sections of the baffle. Tests using this distributor and two conventional inlets, a side and a drop-tube inlets, were conducted. It was concluded by Abu-Hamdan *et al.* (1992) that passive devices of the type used in their experiments or that used by Davis and Bartera (1975) offer no advantage over conventional inlets. More work in this area of distributor design for variable inlet-temperature conditions is still needed.

Theoretical studies of stratified TES tanks were conducted using one-, two- and three-dimensional models. The former two are discussed in the next two sections. Three-dimensional modeling efforts of stratified TES are scarce. The model of Sha and Lin (1978) is a notable model of this kind. Because of computer time limitations three-dimensional models are generally avoided despite their potential for assessing stratified TES tank design concepts.

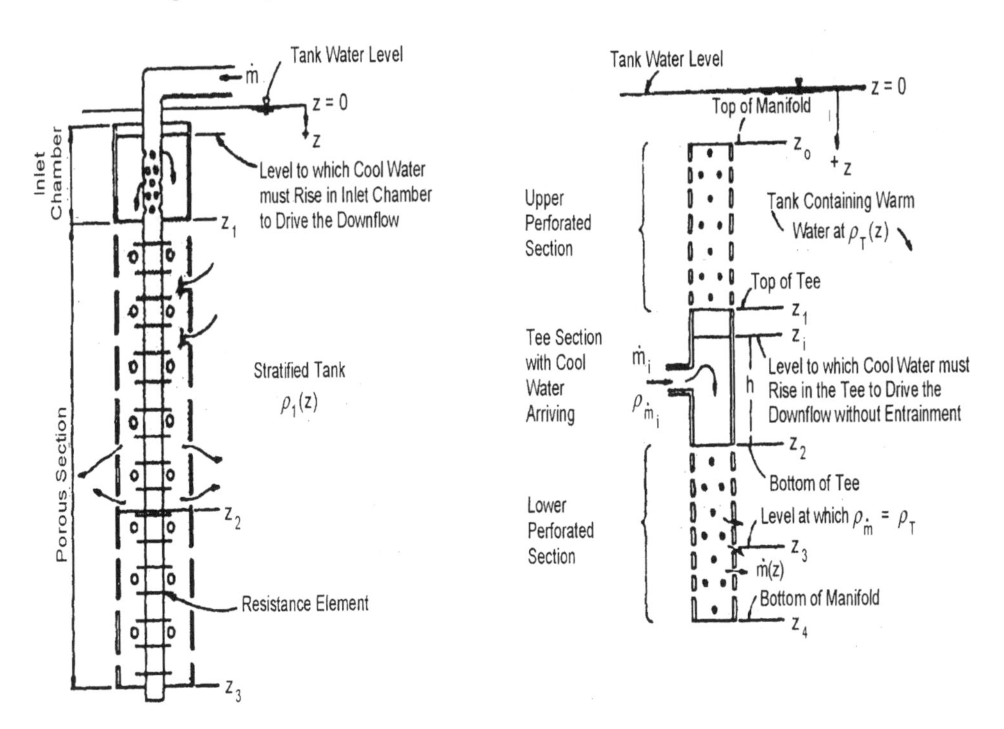

Figure 6.5 Stratified thermal storage tank inlet distributors (Gari *et al.*, 1979; Loehrke *et al.*, 1979).

6.5 One-Dimensional Models

The one-dimensional models of stratified thermal energy storage may be classified into two categories depending on the inlet temperature condition. These are the *stratified* TES tank models in which the inlet temperature is allowed to vary with time, and the *thermocline* TES tank models in which the inlet temperature is maintained constant. The latter condition results in a well-defined thermocline region of relatively small thickness. The former models were first developed for solar energy storage applications characterized by variable inlet temperature condition. In this section we discuss different one-dimensional models developed in the literature for both constant and variable inlet temperature conditions.

The *stratified* TES tank models represent a somewhat realistic condition between two idealized models of extreme cases that have, as discussed in section 6.3, utility in performance evaluation of thermal storage tanks and thermal storage systems. These are the *fully-mixed* and the *fully-stratified* tank models. In the *fully-mixed* model the entire liquid in the tank is assumed to have a uniform temperature which changes with time as a result of net energy addition or withdrawal during the charge or discharge processes or due to the interaction with the surroundings at T_a. The model equation representing this thermal balance may be written as:

$$MC_p \frac{dT}{dt} = \dot{m}_{in} C_p \ (T_{in} - T) - UA(T - T_a) \tag{6.19}$$

Equation 6.19 may be solved numerically for the initial condition $T(o)$ equal to the average temperature of the liquid in the tank. In this equation M is the mass of water in the tank, T_{in} is the inlet temperature, $\dot{m}_{in}$ is the mass flow rate, A is the heat loss surface area, and U is the overall heat transfer coefficient. U may be found experimentally by measuring the rate of decay of initially heated water at uniform temperature in a static tank using Equation 6.21 below rearranged for U or UA. Typical values for hot water storage are U = 0.973 W/m^{2}K (Votsis *et al.*, 1988), UA = 4.57 W/K (Kleinbach *et al.*, 1993), and UA = 2.7 W/K (Davidson *et al.*, 1994). Alternatively, U or UA may be calculated based on forced and free convection correlations depending on the mode of operation, i.e. static or dynamic.

In the fully-stratified model the tank is subdivided into a number of uniform temperature layers and the inflow is assumed to seek its temperature level in the tank without exchanging heat with the adjacent layers or those along its path. The only heat interaction permitted is that with the surroundings at T_a. Under these conditions the energy balance for any layer, J, of mass M_J gives

$$M_J C_p \frac{dT_J}{dt} = -UA(T_J - T_a) \tag{6.20}$$

For an initial temperature $T_{o,J}$, the analytic solution of this equation over a time step, Δt, is

$$(T_J - T_a) = (T_{o,J} - T_a)\exp(-UA\Delta t / M_J C_p) \tag{6.21}$$

The temperature distribution in the storage tank can then be found by applying the solution given by Equation 6.21 to all the layers in the tank. However, during a time step, Δt, the incoming flow fills a finite volume which should be inserted in the tank in between two layers of higher and lower temperatures, respectively. To maintain continuity, the layers in the tank are displaced up or down depending on whether the tank is in the charge or the discharge modes. For example, in the charge mode if the incoming stream is at temperature T_{in}, higher than that of the uppermost layer, then all the layers in the tank are displaced down resulting in the uppermost layer being filled with the incoming stream while the bottom layer exits the tank to the heat source. If T_{in} is somewhere between the two temperatures at the top and bottom layers, the incoming stream is placed in the tank at a location consistent with its temperature. That is, all layers with temperatures greater than T_{in} remain in place while those below are displaced downward. Again the lowest layer exits the tank to the source.

Turning to *stratified* TES tank models, the one-dimensional models of Close (1967), Duffie and Beckman (1980), and Sharp (1978) assume that the flow seeks its temperature level without mixing along its path. In addition, heat loss to the surroundings and mixing with the adjacent layers is allowed. These one-dimensional models were compared by Kuhn *et al.* (1980) by validation with experimental data, and the model of Sharp (1978) was found to give better predictions. Therefore, the latter model is discussed below along with the one-dimensional model of Han and Wu (1978) (known as the *Viscous Entrainment Model*) in which the above assumptions hold except the flow is now assumed to entrain fluid from the layers along its path

In the model of Sharp (1978), which allows for arbitrary variation in fluid inlet temperature, the tank is divided into a number of isothermal, constant volume segments. Energy and mass balance equations are written for each segment while accounting for thermal losses to the surroundings and vertical conduction through the tank walls. Using control functions the inlet fluid is directed to the segment whose temperature most closely matches its temperature without mixing with the segments along its path. However, mixing with the adjacent layers is allowed. Although the model does not explicitly account for turbulence, the number of segments used in the model has an effect on the calculated temperature profile similar to that due to turbulent mixing. This can be seen in Figure 6.6, which shows the temperature profiles calculated for different numbers of segments, N, under a constant inlet temperature. Higher N values result in predictions closer to ideal stratification or, equivalently, less mixing.

Han and Wu's (1978) *Viscous Entrainment* model is a finite-difference model that accounts for heat loss to the surroundings and the mixing effects due to entrainment of the tank fluid by the incoming stream. Also, collector and load flow circuits are incorporated and mass and energy balance equations are derived for both circuits and solved using an implicit finite-difference method. An additional equation describing the rate of viscous entrainment is also provided. A boundary condition parameter, γ, is introduced to account for mixing in the upper and lower regions of the tank due to the introduction of collector flow and load flow, respectively. The calculated transient temperature profiles, at a fixed location in the tank with different values of γ, are shown in Figure 6.7. It is seen that as γ increases the thermocline widens as expected. However, as γ increases beyond 20, unrealistic mixing takes place from one side of the thermocline. That is, the thermocline fronts for different values of γ are at the same location in the tank irrespective of the severity of the mixing at the inlet. Note that the results shown in Figures 6.6 and 6.7 were obtained under constant inlet temperature conditions. It is easier to visualize the validity of these models in this limiting case for the inlet flow condition.

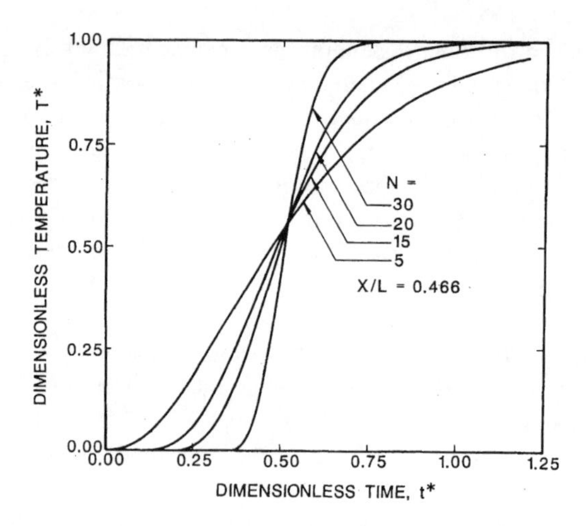

Figure 6.6 Effect of number of segments, *N*, on thermocline predictions using the model of Sharp (1978) (Zurigat *et al.*, 1989).

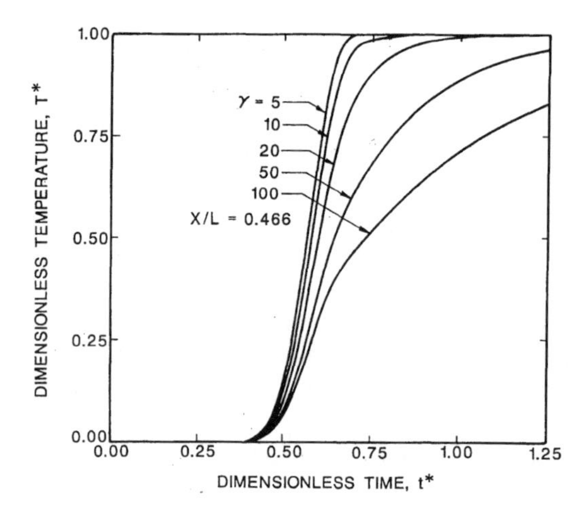

Figure 6.7 Effect of boundary condition mixing parameter, γ, on thermocline predictions using the model of Han and Wu (1978) (Zurigat *et al.*, 1989).

The fact that these models are also applicable to the thermocline TES case makes them more general than the thermocline TES models discussed next. However, in the thermocline TES case, simpler and more accurate models have been developed. The thermocline TES models are based on solving the one-dimensional convection-diffusion equation. This equation can be derived using an energy balance on the control volume shown in Figure 6.8, which represents a fluid region of uniform temperature subject to heat loss to the ambient surroundings through the tank wall and insulation. It is assumed that the flow is one-dimensional, i.e. the velocity is uniform over the tank cross section. In a tank with properly designed inlet and outlet diffusers this assumption of plug or piston-like flow is valid for most of the tank except at and in the vicinity of the diffusers, where the flow may be three-dimensional and turbulent. Furthermore, the thermo-physical properties of the fluid are assumed constant and the thermal inertia of the tank wall and insulation negligible. Justification for the latter assumption has been demonstrated by Gretarsson *et al.* (1994). Then, the resulting energy equation of laminar flow is:

$$\frac{\partial T}{\partial t} + V\frac{\partial T}{\partial x} = \alpha\frac{\partial^2 T}{\partial x^2} + \frac{UP}{A_c \rho C_p}\left(T_a - T\right) \tag{6.22}$$

In this equation A_c is the cross-sectional area of the tank, V is the mean vertical velocity in the tank (based on A_c), U is the overall heat transfer coefficient, and P is the tank perimeter. Although Equation 6.22 seems simple to solve numerically, two potential problems arise. First, the flow experiences mixing at the inlet which is not taken into account in Equation 6.22. Second, unless care is taken, the stability of the numerical solution demands the use of the highly dissipative upstream differencing scheme, thereby compromising the accuracy. These two problems were treated by the model developed by a research group led by Professor Ghajar, which is referred to henceforth as the Ghajar model (Oppel *et al.*, 1986; Zurigat *et al.*, 1988; Zurigat *et al.*, 1991). The turbulent mixing was accounted for by introducing into Equation 6.22 an effective diffusivity factor, ε_{eff}, defined as

$$\varepsilon_{eff} = (\alpha + \varepsilon_H)/\alpha \tag{6.23}$$

Note that for laminar flow the eddy diffusivity for heat $\varepsilon_H = 0$ everywhere and $\varepsilon_{eff} = 1$. For turbulent flow ε_{eff} is much greater than unity. Multiplying the molecular thermal diffusivity in Equation 6.22 by this factor is equivalent to stating that the molecular thermal diffusivity, α, is magnified by turbulence by a factor of ε_{eff}. Equation 6.22 then becomes

$$\frac{\partial T}{\partial t} + V\frac{\partial T}{\partial x} = \alpha\,\varepsilon_{eff}\frac{\partial^2 T}{\partial x^2} + \frac{UP}{A\rho C_p}\left(T_a - T\right) \tag{6.24}$$

A finite-difference solution technique is used to solve Equation 6.24 subject to the appropriate initial and boundary conditions. The heat losses from the top and bottom of the tank are neglected. To ensure stability of the numerical solution of Equation 6.24 under all flow rates, the first space derivative should be descretized using the upwind difference representation. However, this representation produces numerical diffusion which results in inaccurate results. Therefore, to eliminate the numerical diffusion inherent in the upwind differencing of the first space derivative (convective term), Equation 6.24 is separated into two cases. These are the diffusion case

$$\frac{\partial T}{\partial t} = \alpha\,\varepsilon_{eff}\frac{\partial^2 T}{\partial x^2} + \frac{UP}{A_c\rho C_p}\left(T_a - T\right) \tag{6.25}$$

and the convection case

$$\frac{\partial T}{\partial t} + V\frac{\partial T}{\partial x} = 0 \tag{6.26}$$

An implicit finite-difference representation of Equation 6.25 yields the following:

$$\begin{aligned}&\left(-\varepsilon_{eff}\,Fo\right)T'_{n-1} + \left(1 + \phi + 2\varepsilon_{eff}\,Fo\right)T'_n \\ &+ \left(-\varepsilon_{eff}\,Fo\right)T'_{n+1} = \phi T_\theta + T_n\end{aligned} \tag{6.27}$$

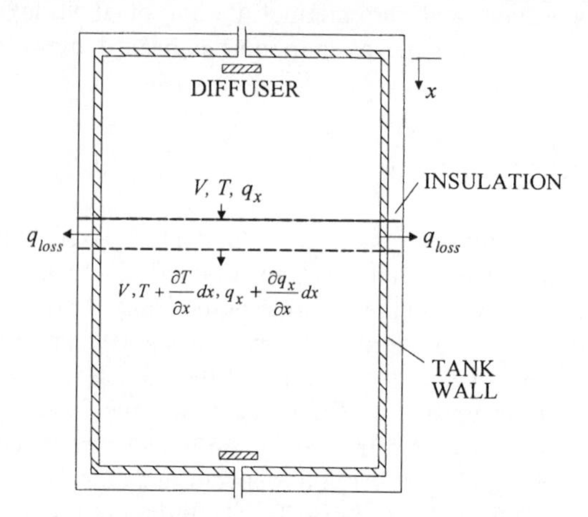

Figure 6.8 Energy balance in thermocline thermal energy storage tank.

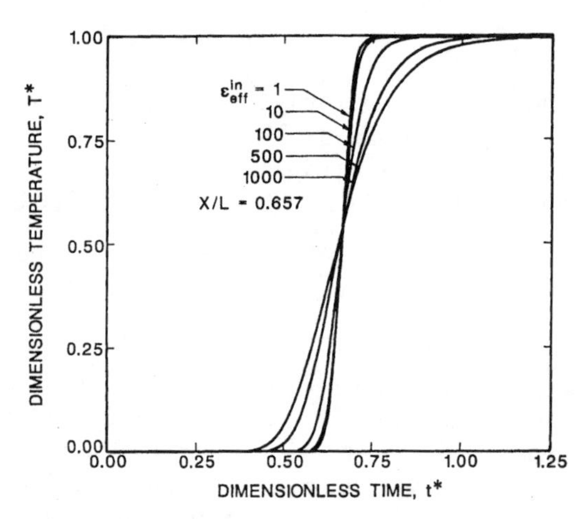

Figure 6.9 Effect of inlet effective diffusivity factor ε_{eff}^{in} on thermocline predictions using the model of Ghajar (Zurigat *et al.*, 1991).

Here, $\phi = UP\Delta t / A_c \rho C_p$ and Fo is the Fourier number; $Fo = \alpha \Delta t/(\Delta x)^2$. An explicit finite-difference representation of Equation 6.26 gives

$$T''_n = T_n - \frac{V\Delta t}{\Delta x}\left(T_n - T_{n-1}\right) \tag{6.28}$$

Here the double prime denotes new values that may be calculated at a time step different from that in Equation 6.27. To obtain the exact solution of Equation 6.28, the Courant number, $C = V\Delta t / \Delta x$, should be equal to unity. This gives

$$T''_n = T_{n-1} \tag{6.29}$$

For the variable flow rate case, C can be different from unity. Therefore, the 'buffer tank' concept (Oppel *et al.*, 1986) is used. This mathematical concept is based on setting the Courant number, C, equal to unity by the choice of Δx or Δt for the maximum expected flow rate, and applying Equation 6.29 only at multiples of Δt for which C is equal to unity for other flow rates. For example if $C = 1/2$, Equation 6.29 would be applied at every other time step.

The variation of ε_{eff} in Equation 6.27 needs to be specified. Since different inlets promote turbulence in varying degrees, it is expected that ε_{eff} will assume different values for different inlets. Based on experiments with fresh-saline water systems (Oppel *et al.*, 1986; Zurigat *et al.*, 1988), ε_{eff} was found to vary spatially from a maximum at the inlet ε_{eff}^{in} to a minimum of unity at the outlet in a decreasing hyperbolic function of the form (other functions were also tried, i.e. linear and exponential)

$$\varepsilon_{eff} = A / N_{sl} + B \tag{6.30}$$

where

$$A = (\varepsilon_{eff}^{in} - 1) / \left(1 - \frac{1}{N_{slt}}\right)$$

$$B = \varepsilon_{eff}^{in} - A$$

Here, N_{sl} is the slab number in increasing order from the inlet and N_{slt} is the total number of slabs, determined by

$$N_{slt} = H / (V\Delta t) \tag{6.31}$$

A problem thus remains in specifying the inlet value of the effective diffusivity factor, ε_{eff}^{in}. Figure 6.9 shows the model's predictions of the transient temperature profiles at a certain location in the tank *(X/L* = 0.657*)* for different values of ε_{eff}^{in} for an arbitrary condition. It is seen that the thermocline gets thicker as ε_{eff}^{in} increases. Despite its artificial nature, the effective diffusivity factor can represent the modifying effect of turbulence caused by the inlet flow. Hence, this factor may be used to characterize the inlet geometry and identify the best inlets. This feature has been demonstrated by Zurigat *et al.* (1988 and 1991). The variation of the inlet effective diffusivity factor as a function of the flow parameters, i.e. the ratio *Re/Ri*, is shown in Figure 6.10 for three different inlet configurations. Here, *Re* is based on the inlet port diameter, while *Ri* is based on the height of the tank, and both are based on the inlet port flow velocity. The three inlets used are the side inlet, the perforated inlet, and the impingement inlet (see Figure 6.4). As expected, the side inlet introduces the highest level of turbulence and mixing. The curves shown in Figure 6.10 are represented by the following correlations:

$$\varepsilon_{eff}^{in} = 0.344\,(\mathrm{Re}/Ri)^{0.894} \quad \text{for the side inlet} \tag{6.32}$$

$$\varepsilon_{eff}^{in} = 3.54\,(\mathrm{Re}/Ri)^{0.586} \quad \text{for the perforated inlet} \tag{6.33}$$

$$\varepsilon_{eff}^{in} = 4.75\,(\mathrm{Re}/Ri)^{0.522} \quad \text{for the impingement inlet} \tag{6.34}$$

The above correlations were incorporated into the Ghajar model which was then confirmed by Zurigat *et al.* (1991) using experimental data from the literature and of their own research. Figures 6.11–6.13 show the predictions compared with the experimental data. The agreement between the predictions and the experiments is good. Thus, the introduction of the effective diffusivity factor has proved to be a powerful tool in modeling the mixing in thermocline TES tanks.

The model of Ghajar described above was compared with five one-dimensional models from the literature by Zurigat *et al.* (1989). The comparison was conducted with respect to experimental data from their own research and the literature for the constant inlet temperature condition, i.e. thermocline TES. The models considered were those of Sharp (1978), Han and Wu (1978), Cole and Bellinger (1982), Cabelli (1977) and Wildin and Truman (1985b). The tank modeled was assumed to be insulated with no wall axial conduction and no thermal inertia for the tank envelope. The first two models are described above and the other three below. The comparison showed varying degrees of agreement with thermocline test data. Generally, the best agreement was obtained by the models of Wildin and Truman (1985b), Cole and Bellinger (1982), and the model of Ghajar. Some adjustment of the parameters is needed in the former two models, while in the latter no adjustment is required in view of the correlations presented in Equations 6.32–6.34.

The model of Cole and Bellinger (1982) is a one-dimensional analytical model with empirically derived parameters. These are the mixing parameter, C, which accounts for mixing due to the introduction of fluid into the tank, the normalized film heat-transfer coefficient, H, which accounts for the fluid-wall thermal interaction, and the capacity ratio, a, which accounts for the effect of wall thermal inertia on stratification. For thin walls $a = 1$. The heat loss to the surroundings is neglected and the model is restricted to a constant flow rate and a constant inlet temperature, as well as a uniform initial temperature distribution. Zurigat *et al.* (1989) found that the parameter C has a significant influence on the thermocline shape, as shown in Figure 6.14, while the parameter H was found to have a negligible effect. Cole and Bellinger (1982) correlated the former parameter with the Fourier and Richardson numbers. The model of Cabelli (1977) is a closed-form solution of the one-dimensional energy equation with heat loss to the environment. No mixing effects are included.

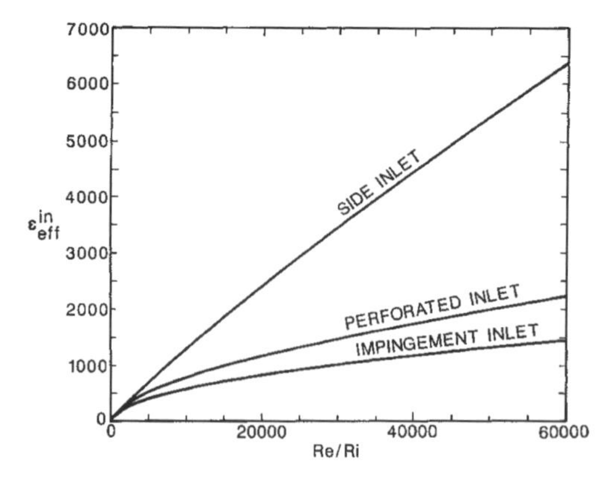

Figure 6.10 The variation of the inlet effective diffusivity factor ε_{eff}^{in} as a function of the ratio *Re/Ri* (Zurigat *et al.*, 1991).

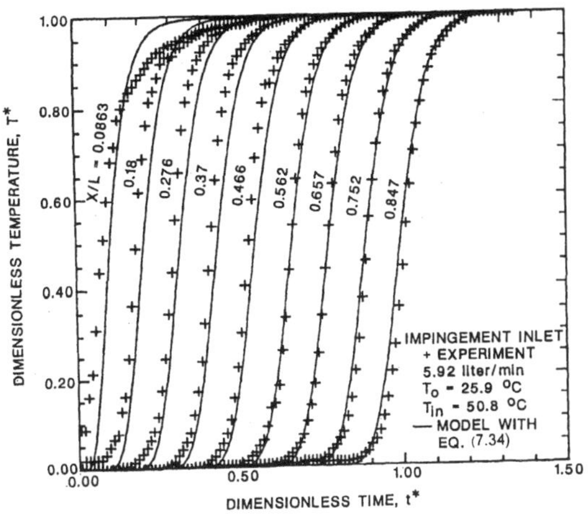

Figure 6.11 Predicted transient temperature profile at different locations in the storage tank using the model of Ghajar for the impingement inlet compared with the experiments (Zurigat *et al.*, 1991).

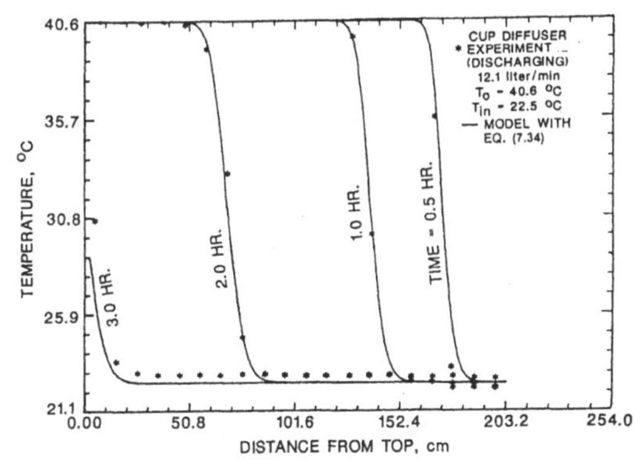

Figure 6.12 Predicted temperature profiles using the Ghajar model compared with the experimental data of Kuhn *et al.* (1980) for the discharge mode using the cup diffuser (Zurigat *et al.*, 1991).

The model of Wildin and Truman (1985b) is a finite-difference model that accounts for mixing at the inlet region, vertical conduction through the tank wall and water, the thermal capacitance of the tank wall and heat exchange with the surroundings. Mixing in the inlet is quantified by averaging the temperatures of a specified number of liquid elements, *NM*, near the inlet. Figure 6.15 shows the effect of the variation of *NM* on the thermocline. Based on experiments with chilled water storage, Wildin and Truman (1985a) concluded that the thickness of the mixing layer for well designed inlet diffusers was no more than 7% of the water depth, up to a maximum of 0.31 m. A well designed diffuser was defined as that which introduces the fluid into the tank in a gravity current flow at the bottom of the tank for cooler water and a surface current flow at the top of the tank for warmer water. A poorly designed diffuser is one which introduces the fluid in a jet-like form. Gretarsson *et al.* (1994) considered the depth of the mixing zone to be 3% of the tank depth for well-designed diffusers and 8% for poorly designed ones. These values are empirically-derived, and thus may differ from one investigator to another.

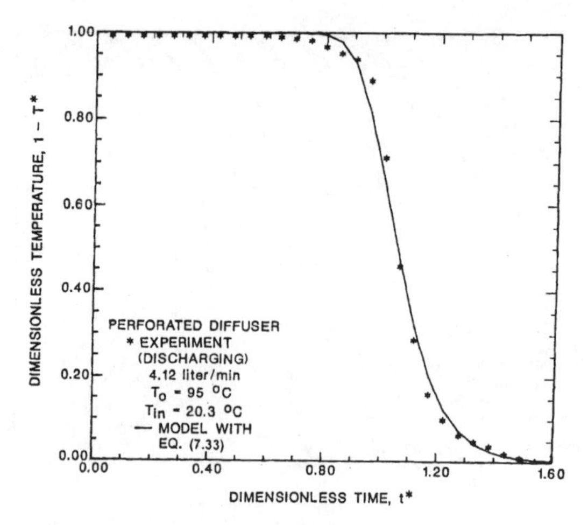

Figure 6.13 Predicted outlet temperature profiles using the model of Ghajar compared with the experimental data of Abdoly (1981) (Zurigat *et al.*, 1991).

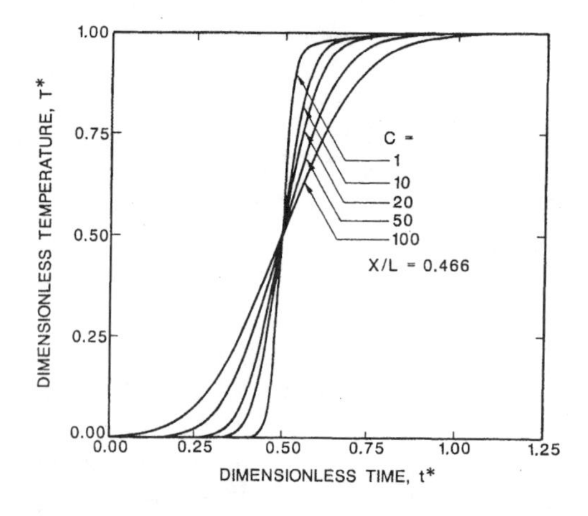

Figure 6.14 Effect of the mixing parameter, *C*, on thermocline predictions using the model of Cole and Bellinger (1982) (Zurigat *et al.*, 1989).

The concept of splitting the governing one-dimensional energy equation used in the Ghajar model was later adopted by other investigators. Votsis *et al.* (1988) assumed a constant eddy diffusivity (or mixing) factor and solved the equations for an insulated tank using an explicit finite difference method. Gretarsson *et al.* (1994) also used the same technique, which they called the 'discrete-time-step' model. The diffusion and heat loss terms in the transient heat conduction equation were discretized for each node and the resulting system of first-order differential equations was solved using the Runge-Kutta method. As with the Ghajar model, the time step and the number of finite fluid elements were determined from the Courant number by setting it to unity. Mixing, however, was introduced in the same way as that used by Wildin and Truman (1985b). That is, a mixing coefficient defined in terms of the percent of the total tank volume was introduced, representing the volume of water in the inlet region that should be mixed with the incoming stream. This coefficient depends on the type of diffuser used. Although the Gretarsson *et al.* (1994) model ignores the tank wall thermal capacitance, they applied their model to concrete tanks of different thickness and compared their results with the predictions of the model of Wildin and Truman (1985b), which accounts for the thermal inertia of the tank walls. The comparison, expressed in terms of the relative difference in the calculated change in internal energy of the water for the two models, showed that ignoring the thermal capacitance introduces a difference of less than 3%. Based on this result, one-dimensional models that ignore the thermal inertia of the tank wall may find justification.

The localized mixing as used by Wildin and Truman (1985b) and Gretarsson *et al.* (1994) was also used in a different way by Nakahara *et al.* (1988), who assumed that the tank consists of two regions: a fully mixed region which increases in extent as filling proceeds, and a plug flow region spanning the rest of the tank. The extent of the fully mixed region was expressed in terms of a dimensionless depth $R = X/H$, where H is the tank height and X is the depth. R was assumed to vary linearly with the dimensionless filling time t^* ($t^* = Vt/H$) as:

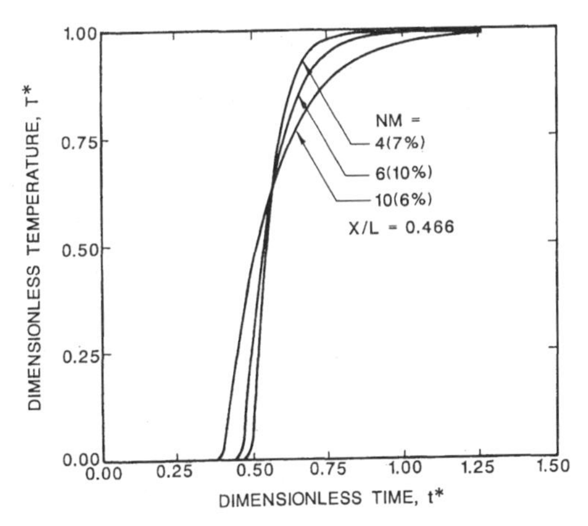

Figure 6.15 Effect of number of mixed segments, *NM*, on thermocline predictions using the model of Wildin and Truman (1985b) (Zurigat *et al.*, 1989).

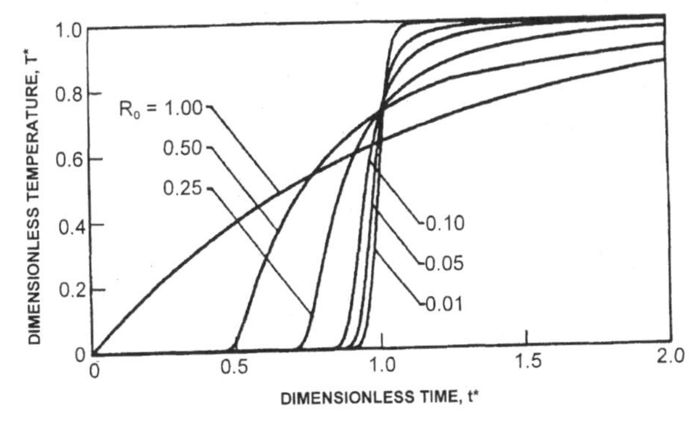

Figure 6.16 The outlet temperature profile predicted by the model of Nakahara *et al.* (1988).

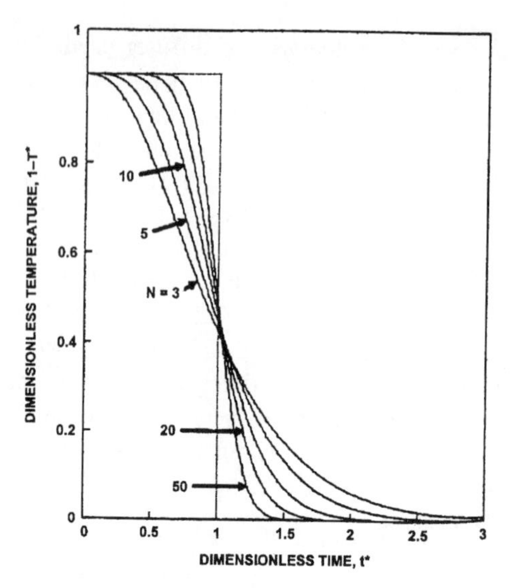

Figure 6.17 Predictions of storage tank outlet temperature profile using the layers-in-series model for different numbers of layers (Mavros *et al.*, 1994).

$$R = R_o + R_k t^* \qquad (6.35)$$

where R_o and R_k are empirical constants. R_k was taken to be 0.4 under all conditions while $R_o = X_o/H$ represents the initial extent of mixing and its value depends on the inlet diffuser type and the inlet flow conditions. Correlations for R_o in terms of Richardson number and the inlet length scale were given for two different inlet types: a pipe inlet and a slot inlet. In the fully mixed region, Equation 6.19 with $U = 0$ is applicable. In terms of the volume of the tank the solution over the time interval Δt becomes:

$$T(t+\Delta t) = T_{in} + (T_o - T_{in})\exp(-\dot{\forall}\Delta t / \forall_T R) \qquad (6.36)$$

In the plug flow region the governing equation is the one-dimensional convection-diffusion equation which is solved numerically. The temperature profiles for different values of R_o are shown in Figure 6.16.

Another thermocline TES model called the *Layers-in-Series* model was developed by Mavros *et al.* (1994). The tank is subdivided into a number of uniform temperature layers and Equation 6.19 without heat loss to the surroundings is written for each layer. The resulting system of first-order ordinary differential equations is then solved in closed form. The inlet temperature of each layer was taken to be that of the upstream layer. Again, the number of layers has an effect on the temperature profile similar to that of mixing and diffusion. Figure 6.17 shows the transient outlet temperature profile predicted by the *Layers-in-Series* model for different numbers of layers. The unrealistic symmetry exhibited by the temperature profile predicted by this model (see Figure 6.17) motivated Mavros *et al.* (1994) to modify their model by adding to each layer a small side tank. Part of the flow entering each layer is diverted into its corresponding side tank and, by continuity, an equal amount is allowed to leave and enter the next layer downstream which experiences a similar flow diversion. To model this flow arrangement two empirical parameters were devised to control the size of the side tank and the flow rate of the stream entering it. Then, energy balance equations are derived for each layer and each side tank, resulting in twice the number of equations used in *Layers-in-Series* model. The resulting system of equations is then solved numerically. As with all one-dimensional models the predictions depend on how well the empirical parameters are correlated with the flow conditions.

The effective diffusivity factor introduced in the model of Ghajar was also used in the same way by Al-Najem and El-Refaee (1997), who assumed it to vary in a spatially decreasing exponential function having a maximum value at the inlet which is correlated with the ratio of Reynolds to Richardson numbers. They solve Equation 6.24 without the heat loss term using a finite-element-based method known as Chapeau-Galarkin method. Homan *et al.* (1996) also used the same approach adopted in the model of Ghajar for quantifying the mixing in the thermocline TES tank. That is, using suitable time and space scales, Equation 6.24 is made dimensionless resulting in the diffusion term being multiplied by the inverse of an effective Peclet number, $Pe_{H,eff} = VH/(\alpha+\varepsilon_H)$. In terms of Peclet number $Pe_{H,eff} = Pe_H/\varepsilon_{eff}$. Then Equation 6.24 without the heat loss term is solved in closed form using Laplace transforms. This solution transposed explicitly for the effective Peclet number gives the functional relation used to determine the ratio ε_H/α (which is one less than ε_{eff}) from experimental data. This method assumes uniform mixing throughout the tank. As pointed out by Musser and Bahnfleth (1998), this assumption oversimplifies. Figure 6.18 gives the ratio ε_H/α as a function of Peclet number, Pe_H, for different inlet configurations. The dimensionless parameter T^*_{max} cited in Figure 6.18 is the maximum outlet temperature allowed by the load, and is given for the discharge mode by $T^*_{max} = (T_{max} - T_o)/(T_{in} - T_o)$. Figure 6.18 shows a good correlation, reinforcing the usefulness of the effective diffusivity factor introduced in the model of Ghajar. The results presented in Figures 6.10 and 6.18 for the effective diffusivity factor differ (in the order of magnitude) because those of Figure 6.10 are for the value at the inlet which is the maximum value while those of Figure 6.18 are the average uniform values.

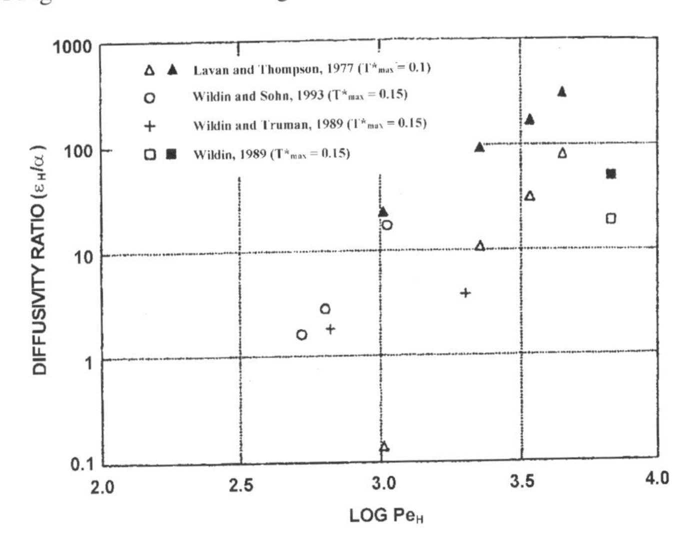

Figure 6.18 Empirical diffusivity ratios for stratified solar and chilled water storage tanks (Homan *et al.*, 1996).

Based on the foregoing, it is seen that the one-dimensional modeling has been the focus of many researchers because of its role in simulations of overall energy systems involving thermal storage. The problem of mixing introduced by the inflow has been treated in different ways, and is empirically based by the nature of one-dimensional modeling.

6.6 Two-Dimensional Models

In these models the conservation equations of flow and heat transfer in stratified thermal storage tanks are the two-dimensional continuity, momentum, and energy equations. In comparison with one-dimensional modeling, two-dimensional modeling involves less assumptions and empiricism, and thus is more realistic and accurate. A wider range of flow thermal and hydrodynamic conditions as well as complex tank geometric parameters may be modeled. Naturally, in addition to the increased computational effort in terms of formulation, programming and output data reduction, more computer time is required. This usually makes two-dimensional models unsuitable when the aim is to study the performance of energy systems involving thermal storage. The two-dimensional modeling of stratified TES has been conducted by several researchers. It should be mentioned at the outset that in solving the governing conservation equations of mass and momentum they are either left in their premitive-variables form or put in the vorticity-stream function form. Retaining the usual variables, the former form requires the solution of the pressure field using different algorithms depending on the numerical method used. For example, the SIMPLE algorithm (Patankar, 1980) and its upgraded versions or the Newton-Raphson pressure-adjustment algorithm (Hirt *et al.*, 1975). The second form has the advantage of identically satisfying the continuity by the stream function and eliminating the pressure by the cross product of the momentum equations. Furthermore, since the problem is of the mixed convection type, the Boussinesq approximation is used by all investigators. In this approximation the density is assumed constant except in the appropriate gravity term in the momentum equations. Also, unless otherwise stated, in the studies discussed below adiabatic conditions are assumed, and no localized resistance terms (baffles) are used. Finally, these numerical studies share the common objective of studying the influence of different design parameters on stratification such as the inlet port type and location, the flow direction, and the Reynolds and Richardson numbers effects.

The general two-dimensional flow and heat transfer model may be derived by assuming two-dimensional turbulent flow with negligible viscous dissipation, and with the Boussinesq approximation invoked. Then, the governing equations written in primitive variables and in conservative form in both Cartesian ($\xi = 0$) and cylindrical coordinates ($\xi = 1, x = r$) reduce as follows:

- Continuity:

$$\frac{\partial u}{\partial x} + \frac{\partial v}{\partial y} + \xi \frac{u}{x} = 0 \tag{6.37}$$

- x- and y-momentum:

$$\frac{\partial u}{\partial t}+\frac{\partial uu}{\partial x}+\frac{\partial uv}{\partial y}+\xi\frac{u^2}{x}=-\frac{1}{\rho_o}\frac{\partial p}{\partial x}+\frac{\partial}{\partial x}\left(\frac{\mu_{eff}}{\rho_o}\frac{\partial u}{\partial x}\right)+\frac{\partial}{\partial y}\left(\frac{\mu_{eff}}{\rho_o}\frac{\partial u}{\partial y}\right)$$
$$+\xi\left[\frac{\mu_{eff}}{x\rho_o}\left(\frac{\partial u}{\partial x}-\frac{u}{x}\right)\right]-\frac{R_x}{\rho_o l_x} \tag{6.38}$$

$$\frac{\partial v}{\partial t}+\frac{\partial uv}{\partial x}+\frac{\partial vv}{\partial y}+\xi\frac{uv}{x}=-\frac{1}{\rho_o}\frac{\partial p}{\partial y}+g_y\beta(T-T_o)+\frac{\partial}{\partial x}\left(\frac{\mu_{eff}}{\rho_o}\frac{\partial v}{\partial x}\right)$$
$$+\frac{\partial}{\partial y}\left(\frac{\mu_{eff}}{\rho_o}\frac{\partial v}{\partial y}\right)+\xi\frac{\mu_{eff}}{x\rho_o}\frac{\partial v}{\partial x}-\frac{R_y}{\rho_o l_y} \tag{6.39}$$

- Energy:

$$\frac{\partial T}{\partial t}+\frac{\partial uT}{\partial x}+\frac{\partial vT}{\partial y}+\frac{\xi}{x}uT=\frac{\partial}{\partial x}\left(\frac{k_{eff}}{\rho_o C_p}\frac{\partial T}{\partial x}\right)+\frac{\partial}{\partial y}\left(\frac{k_{eff}}{\rho_o C_p}\frac{\partial T}{\partial y}\right)+\xi\frac{k_{eff}}{x\rho_o}\frac{\partial T}{\partial x} \tag{6.40}$$

The effective viscosity and conductivity appearing in the governing equations are defined as the sum of the laminar and turbulent contributions, i.e.

$$\mu_{eff}=\mu_\ell+\mu_t$$
$$k_{eff}=k_\ell+k_t \tag{6.41}$$

where μ_t and k_t are the turbulent contributions obtained from a suitable turbulence model. Two simple turbulence models are presented at the end of this section. The resistance terms, R_x and R_y, arise due to the presence of solid or perforated obstructions in the flow field, and are defined by Sha and Lin (1978) as

$$R_x=\frac{1}{2}f\rho|u|u$$
$$R_y=\frac{1}{2}f\rho|v|v \tag{6.42}$$

where the friction factor is calculated based on the diameter and thickness of the perforations. The symbols l_x and l_y appearing in the resistance terms are the appropriate length scales associated with R_x and R_y, respectively. l_x and l_y are taken as the grid sizes in the x and y directions, respectively.

The numerical solution of the above equations, which are highly non-linear and coupled partial differential equations, has been the subject of several studies. These studies, however, differ depending on the assumptions and the numerical methods used. One of the earliest computational studies of stratified thermal storage in dynamic mode was that of Cabelli (1977). The conservation equations in cartesian coordinates and in vorticity-stream function formulation were solved using an implicit finite-difference method. All convective derivatives were discretized using a central-difference representation which is known to lead to instability at high Reynolds numbers. This limited the solution to Reynolds numbers of 200 or less where *Re* (also *Ri*) was based on the inlet port velocity and the height of the tank. The study simulated both vertical and horizontal inflow and outflow. Also, single-flow (charge or discharge) and two-flow (charge and discharge) circuits operating simultaneously were simulated. Although limited in applicability in view of the low *Re* used, the results demonstrated the role of the Richardson number *Ri* in achieving stratification. Increasing *Ri* beyond unity was shown to have a negligible effect on stratification. One has to keep in mind the definition of the Richardson number in light of the reference quantities used. Chan *et al.* (1983) solved the same equations, in primitive variables, using an explicit finite difference method. Different inflow and outflow configurations were simulated. However, their results showed that the flow direction into and from the storage tank has little effect on thermal storage efficiency, contrary to the experimental evidence.

The limitation on Reynolds number values in Cabelli (1977) was later removed by Guo and Wu (1985), who solved the same equations, again in vorticity-stream function formulation, using a finite difference method with a power-law scheme (Patankar, 1980) for the convective derivatives. Thus stability was secured at high values of Reynolds number. However, this method is known to produce numerical diffusion, which compromises the accuracy. As in the study of Cabelli (1977), the Richardson number *Ri* was identified by Guo and Wu (1985) as the important parameter for characterizing the flow pattern and temperature distribution inside the storage tank. For a single-flow circuit, values of $Ri > 1$ were found to provide better stratification. *Ri* is based on the tank height and the inlet velocity. For the case of two-flow circuits, poor stratification was obtained at $Ri = 1$. In an attempt to improve thermal stratification, Han and Wu (1985) incorporated in their numerical study a horizontal baffle located at a distance of one fourth the height of the tank from the bottom and extending slightly beyond the middle of the tank. Simulations of discharge of an initially hot water tank with and without a baffle showed varying degrees of improvement depending on Richardson number. That is, for $Ri < 1.2$ the improvement was excellent and it was good for $Ri = 1.2$, whereas for $Ri > 2.4$ the improvement was marginal. At Richardson numbers of 9.8 or higher the baffle was found to give negligible improvement on stratification. The foregoing results were for laminar flow. Han and Wu (1985) also used a simple eddy viscosity model of mixing length type, and found little difference in the qualitative behavior compared with that predicted by the laminar flow calculations.

Later, Zurigat (1988) (see also Ghajar and Zurigat, 1991) solved the two-dimensional model equations in primitive variables using the numerical solution algorithm of Hirt *et al.* (1975) known as the SOLA algorithm. The model incorporated both cylindrical axisymmetric and cartesian coordinates, laminar and turbulent flows, constant and variable boundary conditions, and localized resistances in the form of perforated and solid baffles of zero thickness. Also, both the weighted upwind and the second-order upwind difference schemes were implemented for all equations. The latter upwind scheme was used to reduce the numerical diffusion inherent in the former. Comparison of the predictions using both schemes with experiments showed that the second-order upwind scheme produces better agreement (Zurigat and Ghajar, 1990). Recognizing that mixing at the inlet is the major source of stratification degradation and that mixing is inlet-design-dependent, the aim was to find if there exist conditions under which the effect of the inlet geometry on stratification vanishes. Therefore, simulations were conducted for the charging of a hot water storage tank with two different inlet configurations, each at Richardson numbers ranging from 5.0 to 46.0, where *Ri* was based on the height of the tank and the tank bulk fluid velocity. The first inlet, a *solid-disk diffuser*, consisted of a circular solid disk of diameter 25.4 mm located 12.7 mm from the inlet pipe, which was of the same diameter and located at the center of the top of the tank. The tank had a diameter of 406.0 mm and a height of 1450 mm. The second inlet, the *perforated inlet*, was formed by adding to the *solid-disk diffuser* a perforated extension which spanned the rest of the tank's cross-sectional area. The Richardson number was modified by varying the temperature difference between the inlet and the initial water while maintaining the same flow rate. Figure 6.19 shows the predictions of thermocline in the storage tank as the thermocline passes through different levels close to the inlet region. The temperature profiles shown were obtained by averaging the temperature field at each horizontal level. It is seen that the addition of the perforated extension to the solid-disk diffuser has resulted in a varying degree of improvement in stratification level in the tank depending on the Richardson number.

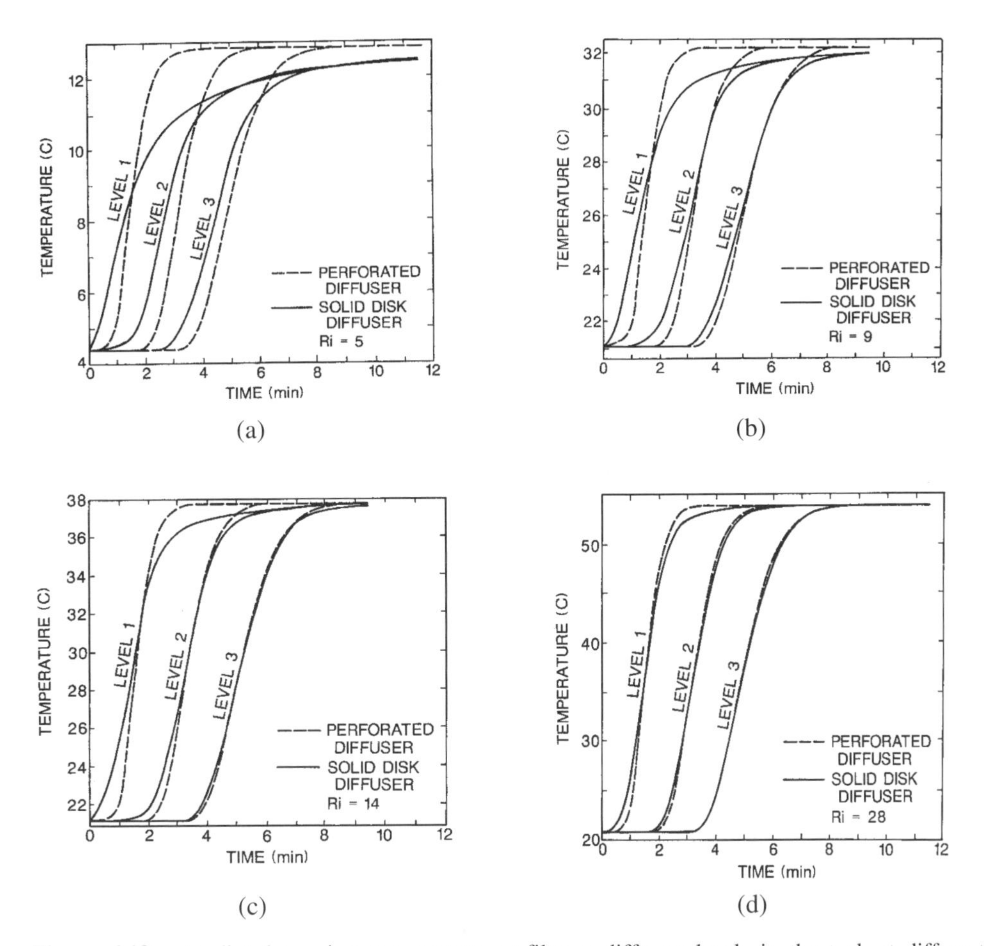

Figure 6.19 Predicted transient temperature profiles at different levels in the tank at different values of Richardson number for two different inlet configurations (Ghajar and Zurigat, 1991).

Consequently, significant improvement is observed at $Ri = 5$. As Ri increases beyond 9.0 the difference in performance of the two inlets becomes insignificant. It was concluded that in thermocline TES tanks, the inlet geometry has a significant effect on stratification for $Ri \leq 5.0$ and a negligible effect for $Ri \geq 10$.

Using the same explicit numerical algorithm, i.e. the SOLA algorithm of Hirt *et al.* (1975), Mo and Miyatake (1996) solved the conservation equations in cartesian coordinates to study the charging of the storage tank with hot water. The inlet and outlet ports were of side slot form, located on the upper and lower corners, respectively, on opposite sides. This study included two new elements in stratified thermal storage work: the use of the QUECKEST discretization scheme of Leonard (1979) which is third-order accurate in both space and time to solve the energy equation and the use of the k-ε turbulence model. Again, comparison with the experimental data showed that the first-order upstream difference scheme produces large numerical diffusion while the QUECKEST scheme produces better agreement with the experiments. The results for different values of the Richardson number ranging from 0.005 to 1.545 at a fixed Reynolds number of 500 show that thermal stratification is maintained at Richardson numbers as low as 0.093. The Reynolds and Grashof numbers, and hence the Richardson number, were based on the slot width and the average velocity at the inlet slot. Mo and Miyatake (1996) concluded that at high Richardson numbers the one-dimensional plug-flow model is justified. At $Ri = 0.005$ severe mixing and short-circuiting of the hot fluid directly to the outlet occurs.

To this end the finite-volume numerical method had not been used in the predictions of flow and heat transfer in stratified thermal storage tanks. Lightstone *et al.* (1989) were probably the first to use this method to solve the stratified thermal storage tank problem. The conservation equations in primitive variables and in axisymmetric cylindrical coordinates were solved. Heat conduction in the tank wall was also included to investigate the effect of wall thermal conductivity. Simulations of tank charging with hot water were carried out for both laminar and turbulent flows. The buoyant jet turbulence model described below in this section was used. Lightstone *et al.* (1989) simulated the experiments of Loehrke and Holzer (1979) in which two different inlet configurations were used: the cup diffuser (see Figure 6.3) and a vertical pipe inlet. Good agreement with the experiments with the cup diffuser was achieved using laminar flow calculations, whereas for the vertical pipe inlet the buoyant jet turbulence model was necessary to achieve good agreement. The results showed that for the vertical pipe inlet the two-dimensional nature of the flow was quite obvious by the large radial temperature variation. For both inlets, mixing and two-dimensional flow behavior were observed to occur, at the early stages of charge, on the tail side of the thermocline while one-dimensional flow behavior dominated the rest of the tank. This is in accord with most experimental results.

Stewart *et al.* (1992) investigated the downward impingement of a cold stream from a slot onto the bottom of a chilled water storage tank filled with warmer water and with no bounding walls. The governing steady two-dimensional flow and energy equations expressed in stream function-vorticity formulation along with the turbulence kinetic energy and length scale (k-ℓ) turbulence model equations were solved by finite difference methods. The flow and geometric parameters investigated were the slot width, W, the distance from the jet exit to the tank bottom, h, the Reynolds number based on jet exit mean velocity, and the difference between the cold and warm water temperatures. Adiabatic boundary conditions were assumed and the initial transients were neglected. The isotherms obtained for different parameters indicated that the temperature difference did not influence the stratification and the other parameters, i.e. Reynolds number, the ratio h/W and W have the dominant effects. Thermal stratification was observed to occur for 50 m < ReW < 200 m at a fixed W/h of unity with W varying between 0.05 and 0.2 m. The steadiness assumed and the absence of side walls are two restrictions that may limit the validity of the results.

In their investigation of the charging of a chilled water storage tank, Cai *et al.* (1993) solved the governing equations in vorticity-stream function formulation and in cartesian coordinates for the turbulent flow case using the k-ℓ turbulence model. An explicit finite-difference method was employed with upwind difference scheme used for the convective terms. The inlet port used was a side slot located at the lower part of the side wall. The results showed that stratification was good at Richardson numbers greater than 5 provided the Reynolds number is less than 10,000 where both are based on the width of the slot and the average velocity at the inlet slot. Doubling the width of the tank while keeping other variables fixed resulted in a thicker thermocline and larger nonuniformity in the horizontal temperature distribution.

Hahne and Chen (1998) studied the charging of a cylindrical hot water storage by solving the laminar flow equations using the vorticity-stream function formulation. The flow inlet and exit were located at the top and bottom of the tank on the centerline. The effects of different storage tank parameters on stratification were studied in terms of the charging efficiency (see Equation 6.10). It was found that for $Ri > 0.25$ the charging efficiency is essentially unaffected by the increase in Peclet number, where both numbers are based on the inlet velocity and the height of the tank. One-dimensional flow behavior was observed at $Ri > 2.5$. Computations were also conducted by Spall (1998) for the charging of a cylindrical chilled water storage tank with a peripheral side slot inlet. The turbulent flow and energy equations were solved using two different turbulence models, the k-ε and the Reynolds Stress Model (RSM). A commercial computer program based on the finite-volume method was used and fairly fine-grid solutions were obtained for different Reynolds and Richardson numbers. The results demonstrated that at fixed values of the Richardson number stratification is independent of the Reynolds number, and for good stratification the Richardson number should be greater than 2. Also, Spall (1998) found that the k-ε model predicted a 50–100% thicker thermocline compared with the RSM predictions. Thus, the k-ε model is more diffusive and the predictions using the RSM model are more accurate. The dependence of stratification on the Richardson number alone arrived at in this study is in accord with that of Mo and Miyatake (1996). However, the experimental data of Wildin (1990) for different diffusers with the same inlet Froude number has revealed that the inlet Reynolds number has a strong influence on thermocline development. This is an issue that needs further study.

Based on the studies presented in this section it is evident that the Richardson number is an important parameter that governs the flow and stratification in stratified thermal storages. The design of stratified TES tanks should be based on maximizing this number within reasonable trade-offs. In heating applications this is easily achieved in view of the high buoyancy differentials available. It was demonstrated that there exists an upper limit on the Richardson number beyond which all current inlet designs in thermocline thermal storage perform equally well. In chilled water storage it is more difficult to achieve high Richardson numbers. In this case, the diffuser design is critical for maintaining stratification. Well designed inlets for chilled water storage are those which introduce the flow at a low velocity in a gravity and surface currents for the charge and discharge modes, respectively.

For the sake of completeness, two simple turbulence models are presented below. These are the *Eddy Viscosity* and the *Vertical Buoyant Jet* turbulence models. In the *Eddy Viscosity* model (Sha *et al.*, 1980) the turbulence viscosity is given by

$$\mu_t = 0.007\, C_\mu \rho U_{\max} \ell \tag{6.43}$$

where $U_{max} = max(u, v)$, $\ell = max(\Delta x, \Delta y)$, and C_μ is a constant given by:

$$C_\mu = \begin{cases} 0.1 & \text{for} \quad \mathrm{Re}_{\max} > 2000 \\ 0.1(0.001\,\mathrm{Re}_{\max} - 1) & \text{for } 1000 \le \mathrm{Re}_{\max} \le 2000 \\ 0 & \text{for} \quad \mathrm{Re}_{\max} < 1000 \end{cases} \tag{6.44}$$

Also, $Re_{max} = max(Re_x, Re_y)$, where $Re_x = \rho u \Delta y/\mu$, $Re_y = \rho v \Delta x/\mu$. The turbulence conductivity is calculated from:

$$k_t = \frac{C_p \mu_t}{\mathrm{Pr}_t} \tag{6.45}$$

where the turbulent Prandtl number Pr_t is evaluated as

$$\mathrm{Pr}_t = 0.8[1 - \exp(-6x10^{-5}\,\mathrm{Re}_{\max}\,\mathrm{Pr}^{1/3})]^{-1} \tag{6.46}$$

In the *Vertical Buoyant Jet* turbulence model (Lightstone *et al.*, 1989) the flow field in the storage tank is divided into three regions, taking into account the nature of the forces acting on the jet issuing from the inlet port and the bulk fluid motion in the tank. These three regions are:

1. The *jet region*: flow here is momentum driven. Axially, this region extends from the inlet port to a distance at which the jet centerline velocity reaches 10% of its initial value. Radially, the jet region extends from the jet centerline to a distance where the jet velocity changes sign. The turbulence viscosity is calculated using the standard jet model:

$$\mu_t = c\rho b v \tag{6.47}$$

   where b and v at the given elevation are the jet half width and the centerline velocity, respectively, and c is a constant (0.0256). Clearly, μ_t does not vary radially if the density remains constant.
2. The *plume region*: the flow in this region is gravity driven in the direction opposite to the jet flow. The plume region extends radially from the outer boundary of the jet region to the locations where the buoyant flow changes direction. In this region the turbulence viscosity is calculated using the same relation given above with $c = 0.068$.
3. The *rest-of-the-tank region:* the flow is unidirectional and constitutes the bulk fluid motion towards the outlet. In this region the flow is assumed laminar.

It must be noted that the boundaries between these regions change with time, and thus should be updated along with μ_t after each time step.

6.7 Conclusions

Stratified thermal energy storage for hot or chilled water fluids has become a common tool in energy conservation and management technology. In this chapter the theoretical and experimental foundations of stratified TES were reviewed. The fluid flow and heat transfer aspects of the subject were introduced and performance measures used by different investigators presented. Design parameters governing the performance of stratified TES were identified. In general, for stratified tanks to perform as expected, several design recommendations must be kept in mind. First, the inlet temperature should be maintained constant whenever possible. Mixing is greatly enhanced by inflows having a variable inlet temperature. The requirement of constant inlet temperature is not difficult to satisfy in chilled water storage systems. However, these systems operate under small density differentials, and thus the inlet diffuser design is critical for maintaining stratification. In solar energy storage applications a variable inlet temperature is more common, and thus maintaining stratification under this condition is difficult. Practical distributors that direct the incoming flow to its temperature level with minimal mixing are not yet reliable. Moreover, they are not flexible in design. More work in this area is still needed. Alternatively, as proposed by some researchers, the solar energy collection strategy may need to be altered to maintain a nearly constant inlet temperature. In thermocline TES tanks the inflow must be introduced at the uppermost and the lowermost levels in the tank for the charging of hot and chilled water storage tanks, respectively. The reverse is true for discharging. Furthermore, the inlet velocity must be maintained at a minimum by using inlet diffusers. In this regard, the diffusers developed for chilled water storage are generally recommended as they will perform even better in hot water storage tanks because of the higher buoyancy differential under which these tanks operate. In thermocline hot water storage there exist flow conditions under which the inlet geometry has a negligible effect on stratification. These conditions are characterized by Ri greater than 10 where the Ri is based on the height of the tank and the water bulk velocity in the tank.

Further recommendations relate to the tank aspect ratio which should be maintained between 3 and 4, and the tank material which should be made of material having a thermal conductivity as close as possible to that of water. This may also be achieved by insulating the interior of the storage tank. This recommendation regarding the tank material is important if the tank is to stay idle while partially charged or discharged. The dimensionless numbers characterizing the flow and heat transfer in stratified TES tanks represent reliable guides for design. However, these numbers should be interpreted in view of the reference velocity and length scales used.

The importance of one-dimensional flow models lies in the fact that they are computationally more efficient than two- or three-dimensional models, which makes them ideal for incorporation into overall energy-system simulation programs. Furthermore, one-dimensional flow should be the target of stratified TES tank designs in practice, as two- or three-dimensional flow would only enhance mixing and degrade the performance. Two- and three-dimensional models, on the other hand, are more capable in accounting for a broad range of hydrodynamic, thermal, and geometric conditions. These models can be used for design assessments and testing of innovative design concepts. The literature on stratified thermal storage is still lacking in the area of three-dimensional modeling. This is understandable in view of the complexity and computational cost involved.

Acknowledgements

We are indebted to the University Center for Energy Research at Oklahoma State University for providing the funding for most of this work. We are also indebted to Ms. Diane Compton for her superb help with the manuscript.

Nomenclature

A	Area
C	Courant number ($C = V\Delta t / \Delta x$)
C_p	specific heat at constant pressure
Ex	exergy
f	friction factor
Fo	Fourier number ($Fo = \alpha\Delta t / (\Delta x)^2$)
Fr	Froude number ($Fr = u_r / \sqrt{g\ell_r}$)
Fr_m	modified Froude number ($Fr_m = u_r / \sqrt{g\beta(T - T_r)\ell_r}$)
Fr_{dm}	densimetric modified Froude number ($Fr_{dm} = u_r / \sqrt{g\,\ell_r(\rho_r - \rho)/\rho_r}$)
g	acceleration of gravity
Gr	Grashof number ($Gr = g\,\beta(T - T_r)\ell_r^3 / \nu^2$)
h	enthalpy in Equation 6.16 and height elsewhere
H	effective height between tank inlet and outlet
k	thermal conductivity
L	length (also the tank height in the ratio X/L when used in the figures)
l	length scales associated with the resistance terms in Equations 6.38 and 6.39
ℓ	turbulence length scale
ℓ_r	reference length
M	mass (or moment in Equations 6.14 and 6.15
$\dot{m}$	mass flow rate
p	pressure
P	perimeter
Pr	Prandtl number ($Pr = \mu C_p/k$)
Pe	Peclet number ($Pe = u_r\ell_r / \alpha$)
q	heat transfer rate
Q	heat transfer
r	radius
Re	Reynolds number ($Re = \rho u_r \ell_r / \mu$)
R_x, R_y	resistance force components in the x and y directions defined by Equations 6.38 and 6.39
Ri	Richardson number ($Ri = g\beta\,\Delta T\,\ell_r / u_r^2$)
Ri_d	densimetric Richardson number ($Ri_d = g(\rho - \rho_r)\,\ell_r / \rho_r u_r^2$)
s	entropy

S	heat loss parameter ($S = UPH^2/A_c k$)
t	time
t^*	dimensionless time ($t^* = V\,t/H$)
T	temperature
T^*	dimensionless temperature ($T^* = (T - T_o)/(T_{in} - T_o)$)
u	velocity in the x- or r-direction
u_r	reference velocity
v	velocity in the y-direction
V	average vertical velocity in the tank
U	overall heat transfer coefficient
x	Cartesian or cylindrical ($x = r$) coordinate in Equations 6.37–6.40 and distance from the inlet in Figure 6.8 and Equations 6.22–6.28
X	distance from the inlet
y	vertical coordinate

Greek Symbols

α	thermal diffusivity
β	coefficient of thermal expansion [$\beta = (\rho_r - \rho)/(\rho_r (T - T_r))$]
ε	turbulence energy dissipation rate
ε_H	Eddy diffusivity for heat
η	efficiency
Δ	designates difference when used as prefix
μ	dynamic viscosity
ν	kinematic viscosity
ρ	density
ξ	index for Cartesian ($\xi = 0$) or cylindrical ($\xi = 1$) coordinates

Special Symbols

$\forall$	volume
$\dot{\forall}$	volume flow rate

Subscripts

a	ambient
c	charge (or cross section when with area A)
d	discharge
E	energy
e	exit
eff	effective
h	high
in	inlet
ℓ	low (when with T) and laminar (with μ or k)
m	mass

r	reference
t	turbulence
th	thermal
T	tank (or total)
o	initial

Superscripts

in	inlet
*	dimensionless quantity

References

Abdoly, M.A. (1981). *Thermal Stratification in Storage Tanks*, PhD Thesis, University of Texas at Dallas.

Abdoly, M.A. and Rapp, D. (1982). Theoretical and experimental studies of stratified thermocline storage of hot water, *Energy Conversion and Management* 22, 275–285.

Abu-Hamdan, M.G., Zurigat, Y.H. and Ghajar, A.J. (1992). An experimental study of a stratified thermal storage under variable inlet temperature for different inlet designs, *International Journal Heat and Mass Transfer* 35, 1927–1934.

Al-Marafie, A., Al-Kandari, A. and Ghaddar, N. (1991). Diffuser design influence on the performance of solar thermal storage tanks, *International Journal of Energy Research* 15, 525–534.

Al-Najem, N.M. (1993). Degradation of a stratified thermocline in a solar storage tank, *International Journal of Energy Research* 17, 183–191.

Al-Najem, N.M., Al-Marafie, A. and Ezuddin, K.Y.(1993). Analytical and experimental investigation of thermal stratification in storage tanks, *International Journal of Energy Research* 17, 77–88.

Al-Najem, N.M. and El-Refaee, M.M. (1997). Numerical study of the prediction of turbulent mixing factor in thermal storage tanks, *Applied Thermal Engineering* 17, 1173–1181.

Baines, W.D., Martin, W.W. and Sinclair, L.A. (1982). On the design of stratified thermal storage tanks, *ASHRAE Transactions* 88, 426–439.

Brumleve, T.D. (1974). *Sensible Heat Storage in Liquids*, Sandia Labs Report, SLL-73-0263.

Cabelli, A. (1977). Storage tanks-A numerical experiment, *Solar Energy* 19, 45–54.

Cai, L., Stewart, W.E. and Sohn, C.W. (1993). Turbulent buoyant flows into a two dimensional storage tank, *International Journal Heat and Mass Transfer* 36, 4247–4256.

Chan, A.M.C., Smereka, P.S. and Guisti, D. (1983). A numerical study of transient mixed convection flows in a thermal storage tank, *ASME-Journal Heat Transfer* 105, 246–253.

Close, D.J. (1967). A design approach for solar processes, *Solar Energy* 11, 112–122.

Cole, R.L. and Bellinger, F.O. (1982). Thermally stratified tanks, *ASHRAE Transactions* 88, Part 2(1), 1005–1017.

Davidson, J.H. and Adams, D.A. (1994). Fabric stratification manifolds for solar water-heating, *Journal of Solar Energy Engineering* 116, 130–136.

Davidson, J.H., Adams, D.A. and Miller, J.A. (1994). A coefficient to characterize mixing in solar water storage tanks, *Journal of Solar Energy Engineering* 116, 94–99.

Davis, E.S. and Bartera, R. (1975). Stratification in solar water heater storage tank, *Proceedings of the Workshop on Solar Energy Storage Subsystems for the Heating and Cooling of Buildings*, Charlottesville, Virginia, pp. 38–42.

Duffie, J.A. and Beckman, W.A. (1980). *Solar Engineering of Thermal Processes*, Wiley, New York.

Gari, H.N., Loehrke, R.I. and Holzer, J.C. (1979). Performance of an inlet manifold for a stratified storage tank, *ASME Paper 79-HT-67, Joint ASME/AICHE 18th National Heat Transfer Conference*, San Diego, California, August.

Ghajar, A.J. and Zurigat, Y.H. (1991). Numerical study of the effect of inlet geometry on mixing in thermocline thermal energy storage, *Numerical Heat Transfer*-Part A 19, 65–83.

Gretarsson, S.P., Pedersen, C.O. and Strand, R.K. (1994), Development of a fundamentally based stratified thermal storage tank model for energy analysis calculations, *ASHRAE Transactions* 100, 1213–1220.

Gross, R.J. (1982), An experimental study of single medium thermocline thermal energy storage, *ASME Paper 82-HT-53*.

Guo, K.L. and Wu, S.T. (1985). Numerical study of flow and temperature stratification in a liquid storage tank, *Journal of Solar Energy Engineering* 107, 15–20.

Hahne, E. and Chen, Y. (1998), Numerical study of flow and heat transfer characteristics in hot-water stores, *Solar Energy* 64, 9–18.

Han, S.M. and Wu, S.T. (1978). Computer simulation of a solar energy system with a viscous-entrainment liquid storage tank model, *Proceedings of the Third Southeastern Conference on Application of Solar Energy*, Edited by Wu, S.T., Christensen, D.L. and Head, R.R., 165–182

Han, S.M. and Wu, S.T. (1985). Enhancement of thermal stratification in a liquid storage tank by a horizontal baffle, In: *Fundamentals of Forced and Mixed Convection,* a collection of papers presented at the 23d National Heat Transfer Conference, Denver, Colorado (Edited by F.A. Kulacki and R.D. Boyd), pp. 197–205.

Hess, C.F. and Miller, C.W. (1982). An experimental and numerical study on the effect of the wall in a cylindrical enclosure-I, *Solar Energy* 28, 145–152.

Hirt, C.W., Nichols, B.D. and Romero, N.C. (1975). *SOLA: A Numerical Solution Algorithm for Transient Fluid Flows*, Report LA-5852, Los Alamos Scientific Laboratory, New Mexico.

Hollands, K.G.T. and Lightstone, M.F. (1989). A review of low-flow, stratified tank solar water heating systems, *Solar Energy* 43, 97–106.

Homan, K.O., Sohn, C.W. and Soo, S.L. (1996). Thermal performance of stratified chilled water storage tanks, *HVAC & R Research* 2, 158–170.

Ismail, K.A.R., Leal, J.F.B. and Zanardi, M.A. (1997). Models of liquid storage tanks, *International Journal of Energy Research* 22, 805–815.

Jaluria, Y. and Gupta, S.K. (1982), Decay of thermal stratification in water body for solar energy storage, *Solar Energy* 28, 137–143.

Kleinback, E.M., Beckman, W.A. and Klein, S.A. (1993). Performance study of one-dimensional models for stratified thermal storage tanks, *Solar Energy* 50, 155–166.

Kuhn, J.K., von Fuchs, G.F. and Zob, A.P. (1980). *Developing and Upgrading of Solar-System Thermal-Energy-Storage Simulation Models*, Final Report, prepared by Boeing Computer Services Company for the Department of Energy, Contract DE-AC02-77CS 34482.

Lavan, Z. and Thompson, J. (1977). Experimental study of thermally stratifed hot water storage tanks, *Solar Energy* 19, 519–524.

Leonard, B.P.(1979). Stable and accurate convective modeling procedure based on quadratic upstream interpolation, *Computational Methods in Applied Mechanical Engineering* 19, 59–98.

Lightstone, M.F., Raithby, G.D. and Hollands, K.G.T. (1989). Numerical simulation of the charging of liquid storage tanks: Comparison with experiment, *Journal of Solar Energy Engineering* 111, 225–231.

Loehrke, R.I., Gari, H.N. and Holzer, J.C. (1979). *Thermal Stratification Enhancement for Solar Energy Applications*, Technical Report HT-TS792, Prepared for the Civil Engineering Laboratory, Naval Construction Battalion Center, Port Hueneme, California.

Loehrke, R.I. and Holzer, J.C. (1979). *Stratified Thermal Storage Experiments*, Technical Report HT-TS793, Mechanical Engineering Department, Colorado State University, Fort Collins, Colorado.

Mavros, P., Belessiotis, V. and Haralambopoulos, D. (1994). Stratified energy storage vessels-characterization of performance and modeling of mixing behaviour, *Solar Energy* 52, 327–336.

Miller, C.W. (1977). Effect of conducting wall on a stratified fluid in a cylinder, *AIAA Paper No. 77-792, AIAA 12th Thermophysics Conference*, Albuquerque, New Mexico.

Mo, Y. and Miyatake, O. (1996). Numerical analysis of the transient turbulent flow field in a thermally stratified thermal storage tank, *Numerical Heat Transfer*-Part A 30, 649–667.

Musser, A. and Bahnfleth, W.P. (1998). Evolution of temperature distribution in a full-scale stratified chilled-water storage tank with radial diffusers, *ASHRAE Transactions* 104, Part 1A, 55–67.

Murthy, S.S., Nelson, J.E.B. and Rao, T.L.S. (1992). Effect of wall conductivity on thermal stratification, *Solar Energy* 49, 273–277.

Nakahara, N., Sagara, K. and Tsujimoto, M. (1988). Water thermal storage tank: Part 2-mixing model and storage model estimation for temperature stratified tanks, *ASHRAE Transactions* 94, Part 2, 371–394.

Nelson, J.E.B., Balakrishnan, A.R. and Murthy, S.S. (1999). Experiments on stratified chilled water tanks, *International Journal of Refrigeration* 22, 216–234.

Oppel, F.J., Ghajar, A.J. and Moretti, P.M. (1986). A numerical and experimental study of stratified thermal storage, *ASHRAE Transactions* 92, 293–309.

Patankar, S.V. (1980). *Numerical Heat Transfer and Fluid Flow*, McGraw-Hill.

Philips, W.F. and Dave, R.N. (1982). Effect of stratification on the performance of liquid-based solar heating systems, *Solar Energy* 29, 111–120.

Rosen, M.A. (1991). On the importance of temperature in performance evaluations for sensible thermal energy storage systems, *Proceedings of the Biennial Congress of the International Solar Energy Society*, Denver, Colorado, USA, 19–23 August. In 1991 Solar World Congress, vol. 2, Part. II (Eds: Arden, M.E., Burley, S.M.A. and Coleman, M.), pp.1931–1935.

Rosen, M.A. and Hooper, F.C. (1991). Evaluation of energy and exergy contents of stratified thermal energy storages for selected storage-fluid temperature distributions, *Proceedings of the Biennial Congress of the International Solar Energy Society*, Denver, Colorado, USA, 19–23 August. In 1991 Solar World Congress, vol. 2, Part. II (Eds: Arden, M.E., Burley, S.M.A. and Coleman, M.), pp. 1961–1966.

Sha, W.T. and Lin, E.I.H. (1978). Three dimensional mathematical model of flow stratification in thermocline storage tanks, In: *Application of Solar Energy* (Edited by S.T. Wu *et al.*), pp.185–202.

Sha, W.T., Lin, E.I.H. Schmitt, R.C., Lin, K.V., Hull, J.R., Oras, J.J. and Domanus, H.M. (1980). *COMMIX-SA-1: A Three-Dimensional Thermodynamic Computer Program for Solar Applications*, Argonne National Laboratory Report, ANL-80-8.

Sharp, M.K. (1978). *Thermal Stratification in Liquid Sensible Heat Storage*, MSc Thesis, Colorado State University, Fort Collins, Colorado.

Sharp, M.K. and Loehrke, R.I. (1979). Stratified thermal storage in residential solar energy applications, *Energy* 3, 106–113.

Sliwinski, B.J., Mech, A.R. and Shih, T.S. (1978). Stratification in thermal storage during charging, *Proceedings of the 6th International Heat Transfer Conference*, Toronto, vol. 4, pp. 149–154.

Spall, R.E. (1998). A numerical study of transient mixed convection in cylindrical thermal storage tanks, *International Journal of Heat and Mass Transfer* 41, 2003–2011.

Stewart Jr., W.E., Becker, B.R., Cai, L. and Sohn, C.W. (1992). Downward impinging flows for stratified chilled water storage, *Proceedings of the 1992 ASME National Heat Transfer Conference*, HTD-vol. 206-2, pp. 131–138.

Tamblyn, R.T. (1980). Thermal storage resisting temperature blending, *ASHRAE Journal* 22, 65–70.

Tran, N., Kreider, J.F. and Brothers, P. (1989). Field measurement of chilled water storage system thermal performance, *ASHRAE Transactions* 95, Part 1, 1106–1112.

Van Koppen, C.W.J., Thomas, J.P.S. and Veltcamp, W.B. (1979). The actual benefits of thermally stratified storage in small and medium size solar system, *Proceedings of ISES Biennial Meeting*, Atlanta, GA, Pergamon Press, New York, vol. 2, pp. 576–580.

Votsis, P.P., Tasson, S.A., Wilson, D.R. and Marquand, C.J. (1988). Experimental and theoretical investigation of mixed and stratified hot water storage tanks, *Proceedings of the Institution of Mechanical Engineers, Part C-Journal of Mechanical Engineering Science* 202, 187–193.

Wuestling, M.D., Klein, S.A. and Duffie, J.A. (1985). Promising control alternatives for solar water heating systems, *Journal of Solar Energy Engineering* 107, 215–221.

Wildin, M.W. (1989). Performance of stratified vertical cylindrical thermal storage tanks, Part II: Prototype tank, *ASHRAE Transactions* 95, Part 1, 1096–1105.

Wildin, M.W. (1990). Diffuser design for naturally stratified thermal storage, *ASHRAE Transactions* 96, Part 1, 1094–1102.

Wildin, M.W. (1991). Flow near the inlet and design parameters for stratified chilled water storage, *National Heat Transfer Conference*, Minneapolis, MN, July 28–31.

Wildin, M.W. (1996). Experimental results from single-pipe diffusers for stratified thermal energy storage, *ASHRAE Transactions* 102, Part 2, 123–132.

Wildin, M.W. and Sohn, C.W. (1993). *Flow and Temperature Distribution in a Naturally Stratified Thermal Storage Tank*, USACERL Technical Report FE-94/01.

Wildin, M.W. and Truman, C.R. (1985a). A summary of experience with stratified chilled water tanks, *ASHRAE Transactions* 92, 956–976.

Wildin, M.W. and Truman, C.R. (1985b). *Evaluation of Stratified Chilled Water Storage Technique*, EPRI Report, EPRI EM-4352, December.

Wildin, M.W. and Truman, C.R. (1989). Performance of stratified vertical cylindrical thermal storage tanks, Part I: Scale model tank, *ASHRAE Transactions* 95, Part 1, 1086–1095.

Wyman, C., Castle, J. and Kreith, F. (1980). A review of collector and energy storage technology for intermediate temperature applications, *Solar Energy* 24, 517–540.

Yoo, J., Wildin, M.W. and Truman, C.R. (1986). Initial formation of a thermocline in stratified thermal storage tanks, *ASHRAE Transactions* 92, Part 2A, 280–291.

Zurigat, Y.H. (1988). *An Experimental and Analytical Examination of Stratified Thermal Storage*, PhD Thesis, Oklahoma State University, Stillwater, Oklahoma, December.

Zurigat, Y.H., Ghajar, A.J. and Moretti, P.M. (1988). Stratified thermal energy storage tank inlet mixing characterization, *Applied Energy* 30, 99–111.

Zurigat, Y.H., Maloney, K.J. and Ghajar, A.J. (1989). A comparison study of one-dimensional models for stratified thermal storage tanks, *Journal of Solar Energy Engineering* 111, 204–210.

Zurigat, Y.H. and Ghajar, A.J. (1990). A comparative study of weighted upwind and second order difference schemes, *Numerical Heat Transfer*-Part B, 18, 61–80.

Zurigat, Y.H., Liche, P.R. and Ghajar, A.J. (1991). Influence of inlet geometry on mixing in thermocline thermal energy storage, *International Journal of Heat and Mass Transfer* 34, 115–125.

7

Modeling of Latent Heat Storage Systems

M. Lacroix

7.1 Introduction

Modeling of solid-liquid phase change heat transfer has received considerable research attention over the last two decades. This is not surprising considering its numerous applications in processes such as the melting of ice and freezing of water, metal casting, welding, coating and purification of metals, crystal growth from melts and solutions, freeze-drying of foodstuffs, nuclear reactor safety, aerodynamic heating of re-entry bodies and thermal spacecraft control. Modeling of solid-liquid phase change is also relevant to the design and development of efficient, cost effective Latent Heat Thermal Energy Storage (LHTES) systems. Indeed, the latent heat-of-fusion energy storage concept, which involves storing and recovering heat through the solid-liquid phase change process, has two undeniable advantages. First, the latent heat of most materials is much higher than their sensible heat, thus requiring a smaller mass of storage medium for storing/recovering a given quantity of thermal energy. Second, the thermal storage process occurs at a nearly constant temperature, which is desirable for efficient operation of most thermal systems. A good understanding of the heat transfer processes involved is essential for accurately predicting the thermal performance of the system and for avoiding costly system overdesign.

Modeling the thermal behavior of LHTES systems is, however, much more complex than the modeling of sensible heat storage systems. There are problems associated with the nonlinear motion of the solid-liquid interface, the possible presence of buoyancy driven flows in the melt, the conjugate heat transfer between the encapsulated Phase Change Material (PCM) and the Heat Transfer Fluid (HTF) in the flow channels and the volume expansion of the PCM upon melting/solidification.

The purpose of this chapter is to provide an overview of the most common modeling techniques for predicting the heat transfer and the thermal behavior of LHTES systems. The review is based essentially on journal articles published in the open literature over the last fifteen years. Virtually hundreds of papers dealing with the mathematical modeling of solid-liquid phase change heat transfer have appeared in the literature during that period of time, and those that are not directly related to latent heat storage devices have been left aside. Among these papers are the works dealing with the melting and solidification of metals and alloys, and articles devoted to the thermodynamics of heat storage in PCMs. Papers on cold storage have also been ignored even though the modeling techniques presented in this review apply equally to these systems. For a general discussion of numerical techniques for the solution of solid-liquid phase change, the reader is referred to the books of Crank (1984) and Alexiades and Solomon (1993), and to the recent review of Voller (1997).

In the following sections, the modeling approaches are classified and reviewed in terms of increasing mathematical complexity and physical insight. First, the modeling of LHTES systems with a porous medium approach is briefly examined. Second, models for which heat transfer in the PCMs is conduction-dominated are presented. Finally, models that take into account the natural convection buoyancy driven flows in the molten PCM are discussed.

7.2 Porous Medium Approach

The main component of a typical packed bed thermal energy storage unit is the insulated vessel containing a large number of PCM capsules (Figure 7.1). This unit may be used for example in solar thermal and waste recovery systems. A heat transfer fluid flows over the capsules for heat storage and heat recovery. Detailed modeling of the heat transfer and fluid flow processes that take place in such a complex arrangement of PCM capsules is not only difficult but also impraticable. The simpler and yet powerful porous medium approach is often preferred (Ismail and Stuginsky, 1999; Adeyibi *et al.*, 1996; Saborio-Aceves *et al.*, 1994; Goncalves and Probert, 1993; Chen, 1992; Beasley *et al.*, 1989; Kondepudi *et al.*, 1988; Torab *et al.*, 1987).

The basic assumption on which porous medium models rest is that the PCM capsules behave as a continuous medium, and not as a medium comprised of independent particles. As a result, coupled heat transfer equations may readily be formulated for both the PCM and the HTF from the application of the first law of thermodynamics.

As an illustrative example, the system depicted in Figure 7.1 is modeled. It is assumed that the thermophysical properties of the PCM and HTF are temperature independent. There are neither heat generation nor chemical reactions within the bed. The fluid flow through the voids in the bed is essentially regarded as plug flow and conditions are uniform in the transverse direction. Consequently, radial dispersion is neglected and the energy conservation equations become one-dimensional in space. Radiation heat transfer is negligible and the thermal gradients within the PCM capsules are ignored. Based on the foregoing assumptions, the energy equations for the HTF and the PCM can be written as

$$\varepsilon(\rho C)_{HTF}\left(\frac{\partial T_{HTF}}{\partial t}+v_{HTF}\frac{\partial T_{HTF}}{\partial x}\right)=\frac{\partial}{\partial x}\left(k_{HTF}\frac{\partial T_{HTF}}{\partial x}\right)+U_{HTF}A_{bed}(T_{PCM}-T_{HTF}) \tag{7.1}$$

$$(1-\varepsilon)(\rho C)_{PCM}\frac{\partial T_{PCM}}{\partial t}=\frac{\partial}{\partial x}\left(k_{PCM}\frac{\partial T_{PCM}}{\partial x}\right)+U_{HTF}A_{bed}(T_{HTF}-T_{PCM}) \tag{7.2}$$

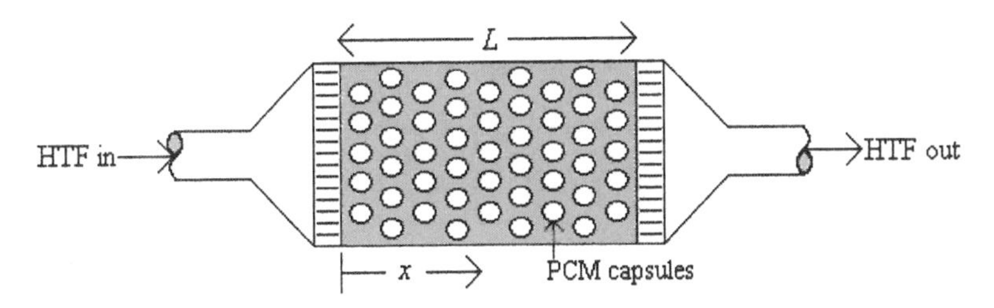

Figure 7.1 Schematic of the thermal energy storage system.

where U_{HTF} is the constant heat transfer coefficient, A_{bed} is the superficial particle area per unit bed volume, v_{HTF} is the mean HTF flow velocity and ε is the void fraction of the bed. The latent heat absorption or release during phase change is taken into account via the relation between the heat capacity of the PCM and the temperature (Goncalves and Probert, 1993). The initial and boundary conditions are specified by

$$T_{HTF}(x,0) = T_{PCM}(x,0) = T_{ini} \tag{7.3}$$

$$T_{HTF}(0,t) = T_{inlet} \tag{7.4}$$

$$\frac{\partial T_{HTF}(L,t)}{\partial x} = 0 \tag{7.5}$$

$$\frac{\partial T_{PCM}(0,t)}{\partial x} = \frac{\partial T_{PCM}(L,t)}{\partial x} = 0 \tag{7.6}$$

The finite-difference equations are obtained upon integrating the governing Equations 7.1–7.2 over the grid-point cluster shown in Figure 7.2. The resulting form of the finite-difference scheme for the HTF is

$$a_P (T_{HTF})_P = a_W (T_{HTF})_W + a_E (T_{HTF})_E + b \tag{7.7}$$

with

$$a_W = \frac{k_{HTF}}{\delta x_w} + \varepsilon(\rho C)_{HTF} v_{HTF} \tag{7.8}$$

$$a_E = \frac{k_{HTF}}{\delta x_e} \tag{7.9}$$

$$a_P = a_W + a_E + a_P^0 + U_{HTF} A_{bed} \Delta x \tag{7.10}$$

$$a_P^0 = \frac{\varepsilon(\rho C)_{HTF} \Delta x}{\Delta t} \tag{7.11}$$

$$b = a_P^0 T_P^0 + U_{HTF} A_{bed} \Delta x (T_{PCM})_P \tag{7.12}$$

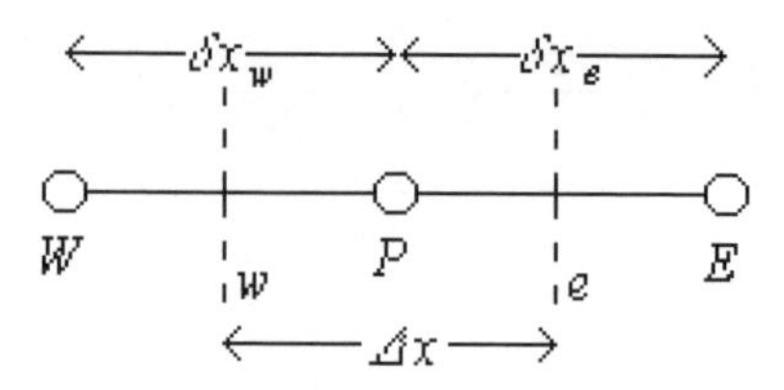

Figure 7.2 Grid-point cluster (1-D).

and for the PCM is

$$a_P(T_{PCM})_P = a_W(T_{PCM})_W + a_E(T_{PCM})_E + b \tag{7.13}$$

with

$$a_W = \frac{k_{PCM}}{\delta x_w} \tag{7.14}$$

$$a_E = \frac{k_{PCM}}{\delta x_e} \tag{7.15}$$

$$a_P = a_W + a_E + a_P^0 + U_{HTF} A_{bed} \Delta x \tag{7.16}$$

$$a_P^0 = \frac{(1-\varepsilon)(\rho C)_{PCM} \Delta x}{\Delta t} \tag{7.17}$$

$$b = a_P^0 T_P^0 + U_{HTF} A_{bed} \Delta x (T_{HTF})_P \tag{7.18}$$

The old (known) values of the HTF and PCM temperatures are denoted by T_{HTF}^0 and T_{PCM}^0, respectively, and the new (unknown) values are denoted by T_{HTF} and T_{PCM}. Since Equations 7.7 and 7.13 are coupled via the source term *b* (Equations 7.12 and 7.18), both are solved iteratively at a given time step, with a Tri-Diagonal Matrix (TDMA) solver.

The above model has been used to predict the thermal behavior of LHTES systems formed by many independent PCM capsules (Adeyibi *et al.*, 1996; Saborio-Aceves *et al.*, 1994). It also allows investigation of the effect of the temperature and mass flow rate of the flue gas, porosity of the bed and dimensions of the vessel on the system performance. Other factors such as the thermal mass of the containment vessel wall have been studied by coupling the model with the heat diffusion equation for the wall. Ismail and Stuginsky (1999) have also extended the above model to account for the thermal diffusion in the radial direction of the bed.

7.3 Conduction-Dominated Phase Change

As seen in the previous section, the porous medium approach fails to account for the thermal gradients that may prevail inside the PCM capsules. Due to the fact that most PCMs employed in commercial LHTES systems have low thermal diffusivities (high Prandtl numbers), the calculated heat transfer rates between the HTF and PCM may be inexact and, as a result, it becomes difficult to predict the thermal behavior of these systems. To overcome this limitation, one must consider the problem of heat transfer within the PCM capsule.

Heat transfer problems involving solid-liquid phase change are difficult to solve primarily because of the moving boundary (or boundaries) that forms between the solid and the liquid phases. The position of this boundary is not known *a priori*, and must be determined as part of the solution, making phase change problems intrinsincally nonlinear in nature. As mentioned in the introduction, although the moving boundary is the most significant problem, it is by no means the only one. Density changes in the PCM produce convective motion in the melt through natural convection, close-contact melting and volume changes. Complications may also occur because of the subcooling/superheating effects, anisotropic phase change or phase change over an extended temperature range.

The present section looks at numerical models of LHTES systems for which it is assumed that the heat transfer in the PCM is conduction dominated. From a mathematical point of view, buoyancy effects and the resulting natural convection motion in the liquid are tacitly ignored because they greatly complicate the analysis of predicting the interface position and heat transfer. From a physical point of view, this assumption is reasonable for PCMs that are gelled in order to avoid incongruent melting. For other PCMs, this assumption amounts to using a maximum capsule thickness S_{max} for which the Rayleigh number, based on the thickness of the capsule, remains smaller than say ≈ 5000

$$S_{max} \leq \left(\frac{5000 \upsilon_l \alpha_l}{g \beta \Delta T} \right)^{1/3} \tag{7.19}$$

where υ_l, α_l and β are the kinematic viscosity, thermal diffusivity and thermal expansion coefficient of the liquid phase, respectively. The symbol g stands for the acceleration of gravity and ΔT is the temperature difference between the heat source and the melting point of the PCM. Smaller values of S_{max} are desirable because of the enhanced heat transfer rates they make possible. For capsule thicknesses larger than S_{max}, natural convection prevails in the melt and it should not be dismissed.

Studies resting on conduction-dominated heat transfer models for rectangular and concentric LHTES systems are summarized in Tables 7.1 and 7.2, respectively. There are a number of points to be made concerning these mathematical models:

1. The LHTES systems are chiefly employed for the management of the supply energy and its utilization (for load leveling of electric power and for the storage of solar energy). The PCMs are paraffins, Glauber's salts, glycol wax, stearic acid, high density polyethylene, lithium fluoride-calcium fluoride and sodium sulphate decahydrate.
2. Rectangular LHTES systems consist of a single PCM layer exposed to its surroundings, or of several parallel layers separated by channels that allow the flow of a HTF. In most concentric LHTES systems, the space between the concentric pipes is filled with the PCM while the HTF flows inside the inner pipe.
3. In the simplest models, heat storage or recovery is achieved by specifying a constant or time-varying wall temperature. Heat storage or recovery may also be achieved by heat conduction through the capsule walls (walls filled with PCMs), by natural or forced convection to the surrounding HTF and/or by surface radiation. Electric resistance

wires are sometimes employed to store heat in the LHTES systems (electric load management). In some cases, the temperature of the HTF (and so the heat transfer rate) is tightly coupled with that of the PCM. This is referred to as 'conjugate heat transfer', and an example of such a model is presented later.
4. Most models can simulate the thermal behavior of the system during the heat storage period (melting) as well as during the heat recovery period (solidification). In some investigations, however, the focus is either on heat storage or on heat recovery.
5. In spite of the fact that natural-convection buoyancy driven flows are sometimes present in the melt (the capsule thickness is larger than the recommended values specified in Equation 7.19), some investigators still rely on heat conduction-based models to predict the thermal behavior of the LHTES system. The effect of natural convection in the melt is, however, taken into account in the model by using an enhanced thermal conductivity (or thermal diffusivity) for the melt (Brousseau and Lacroix, 1998, 1996; Lacroix, 1993a, 1993b; Farid and Hussian, 1990). This enhanced thermal conductivity, called the effective thermal conductivity for the melt k_e, is a function of the Rayleigh number Ra based on the melt layer thickness S_{max}:

$$k_e = k_l \overline{C} Ra^{\bar{n}} \tag{7.20}$$

where

$$Ra = \frac{g\beta \Delta T (S_{max})^3}{\upsilon_l \alpha_l} \tag{7.21}$$

where C and n are constants, depending on the system and are determined experimentally.
6. Finally, models reported in Table 7.1 and 7.2 can be classified into two broad categories according to the approach used for the solution of the heat transfer and phase change problem: the moving-grid temperature-based methods and the fixed-grid enthalpy-based methods.

In moving-grid temperature-based methods, the classical Stefan formulation is invoked. At each time step, the phase front is immobilized and the energy conservation equations for both solid and liquid phases are solved. The temperature gradients at the solid-liquid interface are then calculated, and used in the solid-liquid energy balance equation to predict the next location of the phase front.

In fixed-grid enthalpy-based methods, the need to satisfy explicitly the energy balance at the phase front is eliminated. A fixed grid is applied to the entire physical space and the latent heat is accounted for by using suitable source terms in the energy equation.

An example of the moving-grid temperature-based method is presented in section 7.3.1 for a one-dimensional capsule. In section 7.3.2, the fixed-grid enthalpy-based method is implemented for a two-dimensional capsule.

7.3.1 1-D moving-grid temperature-based method (conduction)

The phase change material is contained in a one-dimensional rectangular or cylindrical capsule (Figure 7.3). Heat is stored or recovered from the PCM via a HTF at the boundary $x = r_i$. The initial temperature of the PCM is T_{ini}. The temperature of the HTF may be time dependent, and is denoted by T_{HTF}. Suddenly, at time $t \geq 0$, heat is transferred between the HTF and the PCM. Heat is conducted through the PCM and eventually melting (heat storage) or solidification (heat recovery) is triggered at the boundary $x = r_i$. The phase front, whose position is denoted by $r(t)$, progresses uniformly from the containment wall. Sensible heat storage in the containment wall is negligible in comparison with the PCM sensible and latent heat. Heat transfer between the PCM and the HTF can be represented by using a heat transfer coefficient U_{HTF}. Superheating and supercooling effects are neglected.

Table 7.1 Rectangular LHTES systems (conduction-dominated heat transfer).

REFERENCE	GEOMETRY	STORAGE	RECOVERY	SOLUTION
Laouadi and Lacroix (1999)	single layer	electric wires	natural convection and radiation	1-D FGE
Amir *et al.* (1999)	single layer	electric wires	natural convection and radiation	1-D FGE
Zuca *et al.* (1999)	single layer	natural convection	natural convection	1-D Perturbation
Brousseau *et al.* (1998, 1996)	multilayer	electric wires	conjugate heat transfer	2-D FGE
Costa *et al.* (1998)	multilayer	electric wires	none	2-D FGE
El Qarnia and Lacroix (1998)	multilayer	none	conjugate heat transfer	3-D FGE
Ismail and Castro (1997)	single layer	conduction through wall	conduction through wall	1-D MGT
Ismail and Henriquez (1997)	single layer	radiation and conduction	radiation and conduction	1-D MGT
Hongjun *et al.* (1996)	multilayer	none	conjugate heat transfer	2-D MGT
Gong and Mujumdar (1996)	single layer	imposed temperature	imposed temperature	1-D FGHC
Rabin *et al.* (1995)	single layer	natural convection	natural convection	1-D FGE
Bansal and Buddhi (1992a)	single layer	forced convection	forced convection	1-D MGT
Hasan *et al.* (1991)	single layer	imposed temperature	imposed temperature	1-D MGT
Ghoneim *et al.* (1991)	single layer	conduction through wall	conduction through wall	1-D FGE
Farid and Hussian (1990)	multilayer	electric wires	forced convection	1-D MGT
Majumdar *et al.* (1990)	multilayer	forced convection	none	1-D FGE
Onyejekwe (1988)	single layer	natural convection	natural convection	1-D FGE

FGE: Fixed-grid enthalpy-based method, MGT: Moving-grid temperature-based method, FGHC: Fixed-grid equivalent heat capacity-based method.

Table 7.2 Concentric LHTES systems (conduction-dominated heat transfer).

REFERENCE	STORAGE	RECOVERY	SOLUTION
Isamail and Goncalves (1999)	conjugate heat transfer	conjugate heat transfer	2-D FGE
Yimer and Senthil (1998)	conjugate heat transfer	conjugate heat transfer	2-D FGE
Hall *et al.* (1998)	radiation	conjugate heat transfer	1-D FGE
Vick *et al.* (1998)	conjugate heat transfer	conjugate heat transfer	1-D MGT
Gong and Mujumdar (1997)	conjugate heat transfer	conjugate heat transfer	2-D FGHC
Velraj *et al.* (1997)	none	forced convection	2-D FGE
Zhang *et al.* (1997)	none	conjugate heat transfer	1-D MGT
Zhang and Faghri (1996)	conjugate heat transfer	none	2-D FGHC
Kurklu *et al.* (1996)	none	conjugate heat transfer	1-D FGE
Lacroix (1993a, 1993b)	conjugate heat transfer	none	2-D FGE
Belleci and Conti (1993)	solar radiation	conjugate heat transfer	2-D FGE
Kerslake and Ibrahim (1993)	conjugate heat transfer	conjugate heat transfer	2-D FGE
Bansal and Buddhi (1992b)	forced convection	forced convection	1-D analytic
Cao *et al.* (1992)	conjugate heat transfer	none	2-D FGHC
Sokolov and Keizman (1991)	forced convection and heat flux	none	1-D MGT
Ghoneim (1989)	conjugate heat transfer	conjugate heat transfer	2-D FGE
Farid and Kanzawa (1989)	conjugate heat transfer	conjugate heat transfer	1-D MGT
Kamimoto *et al.* (1986)	conjugate heat transfer	conjugate heat transfer	1-D FGE

FGE: Fixed-grid enthalpy-based method, MGT: Moving-grid temperature-based method, FGHC: Fixed-grid equivalent heat capacity-based method.

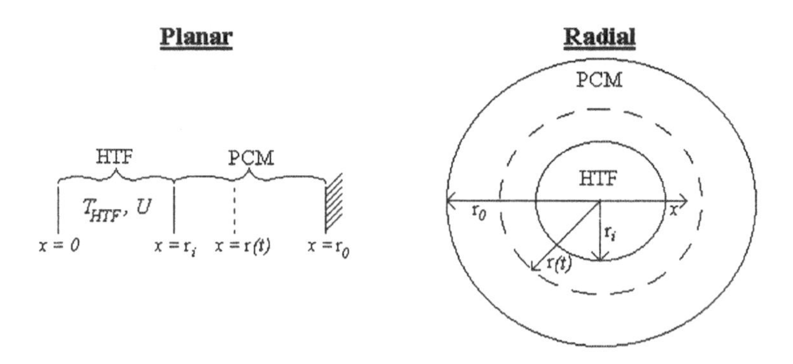

Figure 7.3 Rectangular and cylindrical capsules (1-D).

Based on the above approximations, the temperature distribution in the solid and liquid phases is governed by the following equations:

$$\frac{\partial(x^j T_n)}{\partial t} = \frac{\partial}{\partial x}\left(\alpha_n x^j \frac{\partial T_n}{\partial x}\right) \tag{7.22}$$

The geometric index j takes the value 0 for planar coordinates and 1 for circular coordinates. The subscripts n stands for the liquid phase ($n = l$) or the solid phase ($n = s$). The symbol α_n, is the thermal diffusivity, and it may be different for the solid and liquid phases. The initial and boundary conditions are given by

$$T_s(x,0) = T_l(x,0) = T_{ini} \tag{7.23}$$

$$U_{HTF}(T_{HTF} - T_n(r_i,t)) = -k_n \frac{\partial T_n}{\partial x}(r_i,t) \tag{7.24}$$

$$\frac{\partial T_n}{\partial x}(r_o,t) = 0 \tag{7.25}$$

Note that for melting (heat storage), ($n = l$) in Equation 7.24 and ($n = s$) in Equation 7.25. For solidification (heat recovery), ($n = s$) in Equation 7.24 and ($n = l$) in Equation 7.25. Moreover, the following conditions prevail at the solid-liquid interface $x = r(t)$:

$$T_s(r(t),t) = T_l(r(t),t) = T_m \tag{7.26}$$

$$\rho_s \Delta h_m \frac{dr}{dt} = k_s \frac{\partial T_s(r(t),t)}{\partial x} - k_l \frac{\partial T_l(r(t),t)}{\partial x} \tag{7.27}$$

where T_m is the melting point of the PCM.

The first step in the solution of a solid-liquid phase change problem with a moving grid approach is to cast the Eulerian formulated governing transport Equation 7.22 (fixed grid) to a Eulerian-Lagrangian formulation (moving grid) (Lacroix and Garon, 1992). Performing this straightforward transformation, Equation 7.22 becomes

$$\frac{\partial\left(x^j T_n\right)}{\partial t} - x^j v_{grid}(x) \frac{\partial T_n}{\partial x} = \frac{\partial}{\partial x}\left(\alpha_n x^j \frac{\partial T_n}{\partial x}\right) \tag{7.28}$$

where $v_{grid}(x)$ is the one-dimensional grid deformation velocity. This velocity varies linearly with the x coordinate (Figure 7.4). Its value is zero at the boundaries $x = r_i$ and $x = r_o$ (the boundaries are immobile) and it reaches a maximum at the phase front, i.e. $v_{grid}(x) = dr/dt$, (Equation 7.27).

Equation 7.28 may be solved with a finite-difference method. The difference equation is obtained upon integrating the governing Equation 7.28 over the grid-point cluster shown in Figure 7.2. The resulting finite-difference scheme has the form

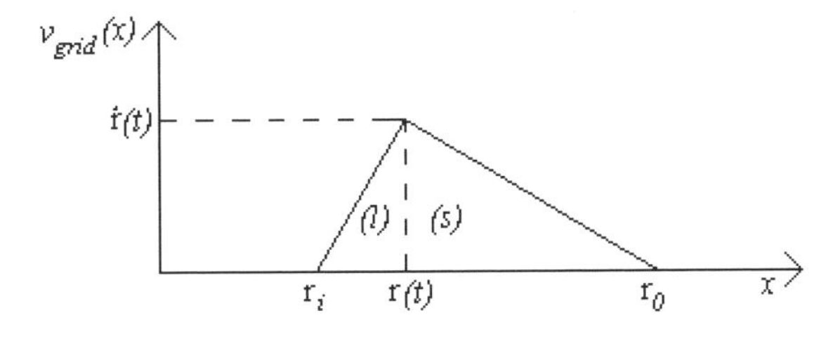

Figure 7.4 Grid deformation velocity.

$$a_P(T_n)_P = a_W(T_n)_W + a_E(T_n)_E + b \tag{7.29}$$

where

$$a_W = \frac{(\alpha_n)x_w^{\ j}}{\delta x_w} - \frac{x_P^{\ j}(v_{grid})_P}{2} \tag{7.30}$$

$$a_E = \frac{(\alpha_n)x_e^{\ j}}{\delta x_e} + \frac{x_P^{\ j}(v_{grid})_P}{2} \tag{7.31}$$

$$a_P = a_W + a_E + a_P^0 \tag{7.32}$$

$$a_P^0 = \frac{x_P^{\ j}\Delta x}{\Delta t} \tag{7.33}$$

$$b = a_P^0(T_n)_P^0 \tag{7.34}$$

Again, the old (known) values of the temperature are denoted by $(T_n)^0$ and the new (unknown) values are denoted by T_n. The terms x_w, x_e and x_p are the radial coordinates at the west face, east face and node P of the control volume, respectively. Solution of Equation 7.29 is obtained from a TDMA. The grid deformation velocity $(v_{grid})_p$ appearing in Equations 7.30 and 7.31 is evaluated from the second-order finite-difference approximation of Equation 7.27.

The numerical solution proceeds through a series of small time intervals during which the solid-liquid interface is assumed to be fixed. For each such time interval, Equation 7.29 is solved in the now-fixed domain. The solution of this equation provides the energy flux at the interface after that time interval. The horizontal displacement of the interface can then be calculated explicitly from Equation 7.27, and a new grid is generated for the next time step.

7.3.2 2-D fixed-grid enthalpy-based method (conduction)

In this case, the PCM is contained in a two-dimensional rectangular capsule of height $(r_o - r_i)$, length L and depth W (direction perpendicular to the plane $x - y$, see Figure 7.5). Or it can also be sandwiched between two concentric tubes of inner and outer radii r_i and r_o, respectively (Figure 7.5). The outside walls of the rectangular cavity or of the cylindrical capsule are adiabatic. Heat is stored or recovered from the PCM via the flow of a HTF at the boundary $x = r_i$. The initial temperature of the PCM, is T_{ini}. Suddenly, at time $t \geq 0$, the HTF starts flowing through the channel and heat is transferred between the HTF and the PCM. Heat is conducted through the PCM and eventually melting (storage) or solidification (recovery) is triggered at the boundary $x = r_i$. Invoking the assumptions made in section 7.3.1, the energy conservation equation for the PCM may be formulated in terms of the sensible enthalpy h:

$$\frac{\partial h}{\partial t} = \frac{1}{x^j}\frac{\partial}{\partial x}\alpha\left(x^j\frac{\partial h}{\partial x}\right) + \frac{\partial}{\partial y}\left(\alpha\frac{\partial h}{\partial y}\right) - \rho_s\Delta h_m\frac{\partial f}{\partial t} \tag{7.35}$$

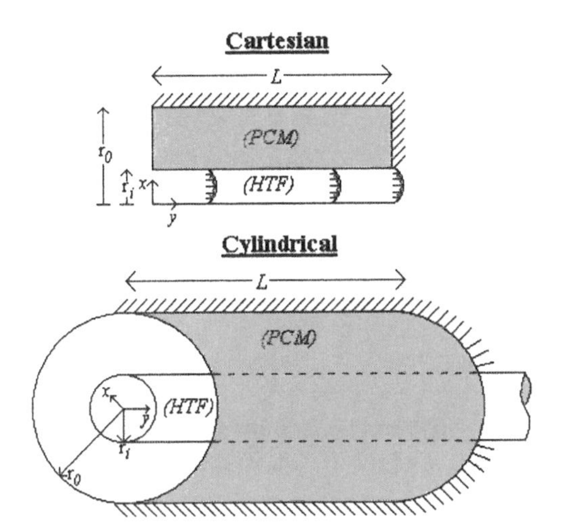

Figure 7.5 Rectangular and cylindrical capsules (2-D).

The superscript j is equal to 0 for cartesian coordinates and 1 for cylindrical coordinates. Here α is the thermal diffusivity, ρ_s is the density of the solid phase and Δh_m is the latent heat of fusion. The third term on the right side of Equation 7.35 stems from Crank's formulation for which the total enthalpy is split into sensible and latent heat components (Crank, 1984):

$$H(T_{PCM}) = h(T_{PCM}) + \rho_s \Delta h_m f \tag{7.36}$$

where

$$h(T_{PCM}) = \int_{T_m}^{T} \rho_k C_k dT_{PCM} \tag{7.37}$$

where ρ_k is the phase density, C_k is the phase specific heat and f is the local liquid fraction. The potential advantage of this formulation is that the enthalpy equation is cast in a standard form, with the problems associated with the phase change isolated in the source term $\rho_s \Delta h_m \partial f/\partial t$.

The initial and boundary conditions for Equation 7.35 are specified by:

$$h(x, y, 0) = h_{ini} \tag{7.38}$$

$$\frac{\partial h}{\partial y}(x, 0, t) = \frac{\partial h}{\partial y}(x, L, t) = \frac{\partial h}{\partial x}(r_o, y, t) = 0 \tag{7.39}$$

$$k_{PCM} \frac{\partial T_{PCM}}{\partial x}(r_i, y, t) = U_{HTF}(T_{PCM} - T_{HTF}(y, t)) \tag{7.40}$$

The temperature of the PCM, T_{PCM}, can be retrieved from the enthalpy definition 7.37. Assuming that heat diffusion in the HTF is negligible with respect to advection, the energy balance for the flow channel yields the following equation:

$$(\rho C)_{HTF} A \frac{\partial T_{HTF}}{\partial t} = PU_{HTF}(T_{PCM} - T_{HTF}) - \dot{m} C_{HTF} \frac{\partial T_{HTF}}{\partial y} \tag{7.41}$$

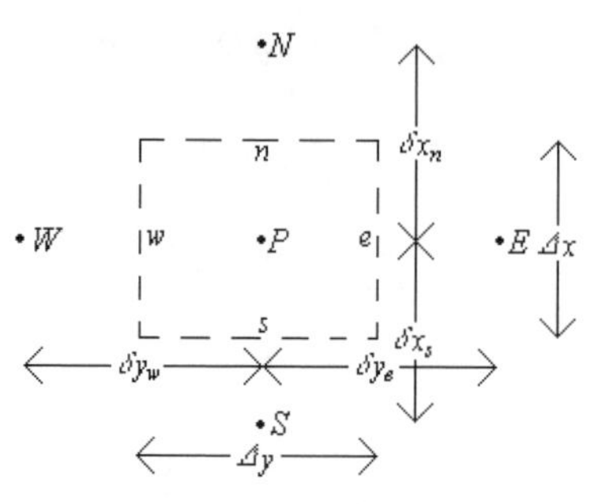

Figure 7.6 Grid-point cluster (2-D).

where $A = Wr_i$ and $P = 2(W + r_i)$ for a rectangular capsule and $A = (\pi r_i)^2$ and $P = 2\pi r_i$ for a cylindrical capsule. Also, $\dot{m}$ is the HTF mass flow rate. The boundary conditions for Equation 7.41 are given by

$$T_{HTF}(0,t) = T_{INLET} \tag{7.42}$$

$$\frac{\partial T_{HTF}}{\partial y}(L,t) = 0 \tag{7.43}$$

Equations 7.35 and 7.41 are solved using a finite-difference solution method. The finite-difference equation for the PCM is obtained upon integrating Equation 7.35 over the control volume in the (x, y) plane depicted in Figure 7.6. The resulting scheme has the form

$$a_P h_P = a_S h_S + a_W h_W + a_E h_E + a_N h_N + b \tag{7.44}$$

where

$$a_S = \alpha \Delta y \cdot \frac{x_s^{\ j}}{\delta x_s} \tag{7.45}$$

$$a_W = \alpha \Delta x \cdot \frac{x_P^{\ j}}{\delta y_w} \tag{7.46}$$

$$a_E = \alpha \Delta x \cdot \frac{x_P^{\ j}}{\delta y_e} \tag{7.47}$$

$$a_N = \alpha \Delta y \cdot \frac{x_n^{\ j}}{\delta x_n} \tag{7.48}$$

$$a_P = a_S + a_W + a_E + a_N + a_p^0 \tag{7.49}$$

$$a_P^0 = x_P^j \frac{\Delta x \Delta y}{\Delta t} \tag{7.50}$$

$$b = a_P^0 h_P^0 + \rho_s \Delta h_m a_P^0 (f_P^0 - f_P^k) \tag{7.51}$$

Equation 7.44 is solved using an alternating direction TDMA solver. The central feature of the present enthalpy fixed grid technique is the source term b, (Equation 7.51). Here, h_P^0 and f_P^0 represent the enthalpy and the local liquid fraction, respectively, from the previous time step. The last term in Equation 7.51 keeps track of the latent heat evolution, and its driving element is the local liquid fraction f. This fraction takes the value of one in fully liquid regions and zero in fully solid regions, and lies in the interval [0,1] in the vicinity of the phase front. Its value is determined iteratively from the solution of the enthalpy equation. Hence, after the $(k+1)^{th}$ numerical solution of the enthalpy equation over the entire computational domain, Equation 7.44 may be rewritten as

$$a_P h_P = a_S h_S + a_W h_W + a_E h_E + a_N h_N + a_P^0 h_P^0 + \rho_s \Delta h_m a_P^0 (f_P^0 - f_P^k) \tag{7.52}$$

If the phase change is occurring about the Pth node, i.e. $1 \geq f \geq 0$, then the kth estimate of the liquid fraction needs to be updated such that the left side of Equation 7.52 is zero, that is,

$$0 = a_S h_S + a_W h_W + a_E h_E + a_N h_N + a_P^0 h_P^0 + \rho_s \Delta h_m a_P^0 (f_P^0 - f_P^{k+1}) \tag{7.53}$$

Subtracting Equation 7.53 from Equation 7.52 leads to the following update for the liquid fraction at nodes where the phase change is taking place:

$$f_P^{k+1} = f_P^k + \frac{a_P h_P}{\rho_s \Delta h_m a_P^0} \tag{7.54}$$

The liquid fraction update is applied at every node after the $(k+1)^{th}$ solution of the linear system (Equation 7.44) for h. Since Equation 7.54 is not adequate for every node, the correction

$$f_P^{k+1} = 0, \text{ if } f_P^{k+1} \leq 0 \quad \text{or} \quad f_P^{k+1} = 1, \text{ if } f_P^{k+1} \geq 1 \tag{7.55}$$

is applied immediately after Equation 7.54. Further details concerning the numerical implementation of the present enthalpy method may be found in the work of Voller (1990).

The finite-difference approximation for the HTF equation (Equation 7.41) is

$$(T_{HTF})_P = \frac{a(T_{HTF})_P^0 + c(T_{HTF})_W + b(T_{PCM})_P}{a + b + c} \tag{7.56}$$

where $a = (\rho C)_{HTF} A / \Delta t$, $b = PU_{HTF}$ and $c = \dot{m} C_{HTF} / \Delta y$. Owing the fact that the PCM enthalpy (Equation 7.44) is tightly coupled with the HTF temperature (Equation 7.56) via the boundary condition (Equation 7.40) at $x = r_i$, both finite-difference equations must be solved iteratively at each time step. At a given time, convergence is usually achieved after few iterations, that is when the change in internal energy of the system is equal to the total energy supplied (storage) or extracted (recovery) at the boundary separating the HTF from the PCM.

7.4 Contact Melting

A phenomenon that should not be overlooked when modeling the thermal behavior of LHTES systems is contact melting. Contact melting occurs when the solid PCM is free to move inside its capsule. If the density of the solid is greater than the density of the liquid, the shrinking solid sinks to the bottom while continuing to experience contact melting by pressing against the capsule wall (if the density of the solid is smaller than that of the liquid, as is the case for ice for example, the expanding solid floats to the top and also experiences contact melting by pressing against the capsule wall). The liquid accumulates in the larger space created above the solid, and melting on the upper surface of the solid is not nearly as intense as on the lower (close-contact) surface, even though natural convection currents may sometimes be present throughout the upper liquid.

In contact melting, a liquid film of finite thickness bridges the temperature gap between the wall and the solid (the melting front). The liquid generated at the melting front is squeezed out from under the solid by the higher pressure maintained in the central section of the film by the weight of the free solid. Therefore, the melting process (film thickness, melting rate) depends on the geometry (shape, size) of the shrinking solid.

The main studies devoted to the modeling of close-contact melting inside capsules are reported in Table 7.3. Note that contact melting has also been the subject of fundamental investigations on sliding, friction and lubrication by Bejan (1989a), Taghavi (1990), Saito *et al.* (1992) and Moallemi *et al.* (1986). The reader is referred to Bejan (1994) for a recent review of this matter.

All models in Table 7.3 are aimed at predicting the time-varying position of the sinking solid PCM. They rely on the conservation equations for the heat transfer and fluid mechanics of the thin film of melt. Their main features are the following:

1. The process is in general quasi-steady, i.e. at every point in time the weight of the solid is balanced by the excess pressure built in the liquid film.
2. Heat transfer in the liquid film is by pure conduction only. Melting on the upper surface of the solid is, in most cases, ignored or modeled as conduction dominated. Saitoh and Kato (1994) and Prasad and Sengupta (1988, 1987) have modeled natural convection dominated melting above the upper surface in a horizontal cylindrical capsule. The effect of contact melting was introduced in their model simply by invoking the results of Bareiss and Beer (1984) for the solid core drop rate due to contact melting and the Nusselt number distribution in the contact zone.
3. The process is symmetric along the central vertical axis, with Poiseuille-type flow in the close contact gap.
4. The pressure at the two openings of the close-contact gap is equal to the hydrostatic pressure in the upper pool of liquid.
5. Finally, the shear stress experienced by the solid as it drops and sweeps the lateral walls is negligible.

As an example of close contact melting, let us briefly examine the case of a parallelepipedic capsule (Figure 7.7). The solid PCM of initial height H, and width $2L$ and $2L_x$ is isothermal at the temperature T_{ini} below the melting point T_m. At time $t = 0$, a constant temperature $T_w = T_m + \Delta T$ is imposed at both the top and bottom surfaces of the cavity while the lateral vertical walls remain adiabatic. Melting is eventually triggered at the top and bottom surfaces of the PCM and, as it proceeds, the solid is assumed to descend vertically while squeezing the melt out of the thin gap separating it from the bottom heated wall.

Table 7.3 Contact melting inside capsules.

REFERENCE	CAPSULE	LIQUID FILM MODEL	LIQUID LAYER MODEL
Lacroix (2001)	parallelepipedic	conduction	conduction
Fomin and Saitoh (1999)	spherical	conduction	none
Chen *et al.* (1995)	rectangular	conduction	conduction
Saitoh and Kato (1994)	cylindrical	conduction (correlation)	convection
Hirata *et al.* (1991)	rectangular	conduction	conduction
Roy and Sengupta (1989)	spherical	conduction	none
Prasad and Sengupta (1987)	cylindrical	conduction (correlation)	convection
Kumar *et al.* (1987)	spherical	conduction	none
Bahrami and Wang (1987)	spherical	conduction	none
Bareiss and Beer (1984)	cylindrical	conduction	convection (correlation)
Moore and Bayazitoglu (1982)	spherical	conduction	none

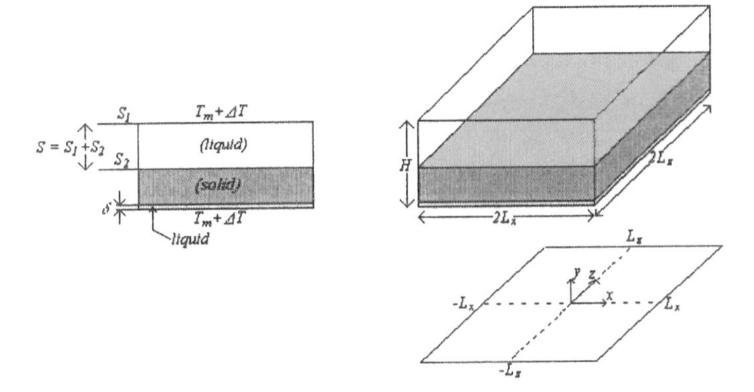

Figure 7.7 Schematic of a parallelepipedic capsule (contact melting).

Using the conduction equation for the solid PCM, the energy and momentum equations for the thin film and the energy equation for the liquid layer above, the simultaneous effect of contact melting at the bottom of the capsule and conduction dominated melting at the top is described by the following set of coupled dimensionless differential equations (Lacroix, 2001):

$$\left(\frac{4C_1}{\rho^* Ar}\left(\frac{dS_1}{dt}\right)\right)^{1/3}\left(\rho^* \Pr\frac{dS_1}{dt}+G\right)=Ste(H-S)^{1/3} \tag{7.57}$$

for contact melting at the bottom and

$$\rho^* \Pr\frac{dS_2}{dt}=\frac{Ste}{S}-G \tag{7.58}$$

for conduction melting at the top. Moreover,

$$S = S_1 + S_2 \tag{7.59}$$

and

$$G = \frac{2k^* \overline{Ste}}{(H - S)} \sum_{n=0}^{\infty} \left((-1)^n - 1\right) \exp\left(\frac{-n^2 \pi^2 \alpha^* t}{\Pr(H - S)^2}\right) \tag{7.60}$$

where *Ar*, *Pr* and *Ste* are the Archimedes, Prandtl and Stefan numbers, respectively. The term *G* accounts for the thermal diffusion in the solid PCM for which the initial degree of subcooling is characterized by $\overline{Ste}$. C_l is a constant which depends on the lateral dimensions of the capsule. k^* and α^* are the dimensionless thermal conductivity and diffusivity, respectively, and ρ^* is the dimensionless density.

The thermal behavior of the capsule during the heat storage period is elucidated from the solution of Equations 7.57–8.59 for the dimensionless drop position *S(t)*. Equations 7.57 and 7.58 are solved with a fourth order Runge-Kutta method.

This model has been used to investigate the effect of the dimensionless numbers *Ste*, *Ar* and $\overline{Ste}$ and of the lateral dimensions L_x and L_z on the performance of the parallelepipedic capsule.

7.5 Convection Dominated Phase Change

Modeling multidimensional phase-change problems with natural convection at the solid-liquid interface is far more difficult, particularly during the melting period. In general, the heat transfer coefficient at the interface is not known and the coupled energy equations in the solid and the liquid, together with the conservation equations of mass and momentum, must be solved. In spite of the fact that the problem of melting in the presence of natural convection in enclosures has attracted considerable research attention (Bejan, 1989b), few investigations are devoted to the modeling of LHTES systems with buoyancy-driven flows in the melt. As exemplified in Tables 7.4 and 7.5, these investigations are chiefly concerned with the heat transfer problem inside rectangular or cylindrical capsules and around tubes. Common features of all these studies are as follows:

1. The PCMs used are pure substances, such as n-octadecane and water.
2. The focus is on the melting process inside the capsule. Only a few investigations have examined cyclic melting-solidification processes.
3. Melting (and solidification) is triggered by imposing a temperature or a heat flux at a boundary. Except for Lacroix (1995), the heat transfer inside the PCM is not coupled with the flow of a HTF.
4. All models reported in Tables 7.4 and 7.5 are two-dimensional. And, as for section 7.3, two different numerical approaches have been used to model the phase change: the moving-grid temperature-based methods and the fixed-grid enthalpy-based methods.

In order to examine the numerical implementation of the two main modeling approaches, the classical convection-dominated melting problem of a pure substance confined to a rectangular enclosure is considered. The physical system is shown in Figure 7.8. The PCM is contained in a two-dimensional cavity of height *H* and width *L*. The walls of the cavity are adiabatic except for the left vertical wall, which allows heat transfer from or to an HTF. Initially, the solid PCM and the surrounding walls are at a uniform temperature $T_m \geq T_{ini}$. Suddenly, at time $t = 0$, the melting process is initiated by imposing a temperature $T_w > T_m$ on the left vertical wall.

It is assumed that the thermophysical properties are constant but may be different for the liquid and the solid phases. Viscous dissipation in the melt is neglected and the Boussinesq approximation is invoked. The liquid phase is Newtonian and incompressible, and the flow is laminar. Furthermore, the Stefan numbers encountered here are small enough so that the normal velocities at the solid-liquid interface resulting from density change upon melting can be neglected. Subject to the previous assumptions, the conservation equations for mass, momentum and energy may be stated as

$$\frac{\partial(\rho_l u)}{\partial x} + \frac{\partial(\rho_l v)}{\partial y} = 0 \tag{7.61}$$

$$\frac{\partial(\rho_l u)}{\partial t} + \frac{\partial(\rho_l uu)}{\partial x} + \frac{\partial(\rho_l vu)}{\partial y} = -\frac{\partial p}{\partial x} + \mu_l\left(\frac{\partial^2 u}{\partial x^2} + \frac{\partial^2 u}{\partial y^2}\right) \tag{7.62}$$

$$\frac{\partial(\rho_l v)}{\partial t} + \frac{\partial(\rho_l uv)}{\partial x} + \frac{\partial(\rho_l vv)}{\partial y} = -\frac{\partial p}{\partial y} + \mu_l\left(\frac{\partial^2 v}{\partial x^2} + \frac{\partial^2 v}{\partial y^2}\right) + \rho_l g\beta_l(T - T_m) \tag{7.63}$$

$$\frac{\partial T}{\partial t} + \frac{\partial(uT)}{\partial x} + \frac{\partial(vT)}{\partial y} = \alpha_n\left(\frac{\partial^2 T}{\partial x^2} + \frac{\partial^2 T}{\partial y^2}\right) \tag{7.64}$$

Table 7.4 Rectangular capsules (natural convection-dominated heat transfer).

REFERENCE	CONDITION	SOLUTION
Bertrand *et al.* (1999)	imposed wall temperature; melting	2-D FGE, MGT
Binet and Lacroix (1998a, 1998b)	electric resistance wires; melting	2-D FGE
Lacroix, Benmadda (1998, 1997)	imposed wall temperature; melting around fins	2-D FGE
Voller *et al.* (1996)	imposed wall temperature; cyclic melting and solidification	2-D FGE
Sasaguchi *et al.* (1996)	imposed heat flux; melting	2-D FGE
Ho and Chu (1996)	imposed time-varying wall temperature; melting-solidification	2-D MGT
Lacroix (1995)	conjugate heat transfer; melting	2-D MGT
Lacroix (1994)	imposed wall temperature, conduction through wall; melting	2-D MGT
Cao and Faghri (1990)	imposed wall temperature; melting	2-D FGE
Benard *et al.* (1986, 1985)	imposed wall temperature; melting	2-D MGT
Gadgil and Gobin (1984)	imposed wall temperature; melting	2-D MGT
Ho and Viskanta (1984a)	imposed wall temperature; melting	2-D MGT
Okada (1984)	imposed wall temperature; melting	2-D MGT

FGE: Fixed-grid enthalpy-based method, MGT: Moving-grid temperature-based method.

Table 7.5 Cylinders (natural convection-dominated heat transfer).

REFERENCE	CONDITION	SOLUTION
Sasaguchi *et al.* (1997)	imposed wall temperature; melting-solidification around two horizontal vertically spaced cylinders	2-D FGE
Wu and Lacroix (1995, 1993a, 1993b)	imposed wall temperature; melting inside vertical cylinder	2-D MGT
Lacroix (1993c)	imposed wall temperature; melting-solidification around two horizontal vertically spaced cylinders	2-D FGE
Prud'Homme *et al.* (1991)	imposed wall temperature; melting inside vertical cylinder	2-D MGT
Wu *et al.* (1991)	imposed wall temperature; melting around vertical cylinder	2-D MGT
Ho and Lin (1986)	imposed wall temperature; outward melting inside cylindrical annulus	2-D MGT
Ho and Viskanta (1984b)	imposed wall temperature; inward melting inside horizontal tube	2-D MGT
Prusa and Yao (1984)	imposed wall heat flux; melting around a horizontal cylinder	2-D MGT
Rieger *et al.* (1983)	imposed wall temperature; melting inside a horizontal tube	2-D MGT
Saitoh and Hirose (1982)	imposed wall temperature; melting inside a horizontal tube	2-D MGT
Rieger *et al.* (1982)	imposed wall temperature; melting around a horizontal cylinder	2-D MGT

FGE : Fixed-grid enthalpy-based method ; MGT : Moving-grid temperature-based method

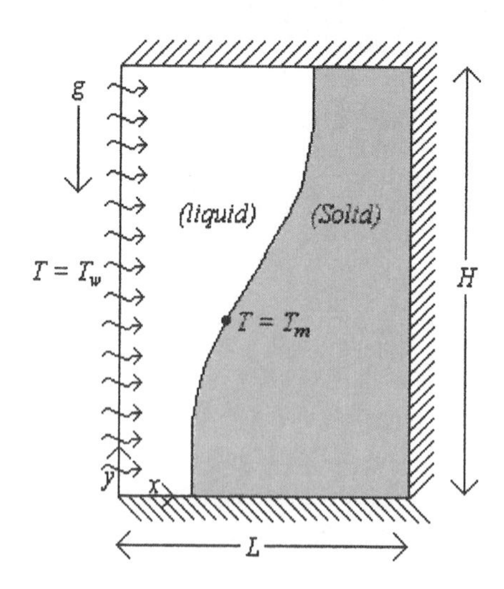

Figure 7.8 Convection-dominated melting inside a rectangular enclosure.

Table 7.6 Variables and parameters of the governing transport Equation 7.65.

Function	ϕ	ρ	Γ	$Q(x,y)$
Temperature liquid	θ_l	1	1	0
Temperature solid	θ_s	1	1	0
Vorticity	ω	1	Pr	$\Pr Ra \frac{\partial \theta_l}{\partial x}$
Stream function	ψ	0	1	ω

Table 7.7 Boundary conditions in the cartesian plane (x, y).

Boundary	Stream Function	Vorticity	Temperature (liquid)	temperature (solid)
Left wall	$\psi = 0$	$\omega = -\frac{\partial^2 \psi}{\partial x^2}$	$\theta_l = 1$	nil
Right wall	nil	nil	nil	$\frac{\partial \theta_s}{\partial x} = 0$
Bottom/top wall	$\psi = 0$	$\omega = -\frac{\partial^2 \psi}{\partial y^2}$	$\frac{\partial \theta_l}{\partial y} = 0$	$\frac{\partial \theta_s}{\partial y} = 0$
Phase front	$\psi = 0$	$\omega = -\frac{\partial^2 \psi}{\partial x^2}$	$\theta_l = 0$	$\theta_s = 0$

The third term on the right side of Equation 7.63 represents the buoyancy force. Of course, in the solid phase, $u = v = 0$, and only the conduction part of Equation 7.64 is solved.

A moving-grid temperature-based methodology is first presented for the solution of the set of Equations 7.61–7.64. Next, a fixed-grid enthalpy-based method is briefly examined.

7.5.1 2-D moving grid temperature-based method (convection)

Due to the fact that in natural convection dominated melting the pressure solution in the melt is of secondary importance, the conservation Equations 7.61–7.64 may be rewritten in terms of the dimensionless stream function ψ, vorticity ω and temperature θ for both phases. The resulting transport equations for the flow property ϕ take the following form in cartesian coordinates:

$$\frac{\partial(\rho\phi)}{\partial t} + \frac{\partial}{\partial x}\left(u\phi - \Gamma\frac{\partial \phi}{\partial x}\right) + \frac{\partial}{\partial y}\left(v\phi - \Gamma\frac{\partial \phi}{\partial y}\right) = Q(x,y) \tag{7.65}$$

where $Q(x,y)$ is a source term, Γ is an exchange coefficient, and ρ is a constant. These parameters and the corresponding dependent variable ϕ are explained in Table 7.6.

The velocity components u and v of the melt are given by

$$u = \frac{\partial \psi}{\partial y} \qquad v = -\frac{\partial \psi}{\partial x} \tag{7.66}$$

Obviously, these velocities are null everywhere in the solid phase of the PCM. The vorticity is defined in the usual manner, i.e.

$$\omega = \frac{\partial v}{\partial x} - \frac{\partial u}{\partial y} \tag{7.67}$$

At time $t = 0$, the conditions are $u = v = \psi = \omega = 0$ and $\theta = \theta_{ini}$. For time $t > 0$, the boundary conditions are specified in Table 7.7. In addition, an energy balance for the solid-liquid PCM interface yields the following dimensionless condition for the moving boundary:

$$-\nabla\theta_l n + k^*\nabla\theta_s n = \frac{\rho^*}{Ste} v_n \tag{7.68}$$

where v_n is the dimensionless local normal interface velocity and $\nabla\theta\, n$ is the local normal heat flux to the solid-liquid interface.

As melting progresses, buoyancy-driven flows in the melt impart heat transfer rates that vary along the moving solid-liquid interface and, as a result, its shape becomes distorted. This curvilinear shape does not, in general, coincide with the grid nodes on a rectangular cartesian mesh, thereby complicating the implementation of the boundary conditions (Figure 7.9). To overcome this problem, the general conservation Equation 7.65 and its boundary conditions (Table 7.7) are converted from the original Eulerian cartesian grid (x,y) to a Eulerian-Lagrangian curvilinear grid (ξ,η). The resulting general conservation equation is slightly more complicated, but its boundary conditions are now specified on straight boundaries, and the computational grid is rectangular and uniformly spaced. Adopting this technique (Lacroix and Garon, 1992), Equation 7.65 becomes, in the transformed coordinate system (ξ,η),

$$\begin{aligned}&\frac{\partial(\rho\phi)}{\partial t} + \frac{1}{J}\left[\frac{\partial x}{\partial \eta}\frac{\partial y}{\partial t} - \frac{\partial y}{\partial \eta}\frac{\partial x}{\partial t}\right]\frac{\partial(\rho\phi)}{\partial \xi} + \frac{1}{J}\left[\frac{\partial y}{\partial \xi}\frac{\partial x}{\partial t} - \frac{\partial x}{\partial \xi}\frac{\partial y}{\partial t}\right]\frac{\partial(\rho\phi)}{\partial \eta}\\ &+\frac{1}{J}\frac{\partial}{\partial \xi}\left[(U\phi) - \frac{\Gamma}{J}\left(\alpha\frac{\partial\phi}{\partial \xi} - \beta\frac{\partial\phi}{\partial \eta}\right)\right] + \frac{1}{J}\frac{\partial}{\partial \eta}\left[(V\phi) - \frac{\Gamma}{J}\left(\gamma\frac{\partial\phi}{\partial \eta} - \beta\frac{\partial\phi}{\partial \xi}\right)\right] = Q(\xi,\eta)\end{aligned} \tag{7.69}$$

The second and third terms on the left side of Equation 7.69 account for the motion of the grid lines. These terms are to the two-dimensional moving grid model what the second term on the left side of Equation 7.28 is to the one-dimensional moving grid model.

The geometric factors U, V, α, β, γ and the Jacobian J of the transformation are given in Appendix A. The source term $Q(\xi,\eta)$ for the vorticity (Table 7.6) becomes

$$Q(\xi,\eta) = \frac{Ra\,\mathrm{Pr}}{J}\left(\frac{\partial y}{\partial \eta}\frac{\partial \theta_l}{\partial \xi} - \frac{\partial y}{\partial \xi}\frac{\partial \theta_l}{\partial \eta}\right) \tag{7.70}$$

The boundary conditions in the computational plane are provided in Table 7.8. The Stefan condition, Equation 7.68, is now rewritten as

$$-\frac{1}{J}\frac{\partial y}{\partial \eta}\frac{\partial \theta_l}{\partial \xi} + \frac{k^*}{J}\frac{\partial y}{\partial \eta}\frac{\partial \theta_S}{\partial \xi} = \frac{\rho^*}{Ste}\frac{dS(y,t)}{dt} \tag{7.71}$$

where $S(y,t)$ represents the time-dependent position of the solid-liquid interface.

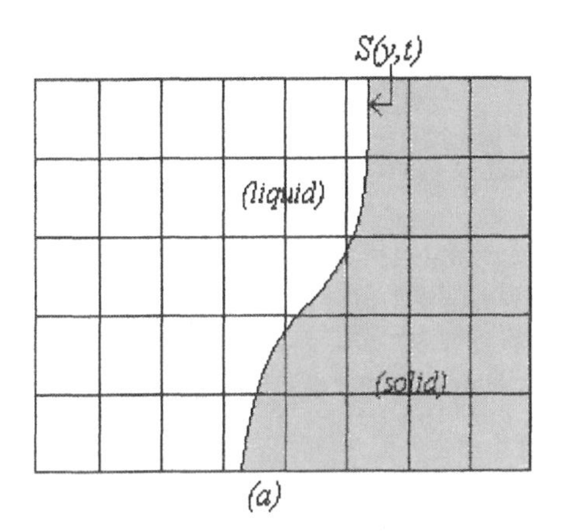

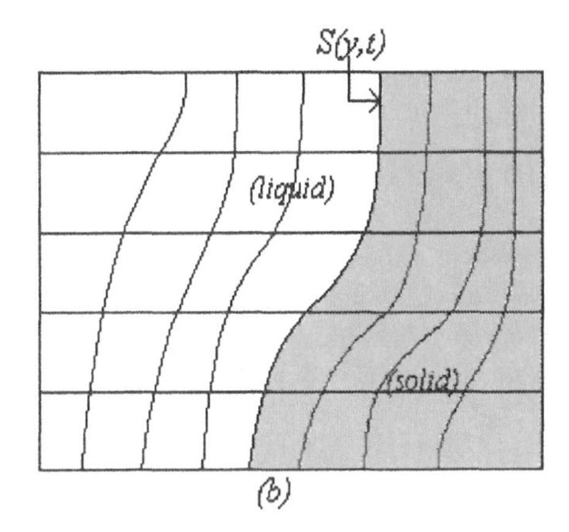

Figure 7.9 Grids. (a) Cartesian, (b) curvilinear.

Table 7.8 Boundary conditions in the computational plane (ξ,η).

Boundary	Stream Function	Vorticity	Temperature (liquid)	Temperature (solid)
Left wall	$\psi = 0$	$\omega = -\frac{\alpha}{J^2}\frac{\partial^2\psi}{\partial\xi^2}$	$\theta_l = 1$	nil
Right wall	nil	nil	nil	$\frac{\partial\theta_s}{\partial\xi} = \frac{\beta}{\alpha}\frac{\partial\theta_s}{\partial\eta}$
Bottom/top wall	$\psi = 0$	$\omega = -\frac{\gamma}{J^2}\frac{\partial^2\psi}{\partial\eta^2}$	$\frac{\partial\theta_l}{\partial\eta} = \frac{\beta}{\gamma}\frac{\partial\theta_l}{\partial\xi}$	$\frac{\partial\theta_s}{\partial\eta} = \frac{\beta}{\gamma}\frac{\partial\theta_s}{\partial\xi}$
Phase front	$\psi = 0$	$\omega = -\frac{\alpha}{J^2}\frac{\partial^2\psi}{\partial\xi^2}$	$\theta_l = 0$	$\theta_s = 0$

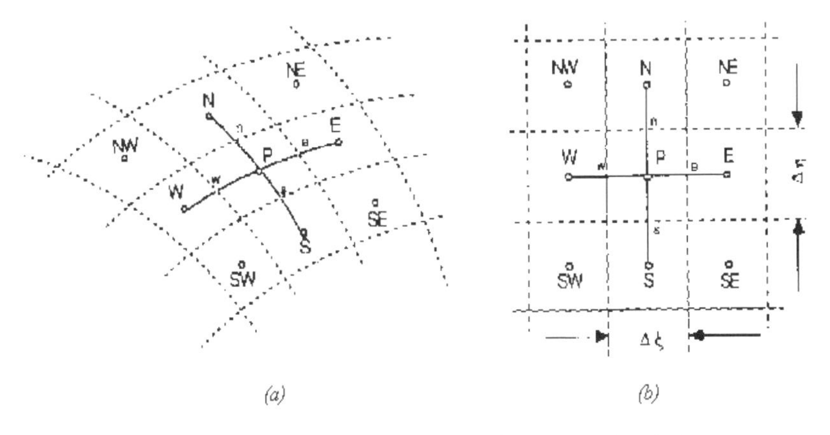

Figure 7.10 Finite-difference grid. (a) Physical plane, and (b) computational plane.

The finite-difference equations are obtained on integrating the general governing Equation 8.69 over each of the control volumes in the (ξ, η) plane (Figure 7.10). The resulting finite-difference scheme has the form

$$a_P\phi_P = a_{SW}\phi_{SW} + a_S\phi_S + a_{SE}\phi_{SE} + a_W\phi_W + a_E\phi_E + a_{NW}\phi_{NW} + a_N\phi_N + a_{NE}\phi_{NE} + b \quad (7.72)$$

Expressions for the coefficients in Equation 7.72 are provided in Appendix B. The linearized equations are solved iteratively for θ, ω and ψ using a line-by-line TDMA. Convergence is achieved, at a given time step, when the largest residual for all difference equations is smaller than a convergence criterion (a small number of the order of 10^{-3}). Its magnitude is set according to the problem to be solved (magnitude of Rayleigh and Stefan numbers, grid size and time step).

The numerical solution proceeds through a series of small time intervals during which the solid-liquid interface is assumed to be fixed. For each such time interval, the field equations are solved implicitly (retaining the second and third terms in Equation 7.69) in the now-fixed computational domain. The solution of the field equations provides the energy fluxes at the interface after that time interval. The horizontal displacement of the interface can then be calculated explicitly from Equation 7.71, and a new computational grid is generated for the next time step. The grid is generated algebraically using a power law clustering function that concentrates grid nodes in the vicinity of the boundaries and the phase front (Lacroix, 1989).

7.5.2 2-D fixed grid enthalpy-based method (convection)

The essential features of the fixed-grid enthalpy-based method have already been discussed in section 7.3.2. This method is extended to convection dominated melting by adding to the conduction model of section 7.3.2 the conservation equations of mass and momentum. Rewriting Equations 7.61–7.64 in terms of the dimensionless velocities u, v pressure p and enthalpy h yields:

$$\frac{\partial u}{\partial x} + \frac{\partial v}{\partial y} = 0 \quad (7.73)$$

$$\frac{\partial u}{\partial t} + \frac{\partial (uu)}{\partial x} + \frac{\partial (vu)}{\partial y} = -\frac{\partial p}{\partial x} + \Pr\left(\frac{\partial^2 u}{\partial x^2} + \frac{\partial^2 u}{\partial y^2}\right) + S_x \quad (7.74)$$

$$\frac{\partial v}{\partial t} + \frac{\partial (uv)}{\partial x} + \frac{\partial (vv)}{\partial y} = -\frac{\partial p}{\partial y} + \Pr\left(\frac{\partial^2 v}{\partial x^2} + \frac{\partial^2 v}{\partial y^2}\right) + S_y \quad (7.75)$$

$$\frac{\partial h}{\partial t} + \frac{\partial (uh)}{\partial x} + \frac{\partial (vh)}{\partial y} = \left(\frac{\partial^2 h}{\partial x^2} + \frac{\partial^2 h}{\partial y^2}\right) + S_h \quad (7.76)$$

The source term S_h in Equation 7.76 is defined as

$$S_h = -\frac{1}{Ste}\frac{\partial f}{\partial t} \quad (7.77)$$

This term keeps track of the latent heat evolution via the liquid fraction f. The liquid fraction f is also used here to drive the velocity components to zero in the solid phase of the PCM via the source terms S_x and S_y in the momentum Equations 7.74–7.75:

$$\begin{aligned} S_x &= -Bfu \\ S_y &= -Bfv + Ra\,\mathrm{Pr}\,h \end{aligned} \tag{7.78}$$

The function B becomes very large when the liquid fraction f is zero, and goes to zero as f tends toward one. A function B based on the Carman-Koseny relation for a porous medium as described by Voller *et al.*, (1987) and Voller and Prakash (1987) is employed:

$$B(f) = \frac{C_2(1-f)^2}{(f^3 + \varepsilon_2)} \tag{7.79}$$

with $C_2 = 1.6\times10^6$ and $\varepsilon_2 = 1.6\times10^6$. This numerical artifact can be viewed as a way of modeling the transition zone between the solid and liquid phases.

The finite difference equations are obtained on integrating the conservation Equations 7.73–7.76 over each of the control volumes in the (x,y) plane (Figure 7.6) using second-order centered differences. The convective terms are discretized, however, with a first order hybrid difference scheme in order to ensure stability. An implicit Euler scheme is used for the time-stepping procedure, and the classical SIMPLE algorithm is adopted for the velocity-pressure coupling (Patankar, 1980). The resulting finite-difference equations are solved iteratively with a line by line TDMA solver by sweeping the cavity from left to right. At a given time, convergence is achieved when the dimensionless residual for the mass conservation equation and the enthalpy conservation equation is less than a given criterion and that the liquid fraction field is stabilized.

7.5.3 Final remarks

The above transformed and fixed-grid methods present some advantages and disadvantages.

The moving-grid temperature-based method is particularly useful for phase change at a fixed temperature, such as that of pure substances. It is found that for single-temperature phase-change problems, the transformed grid method predicts the interface positions and flow structure better than the enthalpy-based formulation. The transformed grid method also needs less computational time than the enthalpy-based method. It has been thoroughly employed to examine the effect of various parameters and phenomena on the melting of pure substances such as the Rayleigh, Stefan and Prandtl numbers, dimensions and aspect ratios of rectangular and cylindrical capsules, surface tension, density inversion of the melt, conjugate heat transfer, etc.

The fixed-grid enthalpy-based method is, on the other hand, suitable for complex substances for which the change of phase occurs over a finite temperature range. It can handle mushy regions with dendrites using the appropriate porous flow equations. The implementation of the enthalpy-based method in existing heat transfer and fluid flow programs such as commercial codes is straightforward, and the method can easily handle multiple phase fronts. The fixed-grid method has been used successfully for the modeling of cyclic melting and solidification problems in which multiple phase fronts coexist, and also for the prediction of phase change in the vicinity of several heated (or cooled) surfaces such as walls with fins, multiple electric resistance wires, vertically spaced horizontal heated and cooled cylinders, etc.

Both methods are valuable for solving phase-change problems. The choice of one method over the other is usually dictated by the problem at hand. As reported by Viswanath and Jaluria (1993), however, more work is needed to make the transformed-grid method more efficient and the fixed-grid method more accurate. In a recent comparison exercise on the simple problem of convection dominated melting of a pure substance from a vertical heated wall (Bertrand *et al.*, 1999), disparities of up to 30% in the predicted Nusselt numbers and up to 20% in the predicted positions of the solid-liquid interface were reported among the various transformed and fixed grid methods. It is clear, therefore, that in the present state of these computational tools, the modeling of convection-dominated melting in LHTES systems should always be conducted very carefully.

7.6 Conclusions

The mathematical complexity of the heat transfer models augments as the emphasis is placed on the physical phenomena occuring inside the PCM. In the porous medium approach, models focus on the overall thermal behavior of the LHTES system. The details of the heat transfer and phase change phenomena inside individual PCM capsules are ignored (or lumped into empirical coefficients), and the resulting mathematical models are very simple. At the other end of the spectrum, models on convection dominated phase change attempt to predict the heat transfer and the buoyancy driven flows inside the PCM capsules. The resulting mathematical models are, in this case, far more elaborate. Both approaches are very useful as they provide complementary information to the designer.

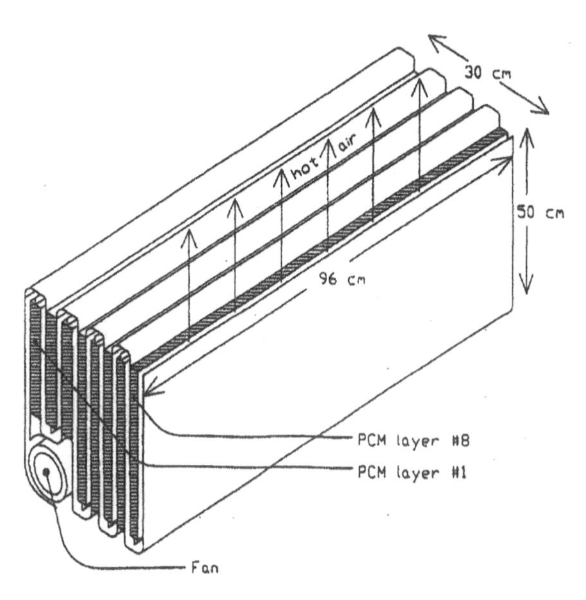

Figure 7.11 Multi-layer storage unit.

As an example, the multi-layer storage unit shown in Figure 7.11 was designed with the help of a two-dimensional heat diffusion model based on the enthalpy method presented in section 7.3.2 (Brousseau and Lacroix, 1998, 1996). This storage unit is used for smoothing the daily electric load profiles. It consists of narrow vertical parallel plates of PCM separated by a rectangular flow passage. Electric resistance wires embedded inside the PCM are used for storage (melting), while the flow of a HTF is employed for heat recovery (solidification). When the prototype was tested in the laboratory, the molten PCM at the top of each layer experienced unacceptably high temperatures that the conduction based model could not have predicted. A model that takes into account the natural convection driven flows inside an individual PCM layer was then developed for predicting the melt temperature in the vicinity of the electric resistance wires (Binet and Lacroix, 1998a, 1998b). With this model, resting on the method presented in section 7.5.2, the optimum shape and size of the PCM layer, together with the optimum distribution of the electric resistance wires, were determined. This information was then fed back to the simpler heat diffusion model for predicting the overall thermal performance of the improved storage unit under different operating scenarios.

In spite of the impressive number of articles published on the subject over the last fifteen years, modeling LHTES systems remains a challenging task. Practical engineering methods for predicting their thermal response and performance, not only of the PCM, but also of a module or of the entire storage unit, in order to facilitate its design are still under development, and are far from the well established methods for the design of sensible heat storage devices and exchangers. There is a need to develop mathematical models that reflect more faithfully the true conditions that prevail inside individual PCM capsules used in LHTES systems. For example, phenomena such as contact and natural convection melting, melting and solidification over a temperature range, superheating and supercooling, thermal expansion upon phase change, etc., should be modeled increasingly as coupled phenomena. Numerical simulations should also be conducted for cyclic melting and solidification processes subject to more realistic boundary conditions. Information resulting from these studies will undoubtedly contribute to improving overall models of LHTES systems.

Acknowledgements

The author wishes to thank the Natural Sciences and Engineering Research Council of Canada for the financial support of this work.

Nomenclature

$a_S, a_W, a_P, a_E, a_N, etc.$	Finite-difference coefficients
A	Surface area
A_{bed}	Particle area per unit volume
Ar	Archimedes $= \dfrac{g(\rho_s - \rho_l)L_x^3}{\rho_s \upsilon_l^2}$
b	Source term

B	Carman-Kozeny relation
C	Heat capacity
C, C_1, C_2	Constants
f	Liquid fraction
g	Acceleration of gravity
h	Sensible enthalpy
H	Total enthalpy (or height)
J	Jacobian
k	Thermal conductivity
$\dot{m}$	Mass flow rate
p	Pressure
P	Perimeter
Pr	Prandtl number $= \dfrac{\upsilon_l}{\alpha_l}$
Q	Source term
r	Coordinate
$r(t)$	Location of phase front (1-D)
Ra	Rayleigh $= \dfrac{g\beta_l \Delta T (S_{\max})^3}{\upsilon_l \alpha_l}$
$S(y,t)$	Location of phase front (2-D)
Ste	Stefan number $= \dfrac{C_l \Delta T}{\Delta h_m}$
$\overline{Ste}$	Subcooling $= \dfrac{C_l (T_{ini} - T_m)}{\Delta h_m}$
$S_{\max}$	Maximum capsule thickness
t	Time
T	Temperature
u, v	Velocity components
U, V	Geometric coefficients
U_{HTF}	Heat transfer coefficient
v_{grid}	Grid deformation velocity (1-D)
x, y	Coordinates
1-D	One-dimensional
2-D	Two-dimensional

Greek letters

α	Geometric coefficient
α_l, α_s	Thermal diffusivity
β	Geometric coefficient
β_l	Thermal expansion coefficient
ε	Void fraction of bed
ϕ	General dependent variable
η, ξ	Transformed coordinates

μ	Dynamic viscosity
θ	Temperature
ρ	Coefficient
ρ_l, ρ_s	Density
υ_l	Kinematic viscosity
ω	Vorticity
ψ	Stream function
$\Delta x, \Delta y, \Delta t$	Space and time increments
Δh_m	Latent heat of fusion
ΔT	Temperature difference $= T_w - T_m$
Γ	Coefficient

Subscripts

e	effective
HTF	heat transfer fluid
ini	initial
l	liquid
m	melting point
n	phase
PCM	phase change material
s	solid
w	wall
s, w, e, n	south, west, east, north faces
SW, S, SE	south west, south and south east nodes
W, P, E	west, central and east grid nodes
NW, N, NE	north west, north and north east nodes

Superscripts

j	= 0 for planar , = 1 for radial coordinate
k	iteration number
$*$	dimensionless parameter
0	previous time step

References

Adebiyi, G.A., Hodge, B.K., Steele, W.G., Jalalzadeh-Aza, A. and Nsofor, E.C. (1996). Computer simulation of a high temperature thermal energy storage system employing multiple families of phase-change storage materials, *Journal of Energy Resources Technology* 118, 102–111.

Alexiades, V. and Solomon, A.D. (1993). *Mathematical Modeling of Melting and Freezing Processes*, Hemisphere Publishing Corporation, Washington, DC.

Amir, M., Lacroix, M. and Galanis, N. (1999). Comportement thermique de dalles chauffantes électriques pour le stockage quotidien, *International Journal of Thermal Sciences* 38, 121–131.

Bahrami, P.A. and Wang, T.G. (1987). Analysis of gravity and conduction-driven melting in a sphere, *Journal of Heat Transfer* 109, 806–809.

Bansal, N.K. and Buddhi, D. (1992a). Performance equations of a collector cum storage system using phase change materials, *Solar Energy* 48(3), 185–194.

Bansal, N.K. and Buddhi, D. (1992b). An analytical study of a latent heat storage system in a cylinder, *Energy Conversion and Management* 33(4), 235–242.

Bareiss, M. and Beer, H. (1984). An analytical solution of the heat transfer process during melting of an unfixed solid phase change material inside a horizontal tube, *International Journal of Heat and Mass Transfer* 27(5), 739–746.

Beasley, D.E., Ramanarayanan, C. and Torab, H. (1989). Thermal response of a packed bed of spheres containing a phase-change material, *International Journal of Energy Research* 13, 253–265.

Bejan, A. (1994). Contact melting heat transfer and lubrication, *Advances in Heat Transfer* 24, 1–38.

Bejan, A. (1989a). The fundamentals of sliding contact melting and friction, *Journal of Heat Transfer* 111, 13–20.

Bejan, A. (1989b). Analysis of melting by natural convection in an enclosure, *International Journal of Heat and Fluid Flow* 10(3), 245–252.

Belleci, C. and Conti, M. (1993). Transient behaviour analysis of a latent heat thermal storage module, *International Journal of Heat and Mass Transfer* 36(15), 3851–3857.

Benard, C., Gobin, D. and Martinez, F. (1985). Melting in rectangular enclosures: experiments and numerical simulations, *Journal of Heat Transfer* 107, 794-803.

Benard, C., Gobin, D. and Zanoli, A. (1986). Moving boundary problem: heat conduction in the solid phase of a phase-change material during melting driven by natural convection in the liquid, *International Journal of Heat and Mass Transfer* 29(11), 1669–1681.

Bertrand, O., Binet, B., Combeau, H., Couturier, S., Delannoy, Y., Gobin, D., Lacroix, M., Le Quéré, P., Médale, M., Mencinger, J., Sadat, H. and Vieira, G. (1999). Melting driven by natural convection: a comparison exercise, *International Journal of Thermal Sciences* 38, 5–26.

Binet, B. and Lacroix, M. (1998a). Numerical study of natural-convection dominated melting inside uniformly and discretely heated rectangular cavities, *Journal of Numerical Heat Transfer-Part A* 33, 207–224.

Binet, B. and Lacroix, M. (1998b). Étude numérique de la fusion dans des enceintes rectangulaires chauffées uniformément ou discrètement par les parois latérales conductrices, *International Journal of Thermal Sciences* 37, 607–620.

Brousseau, P. and Lacroix, M. (1998). Numerical simulation of a multi-layer latent heat thermal energy storage system, *International Journal of Energy Research* 22, 1–15.

Brousseau, P. and Lacroix, M. (1996). Study of the thermal performance of a multi-layer pcm storage unit, *Energy Conversion and Management* 37(5), 599–609.

Cao, Y. and Faghri, A. (1992). A study of thermal energy storage systems with conjugate turbulent forced convection, *Journal of Heat Transfer* 114, 1019–1027.

Cao, Y. and Faghri, A. (1990). A numerical analysis of phase-change problems including natural convection, *Journal of Heat Transfer* 112, 812–816.

Chen, S.L. (1992). One-dimensional analysis of energy storage in packed capsules, *Journal of Solar Energy Engineering* 114, 127–130.

Chen, W.Z., Cheng, S.M., Luo, Z. and Gu, W.M. (1995). Analysis of contact melting of phase change materials inside a heated rectangular capsule, *International Journal of Energy Research* 19, 337–345.

Costa, M., Buddhi, D. and Oliva, A. (1998). Numerical simulation of a latent heat thermal energy storage system with enhanced heat conduction, *Energy Conversion and Management* 39(3/4), 319–330.

Crank, J. (1984). *Free and Moving Boundary Problems*, Clarendon Press, Oxford.

El Qarnia, H. and Lacroix, M. (1998). Modélisation d'un échangeur de chaleur compact à courants croisés séparés par des couches de matériau à changement de phase, *International Journal of Thermal Sciences* 37, 514–524.

Farid, M. and Husian, R. (1990). an electrical storage heater using the phase-change method of heat storage, *Energy Conversion and Management* 30(3), 219–230.

Farid, M. and Kanzawa, A. (1989). Thermal performance of a heat storage module using pcm's with different melting temperatures: mathematical modeling, *Journal of Solar Energy Engineering* 111, 152–157.

Fomin, S.A. and Saitoh, T.S. (1999). Melting of unfixed material in spherical capsule with non-isothermal wall, *International Journal of Heat and Mass Transfer* 42, 4197–4205.

Gadgil, A. and Gobin, D. (1984). Analysis of two-dimensional melting in rectangular enclosures in presence of convection, *Journal of Heat Transfer* 106, 20–26.

Ghoneim, A.A., Klein, S.A. and Duffie, J.A. (1991). Analysis of collector-storage buildings walls using phase-change materials, *Solar Energy* 47(3), 237–242.

Ghoneim, A.A. (1989). Comparison of theoretical models of phase-change and sensible heat storage for air and water-based solar heating systems, *Solar Energy* 42(3), 209–220.

Goncalves, L.C.C. and Probert, S.D., (1993). Thermal-energy storage: dynamic performance characteristics of cans each containing a phase-change material, assembled as a packed-bed, *Applied Energy* 45, 117–155.

Gong, Z. and Mujumdar, A.S. (1997). Finite-element analysis of cyclic heat transfer in a shell-and-tube latent heat energy storage exchanger, *Applied Thermal Engineering* 17(6), 583–591.

Gong, Z. and Mujumdar, A.S. (1996). Enhancement of energy charge-discharge rates in composite slabs of different phase change materials, *International Journal of Heat and Mass Transfer* 39(4), 725–733.

Hall, C.A., Glapke, E.K. and Cannon, J.N. (1998). Modeling cyclic phase change and energy storage in solar heat receivers, *Journal of Thermophysics and Heat Transfer* 12(3), 406–413.

Hasan, M., Mujumdar, A.S. and Weber, M.E. (1991). Cyclic melting and freezing, *Chemical Engineering Science* 46(7), 1573–1587.

Hirata, T., Makino, Y. and Kaneko, Y. (1991). Analysis of close-contact melting for octadecane and ice inside isothermally heated horizontal rectangular capsule, *International Journal of Heat and Mass Transfer* 34(12), 3097–3106.

Ho, C.J. and Chu, C.H. (1996). Numerical simulation of heat penetration through a vertical rectangular phase change material/air composite cell, *International Journal of Heat and Mass Transfer* 39(9), 1785–1795.

Ho, C.J. and Lin, H. (1986). Outward melting in a cylindrical annulus, *Journal of Energy Resources Technology* 108, 240–245.

Ho, C.J. and Viskanta, R. (1984a). Heat transfer during melting from an isothermal vertical wall, *Journal of Heat Transfer* 106, 12–19.

Ho, C.J. and Viskanta, R. (1984b). Heat transfer during inward melting in a horizontal tube, *International Journal of Heat and Mass Transfer* 27(5), 705–716.

Ismail, K.A.R. and Goncalves, M.M. (1999). Thermal performance of a pcm storage unit, *Energy Conversion and Management* 40, 115–138.

Ismail, K.A.R. and Stuginsky, R. (1999). A parametric study on possible fixed bed models for pcm and sensible heat storage. *Applied Thermal Engineering* 19, 757–788.

Ismail, K.A.R. and Castro, J.N.C. (1997). PCM thermal insulation in buildings. *International Journal of Energy Research* 21, 1281–1296.

Ismail, K.A.R. and Henriquez, J.R. (1997). PCM glazing systems. *International Journal of Energy Research* 21, 1241–1255.

Kamimoto, M., Abe, Y., Sawata, S., Tani, T. and Ozawa, T. (1986). Latent thermal storage unit using form-stable high density polyethylene; part ii: numerical analysis of heat transfer. *Journal of Solar Energy Engineering* 23, 290–297.

Kerslake, T. and Ibrahim, M.B. (1993). Analysis of thermal energy storage material with change-of-phase volumetric effects. *Journal of Solar Energy Engineering* 115, 22–31.

Kondepudi, S., Somasundaram, S. and Anand, N.K. (1988). A simplified model for the analysis of a phase-change material-based, thermal energy storage system. *Heat Recovery Systems & CHP* 8(3), 247–254.

Kumar, A., Prasad, A. and Upadhaya, S.N. (1987). Spherical-phase change energy storage with constant temperature heat injection. *Journal of Energy Resources Technology* 109, 101–104.

Kurklu, A., Wheldon, A. and Hadley, P. (1996). Mathematical modeling of the thermal performance of a phase-change material (PCM) store: cooling cycle. *Applied Thermal Engineering* 16(7), 613–623.

Lacroix, M. (2001). Contact melting of a phase change material in a heated parallelepipedic capsule. *Energy Conversion and Management* 42, 35–47.

Lacroix, M. (1995). Numerical study of natural convection dominated melting of a pcm with conjugate forced convection. *Transactions of the Canadian Society for Mechanical Engineering* 19, 455–470.

Lacroix, M. (1994). Coupling of wall conduction with natural convection dominated melting of a phase change material. *Numerical Heat Transfer-Part A* 26, 483–498.

Lacroix, M. (1993a). Numerical simulation of a shell-and-tube latent heat thermal energy storage unit. *Solar Energy* 50(4), 357–367.

Lacroix, M. (1993b). Study of the heat transfer behavior of a latent heat thermal energy storage unit with a finned tube. *International Journal of Heat and Mass Transfer* 36(8), 2083–2092.

Lacroix, M. (1993c). Numerical simulation of melting and resolidification of a phase change material around two cylindrical heat exchangers. *Numerical Heat Transfer-Part A* 24, 143–160.

Lacroix, M. (1989). Computation of heat transfer during melting of a pure substance from an isothermal wall. *Numerical Heat Transfer-Part B* 15, 191–210.

Lacroix, M. and Benmadda, M. (1998). Analysis of natural convection melting from a heated wall with vertically oriented fins, *International Journal of Numerical Methods for Heat and Fluid Flow* 8(4), 465–478.

Lacroix, M. and Benmadda, M. (1997). Numerical simulation of natural convection dominated melting and solidification from a finned vertical wall. *Numerical Heat Transfer-Part A* 31, 71–86.

Lacroix, M. and Garon, A. (1992). Numerical solution of phase change problems: an eulerian-lagrangian approach. *Numerical Heat Transfer-Part B* 19, 57–78.

Laouadi, A. and Lacroix, M. (1999). Thermal performance of a latent heat energy storage ventilated panel for electric load management. *International Journal of Heat and Mass Transfer* 42, 275–286.

Li, H., Hsieh, C.K. and Goswami, D.Y. (1996). Conjugate heat transfer analysis of fluid flow in a phase change energy storage unit. *International Journal of Numerical Methods for Heat and Fluid Flow* 6, 77–90.

Majumdar, P. and Saidbakhsh, A. (1990). A heat transfer model for phase change thermal energy storage, *Heat Recovery Systems & CHP* 10(5/6), 457–468.

Moellemi, M.K., Webb, B.W. and Viskanta, R. (1986). An experimental and analytical study of close-contact melting, *Journal of Heat Transfer* 108, 894–899.

Okada, M. (1984). Analysis of heat transfer during melting from a vertical wall, *International Journal of Heat and Mass Transfer* 27(11), 2057–2066.

Onyejekwe, D.C. (1988). A generalized model for low temperature energy storage using liquid/solid PCM, Journal of Energy Resources 28(4), 281–285.

Patankar, S.V. (1980). *Numerical Heat Transfer and Fluid Flow*, McGraw-Hill, Washington DC.

Prasad, A. and Sengupta, S. (1987). Numerical investigation of melting inside a horizontal cylinder including the effects of natural convection, *Journal of Heat Transfer* 109, 803–809.

Prud'Homme, M., Nguyen, T.H. and Wu, Y.K. (1991). Simulation numérique de la fusion à l'intérieur d'un cylindre adiabatique chauffé par le bas, *International Journal of Heat and Mass Transfer* 34(9), 2275–2286.

Prusa, J. and Yao, L.S. (1984). Melting around a horizontal heated cylinder: part i- perturbation and numerical solutions for constant heat flux boundary condition, *Journal of Heat Transfer* 106, 376–384.

Rabin, Y., Bar-Niv, I., Korin, E. and Mikic, B. (1995). Integrated solar collector storage system based on a salt-hydrate phase-change material, *Solar Energy* 55(6), 435–444.

Rieger, H., Projahn, U., Bareiss, M. and Beer, H. (1983). Heat transfer during melting inside a horizontal tube, *Journal of Heat Transfer* 105, 226–234.

Rieger, H., Projahn, U., Bareiss, M. and Beer, H. (1982). Analysis of the heat transport mechanisms during melting around a horizontal circular cylinder, *International Journal of Heat and Mass Transfer* 25(1), 137–147.

Roy, S.K. and Sengupta, S. (1989). Melting of a free solid in a spherical enclosure : effects of subcooling, *Journal of Solar Energy Engineering* 111, 32–36.

Saborio-Aceves, S., Nakamura, H. and Reistad, G.M. (1994). Optimum efficiencies and phase change temperature in latent heat storage systems, *Journal of Energy Resources Technology* 116, 79–86.

Saito, A., Hong, H. and Hirokane, O. (1992). Heat transfer enhancement in the direct contact melting process, *International Journal of Heat and Mass Transfer* 35(2), 295–305.

Saitoh, T. and Hirose, K. (1982). High Rayleigh number solutions to problems of latent heat thermal energy storage in a horizontal cylinder capsule, *Journal of Heat Transfer* 104, 545–553.

Saitoh, T.S. and Kato, H. (1994). Numerical analysis for combined natural-convection and close-contact melting in a horizontal cylindrical capsule, *Heat Transfer-Japanese Research* 23(2), 198–213.

Sasaguchi, K., Ishihara, A. and Zhang, H. (1996). Numerical study on utilization of melting of phase change material for cooling of a heated surface at a constant rate, *Numerical Heat Transfer-Part A* 29, 19–31.

Sasaguchi, A., Kusano, K. and Viskanta, R. (1997). A numerical analysis of solid-liquid phase change heat transfer around a single and two horizontal, vertically spaced cylinders in a rectangular cavity, *International Journal of Heat and Mass Transfer* 40(6), 1343–1354.

Sokolov, M. and Keizman, Y. (1991). Performance indicators for solar pipes with phase change storage, Solar Energy 47(5), 339–346.

Taghavi, K. (1990). Analysis of direct-contact melting under rotation, *Journal of Heat Transfer* 112, 137–143.

Torab, H. and Beasley, D.E. (1987). Optimization of a packed bed thermal energy sorage unit, *Journal of Solar Energy Engineering* 109, 170–175.

Velraj, R., Seeniraj, R.V., Hafner, B., Faber, C. and Schwarzer, K. (1997). Experimental analysis and numerical modeling of inward solidification on a finned vertical tube for a latent heat storage unit, *Solar Energy* 60(5), 281–290.

Vick, B., Nelson, D.J. and Yu, X. (1998). Freezing and melting with multiple phase fronts along the outside of a tube, *Journal of Heat Transfer* 120, 422–429.

Viswanath, R. and Jaluria, Y. (1993). A comparison of different solution methodologies for melting and solidification problems in enclosures, *Numerical Heat Transfer-Part A* 24, 77–105.

Voller, V.R. (1997). An overview of numerical methods for solving phase change problems, Chapter 9, pp. 341-380, in *Advances in Numerical Heat Transfer* (Eds. W.J. Minkowycz and E.M. Sparrow), Taylor & Francis, New York.

Voller, V.R. (1990). Fast implicit finit-difference method for the analysis of phase change problems, *Numerical Heat Transfer-Part B* 17, 155–169.

Voller, V.R., Cross, M. and Markatos, N.C. (1987). An enthalpy method for convection/diffusion phase change, *International Journal for Numerical Methods in Engineering* 24, 271–284.

Voller, V.R., Felix, P. and Swaminathan, C.R. (1996). cyclic phase change with fluid flow, *International Journal of Numerical Methods for Heat and Fluid Flow* 6(4), 57–64.

Voller, V.R. and Prakash, C. (1987). A fixed grid numerical modeling methodology for convection/diffusion mushy region phase-change problems, *International Journal of Heat and Mass Transfer* 30(8), 1709–1719.

Wu, Y.K. and Lacroix, M. (1995). Melting of a PCM inside a vertical cylindrical capsule, *International Journal for Numerical Methods in Fluids* 20, 559–572.

Wu, Y.K. and Lacroix, M. (1993a). Numerical study of natural convection dominated melting within an isothermal vertical cylinder, *Transactions of the Canadian Society for Mechanical Engineering* 17(3), 281–296.

Wu, Y.K. and Lacroix, M. (1993b). Analysis of natural convection melting of a vertical ice cylinder involving density anomaly, *International Journal of Numerical Methods for Heat and Fluid Flow* 3, 445–456.

Wu, Y.K., Prud'Homme, M. and Nguyen, T.H. (1989). Étude numérique de la fusion autour d'un cylindre vertical soumis à deux types de conditions limites, *International Journal of Heat and Mass Transfer* 32(10), 1927–1938.

Yimer, B. and Senthil, K. (1998). Experimental and analytical phase change heat transfer, *Energy Conversion and Management* 39(9), 889–897.

Zhang, Y., Chen, Z. and Faghri, A. (1997). Heat transfer during solidification around a horizontal tube with internal convective cooling, *Journal of Solar Energy Engineering* 119, 44–47.

Zhang, Y. and Faghri, A. (1996). Heat transfer enhancement in latent heat thermal energy storage system by using the internally finned tube, *International Journal of Heat and Mass Transfer* 39(15), 3165–3173.

Zuca, S., Pavel, P.M. and Constantinescu, M. (1999). Study of one dimensional solidification with free convection in an infinite plate geometry, *Energy Conversion and Management* 40, 261–271.

Appendix A

The geometric factors U, V, α, β, γ and the Jacobian J of the transformation (Equation 7.69) are defined as

$$U = \frac{\partial \psi}{\partial \eta} \; ; \; V = -\frac{\partial \psi}{\partial \xi}$$

$$\alpha = \left(\frac{\partial x}{\partial \eta}\right)^2 + \left(\frac{\partial y}{\partial \eta}\right)^2 \; ; \; \beta = \frac{\partial x}{\partial \xi}\frac{\partial x}{\partial \eta} + \frac{\partial y}{\partial \xi}\frac{\partial y}{\partial \eta}$$

$$\gamma = \left(\frac{\partial x}{\partial \xi}\right)^2 + \left(\frac{\partial y}{\partial \xi}\right)^2 \; ; \; J = \frac{\partial x}{\partial \xi}\frac{\partial y}{\partial \eta} - \frac{\partial x}{\partial \eta}\frac{\partial y}{\partial \xi}$$

Appendix B

The coefficients for the finite-difference Equation 7.72 are specified by:

$$a_{SW} = -(E_w + E_s) \; ; \; a_S = -\frac{\rho}{2}\left(\frac{\partial x}{\partial \xi}\frac{\partial y}{\partial t} - \frac{\partial y}{\partial \xi}\frac{\partial x}{\partial t}\right)_P + A_s + E_e - E_w \; ; \; a_{SE} = (E_e + E_s)$$

$$a_W = \frac{\rho}{2}\left(\frac{\partial x}{\partial \eta}\frac{\partial y}{\partial t} - \frac{\partial y}{\partial \eta}\frac{\partial x}{\partial t}\right)_P + A_w + E_n - E_s \; ; \; a_E = -\frac{\rho}{2}\left(\frac{\partial x}{\partial \eta}\frac{\partial y}{\partial t} - \frac{\partial y}{\partial \eta}\frac{\partial x}{\partial t}\right)_P + A_e - E_n + E_s$$

$$a_{NW} = (E_w + E_n) \; ; \; a_N = \frac{\rho}{2}\left(\frac{\partial x}{\partial \xi}\frac{\partial y}{\partial t} - \frac{\partial y}{\partial \xi}\frac{\partial x}{\partial t}\right)_P + A_n - E_e + E_w$$

$$a_{NE} = -(E_e + E_n) \; ; \; a_P = a_S + a_W + a_E + a_N + \left(\frac{\rho J}{\Delta t}\right)_P \; ; \; b = \left(\frac{\rho J}{\Delta t}\right)_P \phi_P^0 + (JQ)_P$$

where

$$A_s = D_s + \frac{C_s}{2} \; ; \; A_w = D_w + \frac{C_w}{2} \; ; \; A_e = D_e - \frac{C_e}{2} \; ; \; A_n = D_n - \frac{C_n}{2}$$

The coefficients C, D and E are defined as

$$C_w = (\rho U)_w \; ; \; C_e = (\rho U)_e \; ; \; C_s = (\rho V)_s \; ; \; C_n = (\rho V)_n$$

$$D_w = \left(\frac{\Gamma}{J}\alpha\right)_w \; ; \; D_e = \left(\frac{\Gamma}{J}\alpha\right)_e \; ; \; D_s = \left(\frac{\Gamma}{J}\gamma\right)_s \; ; \; D_n = \left(\frac{\Gamma}{J}\gamma\right)_n$$

$$E_w = \left(\frac{\beta\Gamma}{4J}\right)_w \; ; \; E_e = \left(\frac{\beta\Gamma}{4J}\right)_e \; ; \; E_s = \left(\frac{\beta\Gamma}{4J}\right)_s \; ; \; E_n = \left(\frac{\beta\Gamma}{4J}\right)_n$$

The space increments in the computational plane (ξ, η) are set to $\Delta\xi = \Delta\eta = 1$.

8

Heat Transfer with Phase Change in Simple and Complex Geometries

K.A.R. Ismail

8.1 Introduction

The subject of heat transfer with phase change around solid bodies is receiving continuously increasing interest because of its importance in the thermal science area, and also because of the large number of applications in engineering, bioengineering, materials science and many other areas. Important applications can be found in engineering applications as in the energy storage applications, thermal insulation of walls, roofs and windows, and in the food processing industries. Other applications include water flow, freezing of rivers, lakes and plantation fields. Thermal energy storage as a means of managing energy demand and utilization is now accepted, and many countries are not sparing efforts to encourage their societies to adopt this solution in their energy systems. Available and promising thermal storage systems include sensible heat energy storage in liquids or solids, or in both. Energy storage in the form of latent heat of phase transition is becoming popular because of its compact size and the nearly constant temperature of operation. These storage units are often preferred to other systems, and many installations of large, medium and small capacities are available and, in some cases, in operation.

Cool thermal storage systems have been intensively studied for their ability to use efficiently stored cooling produced during off-peak hours to provide the energy necessary for daytime applications. Basically, there are two main methods of storing cool thermal energy, either in the form of sensible heat or alternatively in the form of latent heat of phase transformation. The chilled water sensible heat storage concept, with or without thermal stratification, is relatively simple and is considered a well-developed technology. The small differences in temperature, the fluid flow turbulence, technical problems associated with the fluid entry and exit manifolds geometry, and finally the large volume usually required represent difficulties for their use as anything more than a thermal buffer.

Homan and Soo (1997) reported the results of a model of transient stratified flow into a chilled-water tank, while Ismail *et al.* (1997) presented a two dimensional model for stratified storage tank based upon the continuity, momentum and energy equations. As a special case, they simplified the two-dimensional model to obtain the one-dimensional pure conduction model. The results from the two models were compared with available numerical and experimental results.

Cool thermal storage using the latent heat concept as an alternative to sensible heat storage offers a good option because of its high storage density and the nearly constant-temperature heat removal characteristics during the discharge cycle. Since the principle of latent cool storage necessarily involves a change of state, heat transfer must occur in different modes depending on the state of charging or discharging. During the periods of low cooling demand the system removes heat from the thermal storage medium (water, PCM, etc.) to be used later to meet the air conditioning or process cooling load demand. The use of this concept for air conditioning applications is increasing due to the need to reduce peak power requirements. This peak can be shifted to periods of low power demand and the chilling plant can then operate close to its maximum efficiency range while the peak electric demand can be reduced significantly, reducing operational costs.

The basic operational principles of an ice bank are presented in detail in Grumman and Butkus (1988). Silver *et al.* (1989) developed mathematical models for the components of the ice banks with the objective of simulating the thermal performance of such equipment, performing complete energy analysis and testing control strategies. Chen and Yue (1991) investigated the thermal performance of an ice-water cold storage system both theoretically and experimentally. They used a general lumped model to determine the thermal storage characteristics. The solution of the system of equations is found using the Laplace transformation method. Effects of several operational parameters on the charge characteristics are quantitatively clarified.

Somasumdaram *et al.* (1993) analyzed two basic principles for ice formation, that is, ice building and ice harvesting. In the first type the ice is constantly formed over the evaporator, while in the second type the ice is formed periodically and removed by local instantaneous fusion. One of the available commercial ice bank storage systems is the ice-on-coil type equipment. In this system the chilled water leaving the chiller passes through a large number of coiled tubes in the tank and effectively transfers heat from the storage medium (ice/water) to the transport fluid (brine). During the charging period, cold brine is circulated through the tubes and ice builds up on the outside of the coiled tubes, and during the discharging period, relatively warm brine is circulated through the same tubes and ice is melted. Jekel *et al.* (1993) developed a model to simulate the time-dependent performance of a static ice-on-coil storage tank. Basic heat transfer and thermodynamic relations were employed to solve for the rate of heat transfer from the brine to the tank during both charging and discharging. The performance predicted by the model was found to compare well with the manufacturer's data. Vargas and Bejan (1995) showed that the production of ice by convection cooling followed by contact melting can be maximized by properly selecting the frequency of the intermittent freezing and removal cycle.

Knebel (1995) evaluated an ice harvesting system composed of plane evaporators installed above the storage tank. During the charging period, water under low pressure is distributed uniformly over the plane surface of the evaporators and is solidified in relatively thin layers varying between 5 mm and 9 mm in thickness. Periodically an instantaneous fusion process removes these layers. Lee and Jones (1996), Vick *et al.* (1996) and Neto and Krarti (1997) reported the results of their numerical and experimental studies on ice-on-coil thermal energy storage systems.

Bedecarrats *et al.* (1996) presented a theoretical and experimental study on a commercial system, which uses phase change material encapsulated in spherical containers stacked in the storage tank. They developed a simulation program that considers aspects of both the surrounding heat transfer fluid and the phase change material during the charging and discharging periods. The simulation results were found to compare well with the experimental observations. Ismail and Gonçalves (1997) developed a two-dimensional conduction-based model for solving the phase change heat transfer problem around a vertical cylinder submersed in PCM (water). The energy equation is coupled to the flow problem by an energy balance. The system of equations was solved numerically by using an average control volume technique and the ADI approach. The results show the effects of the variation of the Biot number, the Stefan number, the inlet fluid temperature and the ratio of the outer to the inner tube radius on the solidified mass fraction, NTU, effectiveness and the time for complete solidification.

The literature on solutions of problems of phase change with or without thermal free convection and in finned or finless geometry is vast and extremely rich. Some of these results and applications can be found in Rao and Sastri (1984), Sparrow and Chuck (1984), Voller (1985), Voller and Prakash (1987), Kim and Kaviany (1990), Schneider (1990), Voller (1990), Date (1991) and Clavier *et al.* (1994). Numerous experimental and theoretical studies have been realized at The Thermal Storage and Heat Pipes Laboratory, Faculty of Mechanical Engineering, The State University of Campinas, Brazil, on latent heat thermal storage, phase change problems and applications in thermal insulation, thermal windows, PCM walls and roofs, as described in Ismail (1998,1999), Castro (1991), Gonçalves (1996), Neto (1996), Quispe (1996), Henríquez (1996) and Ismail and Alves (1985, 1986, 1989).

This chapter presents the models and results of some studies on heat transfer with phase change using simple and finned geometries relevant to latent heat thermal storage. The models presented were solved using different approaches illustrating the different methods. In the case of simple cylindrical geometry, the immobilization method was used and the results were compared with available predictions to validate the results. The subsequent sections handle the case of a spherical capsule and discuss the effects of the geometrical and operational parameters, and also treat the case of plane geometry with and without fins. The last section presents the cases of axial, radial and alternating finned geometries suitable for vertical and horizontal latent heat storage systems.

8.2 Phase Change around Cylindrical Geometry

Landau (1950) suggested the boundary immobilization method, referring to a transformation of coordinates, such that in the new coordinate system the moving interface can be immobilized, and the solution can be realized in the fixed domain. Since then the method has become a technique for handling phase-change heat transfer problems. The Landau transformation of coordinates was extended and applied to different cases by Crank (1957), Lotkin (1960), Murray and Landis (1959), Duda *et al.* (1975), Saitoh (1978),

Sparrow *et al.* (1978b), Sparrow and Chuck (1984), Lacroix (1989), Lacroix and Voller (1990) and Kim and Kaviany (1990, 1992).

The present model for the problem of solidification around a cylinder submersed in a phase change material is based upon the conservation equation of energy, as well as the usual boundary conditions for phase-change problems. It is assumed that the phase-change material is initially in the liquid phase and at a temperature slightly above or equal to the phase-change temperature. Also it is assumed that the problem is axisymmetric, and hence two-dimensional. The external bounding surfaces are adiabatic and the working fluid is assumed to have a uniform and specified temperature at entry and being hydrodynamically fully developed. The thermophysical properties of the phase change material, tube material and working fluid are uniform and independent of the temperature. The thermophysical properties of the phase-change material can be considered different for the two phases, while density is considered the same for both phases. Finally, as the solidification problem is being treated and the initial temperature of the liquid phase is not very high, the mechanism of heat transfer in the PCM is considered to be pure conduction, and hence there are no convection effects in the liquid phase. The inclusion of the liquid phase allowed the development of a more general formulation compared with that presented by Hsu *et al.* (1981).

The problem is illustrated in Figure 8.1, which can be subdivided into three regions: the interior of the inner cylinder, the wall of the inner tube and the annular space with the PCM phases.

The immobilization of the frontiers is achieved by a coordinate transformation in such a way that the moving domain is changed to a fixed region. If one uses a new set of dimensionless coordinates initially proposed by Landau (1950), of the form

$$\eta_s = \frac{r - r_w}{\delta_s(z,t)} \; ; \; \eta_l = \frac{r - r_\delta(z,t)}{\delta_l(z,t)} \; ; \; \xi = \frac{z}{r_w} \tag{8.1}$$

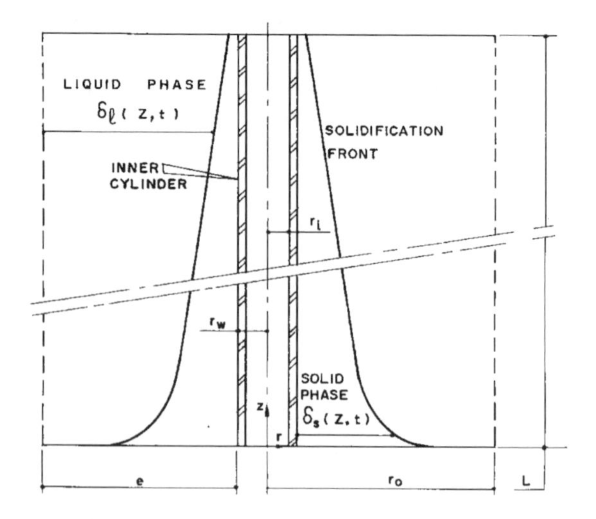

Figure 8.1 Layout of the problem.

one can describe the annular space in terms of the new variables and, as a result, the moving boundary becomes fixed and parallel to the other boundaries of the problem. Consequently, the solid and liquid phases are confined in the region limited by $0 \le \eta_{s,l} \le 1$; $0 \le \xi \le L/r_w$. The method to be used in the solution of the heat transfer problem in the three regions of the transformed domain is the finite volume method due to Patankar (1980). For the PCM region the same formulation, developed by Hsu *et al.* (1981), is used with an extension for inclusion of the liquid phase. The transformed domain is divided in a grid of control volumes and Figure 8.2 shows the grid in the transformed region and in the physical domain.

Considering the physical grid, it is essential that the control-volumes move along time to follow the solidification front. In this manner, the integral form of the energy equation for the control volumes in the phase-change material must be obtained considering the control surface velocity. Using the Leibinitz rule, the integral form of the energy equation for the control volumes in the phase-change region in the physical domain can be written as

$$\frac{d}{dt}\left[\int_{V.C.} \rho c_{ps,l} T_{s,l}\, dV\right] = \int_{V.C.} \rho c_{ps,l} \frac{\partial T_{s,l}}{\partial t} dV + \int_{S.C.} \rho c_{ps,l} T_{s,l}\left(\vec{u}_b \cdot \vec{n}\right) dS \tag{8.2}$$

where $T_{s,l}$ is the temperature, ρ is the density and $c_{ps,l}$ is the specific heat. The terms $\vec{u}_b$ and $\vec{n}$ are the velocity of the control surface and the vector normal to this surface, respectively.

Let us define a dimensionless temperature as

$$\theta_{s,l} = \frac{T_{s,l} - T_m}{T_m - T_{in}} \tag{8.3}$$

where T_m is the phase-change temperature and T_{in} is the entry temperature of the working fluid. The subscripts 's' and 'l' refer to the solid and liquid phases, respectively.

Considering that thermophysical properties in each phase are uniform and independent of time, Equation 8.2 can be written as

$$\frac{d}{dt}\left[\int_{V.C.} \theta_{s,l}\, dV\right] = \int_{V.C.} \frac{\partial \theta_{s,l}}{\partial t} dV + \int_{S.C.} \theta_{s,l}\left(\vec{u}_b \cdot \vec{n}\right) dS \tag{8.4}$$

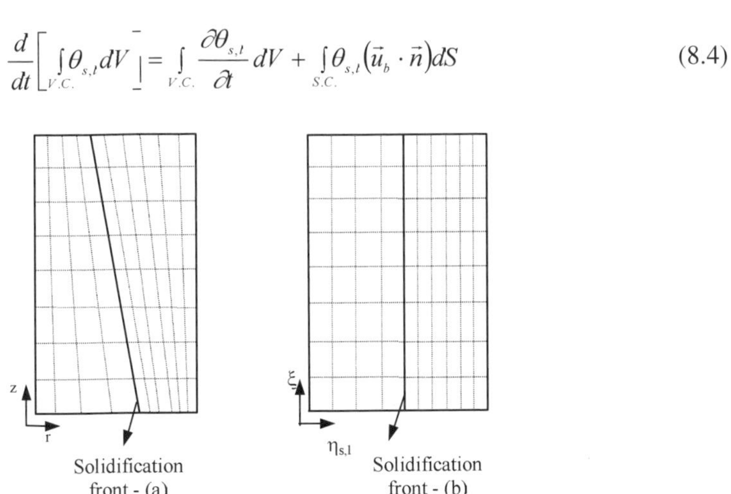

Figure 8.2 Control-volumes grid in the (a) physical and (b) transformed domains.

If we assume that the heat transfer in the PCM is by thermal conduction, one can write

$$\frac{\partial \theta_{s,l}}{\partial t} = \alpha_{s,l} \nabla^2 \theta_{s,l} \tag{8.5}$$

By virtue of Equation 8.5 and Gauss theorem, Equation 8.4 can be written as

$$\frac{d}{dt}\left[\int_{V.C.} \theta_{s,l} dV\right] = \int_{S.C.} \left(\alpha_{s,l} \nabla \theta_{s,l} + \theta_{s,l} \vec{u}_b\right) \cdot \vec{n} dS \tag{8.6}$$

Equation 8.6 represents the energy integral equation for a PCM control volume in the physical domain. In order to express Equation 8.6 in the new coordinates $\eta_{s,l}(r,z,t)$, $\xi(z)$ and $\tau(t)$, where $\tau(t)$ is a dimensionless time, we need some transformation equations for the derivatives in terms of (r,z,t) to be expressed in terms of $(\eta_{s,l},\xi,\tau)$, and these can be written as

$$\left.\frac{\partial}{\partial r}\right|_{z,t} = \frac{\partial \eta_{s,l}}{\partial r}\left.\frac{\partial}{\partial \eta_{s,l}}\right|_{\xi,\tau} = \frac{1}{\delta_{s,l}(z,t)}\left.\frac{\partial}{\partial \eta_{s,l}}\right|_{\xi,\tau} \tag{8.7a}$$

$$\left.\frac{\partial}{\partial z}\right|_{r,t} = \frac{-\beta_{s,l}}{\Delta_{s,l} r_w}\left.\frac{\partial}{\partial \eta_{s,l}}\right|_{\xi,\tau} + \frac{1}{r_w}\left.\frac{\partial}{\partial \xi}\right|_{\eta_{s,l},\tau} \tag{8.7b}$$

$$\left.\frac{\partial}{\partial t}\right|_{r,z} = \frac{\partial \eta_{s,l}}{\partial t}\left.\frac{\partial}{\partial \eta_{s,l}}\right|_{\xi,\tau} + \frac{\partial \tau}{\partial t}\left.\frac{\partial}{\partial \tau}\right|_{\eta_{s,l},\xi} = \frac{\alpha_s}{r_w^2} Ste\left[-\frac{1}{\Delta_{s,l}}\gamma_{s,l}\left.\frac{\partial}{\partial \eta_{s,l}}\right|_{\xi,\tau} + \left.\frac{\partial}{\partial \tau}\right|_{\eta_{s,l},\xi}\right] \tag{8.7c}$$

where

$$\Delta_{s,l} = \frac{\delta_{s,l}}{r_w} : \beta_s = \eta_s \frac{\partial \Delta_s}{\partial \xi} : \beta_l = (\eta_l - 1)\frac{\partial \Delta_l}{\partial \xi} \tag{8.7d}$$

$$\gamma_s = \eta_s \frac{\partial \Delta_s}{\partial \tau} : \gamma_l = (\eta_l - 1)\frac{\partial \Delta_l}{\partial \tau} \tag{8.7e}$$

The transformation relations of Equations 8.7 enable us to express the energy equation, Equation 8.6, in terms of the new variables. Figure 8.3 illustrates a section of a typical control volume.

The faces S1, S2, S3 and S4 are the divisions of the control surface used to calculate the integrals in Equation 8.6. Note that S1 and S3 move in the radial direction, while S2 and S4 are fixed. The differential element for S1 can be written as

$$dS1 = 2\pi r dz\left[1 + \left(\frac{dr}{dz}\right)^2\right]^{-1/2} \tag{8.8}$$

which, when expressed in terms of the new coordinates, can be written as

$$dS1 = 2\pi r_w^2 R_{s,l} \chi_{s,l}^{1/2} d\xi \;; \quad R_{s,l} = \frac{r}{r_w} : \chi_{s,l} = 1 + \beta_{s,l}^2 \tag{8.9}$$

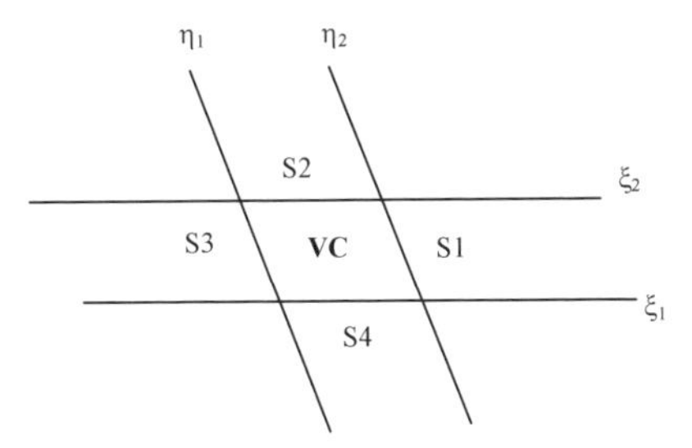

Figure 8.3 Typical control-volume.

One can easily show that the unit normal vector in both the liquid and solid phases can be written as

$$\vec{n}_{S1(s,l)} = \frac{\nabla \eta_{s,l}}{\left|\nabla \eta_{s,l}\right|} = \frac{\left(\vec{i}_r - \beta_{s,l}\,\vec{i}_z\right)}{\chi_{s,l}^{1/2}} \tag{8.10}$$

while the velocity across the surface S1 is

$$\vec{u}_{bS1(s,l)} = \left.\frac{\partial r}{\partial t}\right|_{S1(s,l)} \vec{i}_r = \left(\frac{\alpha_s Ste}{r_w}\right)\gamma_{s,l}\,\vec{i}_r \tag{8.11}$$

If we substitute the terms expressed in the new variables in Equation 8.6 and integrate one obtains

$$\int_{S1}\left(\alpha_{s,l}\nabla\theta_{s,l} + \theta_{s,l}\vec{u}_b\right)\cdot\vec{n}\,dS = 2\pi r_w \alpha_s \int_{\xi_1}^{\xi_2}\left[\frac{\alpha_{s,l}}{\alpha_s}\frac{\chi_{s,l}}{\Delta_{s,l}}\frac{\partial\theta_{s,l}}{\partial\eta_{s,l}} - \frac{\alpha_{s,l}}{\alpha_s}\beta_{s,l}\frac{\partial\theta_{s,l}}{\partial\xi} + \theta_{s,l} Ste\,\gamma_{s,l}\right]R_{s,l}\,d\xi \tag{8.12}$$

where ξ_1 and ξ_2 are the coordinates of the lower and upper faces of the control volume. A similar procedure is used to handle the integral across the surface S3, except that the terms are multiplied by (−1).

The evaluation of the integrals across the surfaces S2 and S4 follows a similar procedure. Consequently, the integral across the surface S2 for each phase can be written as

$$\int_{S2}\left(\alpha_{s,l}\nabla\theta_{s,l} + \theta_{s,l}\vec{u}_b\right)\cdot\vec{n}\,dS = 2\pi r_w \alpha_s \frac{\alpha_{s,l}}{\alpha_s}\int_{\eta_1}^{\eta_2}\left[-\frac{\beta_{s,l}}{\Delta_{s,l}}\frac{\partial\theta_{s,l}}{\partial\eta_{s,l}} + \frac{\partial\theta_{s,l}}{\partial\xi}\right]R_{s,l}\Delta_{s,l}\,d\eta_{s,l} \tag{8.13}$$

A similar treatment is carried out in the case of S4 except that the result is multiplied by (−1). To treat the left-hand side of Equation 8.6, the volume differential element can be written as

$$dV_{s,l} = 2\pi r\,dr\,dz = 2\pi r_{s,l}\delta_{s,l} r_w\,d\eta_{s,l}\,d\xi = 2\pi r_w^3 R_{s,l}\Delta_{s,l}\,d\eta_{s,l}\,d\xi \tag{8.14}$$

and the volume integral in the transformed domain as

$$\frac{d}{dt}\left[\int_{VC}\theta_{s,l}\,dV\right] = 2\pi\alpha_s r_w Ste\frac{d}{d\tau}\left[\int_{VC}R_{s,l}\Delta_{s,l}\theta_{s,l}\,d\xi\,d\eta_{s,l}\right] \tag{8.15}$$

Consequently, the energy equation, Equation 8.6, can be written as

$$\frac{d}{d\tau}\left[\int_{VC} R_{s,l}\Delta_{s,l}\theta_{s,l}\,d\xi d\eta_{s,l}\right]Ste = \int_{\xi 1}^{\xi 2}\left(\Omega_{s,l}+\Lambda_{s,l}\right)_{S3}d\xi - \int_{\xi 1}^{\xi 2}\left(\Omega_{s,l}+\Lambda_{s,l}\right)_{S1}d\xi$$
$$+\int_{\eta 1}^{\eta 2}\left(\Gamma_{s,l}+\Psi_{s,l}\right)_{S4}d\eta_{s,l} - \int_{\eta 1}^{\eta 2}\left(\Gamma_{s,l}+\Psi_{s,l}\right)_{S2}d\eta_{s,l} \tag{8.16}$$

where

$$\Omega_s = R_s\left[-\frac{\chi_s}{\Delta_s}\frac{\partial\theta_s}{\partial\eta_s} - \eta_s\theta_s Ste\frac{\partial\Delta_s}{\partial\tau}\right];\Omega_l = R_l\left[-\frac{\alpha_l}{\alpha_s}\frac{\chi_l}{\Delta_l}\frac{\partial\theta_l}{\partial\eta_l} - \theta_l Ste\left(\frac{\partial\Delta_s}{\partial\tau}+\eta_l\frac{\partial\Delta_l}{\partial\tau}\right)\right] \tag{8.17a}$$

$$\Lambda_{s,l} = \beta_{s,l}R_{s,l}\frac{\alpha_{s,l}}{\alpha_s}\frac{\partial\theta_{s,l}}{\partial\xi} \tag{8.17b}$$

$$\Gamma_{s,l} = -R_{s,l}\Delta_{s,l}\frac{\alpha_{s,l}}{\alpha_s}\frac{\partial\theta_{s,l}}{\partial\xi} \tag{8.17c}$$

$$\Psi_{s,l} = \beta_{s,l}R_{s,l}\frac{\alpha_{s,l}}{\alpha_s}\frac{\partial\theta_{s,l}}{\partial\eta_l} \tag{8.17d}$$

Equation 8.16 is the energy integral equation to be solved for each PCM control volume in the transformed domain subject to the appropriate boundary conditions. The terms Ω_s and Ω_l in Equation 8.17a are called *pseudo-convective* because they have a transport component associated with the phase change front velocity. Their discretization is discussed in the numerical formulation. The terms Λ_s, Λ_l, Ψ_s and Ψ_l in Equations 8.17b and 8.17d represent mixed derivatives terms in the differential formulation, and are generally associated with an anisotropic behavior. Here they appear as a consequence of using a non-orthogonal coordinate system. Relating the heat conduction in the PCM across the wall of the inner tube with the convection in the interior of the cylinder does the coupling condition between the above problem and the flow problem.

The heat transfer across the wall of the cylinder can be described by

$$\rho_w c_{pw}\frac{\partial T_w}{\partial t} = \frac{1}{r}\frac{\partial}{\partial r}\left(k_w r\frac{\partial T_w}{\partial r}\right)+\frac{\partial}{\partial z}\left(k_w\frac{\partial T_w}{\partial z}\right) \tag{8.18}$$

valid in the domain $r_i \le r \le r_w; 0 \le z \le L$, which, upon substituting the dimensionless variables, can be written as

$$\frac{\alpha_s}{\alpha_w}Ste\frac{\partial\theta_w}{\partial\tau} = \frac{1}{R}\frac{\partial}{\partial R}\left(R\frac{\partial\theta_w}{\partial R}\right)+\frac{\partial}{\partial\xi}\left(\frac{\partial\theta_w}{\partial\xi}\right) \tag{8.19}$$

The coupling with Equation 8.16 is realized by equalizing the heat fluxes in the faces of the control volume in the phase change material and the wall of the tube.

When performing an energy balance in the tube, the equation for the fluid is obtained as

$$\frac{1}{\alpha_f}\left(\frac{\partial T_f}{\partial t}+U_m\frac{\partial T_f}{\partial z}\right) = \frac{\partial^2 T_f}{\partial z^2}+\frac{4h}{Dk_f}\left(T_{wi}-T_f\right) \tag{8.20}$$

where T_f is the bulk temperature, U_m is the mean fluid velocity, h is the local convection coefficient and T_{wi} the inner wall temperature. In terms of the new variables, Equation 8.20 can be written as

$$\frac{\alpha_s}{\alpha_f} Ste \frac{\partial \theta_f}{\partial \tau} + \mathrm{Re}\,\mathrm{Pr}\left(\frac{r_w}{D}\right)\frac{\partial \theta_f}{\partial \xi} = \frac{\partial^2 \theta_f}{\partial \xi^2} + 4\left(\frac{r_w}{D}\right)^2 Nu\left(\theta_{wi} - \theta_f\right) \tag{8.21}$$

In terms of the Biot and Stanton numbers and neglecting axial conduction, Equation 8.20 can be written as

$$\frac{(\rho c_p)_f}{(\rho c_p)_s}\frac{Ste}{Bi}\frac{\partial \theta_f}{\partial \tau} + \frac{1}{St}\frac{\partial \theta_f}{\partial \xi} = 2\left(\theta_{wi} - \theta_f\right) \tag{8.22}$$

The end caps of the system and the external cylinder wall are considered thermally insulated. This condition is equally extended to the phase-change material solid and liquid, as well as the wall of the internal tube adjacent to the end caps and the external surface. The boundary conditions expressed in terms of the new variables can be written as

$$\left.\frac{\partial \theta_w}{\partial \xi}\right|_{\xi=0,L/r_w} = 0\,;\quad \left.\frac{\partial \theta_{s,l}}{\partial \xi}\right|_{\xi=0,L/r_w} = 0\,;\quad \left.\frac{\partial \theta_{s,l}}{\partial \eta_{s,l}}\right|_{\eta_{s,l}=1} = 0 \tag{8.23}$$

If the tube wall is neglected, the boundary condition between the refrigeration fluid and the solid phase can be written as

$$\frac{hr_i}{k_s}\left(\theta_{wi} - \theta_f\right) = \frac{1}{\Delta_s}\left.\frac{\partial \theta_s}{\partial \eta_s}\right|_{\eta_s=0} \quad \text{or} \quad Bi\left(\theta_{wi} - \theta_f\right) = \frac{1}{\Delta_s}\left.\frac{\partial \theta_s}{\partial \eta_s}\right|_{\eta_s=0} \tag{8.24}$$

The boundary conditions involving the inner tube wall are

$$\frac{k_w}{k_s}\left.\frac{\partial \theta_w}{\partial R}\right|_{R=1} = \frac{1}{\Delta_s}\left.\frac{\partial \theta_s}{\partial \eta_s}\right|_{\eta_s=0} \tag{8.25}$$

and

$$\frac{k_w}{k_f}\left.\frac{\partial \theta_w}{\partial R}\right|_{R=r_i/r_w} = Nu\left(\frac{r_w}{D}\right)\left(\theta_{wi} - \theta_f\right) \tag{8.26}$$

The boundary conditions at the solid-liquid interface can be expressed in the form

$$\theta_s(\eta_s = 1, \xi, \tau) = \theta_l(\eta_l = 0, \xi, \tau) = 0 \tag{8.27}$$

while the energy balance at the interface can be expressed as

$$\left[1 + \left(\frac{\partial \Delta_s}{\partial \xi}\right)^2\right]\left[\frac{1}{\Delta_s}\left.\frac{\partial \theta_s}{\partial \eta_s}\right|_{\eta_s=1} - \frac{k_l}{k_s}\frac{1}{\Delta_l}\left.\frac{\partial \theta_l}{\partial \eta_l}\right|_{\eta_l=0}\right] = \frac{\partial \Delta_s}{\partial \tau} \tag{8.28}$$

The last equation is used to determine the position of the solidification front using the temperature distribution obtained from the solution of the energy equation in the phase change material. At this stage we have the basic equations and the associated boundary conditions to solve the system of equations which permit determination of the temperature distribution in each region, as well as the position of the solidification front in terms of time. The numerical solution of the equations was performed using the implicit-explicit formulation as suggested by Sparrow and Chuck (1984). The temperature equations were discretized in the implicit form, while the thermal energy balance condition at the interface, Equation 8.28, was discretized explicitly. This procedure avoids the need for iterations between the temperature distribution and the interface location values. The discretization of the PCM equations takes into account the existence of pseudo-convective terms, Equation 8.17a, and pseudo-anisotropic terms, Equations 8.17b and 8.17d. In treating the pseudo-convective terms, central-differences and the power-law scheme were used in a formulation similar to those presented in Hsu *et al.* (1981). The pseudo-anisotropic terms were evaluated explicitly and updated during the iterative solution of the linear system. Singularities in the beginning of the process were avoided by starting the method with a semi-analytical solution based on quasi-steady approximation.

Numerical experiments were performed to check the influence of the grid size and time step in the results. In the comparisons presented below, the grid adopted had, in the radial direction, 1 control volume for the refrigerant fluid, 1 control volume for the tube wall, 10 control volumes for the solid phase and 10 control volumes for the liquid phase. In the axial direction there were 50 control volumes for each region. The dimensionless time step used was 0.01 or 0.001, depending on the time for complete solidification of the system.

Figure 8.4 shows a comparison between the results due to Sparrow and Hsu (1981) and the present predictions using the power-law scheme and the central-differences discretization of the pseudo-convective terms. As can be seen, the agreement is good. Bellecci and Conti (1993) and Zhang and Faghri (1996) solved problems similar to what is being considered here, but using different methodologies for treating the phase-change problem. In their works the discussion about the evaluation of the Nusselt number in the heat transfer fluid was raised. Bellecci and Conti (1993) used a steady-state correlation for the constant wall heat flux condition, while Zhang and Faghri (1996) argued that this procedure would lead to errors because the phase-change problem is an intrinsically transient one, especially for fluids with moderate Prandtl numbers. They solved the problem of fusion of paraffin, initially at the phase-change temperature and in annular space, by an approximate analytical method (the integral method). The fluid flow was assumed laminar and the Nusselt number calculated by an iterative method considering the variation of the tube wall temperature. Figure 8.5 shows results for the evolution of the phase change front using both correlations of constant wall temperature and constant wall heat flux for determining the Nusselt number, and comparing with the results of Zhang and Faghri (1996). The comparison shows that differences in the interface predictions are small for the three methods, and the results due to Zhang and Faghri (1996) seem to occupy an intermediate position between the steady-state correlation's results. Based on Figure 8.5, one can conclude that it appears to be possible to use a simplified approach to evaluate the Nusselt number in the heat transfer fluid.

One experimental study available was performed by Sinha and Gupta (1982), where they studied the solidification of water surrounding a horizontal copper cylinder cooled by the flow of refrigerant inside the tube to keep it in an isothermal condition. The comparison with their experimental results is shown in Figure 8.6. In this experiment the initial water temperature is 25°C and the surface temperature of the inner tube –14.7°C. The solidification process started at about $\tau = 0.185$, approximately 25 minutes; this instant was assumed to be $\tau = 0$ in the numerical analysis. This late start of solidification in their experiments is due to the fact that the initial water temperature was high.

Sablani *et al.* (1990) solved the problem of complete solidification in annular space by the immobilization technique using a fully implicit finite-difference scheme. They developed a correlation of the dimensionless time for the complete solidification in terms of the Biot number and the geometry of the annular space. Their correlation is

$$\tau_c = 1.598 Bi^{-0.639} \left(L / r_w \right)^{0.153} E^{1.533}. \tag{8.29}$$

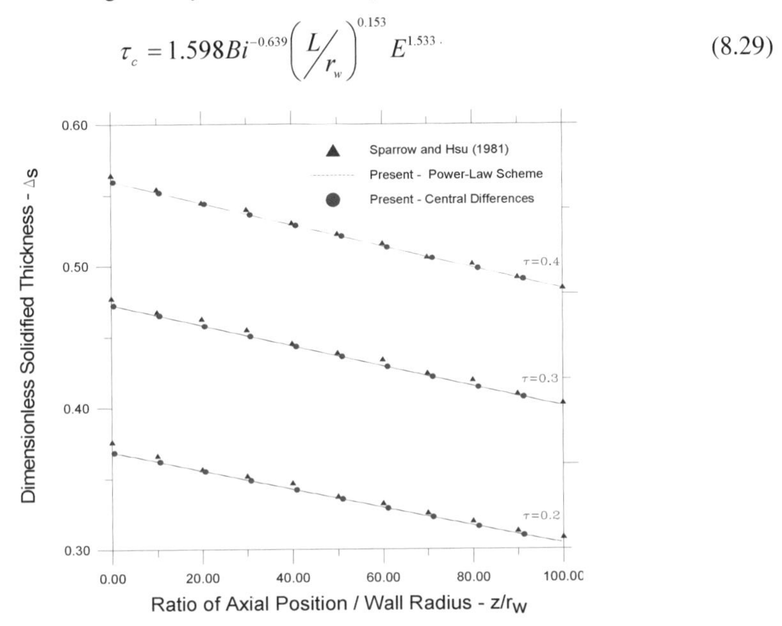

Figure 8.4 Comparison of different discretization schemes and with the results of Sparrow and Hsu (1981). (Ste = 1.0; St = 0.003; Bi = 5.0; L/r_w = 100).

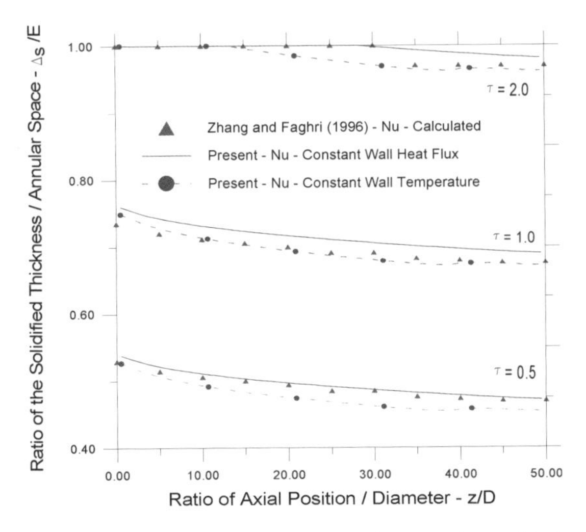

Figure 8.5 Demonstration of the Nusselt number effect and comparison with the results of Zhang and Faghri (1996). (Ste = 0.5; Pe = 4000; L/D = 50; r_w/D = 0.575; E = 1.304).

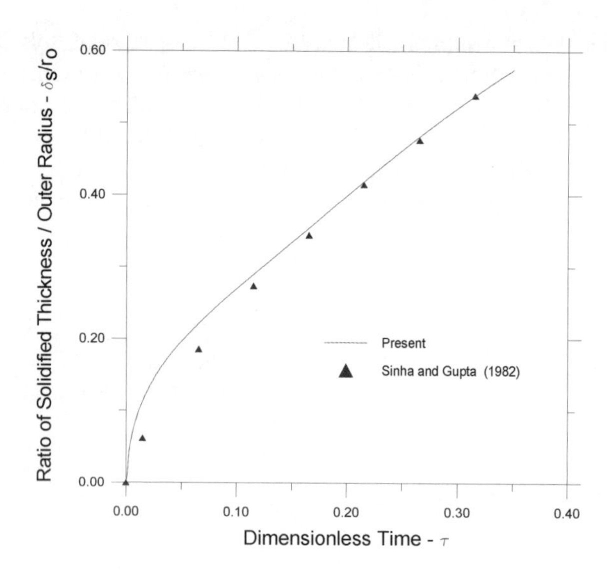

Figure 8.6 Comparisons of the present predictions with the experimental results of Sinha and Gupta (1982). (Ste = 0.09; L/D = 80; E = 3.08).

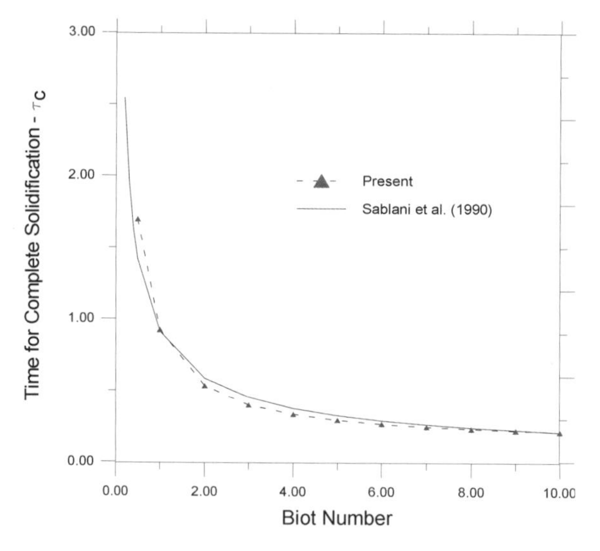

Figure 8.7 Comparisons with the correlation due to Sablani *et al.* (1990). (Ste = 0.1; θ_{in}= 0.5; St = 0.003; r_w/D = 0.5; L/r_w = 60; E = 0.46).

Figure 8.7 compares their results with the present numerical predictions under the same conditions. As can be seen, the time for the complete solidification seems to agree well.

Cao and Faghri (1991) solved the problem of fusion of paraffin around a cylinder inside which flows a working fluid of small Prandtl number (liquid metal). The equations of movement and energy were solved simultaneously with the heat conduction equation for the paraffin and tube wall. The paraffin was initially at a temperature higher than the fusion temperature. In their study they used the enthalpy method, in contrast to the present study where we used the immobilization technique. The results obtained are presented in Figures 8.8 and 8.9. Figure 8.8 shows a good agreement between their results and the present predictions. The differences in the fluid temperature values occur because they evaluate local temperature while this work calculates bulk temperature. Figure 8.9 shows the comparative results for the interface positions. One can observe a relatively big difference in the initial stages, due to the fact that the immobilization starts with an infinitesimally small layer solidified, while this artifice is not necessary in case of the enthalpy method. In the subsequent time the results seem to be in good agreement.

As a demonstration one can consider the case of a system of latent heat storage consisting of an array of tubes immersed in PCM initially solid or liquid. The tubes carry a heat transfer fluid inside, which causes the phase change, melting or solidification, depending on the fluid and PCM conditions. In general in defining the layout of the tubes in the storage tank, care is taken to allow for the simultaneous phase-change process around all the tubes. Generally, a radial distance is allowed to permit an efficient solidification/fusion process without overlapping of the interfaces. This limiting radius can be considered as a symmetry surface beyond which there is no effective change in the interface position. The existence of a symmetry circle around the tubes is the main reason for the use of the single-tube model with the PCM in the annular region between the tube and an isolated external surface. The radius of the external surface is the symmetry radius of the tube array. The case study under consideration has the following dimensions: tube external radius: $r_w = 0.0127$ m, tube wall thickness: $\delta r = 0.0015$ m, tube inner diameter: $D = 0.0224$ m, and tube length: $L = 1$ m. The PCM chosen was water, since ice-bank applications and low-temperature thermal energy storage are the main concerns. The heat transfer fluid was a mixture 70% water and 30% ethylene glycol, commonly used as a secondary refrigerant in ice-banks.

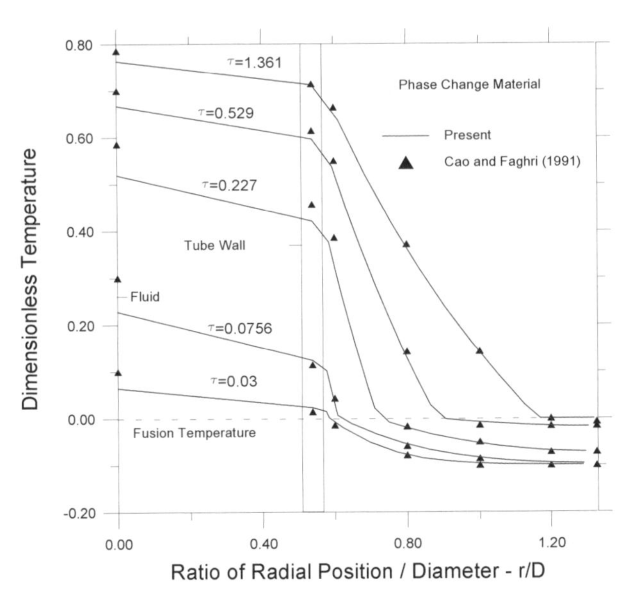

Figure 8.8 Temperature distribution. Comparison with the results of Cao and Faghri (1991).

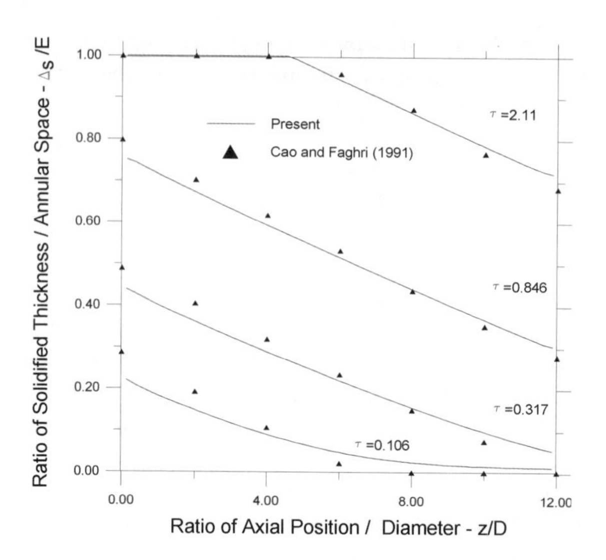

Figure 8.9 Interface position. Comparison with the results of Cao and Faghri (1991).

The initial temperature of the liquid phase was assumed at 5°C, in order to minimize natural convection effects. The fluid entry temperature was varied in a range from –20°C to –5°C. The symmetry radius of the system was not initially specified because one of the desired results is the time for complete solidification of the system as a function of the annular space and the fluid inlet temperature.

The definition of an adequate value for the annular space thickness was done by investigating the effect of the increase of the external radius on the time for complete solidification. The effect of the annular spacing is shown in Figure 8.10 for an initial liquid temperature of 5°C and different fluid entry temperatures varying from –5°C to –20°C. Figure 8.11 shows similar results as those of Figure 8.10, but with the time for complete solidification expressed as a function of the fluid entry temperature, for different values of the annular space thickness. From both figures it is possible to notice the strong influence of increasing the external radius of the symmetry circle and also the dominant effect of the refrigerant entry temperature on the time for complete solidification. This is basically caused by the continuous increase in the thermal resistance of the formed solid phase of low thermal conductivity, as its thickness grows. In order to define some values for the present analysis we considered the case of ice-bank application and night coolness generation with a solidification time of eight hours, hence we found a value of $T_{in} = -10$°C and $E = 3.0$ or symmetry circle external radius of $r_o = 0.0508$ m.

Having all the dimensions of the system defined, one can present the results of the parametric study. Figure 8.12 shows the effect of varying the initial temperature of the liquid phase on the thickness distribution of the solidified mass along the axial direction. One can observe that for an initial temperature of the liquid nearer to the phase change temperature, the evolution of thickness of the solidified mass is a little faster.

Figure 8.13 shows the effect of the material of the inner tube on the PCM solidification front growth. It is found that for metallic conducting materials, with thermal-conductivity of the same order, the differences are relatively small. When compared to the PVC results the differences are significant, because the thermal resistance of the tube wall is much higher and even greater than that of the solid phase. This result seems to be important in the design of ice storage systems usually fabricated using plastic tubing for their reasonable cost and ease of the manufacturing process.

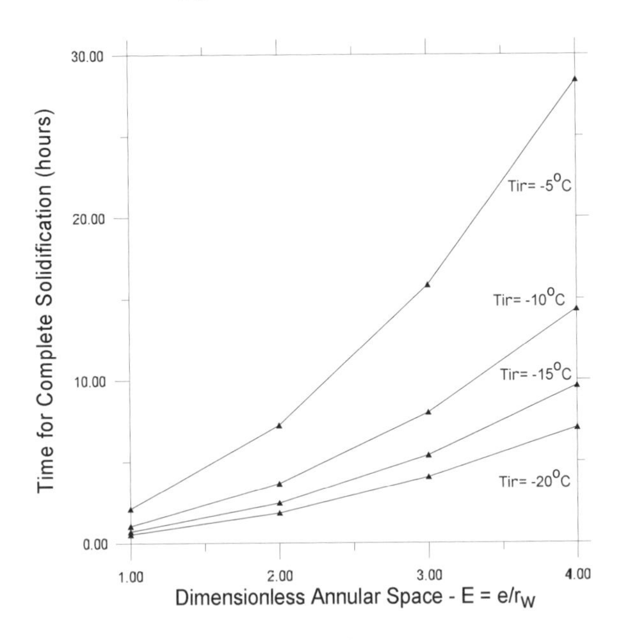

Figure 8.10 Time for complete solidification vs. annular space.

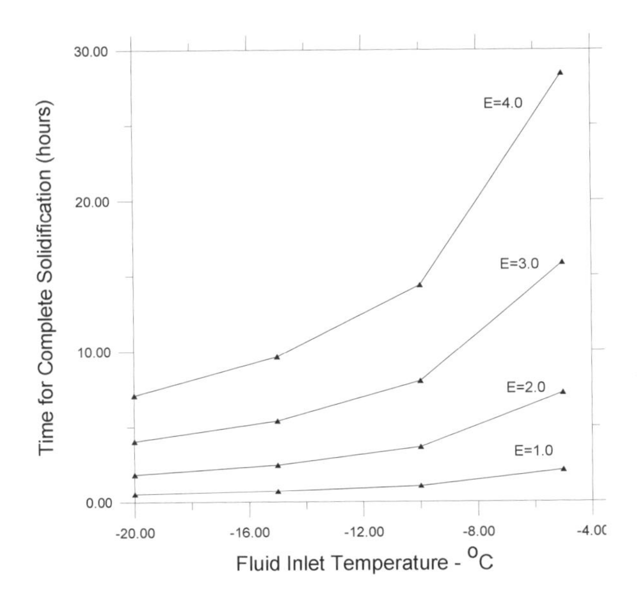

Figure 8.11 Time for complete solidification vs. inlet fluid temperature.

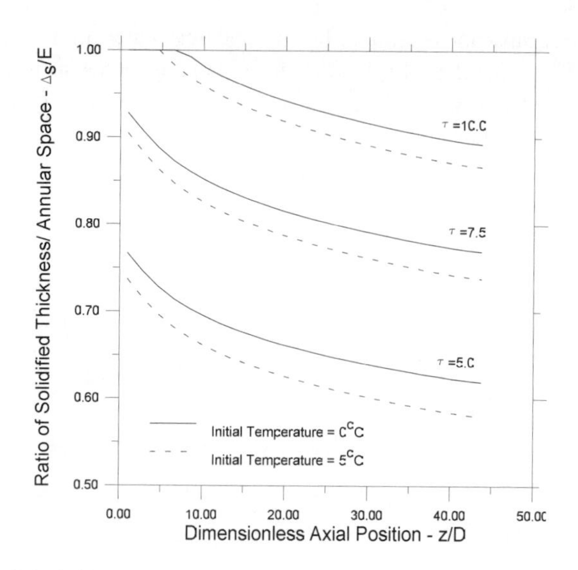

Figure 8.12 Effect of the initial liquid temperature on the solidification front evolution.

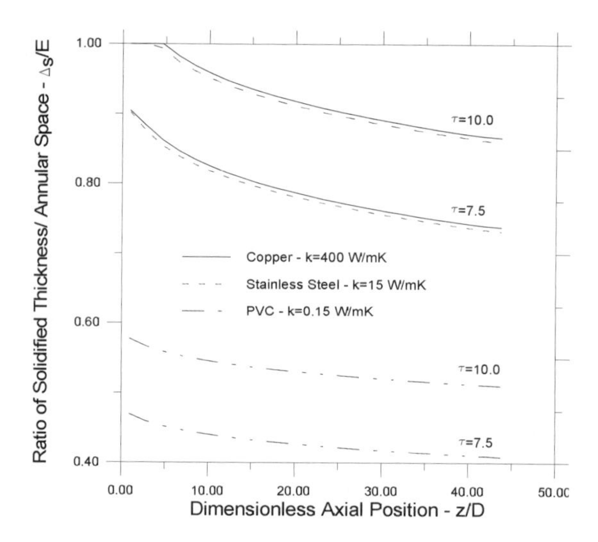

Figure 8.13 Effect of tube wall thermal conductivity on the solidification front evolution.

8.2.1 Concluding remarks

The method of immobilization was presented and applied to the solution of the case of PCM phase change in an annular space representing the main section of a latent heat thermal energy storage system. As can be seen, the method is straightforward and produces precise results, as confirmed by the different comparisons realized to validate the method. As a demonstration the method was used to investigate some of the important geometrical parameters usually calculated in latent heat thermal storage systems.

8.3 Solidification of PCM inside a Spherical Capsule

The process of solidification and fusion in spherical geometries is simple if formulated as a one-dimensional heat transfer problem with phase change, although then the model and results do not represent the real situation and big differences can be found due to the presence of natural convection. Here we present a simple model based upon pure conduction inside a spherical shell. Details on other models can be found in Ismail (1998, 1999).

Tao (1967) presented a method for the analysis of solidification in cylindrical and spherical geometries. The method analyzes the solidification process in cylinders and spheres considering an average heat transfer coefficient between the external surface of the solidified mass and the cooling fluid. He used a fixed grid approach to solve the mathematical models developed. Megerlin (1968) obtained a series solution for the heat conduction equation with phase change in spherical geometry. He adopted a polynomial approximation for the temperature profile. Cho and Sunderland (1974), Shih and Chou (1971), Pedroso and Domoto (1973), Riley *et al.* (1974), Kern and Wells (1977), Moore and Bayazitoglu (1982), Hill and Kucera (1983), Milanez and Ismail (1984), Prud'homme *et al.* (1989), Gillessen *et al.* (1988), Gibbs and Hasnain (1995), Feng *et al.* (1996) and Folio and Lacour (1996), among many others, reported the results of different numerical and experimental studies on the process of phase change within spherical shells.

Consider a sphere of external radius r_e initially filled with liquid PCM at temperature T_m, which is the phase change temperature. At time $t = 0$ the external surface of the sphere is subject to convective cooling, while the ambient temperature is kept constant at $T = T_\infty$. Considering a pure conduction model for the solidification process, one can write the governing equations with the respective boundary and initial conditions as:

$$\frac{\partial T}{\partial t} = \frac{k}{\rho c_{PS}}\left[\frac{2}{r}\frac{\partial T}{\partial r} + \frac{\partial^2 T}{\partial r^2}\right] \qquad r_m < r < r_e \tag{8.30}$$

$$k\frac{\partial T}{\partial r} = \rho\lambda\frac{ds}{dt} \qquad \text{at } r = r_m \tag{8.31}$$

$$-k\frac{\partial T}{\partial r} = h(T - T_\infty) \qquad \text{at } r = r_e \tag{8.32}$$

$$T = T_m \qquad \text{at } 0 \le r \le r_m \tag{8.33}$$

In order to facilitate the numerical calculations we adopt the following dimensionless variables:

$$R^* = 1 - \frac{r}{r_e}, \quad S = 1 - \frac{r_m}{r_e}, \quad \tau = \frac{\alpha t}{r_e^2} \tag{8.34}$$

$$\theta = \frac{T - T_\infty}{T_m - T_\infty}, \quad Ste = \frac{c_{PS}(T_m - T_\infty)}{\lambda}, \quad Bi = \frac{hr_e}{k} \tag{8.35}$$

In terms of the new dimensionless variables, the model equations can be written as:

$$\frac{\partial\theta}{\partial t} = \frac{2}{R^* - 1}\frac{\partial\theta}{\partial R^*} + \frac{\partial^2\theta}{\partial R^{*2}}, \quad 0 < R^* < S \tag{8.36}$$

$$Ste\frac{\partial\theta}{\partial R^*} = \frac{dS}{d\tau}, \quad R^* = S \tag{8.37}$$

$$\frac{\partial\theta}{\partial R^*} = Bi\theta, \quad R^* = 0 \tag{8.38}$$

$$\theta = 1 \qquad \begin{cases} \text{for} \quad \tau = 0 \quad \text{at} \quad 0 < R^* < 1 \\ \text{for} \quad \tau > 0 \quad \text{at} \quad S < R^* < 1 \end{cases} \tag{8.39}$$

The numerical solution is realized using the finite difference approximation and the moving grid approach, where the point $n = 0$ at $R^* = 0$ and the point $n = N$ is at $R^* = S$.

The stability condition adopted is

$$\frac{\Delta\tau N^2}{R_N^{*2}} \le \frac{1}{2} \tag{8.40}$$

In the numerical procedure the initial layer of the solidified mass was calculated using the simplified model due to London and Seban (1943). Several numerical attempts were realized to establish the best grid for the space and time variables. It is found that those 21 grid points in the radial direction while the time step were always calculated according to the stability criterion.

In order to validate the numerical code, comparisons were realized with available results. Figure 8.14 shows a comparison between the present model and the results of Shih and Chou (1971) and Hill and Kucera (1983). As can be seen, the present prediction lies between the two results.

Pedroso and Domoto (1973) used the perturbation method to obtain a solution for the case of constant temperature on the external surface. They obtained a relation for the interface position in terms of time. London and Seban (1943) obtained a solution for the solidification of spherical mass using the method of quasi-steady state approximation with a convective boundary condition on the external surface. These two models are compared with the present predictions under the same conditions. The results are shown in Figure 8.15. As can be noticed, the present prediction shows better agreement with the results of London and Seban (1943). Another comparison with London and Seban (1943) for the case of heat convection on the external surface of the sphere is shown in Figure 8.16. As can be verified, the agreement is good during the initial period up to four hours, and then deviates for longer time intervals.

Having validated the model and the numerical results, one can study the parameters of importance such as the sphere radius, the thickness of the spherical shell, the material of the sphere and the temperature of the external surface, as well as the Biot number, and their influence on the time for complete solidification and solidified mass fraction. These parameters are extremely important in the design and analysis of thermal storage systems with spherical capsules containing the PCM. In the present analysis, the solidified mass fraction is defined as

$$M_{fr} = \frac{M_{Solid}}{M_{Total}} \tag{8.41}$$

The effect of the material of the spherical capsule on the solidified mass fraction is shown in Figure 8.17. The materials tested are copper, stainless steel, Teflon and PVC. In the case of the metallic materials, the solidification velocities are coincidence. Comparing the results for Teflon and PVC, Teflon is found to show better performance than PVC. Teflon and PVC are promising candidates for their low cost and ease of manufacturing.

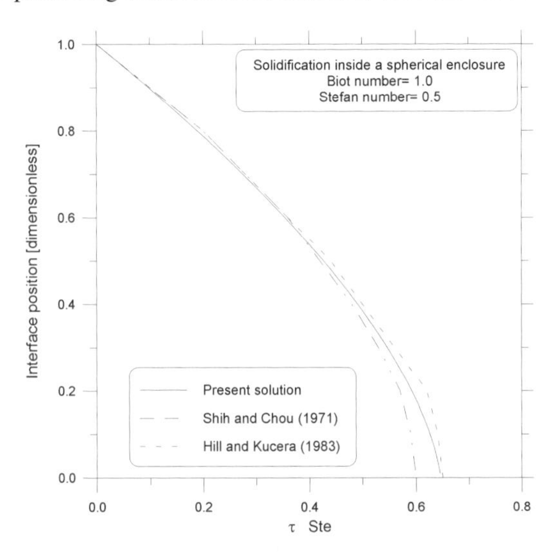

Figure 8.14 Comparison of the present solution with the results due to Shih and Chou (1971) and Hill and Kucera (1983).

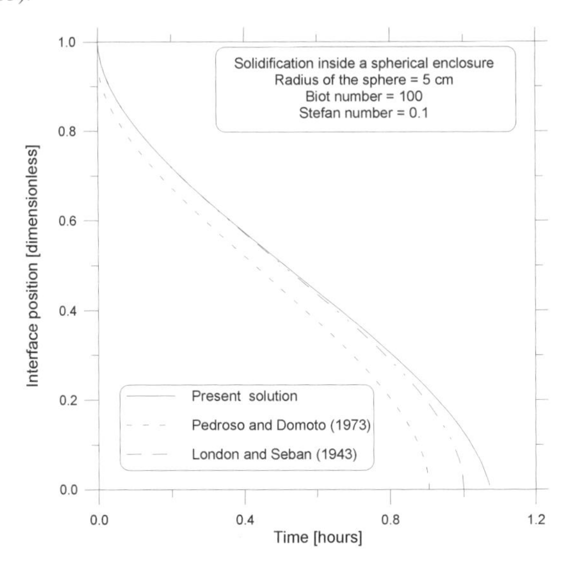

Figure 8.15 Comparison of the present solution with the results due to Pedroso and Domoto (1973) and London and Seban (1943).

Figure 8.18 shows the effect of the sphere size on the time for complete solidification. The simulated capsules have a diameter of 20, 40, 100, 140, 200, 300 and 400 mm. A general correlation for the effect of sphere size on the time for complete solidification for different initial working fluid temperature is

$$t_c = m r_e^{2.036} \tag{8.42}$$

where

$$m = \begin{cases} 4259.41 & \text{for} \quad T_\infty = -5\ ^\circ C \\ 2174.52 & \text{for} \quad T_\infty = -10\ ^\circ C \\ 1127.27 & \text{for} \quad T_\infty = -20\ ^\circ C \end{cases} \tag{8.43}$$

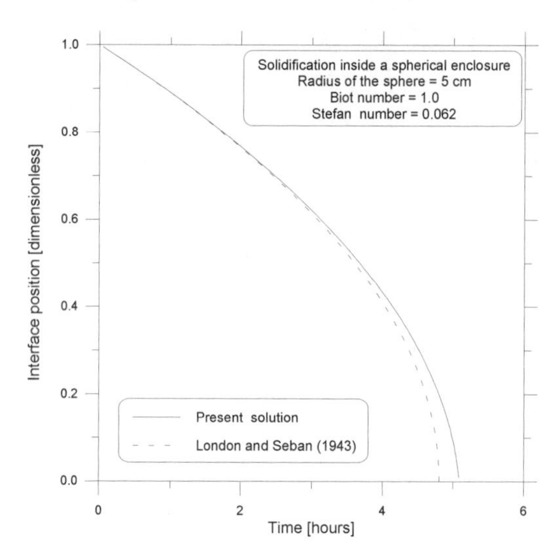

Figure 8.16 Comparison of present solution with the results due to London and Seban (1943).

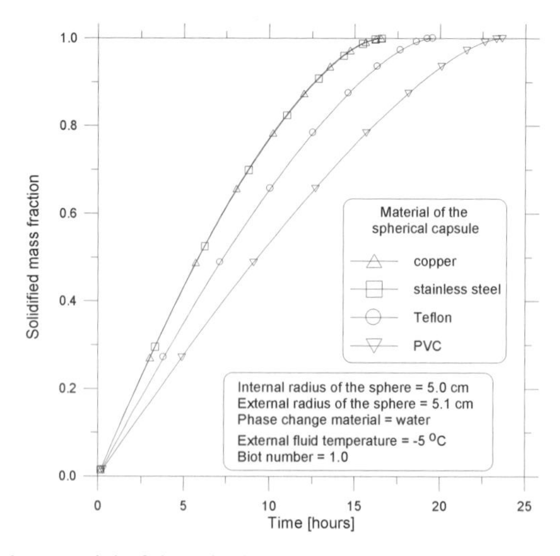

Figure 8.17 Effect of the material of the spherical capsule on the solidified mass fraction.

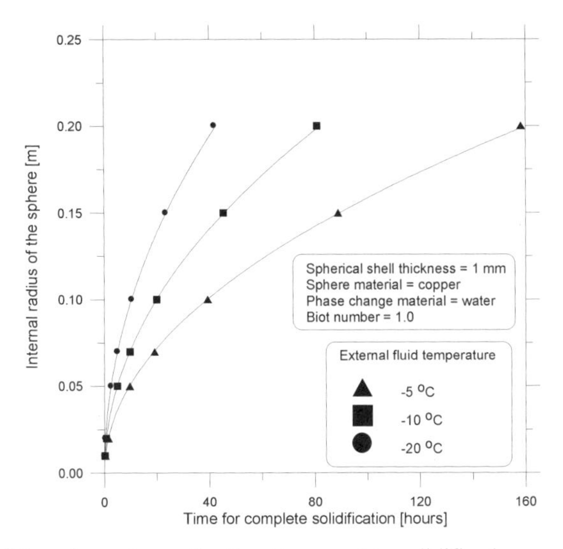

Figure 8.18 Effect of the sphere size on the time for complete solidification.

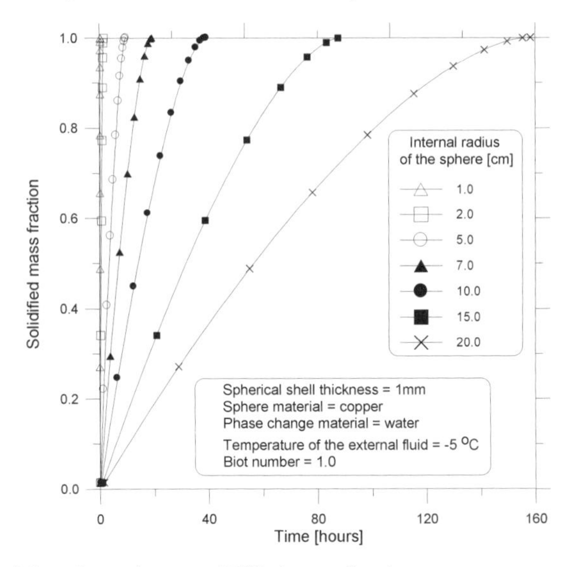

Figure 8.19 Effect of the sphere size on solidified mass fraction.

As can be seen, low surface temperatures reduce the time for complete solidification. Also one can observe that the increase of the size of the sphere leads to increasing the time for complete solidification, as can be verified in Figures 8.18 and 8.19.

Figure 8.20 shows the effect of varying the temperature of the external fluid. As can verified, the decrease of the external fluid temperature leads to the reduction of the time for the complete solidification. Figure 8.21 shows the effect of varying this temperature on the solidified mass fraction. A numerical correlation relating the working fluid temperature to the time for the complete solidification is

$$t_c = 1.52977 \cdot 10^{-4}\, T_\infty^4 - 0.156049 T_\infty^3 + 59.6937988 T_\infty^2 - 10148.7 T_\infty + 647014 \qquad (8.44)$$

The effect of varying the Biot number is shown in Figures 8.22 and 8.23. As can be seen, the increase of the Biot number leads to a reduction of the time for complete solidification. Figure 8.23 shows the effect of the Biot number on the solidified mass fraction. Again, a numerical correlation relating the time for the complete solidification to the Biot number is

$$t_c = 13.526\,Bi^{-0.5685} \tag{8.45}$$

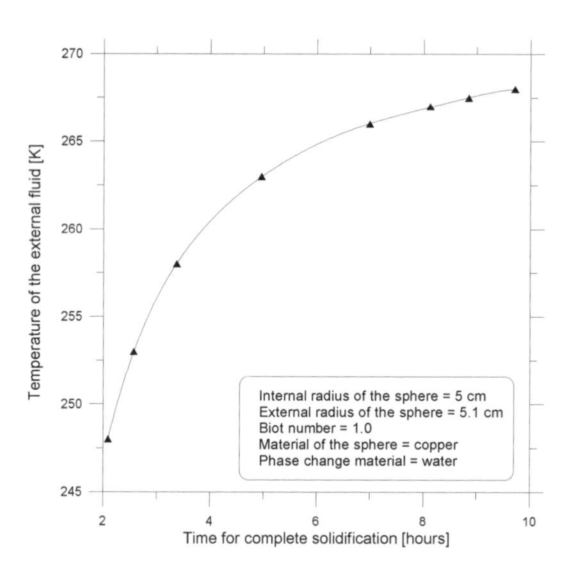

Figure 8.20 Effect of the external fluid temperature on the time for complete solidification.

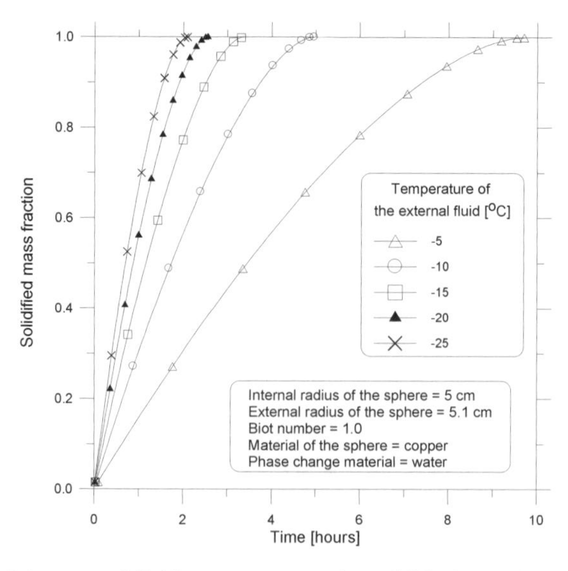

Figure 8.21 Effect of the external fluid temperature on the solidified mass fraction.

8.3.1 Concluding remarks

Phase change heat transfer within spherical capsules was treated by using a one-dimensional heat conduction model in which the natural convection effects were ignored in order to obtain a straight forward model easy to handle and implement as a subroutine for numerical calculations of storage systems. The model was validated by comparisons with available results. Parameters of interest in the design and calculation of latent heat storage systems such as the diameter of the spherical shell, material of the shell, and the Biot number were evaluated and discussed.

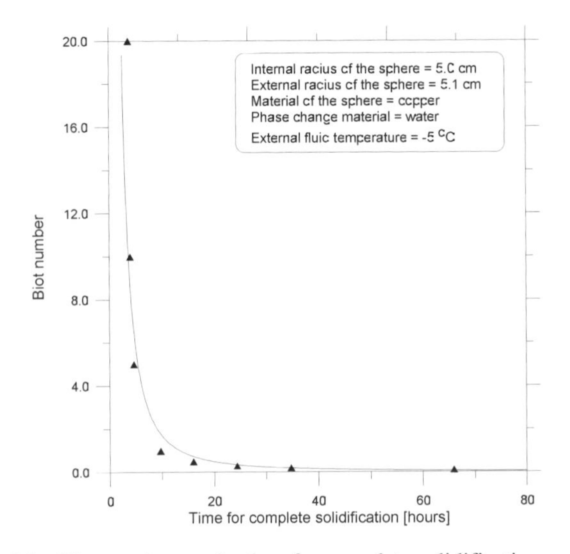

Figure 8.22 Effect of the Biot number on the time for complete solidification.

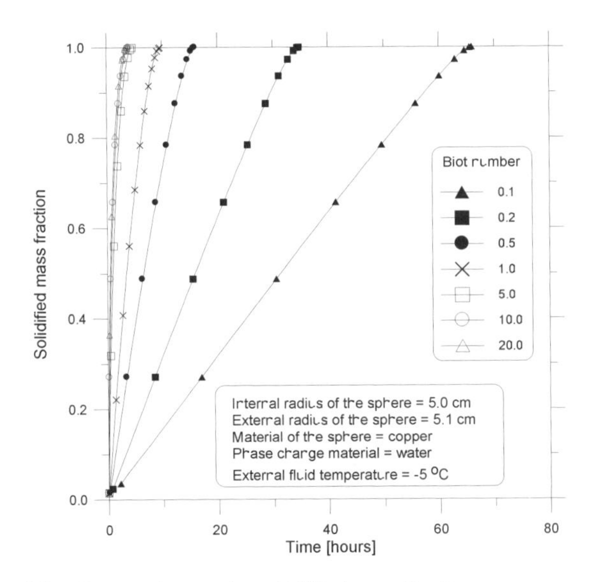

Figure 8.23 Effect of the Biot number on the solidified mass fraction.

8.4 Plane Finned Geometries For Latent Heat Storage Applications

A series of numerical and experimental studies were realized on parallel plate finned and finless geometries with the objective of developing thermal models to use in energy storage systems, and particularly ice storage systems for refrigeration and air conditioning applications. The first model is based upon pure conduction formulation and solved numerically by a fixed grid scheme. The results are omitted here for brevity, and more details can be found in Ismail (1999). The second model allows for natural convection in the PCM liquid phase, and was formulated and solved by the finite element approach. The classical problem of convection in a cavity was solved and compared with available data to establish the validity of the model. Further, the problem of a rectangular cavity with internal fins filled with PCM initially at a temperature different from the phase change material was solved using the same numerical approach. Traditionally, a great deal of complicated combined thermal fluid flow problems is solved by the control volume method based upon the finite difference approximation. The use of the co-localized (or equal order methods) changed this situation, and more frequently, methods based upon finite element approaches are preferred by research workers in the heat and fluid flow area.

One of the pioneer studies in this area is the work of Marshall *et al.* (1978) in which a formulation based upon a 'penalty function' was used to obtain a solution for a convective problem in a square cavity and presented results for different Rayleigh numbers (10^4, 10^5, 10^6 and 10^7).

Different possible solutions suitable for handling fluid movement by using co-localized grids were compiled and presented by Schneider *et al.* (1978). They examined and incorporated into their work solutions of several test problems using the penalty function method, the false compressibility method and the velocity correction procedure. They concluded that the method of correcting the velocity rendered the best results. The most common methods can be found in Gresho (1990) and Gresho and Chan (1990). A complete analysis of the alternative methods can be found in Saabas and Baliga (1994).

De Vahl Davis and Jones (1983) presented results of numerous methods of solution applied to a test problem of free convection in a cavity for different Rayleigh numbers. Later de Vahl Davis (1983) presented a solution for the proposed free convection problem. He obtained results for different grids extrapolated in such a way to obtain a solution with minimum possible error.

Here we present a solution of the free convection problem with phase change based on extending the finite element model proposed by Rice and Schnipke (1986). Their model was extended by including the necessary terms for the treatment of free convection flows and their correct manipulation. An advantage of the present method is the use of primitive variables which facilitates the application of the boundary conditions, and also the utilization of the false transient scheme which permits better convergence levels. The results are analyzed and compared with de Vahl Davis's model (1983) to validate the present method of solution. After validating the model and the numerical scheme of solution, a model for a cavity filled with PCM is solved by the same numerical procedure, the results were compared with other available numerical and approximate solutions and good agreement was achieved. Having validated the model for phase change test problems used in the comparison a cavity having internal fins filled with PCM initially at a temperature different from the phase change temperature and subject a variety of boundary conditions was formulated, solved numerically, and different operational and geometrical conditions were investigated and discussed.

8.4.1 Formulation of the Problem

With the reference to Figure 8.24, one can write the two-dimensional Navier Stokes equation for the incompressible fluid with the energy equation in the form

$$\left.\begin{aligned}
\rho\frac{\partial u}{\partial t} + \rho u\frac{\partial u}{\partial x} + \rho v\frac{\partial u}{\partial y} &= -\frac{\partial p}{\partial x} + \mu\left(\frac{\partial^2 u}{\partial x^2} + \frac{\partial^2 u}{\partial y^2}\right) + S_x \\
\rho\frac{\partial v}{\partial t} + \rho u\frac{\partial v}{\partial x} + \rho v\frac{\partial v}{\partial y} &= -\frac{\partial p}{\partial y} + \mu\left(\frac{\partial^2 v}{\partial x^2} + \frac{\partial^2 v}{\partial y^2}\right) + S_y \\
\rho c_p\frac{\partial T}{\partial t} + \rho u c_p\frac{\partial T}{\partial x} + \rho v c_p\frac{\partial T}{\partial y} &= k\left(\frac{\partial^2 T}{\partial x^2} + \frac{\partial^2 T}{\partial y^2}\right) + S_e
\end{aligned}\right\} \tag{8.46}$$

where the source term S of the momentum equation includes the gravitational effects. The continuity equation can be written as

$$\frac{\partial u}{\partial x} + \frac{\partial v}{\partial y} = 0 \tag{8.47}$$

Notice that the momentum equations are identical to the energy equations, except for the pressure term, which can be included in the source term. In this manner, applying the weighted residuals formulation to the weak form of the equation, the resulting linear system can be presented by

$$\left.\begin{aligned}
\left(\frac{1}{\Delta t}[M] + [C] + v[K]\right)\{u\} &= \frac{1}{\Delta t}[M]\{u^0\} + \frac{1}{\rho}\{S_x\} - \frac{1}{\rho}[M]\frac{\partial p}{\partial x} + v\int_\Gamma N_i\frac{\partial u}{\partial n}d\Gamma \\
\left(\frac{1}{\Delta t}[M] + [C] + v[K]\right)\{v\} &= \frac{1}{\Delta t}[M]\{v^0\} + \frac{1}{\rho}\{S_y\} - \frac{1}{\rho}[M]\frac{\partial p}{\partial y} + v\int_\Gamma N_i\frac{\partial v}{\partial n}d\Gamma \\
\left(\frac{1}{\Delta t}[M] + [C] + \alpha[K]\right)\{T\} &= \frac{1}{\Delta t}[M]\{T^0\} + \frac{1}{\rho c}\{S_e\} + \alpha\int_\Gamma N_i\frac{\partial T}{\partial n}d\Gamma
\end{aligned}\right\} \tag{8.48}$$

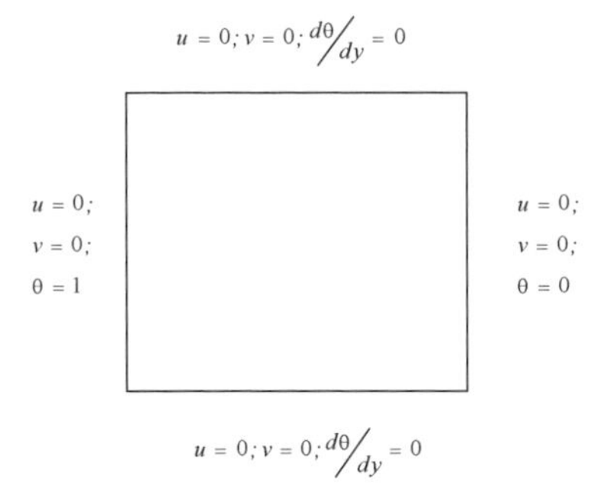

Figure 8.24 Layout of the problem.

where the values of the matrices [M], [C], [K] and {S} are given by

$$[M] = \int_{\Omega} N_i \cdot d\Omega \tag{8.49}$$

$$[C] = \int_{\Omega} \left(N_k u_k N_i \frac{\partial N_j}{\partial x} + N_k v_k N_i \frac{\partial N_j}{\partial y} \right) \cdot d\Omega \tag{8.50}$$

$$[K] = \int_{\Omega} \left(\frac{\partial N_i}{\partial x} \frac{\partial N_j}{\partial x} + \frac{\partial N_i}{\partial y} \frac{\partial N_j}{\partial y} \right) \cdot d\Omega \tag{8.51}$$

$$\{S\} = \int_{\Omega} N_i \cdot d\Omega \cdot \vec{S} \tag{8.52}$$

If the indicated operations are performed on the matrices, one can obtain a linear system of the form

$$[\overline{A}_{\phi}] \cdot \{\phi\} = \{\overline{B}_{\phi}\} \tag{8.53}$$

where [$\overline{A}_{\phi}$] and { $\overline{B}_{\phi}$ } are identified with the respective variables: *u*, *v* or *T*. In this solution the temperature field is the first to be solved, since it is necessary to determine the values of the source terms of momentum equations. This system of equations will be used later in the discretization of the continuity equation.

In the present analysis a better approximation of the pressure field is adopted and the expression, which appears in the general equation, can be approximated by

$$\int_{\Omega} W \frac{\partial p}{\partial z} \cdot d\Omega = \int_{\Omega} W \frac{\partial N_j}{\partial z} \cdot d\Omega \cdot p_j = \int_{\Omega} N_i \frac{\partial N_j}{\partial z} \cdot d\Omega \cdot p_j$$

where z is an arbitrary direction. In this manner a new matrix can be defined as

$$[P'_z] = \int_{\Omega} N_i \frac{\partial N_j}{\partial z} \cdot d\Omega \tag{8.54}$$

Hence, the general two-dimensional expression for the momentum equation can be written as

$$\left(\frac{1}{\Delta t}[M] + [C] + \nu[K] \right)\{u\} = \frac{1}{\Delta t}[M]\{u^0\} + \frac{1}{\rho}\{S_x\} - \frac{1}{\rho}[P'_x]\{p\} + \nu \int_{\Gamma} N_i \frac{\partial u}{\partial n} d\Gamma \tag{8.55}$$

$$\left(\frac{1}{\Delta t}[M] + [C] + \nu[K] \right)\{v\} = \frac{1}{\Delta t}[M]\{v^0\} + \frac{1}{\rho}\{S_y\} - \frac{1}{\rho}[P'_y]\{p\} + \nu \int_{\Gamma} N_i \frac{\partial v}{\partial n} d\Gamma \tag{8.56}$$

Similarly, treating the continuity equation by using the method of weighted residuals, it can be written in the form

$$\varepsilon = \int_{\Omega} W \left(\frac{\partial u}{\partial x} + \frac{\partial v}{\partial y} \right) \cdot d\Omega$$

which in its weak form becomes

$$\int_{\Omega}\left(\frac{\partial W}{\partial x}u+\frac{\partial W}{\partial y}v\right)\cdot d\Omega=\int_{\Gamma} Wu_n\cdot d\Gamma \tag{8.57}$$

where u_n is the velocity normal to the element face.

One can observe that the discretized equations establish a relation between the velocity field and the pressure gradients. In matrix form, one can write the momentum equations in the form

$$a_{i,i}u_i=-\sum_{j\neq i}a_{i,j}u_j-b_i\frac{\partial p}{\partial x}+s_i^{p,x} \tag{8.58}$$

If we define a set of new variables in the form

$$\hat{u}_i=-\frac{\sum_{j\neq i}a_{i,j}u_j}{a_{i,i}};\quad \hat{v}_i=-\frac{\sum_{j\neq i}a_{i,j}v_j}{a_{i,i}};\quad K_{p,i}=\frac{b_i}{a_{i,i}};\quad S_i^{p,z}=\frac{s_i^{p,z}}{a_{i,i}} \tag{8.59}$$

one can write Equation 8.86 in the form

$$u_i=\hat{u}_i+S_i^{p,x}-K_{p,i}\frac{\partial p}{\partial x} \tag{8.60}$$

$$v_i=\hat{v}_i+S_i^{p,y}-K_{p,i}\frac{\partial p}{\partial y} \tag{8.61}$$

Since the values of u and v can be expressed in terms of the interpolation functions N_j as

$$u=N_ju_j=N_j\left(\hat{u}_j+S_j^{p,x}-K_{p,j}\frac{\partial p}{\partial x}\right)\quad\text{and}\quad v=N_jv_j=N_j\left(\hat{v}_j+S_j^{p,y}-K_{p,j}\frac{\partial p}{\partial y}\right)$$

These expressions can be substituted into the weak form of the continuity equation to obtain

$$\int_{\Omega}\left[\frac{\partial N_i}{\partial x}N_j\left(\hat{u}_j+S_j^{p,x}-K_{p,j}\frac{\partial p}{\partial x}\right)+\frac{\partial N_i}{\partial y}N_j\left(\hat{v}_j+S_j^{p,y}-K_{p,j}\frac{\partial p}{\partial y}\right)\right]\cdot d\Omega=\int_{\Gamma} Wu_n\cdot d\Gamma \tag{8.62}$$

Noting that the pressure term in the element can be also expressed in terms of the interpolation function as

$$p=N_i(x,y)P_i\Rightarrow\begin{cases}\dfrac{\partial p}{\partial x}=\dfrac{\partial N_i(x,y)}{\partial x}p_i\\[2ex]\dfrac{\partial p}{\partial y}=\dfrac{\partial N_i(x,y)}{\partial y}p_i\end{cases}$$

which when substituted into Equation 8.62 to obtain the general expression for the continuity equation in the form

$$\begin{aligned}&\int_{\Omega}\left[\frac{\partial N_i}{\partial x}\left(N_jK_{p,j}\right)\frac{\partial N_k}{\partial x}+\frac{\partial N_i}{\partial y}\left(N_jK_{p,j}\right)\frac{\partial N_k}{\partial y}\right]\cdot d\Omega\cdot p_k=\\&\int_{\Omega}\left[\frac{\partial N_i}{\partial x}N_j\left(\hat{u}_j+S_j^{p,x}\right)+\frac{\partial N_i}{\partial y}N_j\left(\hat{v}_j+S_j^{p,y}\right)\right]\cdot d\Omega-\int_{\Gamma} Wu_n\cdot d\Gamma\end{aligned} \tag{8.63}$$

Considering that the vectors P_i and the $K_{p,j}$ are constants for a given element, and consequently can be put outside the integral sign, the problem is reduced to a system of equations of the form

$$[\overline{A}_p]\cdot\{p\}=\{\overline{B}_p\} \tag{8.64}$$

The terms $[\overline{A}_p]$ and $\{\overline{B}_p\}$ are given by the relations below:

$$[\overline{A}_p]=\int_\Omega\left[\frac{\partial N_i}{\partial x}\frac{\partial N_j}{\partial x}+\frac{\partial N_i}{\partial y}\frac{\partial N_j}{\partial y}\right]\left(N_t K_{p,t}\right)\cdot d\Omega$$

$$\{\overline{B}_p\}=\int_\Omega\frac{\partial N_i}{\partial x}N_j\cdot d\Omega\cdot\left(\hat{u}_j+S_j^{p,x}\right)+\int_\Omega\frac{\partial N_i}{\partial y}N_j\cdot d\Omega\cdot\left(\hat{v}_j+S_j^{p,y}\right)-\int_\Gamma Wu_n\cdot d\Gamma$$

When the pressure field is determined one can correct the velocity field before calculating the new coefficients. This can be done as follows:

$$u_j=\hat{u}_j+S_j^{p,x}-\frac{1}{\rho\cdot a_{i,i}}\int_\Omega W\frac{\partial p}{\partial x}\cdot d\Omega \tag{8.65}$$

$$v_j=\hat{v}_j+S_j^{p,y}-\frac{1}{\rho\cdot a_{i,i}}\int_\Omega W\frac{\partial p}{\partial y}\cdot d\Omega \tag{8.66}$$

Alternatively, one can interpolate the pressure field and use the Galerkin approximation in the same way as used in the case of the momentum equation to obtain the following relations:

$$u_j=\hat{u}_j+S_j^{p,x}-\frac{1}{\rho\cdot a_{i,i}}[P'_x]\{p\} \tag{8.67}$$

$$v_j=\hat{v}_j+S_j^{p,y}-\frac{1}{\rho\cdot a_{i,i}}[P'_y]\{p\} \tag{8.68}$$

Having defined the system of equations, it is necessary to establish the boundary conditions of the problem. In case of the momentum equation one can use the no-slip condition or the condition of known velocity. Both can be applied simply by changing the corresponding lines in the matrix to the value of interest. The boundary conditions at exit are more difficult to establish because the velocity profile is not known and the most used condition is that $d\bar{u}_n/d\bar{n}=0$, because it does not need any further correction.

In case of the pressure equation, the boundary conditions are more difficult. It was demonstrated that to obtain the weak form of the continuity equation, the term $\int_S W\bar{u}_n dS$ has to be evaluated in terms of the predetermined velocity distribution. It must be mentioned that this term represents the normal velocity, and hence need to be evaluated at the frontiers where mass influx or efflux occurs. As mentioned earlier, the method of solution is validated by comparing predictions from the present method with the results of de Vahl Davis and Jones (1983) and de Vahl Davis (1983), as shown in Figure 8.24. De Vahl Davis solved the problem of natural convection using a formulation based upon the vorticity and stream function method and obtained solutions for different size grids. From these results de Vahl Davis obtained his solution (considered as a standard solution) by extrapolating the grids until achieving precise results. The present problem is a closed cavity with the two vertical walls maintained at different temperatures, as shown in Figure 8.25, while the top and bottom surfaces are thermally insulated. The numerical predictions were obtained by using a grid of 21×21 points and the results were further interpolated by a cubic spline. The grid size was chosen in order to be able to compare with de Vahl Davis's results under the same conditions. Numerical trials were performed to establish the grid size most suitable for the present study. No further attempts were realized since the 21×21 grid (the same used by de Vahl Davis) gave satisfactory results. Obviously for high values of the Rayleigh number the errors encountered are appreciable, and hence it would be necessary to perform grid size testing to establish a suitable grid size.

The four schemes investigated in the calculations are presented below:

- Formulation 1: as proposed by Morgan *et al.* (1978) except that the thermal capacity in terms of time is corrected three times, which was not done in Morgan's original work.
- Formulation 2: as proposed by Comini and Saro (1990).
- Formulation 3: as proposed by Pham (1986).
- Formulation 4: similar to Pham's but does not utilize correction of the enthalpy at the end of each iteration. The tests were realized using IBM Risc 6000 workstation and the results are shown below:

Method	Dimensionless time for complete solidification
Formulation 1	0.5118
Formulation 2	0.5491
Formulation 3	0.5485
Formulation 4	0.5485
Analytical Solution	0.5484

Figure 8.26 shows the results for the interface position evolution at x = 0.1. One can find that, except for Morgan's formulation, all the schemes seem to produce similar results.

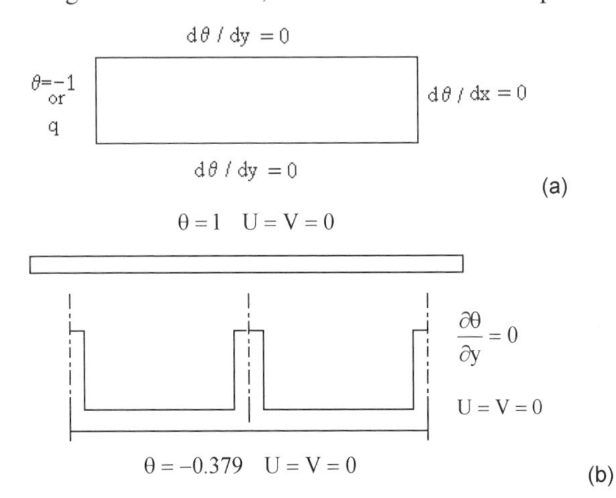

Figure 8.25 Sketch of the finned cavity filled with PCM. (a) Layout of the problem. (b) Geometry of the finned cavity.

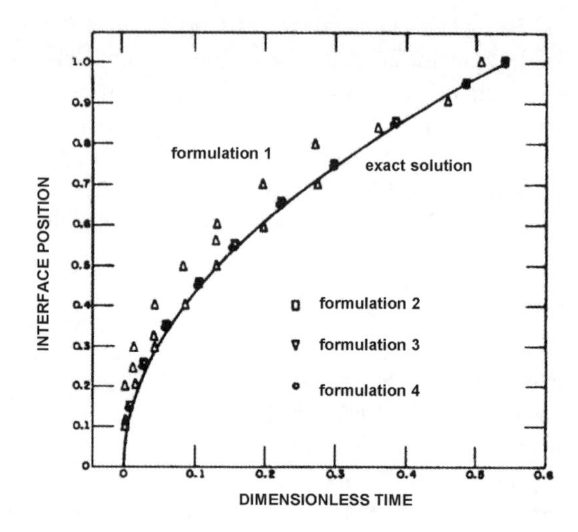

Figure 8.26 Comparison between different formulations schemes.

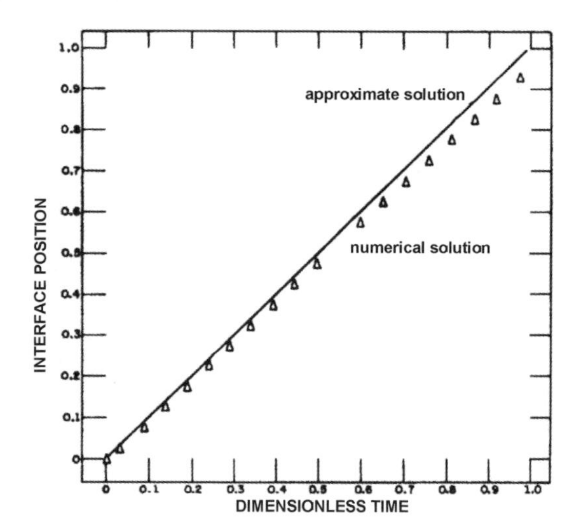

Figure 8.27 Evolution of the solidification front.

In the present analysis we adopted formulation 4. In the case of a heat flux boundary condition, the results obtained were compared with an approximate solution due to Alexiades and Solomon (1993), as shown in Figure 8.27. As can be seen, the agreement is reasonably good confirming the validity of the proposed model and method of solution. The same formulation and method of solution were applied to a cavity filled with PCM initially at a temperature different from the phase change temperature. The boundaries of the cavity may be subject to different boundary conditions and may have extended surfaces in between the cavity walls. The details of the model, the numerical solution and additional results can be found in Scalon (1998). Figure 8.28 shows that when the initial temperature is close to the phase change temperature there is practically no effect on the complete fusion time. Initial temperature much below the phase change temperature seems to have a strong effect on the fusion rate, and consequently on the time for nearly complete fusion. The effects of varying the Stefan number Ste in the range of great interest for heat storage applications are shown in Figure 8.29. One can observe that during the initial instants the process is dominated by conduction and the Stefan number does not affect the phase change process. After this period, convection starts to be more dominant and the Stefan number starts to influence the phase change process.

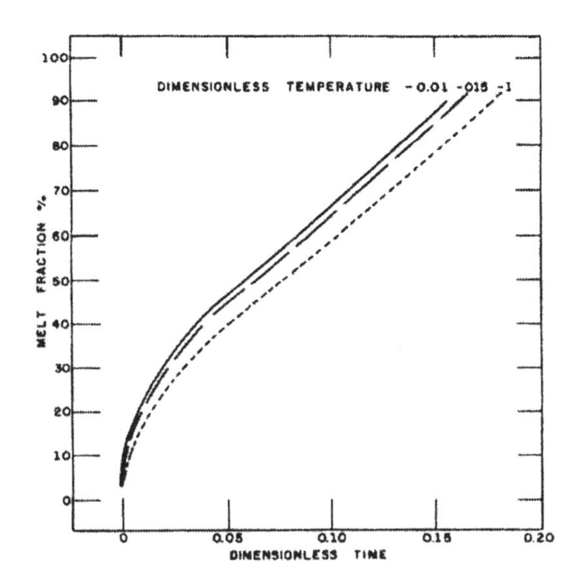

Figure 8.28 Effect of the initial solid phase PCM temperature on the melt fraction.

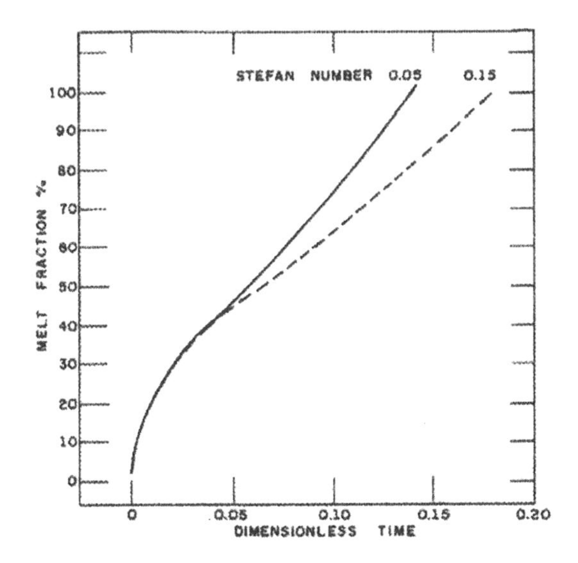

Figure 8.29 Effects of the Stefan number on the melt fraction.

The effect of the Rayleigh number is shown in Figure 8.30. One can observe that during the initial instants, the process is dominated by conduction, and consequently the Rayleigh number does not influence the heat transfer process and the curves seem to be coincident during this interval of time. When the convection process becomes more dominant, the corresponding Rayleigh number influences the phase change process as shown in Figure 8.30. The effect of the geometry of the fin is shown in Figure 8.31. As can be noticed, the increase of the fin length causes the melt mass fraction and the time for fusion to increase.

8.4.2 Concluding remarks

The case of plane finned geometry suitable for latent heat storage applications was presented and discussed. The thermal model takes into account possible convection effects and was treated by the finite element approach not widely used in this type of problem. Effects of the fin length, the Stefan and the Rayleigh numbers on the solidified mass fraction were presented and discussed.

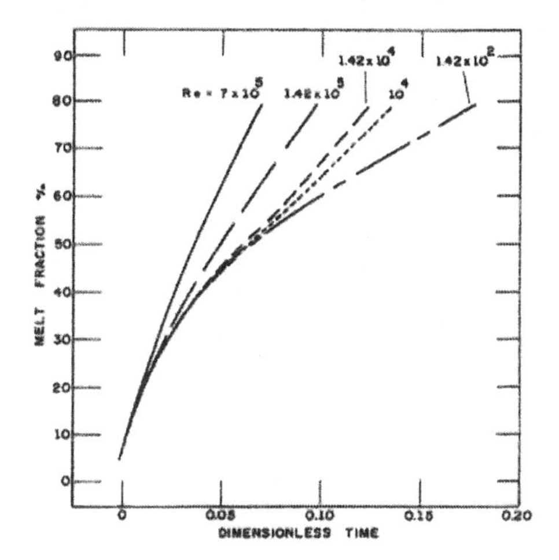

Figure 8.30 Effects of the Rayleigh number on the melt fraction.

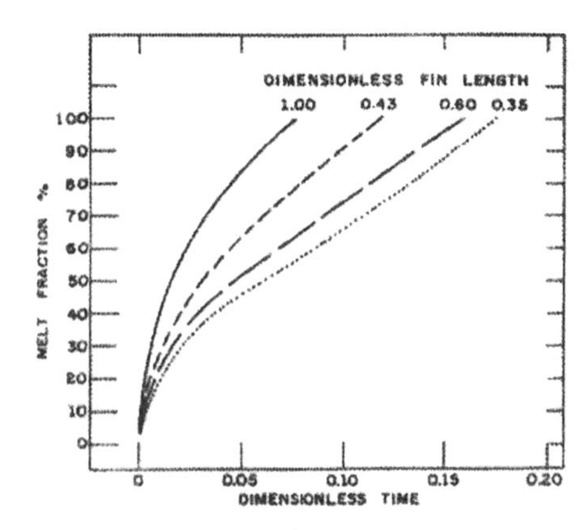

Figure 8.31 Effects of the fin length on the melt fraction.

8.5 Phase Change around Isothermal Finned Cylinder

In this section we present the models and results of numerical and experimental studies realized on finned geometries of potential use in thermal energy storage system and cold storage applications. Three forms of finned geometries were investigated, namely axially, radially and alternating finned cylinders. These fins help in reducing the thermal resistance problem in storage systems, and alleviating the non-desirable effects of natural convection within the PCM.

8.5.1 Axially finned cylinder

First we present a model for simulating heat transfer with phase change around an axially finned tube. Sparrow *et al.* (1978a) investigated experimentally a finned tube with four fins. In their work they observed a presence of natural convection in the liquid phase which can delay or even interrupt completely the solidification process. They also noticed that the use of fins could delay the domination of natural convection on the heat conduction process during the solidification process. How the number of fins and their geometry can affect the process was not fully investigated. In their experiments, the temperature of the tube was kept constant and at a value higher than the phase change temperature, conditions which always permit the presence of natural convection. Later Sparrow and Hsu (1981) studied the transition which occurs during a solidification process controlled by natural convection to a process dominated by conduction, and observed that the transition is a function of the temperature difference between the tube wall temperature and the initial temperature of the superheated liquid phase of the PCM. They noticed that natural convection dominates the process during the first instances, and consequently delays the solidification process. The use of finned tubes can accelerate the transition of natural convection to pure conduction. This aspect of problem was not treated in the literature.

Zhang and Faghri (1996) used the approximate analytical method (integral energy method) to solve the fusion problem in an annular space with water as the working fluid flowing under a laminar flow regime, while the phase change material used is paraffin initially at the fusion temperature. The integral method was used in the PCM while the temperature, the interface position and Nusselt number were calculated in each time increment by an iterative procedure. Choi and Kim (1992) analyzed experimentally the performance of latent heat storage with finned and finless tubes. Figure 8.32 shows the finned tube and the symmetry region. The external surface can be a thermally insulated tube or a symmetry circle in a system composed of multi-tubes immersed in the storage tank filled with PCM.

Figure 8.33 shows details of the geometry of the symmetrical region under study, where the sides are thermally insulated while the tube wall is maintained under constant temperature.

Considering that the heat transfer process is controlled by pure conduction, one can write the conduction equation for the solid and liquid phases, respectively, as

$$\rho_s c_{p_s} \frac{\partial T_s}{\partial t} = \frac{1}{r}\frac{\partial}{\partial r}\left(r k_s \frac{\partial T_s}{\partial r} \right) + \frac{1}{r}\frac{\partial}{\partial \phi}\left(\frac{k_s}{r}\frac{\partial T_s}{\partial \phi} \right) \tag{8.69}$$

$$\rho_l c_{pl}\frac{\partial T_l}{\partial t}=\frac{1}{r}\frac{\partial}{\partial r}\left(rk_l\frac{\partial T_l}{\partial r}\right)+\frac{1}{r}\frac{\partial}{\partial \phi}\left(\frac{k_l}{r}\frac{\partial T_l}{\partial \phi}\right) \tag{8.70}$$

and the energy balance equation at the solid-liquid interface can be written as

$$\left(k_s\frac{\partial T_s}{\partial r}-k_l\frac{\partial T_l}{\partial r}\right)\left(1+\frac{1}{r_s^2}\left(\frac{\partial r_s}{\partial \phi}\right)^2\right)=\rho_s\lambda\frac{\partial r_s}{\partial t} \tag{8.71}$$

Adopting the new variables in the following form, we obtain Equations 8.72 and 8.73:

$$\theta=\frac{T_m-T}{T_m-T_w},\quad R=\frac{r}{r_i},\quad \tau=\frac{k_s t}{c_{ps} r_i^2},\quad \zeta=\frac{\Delta T}{T_m-T_w}$$

$$\bar{C}(\theta)\frac{\partial \theta}{\partial \tau}=\frac{1}{R}\frac{\partial}{\partial R}\left(R\bar{k}(\theta)\frac{\partial \theta}{\partial R}\right)+\frac{1}{R}\frac{\partial}{\partial \phi}\left(\frac{\bar{k}(\theta)}{R}\frac{\partial \theta}{\partial \phi}\right) \tag{8.72}$$

$$\theta=1,\ \text{for}\ R=1\qquad \frac{\partial \theta}{\partial R}=0\ \text{for R}=\frac{\text{r}_\text{m}}{\text{r}_\text{i}}\qquad \frac{\partial \theta}{\partial \phi}=0\ \text{for}\ \phi=0$$

$$\frac{\partial \theta}{\partial \phi}=0,\quad \text{for}\ \phi=\phi_\text{m} \tag{8.73}$$

The integration of Equation 8.72 is realized by using the finite control volume technique; the domain of the problem is divided into a convenient number of control volumes and Equation 8.72 is integrated over R, ϕ and τ to obtain, after some mathematical manipulations, the required algebraic equations. The same procedure is applied with respect to the boundary conditions of the problem. Details of the mathematical manipulations are omitted here for the sake of brevity. The present solution is based on the Alternating Direction Implicit formulation because it is unconditionally stable. More details about the numerical solution can be found in Gonçalves (1996).

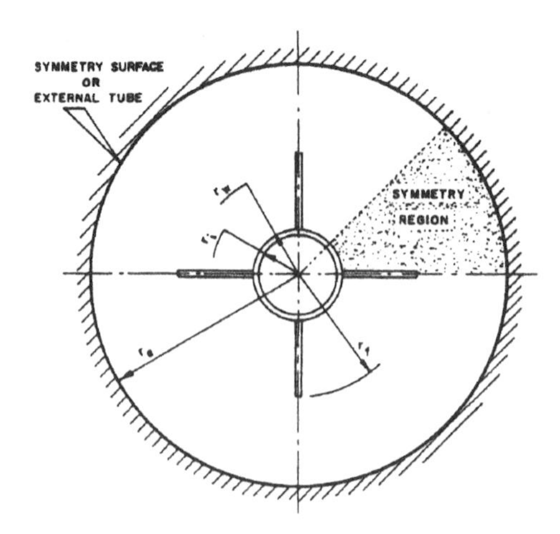

Figure 8.32 General layout of the problem.

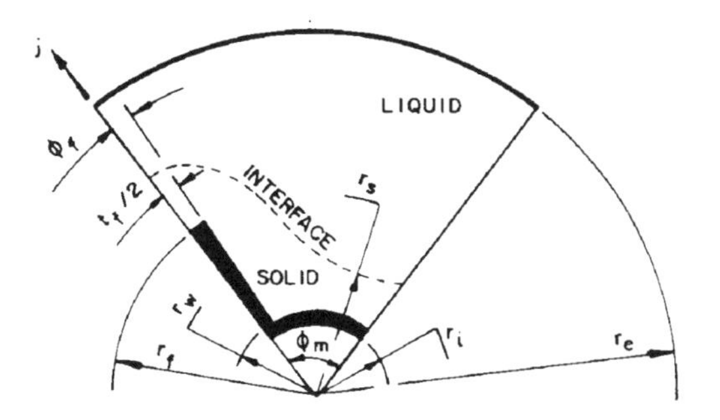

Figure 8.33 Details of the symmetry region.

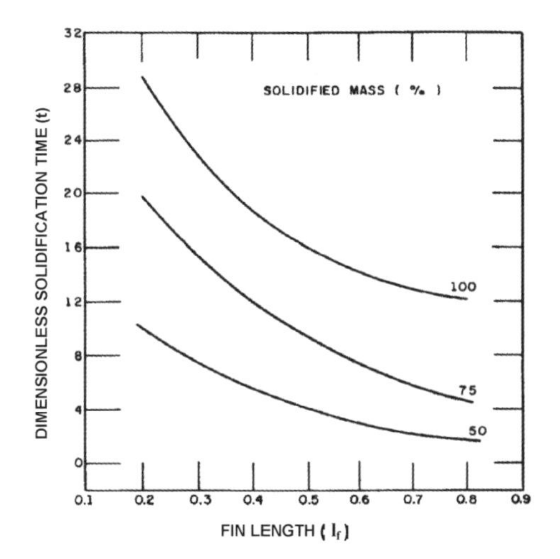

Figure 8.34 Variation of the solidification time with the fin length.

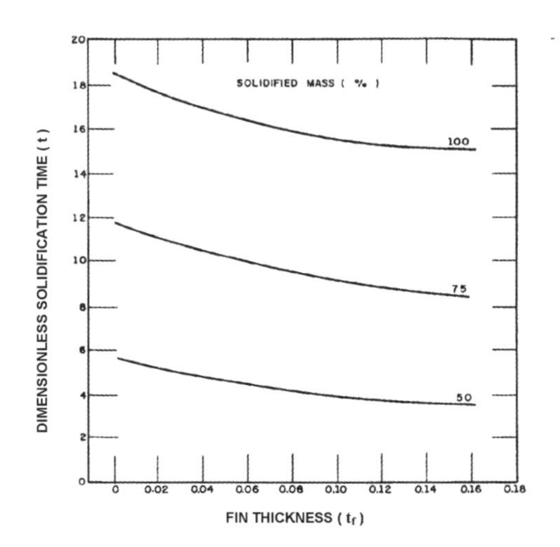

Figure 8.35 Effect of the fin thickness on the solidification time.

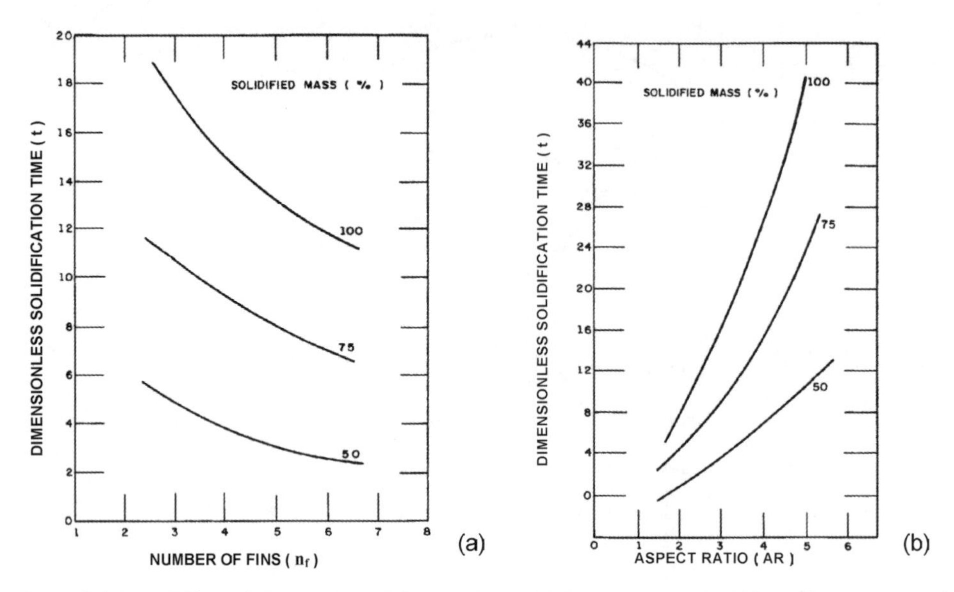

Figure 8.36 (a) Effect of the number of fins on the solidification time. (b) Effect of the aspect ratio of annular space on the solidification time.

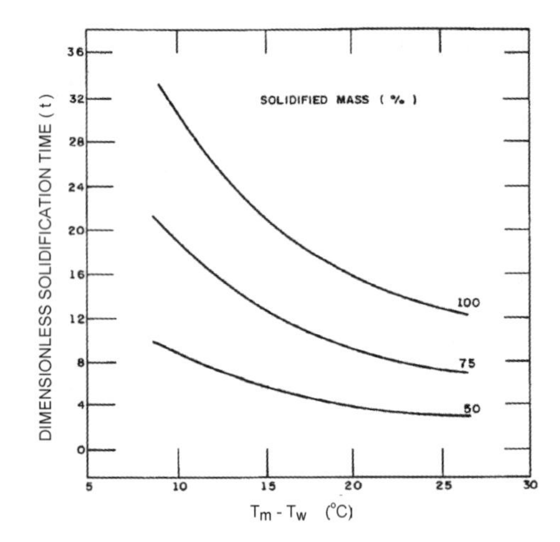

Figure 8.37 Effect of the super heating on the solidification time.

Figure 8.34 shows the influence of the fin length on the time for complete solidification. As can be seen, the increase of the fin length reduces the time for complete solidification. The effect of the fin thickness is found to be relatively small, and the reduction in the time for complete solidification is found to be marginal as can be verified from Figure 8.35. The influence of the number of fins on the time for complete solidification is shown in Figure 8.36a. As can be seen, increasing the number of the fins leads to a noticeable reduction in the time for complete solidification. Figure 8.36b shows the effect of the aspect ratio on the time for complete solidification. One can observe the strong adverse influence of the aspect ratio on the time for complete solidification. The influence of the degree of superheat of the PCM on the time for complete solidification is also presented in Figure 8.37.

Concluding remarks

The parameters analyzed numerically include the fin length, fin thickness, number of fins, the aspect ratio of the annular space and the temperature difference between the phase change temperature and the tube wall temperature. The results indicate that these parameters have significant effects on the time for complete solidification. From the results presented, one can observe that the fin thickness has a relatively small influence on the solidification time and that the fin length affects strongly the time for complete solidification and the solidification rate. Also the number of fins has a strong influence on the time for complete solidification and the rate of mass solidification. The aspect ratio of the annular space has a strong effect on the time for solidification of a certain mass, and increasing this ratio leads to increasing the time for complete solidification. The temperature difference has an opposite effect on the solidification of a certain mass of PCM, and the time for the complete solidification seems to decrease with the increase of this temperature difference.

The thermal model and the results of the effects of the main parameters of the problem of axially finned geometry should be of great help in understanding and appreciating the advantages of using finned geometries in latent heat storage systems. As can be seen, the model is easy to incorporate within a program for thermal storage analysis and design.

8.5.2 Radial finned cylinders

In a similar manner, using the same approach as in the previous case solves the problem of radial finned tube submerged in a PCM. A model for the solidification around a radially finned tube with constant wall temperature is developed and solved numerically. Figure 8.38 shows a typical finned tube arrangement inside a storage tank. The symmetry circle is the limiting boundary for phase change around the tube, and hence one tube is considered here as representing the phase change problem. The model is based upon pure heat conduction formulation, and the enthalpy method due to Bonacina *et al.* (1973) was adopted. The finite difference approach and the Alternating Direction implicit scheme were used to discretize the system of equations and the associated boundary, initial and final conditions.

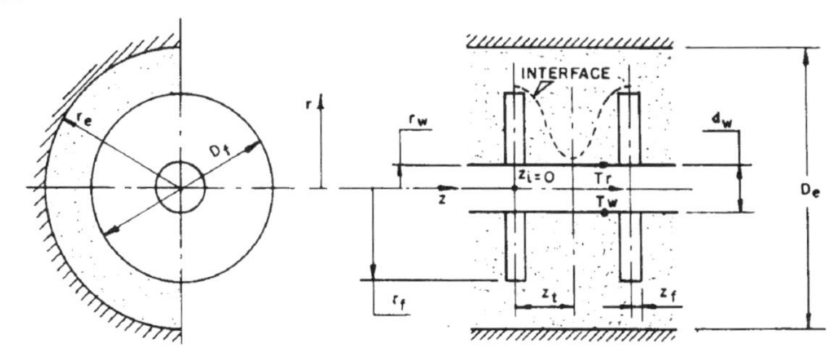

Figure 8.38 Layout of the problem.

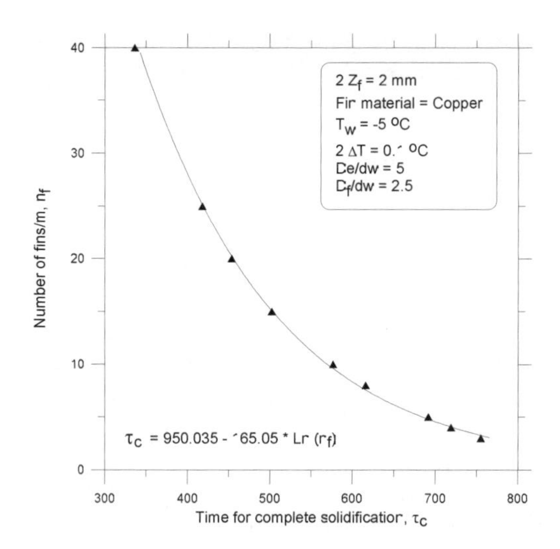

Figure 8.39 Effect of the number of fins/m on the time for complete solidification.

The problem under consideration is presented in Figure 8.38, and one can identify the symmetry circle and the representative domain of the problem under consideration. In the present formulation we consider that the phenomenon of solidification is controlled by pure conduction, and at any instant one can write the governing equations and the boundary conditions of the problem as below.

The energy equation for the PCM solid phase is

$$\rho_s c_{p_s} \frac{\partial T_s}{\partial t} = \frac{1}{r}\frac{\partial}{\partial r}\left(r k_s \frac{\partial T_s}{\partial r}\right) + \frac{\partial}{\partial z}\left(k_s \frac{\partial T_s}{\partial z}\right) \tag{8.74}$$

The energy equation for the PCM liquid phase is

$$\rho_l c_{p_l} \frac{\partial T_l}{\partial t} = \frac{1}{r}\frac{\partial}{\partial r}\left(r k_l \frac{\partial T_l}{\partial r}\right) + \frac{\partial}{\partial z}\left(k_l \frac{\partial T_l}{\partial z}\right) \tag{8.75}$$

The boundary conditions at the interface can be written as

$$\left(k_s \frac{\partial T_s}{\partial r} - k_l \frac{\partial T_l}{\partial r}\right)\left(1 + \left(\frac{\partial s}{\partial z}\right)^2\right) = \rho_s \lambda \frac{\partial s}{\partial t};\ \mathrm{r} = \mathrm{s(t)} \tag{8.76}$$

$$\begin{array}{lll} T_s = T_l = T_m & & r = s(t) \\ At & \mathrm{r} = \mathrm{r_w}: & T = T_w \\ At & \mathrm{r} = \mathrm{r_e}: & \dfrac{\partial T}{\partial r} = 0 \\ At & \mathrm{z} = \mathrm{z_i} = 0: & \dfrac{\partial T}{\partial z} = 0 \\ At & \mathrm{z} = \mathrm{z_t}: & \dfrac{\partial T}{\partial z} = 0 \end{array} \tag{8.77}$$

The initial and final conditions can be written as

$$\begin{aligned} T(r,z,t=0) &= T_m + \Delta T \\ T(r,z,t_f) &= T_m - \Delta T \end{aligned} \tag{8.78}$$

where ΔT is half the phase change temperature range. Following Bonacina *et al.* (1973), and the same approach as in the previous case, one can write the governing equations and the associated boundary conditions as

$$\bar{C}(\theta)\frac{\partial \theta}{\partial \tau} = \frac{1}{R}\frac{\partial}{\partial R}\left(R\bar{k}(\theta)\frac{\partial \theta}{\partial R}\right) + \frac{\partial}{\partial Z}\left(\bar{k}(\theta)\frac{\partial \theta}{\partial Z}\right) \tag{8.79}$$

The boundary conditions expressed in terms of the new variables can be written as:

$$\left.\begin{aligned} &\phi_w = 0 && R = 1 \\ &\frac{\partial \theta}{\partial R} = 0 && R = R_e \\ &\frac{\partial \theta}{\partial Z} = 0 && Z = Z_i = 0 \\ &\frac{\partial \theta}{\partial Z} = 0 && Z = Z_t \end{aligned}\right\} \tag{8.80}$$

$$\left.\begin{aligned} &\text{The initial condition}: \theta = 1 \\ &\text{The final condition}: \theta = 1 - 2\zeta \end{aligned}\right\} \tag{8.81}$$

where the new dimensionless variables are as below:

$$\begin{aligned} &\theta = \frac{T_m - T}{T_m - T_w}; \quad R = \frac{r}{r_w}; \quad Z = \frac{z}{r_w}; \\ &\bar{k} = \frac{k}{k_s}; \quad \bar{C} = \frac{c_p}{c_{p_s}}; \quad \zeta = \frac{\Delta T}{T_m - T_w} \\ &\tau = \frac{k_s t}{c_{p_s} r_w^2}; \quad Ste_{\Delta T} = \frac{c_{p_s}\Delta T}{\rho_s \lambda} \end{aligned} \tag{8.82}$$

The present model enables the prediction of the solidified mass fraction, the time for complete solidification, the interface velocity and the interface positions. Also it enables investigation of the parameters of interest, e.g. the number of fins, fin radius, fin thickness, fin material, effect of the annular aspect ratio and the phase change temperature range.

Equation 8.79 is transformed into a system of algebraic equations by using the control volume technique. The domain of the problem is divided into a convenient number of control volumes, and Equation 8.79 is integrated over R, Z and τ to obtain the algebraic equations. The same procedure is done with respect to the boundary conditions, the initial and final conditions. Having obtained the system of algebraic equations, it is possible to proceed with the numerical solution using a variety of numerical options such as implicit, explicit and alternating direction implicit among many others. In this study we adopted the alternating direction implicit method because of its unconditional stability. More details are available in Gonçalves (1996) and Ismail (1999).

Figure 8.39 shows the effects of the number of fins on the time for complete solidification. The time for complete solidification decreases with the increase of the number of fins. A general equation relating the number of fins to the time for complete solidification is obtained. For the case of a tube of 5–25 mm diameters, fin thickness of 2 mm, wall temperature of –5°C and ratio of diameter of fin to diameter of tube of 2.5, the equation can be written in the form

$$\tau_c = 950.035 - 165.05\ Ln(n_f) \tag{8.83}$$

where τ_c is the time for complete solidification. The dimensionless time can be converted in dimensional time by the equation $t = 132.23\ \tau$.

The influence of the dimensionless fin length D_f/d_w on the time for complete solidification is shown in Figure 8.40, and the corresponding correlation is

$$\tau_c = 1057.468\left(\frac{D_f}{d_w}\right)^{-0.69956} \tag{8.84}$$

The influence of the wall temperature on the solidified mass fraction is investigated varying the wall temperature from –5 to –20°C. As can be seen in Figure 8.41, reducing the wall temperature leads to a big reduction in the solidification time for a fixed solidified mass fraction, and its variation as a function of the wall temperature can be given by the equation

$$\tau_c = 879.855\ e^{-0.1002163\ Tw} \tag{8.85}$$

In the preceding analysis we considered that the range of phase change temperature is 0.1°C. This parameter affects the solidification characteristics as predicted by the present model. The effect of the phase change temperature range was investigated by varying the value of the range from 0.1 to 0.5°C. One can notice that the effect of the phase change temperature range is rather strong and increases the time for complete solidification, as can be observed from Figure 8.42.

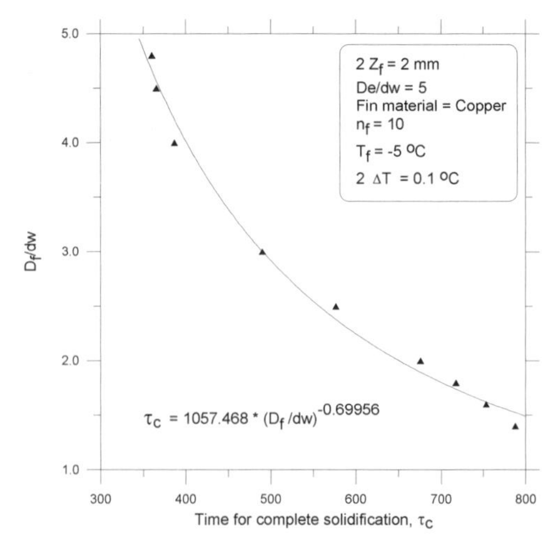

Figure 8.40 Effect of the fin length on the time for complete solidified.

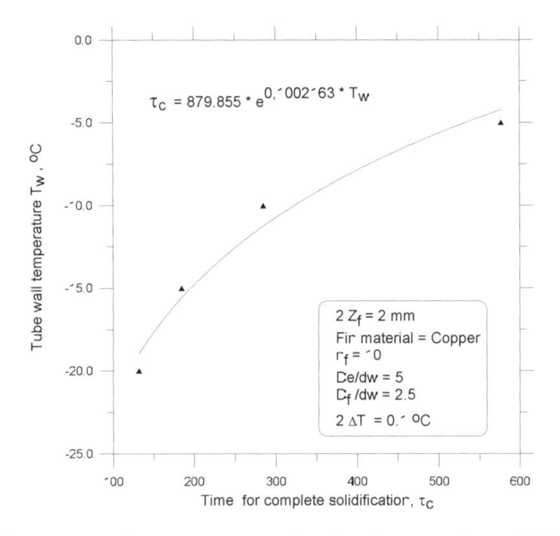

Figure 8.41 Effect of the tube wall temperature on the time for complete solidification.

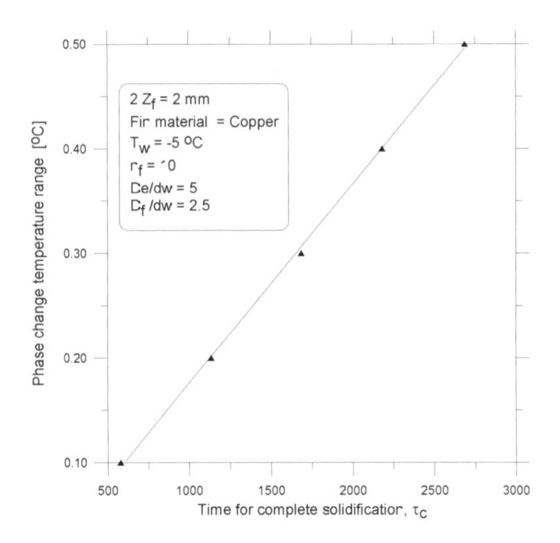

Figure 8.42 Effect of the phase change temperature range on the time for complete solidification.

The variation of the time for complete solidification with the range of the phase change temperature can be predicted by the equation

$$\tau_c = 5269.148\, \Delta T + 72.619 \tag{8.86}$$

Figure 8.43 shows the effects of the fin thickness on the solidified mass fraction and the time for the complete solidification. As it can be noticed the thickness has little effect on the time for complete solidification.

Figure 8.44 presents the effect of the aspect ratio D_e/d_w on the solidified mass fraction and the time for complete solidification As can be seen, the time for complete solidification increases with the increase of the aspect ratio D_e/d_w.

The equation relating the aspect ratio to the time for complete solidification for the given fixed conditions can be written as

$$\tau_c = 13.7864\, e^{0.754808\,(De/d_w)} \tag{8.87}$$

The material of the refrigerant carrying tube has a strong influence on the solidified mass fraction and the time for complete solidification, as seen in Figure 8.45. As expected, a good thermal conducting material produces quick solidification and hence shorter periods to obtain complete solidification due to the overall low thermal resistance.

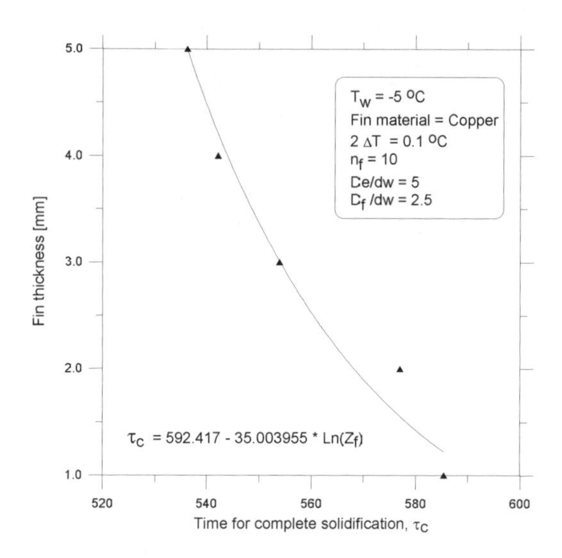

Figure 8.43 Effect of the fin thickness on the time for complete solidification.

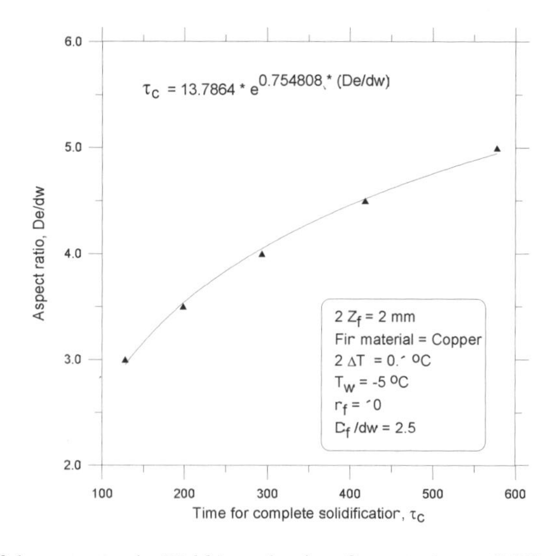

Figure 8.44 Effect of the aspect ratio (D_e/d_w) on the time for complete solidification.

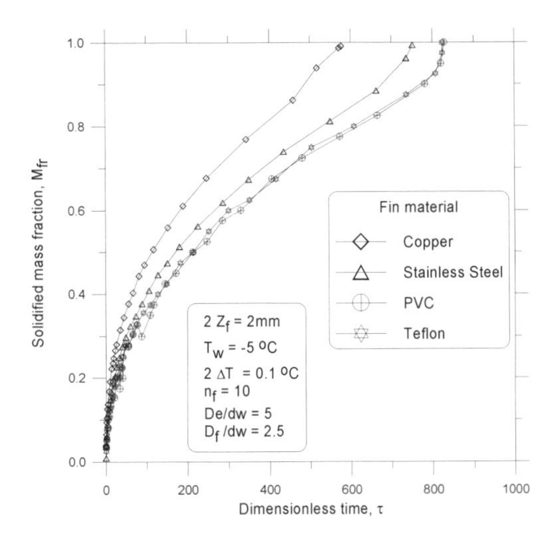

Figure 8.45 Effect of the fin material on the solidified mass fraction.

Concluding remarks

One can conclude from the results of the present analysis that the model is able to predict the thermal performance of a radially finned tube. The study shows that the number of fins, fin length, fin thickness and the aspect ratio of the finned tube arrangement have a strong influence on the time for complete solidification and the solidification rate. Also, the tube material as well as the tube wall temperature seem to exert a strong effect on the time for complete solidification. The influence equations presented enable the prediction of the performance under different operational conditions.

8.6 Conclusion

This chapter deals with the problems of heat transfer with phase change in simple and complex geometries of latent heat storage systems. In addition to a comprehensive literature review of the state of the art, different simple and complex geometries are analyzed using various effective analytical and numerical methods. The results are presented and validated with actual and existing data. These methods are essential to determine the thermal performance of latent heat storage systems.

Acknowledgements

The author of this chapter wishes to express his gratitude to the Brazilian National Research Council (CNPQ) for the financial support for research projects in the area of energy storage, and also for the research scholarship to the author and the study scholarships for his post graduate students. Acknowledgements are also due to CAPES for the scholarships, and to FAPESP for the financial support to research projects in the area of cold storage applications.

Nomenclature

Bi	Biot number
c_p	specific heat (J/kg K)
C	heat capacity per unit volume (= ρc_p) (J m^{-3} K)
$C(T)$	dimensionless heat capacity per unit volume including the phase change
C^*	ratio of the thermal conductivity of the PCM at the initial stage to that of the fin material
CA	dimensionless fin length ($=$(fin length$/r$))
d_w	tube diameter (m)
D	internal diameter (m)
D_e	external or simmetry diameter (m)
D_f	fin diameter (m)
E	dimensionless width of the annular space ($= e/r_w$)
e	width of the annular space (m)
h	convection heat transfer coefficient (W/m^2 K)
$\vec{i}$	unit vector
k	thermal conductivity (W/m K)
$k(T)$	dimensionless thermal conductivity including the phase change
L	tube length (m)
l_f	dimensionless fin length ($= (r_f - r_w)/(r_e - r_w)$)
$\vec{n}$	normal unit vector to the control surface
n_f	number of fins
N	number of grid points
N_i	interpolation function
Nu	Nusselt number ($= hL/k$)
M_{Solid}	solidified mass (kg)
M_{fr}	solidified mass fraction
M_{Total}	total mass (kg)
p	pressure (Pa)
P	dimensionless pressure ($= p/\rho(L/\alpha)^2$)
Pe	Peclet number (= $U_m D/\alpha_f$)
Pr	Prandtl number (= ν/α)
q	wall heat flux
Q	thermal energy stored (kJ)
r	radius or radial coordinate (m)
r_e	radius of external cylinder or radius of the symmetry circle (m)
r_f	fin extreme radius (m)
r_i	internal radius of the tube (m)
r_m	interface position (m)
r_o	symmetry (external) radius (m)
r_s	radial position of the solid/liquid interface (m)
r_w	external radius of the tube (m)
r_δ	solid-liquid interface radius (m)
R	dimensionless radial coordinate
R^*	dimensionless radial coordinate

Ra	Rayleigh number $\left(= g\,\beta\,\Delta T\,L^3/\nu^2\,\mathrm{Pr}\right)$
Re	Reynolds number $(=U_m D/\nu_f)$
$s(t)$	solid liquid interface
S	dimensionless interface position
$S1, S2, S3, S4$	control volume faces
$S^{p,z}$	source term associated to the pressure
S_x, S_y, S_z	source terms
St	Stanton number $(= (h/U_m \rho c_p))$
Ste	Stefan number $(= c_{pS}(T_m - T_{in})/\lambda)$
t	time (s)
t_c	time of complete solidification (s)
T	temperature (K)
T_f	phase change temperature (K)
T_i	initial temperature of the PCM (K)
T_{in}	inlet fluid temperature or initial temperature (K)
T_m	phase change temperature (K)
T_p	wall temperature of the cold plate (K)
T_w	tube wall temperature (K)
T_∞	coolant temperature (K)
u	velocity in the x direction (m/s)
U	dimensionless velocity in the x direction $(= uL/\alpha)$
$\vec{u}_b$	control surface velocity (m/s)
U_m	mean flow velocity (m/s)
v	velocity in the y direction (m/s)
V	dimensionless velocity in the y direction $(= vL/\alpha)$
W	arbitrary weight function
W^*	dimensionless fin thickness (= (fin width / r_i))
x	X-coordinate (m)
X	dimensionless coordinate $(= x/L)$
Δx	increment in the x-direction
ΔT	half the phase change temperature range (K)
z, Z	axial coordinates (m)
z_t	half the spacing between radial fins
$[C]$	matrix of the convective terms
$[K]$	conductance matrix
$[M]$	global mass matrix
$[P_z]$	matrix of the pression terms in z direction

Greek Symbols

α	thermal diffusivity (m^2/s)
β	auxiliary variable
δ	thickness of the solid or liquid layer (m)
Δ	dimensionless thickness of the solid or liquid layer $(= \delta/r_w)$

γ	auxiliary variable
ν	kinematic viscosity (m^2/s)
μ	dynamic viscosity (kg/m s)
η	transformed coordinate
τ	dimensionless time
λ	latent heat of fusion or solidification (J/Kg)
ξ	transformed coordinate
ζ	dimensionless phase change temperature range ($= \Delta T/(T_m - T_w)$)
ψ	streamline function
$\psi(0)$	dimensionless superheating parameter ($= (T_i(0) - T_m)/(T_i(0) - T_w)$)
θ	dimensionless temperature ($= (T_m - T)/(T_m - T_w)$)
ρ	density (kg/m^3)
ϕ	tangential coordinate (°)
ϕ_f	half fin angle (°)
ϕ_m	half angle between two successive fins (°)
Γ	auxiliary variable
Λ	auxiliary variable
Ω	auxiliary variable or elementary domain
Ψ, χ	auxiliary variables

Subscripts

c	complete solidification
e	external
f	refrigerating fluid, fin or final
i	internal
l	liquid phase
r	radial direction
s	solid phase
w	wall

Abbreviations

CVFEM	Control Volume Finite Element Method
PCM	Phase Change Material
ADI	Alternating Direction Implicit
AR	Aspect Ratio ($= (r_e - r_w)/r_w$)

References

Alexiades, V. and Solomon, A.D. (1993). *Mathematical Modeling of Melting and Freezing Processes*, Hemisphere Publishing Co.

Bedecarrats, J. P., Strub, F., Falcon, B. and Dumas. J.P. (1996). Phase change thermal energy storage using spherical capsules: performance of a test plant, *International Journal of Refrigeration* 19(3), 187-196.

Bellecci, C. and Conti, M. (1993). Phase change thermal storage: transient behavior analysis of a solarreceiver/storage module using the enthalpy method, *International Journal of Heat and Mass Transfer* 36, 2157–2163.

Bonacina, C., Fasano, A. and Primicerio, M. (1973). Numerical solutions of phase change problems, *International Journal of Heat and Mass Transfer* 16, 1825–1832.

Cao, Y. and Faghri, A. (1991). Performance characteristics of a thermal energy storage module: a transient PCM/forced convection conjugate analysis, *International Journal of Heat and Mass Transfer* 34, 93–101.

Castro, J.N.C. (1991). *Paredes Termicas*, PhD Thesis, Faculty of Mechanical Engineering, The State University of Campinas, Brazil.

Chen, S. L. and. Yue, J. S. (1991). Thermal performance of cool storage in packed capsules for air conditioning, *Heat Recovery System and CHP* 11(6), 551–561.

Cho, S. H. and Sunderland, J. E. (1974). Phase change problems with temperature-dependent thermal conductivity, *Journal of Heat Transfer* 96(2), 214–217.

Choi, J. C. and Kim, S. D. (1992). Heat transfer characteristics of a latent heat storage system using $MgCl_2 6H_2O$, *Energy* 17, 1153–1162.

Clavier, L., Arquis, E. and Caltagirone, J.P. (1994). A fixed grid method for the numerical solution of phase change problems, *International Journal for Numerical Methods in Engineering* 37, 4247–4261.

Comini, G and Saro, O. (1990). A conservative algorith for multidimensional conduction phase change, *International Journal for Numerical Methods in Engineering* 30, 697–709.

Crank, J. (1957). Two methods for the numerical solution of moving-boundary problems in diffusion and heat flow, *Quart. Journal of Mech. and Applied Mathematics* 10, 220–231.

Date, A.W. (1991). A strong enthalpy formulation for the Stefan problems, *International Journal Heat and Mass Transfer* 34, 2231–2235.

De Vahl Davis, G. and Jones, I. (1983). Natural convection in a square cavity: a comparison exercise, *International Journal for Numerical Methods in Fluids* 3, 227–248.

De Vahl Davis, G. (1983). Natural convection of air in a square cavity: a Benchmark numerical solution, *International Journal for Numerical Methods in Fluids* 3, 249–264.

Duda, J.L.. Malone, M. F., Notter, R.H. and Vrentas, J.S. (1975). Analysis of two-dimensional diffusion-controlled moving boundary problems, *International Journal of Heat and Mass Transfer* 8, 901–910.

Feng, Z.G., Michaelides, E.E. and Scibilia, M.F. (1996). The energy equation of a sphere in an unsteady and non-uniform temperature field, *Rev. Gén. Therm.* 35, 5–13.

Folio, F. and Lacour, A. (1996). Heat transfer by conduction and convection between a spherical droplet particle and a two-phase fluid, *International Journal of Rapid Solidification* 9(2), 75–89.

Gibbs, B.M. and Hasnain, S.M. (1995). DSC study of technical grade phase change heat storage materials for solar heating applications, *Journal of Solar Engineering* 2, 1053–1062.

Gillessen, F., Herlach, D.M. and Feuerbacher, B. (1988). Glass formation by containerless solidification of metallic droplets in drop tube experiments, *Journal of the Less-Common Metals* 145, 145–152.

Gonçalves, M.M. (1996). *Armazenamento de Calor Latente Anular Com Aletas Alternadas*, PhD Thesis, Faculty of Mechanical Engineering, The State University of Campinas, Brazil.

Gresho, P. and Chan, S.T. (1990). On the theory of semi-implicit projection methods for viscous incompressible flow and its implementation via finite element method that also introduces a nearly consistent mass matrix, *International Journal for Numerical Methods in Fluids* 11, 621–659.

Gresho, P. (1990). On the theory of semi-implicit projection methods for viscous incompressible flow and its implementation via finite element method that also introduces a nearly consistent mass matrix, *International Journal for Numerical Methods in Fluids* 11, 587–620.

Grumman, D.L. and Butkus, A.S. (1988). The ice storage option, *ASHRAE Journal* 30 (5), 20–26.

Henríquez, J.R. (1996). *Estudo Numérico e Experimental Sobre Vidros Térmicos*, MSc Thesis, Faculty of Mechanical Engineering, The State University of Campinas, Brazil.

Hill J.M. and Kucera A. (1983). Freezing a saturated liquid inside a sphere, *International Journal Heat and Mass Transfer* 26(11), 1631–1637.

Homan, K.O. and Soo, S.L. (1997). Model for the transient stratification into a chilled-water storage tank, *International Journal Heat and Mass Transfer* 40(18), 4367–4377.

Hsu, C.F., Sparrow, E.M. and Patankar, S.V. (1981). Numerical solution of moving boundary problems by boundary immobilization and control-volume based finite-difference scheme, *International Journal of Heat and Mass Transfer* 24, 1335–1343.

Ismail, K.A.R. and Alves, C.L. (1985). Modeling and analysis of phase change thermal energy storage in shell and circular tube annulus configuration, *Proceedings of the 7th Miami International Conference on Alternative Energy Sources*, Miami, Florida, pp. 321–334.

Ismail, K.A.R. and Alves, C.L. (1986). Transient analysis of latent heat storage systems of the shell tube type, *Proceedings of the World Congress III of Chem. Eng.,* Tokyo, Japan, vol. 1, pp. 668–671.

Ismail, K.A.R. and Alves, C.L. (1989). Numerical solution of finned geometries immersed in phase change material, *ASME-HTD* 109, pp. 31–36.

Ismail, K.A.R., Leal, J.F.B. and Zanardi, M.A. (1997). Models of liquid storage tanks, *Energy* 22, 805–815.

Ismail, K.A.R. and Gonçalves, M.M. (1997). Analysis of a latent heat cold storage unit, *International Journal of Energy Research* 21, 1223–1239.

Ismail, K.A.R. (1998). *Cold Storage Systems: Fundamentals and Modeling*, State University of Campinas, Campinas-SP-Brazil (in Portuguese).

Ismail, K.A.R. (1999). *Modeling of Thermal Processes; Fusion and Solidification*, State University of Campinas, Campinas-SP-Brazil (in Portuguese).

Jekel, T. B., Mitchell, J. W. and Klein, S. A. (1993). Modeling of ice-storage tanks, *ASHRAE Transactions* 99 (1), 1016-1048.

Kern, J. and Wells, G.L. (1977). Simple analysis and working equations for the solidification of cylinders and spheres, *Metallurgical Transaction* 8(b), 99.

Kim, C.J. and Kaviany, M. (1990). A numerical method for phase-change problems, *International Journal of Heat and Mass Transfer* 33, 2721–2734.

Kim, C.J. and Kaviany, M. (1992). A numerical method for phase-change problems with convection and diffusion, *International Journal of Heat and Mass Transfer* 35, 457–467.

Knebel, D. (1995). Predicting and evaluating the performance of ice harvesting thermal energy storage systems, *ASHRAE Journal* 37, 22–30.

Lacroix, M. and Voller, V.R. (1990). Finite difference solutions of solidification phase change problems transformed versus fixed grids, *Numerical Heat Transfer-Part B* 17, 24–41.

Lacroix, M. (1989). Computation of heat transfer during melting of a pure substance from an isothermal wall, *Numerical Heat Transfer* 15, 191–210.

Landau, H.G. (1950) Heat conduction in a melting solid, *Quart. of Applied Mathematics* 8, 81–94.

Lee, A.H.W. and Jones, J.W. (1996). Laboratory performance of an ice-on-coil thermal energy storage system for residential and light commercial applications, *Energy* 21(2), 115–130.

London, A.L and Seban, R.A. (1943). Rate of ice formation, *Transactions of ASME* 65, 771–778.

Lotkin, M. (1960). The calculation of heat flow in melting solids, *Quart. of Applied Mathematics* 18, 79–85.

Marshall, R., Heinrich, J. and Zienkiewics, O. (1978). Natural covection in square enclosure by a finite element, penalty function method using primitive fluid variables, *Numerical Heat Transfer* 1, 331–349.

Megerlin, F. (1968) Geometrisch eindimensionale warmeleitung beim schmelzen und erstarrem, forsch., *Ing. Wes*, 34, 40.

Milanez, L.F. and Ismail K.A.R. (1984). Solidification in spheres, theoretical and experimental investigation, Presented at *the 3rd Int. Conference on Multi-Phase and Heat Transfer*, Miami.

Moore, F.E. and Bayazitoglu, Y. (1982). Melting within a spherical enclosure, *Journal of Heat Transfer* 104, 19–23.

Morgan, K., Lewis,R. and Zienkiewicz, O. (1978). An improved algorithm for heat conduction problems with phase change, *International Journal for Numerical Methods in Engineering* 12, 1191–1195.

Murray W. D. and Landis, F. (1959). Numerical and machine solutions of transient heat-conduction problems involving melting or freezing: part. I-method of analysis and sample solutions, *Journal of Heat Transfer* 81, 106–112.

Neto J.H.M. and Krarti, M. (1997). Experimental validation of a numerical model for an internal melt ice-on-coil thermal storage tank, *ASHRAE Transactions* 103(1), 125–138.

Neto, F.M. (1996). *Estudo Parametrico de um Armazenador Termico Tipo Gelo Sobre Serpentina*, Ph.D Thesis, Faculty of Mechanical Engineering, The State University of Campinas, Brazil.

Neto, J.H.M. and Krarti, M. (1997). Deterministic model for an internal melt ice-on-coil thermal storage tank, *ASHRAE Transactions* 103(1), 113–124.

Patankar, S.V. (1980). *Numerical Heat Transfer and Fluid Flow*, Hemisphere, New York.

Pedroso R.I. and Domoto G.A. (1973). Perturbation solutions for spherical solidification of saturated liquids, *Journal of Heat Transfer* 95(1), 42–46.

Pham, Q. (1986). The use of lumped capacitance in the finite element solution of heat conduction problems with phase change, *International Journal of Heat and Mass Transfer* 29, 285–291.

Prud'homme, M., Nguyen, T.H. and Nguyen, D.L. (1989). A heat transfer analysis for solidification of slabs, cylinders, and spheres, *Journal of Heat Transfer* 111, 699–705.

Quispe, O.C. (1996). *Avaliação Numérica Experimental do Conceito De Banco de Gelo de Placas Paralelas*, MSc Thesis, Faculty of Mechanical Engineering, The State University of Campinas, Brazil.

Rao, P.R.and Sastri, V.M.K. (1984). Efficient numerical method for two dimensional phase change problems, *International Journal Heat and Mass Transfer* 27(11), 2077–2084.

Rice, J. and Schnipke, R. (1986). An equal-order velocity pressure formulation that does not exhibit spurious pressure modes, *Comp. Meth. App. Mech. Eng.*, 58, 135–149.

Riley, D.S., Smith, S.T. and Poots, G. (1974). The inward solidification of spheres and circular cylinders, *International Journal Heat and Mass Transfer* 17, 1507–1516.

Saabas, H. and Baliga, B. (1994). Co-located equal order control-volume finite element method for multidimensional incompressible flow, part I: formulation, *Numerical Heat Transfer* 26, 381–407.

Sablani, S., Sastri, S.M.K. and Venkateshan, S.P. (1990). Numerical study of two-dimensional freezing in an annulus, *Journal of Thermophysics and Heat Transfer* 4, 398–400.

Saitoh, T. (1978). Numerical method for multi-dimensional freezing problems in arbitrary domains, *Journal of Heat Transfer*100, 294–299.

Scalon, V.L. (1998). *Estudo da Fusão de um PCM em Geometria Plana Aletada Incluindo os Efeitos Convectivos*, MSc Thesis, The State University of Campinas, Brazil.

Schneider, G., Raithby, G., and Yvanovich, M. (1978). Finite-element solution procedures for solving the incompressible, Navier-Stokes equations using equal order interpolation, *Numerical Heat Transfer* 1, 433–451.

Schneider, G.E. (1990). Computation of solid –liquid phase change including free convection-comparison with data, *Journal of Thermophysics* 4(3), 366–373.

Shih, Y.P. and Chou, T.C. (1971). Analytical solution for freezing a saturated liquid inside or outside spheres, *Chem. Engng. Sci.*, 26, 1787–1793.

Silver, S. C. Milbitz, A. Jones, J. W. Peterson, J. L. and Hunn, B. D. (1989). Component models for computer simulation of ice storage tanks, *ASHRAE Transactions* 95, 1214–1226.

Sinha, T.K. and Gupta, J.P. (1982). Solidification in an annulus, *International Journal of Heat and Mass Transfer* 25, 1771–1773.

Somasundaram, S., Drost, M.K., Brown, D.R.and Antoniak, Z.I. (1993). Integrating thermal energy storage in power plants, *Mechanical Engineering* 115(9), 84–90.

Sparrow, E.M. and Chuck, W. (1984). An implicit/explicit numerical solution scheme for phase-change problems, *Numerical Heat Transfer* 7, 1–15.

Sparrow, E.M., Ramadhyani, S. and Patankar, S.V. (1978b). Effect of subcooling on cylindrical melting, *Journal of Heat Transfer* 100, 395–402.

Sparrow, E.M., Schmidt, R.R. and Ramsey, J.W. (1978a) Experiments on the role of natural convection in the melting of solids, *Journal of Heat Transfer* 100, 11–16.

Sparrow, E.M. and Chuck, W. (1984) Planar or radial freezing with solid-liquid transitions and convective heating at the solid liquid interface, *Numerical Heat Transfer* 7, 17–38.

Sparrow, E.M. and Hsu, C.F. (1981). Analysis of two-dimensional freezing on the outside of a coolant-carrying tube, *International Journal of Heat and Mass Transfer* 24, 1345–1357.

Tao, L.C. (1967). Generalized numerical solutions of freezing a saturated liquid in cylinders and spheres, *AIChE Journal* 13, 165–169.

Vargas, J.V.C. and Bejan, A. (1995) Fundamentals of ice making by convection cooling followed by contact melting, *International Journal of Heat and Mass Transfer* 38(15), 2833–2834.

Vick, B., Nelson, D.J. and Yu, X. (1996). Validation of the algorithm for ice-on-pipe brine thermal storage systems, *ASHRAE Transactions* 102, 55–62.

Vick, B., Nelson, D. J. and Yu, X. (1996). Model of an ice-on-Pipe brine thermal storage component, *ASHRAE Transactions* 102, 45–54.

Voller V.R.and Prakash, C. (1987). A fixed grid numerical modeling methodology for convection-diffusion mush region phase change problems, *International Communications in Heat Mass Transfer* 30(8), 1709–1719.

Voller, V.R. (1990). Fast implicit finite difference method for the analysis of phase change problems, *Numerical Heat Transfer* 17, 155–169.

Voller, V.R. (1985). Implicit finite-difference solutions of the enthalpy formulation on Stefan problems, *IMA-Journal of Numerical Analysis* 5, 201–214.

Zhang, Y. and Faghri, A. (1996). Semi-analytical solution of thermal energy storage system with conjugate laminar forced convection, *International Journal of Heat and Mass Transfer* 39, 717–724.

9

Thermodynamic Optimization of Thermal Energy Storage Systems

A. Bejan

9.1 Introduction – Optimal Distribution of Imperfection: The Generation of System Flow Structure

In this chapter we illustrate the most fundamental aspects of the application of the method of thermodynamic optimization (e.g. Bejan, 1996, 1997) to the conceptual design of systems for thermal energy storage. Two classes of storage systems are reviewed: systems with energy storage as sensible-heat, and systems with latent-heat storing materials. The objective of the method is to identify the ways (features, procedures) in which the system fulfils its functions while performing at the highest thermodynamic level possible. We seek designs that operate least irreversibly, i.e. with minimum generation of entropy.

Designs are destined to remain imperfect, because finite-size constrains such as specified heat transfer surfaces will always force currents to flow against resistances. Designs will always operate irreversibly. The challenge is to do the 'best possible' under the constraints. This is achieved by spreading the imperfections through the system in optimal ways. Optimal spreading means the generation of physical structure—the actual being of the engineered system. In the applications illustrated in this chapter we show the emergence of structures of two types, structure in time (section 9.2) and structure in space and on the temperature scale (sections 9.3–9.6).

'Optimal distribution of imperfection' is the principle that generates flow topology (structure, configuration, construction) in flow systems, i.e. systems far from thermodynamic equilibrium. This 'constructal' principle was identified in engineering (Bejan, 2000). It was shown that the principle governs the generation of morphology not only in engineered flow systems, but also in naturally occurring flow systems. The thought that the constructal principle is a law of physics that accounts for the generation of flow structure everywhere is 'constructal theory'.

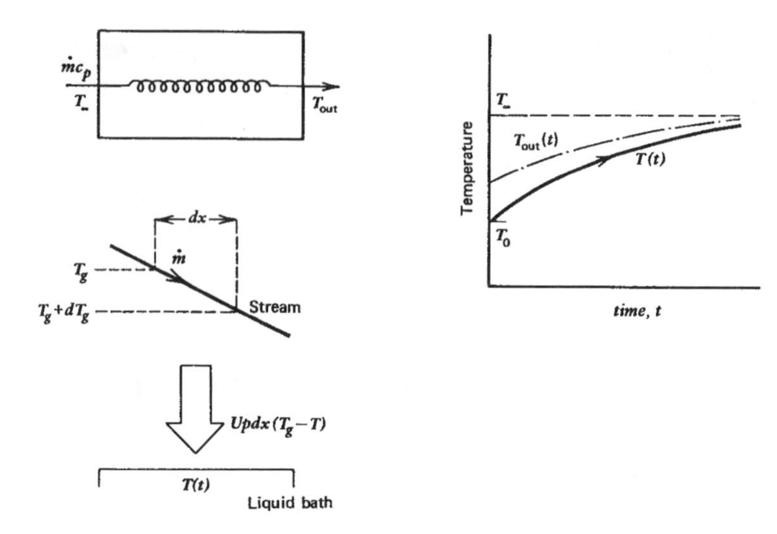

Figure 9.1 Liquid bath element for sensible heat storage (Bejan, 1978).

Important to design engineers who contemplate the application of the constructal principle is the reminder that in the beginning the physical configuration of the system is not known. The configuration is the unknown, and this unknown is the path to better performance. The system is free to 'morph' into better and better structures. The designer should view the system as a changing structure, not as a frozen object the operation of which might be improved if some coincidental flow parameters are adjusted.

9.2 Sensible Heat Storage

Consider the operation of the sensible heat storage system shown schematically in Figure 9.1. The system consists of a large liquid bath of mass M and specific heat C placed in an insulated vessel. Hot gas enters the system through one port, is cooled while flowing through a heat exchanger immersed in the bath, and is eventually discharged into the atmosphere. Gradually, the bath temperature T, as well as the gas outlet temperature T_{out}, rises, approaching the hot gas inlet temperature T_∞. The bath is filled with an incompressible liquid such as water or air. The bath liquid is thermally well mixed so that at any given time its temperature is uniform, T(t). It is assumed that initially the bath temperature equals the environment temperature T_0.

The time dependence of the bath temperature T(t) and gas outlet temperature $T_{out}(t)$ can be derived analytically using the drawing that appears on the left side of Figure 9.1. Locally, the gas-liquid heat transfer is balanced by the enthalpy change in the gas stream,

$$Up\,dx\left(T_g - T\right) + \dot{m}c_p dT_g = 0, \tag{9.1}$$

where p is the local heat transfer area for unit length (wetted perimeter). Integrating over the gas flow path, from x = 0 ($T_g = T_\infty$) to x = L ($T_g = T_{out}$), we obtain

$$\frac{T_{out}(t) - T(t)}{T_\infty - T(t)} = \exp(-N_{tu}) \tag{9.2}$$

with the usual notation for the number of heat transfer units, $N_{tu} = UpL/(\dot{m} c_p)$. Applying the first law of thermodynamics to the liquid bath as a system, we write

$$MC\frac{dT}{dt} = \dot{m}c_p\left(T_\infty - T_{out}\right) \tag{9.3}$$

Note that both T and T_{out} are functions of time. Combining Equations 9.2 and 9.3 and integrating in time from t = 0 (T = T_0) to any time t(T, T_{out}) yields

$$\frac{T(t)-T_0}{T_\infty - T_0} = 1-\exp(-y\theta) \tag{9.4}$$

$$\frac{T_{out}(t)-T(t)}{T_\infty - T(t)} = 1-y\exp(-y\theta) \tag{9.5}$$

where, for brevity, we used the dimensionless groups

$$y = 1-\exp(-N_{tu}) \tag{9.6}$$

$$\theta = \frac{\dot{m}c_p}{MC}t \tag{9.7}$$

Both T and T_{out} approach T_∞ asymptotically—the faster, the higher the N_{tu}. The ability to store energy increases with increasing the charging time θ and the number of heat transfer units N_{tu}. It is shown next that an optimal charging time exists for which the element of Figure 9.1 stores the most exergy per unit of exergy drawn from the high-temperature gas source T_∞.

As shown in Figure 9.2, the batch heating process is accompanied by two irreversibilities. First, the heat transfer between the hot stream and the cold bath always takes place across a finite ΔT. Secondly, the gas stream exhausted into the atmosphere is eventually cooled down to T_0, again by heat transfer across a finite ΔT. A third irreversibility source, neglected in the present analysis, but treated in Bejan (1978, 1996), is the frictional pressure drop on the gas side of the heat exchanger. The combined effect of these irreversibilities is a basic characteristic of all sensible heat storage systems: only a fraction of the exergy brought in by the hot stream is stored in the liquid bath.

The entropy generation rate in the system defined by the dashed boundary in Figure 9.2 is

$$\dot{S}_{gen} = \dot{m}c_p\ln\frac{T_0}{T_\infty} + \frac{Q_0}{T_0} + \frac{d}{dt}\left(MC\ln T\right) \tag{9.8}$$

where $Q_0 = \dot{m} c_p (T_{out} - T_0)$. We are interested in the entropy generated during the time interval 0 – t; this quantity can be calculated by integrating Equation 9.8 and using the $T_{out}(t)$ expression

$$\frac{1}{MC}\int_0^t \dot{S}_{gen}dt = \theta\left[\ln\frac{T_0}{T_\infty} + \frac{T_\infty - T_0}{T_0}\right] + \ln\left[1 + \frac{T_\infty - T_0}{T_0}\left(1-e^{-y\theta}\right)\right] + \frac{T_\infty - T_0}{T_0}(1-e^{-y\theta}) \tag{9.9}$$

This result is more instructive if we express it as the ratio of destroyed exergy, $T_0 \int_0^t \dot{S}_{gen} dt$, divided by the total exergy content of the gas drawn from the hot gas supply:

$$E_x = \dot{m}tc_p \left(T_\infty - T_0 - T_0 \ln \frac{T_\infty}{T_0} \right) \tag{9.10}$$

We define the entropy generation number

$$N_S = \frac{T_0 \int_0^t \dot{S}_{gen} dt}{E_x} = 1 - \frac{\tau\left(1-e^{-y\theta}\right) - \ln\left[1+\tau\left(1-e^{-y\theta}\right)\right]}{\theta\left[\tau - \ln\left(1+\tau\right)\right]} \tag{9.11}$$

where $\tau = (T_\infty - T_0)/T_0$. As shown in Figure 9.3, N_s depends on the charging time θ, the heat transfer area (N_{tu}), and the dimensionless temperature τ. It is clear that for any given τ and N_{tu} an optimal time θ_{opt} exists when the fraction of accumulated irreversibility N_s reaches its minimum. Away from this minimum, N_s approaches unity. In the $\theta \to 0$ limit the entire exergy content of the hot stream is destroyed by heat transfer to the liquid bath, which is initially at atmospheric temperature T_0. In the $\theta \to \infty$ limit the irreversibility shifts outside the liquid bath. The gas stream exits from the heat exchanger as hot as it enters (T_∞), hence its exergy content is destroyed entirely by direct heat transfer to T_0.

The optimal charging time can be calculated explicitly in the limit $\tau \to 0$, where the entropy generation number, Equation 9.11, becomes

$$N_S = 1 - \frac{1}{\theta}\left(1-e^{-y\theta}\right)^2 \tag{9.12}$$

Solving $\partial N_s/\partial\theta = 0$ we find:

$$\theta_{opt} = \frac{1.256}{y} = \frac{1.256}{1-\exp\left(-N_{tu}\right)} \tag{9.13}$$

This result suggests that for all heat exchangers with $N_{tu} >> 1$, the optimal charging time t_{opt} is of the same order of magnitude as $MC/(\dot{m}c_p)$. In other words, we must terminate the heating process when the thermal inertia of the hot gas used ($\dot{m}c_pt$) is comparable to the thermal inertia of the liquid bath.

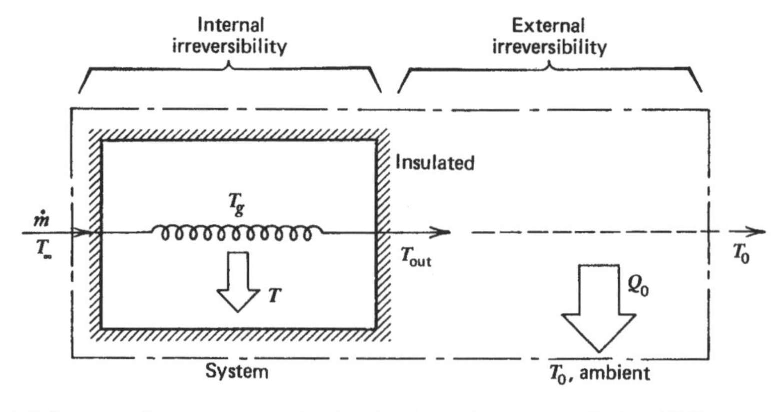

Figure 9.2 Sources of entropy generation in a batch heating process (Bejan, 1978).

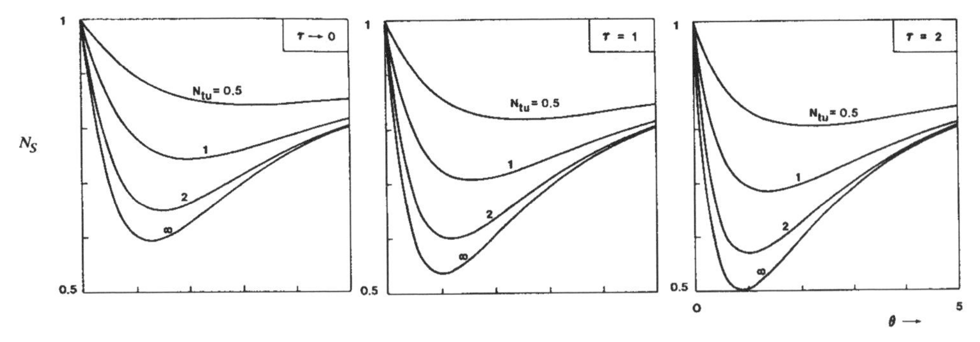

Figure 9.3 Fraction of exergy destroyed during the storage process (Bejan, 1978).

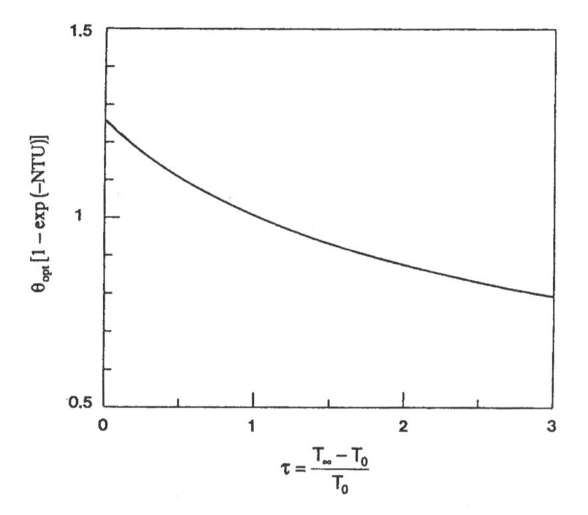

Figure 9.4 Optimal charging time for minimum entropy generation during the sensible heat storage process (Bejan, 1978).

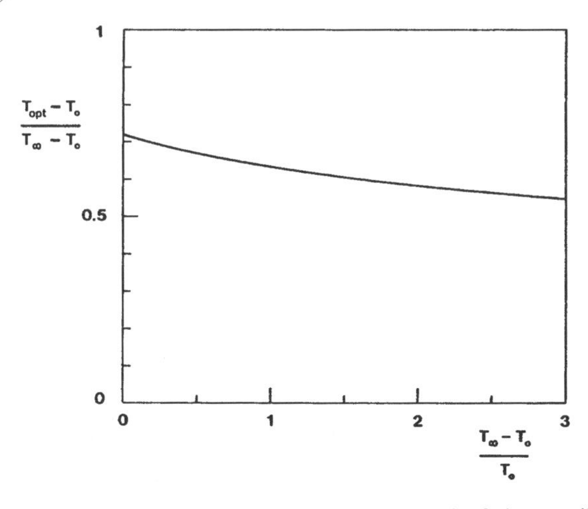

Figure 9.5 The storage element temperature at the optimal end of the sensible heating process (Bejan, 1978).

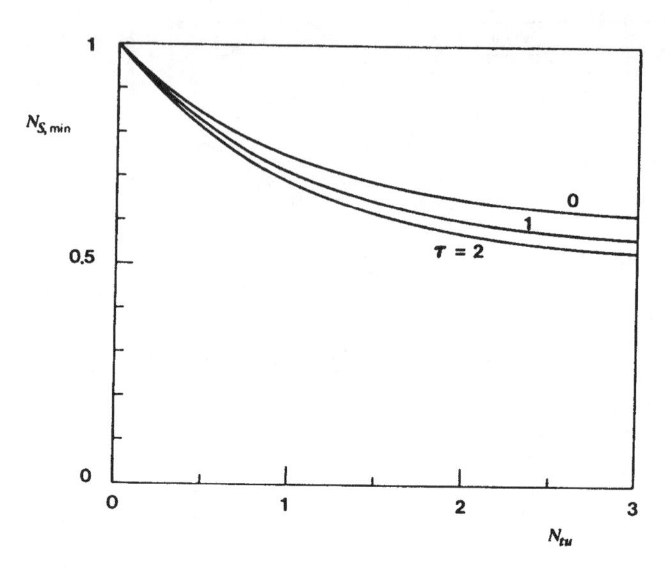

Figure 9.6 Minimum entropy generation during the sensible heating process (Bejan, 1978).

In the general case when τ is finite, the optimal charging time depends on N_{tu} and τ. Figure 9.4 shows this relationship, which was obtained by minimizing N_s numerically. The temperature of the liquid bath at the optimal end of the heating process is shown in Figure 9.5. The thermal energy stored as sensible heat at the end of the process, $MC(T_{opt} - T_0)$, is considerably less than the maximum energy storage capability of the liquid bath, $MC(T_\infty - T_0)$. Moreover, as shown in Figure 9.3, if one seeks to heat the bath until the stored thermal energy reaches its ceiling value ($T = T_\infty$), one runs the risk of storing a very small fraction of the exergy drawn from the high temperature source.

Finally, in Figure 9.6 we see the minimum N_s corresponding to a heating process terminated at θ_{opt}. The fraction of destroyed exergy is at least as large as 50% for most cases (τ, N_{tu}) of practical interest. In the limit $\tau \rightarrow 0$ considered earlier in Equations 9.12 and 9.13, the minimum entropy generation number is given by $N_{s,min} = 0.593 + 0.407\exp(-N_{tu})$ accordingly.

9.3 Series of Sensible-Heat Storage Units

The engineering message of the preceding analysis and optimization is that the storage process of Figure 9.1 is intrinsically inefficient. We saw that even under the best thermodynamic operating conditions, the process destroys approximately as much exergy as it stores (Figure 9.6). We must ask the question, what can be done to make the storage process more efficient? Clearly, we must seek to reduce the internal and external heat transfer irreversibilities discussed in connection with Figure 9.2.

The reduction of these irreversibilities hinges on the ability to minimize the two temperature gaps, the gap between hot inlet (T_∞) and storage unit (T) and the temperature gap between exhaust (T_{out}) and ambient (T_0). The ambient temperature and—at least in the initial phase of the process—the temperature of the storage material are fixed (equal to T_0). Thus, we must concentrate on bringing the stream inlet temperature closer to T and also on bringing the exhaust temperature T_{out} closer to T_0. This task is accomplished in strikingly simple form by the use of a number of storage units positioned in series, as shown in Figure 9.7. The stream exhausted by unit (i – 1) becomes the exergy source stream for unit (i). Next, the stream exhausted by unit (i) is not rejected to ambient temperature, but to a higher temperature, the temperature of unit (i + 1).

It is shown schematically in Figure 9.7(a) that in such an arrangement the temperature of the storage elements decreases monotonically in the direction of flow. The temperature gaps, stream-storage material and exhaust-ambient, are now considerably smaller. Figure 9.7(b) shows the temperature distribution in the series arrangement at one instant during the exergy recovery phase. In this second phase the cold stream collects exergy from the storage units by running in the reverse direction.

This proposal was investigated in great detail by Taylor *et al.* (1990), based on a solid *distributed storage element* model, in which the storage material temperature varied continuously along the stream. Taylor *et al.* showed that the longitudinal conduction of heat through the storage material during the periodic operation of the heat exchanger can have a major impact on the overall irreversibility of the installation. The overall irreversibility figure N_s is again a strong function of the time interval required by the storage part of the cycle: the identification of the optimal storage time interval is critical. The overall N_s is also affected by the geometric aspect ratio of the storage material. The numerical examples given in Taylor *et al.*'s study reveal N_s values that cover the range 0.2 to 0.8. The series of storage units was optimized further by Sekulic and Krane (1992).

The cyclical storage and removal of exergy from a continuous one-dimensional stretch of storage material was also studied by Mathiprakasam and Beeson (1983). One interesting effect illustrated by them is that of the direction of flow during the removal phase. They found that the second law efficiency $(1 - N_s)$ is always lower if the exergy-removal stream flows in the same direction as the original exergy-supply stream (i.e. in parallel), lower than in the counterflow arrangement discussed in the preceding paragraph.

Closely related to the continuous one-dimensional storage scheme with periodic counterflow circulation is the class of periodic heat exchangers recognized as regenerators. The design of this type of heat exchanger was approached on the basis of entropy generation minimization by San *et al.* (1987). Their model consists of two-dimensional parallel-plate channels sandwiched between slabs of storage material. The longitudinal conduction of heat through the storage material is neglected. An important difference between this regenerator model and the continuous storage system analyzed by Taylor *et al.* (1990) is that in the case of the regenerator, the stream exhausted during the storage phase is not dumped into the atmosphere. That stream and the exergy still in it are considered usable. Therefore, the total entropy generation figure of one full cycle in the operation of the regenerator is due to four contributions, namely, the ΔT- and ΔP-inspired irreversibilities of the storage part and the removal part of the cycle.

The minimization of entropy generation in periodic-flow regenerative heat exchangers was also studied by Hutchinson and Lyke (1987) and Shen and Worek (1993). Regenerators for cryogenic applications were optimized by Matsumoto and Shiino (1989), Das and Sahoo (1991), and Sahoo and Das (1994). The thermodynamic arguments of this section were combined with overall cost minimization arguments into a thermoeconomic extension of the sensible heat storage process by Badar *et al.* (1993) and Kotas and Jassim (1993).

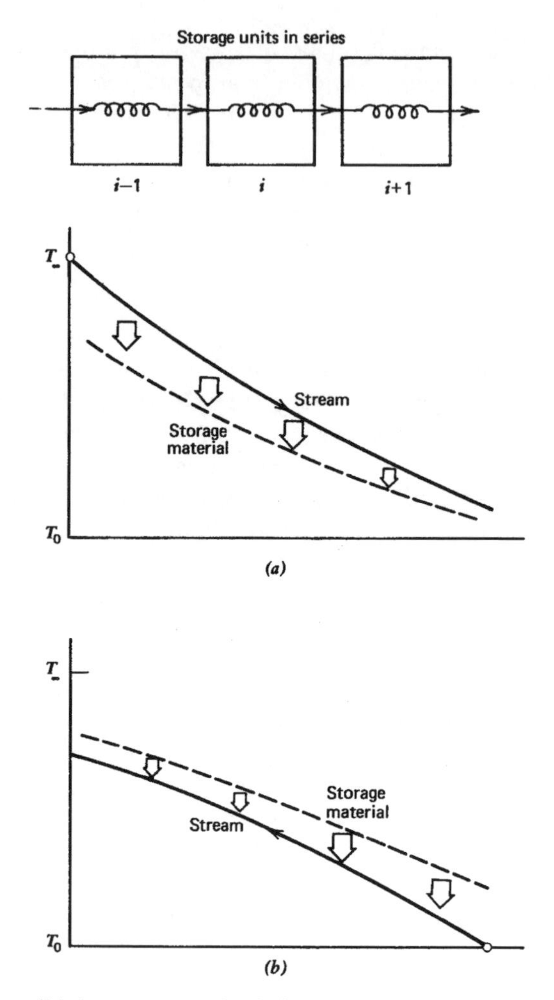

Figure 9.7. Series of sensible heat storage units during (a) charging and (b) discharging.

9.4 Storage Followed by Removal of Sensible Heat

The minimization of entropy generation during a complete cycle was described by Krane (1987). Figure 9.8 shows qualitatively the evolution of the liquid bath temperature during the storage period and during the exergy removal period that immediately follows. The liquid temperature varies periodically without ever reaching the limiting temperature levels T_0 and T_∞.

Insight into the irreversibility composition of the storage and removal cycle is provided by Figure 9.9. The only parameter that varies in this example is the duration of the storage part of the cycle, θ. The storage part is accompanied by the irreversibilities already mentioned, namely, the contributions due to heat exchanger ΔT, heat exchanger ΔP, and the dumping of the used stream into the atmosphere. The exergy removal part of the cycle is plagued by irreversibilities due only to heat exchanger ΔT and ΔP. The gas stream $\dot{m}_r$, heated by the liquid pool during the removal phase—the fruit of the entire scheme—is delivered to a power cycle that can use its exergy content. In Figure 9.9 the two pressure drop effects (during storage and removal) are shown added under the same curve. The conclusions reached during the study of the storage phase alone (section 9.2) are reinforced by Krane's study of the complete storage and removal cycle. The optimal storage time interval, for example, is such that in dimensionless terms it emerges once more as a number of order 1.

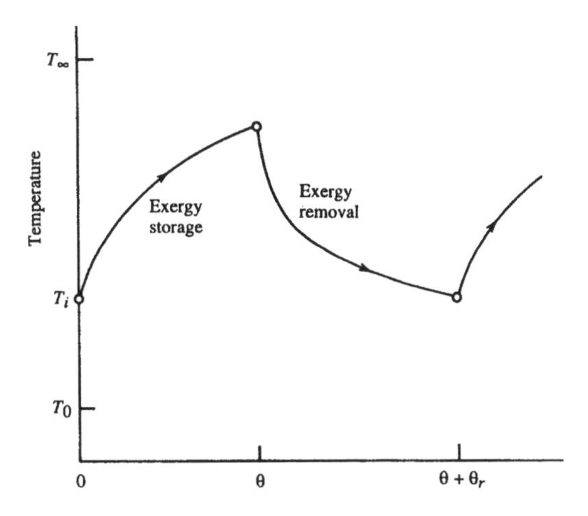

Figure 9.8. The bath system temperature evolution during a complete exergy storage and exergy removal cycle (Krane, 1987).

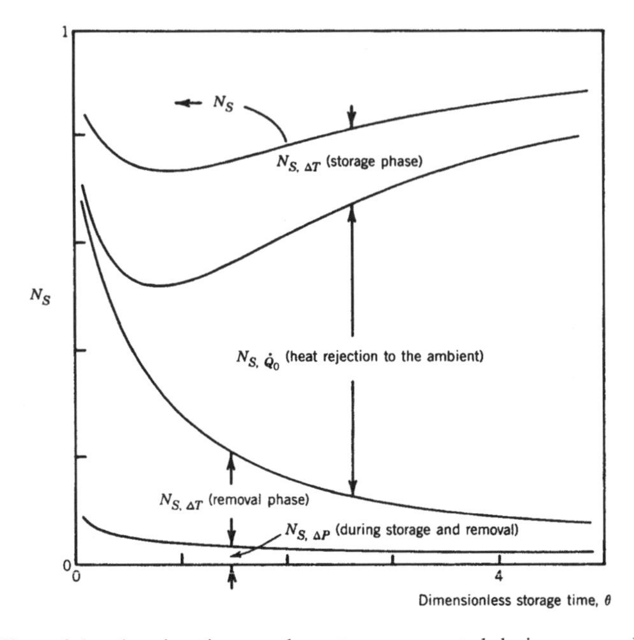

Figure 9.9 The effect of the charging time on the entropy generated during a complete storage and removal cycle (Krane, 1987).

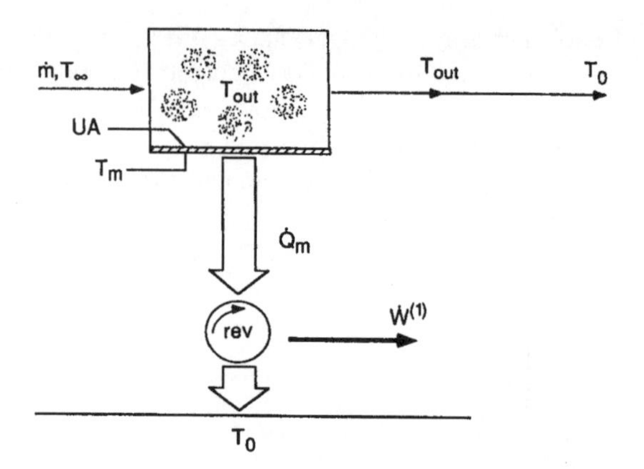

Figure 9.10 The steady production of power using a single phase-change material and a mixed stream (Lim *et al.*, 1992).

A review of the thermodynamic optimization of sensible heat storage methods was presented by Rosen *et al.* (1988). Treated were the individual storage and removal processes as well as the complete cycle. The emphasis was on the development of simple and consistent ways (conventions) to evaluate and compare the performance of competing designs. This work was continued by Rosen (1992), who showed that different efficiency measures are suitable for different applications, and that it is important to agree on a common efficiency definition before comparing designs of the same class.

Another fundamental problem is how to cool or heat a system to a desired temperature by using the minimum quantity of precious fluid (cooling or heating agent). Bejan and Schultz (1982) treated this question relative to the cooldown (refrigeration storage) process, because the analytical treatment of the heating process is identical. Note that minimizing the amount of cooling agent (cryogen) used is equivalent to minimizing the refrigerator work needed to produce (liquefy) the cryogen, or to minimizing the entropy generated in the cold space during the cooldown process.

The minimization of entropy generation during cold storage has received considerable coverage in the recent literature. Sahoo (1989) optimized refrigeration storage in units plagued by heat leak from room temperature. Pedinelli *et al.* (1993) optimized cold storage systems that operate in cycles (charging and discharging), while the storage material experiences sensible cooling or phase change. They also accounted for the effects of stratification. Dumas *et al.* (1992) optimized cold storage units for air conditioning and refrigeration, which are based on freezing and melting processes. The next section focuses on the optimization of this important class of time-dependent process.

9.5 Latent Heat Storage

We now turn our attention to the storage process in which the heated material melts, i.e. to latent heat storage as opposed to sensible heat storage. The simplest way to perform the thermodynamic optimization of the latent heat storage process is shown in Figure 9.10 (Lim *et al.*, 1992). The hot stream of initial temperature T_∞ comes in contact with a single phase-change material through a finite thermal conductance UA, assumed known, where A is the contact area between the melting material and the stream, and U is the overall heat transfer coefficient based on A. The phase-change material (solid or liquid) is at the melting point T_m. The stream is well mixed at the temperature T_{out}, which is also the temperature of the stream exhausted into the atmosphere (T_0).

The steady operation of the installation described in Figure 9.10 accounts for the cyclic operation in which every infinitesimally short storage (melting) stroke is followed by a short energy retrieval (solidification) stroke. During the solidification stroke the flow $\dot{m}$ is stopped, and the recently melted phase-change material is solidified to its original state by the cooling effect provided by the heat engine positioned between T_m and T_0. In this way the steady-state model of Figure 9.10 represents the *complete cycle*, i.e. storage followed by retrieval.

The steady cooling effect of the power plant can be expressed in two ways:

$$\dot{Q}_m = UA\left(T_{out} - T_m\right) \tag{9.14}$$

$$\dot{Q}_m = \dot{m}c_p\left(T_\infty - T_{out}\right) \tag{9.15}$$

By eliminating T_{out} between these two equations, we obtain

$$\dot{Q}_m = \dot{m}c_p \frac{N_{tu}}{1+N_{tu}}\left(T_\infty - T_{out}\right) \tag{9.16}$$

where $N_{tu} = UA/\,\dot{m}c_p$.

Of interest here is the maximum rate of exergy, or useful work ($\dot{W}$ in Figure 9.10), that can be extracted from the phase-change material. For this, we model as reversible the cycle executed by the working fluid between T_m and T_0:

$$\dot{W} = \dot{Q}_m\left(1 - \frac{T_0}{T_m}\right) \tag{9.17}$$

and, after combining with Equation 9.16, we obtain

$$\dot{W} = \dot{m}c_p \frac{N_{tu}}{1+N_{tu}}\left(T_\infty - T_m\right)\left(1 - \frac{T_0}{T_m}\right) \tag{9.18}$$

By maximizing $\dot{W}$ with respect to T_m, i.e. with respect to the type of phase-change material, we obtain the optimal melting and solidification temperature

$$T_{m,opt} = (T_\infty T_0)^{1/2} \tag{9.19}$$

The maximum power output that corresponds to this optimal choice of phase-change material is

$$\dot{W}_{max} = \dot{m}c_p T_\infty \frac{N_{tu}}{1+N_{tu}}\left[1 - \left(\frac{T_0}{T_\infty}\right)^{1/2}\right]^{-2} \tag{9.20}$$

The same results, Equations 9.19 and 9.20 can be obtained by minimizing the total rate of entropy generation. Equation 9.19 was also obtained by Bjurstrom and Carlsson (1985) and

Adebiyi and Russell (1987), who analyzed the heating (melting) portion of the process based on a lumped model and the entropy generation minimization approach used in Bejan (1978) for sensible heat storage.

The details of the actual melting and solidification heat transfer process were taken into account by De Lucia and Bejan (1990, 1991). The effect of temperature distribution in the melt layer, or the effect of liquid superheating during melting, was analyzed by De Lucia and Bejan (1991) using the unidirectional time-dependent conduction process shown in Figure 9.11. The degree of liquid superheating is expressed in dimensionless terms by the Stefan number

$$Ste = \frac{c_p(T_\infty - T_m)}{h_{sf}} \tag{9.21}$$

where c_p is the specific heat of the heating agent, and h_{sf} is the latent heat of melting of the storage material. De Lucia and Bejan (1990) showed that in the limit Ste << 1 the optimal melting temperature is the same as the constant indicated by Equation 9.19. When Ste is of order 1 or greater, $T_{m,opt}$ is no longer a constant but depends on Ste, the absolute temperature ratio T_∞/T_0, the number of heat transfer units during melting N_{tu}, and the duration (t) of the melting process:

$$\tau = \frac{tU^2\left(T_\infty - T_m\right)}{k\rho h_{sf}} \tag{9.22}$$

where k is the thermal conductivity of the liquid phase, and ρ is the density of the liquid and solid phases. Figure 9.12 shows that the effects of Ste and τ are relatively weak, and that Equation 9.19 is a good approximation even when liquid superheating is significant.

The complete cycle of melting during the finite time τ, Equation 9.22, followed by solidification during the next time interval $\tilde{\tau}$ was optimized by De Lucia and Bejan (1991). The solidification dimensionless parameters are defined by

$$\tilde{\tau} = \frac{t_* U_*^2\left(T_m - T_0\right)}{k_s \rho h_{sf}} \qquad \tilde{N}_{tu} = \frac{U_* A}{(\dot{m}c_p)_*} \tag{9.23}$$

where the asterisks indicate the corresponding physical parameters of the installation used during the solidification process, and k_s is the thermal conductivity of the solid phase. The melting and solidification heat transfer problems were formulated as unidirectional time-dependent conduction processes. The liquid superheating during melting and the solid subcooling during solidification were assumed small enough such that their respective Stefan numbers were less than 1. De Lucia and Bejan (1991) showed that the entropy generated during the melting and solidification cycle ($t + t_*$) depends on seven dimensionless parameters: τ, N_{tu}, T_m/T_0, T_∞/T_0, $\tilde{\tau}$, $\tilde{N}_{tu}$ and $k_sU/(kU_*)$. The melting temperature $T_{m,opt}$ that minimizes the time-integrated entropy generation depends on the remaining six parameters.

Figure 9.13 shows the optimal melting temperature in the special case in which $N_{tu} = 1$, $\tilde{N}_{tu} = 1$, $\tau = \tilde{\tau}$, and $k_sU = kU_*$. Once again, the $T_{m,opt}$ constant furnished by Equation 9.19 is a fairly good approximation, although, as expected, the entropy generated during the complete melting and solidification cycle is significantly greater than during the melting process alone.

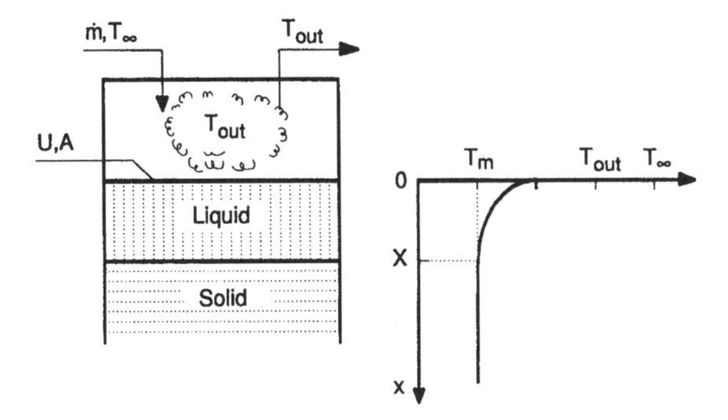

Figure 9.11 Unidirectional time-dependent heating of a layer of phase-change material (De Lucia and Bejan, 1991).

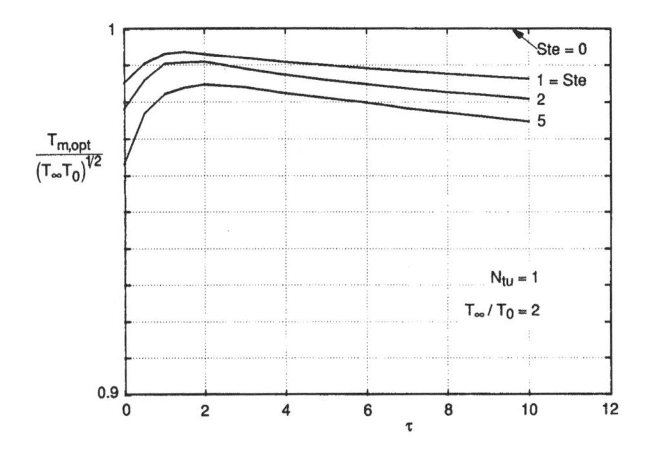

Figure 9.12. The effects of liquid superheating (Ste) and the duration of the melting process (τ) on the optimal melting temperature (De Lucia and Bejan, 1991).

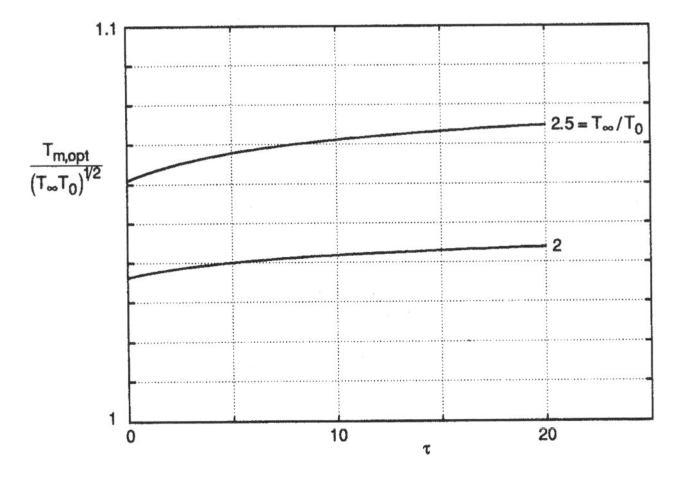

Figure 9.13 The optimal melting temperature for minimum entropy generation during a complete melting and solidification cycle for $N_{tu} = 1$, $N_{tu} = 1$, $\tau = \tilde{\tau}$, and $k_s U = kU_*$.

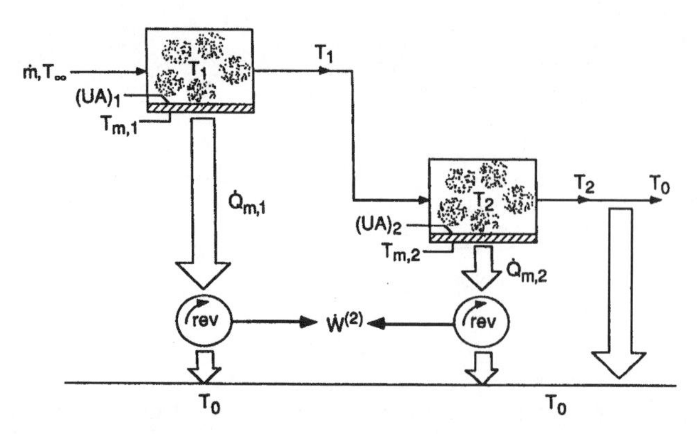

Figure 9.14 Power production based on melting and solidification in two phase-change materials placed in series (Lim *et al.*, 1992).

9.6 Series of Latent-Heat Storage Units

One way to improve the power output of the single-element installation of Figure 9.10 is by placing the exhaust in contact with a second phase-change element of a lower temperature (Lim *et al.*, 1992). This method is illustrated in Figure 9.14, in which, individually, each phase-change element has the features of the element described in Figure 9.10. In general, these two elements contain different phase-change materials ($T_{m,1}$, $T_{m,2}$), and their heat exchanger surfaces are not identical [$(UA)_1$, $(UA)_2$]. The well-mixed gas temperatures above each heat exchanger surface are T_1 and T_2. The total power output from the arrangement of Figure 9.14 is

$$\dot{W}^{(2)} = \dot{Q}_{m,1}\left(1-\frac{T_0}{T_{m,1}}\right)+\dot{Q}_{m,2}\left(1-\frac{T_0}{T_{m,2}}\right) = \dot{m}c_p\left(T_\infty - T_1\right)\left(1-\frac{T_0}{T_{m,1}}\right)+\dot{m}c_p\left(T_1-T_2\right)\left(1-\frac{T_0}{T_{m,2}}\right) \tag{9.24}$$

An analysis equivalent to that contained between Equations 9.14 and 9.16 yields

$$T_1 = \frac{T_\infty + N_{tu,1}T_{m,1}}{1+N_{tu,1}} \qquad T_2 = \frac{T_1 + N_{tu,2}T_{m,2}}{1+N_{tu,2}} \tag{9.25}$$

where the respective numbers of heat transfer units of the two elements are

$$N_{tu,1} = \frac{(UA)_1}{\dot{m}c_p} \qquad N_{tu,2} = \frac{(UA)_2}{\dot{m}c_p} \tag{9.26}$$

For a meaningful comparison of the performance of the two-element system ($N_{tu,1}$, $N_{tu,2}$) relative to the reference single-element system (N_{tu}), it is reasonable to adopt the overall heat exchanger size constraint

$$N_{tu,1} + N_{tu,2} = N_{tu} \tag{9.27}$$

This constraint can be written in terms of a thermal conductance allocation ratio x

$$N_{tu,1} = x\,N_{tu} \qquad N_{tu,2} = (1-x)\,N_{tu} \tag{9.28}$$

The resulting expression for the total power output $\dot{W}^{(2)}$ can be nondimensionalized by dividing $\dot{W}^{(2)}$ by the single-element power maximum derived in Equation 9.20:

$$\widetilde{W} = \frac{\dot{W}^{(2)}}{\dot{W}^{(1)}_{max}} \tag{9.29}$$

where the superscript (1) indicates the use of a single phase-change material. The resulting expression for $\tilde{W}$ depends on the new dimensionless variables

$$\tau_1 = \frac{T_{m,1}}{(T_\infty T_0)} \qquad \tau_2 = \frac{T_{m,2}}{(T_\infty T_0)^{1/2}} \qquad y = \left(\frac{T_0}{T_\infty}\right)^{1/2} \tag{9.30}$$

In sum, the total power output $\tilde{W}$ is a function of five parameters, N_{tu}, x, y, τ_1, and τ_2, of which only three (x, τ_1, τ_2) are degrees of freedom in the design of the two-element installation of Figure 9.14. The total number of heat transfer units (N_{tu}) is constrained by economic considerations, while the temperature parameter y is fixed by the given initial temperature of the stream, T_∞. It was shown that x, τ_1, and τ_2 can be selected so that the overall power output $\tilde{W}$ is maximized.

Much simpler and easier to see is the class of designs in which x is fixed, and only the two melting temperatures (τ_1, τ_2) can vary. Let us begin with the case x = 1/2, in which the heat exchanger inventory is divided equally between the two elements. The optimal melting temperatures $\tau_{1,opt}$, $\tau_{2,opt}$) can be determined by solving the system ($\partial\tilde{W}/\partial\tau_1 = 0$, $\partial\tilde{W}/\partial\tau_2 = 0$), after substituting x = 1/2 into the $\tilde{W}$ expression. This system of two equations becomes

$$\frac{1}{\tau_{1,opt}^2} + \frac{N_{tu}}{2+N_{tu}}\left(1 - \frac{y}{\tau_{2,opt}}\right) = 1 \tag{9.31}$$

$$\frac{1}{\tau_{2,opt}^2} = 1 + \frac{N_{tu}}{2+N_{tu}}(y\tau_{1,opt} - 1) \tag{9.32}$$

The solution for $\tau_{1,opt}$ ($N_{tu,y}$) and $\tau_{2,opt}$ ($N_{tu,y}$) is shown in Figures 9.15 and 9.16. The fact that $\tau_{1,opt} > 1$ and $\tau_{2,opt} < 1$ indicates that the optimal melting temperature of the upstream element must be higher than the single value recommended by Equation 9.19, and, at the same time, the optimal melting temperature of the downstream element must be lower than $(T_\infty T_0)^{1/2}$. The departure of both $\tau_{1,opt}$ and $\tau_{2,opt}$ from the value 1 (i.e. from Equation 9.19) is more accentuated as N_{tu} increases and as y decreases. This means that as the total area A and the inlet temperature T_∞ increase, the optimization of the two-element system gains in importance relative to simply using Equation 9.19 for the selection of both materials.

The total relative power output that corresponds to the optimal temperatures calculated in Figures 9.15 and 9.16 is plotted in Figure 9.17. The subscript of $\tilde{W}_{max,max}$ on the ordinate is a reminder that $\tilde{W}$ was maximized twice, with respect to τ_1 and τ_2. Figure 9.17 shows that the relative gain in power output increases as N_{tu} and $y = (T_0/T_\infty)^{1/2}$ increase. For example, when $T_\infty/T_0 = 2$ and $N_{tu} >> 1$, the temperature span parameter is y = 0.707, and $\tilde{W}_{max,max} = 1.282$. In other words, the optimal two-element design promises to produce

28.2% more power than the optimal single-element design. The ceiling value of 4/3 reached by $\tilde{W}_{max,max}$ at y = 1 is deceptive, because in the limit y→1 the power $\tilde{W}$ produced by either the scheme of Figure 9.10 or the scheme of Figure 9.14 approaches zero.

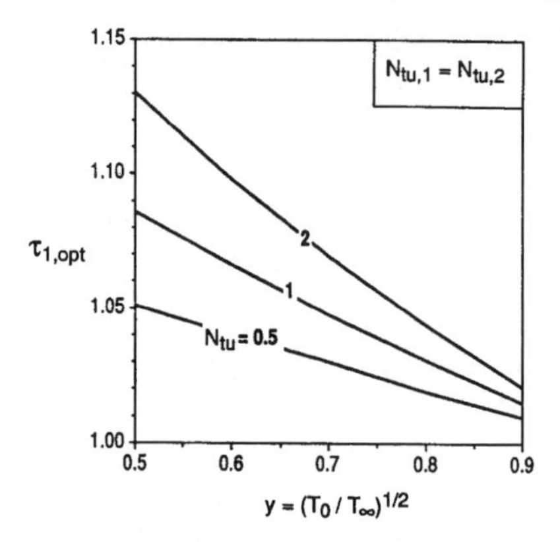

Figure 9.15 The optimal melting temperature of the phase-change material used in the upstream element of Figure 9.14 (Lim *et al.*, 1992).

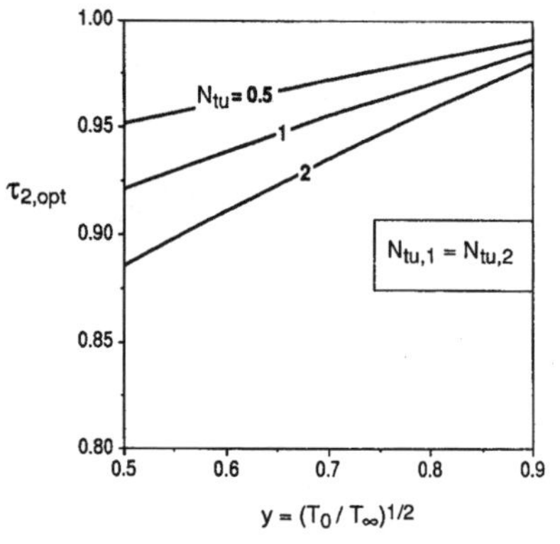

Figure 9.16 The optimal melting temperature of the phase-change material used in the downstream element of Figure 9.14 (Lim *et al.*, 1992).

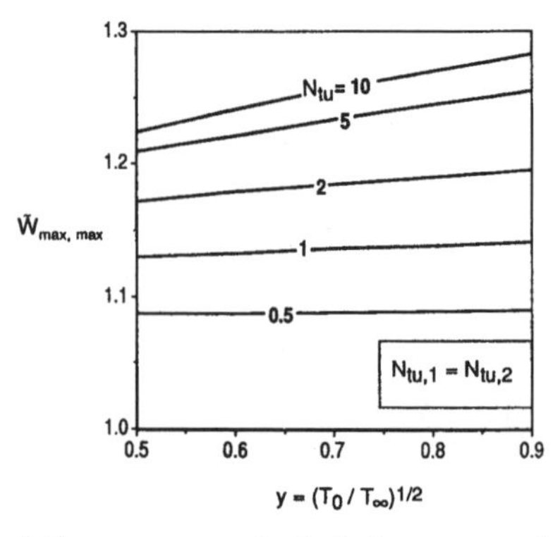

Figure 9.17 The maximum relative power output that corresponds to the optimal melting temperatures reported in Figures 9.15 and 9.16 (Lim *et al.*, 1992).

In the analysis reported in this section the two heat exchangers were assumed to have the same size ($N_{tu,1} = N_{tu,2}$ or x = 1/2), and the relative power function $\tilde{W}$ was maximized only with respect to τ_1 and τ_2. In a subsequent effort Lim *et al.* (1992) relaxed the assumption that x is fixed, and maximize $\tilde{W}$ with respect to x, τ_1, and τ_2 simultaneously. They carried out this work numerically while searching for x_{opt}, $\tau_{1,opt}$, and $\tau_{2,opt}$ as functions of N_{tu} and y. In the domain represented by $0.5 \leq N_{tu} \leq 10$ and $0.5 \leq y \leq 0.9$ they found that the optimal value of x varies about 0.5±0.0001, i.e. that $\tau_{1,opt}$, $\tau_{2,opt}$, and the three times maximized $\tilde{W}$ have practically the same values as in Figures 9.15–9.17.

In conclusion, the x = 1/2 value assumed at the start of this section is a very good approximation for the optimal value that maximizes $\tilde{W}$. For minimum entropy generation, then, the total heat exchanger inventory N_{tu} must be divided equally between the two phase-change elements.

In view of the improvement in thermodynamic performance registered in going from the single-element scheme (Figure 9.10) to the two-element scheme (Figure 9.14), it is reasonable to think of the limit in which we employ not two but an infinite number of elements (materials) in series (Lim *et al.*, 1992). The melting points of these elements vary infinitesimally from one element to the next, so that the longitudinal distribution of melting points is represented by the function $T_m(x)$. This new arrangement is analyzed in Figure 9.18, where x indicates the position of each element along the stream $\dot{m}$.

As a preliminary step in the analysis of Figure 9.18, it is useful to consider the performance of the simpler scheme shown in Figure 9.19. Here, the melting temperature is unique; in other words, from the inlet to the outlet of the hot stream the melting and solidification processes occur in a layer of the same phase-change material. The essential difference between Figures 9.10 and 9.19 is that in Figure 9.19, the hot stream is not mixed: its temperature varies smoothly along the layer of phase-change material.

The power that can be extracted from the scheme of Figure 9.19 can be calculated with Equation 9.17, in which the heating rate is now given by

$$\dot{Q}_m = \dot{m}c_p\left(T_\infty - T_m\right)\left(1 - e^{-N_{tu}}\right) \tag{9.33}$$

Finally, by combining Equations 9.17 and 9.33 we obtain

$$\dot{W}_{Fig.9.19} = \dot{m}c_p\left(1 - e^{-N_{tu}}\right)\left(T_\infty - T_m\right)\left(1 - \frac{T_0}{T_m}\right) \tag{9.34}$$

By solving $\partial\widetilde{W}_{Fig.9.19} / \partial T_m = 0$, we find that the maximum power output occurs at the same optimal melting point as in Equation 9.19. The power output that corresponds to this optimum is

$$\widetilde{W}_{Fig.9.19,max} = \dot{m}c_p T_\infty\left(1 - e^{-N_{tu}}\right)\left[1 - \left(\frac{T_0}{T_\infty}\right)^{1/2}\right]^{2} \tag{9.35}$$

Now we turn our attention to the total power output, which is the sum of the power produced by each slice of infinitesimal thickness *dx* in Figure 9.18:

$$d\dot{W} = \left(hp\,dx\right)\left(T - T_m\right)\left(1 - \frac{T_0}{T_m}\right) \tag{9.36}$$

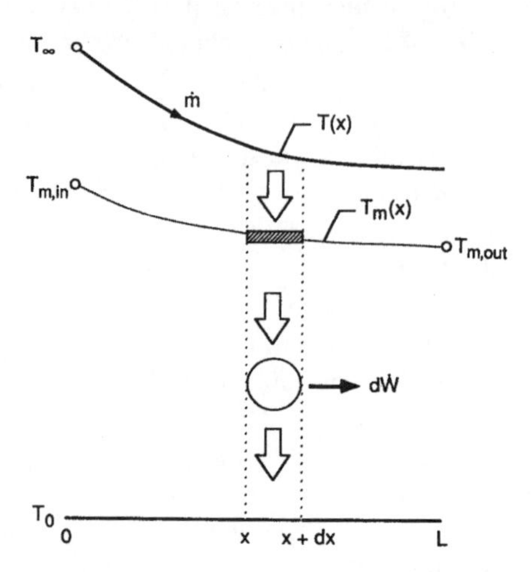

Figure 9.18 Series of infinitesimal phase-change elements in contact with an unmixed stream (Lim *et al.*, 1992).

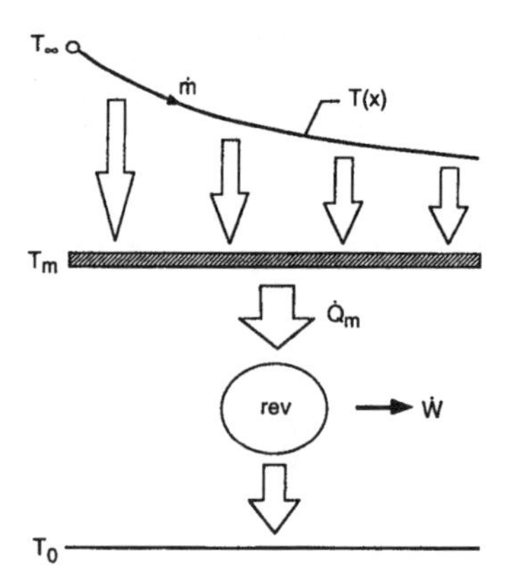

Figure 9.19 Single phase-change material in contact with an unmixed stream (Lim *et al.*, 1992).

In this expression p is the heat transfer area per unit length, while T(x) is the bulk temperature of the hot stream entering the slice (control volume) of thickness dx. The dx-thin control volume is analogous to the system of Figure 9.19. They are analogous in the sense that each of these systems has a unique T_m value for the phase-change material. The only difference between them is the length of the area of heat exchange, namely dx in Figure 9.18 vs. L = A/p in Figure 9.19. While maximizing Equation 9.34, we found that the optimal melting temperature $T_{m,opt} = (T_\infty T_0)^{1/2}$ is not a function of the length of thermal contact. This means that the optimal local melting point for the dx-thin element of Figure 9.18 is simply

$$T_{m,opt}(x) = \left[T(x) \cdot T_0\right]^{1/2} \tag{9.37}$$

The two temperature distributions T(x) and $T_{m,opt}(x)$ can be obtained by combining Equation 9.37 with the analysis of the heat exchange between T and T_m:

$$\dot{m}c_p\left(-\frac{dT}{dx}\right)=hp\left(T-T_m\right) \tag{9.38}$$

By eliminating T(x) between Equations 9.37 and 9.38, and by integrating from x = 0 to x = L, we obtain the relationship between the temperatures of the two ends of the string of phase-change elements:

$$\frac{T_{m,in}-T_0}{T_{m,out}-T_0}=\exp\left(\frac{1}{2}N_{tu}\right) \tag{9.39}$$

In the above equation we set $T_{m,in} = (T_\infty T_0)^{1/2}$ in accordance with Equation 9.37. The total power delivered by the scheme of Figure 9.18 is the result of integrating $d\dot{W}$ from x = 0 to x = L. In that integral we replace T(x) with $T_{m,opt}/T_0$ and after some algebra we obtain

$$\dot{W}_{max}^{(\infty)}=2\dot{m}c_p\int_{T_{m,out}}^{T_{m,in}}\left(\frac{T_m}{T_0}-1\right)dT=2\dot{m}c_p\left[\frac{1}{2T_0}\left(T_{m,in}^2-T_{m,out}^2\right)-T_{m,in}+T_{m,out}\right] \tag{9.40}$$

The superscript (∞) indicates that this maximum power output is based on an infinite number of elements in series. Finally, by writing $T_{m,in} = (T_\infty T_0)^{1/2}$, and by eliminating $T_{m,out}$ between Equations 9.39 and 9.40, it is possible to show that

$$\dot{W}_{max}^{(\infty)}=\dot{m}c_pT_\infty\left(1-e^{-N_{tu}}\right)\left[1-\left(\frac{T_0}{T_\infty}\right)^{1/2}\right]^2 \tag{9.41}$$

Equations 9.41 and 9.35 are identical, i.e. the maximum power from the scheme of Figure 9.18 is the same as that produced by the scheme of Figure 9.19. How this maximum power output compares to that of the single-element arrangement (Figure 9.10) is illustrated in Figure 9.20. Plotted on the ordinate is the ratio

$$\frac{\dot{W}_{max}^{(\infty)}}{\dot{W}_{max}^{(1)}}=\left(\frac{1}{N_{tu}}+1\right)\left(1-e^{-N_{tu}}\right) \tag{9.42}$$

which depends only on the total number of heat transfer units. Its maximum value, 1.298, occurs at N_{tu} = 1.793, indicating that relative to the arrangement of Figure 9.10 the optimal designs in Figures 9.18 and 9.19 can produce up to 30% more power.

9.7 Other Configurations and Storage Models

Charach's (1994) review shows that the thermodynamic optimization of storage installations with melting and solidification has attracted a lot of interest. For example, Adebiyi (1991) considered a bed with particles with several ratios of latent heat to sensible heat storage capability. He modeled the heat conduction as one-dimensional in the cylindrical pellet geometry. For the complete storage and removal cycle, he found that the optimal phase-change temperature is equal to the arithmetic average of the heat source and ambient temperatures.

Charach and Zemel (1992a) optimized the thermodynamic performance of latent heat storage in the shell of a shell-and-tube heat exchanger. The focus was on the effect of two-dimensional heat transfer. Charach and Zemel (1992a) also considered the effect of pressure drop on the stream side of the heat exchanger. Their analysis was extended by Charach and Zemel (1992b) and Charach (1993) to the complete melting and solidification cycle by using a quasi-steady treatment of the phase change process occurring in the shell. These latest studies showed that the optimal phase-change temperature is bounded from above and from below by the arithmetic and, respectively, geometric averages of the source and ambient temperatures.

Another interesting direction is highlighted by the study of latent heat storage units coupled in series with a power plant, and optimized over the entire storage and removal cycle. Bellecci and Conti (1994a,b) showed that minimization of entropy generation and operation stability are two competing criteria in the optimization of the aggregate installation. The arithmetic average of the extreme temperatures again emerges as the optimal phase-change temperature. Detailed modeling and numerical simulations of the heat transfer behavior of the shell-and-tube phase-change heat exchanger were performed by Bellecci and Conti (1993a,b).

Aceves-Saborio *et al.* (1994) developed a systematic way of modeling phase-change storage systems by applying the lumped model to many independent elements. They also considered the more general case in which the phase-change material melts over a finite temperature range. Aceves-Saborio *et al.* (1993) optimized a single latent-heat cell by first simulating numerically the phase-change process. This is an important and difficult task because the natural convection currents that occur in the liquid control the shape and movement of the liquid-solid interface (e.g. section 9.4 in Bejan, 1995). It was shown by De Lucia and Bejan (1990) that when the melting process is controlled by natural convection, the optimal melting temperature is equal to the geometric average shown in Equation 9.19.

Adebiyi *et al.* (1992) constructed a numerical model for simulating and then optimizing the performance of storage systems with multiple phase-change materials. They found that the second law efficiency of systems with multiple materials can exceed by 13–26% the efficiency of systems employing a single material. Fundamental studies of entropy generation minimization in time-dependent unidirectional heat conduction were conducted by Charach and Rubinstein (1989) and Gordon *et al.* (1990).

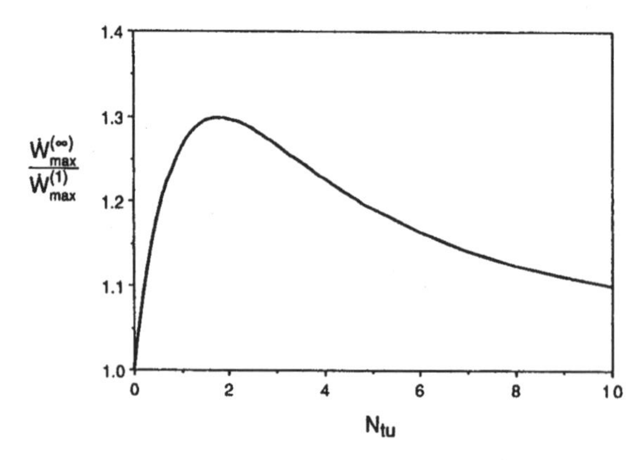

Figure 9.20 The relative performance of the optimized arrangements of Figures 9.19 and 9.10 (Lim *et al.*, 1992).

In this chapter we reviewed the early work on the fundamental thermodynamic problem of how to cool or heat a body in the least irreversible manner possible while subject to size and time constraints. We showed that this fundamental problem triggered a significant amount of work in the 1980s and 1990s in engineering. More recently the same basic question was subjected to study in physics (Andresen and Gordon, 1992a,b). These authors minimized the generation of entropy instead of the amount of heating or cooling agent used. As heat source or heat sink they assumed a heat reservoir of temperature that can be varied at will. Between the heat reservoir and the thermal inertia they assumed several heat transfer rate laws, for example, convection, with a constant heat transfer coefficient, and radiation, with constant (temperature independent) emissivities. In a companion paper, Andresen and Gordon (1992b) considered a related problem—the heating of the body of interest is effected by a stream, while the heat reservoir temperature may change or remain constant. The result of the entropy generation minimization procedure is again an optimal flow rate of heating agent. In both papers the effect of pressure drop was neglected.

The optimization techniques displayed in this chapter are employed further in Bejan (2000), where it is shown that the temporal and spatial structure of natural flow systems can also be reasoned based on global performance maximization (constructal theory).

Nomenclature

A	area, m^2
c_p	specific heat at constant pressure, $J\ kg^{-1}\ K^{-1}$
C	specific heat, $J\ kg^{-1}\ K^{-1}$
E_x	exergy, J
h	heat transfer coefficient, $W\ m^{-2}K^{-1}$
h_{sf}	latent heat of melting, $J\ kg^{-1}$
i	index, Figure 9.7
k	thermal conductivity, $W\ m^{-1}K^{-1}$
L	length, m
$\dot{m}$	mass flow rate, $kg\ s^{-1}$
M	mass, kg
N_s	entropy generation number, Equation 9.11
N_{tu}	number of heat transfer units, Equations 9.2 and 9.23
p	perimeter of contact, m
Q	heat transfer, W
S_{gen}	entropy generation rate, $W\ K^{-1}$
Ste	Stefan number, Equation 9.21
t	time, s
T	temperature, K
U, U_*	overall heat transfer coefficient, $W\ m^{-2}K^{-1}$
$\dot{W}$	power, W
$\tilde{W}$	dimensionless power, Equation 9.29
x	longitudinal coordinate, m; conductance allocation ratio, Equation 9.28
y	function, Equation 9.6

Greek Symbols

ΔP	pressure drop, Pa
ΔT	temperature drop, k
θ	dimensionless time, Equation 9.7
ρ	density, kg m^{-3}
τ	dimensionless temperature difference, Figure 9.4
τ, $\tilde{\tau}$	dimensionless time, Equations 9.22 and 9.23

Subscripts

f	fluid
g	gas
in	inlet
m	melting
max	maximum
min	minimum
opt	optimum
out	outlet
r	removal
s	solid

References

Aceves-Saborio, S., Hernandez-Guerrero, A. and Nakamura, H. (1993). Heat transfer and exergy analysis of the charge process of a heat storage cell, *ASME HTD* 266, 73–80.

Aceves-Saborio, S., Nakamura, H. and Reistad, G.M. (1994). Optimum efficiencies and phase-change temperatures in latent heat storage systems, *J. Energy Resources Technol.*, 116, 79–86.

Adebiyi, G. (1991). A second law study of packed bed energy storage systems utilizing phase-change materials, *J. Solar Energy Eng.*, 113, 146–156.

Adebiyi, G. and Russell, L.D. (1987). A second law analysis of phase-change thermal energy storage systems, *ASME HTD* 80, 9–20.

Adebiyi, G.A., Hodge, B.K., Steele, W.G., Jalalzadeh, A. and Nsofor, E.C. (1992). Computer simulation of high temperature thermal energy storage system employing multiple families of phase-change materials, *ASME AES* 27, 1–11.

Andresen, G. and Gordon, J.M. (1992). Optimal paths for minimizing entropy generation in a common class of finite-time heating and cooling processes, *Int. J. Heat Fluid Flow*, 13, 294–299.

Andresen, G. and Gordon, J.M. (1992b). Optimal heating and cooling strategies for heat exchanger design, *J. Appl. Phys*. 71, 76–79.

Badar, M.A., Zubair, S.M. and Al-Farayedhi, A.A. (1993). Second-law-based thermoeconomic optimization of a sensible heat thermal energy storage system, *Energy* 18, 641–649.

Bejan, A. (1978). Two thermodynamic optima in the design of sensible heat units for energy storage, *J. Heat Transfer*, 100, 708–712.

Bejan, A., (1995). *Convection Heat Transfer*, 2nd ed. Wiley, New York.

Bejan, A. (1996). *Entropy Generation Minimization.* CRC Press, Boca Raton, FL.

Bejan, A. (1997). *Advanced Engineering Thermodynamics*, Wiley, New York.

Bejan, A. (2000). *Shape and Structure, from Engineering to Nature*, Cambridge University Press, Cambridge, UK.

Bejan, A. and Schultz, W. (1982). Optimum flowrate history for cooldown and energy storage processes, *Int. J. Heat Mass Transfer* 25, 1087–1092.

Bellecci, C. and Conti, M. (1993a). Latent heat thermal storage for solar dynamic power generation, *Solar Energy* 51, 169–173.

Bellecci, C. and Conti, M. (1993b). Phase change thermal storage: transient behavior analysis of a solar receiver/storage module using enthalpy method, *Int. J. Heat Mass Transfer* 36, 2157–2163.

Bellecci, C. and Conti, M. (1994a). Thermal storage for solar dynamic power generation: performance indicators in a second law perspective, *SOLCOM-1 Int. Conf. Comparative Assessment of Solar Power Technologies*, Program and Abstracts, p.3. A Roy, Ed. Israel Ministry of Science and Technology, Jerusalem.

Bellecci, C. and Conti, M. (1994b). Phase change energy storage: entropy production, irreversibility and second law efficiency, *Solar Energy* 53, 163–170.

Bjurstrom, H. and Carlsson, B. (1985). An exergy analysis of sensible and latent heat storage, *Heat Recovery Syst.* 5, 233–250.

Charach, Ch. (1993). Second-law efficiency of an energy storage-removal cycle in a phase-change material shell-and-tube exchanger, *J. Solar Energy Eng.* 115, 240–243.

Charach, Ch. (1994). On the second law efficiency of thermal energy storage. *Proc. Int. Conf. Comparative Assessments of Solar Power Technologies*, A. Roy and W. Grasse, Eds. Muller Verlag, Karlsruhe.

Charach, Ch. and Rubinstein, I. (1989). On entropy generation in phase-change heat conduction, *J. Appl. Phys.* 66, 4053–4061.

Charach, Ch. and Zemel, A. (1992a). Thermodynamic analysis of latent heat storage in a shell-and-tube heat exchanger, *J. Solar Energy Eng.* 114, 93–99.

Charach, Ch. and Zemel, A. (1992b). Irreversible thermodynamics of phase-change heat transfer: basic principles and applications latent heat storage, *Open Syst. Inform. Dyn.* 1, 423–458.

Das, S. K. and Sahoo, R. K. (1991). Thermodynamic optimization of regenerators, *Cryogenics* 31, 862–868.

DeLucia, M. and Bejan, A. (1990). Thermodynamics of energy storage by melting due to conduction or natural convection, *J. Solar Energy Eng.* 112, 110–116.

De Lucia, M. and Bejan, A. (1991). Thermodynamics of phase-change energy storage: the effects of liquid superheating during melting, and irreversibility during solidification, *J. Solar Energy Eng.* 114, 2–10.

Dumas, J.P., Strub, F., Bedecarrats, J.P., Zeraouli, Y., Broto, F., and Lenotre, Ch. (1992). Influence of undercooling on cold storage systems, *Proc. Int. Symp. Efficiency, Costs Optimization and Simulation of Energy Systems* (ECOS'92), pp. 571–575. Zaragoza, Spain.

Gordon, J.M., Rubinstein, I., and Zarmi, Y. (1990). On optimal heating and cooling strategies for melting and freezing, *J. Appl. Phys.* 67, 81–84.

Hutchinson, R.A. and Lyke, S.E. (1987). Microcomputer analysis of regenerative heat exchangers for oscillating flows, *Proc. 1987 ASME/JSME Thermal Eng. Joint Conf.* 2, 653, New York.

Kotas, T.J. and Jassim, R.K. (1993). Costing of exergy flows in the thermoeconomic optimization of the geometry of rotary regenerators, *Proc. Int. Conf. Energy Systems and Ecology (ENSEC'93)*, pp. 313–322. Cracow, Poland.

Krane, R.J. (1987). A second law analysis of the optimum design and operation of thermal energy storage systems, *Int. J. Heat Mass Transfer* 30, 43–57.

Lim, J.S., Bejan, A. and Kim, J.H. (1992). Thermodynamic optimization of phase-change energy storage using two or more materials, *J. Energy Resources Technol.* 114, 84–90.

Mathiprakasam, B. and Beeson, J. (1983). Second law analysis of thermal energy storage devices, *Proc. AIChE Symp. Ser., National Heat Transfer Conf.*, p. 161–168.

Matsumoto, K. and Shiino, M. (1989). Thermal regenerator analysis: analytical solution for effectiveness and entropy production in regenerative process, *Cryogenics* 29, 888–894.

Pedinelli, N., Rosen, M.A. and Hooper, F.C. (1993). Thermodynamic assessment of cold capacity thermal energy storage systems, *Proc. Int. Conf. Energy Systems and Ecology (ENSEC'93)*, pp. 705–712, Cracow, Poland.

Rosen, M.A. (1992). Appropriate thermodynamic performance measures for closed systems for thermal energy storage, *J. Solar Energy Eng.* 114, 100–105.

Rosen, M.A., Hooper, F.C. and Barbaris, L.N. (1988). Exergy analysis for the evaluation of closed thermal energy storage systems, *J. Solar Energy Eng.* 110, 255–261.

Sahoo, R.K. (1989). Exergy maximization in refrigeration storage units with heat leak, *Cryogenics*, 29, 59–64.

Sahoo, R.K. and Das, S.K. (1994). Exergy maximization in cryogenic regenerators, *Cryogenics* 34, 475–482.

San, J.Y., Worek, W.M. and Lavan, Z. (1987). Second-law analysis of a two-dimensional regenerator, *Energy* 12, 485–496.

Sekulic, D.P. and Krane, R.J. (1992). The use of multiple storage elements to improve the second law efficiency of a thermal energy storage system. Parts I and II. *Proc. Int. Symp. Efficiency, Costs, Optimization and Simulation of Energy Systems (ECOS'92)*, pp. 61-72, Zaragoza, Spain.

Shen, C. M. and Worek, W. M. (1993). Second-law optimization of regenerative heat exchangers, including the effect of matrix heat conduction, *Energy* 18, 355-363.

Taylor, M. J., Krane, R. J. and Parsons, J. R. (1990). Second law optimization of a sensible heat thermal energy storage system with a distributed storage element. Parts 1 and 2. In: *A Future for Energy: Proc. Florence World Energy Symp.* S.S. Stecco and M.J. Moran, Eds., pp. 885-908. Pergamon Press, Oxford.

10

Energy and Exergy Analyses of Thermal Energy Storage Systems

M.A. Rosen and I. Dincer

10.1 Introduction

Thermal Energy Storage (TES) systems for heating or cooling capacity are often utilized in applications where the occurrence of a demand for energy and that of the economically most favorable supply of energy are not coincident. Thermal storages are an essential element of many energy conservation programs, in industry, in commercial building, and in solar energy utilization. Numerous reports on TES applications and studies have been published (Hahne, 1986; Kleinbach *et al.*, 1993; Dincer *et al.*, 1997a, 1997b; Dincer, 1999; Beckman and Gilli, 1984; Bejan, 1982, 1995; Jansen and Sorensen, 1984).

Many types of TES systems exist for storing either heating or cooling capacity. The storage medium often remains in a single phase during the storing cycle (so that only sensible heat is stored), but sometimes undergoes phase change (so that some of the energy is stored as latent heat). Sensible TESs (e.g. liquid water systems) exhibit changes in temperature in the store as heat is added or removed (Ismail and Stuginsky, 1999; Ismail *et al.*, 1997). In latent TESs (e.g. liquid water/ice systems and eutectic salt systems), the storage temperature remains fixed during the phase-change portion of the storage cycle (Adebiyi *et al.,* 1996; Laouadi and Lacroix, 1999; Brousseau and Lacroix, 1999; Costa *et al.*, 1998). The storage medium can be located in storage 'containers' of various types and sizes, including storage tanks, ponds, caverns and underground aquifers.

The authors have investigated methodologies for TES evaluation and comparison (e.g. Rosen, 1991, 1992b, 1999b; Rosen and Dincer, 1999a), and concluded that, while many technically and economically successful thermal storages are in operation, no generally valid basis for comparing the achieved performance of one storage with that of another operating under different conditions has found broad acceptance. The energy efficiency of a TES system, the ratio of the energy recovered from storage to that originally input, is conventionally used to measure TES performance. The energy efficiency, however, is an inadequate measure because it does not take into account all the considerations necessary in TES evaluation (e.g. how nearly the performance of the system approaches the ideal, the storage duration, the temperatures of the supplied and recovered thermal energy and of the surroundings).

Exergy analysis is a thermodynamic analysis technique based on the Second Law of Thermodynamics which provides an alternative and illuminating means of assessing and comparing TES systems rationally and meaningfully. In particular, exergy analysis yields efficiencies which provide a true measure of how nearly actual performance approaches the ideal, and identifies more clearly than energy analysis the causes and locations of thermodynamic losses. Consequently, exergy analysis can assist in improving and optimizing TES designs. Increasing application and recognition of the usefulness of exergy methods by those in industry, government and academia has been observed in recent years (Moran, 1989, 1990; Kotas, 1995; Edgerton, 1982; Ahern, 1980; Morris *et al.*, 1988). The present authors, for instance, have examined exergy analysis methodologies (Rosen and Horazak, 1995; Rosen, 1999a) and applied them to industrial systems (Rosen and Scott, 1998; Rosen and Horazak, 1995), countries (Rosen and Dincer, 1997b; Rosen, 1992a), environmental impact assessments (Crane *et al.*, 1992; Rosen and Dincer, 1997a, 1999b; Gunnewiek and Rosen, 1998) and TES. To date, however, few exergy analyses of TES systems have been performed (Bader *et al.*, 1993; Bascetincelik *et al.*, 1998; Bejan, 1978, 1982, 1995; Bjurstrom and Carlsson, 1985; Hahne *et al.*, 1989; Krane, 1985, 1987; Krane and Krane, 1991; Mathiprakasam and Beeson, 1983; Moran and Keyhani, 1982; Rosenblad, 1985; Taylor, 1986). Exergy analysis is described extensively elsewhere (e.g. Gaggioli, 1983; Maloney and Burton, 1980; Moran and Shapiro, 2000; Rosen, 1999a; Kestin, 1980), and in the next section background material is provided on energy and exergy analyses that is relevant to applications to TES systems.

The main objectives of this chapter are to describe how energy and exergy analyses of TES systems are performed, and to demonstrate the usefulness of such analyses in providing insights into TES behavior and performance. In the presentation, the thermodynamic considerations and challenges involved in the evaluation of TES systems are discussed, and recent advances in addressing these challenges are described in the hope that standardized exergy-based methodologies can evolve for TES evaluation and comparison.

The topics covered in this chapter can be summarized as follows. First, theoretical and practical aspects of energy and exergy analyses are described (section 10.2) and general thermodynamic considerations in TES evaluation are discussed (section 10.3). Then, the use of exergy in evaluating a closed TES system is detailed (section 10.4), with in-depth discussions of two critical evaluation factors: appropriate TES efficiency measures (section 10.5) and the importance of temperature in performance evaluations for sensible TES systems (section 10.6). Next, applications of exergy analysis to a wide range of TESs are considered, including aquifer TES systems (section 10.7), thermally stratified storage systems (section 10.8) and cold TES systems (section 10.9). Finally, uses of exergy analysis in optimization and design activities are illustrated by examining exergy-based optimal discharge periods for closed TES systems (section 10.10). Many illustrative examples are given in all above listed sections.

10.2 Theory: Energy and Exergy Analyses

This section reviews aspects of thermodynamics most relevant to energy and exergy analyses. Fundamental principles and such related issues as reference-environment selection, efficiency definition, and material-properties acquisition are discussed. Others also discuss these issues (e.g. Moran and Shapiro, 2000; Moran, 1989, 1990; Kotas, 1995; Gaggioli, 1983). General implications of exergy analysis results are discussed, and a step-by-step procedure for energy and exergy analyses is given.

A note on terminology is in order here. A relatively standard terminology and nomenclature has evolved for conventional classical thermodynamics. However, there is at present no generally agreed upon terminology and nomenclature for exergy analysis. A diversity of symbols and names exist for basic and derived quantities (Kotas *et al.*, 1987; Lucca, 1990). For example, exergy is often called available energy, availability, work capability, essergy, etc.; and exergy consumption is often called irreversibility, lost work, dissipated work, dissipation, etc. The exergy analysis nomenclature used here follows that proposed by Kotas *et al.* (1987) as a standard exergy-analysis nomenclature. For the reader unfamiliar with exergy, a glossary of selected exergy terminology is included (see Appendix 10A).

10.2.1 Motivation for energy and exergy analyses

Thermodynamics permits the behavior, performance and efficiency to be described for systems for the conversion of energy from one form to another. Conventional thermodynamic analysis is based primarily upon the First Law of Thermodynamics, which states the principle of conservation of energy. An energy analysis of an energy-conversion system is essentially an accounting of the energies entering and exiting. The exiting energy can be broken down into products and wastes. Efficiencies are often evaluated as ratios of energy quantities, and are often used to assess and compare various systems. Conventional thermal storages, for example, are often compared based on their energy efficiencies.

However, energy efficiencies are often misleading in that they do not always provide a measure of how nearly the performance of a system approaches ideality. Further, the thermodynamic losses which occur within a system (i.e. those factors which cause performance to deviate from ideality) often are not accurately identified and assessed with energy analysis. The results of energy analysis can indicate the main inefficiencies to be within the wrong sections of the system, and a state of technological efficiency different than actually exists.

Exergy analysis permits many of the shortcomings of energy analysis to be overcome. Exergy analysis is based on the Second Law of Thermodynamics, and is useful in identifying the causes, locations, and magnitudes of process inefficiencies. The exergy associated with an energy quantity is a quantitative assessment of its usefulness or quality. Exergy analysis acknowledges that, although energy cannot be created or destroyed, it can be degraded in quality, eventually reaching a state in which it is in complete equilibrium with the surroundings and hence of no further use for performing tasks. This statement is of particular importance to TES systems in that, from a thermodynamic perspective, one wishes to recover as much thermal energy as is reasonably possible after the input energy is stored, with little or no degradation of temperature towards the environmental state.

For TES systems, exergy analysis allows one to determine the maximum potential associated with the incoming thermal energy. This maximum is retained and recovered only if the thermal energy undergoes processes in a reversible manner. No further useful thermal energy or exergy can be extracted by allowing a system and its environment to interact if they are in equilibrium. Losses in the potential for exergy recovery occur in the real world because actual processes are always irreversible.

The exergy flow rate of a flowing commodity is the maximum rate that work may be obtained from it as it passes reversibly to the environmental state, exchanging heat and materials only with the surroundings. In essence, exergy analysis states the theoretical limitations imposed upon a TES system, clearly pointing out that no real system can conserve thermal exergy, and that only a portion of the input thermal exergy can be recovered. Also, exergy analysis quantitatively specifies practical TES limitations by providing losses in a form in which they are a direct measure of lost thermal exergy.

10.2.2 Conceptual balance equations for mass, energy and entropy

A general balance for a quantity in a system may be written as

$$\textit{Input} + \textit{Generation} - \textit{Output} - \textit{Consumption} = \textit{Accumulation} \tag{10.1}$$

Input and output refer respectively to quantities entering and exiting through system boundaries. Generation and consumption refer respectively to quantities produced and consumed within the system. Accumulation refers to build-up (either positive or negative) of the quantity within the system.

Versions of the general balance equation above may be written for mass, energy, entropy and exergy. Mass and energy, being subject to conservation laws (neglecting nuclear reactions), can be neither generated nor consumed. Consequently, the general balance (Equation 10.1) written for each of these quantities becomes

$$\textit{Mass input} - \textit{Mass output} = \textit{Mass accumulation} \tag{10.2}$$

$$\textit{Energy input} - \textit{Energy output} = \textit{Enegy accumulation} \tag{10.3}$$

Before giving the balance equation for exergy, it is useful to examine that for entropy:

$$\textit{Entropy input} + \textit{Entropy generation} - \textit{Entropy output} = \textit{Entropy accumulation} \tag{10.4}$$

Entropy is created during a process due to irreversibilities, but cannot be consumed. By combining the conservation law for energy and non-conservation law for entropy, the exergy balance can be obtained:

$$\textit{Exergy input} - \textit{Exergy output} - \textit{Exergy consumption} = \textit{Exergy accumulation} \tag{10.5}$$

Exergy is consumed due to irreversibilities. Exergy consumption is proportional to entropy creation. Equations 10.3 and 10.5 demonstrate an important main difference between energy and exergy: energy is conserved while exergy, a measure of energy quality or work potential, can be consumed.

These balances describe what is happening in a system between two instants of time. For a complete cyclic process where the initial and final states of the system are identical, the accumulation terms in all the balances are zero.

10.2.3 Detailed balance equations for mass, energy and entropy

Two types of systems are normally considered: open (flow) and closed (non-flow). In general, open systems have mass, heat and work interactions, and closed systems heat and work interactions. Mass flow 'into,' heat transfer 'into' and work transfer 'out of' the system are defined to be positive. Mathematical formulations of the principles of mass and energy conservation and entropy non-conservation can be written for any system, following the general physical interpretations in Equations 10.2–10.4.

Consider a non-steady flow process in a time interval t_1 to t_2. Balances of mass, energy and entropy, respectively, can be written for a control volume as

$$\sum_i m_i - \sum_e m_e = m_2 - m_1 \tag{10.2a}$$

$$\sum_i (e + Pv)_i m_i - \sum_e (e + Pv)_e m_e + \sum_r (Q_r)_{1,2} - (W')_{1,2} = E_2 - E_1 \tag{10.3a}$$

$$\sum_i s_i m_i - \sum_e s_e m_e + \sum_r (Q_r / T_r)_{1,2} + \Pi_{1,2} = S_2 - S_1 \tag{10.4a}$$

Here, m_i and m_e denote respectively the amounts of mass input across port i and exiting across port e; $(Q_r)_{1,2}$ denotes the amount of heat transferred into the control volume across region r on the control surface; $(W')_{1,2}$ denotes the amount of work transferred out of the control volume; $\Pi_{1,2}$ denotes the amount of entropy created in the control volume; m_1, E_1 and S_1 denote respectively the amounts of mass, energy and entropy in the control volume at time t_1 and m_2, E_2 and S_2 denote respectively the same quantities at time t_2; and e, s, P, T and v denote specific energy, specific entropy, absolute pressure, absolute temperature and specific volume, respectively. The total work W' done by a system excludes flow work, and can be written as

$$W' = W + W_x \tag{10.6}$$

where W is the work done by a system due to change in its volume and W_x is the shaft work done by the system. The term 'shaft work' includes all forms of work that can be used to raise a weight (i.e. mechanical work, electrical work, etc.), but excludes work done by a system due to change in its volume. The specific energy e is given by

$$e = u + ke + pe \tag{10.7}$$

where u, ke and pe denote respectively specific internal, kinetic and potential (due to conservative force fields) energies. For irreversible processes $\Pi_{1,2} > 0$, and for reversible processes $\Pi_{1,2} = 0$.

The left sides of Equations 10.2a, 10.3a and 10.4a represent the net amounts of mass, energy and entropy transferred into (and in the case of entropy created within) the control volume, while the right sides represent the amounts of these quantities accumulated within the control volume.

For the mass flow m_j across port j,

$$m_j = \int_{t_1}^{t_2} \left[\int_j (\rho V_n dA)_j \right] dt \tag{10.8}$$

Here, ρ is the density of matter crossing an area element dA on the control surface in time interval t_1 to t_2 and V_n is the velocity component of the matter flow normal to dA. The integration is performed over port j on the control surface. One-dimensional flow (i.e. flow in which the velocity and other intensive properties do not vary with position across the port) is often assumed. Then the previous equation becomes

$$m_j = \int_{t_1}^{t_2} (\rho V_n A)_j dt \tag{10.8a}$$

It has been assumed that heat transfers occur at discrete regions on the control surface and the temperature across these regions is constant. If the temperature varies across a region of heat transfer,

$$(Q_r)_{1,2} = \int_{t_1}^{t_2} \left[\int_r (qdA)_r \right] dt \tag{10.9}$$

and

$$(Q_r / T_r)_{1,2} = \int_{t_1}^{t_2} \left[\int_r (q / T)_r dA_r \right] dt \tag{10.10}$$

where T_r is the temperature at the point on the control surface where the heat flux is q_r. The integral is performed over the surface area of region A_r.

The quantities of mass, energy and entropy in the control volume (denoted by m, E and S) on the right sides of Equations 10.2a, 10.3a and 10.4a, respectively, are given more generally by

$$m = \int \rho dV \tag{10.11}$$

$$E = \int \rho e dV \tag{10.12}$$

$$S = \int \rho s dV \tag{10.13}$$

where the integrals are over the control volume.

For a closed system, $m_i = m_e = 0$ and Equations 10.2a, 10.3a and 10.4a become

$$0 = m_2 - m_1 \tag{10.2b}$$

$$\sum_r (Q_r)_{1,2} - (W')_{1,2} = E_2 - E_1 \tag{10.3b}$$

$$\sum_r (Q_r / T_r)_{1,2} + \Pi_{1,2} = S_2 - S_1 \tag{10.4b}$$

10.2.4 Basic quantities for exergy analysis

Several quantities related to the conceptual exergy balance are described here, following the presentations by Moran (1989) and Kotas (1995).

Exergy of a closed system

The exergy Ξ of a closed system of mass m, or the non-flow exergy, can be expressed as

$$\Xi = \Xi_{ph} + \Xi_o + \Xi_{kin} + \Xi_{pot} \tag{10.14}$$

where

$$\Xi_{pot} = PE \tag{10.15}$$

$$\Xi_{kin} = KE \tag{10.16}$$

$$\Xi_o = \sum_i (\mu_{io} - \mu_{ioo}) N_i \tag{10.17}$$

$$\Xi_{ph} = (U - U_o) + P_o(V - V_o) - T_o(S - S_o) \tag{10.18}$$

where the system has a temperature T, pressure P, chemical potential μ_i for species i, entropy S, energy E, volume V and number of moles N_i of species i. The system is within a conceptual environment in an equilibrium state with intensive properties T_o, P_o and μ_{ioo}. The quantity μ_{io} denotes the value of μ at the environmental state (i.e. at T_o and P_o). The terms on the right side of Equation 10.14 represent respectively physical, chemical, kinetic and potential components of the non-flow exergy of the system.

The exergy Ξ is a property of the system and conceptual environment, combining the extensive properties of the system with the intensive properties of the environment.

Physical non-flow exergy is the maximum work obtainable from a system as it is brought to the environmental state (i.e. to thermal and mechanical equilibrium with the environment), and chemical non-flow exergy is the maximum work obtainable from a system as it is brought from the environmental state to the dead state (i.e. to complete equilibrium with the environment).

Exergy of a flowing stream of matter

The exergy of a flowing stream of matter $\in$ is the sum of non-flow exergy and the exergy associated with the flow work of the stream (with reference to P_o), i.e.

$$\in = \Xi + (P - P_o)V \tag{10.19}$$

Alternatively, $\in$ can be expressed following Equation 10.14 in terms of physical, chemical, kinetic and potential components:

$$\in = \in_{ph} + \in_o + \in_{kin} + \in_{pot} \tag{10.20}$$

where

$$\in_{pot} = PE \tag{10.21}$$

$$\in_{kin} = KE \tag{10.22}$$

$$\in_o = \Xi_o = \sum_i (\mu_{io} - \mu_{ioo}) N_i \tag{10.23}$$

$$\in_{ph} = (H - H_o) - T_o(S - S_o) \tag{10.24}$$

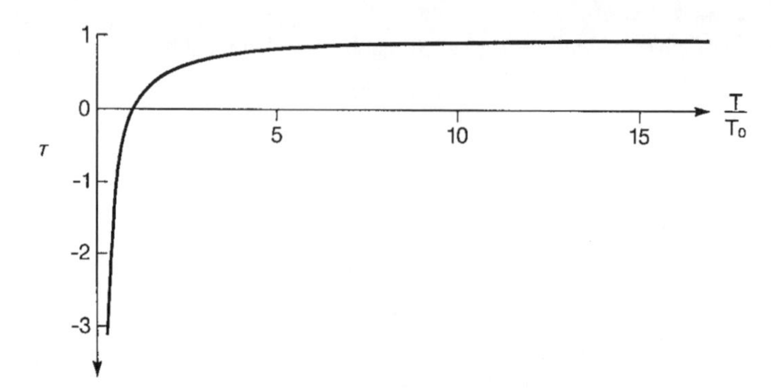

Figure 10.1 The relation between the exergetic temperature factor τ and the absolute temperature ratio T/T_o. The factor τ is equal to zero when $T = T_o$. For heat transfer at above-environment temperatures (i.e. $T > T_o$), $0 < \tau \leq 1$. For heat transfer at sub-environment temperatures (i.e. $T < T_o$), $\tau < 0$, implying that exergy and energy flow in opposite directions in such cases. Note that the magnitude of exergy flow exceeds that of the energy flow when $\tau < -1$, which corresponds to $T < T_o/2$.

Exergy of thermal energy

Consider a control mass, initially at the dead state, being heated or cooled at constant volume in an interaction with some other system. The heat transfer experienced by the control mass is Q. The flow of exergy associated with the heat transfer Q is denoted by X, and can be expressed as

$$X = \int_i^f (1 - T_o/T)\delta Q \tag{10.25}$$

where δQ is an incremental heat transfer, and the integral is from the initial state (i) to the final state (f). This 'thermal exergy' is the minimum work required by the combined system of the control mass and the environment in bringing the control mass to the final state from the dead state.

Often the dimensionless quantity in parentheses in this expression is called the 'exergetic temperature factor' and denoted τ:

$$\tau = 1 - T_o/T \tag{10.26}$$

The relation between τ and the temperature ratio T/T_o is illustrated in Figure 10.1.

If the temperature T of the control mass is constant, the thermal exergy transfer associated with a heat transfer is

$$X = (1 - T_o/T)Q = \tau Q \tag{10.25a}$$

For heat transfer across a region r on a control surface for which the temperature may vary,

$$X = \int_r [q(1 - T_o/T)dA]_r \tag{10.25b}$$

where q_r is the heat flow per unit area at a region on the control surface at which the temperature is T_r.

Exergy of work and electricity

Equation 10.6 separates total work W' into two components: W_x and W. The exergy associated with shaft work W_x is by definition W_x. Similarly, the exergy associated with electricity is equal to the energy.

The exergy transfer associated with work done by a system due to volume change is the net usable work due to the volume change, and is denoted by W_{NET}. Thus for a process in time interval t_1 to t_2,

$$\left(W_{NET}\right)_{1,2} = W_{1,2} - P_o\left(V_2 - V_1\right) \tag{10.27}$$

where $W_{1,2}$ is the work done by the system due to volume change ($V_2 - V_1$). The term $P_o(V_2 - V_1)$ is the displacement work necessary to change the volume against the constant pressure P_o exerted by the environment.

Exergy consumption

For a process occuring in a system, the difference between the total exergy flows into and out of the system, less the exergy accumulation in the system, is the exergy consumption I, expressible as

$$I = T_o \Pi \tag{10.28}$$

Equation 10.28 points out that exergy consumption is proportional to entropy creation, and is known as the Gouy-Stodola relation.

10.2.5 Detailed exergy balance

An analogous balance to those given in Equations 10.2a, 10.3a and 10.4a can be written for exergy, following the physical interpretation of Equation 10.5. For a non-steady flow process during time interval t_1 to t_2

$$\sum_i \varepsilon_i m_i - \sum_e \varepsilon_e m_e + \sum_r \left(X_r\right)_{1,2} - \left(W_x\right)_{1,2} - \left(W_{NET}\right)_{1,2} - I_{1,2} = \Xi_2 - \Xi_1 \tag{10.5a}$$

where $(W_{NET})_{1,2}$ is given by Equation 10.27 and

$$\left(X_r\right)_{1,2} = \int_{t_1}^{t_2}\left[\int_r \left(1 - T_o/T_r\right) q_r dA_r\right] dt \tag{10.29}$$

$$I_{1,2} = T_o \Pi_{1,2} \tag{10.30}$$

$$\Xi = \int \rho \xi dV \tag{10.31}$$

Here, I and Π respectively denote exergy consumption and entropy creation, Ξ denotes specific non-flow exergy, and the integral for Ξ is performed over the control volume. The first two terms on the left side of Equation 10.5a represent the net input of exergy associated with matter, the third term the net input of exergy associated with heat, the fourth and fifth terms the net input of exergy associated with work, and the sixth term the exergy consumption. The right side of Equation 10.5a shows the accumulation of exergy.

For a closed system, Equation 10.5a simplifies to

$$\sum_r \left(X_r\right)_{1,2} - \left(W_x\right)_{1,2} - \left(W_{NET}\right)_{1,2} - I_{1,2} = \Xi_2 - \Xi_1 \tag{10.5b}$$

When volume is fixed, $(W_{NET})_{1,2} = 0$ in Equations 10.5a and 10.5b. Also, when the initial and final states are identical as in a complete cycle, the right sides of Equations 10.5a and 10.5b are zero.

10.2.6 The reference environment

Exergy is evaluated with respect to a reference environment. The intensive properties of the reference environment determine the exergy of a stream or system. The exergy of the reference environment is zero. The exergy of a stream or system is zero when it is in equilibrium with the reference environment. The reference environment is in stable equilibrium, with all parts at rest relative to one another. No chemical reactions can occur between the environmental components. The reference environment acts as an infinite system, and is a sink and source for heat and materials. It experiences only internally reversible processes in which its intensive state remains unaltered (i.e. its temperature T_o, pressure P_o and the chemical potentials μ_{ioo} for each of the i components present remain constant).

The natural environment does not have the theoretical characteristics of a reference environment. The natural environment is not in equilibrium, and its intensive properties exhibit spatial and temporal variations. Many chemical reactions in the natural environment are blocked because the transport mechanisms necessary to reach equilibrium are too slow at ambient conditions. Thus, the exergy of the natural environment is not zero; work could be obtained if it were to come to equilibrium. Consequently, models for the reference environment are used which try to achieve a compromise between the theoretical requirements of the reference environment and the actual behavior of the natural environment.

One important class of reference-environment models is the natural-environment-subsystem type. These models attempt to simulate realistically subsystems of the natural environment. One such model consisting of saturated moist air and liquid water in phase equilibrium was proposed by Baehr and Schmidt (1963). An extension of the above model which allowed sulphur-containing materials to be analyzed was proposed by Gaggioli and Petit (1977) and Rodriguez (1980). The temperature and pressure of this reference environment (see Table 10.1) are normally taken to be 25°C and 1 atm, respectively, and the chemical composition is taken to consist of air saturated with water vapor, and the following condensed phases at 25°C and 1 atm: water (H_2O), gypsum ($CaSO_4{\cdot}2H_2O$), and limestone ($CaCO_3$). The stable configurations of C, O and N, respectively, are taken to be those of CO_2, O_2 and N_2 as they exist in air saturated with liquid water at T_o and P_o; of hydrogen is taken to be in the liquid phase of water saturated with air at T_o and P_o; and of S and Ca, respectively, are taken to be those of $CaSO_4{\cdot}2H_2O$ and $CaCO_3$ at T_o and P_o.

The analyses in this chapter use the natural-environment-subsystem model described in Table 10.1, but with a temperature modified to reflect the approximate mean ambient temperature of the location of the TES for the time period under consideration (e.g. annual, seasonal, monthly).

Table 10.1 A reference-environment model.

Temperature:	$T_o = 298.15$ K	
Pressure:	$P_o = 1$ atm	
Composition:	(i) Atmospheric air saturated with H_2O at T_o and P_o, having the following composition:	
	Air constituents	Mole fraction
	N_2	0.7567
	O_2	0.2035
	H_2O	0.0303
	Ar	0.0091
	CO_2	0.0003
	H_2	0.0001
	(ii) The following condensed phases at T_o and P_o:	
	Water (H_2O)	
	Limestone ($CaCO_3$)	
	Gypsum ($CaSO_4 \cdot 2H_2O$)	

Source: Adapted from Gaggioli and Petit (1977).

Other classes of reference-environment models have been proposed:

- *Reference-substance models.* These are in which a 'reference substance' is selected and assigned zero exergy for every chemical element. One such model in which the reference substances were selected as the most valueless substances found in abundance in the natural environment was proposed by Szargut (1967). The criteria for selecting such reference substances are consistent with the notion of simulating the natural environment, but are primarily economic in nature, and are vague and arbitrary with respect to the selection of reference substances. Part of this environment is the composition of moist air, including N_2, O_2, CO_2, H_2O and the noble gases; gypsum (for sulphur) and limestone (for calcium). Another model in this class, in which reference substances are selected arbitrarily, was proposed by Sussman (1980, 1981). This model is not similar to the natural environment, consequently absolute exergies evaluated with this model do not relate to the natural environment, and cannot be used rationally to evaluate efficiencies. Since exergy-consumption values are independent of the choice of reference substances, they can be rationally used in analyses.
- *Equilibrium models.* A model in which all the materials present in the atmosphere, oceans and a layer of the crust of the earth are pooled together and an equilibrium composition is calculated for a given temperature was proposed by Ahrendts (1980). The selection of the thickness of crust considered is subjective, and is intended to include all materials accessible to technical processes. Ahrendts considered thicknesses varying from 1 m to 1000 m, and a temperature of 25°C. For all thicknesses, Ahrendts found that the model differed significantly from the natural environment. Exergy values obtained using these environments are significantly dependent on the thickness

of crust considered, and represent the absolute maximum amount of work obtainable from a material. Since there is no technical process available which can obtain this work from materials, Ahrendts' equilibrium model does not give meaningful exergy values when applied to the analysis of real processes.

- *Constrained-equilibrium models.* Ahrendts (1980) also proposed a modified version of his equilibrium environment in which the calculation of an equilibrium composition excludes the posibility of the formation of nitric acid (HNO_3) and its compounds. That is, all chemical reactions in which these substances are formed are in constrained equilibrium, and all other reactions are in unconstrained equilibrium. When a thickness of crust of 1 m and temperature of 25°C were used, the model was similar to the natural environment.
- *Process-dependent models.* A model which contains only components that participate in the process being examined in a stable equilibrium composition at the temperature and total pressure of the natural environment was proposed by Bosnjakovic (1963). This model is dependent on the process examined, and is not general. Exergies evaluated for a specific process-dependent model are relevant only to the process; they cannot rationally be compared with exergies evaluated for other process-dependent models.

Many researchers have examined the characteristics of and models for reference environments (Wepfer and Gaggioli, 1980; Sussman, 1981; Ahrendts, 1980), and the sensitivities of exergy values to different reference-environment models (Rosen and Scott, 1987).

10.2.7 Efficiencies

General efficiency concepts

Efficiency has always been an important consideration in decision making regarding resource utilization. Efficiency is defined as 'the ability to produce a desired effect without waste of, or with minimum use of, energy, time, resources, etc.,' and is used by people to mean the effectiveness with which something is used to produce something else, or the degree to which the ideal is approached in performing a task.

For general engineering systems, non-dimensional ratios of quantities are typically used to determine efficiencies. Ratios of energy are conventionally used to determine efficiencies of engineering systems whose primary purpose is the transformation of energy. These efficiencies are based on the First Law of Thermodynamics. A process has maximum efficiency according to the First Law if energy input equals recoverable energy output (i.e. if no 'energy losses' occur). However, efficiencies determined using energy are misleading, because in general they are not measures of 'an approach to an ideal.'

To determine more meaningful efficiencies, a quantity is required for which ratios can be established which do provide a measure of an approach to an ideal. Thus, the Second Law must be involved, as this law states that maximum efficiency is attained (i.e. ideality is achieved) for a reversible process. However, the Second Law must be quantified before efficiencies can be defined.

The 'increase of entropy principle,' which states that entropy is created due to irreversibilities, quantifies the Second Law. From the viewpoint of entropy, maximum efficiency is attained for a process in which entropy is conserved. Entropy is created for non-ideal processes. The magnitude of entropy creation is a measure of the non-ideality or irreversibility of a process. In general, however, ratios of entropy do not provide a measure of an approach to an ideal.

A quantity which has been discussed in the context of meaningful measures of efficiency is negentropy (Hafele, 1981). Negentropy is defined such that the negentropy consumption due to irreversibilities is equal to the entropy creation due to irreversibilities. As a consequence of the 'increase of entropy principle,' maximum efficiency is attained from the viewpoint of negentropy for a process in which negentropy is conserved. Negentropy is consumed for non-ideal processes. Negentropy is a measure of order. Consumptions of negentropy are therefore equivalent to degradations of order. Since the abstract property of order is what is valued and useful, it is logical to attempt to use negentropy in developing efficiencies. However, general efficiencies cannot be determined based on negentropy because its absolute magnitude is not defined.

Negentropy can be further quantified through the ability to perform work. Then, maximum efficiency is attainable only if, at the completion of a process, the sum of all energy involved has an ability to do work equal to the sum before the process occurred. Exergy is a measure of the ability to perform work and, from the viewpoint of exergy, maximum efficiency is attained for a process in which exergy is conserved. Efficiencies determined using ratios of exergy do provide a measure of an approach to an ideal. Exergy efficiencies are often more intuitively rational than energy efficiencies, because efficiencies between 0% and 100% are always obtained. Measures which can be greater than 100% when energy is considered, such as coefficient of performance, are normally between 0% and 100% when exergy is considered. In fact, some researchers (e.g. Gaggioli, 1983) call exergy efficiencies 'real' or 'true' efficiencies, while calling energy efficiencies 'approximations to real' efficiencies.

Energy and exergy efficiencies

Many researchers (Sussman, 1981; Hevert and Hevert, 1980; Alefeld, 1990) have examined efficiencies and other measures of performance. Different efficiency definitions generally answer different questions.

Energy (η) and exergy (ψ) efficiencies are often written for steady-state processes occurring in systems as

$$\eta = \frac{Energy\ in\ product\ outputs}{Energy\ in\ inputs} = 1 - \frac{Energy\ loss}{Energy\ in\ inputs} \tag{10.32}$$

$$\psi = \frac{Exergy\ in\ product\ outputs}{Exergy\ in\ inputs} = 1 - \frac{Exergy\ loss\ plus\ consumption}{Exergy\ in\ inputs} \tag{10.33}$$

Two other common exergy-based efficiencies for steady-state devices are as follows:

$$Rational\ efficiency = \frac{Total\ exergy\ output}{Total\ exergy\ input} = 1 - \frac{Exergy\ consumption}{Total\ exergy\ input} \tag{10.34}$$

Table 10.2 Base enthalpy and chemical exergy values of selected species.

Species	Specific base enthalpy (kJ/g-mol)	Specific chemical exergy* (kJ/g-mol)
Ammonia (NH_3)	382.585	2.478907 ln y + 337.861
Carbon (graphite) (C)	393.505	410.535
Carbon dioxide (CO_2)	0.000	2.478907 ln y + 20.108
Carbon monoxide (CO)	282.964	2.478907 ln y + 275.224
Ethane (C_2H_6)	1,564.080	2.478907 ln y + 1,484.952
Hydrogen (H_2)	285.851	2.478907 ln y + 235.153
Methane (CH_4)	890.359	2.478907 ln y + 830.212
Nitrogen (N_2)	0.000	2.478907 ln y + 0.693
Oxygen (O_2)	0.000	2.478907 ln y + 3.948
Sulphur (rhombic) (S)	636.052	608.967
Sulphur dioxide (SO_2)	339.155	2.478907 ln y + 295.736
Water (H_2O)	44.001	2.478907 ln y + 8.595

* y represents the molal fraction for each of the respective species.

Source: Compiled from data in Rodriguez (1980) and Gaggioli and Petit (1977).

$$\textit{Task efficiency} = \frac{\textit{Theoretical minimum exergy input required}}{\textit{Actual exergy input}} \tag{10.35}$$

Exergy efficiencies often give more illuminating insights into process performance than energy efficiencies because (i) they weigh energy flows according to their exergy contents, and (ii) they separate inefficiencies into those associated with effluent losses and those due to irreversibilities. In general, exergy efficiencies provide a measure of potential for improvement.

10.2.8 Properties for energy and exergy analyses

Many material properties are needed for energy and exergy analyses of processes. Sources of conventional property data are abundant for many substances (e.g. steam, air and combustion gases (Keenan *et al.*, 1992) and chemical substances (Chase *et al.*, 1985)).

Energy values of heat and work flows are absolute, while the energy values of material flows are relative. Enthalpies are evaluated relative to a reference level. Since energy analyses are typically concerned only with energy differences, the reference level used for enthalpy calculations can be arbitrary. For the determination of some energy efficiencies, however, the enthalpies must be evaluated relative to specific reference levels (e.g. for energy-conversion processes, the reference level is often selected so that the enthalpy of a material equals its higher heating value (*HHV*).

If, however, the results from energy and exergy analyses are to be compared, it is necessary to specify reference levels for enthalpy calculations such that the enthalpy of a compound is evaluated relative to the stable components of the reference environment. Thus, a compound which exists as a stable component of the reference environment is defined to have an enthalpy of zero at T_o and P_o. Enthalpies calculated with respect to such conditions are referred to as 'base enthalpies' (Rodriguez, 1980). The base enthalpy is similar to the enthalpy of formation. While the latter is the enthalpy of a compound (at T_o and P_o) relative to the elements (at T_o and P_o) from which it would be formed, the former is the enthalpy of a component (at T_o and P_o) relative to the stable components of the environment (at T_o and P_o). For many environment models, the base enthalpies of material fuels are equal to their *HHV*s.

Base enthalpies for many substances, corresponding to the reference-environment model in Table 10.1, are listed in Table 10.2 (Rodriguez, 1980). It is required for chemical exergy values to be determined for exergy analysis. Many researchers have developed methods for evaluating chemical exergies, and tabulated values (e.g. Rodriguez, 1980; Szargut, 1967; Sussman, 1980). Included are methods for evaluating the chemical exergies of solids, liquids and gases. For complex materials (e.g. coal, tar, ash), approximation methods have been developed. By considering environmental air and gaseous process streams as ideal gas mixtures, chemical exergy can be calculated for gaseous streams using component chemical exergy values, i.e. values of $(\mu_{io} - \mu_{ioo})$ listed in Table 10.2.

10.2.9 Implications of results of exergy analyses

The results of exergy analyses of TES systems have direct implications on application decisions and on research and development (R&D) directions.

Further, exergy analyses more than energy analyses provide insights into the 'best' directions for R&D effort. Here, 'best' is loosely taken to mean 'most promising for significant efficiency gains.' There are two main reasons for this statement:

- Exergy losses represent true losses of the potential that exists to generate the desired product (recovered thermal energy with little temperature degradation, here) from the given driving input (input thermal energy and electricity, here). This is not true in general for energy losses. Thus, if the objective is to increase TES efficiency while accounting for temperature degradation, focusing on exergy losses permits R&D to focus on reducing losses that will affect the objective.
- Exergy efficiencies always provide a measure of how nearly the operation of a system approaches the ideal, or theoretical upper limit. This is not in general true for energy efficiencies. By focusing R&D effort on those plant sections or processes with the lowest exergy efficiencies, the effort is being directed to those areas which inherently have the largest margins for efficiency improvement. By focusing on energy efficiencies, on the other hand, one can expend R&D effort on topics for which little margins for improvement, even theoretically, exist.

Exergy analysis results typically suggest that R&D efforts should concentrate more on internal rather than external exergy losses, based on thermodynamic considerations, with a higher priority for the processes having larger exergy losses. Although this statement suggests focusing on those areas for which margins for improvement are greatest, it does not indicate that R&D should not be devoted to those processes having low exergy losses, as simple and cost-effective ways to increase efficiency by reducing small exergy losses should certainly be considered when identified.

More generally, it is noted that application and R&D allocation decisions should not be based exclusively on the results of energy and exergy analyses, even though these results provide useful information to assist in such decision making. Other factors must also be considered, such as economics, environmental impact, safety, and social and political implications.

10.2.10 Steps for energy and exergy analyses

A simple procedure for performing energy and exergy analyses involves the following steps:

- Subdivide the process under consideration into as many sections as desired, depending on the depth of detail and understanding desired from the analysis.
- Perform conventional mass and energy balances on the process, and determine all basic quantities (e.g. work, heat) and properties (e.g. temperature, pressure) (section 10.2.3).
- Based on the nature of the process, the acceptable degree of analysis complexity and accuracy, and the questions for which answers are sought, select a reference-environment model (section 10.2.6).
- Evaluate energy and exergy values, relative to the selected reference-environment model (sections 10.2.4 and 10.2.8).
- Perform exergy balances, including the determination of exergy consumptions (section 10.2.5).
- Select efficiency definitions, depending on the measures of merit desired, and evaluate values for the efficiencies (section 10.2.7).
- Interpret the results, and draw appropriate conclusions and recommendations, relating to such issues as design changes, retrofit plant modifications, etc. (section 10.2.9).

10.3 Thermodynamic Considerations in TES Evaluation

Several of the more important thermodynamic factors to be considered during the evaluation and comparison of TES systems are discussed in this section (Rosen and Dincer, 1999a).

10.3.1 Determining important analysis quantities

The two most significant quantities to consider when evaluating TES systems are energy and exergy. Exergy analysis involves the examination of the exergy at different points in a series of energy conversion steps, and the determination of meaningful efficiencies and of the steps having the largest losses (i.e. the largest margin for improvement). The authors and others feel that the use of exergy analysis circumvents many of the problems associated with conventional TES evaluation and comparison methodologies by providing a more rational basis.

Exergy analysis permits more rational and convenient evaluation than energy analysis of TES systems for cooling capacity as well as heating capacity. Exergy analysis applies equally well to systems for storing thermal energy at temperatures above and below the temperature of the environment, because the exergy associated with such energy is always greater than or equal to zero. Energy analysis is more difficult to apply to such storage systems because efficiency definitions have to be carefully modified when cooling capacity, instead of heating capacity, is stored, or when both warm and cool reservoirs are included.

10.3.2 Obtaining appropriate measures of efficiency

The evaluation of a TES system requires a measure of performance which is rational, meaningful and practical. The conventional energy storage efficiency, as pointed out earlier, is an inadequate measure. A more perceptive basis for comparison is apparently needed if the true usefulness of thermal storages is to be assessed, and so permit maximization of their economic benefit. Efficiencies based on ratios of exergy do provide rational measures of performance, since they can measure the approach of the performance of a system to the ideal.

That the energy efficiency is an inappropriate measure of thermal storage performance can best be appreciated through a simple example. Consider a perfectly insulated thermal storage containing 1000 kg of water, initially at 40°C. The ambient temperature is 20°C.

A quantity of 4200 kJ of heat is transferred to the storage through a heat exchanger from an external body of 100 kg of water cooling from 100°C to 90°C (i.e. with Equation 10.168, (100 kg)(4.2 kJ/kg K)(100 – 90)°C = 4200 kJ). This heat addition raises the storage temperature 1.0°C, to 41°C (i.e. with Equation 10.181, (4200 kJ)/((1000 kg)(4.2 kJ/kg K)) = 1.0°C). After a period of storage, 4200 kJ of heat are recovered from the storage through a heat exchanger which delivers it to an external body of 100 kg of water, raising the temperature of that water from 20°C to 30°C (i.e. with Equation 10.168, ΔT = (4200 kJ)/((100 kg)(4.2 kJ/kg K)) = 10°C). The storage is returned to its initial state at 40°C.

For this storage cycle the energy efficiency, the ratio of the heat recovered from the storage to the heat injected, is 4200 kJ/4200 kJ = 1, or 100%. But the recovered heat is at only 30°C, and of little use, having been degraded even though the storage energy efficiency was 100%. With Equation 10.170a, the exergy recovered in this example is evaluated as (100 kg)(4.2 kJ/kg K)[(30 – 20)°C – (293 K) ln (303/293)] = 70 kJ, and the exergy supplied as (100 kg)(4.2 kJ/kg K)[(100 – 90)°C – (293 K) ln (373/363)] = 856 kJ. Thus the exergy efficiency, the ratio of the thermal exergy recovered from storage to that injected, is 70/856 = 0.082, or 8.2%, a much more meaningful expression of the achieved performance of the storage cycle.

In most TES investigations, the energy and exergy efficiency definitions in Equations 10.32 and 10.33 are used. These efficiency definitions are dependent on what quantities are considered to be products and inputs. Two possible sets of efficiency definitions are presented in Table 10.3. The energy or exergy initially in the store is neglected in the first definition, and considered to be an 'input' in the second definition. Depending on the particular circumstances, the energy efficiency definitions in Table 10.3 can yield values which are identical or radically different. The same statement can be made for the exergy efficiencies. Regardless of definition, however, the authors feel that the use of exergy analysis is necessary for evaluating TESs.

Table 10.3 Two overall energy (η) and exergy (ψ) TES efficiencies.

Efficiency	Definition 1	Definition 2
η	$\dfrac{\textit{Energy recovered from TES}}{\textit{Energy input to TES}}$	$\dfrac{\textit{Energy recovered from and remaining in TES}}{\textit{Energy input to and originally in TES}}$
ψ	$\dfrac{\textit{Exergy recovered from TES}}{\textit{Exergy input to TES}}$	$\dfrac{\textit{Exergy recovered from and remaining in TES}}{\textit{Exergy input to and originally in TES}}$

10.3.3 Pinpointing losses

With energy analysis, all losses are attributable to energy releases across system boundaries. With exergy analysis, losses are divided into two types: those associated with releases of exergy from the system and those associated with internal consumptions of exergy (Alefeld, 1990). For a TES system, the total exergy loss is the sum of the exergy associated with heat loss to the surroundings and the exergy loss due to internal exergy consumptions, such as by reductions in availability of the stored heat through mixing of warm and cool fluids. The division of losses associated with exergy analysis allows the causes of inefficiencies to be more accurately identified than does energy analysis, and R&D effort to be more effectively allocated. The analysis of the heat flows from or into TESs is often investigated (e.g. Rosen, 1990, 1998a, 1998b), and is discussed elsewhere in this book.

10.3.4 Assessing the effects of stratification

Water tanks are one of the most economic devices for TES. For many TES applications, performance is strongly dependent on the temperature required to meet the thermal-energy demand, and stratification within the tank can play a significant role (Yoo *et al.*, 1998; Holman *et al.*, 1996; Mavros *et al.*, 1994; Nelson *et al.*, 1999; Gretarsson *et al.*, 1994). In most cases in practice, a vertical cylindrical tank with a hot water inlet (outlet) at the top and a cold water inlet (outlet) at the bottom is used. The hot and cold water in the tank are usually stratified initially into two layers, with a mixing layer in between. The degree of stratification is dependent on the volume and configuration of the tank, the size, location and design of the inlets and outlets, the flow rates of the entering and exiting streams, and the durations of the charging, storing and discharging periods.

Four primary factors contribute to the loss of stratification and hence the degradation of the stored energy:

- heat losses to (or leakages from) the surrounding environment,
- heat conduction from the hot portions of the storage fluid to the colder portions,
- vertical conduction in the tank wall, and
- mixing during charging and discharging periods.

Among these, the last item generally is the major cause of loss of stratification, with particularly significant mixing losses occurring during lengthy storing periods. Improving stratification often leads to substantial improvement in TES efficiency relative to a system incorporating a thermally mixed storage tank.

TES evaluation methodologies must quantitatively and clearly assess the effects of stratification on system performance. The effects of stratification are more clearly assessed with exergy analysis than with energy analysis due to the internal spatial temperature variations stratified storages exhibit (Hahne *et al.*, 1989; Krane and Krane, 1991; Rosen and Tang, 1997). Through carefully managing the injection, recovery and holding of heat (or cold) during a storage cycle so that temperature degradation is minimized, better storage-cycle performance can be achieved (as measured by better thermal energy recovery and temperature retention). This improved performance is accounted for explicitly with exergy analysis through increased exergy efficiencies. Some relatively simple approaches for stratified TES evaluation have been developed (Hooper *et al.*, 1988; Rosen and Hooper, 1991a, 1991b, 1992, 1994), which take into consideration Second-Law concerns, and which assess the performance of real stratified storages with an accuracy acceptable for most purposes, and certainly superior to that from assessments based only on the First Law.

10.3.5 Accounting for time duration of storage

Rational measures of merit for the evaluation and comparison of TES systems must account for the length of time thermal energy is in storage. The length of time that thermal energy is retained in a TES does not enter into the expressions for thermal efficiency and exergy efficiency for thermal storages, although it is clearly a dominant consideration in the overall effectiveness for such systems. The authors have examined the relation between the length of time thermal energy is held in storage and storage effectiveness, and have developed an approach for comparing TESs using a time parameter (Barbaris *et al.*, 1988).

10.3.6 Accounting for variations in reference-environment temperature

Over the time periods involved in some TES cycles (up to six months for seasonal systems), the value of the reference environment temperature T_o varies with time. The value of T_o also varies with location. Since the results of TES evaluations based on energy and exergy analyses depend on the value of T_o, the temporal and spatial dependences must be considered in such evaluations.

The value of $T_o(t)$ can often be assumed to be the same as the ambient temperature variation with time, $T_{amb}(t)$. On an annual basis, the ambient temperature varies approximately sinusoidally with time t about the annual mean:

$$T_{amb}(t) = \overline{T}_{amb} + \Delta T_{amb}\left[\sin\frac{2\pi t}{period} + (phase\ shift)\right] \tag{10.36}$$

where T_{amb} is the mean annual ambient temperature and ΔT_{amb} is the maximum temperature deviation from the annual mean. The values of the parameters in Equation 10.36 vary spatially and the period is one year.

Table 10.4 Some exergy-related thermodynamic considerations in TES system evaluations.

- Determining important analysis quantities
- Evaluating storages for cooling as well as heating capacity
- Obtaining appropriate measures of efficiency
- Pinpointing losses
- Assessing the effects of stratification
- Assessing the performance of subprocesses
- Accounting for temporal and spatial variations in T_o
- Accounting for the time duration of storage

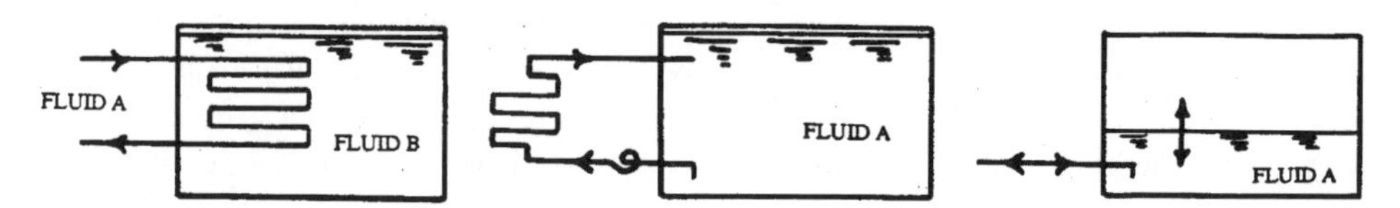

Figure 10.2 Three basic types of thermal storage. Top: a closed system in which heat transfer occurs between the transport fluid A and the storage fluid B. Center: an open system of constant mass in which the same fluid is used for both transport and storage of heat. Bottom: an open system of variable mass.

In evaluating the performance of most storages (particularly of long-term storages), a constant value of T_o can be assumed. Some possible values for T_o are:

- the appropriate seasonal mean value of the temperature of the atmosphere;
- the appropriate annual mean value of the temperature of the atmosphere;
- the temperature of soil far enough below the surface that the temperature remains approximately constant throughout the year, i.e. near the water table (this temperature is usually near to that specified in the previous point); and
- the lowest value of the atmosphere temperature during the year, for heat storage processes, and the highest value of the atmosphere temperature during the year, for cooling capacity storage processes.

10.3.7 Closure

The factors discussed here that significantly impact on the evaluation and comparison of TES systems are summarized in Table 10.4.

10.4 Exergy Evaluation of a Closed TES System

The use of exergy analysis in the evaluation of a specific TES system (a simple closed tank storage with heat transfers by heat exchanger) is described in this section, following an earlier report (Rosen *et al.*, 1988). A complete storing cycle, as well as the individual charging, storing and discharging periods, are considered. A numerical example for a simple case is given (see section 10.4.6). This application highlights the fact that, although energy is conserved in an adiabatic system, mixing of the high- and low-temperature portions of the storage medium causes a consumption (or destruction) of exergy, which is conserved only in fully reversible processes.

A clear understanding of the TES type under consideration and a clear declaration of the assumptions used are important in establishing a consistent basis for TES analysis and comparison. Three basic sensible heat storage types are shown in Figure 10.2. The first type, which is considered here, represents a closed system storing heat in a fixed amount of storage fluid B, to or from which heat is transferred through a heat exchanger by means of a heat transport fluid A. The second type is also of fixed mass, but is open, and the transport fluid and the storage fluid are the same substance. No heat exchanger is involved. The third type uses an open system storing a variable amount of the combined heat transport and storage fluid. For each of these storage types, further characteristics can yield additional cases and conditions (e.g. adiabatic or non-adiabatic boundaries, complete or incomplete cycles, fully mixed or stratified storage fluid, steady or intermittent fluid flows, steady or variable ambient conditions, short or long storage periods, constant or variable physical properties, and one-, two- or three-dimensional heat flows).

10.4.1 Description of the case considered

A specific, simple case is considered in which a TES system undergoes a complete storage cycle, ending with the final state identical to the initial state. Figure 10.3 illustrates the three periods in the overall storage process considered. The TES may be stratified. Other characteristics of the considered case are:

- non-adiabatic storage boundaries,
- finite charging, storing and discharging time periods,
- surroundings at constant temperature and pressure,
- constant storage volume,
- negligible work interactions (e.g. pump work), and
- negligible kinetic and potential energy terms.

The operation of the heat exchangers is simplified by assuming that there are no heat losses to the surrounding environment from the charging and discharging fluids. That is, it is assumed for the charging period that all heat removed from the charging fluid is added to the storage medium, and for the discharging period that all heat added to the discharging fluid originated in the storage medium. This assumption is valid if heat losses from the charging and discharging fluids are small compared with heat losses from the storage medium. This assumption can be extended by lumping actual heat losses from the charging and discharging fluids together with heat losses from the TES. Also, as is often done for practical systems, the charging and discharging fluid flows are considered steady and with time-independent properties, and modeled as one-dimensional.

10.4.2 Analysis of the overall process

For the cases considered (Figure 10.3), energy and exergy balances and efficiencies are provided for the overall process.

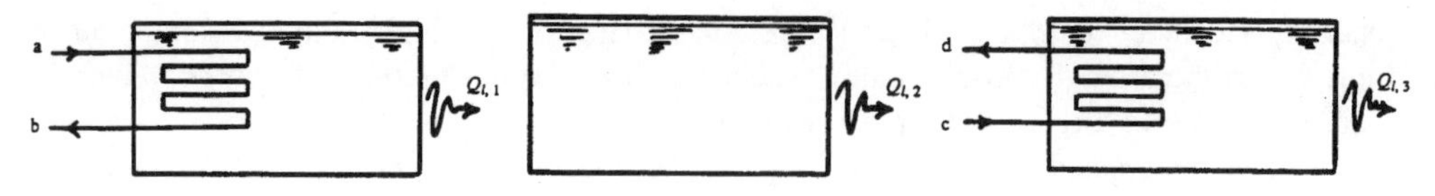

Figure 10.3 The three stages in a simple heat storage process: charging period (left), storing period (center) and discharging period (right).

Overall energy balance

Following Equations 10.3 and 10.3a, an energy balance for the overall storage process can be written as

$$Energy\ input - [\,Energy\ recovered + Energy\ loss\,] = Energy\ accumulation \tag{10.37}$$

or

$$(H_a - H_b) - [(H_d - H_c) + Q_l] = \Delta E \tag{10.37a}$$

where H_a, H_b, H_c and H_d are the total enthalpies of the flows at states a, b, c and d, respectively; Q_l denotes the heat losses during the process and ΔE the accumulation of energy in the TES. In Equation 10.37a, $(H_a - H_b)$ represents the net heat delivered to the TES and $(H_d - H_c)$ the net heat recovered from the TES. The quantity in square brackets represents the net energy output from the system. The terms ΔE and Q_l are given by

$$\Delta E = E_f - E_i \tag{10.38}$$

$$Q_l = \sum_{j=1}^{3} Q_{l,j} \tag{10.39}$$

Here E_i and E_f denote the initial and final energy contents of the storage, and $Q_{l,j}$ denotes the heat losses during the period j, where j = 1, 2, 3 correspond to the charging, storing and discharging periods, respectively. In the case of identical initial and final states, $\Delta E = 0$ and the overall energy balance simplifies.

Overall exergy balance

Following Equations 10.5 and 10.5a, an overall exergy balance can be written as

$$\begin{aligned} &Exergy\ input - [\,Exergy\ recovered + Exergy\ loss\,] - Exergy\ consumption = \\ &Exergy\ accumulation \end{aligned} \tag{10.40}$$

or

$$(\epsilon_a - \epsilon_b) - [(\epsilon_d - \epsilon_c) + X_l] - I = \Delta\Xi \tag{10.40a}$$

where $\epsilon_a, \epsilon_b, \epsilon_c$ and ϵ_d are the exergies of the flows at states a, b, c and d, respectively; and X_l denotes the exergy loss associated with Q_l; I is the exergy consumption; and $\Delta\Xi$ is the exergy accumulation. In Equation 10.40a, $(\epsilon_a - \epsilon_b)$ represents the net exergy input and $(\epsilon_d - \epsilon_c)$ is the net exergy recovered. The quantity in square brackets represents the net exergy output from the system. The terms I, X_l and $\Delta\Xi$ are given respectively by

$$I = \sum_{j=1}^{3} I_j \tag{10.41}$$

$$X_l = \sum_{j=1}^{3} X_{l,j} \tag{10.42}$$

$$\Delta\Xi = \Xi_f - \Xi_i \tag{10.43}$$

Here, I_1, I_2 and I_3 denote respectively the consumptions of exergy during the charging, storing and discharging periods; $X_{l,1}$, $X_{l,2}$ and $X_{l,3}$ denote the exergy losses associated with heat losses during the same periods; and Ξ_i and Ξ_f denote the initial and final exergy contents of the storage. When the initial and final states are identical, $\Delta\Xi = 0$.

The exergy content of the flow at the states $k = a, b, c, d$ is evaluated as

$$\in_k = (H_k - H_o) - T_o(S_k - S_o) \tag{10.44}$$

where $\in_k$, H_k and S_k denote the exergy, enthalpy and entropy of state k, respectively, and H_o and S_o the enthalpy and the entropy at the temperature T_o and pressure P_o of the reference environment. The exergy expression in Equation 10.44 only includes physical (or thermomechanical) exergy. Potential and kinetic exergy components are, as pointed out earlier, considered negligible for the devices under consideration. The chemical component of exergy is neglected because it does not contribute to the exergy flows for sensible TES systems. Thus, the exergy differences between the inlet and outlet for the charging and discharging periods are, respectively:

$$\in_a - \in_b = (H_a - H_b) - T_o(S_a - S_b) \tag{10.45}$$

and

$$\in_d - \in_c = (H_d - H_c) - T_o(S_d - S_c) \tag{10.46}$$

Here it has been assumed that T_o and P_o are constant, so that H_o and S_o are constant at states a and b, and at states c and d.

For a fully mixed tank, the exergy losses associated with heat losses to the surroundings are evaluated following Equation 10.25 as

$$X_{l,j} = \int_i^f \left(1 - \frac{T_o}{T_j}\right) dQ_{l,j} \quad \text{for } j = 1, 2, 3 \tag{10.47}$$

where j represents the particular period. If T_1, T_2 and T_3 are constant during the respective charging, storing and discharging periods, then $X_{l,j}$ may be written with Equation 10.25a as follows:

$$X_{l,j} = \left(1 - \frac{T_o}{T_j}\right) Q_{l,j} \tag{10.47a}$$

Sometimes when applying Equation 10.47a to TES systems, T_j represents a mean temperature within the tank for period j.

Overall energy and exergy efficiencies

Following Equations 10.32 and 10.33, the energy efficiency η can be defined as

$$\eta = \frac{\textit{Energy recovered from TES during discharging}}{\textit{Energy input to TES during charging}} = \frac{H_d - H_c}{H_a - H_b} = 1 - \frac{Q_l}{H_a - H_b} \tag{10.48}$$

and the exergy efficiency ψ as

$$\psi = \frac{\textit{Exergy recovered from TES during discharging}}{\textit{Exergy input to during charging}} = \frac{\in_d - \in_c}{\in_a - \in_b} = 1 - \frac{X_l + I}{\in_a - \in_b} \quad (10.49)$$

The efficiency expressions in Equations 10.48 and 10.49 do not depend on the initial energy and exergy contents of the TES.

If the TES is adiabatic, $Q_{l,j} = X_{l,j} = 0$ for all j. Then the energy efficiency is fixed at unity and the exergy efficiency simplifies to

$$\psi = 1 - \frac{I}{\in_a - \in_b} \quad (10.49a)$$

emphasizing the point that even when TES boundaries are adiabatic and there are therefore no energy losses, the exergy efficiency is less than unity due to internal irreversibilities.

10.4.3 Analysis of subprocesses

Many different efficiencies based on energy and exergy can be defined for the charging, storing and discharging periods (as discussed in section 10.5). In the present analysis of the subprocesses (see Figure 10.3), only one set of efficiencies is considered.

Analysis of charging period

An energy balance for the charging period can be written as follows:

$$\textit{Energy input} - \textit{Energy loss} = \textit{Energy accumulation} \quad (10.50)$$

$$(H_a - H_b) - Q_{l,1} = \Delta E_1 \quad (10.50a)$$

Here,

$$\Delta E_1 = E_{f,1} - E_{i,1} \quad (10.51)$$

and $E_{i,1}$ and $E_{f,1}$ denote the initial and the final energy of the TES for the charging period. Note that $E_{i,1}$ η E_i (see Equation 10.38). A charging-period energy efficiency can be defined as

$$\eta_1 = \frac{\textit{Energy accumulation in TES during charging}}{\textit{Energy input to TES during charging}} = \frac{\Delta E_1}{H_a - H_b} \quad (10.52)$$

An exergy balance for the charging period can be written as

$$\textit{Exergy input} - \textit{Exergy loss} - \textit{Exergy consumption} = \textit{Exergy accumulation} \quad (10.53)$$

$$(\in_a - \in_b) - X_{l,1} - I_1 = \Delta\Xi_1 \quad (10.53a)$$

Here,

$$\Delta\Xi_1 = \Xi_{f,1} - \Xi_{i,1} \quad (10.54)$$

and $\Xi_{i,1}$ and $\Xi_{f,1}$ are the initial and the final exergy of the TES for the charging period. Note that $\Xi_{i,1} \equiv \Xi_i$ (see Equation 10.43). A charging period exergy efficiency can be defined as

$$\psi_1 = \frac{\text{Exergy accumulation in TES during charging}}{\text{Exergy input to TES during charging}} = \frac{\Delta\Xi_1}{\in_a - \in_b} \tag{10.55}$$

The charging efficiencies in Equations 10.52 and 10.55 indicate the fraction of the input energy/exergy which is accumulated in the store during the charging period.

Analysis of storing period

An energy balance for the storing period can be written as

$$-\text{Energy loss} = \text{Energy accumulation} \tag{10.56}$$

$$-Q_{l,2} = \Delta E_2 \tag{10.56a}$$

Here,

$$\Delta E_2 = E_{f,2} - E_{i,2} \tag{10.57}$$

and $E_{i,2}$ ($\equiv E_{f,1}$) and $E_{f,2}$ denote the initial and final energy contents of the TES for the storing period. An energy efficiency for the storing period can be defined as

$$\eta_2 = \frac{\text{Energy accumulation in TES during charging and storing}}{\text{Energy accumulation in TES during charging}} = \frac{\Delta E_1 + \Delta E_2}{\Delta E_1} \tag{10.58}$$

Using Equation 10.56a, the energy efficiency can be rewritten as

$$\eta_2 = \frac{\Delta E_1 - Q_{l,2}}{\Delta E_1} \tag{10.58a}$$

An exergy balance for the storing period can be written as

$$-\text{Energy loss} - \text{Exergy consumption} = \text{Exergy accumulation} \tag{10.59}$$

$$-X_{l,2} - I_2 = \Delta\Xi_2 \tag{10.59a}$$

Here,

$$\Delta\Xi_2 = \Xi_{f,2} - \Xi_{i,2} \tag{10.60}$$

and $\Xi_{i,2}(\equiv \Xi_{f,1})$ and $\Xi_{f,2}$ denote the initial and the final exergies of the system for the storing period. An exergy efficiency for the storing period can be defined as

$$\psi_2 = \frac{\text{Exergy accumulation in TES during charging and storing}}{\text{Exergy accumulation in TES during charging}} = \frac{\Delta\Xi_1 + \Delta\Xi_2}{\Delta\Xi_1} \tag{10.61}$$

Using Equation 10.59a, the exergy efficiency can be rewritten as

$$\psi_2 = \frac{\Delta\Xi_1 - (X_{l,2} + I_2)}{\Delta\Xi_1} \tag{10.61a}$$

The storing efficiencies in Equations 10.58 and 10.61 indicate the fraction of the energy/exergy accumulated during charging which is still retained in the store at the end of the storing period.

Analysis of discharging period

An energy balance for the discharging period can be written as

$$-[Energy\ recovered + Energy\ loss] = Energy\ accumulation \tag{10.62}$$

$$-[(H_d - H_c) + Q_{l,3}] = \Delta E_3 \tag{10.62a}$$

Here,

$$\Delta E_3 = E_{f,3} - E_{i,3} \tag{10.63}$$

and $E_{i,3}$ (= $E_{f,2}$) and $E_{f,3}$ (= E_f in Equation 10.38) denote the initial and final energies of the store for the discharging period. The quantity in square brackets represents the energy output during discharging. An energy efficiency for the discharging period can be defined as

$$\eta_3 = \frac{Energy\ recovered\ from\ TES\ during\ discharging}{Energy\ accumulation\ in\ TES\ during\ charging\ and\ storing} = \frac{H_d - H_c}{\Delta E_1 + \Delta E_2} \tag{10.64}$$

Using Equation 10.56a, the energy efficiency can be rewritten as

$$\eta_3 = \frac{H_d - H_c}{\Delta E_1 - Q_{l,2}} \tag{10.64a}$$

An exergy balance for the discharging period can be written as follows:

$$-[Exergy\ recovered + Exergy\ loss] - Exergy\ consumption = Exergy\ accumulation \tag{10.65}$$

$$-[(\in_d - \in_c) + X_{l,3}] - I_3 = \Delta\Xi_3 \tag{10.65a}$$

Here,

$$\Delta\Xi_3 = \Xi_{f,3} - \Xi_{i,3} \tag{10.66}$$

and $\Xi_{i,3}$ (= $\Xi_{f,2}$) and $\Xi_{f,3}$ (= Ξ_f in Equation 10.43) denote the initial and final exergies of the store for the discharging period. The quantity in square brackets represents the exergy output during discharging. An exergy efficiency for the discharging period can be defined as

$$\psi_3 = \frac{Exergy\ recovered\ from\ TES\ during\ discharging}{Exergy\ accumulation\ in\ TES\ during\ charging\ and\ storing} = \frac{\in_d - \in_c}{\Delta\Xi_1 + \Delta\Xi_2} \tag{10.67}$$

Using Equation 10.59a, the exergy efficiency can be rewritten as

$$\psi_3 = \frac{\in_d - \in_c}{\Delta\Xi_1 - (X_{l,2} + I_2)} \tag{10.67a}$$

The discharging efficiencies in Equations 10.64 and 10.67 indicate the fraction of the energy/exergy input during charging and still retained at the end of storing which is recovered during discharging.

10.4.4 *Alternative formulations of subprocess efficiencies*

Several alternative subprocess efficiency formulations, that can be useful depending upon the inclination of the analyst and the application addressed, are given here.

In the expressions for the subprocess energy and exergy efficiencies, i.e. in Equations 10.52, 10.55, 10.58a, 10.61a, 10.64a and 10.67a, the terms ΔE_1 and $\Delta\Xi_1$ can be eliminated using Equations 10.50a and 10.53a, respectively. After substitutions for ΔE_1 and $\Delta\Xi_1$ and minor re-arrangement of the equations, the following are obtained:

$$\eta_1 = \frac{(H_a - H_b) - Q_{l,1}}{(H_a - H_b)} = 1 - \frac{Q_{l,1}}{H_a - H_b} \tag{10.52}$$

$$\psi_1 = \frac{(\epsilon_a - \epsilon_b) - (X_{t,1} + I_1)}{(\epsilon_a - \epsilon_b)} = 1 - \frac{X_{t,1} + I_1}{\epsilon_a - \epsilon_b} \tag{1055a}$$

$$\eta_2 = \frac{[(H_a - H_b) - Q_{l,1}] - Q_{l,2}}{(H_a - H_b) - Q_{l,1}} = \frac{(H_a - H_b) - \sum_{j=1}^{2} Q_{l,j}}{(H_a - H_b) - Q_{l,1}} = 1 - \frac{Q_{l,2}}{(H_a - H_b) - Q_{l,1}} \tag{10.58b}$$

$$\psi_2 = \frac{[(\epsilon_a - \epsilon_b) - (X_{t,1} + I_1)] - [X_{t,2} + I_2]}{(\epsilon_a - \epsilon_b) - (X_{t,1} + I_1)} = \frac{(\epsilon_a - \epsilon_b) - \sum_{j=1} (X_{t,j} + I_j)}{(\epsilon_a - \epsilon_b) - (X_{t,1} + I_1)} = 1 - \frac{X_{t,2} + I_2}{(\epsilon_a - \epsilon_b) - (X_{t,1} + I_1)} \tag{10.61b}$$

$$\eta_3 = \frac{H_d - H_c}{[(H_a - H_b) - Q_{l,1}] - Q_{l,2}} = \frac{H_d - H_c}{(H_a - H_b) - \sum_{j=1}^{2} Q_{l,j}} \tag{10.64b}$$

$$\psi_3 = \frac{\epsilon_d - \epsilon_c}{[(\epsilon_a - \epsilon_b) - (X_{t,1} + I_1)] - (X_{t,2} + I_2)} = \frac{\epsilon_d - \epsilon_c}{(\epsilon_a - \epsilon_b) - \sum_{j=1}^{2} (X_{t,j} + I_j)} \tag{10.67b}$$

The above equations for η_3 and ψ_3 can be further modified if the terms in the numerators are eliminated using Equations 10.37a and 10.40a with $\Delta E = \Delta\Xi = 0$. Then,

$$\eta_3 = \frac{(H_a - H_b) - Q_l}{(H_a - H_b) - \sum_{j=1}^{2} Q_{l,j}} = \frac{(H_a - H_b) - \sum_{j=1}^{3} Q_{l,j}}{(H_a - H_b) - \sum_{j=1}^{2} Q_{l,j}} = 1 - \frac{Q_{l,3}}{(H_a - H_b) - \sum_{j=1}^{2} Q_{l,j}} \tag{10.64c}$$

$$\psi_3 = \frac{(\epsilon_a - \epsilon_b) - (X_l + I)}{(\epsilon_a - \epsilon_b) - \sum_{j=1}^{2} (X_{l,j} + I_j)} = \frac{(\epsilon_a - \epsilon_b) - \sum_{j=1}^{3} (X_{l,j} + I_j)}{(\epsilon_a - \epsilon_b) - \sum_{j=1}^{2} (X_{l,j} + I_j)} = 1 - \frac{X_{l,3} + I_3}{(\epsilon_a - \epsilon_b) - \sum_{j=1}^{2} (X_{l,j} + I_j)} \tag{10.67c}$$

10.4.5 *Relations between performance of subprocesses and overall process*

The total energy and exergy efficiencies can be written as the products of the energy and exergy efficiencies of the charging, storing and discharging periods. That is,

$$\eta = \prod_{j=1}^{3} \eta_j \tag{10.68}$$

$$\psi = \prod_{j=1}^{3} \psi_j \tag{10.69}$$

The energy efficiency relationship can be verified by multiplying together Equations 10.52, 10.58a and 10.64a and comparing the result to Equation 10.48. Similarly, the exergy efficiency relationship can be verified by comparing the product of Equations 10.55, 10.61a and 10.67a to Equation 10.49a. Equations 10.68 and 10.69 also can be shown to hold when using the alternative formulations of the subprocess efficiencies.

In addition, it can be shown that the summations of the energy or exergy balance equations respectively for the three subprocesses give the energy or exergy balance equations for the overall process. This statement can be verified by noting that

$$\sum_{j=1}^{3} \Delta E_j = E_{3,f} - E_{1,i} = E_f - E_i = \Delta E \tag{10.70}$$

$$\sum_{j=1}^{3} \Delta \Xi_j = \Xi_{3,f} - \Xi_{t,i} = \Xi_f - \Xi_i = \Delta \Xi \tag{10.71}$$

and by comparing the sum of Equations 10.50a, 10.56a and 10.62a with Equation 10.37a for energy, and by comparing the sum of Equations 10.53a, 10.59a and 10.65a with Equation 10.40a for exergy. In writing Equations 10.70 and 10.71, it has been noted for period j that

$$\Delta E_j = E_{f,j} - E_{i,j} \tag{10.72}$$

$$\Delta \Xi_j = \Xi_{f,j} - \Xi_{i,j} \tag{10.73}$$

and that $E_{i,1} = E_i$, $E_{f,3} = E_f$, and $E_{i,j+1} = E_{f,j}$ for j = 1,2, while analogous expressions hold for the Ξ terms.

10.4.6 Example

Consider two different thermal storages, each of which undergoes a similar charging process. In each charging operation heat is transferred to a closed thermal storage from a stream of 1000 kg of water which enters at 85°C and leaves at 25°C (see Figure 10.4). Consider Cases A and B, representing two different modes of operation. For Case A, heat is recovered from the storage after one day by a stream of 5000 kg of water entering at 25°C and leaving at 35°C. For Case B, heat is recovered from the storage after 100 days by a stream of 1000 kg of water entering at 25°C and leaving at 75°C.

Energy and exergy analyses of the overall processes are performed for both cases, using superscripts A and B to denote Cases A and B, respectively. In both cases the temperature of the surroundings remains constant at 20°C, and the final state of the storage is the same as the initial state. Water is taken to be an incompressible fluid having a specific heat at constant pressure of $c_p = 4.18$ kJ/kg K, and heat exchanges during charging and discharging are assumed to occur at constant pressure.

The numerical values used in the example were selected to illustrate the concepts discussed in this section, and to resemble values for possible practical system configurations. Several physical implications of the selected numerical values are as follows. First, the inlet and outlet temperatures for the charging and discharging fluids imply that a stratified temperature profile exists in the TES after charging. Secondly, the higher discharging fluid temperature for Case B implies that a greater degree of stratification is maintained during the storing period for Case B (or that greater internal mixing occurs for Case A). Thirdly, the quantities of discharging fluid and the associated temperatures imply that the discharging fluid is circulated through the TES at a greater rate for Case A than for Case B.

Energy analysis for the overall process

The net heat input to the storage during the charging period for each case is

$$H_a - H_b = m_1 c_p (T_a - T_b) = 1000\,\text{kg} \times 4.18\,\text{kJ/kg K} \times (85 - 25)\,\text{K} = 250{,}800\,\text{kJ}$$

For Case A, the heat recovered during the discharging period is

$$(H_d - H_c)^A = 5000\,\text{kg} \times 4.18\,\text{kJ/kg K} \times (35 - 25)\text{K} = 209{,}000\,\text{kJ}$$

The energy efficiency of storage is (see Equation 10.48)

$$\eta^A = \frac{\textit{Heat recovered}}{\textit{Heat input}} = \frac{(H_d - H_c)^A}{H_a - H_b} = \frac{209{,}000\,\text{kJ}}{250{,}800\,\text{kJ}} = 0.833$$

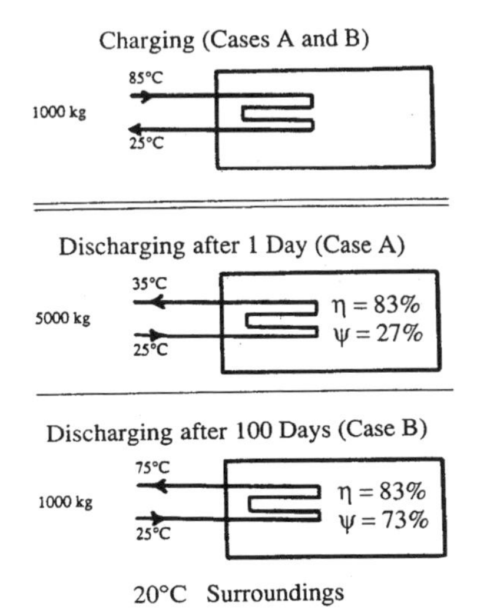

Figure 10.4 An example in which two cases are considered. Shown are the charging process, which is identical for Cases A and B (top), the discharging process for Case A (center), and the discharging process for Case B (bottom).

The heat lost to the surroundings during storage is (see Equation 10.37a with $\Delta E = 0$)

$$Q_l^A = (H_a - H_b) - (H_c - H_d)^A = 250{,}000\,\text{kJ} - 209{,}000\,\text{kJ} = 41{,}800\,\text{kJ}$$

For Case B, the heat recovered during discharging, the energy efficiency and the heat lost to the surroundings can be evaluated similarly:

$$(H_d - H_c)^B = 1000\,\text{kg} \times 4.18\,\text{kJ/kg K} \times (75-25)\text{K} = 209{,}000\,\text{kJ}$$

$$\eta^B = \frac{209{,}000\,\text{kJ}}{250{,}800\,\text{kJ}} = 0.833$$

$$Q_l^B = 250{,}800\,\text{kJ} - 209{,}000\,\text{kJ} = 41{,}800\,\text{kJ}$$

The values of the three parameters evaluated above for Case B are the same as the corresponding values for Case A.

Exergy analysis for the overall process

The net exergy input during the charging period ($\in_a - \in_b$) can be evaluated with Equation 10.45. In that expression, the quantity ($H_a - H_b$) represents the net energy input to the store during charging, evaluated as 250,800 kJ in the previous subsection. Noting that the difference in specific entropy can be written assuming incompressible substances having a constant specific heat as

$$s_a - s_b = c_p \ln\frac{T_a}{T_b} = 4.18\frac{\text{kJ}}{\text{kg K}} \times \ln\frac{358\,\text{K}}{298\,\text{K}} = 0.7667\frac{\text{kJ}}{\text{kg K}}$$

the quantity $T_o(S_a - S_b)$, which represents the unavailable part of the input heat, is

$$T_o(S_a - S_b) = T_o m_1 (s_a - s_b) = 293\,\text{K} \times 1000\,\text{kg} \times 0.7667\,\text{kJ/kg K} = 224{,}643\,\text{kJ}$$

where m_1 denotes the mass of the transport fluid cooled during the charging period. Then, the net exergy input is

$$\in_a - \in_b = 250{,}800\,\text{kJ} - 224{,}643\,\text{kJ} = 26{,}157\,\text{kJ}$$

The net exergy output during the discharging period ($\in_d - \in_c$) can be evaluated using Equation 10.46 and, denoting the mass of the transport fluid circulated during the discharging period as m_3, in a similar three-step fashion for Cases A and B. For Case A,

$$(s_d - s_c)^A = c_p \ln\frac{T_d^A}{T_c^A} = 4.18\frac{\text{kJ}}{\text{kg K}} \times \ln\frac{308\,\text{K}}{298\,\text{K}} = 0.1379\frac{\text{kJ}}{\text{kg K}}$$

$$T_o(S_d - S_c)^A = T_o m_3^A (s_d - s_c)^A = 293\,\text{K} \times 5000\,\text{kg} \times 0.1379\,\text{kJ/kg K} = 202{,}023\,\text{kJ}$$

$$(\in_d - \in_c)^A = 209{,}000\,\text{kJ} - 202{,}023\,\text{kJ} = 6{,}977\,\text{kJ}$$

For Case B,

$$(s_d - s_c)^B = c_p \ln\frac{T_d^B}{T_c^B} = 4.18\frac{\text{kJ}}{\text{kg K}} \times \ln\frac{348\,\text{K}}{298\,\text{K}} = 0.6483\frac{\text{kJ}}{\text{kg K}}$$

$$T_o(S_d - S_c)^B = T_o m_3^B (s_d - s_c)^B = 293\,\text{K} \times 1000\,\text{kg} \times 0.6483\,\text{kJ/kg K} = 189{,}950\,\text{kJ}$$

$$(\epsilon_d - \epsilon_c)^B = 209{,}000\,\text{kJ} - 189{,}950\,\text{kJ} = 19{,}050\,\text{kJ}$$

Thus, the exergy efficiency (see Equation 10.49) for Case A is

$$\psi^A = \frac{(\epsilon_d - \epsilon_c)^A}{\epsilon_a - \epsilon_b} = \frac{6977\ \text{kJ}}{26{,}157\ \text{kJ}} = 0.267$$

and for Case B

$$\psi^B = \frac{(\epsilon_d - \epsilon_c)^B}{\epsilon_a - \epsilon_b} = \frac{19{,}050\ \text{kJ}}{26{,}157\ \text{kJ}} = 0.728$$

which is considerably higher than for Case A.

The exergy losses (total) can be evaluated with Equation 10.40a (with $\Delta\Xi = 0$) as the sum of the exergy loss associated with heat loss to the surroundings and the exergy loss due to internal exergy consumptions:

$$(X_l + I)^A = (\epsilon_a - \epsilon_b) - (\epsilon_d - \epsilon_c)^A = 26{,}157\,\text{kJ} - 6{,}977\,\text{kJ} = 19{,}180\,\text{kJ}$$

$$(X_l + I)^B = (\epsilon_a - \epsilon_b) - (\epsilon_d - \epsilon_c)^B = 26{,}157\,\text{kJ} - 19{,}050\,\text{kJ} = 7{,}107\,\text{kJ}$$

Here, no attempt has been made to evaluate the individual values of the two exergy loss parameters. If it is assumed that heat is transferred at T_o to the surroundings, then $X_l^A = X_l^B = 0$, and all the exergy losses are internal consumptions.

Table 10.5 Comparison of the performance of a TES for two cases.

	Case A	Case B
General parameters		
Storing period (days)	1	100
Charging-fluid temperatures (in/out) (°C)	85/25	85/25
Discharging-fluid temperatures (in/out) (°C)	25/35	25/75
Energy parameters		
Energy input (kJ)	250,800	250,800
Energy recovered (kJ)	209,000	209,000
Energy loss (kJ)	41,800	41,800
Energy efficiency (%)	83.3	83.3
Exergy parameters		
Exergy input (kJ)	26,157	26,157
Exergy recovered (kJ)	6,977	19,050
Exergy loss (kJ)	19,180	7,107
Exergy efficiency (%)	26.7	72.8

Comparative summary

The two cases are summarized and compared in Table 10.5. Although the same quantity of energy is discharged for Cases A and B, a greater quantity of exergy is discharged for Case B. In addition, Case B stores the energy and exergy for a greater duration of time.

10.4.7 Closure

This section demonstrates the application of exergy analysis to closed TES systems. The use of exergy analysis clearly takes into account the external and temperature losses in TES operations, and hence it more correctly reflects their thermodynamic behavior. Exergy and energy analyses do not quantitatively assess the value associated with the length of time the heat is held in storage, so this factor must be considered separately. Other TES types are considered in subsequent sections.

10.5 Appropriate Efficiency Measures for Closed TES Systems

Generally accepted standards have not been established for TES evaluation and comparison for at least two reasons. First, many different but valid efficiency measures can be defined. As shown in Table 10.3, for example, TES energy efficiency can be defined as the ratio of energy recovered to energy input, or as the ratio of energy recovered and remaining in a TES to energy input and originally in the TES; both definitions are reasonable but can yield different values in some circumstances. Secondly, while most TES efficiency definitions are based on energy, more meaningful efficiencies can be defined based on exergy, which takes into account temperature level (and hence quality) of the energy transferred.

In this section, several categories of TES energy and exergy efficiencies are discussed, following an earlier analysis (Rosen, 1992b). The overall storage process and the charging, storing and discharging subprocesses are considered. An illustrative example is presented. This section is an extension of the previous one where a limited set of TES efficiencies are considered, and helps determine which efficiencies are most appropriate in different circumstances.

10.5.1 TES model considered

As in Section 10.4, a simple TES model is considered which undergoes a storage process (Figure 10.5) involving distinct charging, storing and discharging periods (Figure 10.6). The model is described in section 10.4.1. Heat is transferred at specified temperatures to and from the TES. For simplicity, only the net thermal energy and thermal exergy transfers associated with material flows are considered here, rather than the energy and exergy values of the material flows themselves.

10.5.2 Energy and exergy balances

Energy and exergy balances for the overall storage process (Figure 10.5) can be expressed, following Equations 10.37 and 10.40, respectively, as

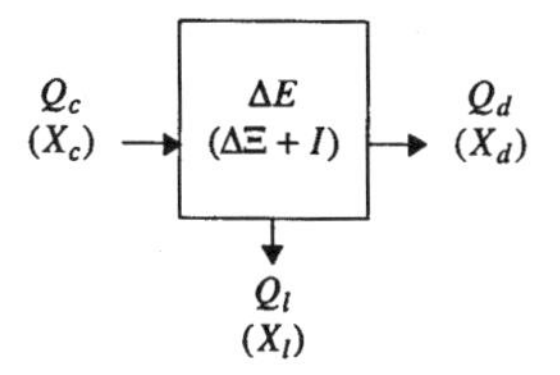

Figure 10.5 The overall heat storage process, showing energy parameters (terms not in parentheses) and exergy parameters (terms in parentheses).

$$Q_c - [Q_d + Q_l] = \Delta E \tag{10.74}$$

$$X_c - [X_d + X_l] - I = \Delta\Xi \tag{10.75}$$

where Q_c, Q_d and Q_l denote respectively the heat input during charging, recovered during discharging and lost during the entire process; and X_c, X_d and X_l denote respectively the exergy transfers associated with Q_c, Q_d and Q_l. For the case of a complete cycle (i.e. a process with identical initial and final states), $\Delta E = \Delta\Xi = 0$. The exergy Ξ can be written with Equation 10.18 which, for a tank of constant volume and a working fluid which behaves ideally, becomes

$$\Xi = mc_v[(T - T_o) - T_o \ln(T/T_o)] \tag{10.76}$$

where T, c_v and m respectively denote for the TES temperature, specific heat at constant volume and mass. The terms Q_l, X_l, I, ΔE and $\Delta\Xi$ can be written in terms of subprocess values as in Equations 10.39, 10.42, 10.70 and 10.71, respectively.

Energy and exergy balances, respectively, can be written for the charging period as in Equations 10.50a (with Q_c replacing $H_a - H_b$) and 10.53 (with X_c replacing $\epsilon_a - \epsilon_b$), for the storing period as in Equations 10.56a and 10.59a, and for the discharging period as in Equations 10.62a (with Q_d replacing $H_d - H_c$) and 10.65a (with X_d replacing $\epsilon_d - \epsilon_c$).

By introducing the definition $E' \equiv E - E_o$, where E' represents the TES energy content, Equations 10.70 and 10.72 can be written as

$$\Delta E = (E_f - E_o) - (E_i - E_o) = E'_f - E'_i \tag{10.70a}$$

$$\Delta E_j = (E_{f,j} - E_o) - (E_{i,j} - E_o) = E'_{f,j} - E'_{i,j} \tag{10.72a}$$

The terms Ξ and E' are analogous (e.g. both are equal to zero at the dead state, i.e. $\Xi_o = E'_o = 0$).

10.5.3 Energy and exergy efficiencies

Since several quantities can be considered to be products and inputs for TES systems, various efficiency definitions are possible. For each subprocess and the overall storage process, energy and exergy efficiencies are evaluated here based on the quantities of heat transferred to and from a TES, the temperatures at which the heat transfers occur, and the initial and final TES states. Furthermore, for each energy and exergy efficiency, four cases are considered (where possible). The initial energy or exergy in the store is neglected in Cases *A* and *B*, and accounted for in Cases *C* and *D*, while the net accumulation of energy or exergy in the store is treated as a loss for Cases *A* and *C*, and as a product for Cases *B* and *D*. Here, efficiency cases are denoted by superscripts and, following the notation introduced earlier, efficiencies with subscripts denote period efficiencies while efficiencies without subscripts denote overall efficiencies. Other efficiency cases are possible but not considered here.

10.5.4 Overall efficiencies

The following four overall energy efficiency η definitions are considered:

$$\eta^A = \frac{\textit{Energy recovered}}{\textit{Energy input}} = 1 - \frac{\textit{Energy loss} + \textit{Energy accumulation}}{\textit{Energy input}}$$

$$\eta^A = \frac{Q_d}{Q_c} = 1 - \frac{Q_l + \Delta E}{Q_c} \tag{10.77}$$

$$\eta^B = \frac{\textit{Energy recovered} + \textit{Energy accumulation}}{\textit{Energy input}} = 1 - \frac{\textit{Energy loss}}{\textit{Energy input}}$$

$$\eta^B = \frac{Q_d + \Delta E}{Q_c} = 1 - \frac{Q_l}{Q_c} \tag{10.77a}$$

$$\eta^C = \frac{\textit{Energy recovered}}{\textit{Energy input} + \textit{Initial energy in store}} = 1 - \frac{\textit{Energy loss} + \textit{Final energy in store}}{\textit{Energy input} + \textit{Initial energy in store}}$$

$$\eta^C = \frac{Q_d}{Q_c + E'_i} = 1 - \frac{Q_l + E'_f}{Q_c + E'_i} \tag{10.77b}$$

$$\eta^D = \frac{\textit{Energy recovered} + \textit{Final energy in store}}{\textit{Energy input} + \textit{Initial energy in store}} = 1 - \frac{\textit{Energy loss}}{\textit{Energy input} + \textit{Initial energy in store}}$$

$$\eta^D = \frac{Q_d + E'_f}{Q_c + E'_i} = 1 - \frac{Q_l}{Q_c + E'_i} \tag{10.77c}$$

Four overall exergy efficiencies, analogous to those in Equations 10.77–10.77c, respectively, are defined:

$$\psi^A = \frac{X_d}{X_c} = 1 - \frac{X_l + I + \Delta\Xi}{X_c} \tag{10.78}$$

$$\psi^B = \frac{X_d + \Delta\Xi}{X_c} = 1 - \frac{X_l + I}{X_c} \tag{10.78a}$$

$$\psi^C = \frac{X_d}{X_c + \Xi_i} = 1 - \frac{X_l + I + \Xi_f}{X_c + \Xi_i} \tag{10.78b}$$

$$\psi^D = \frac{X_d + \Xi_f}{X_c + \Xi_i} = 1 - \frac{X_l + I}{X_c + \Xi_i} \qquad (10.78c)$$

Note that $\eta^A = \eta^B = \eta^C = \eta^D$ if $E_i' = E_f' = 0$, and $\psi^A = \psi^B = \psi^C = \psi^D$ if $\Xi_i = \Xi_f = 0$, while $\eta^A = \eta^B$ if $\Delta E = 0$, and $\psi^A = \psi^B$ if $\Delta\Xi = 0$. Also, if $\Delta E < 0$ and $\Delta\Xi < 0$, the definitions for Cases *A* and *B* do not provide rational measures of performance for normal applications, while those for Cases *C* and *D* do. Furthermore if the store is adiabatic, $Q_{l,j} = X_{l,j} = 0$, and all the efficiencies simplify (e.g. the energy efficiencies in Equations 10.77a and 10.77c become 100%).

10.5.5 *Charging-period efficiencies*

The following two energy efficiency definitions for the charging period are considered:

$$\eta_1^B = \frac{\textit{Energy accumulation}}{\textit{Energy input}} = 1 - \frac{\textit{Energy loss}}{\textit{Energy input}}$$

$$\eta_1^B = \frac{\Delta E_1}{Q_c} = 1 - \frac{Q_{l,1}}{Q_c} \qquad (10.79)$$

$$\eta_1^D = \frac{\textit{Final energy in store}}{\textit{Energy input} + \textit{Initial energy in store}} = 1 - \frac{\textit{Energy loss}}{\textit{Energy input} + \textit{Initial energy in store}}$$

$$\eta_1^D = \frac{E'_{f,1}}{Q_c + E'_{i,1}} = 1 - \frac{Q_{l,1}}{Q_c + E'_{i,1}} \qquad (10.79a)$$

Two analagous exergy efficiencies are defined:

$$\psi_1^B = \frac{\Delta\Xi_1}{X_c} = 1 - \frac{X_{l,1} + I_1}{X_c} \qquad (10.80)$$

$$\psi_1^D = \frac{\Xi_{F,1}}{X_c + \Xi_{I,1}} = 1 - \frac{X_{L,1} + I_1}{X_c + \Xi_{I,1}} \qquad (10.80a)$$

Meaningful charging-period efficiencies can not be defined for Cases *A* and *C*, since accumulations are treated as losses for these cases, resulting in the efficiencies reducing to zero.

10.5.6 *Storing-period efficiencies*

As for the charging period, only two sets of storing-period efficiencies can be defined. The two energy efficiencies are

$$\eta_2^B = \frac{\textit{Energy accumulation during charging and storing}}{\textit{Energy accumulation during charging}} = 1 - \frac{\textit{Energy loss during storing}}{\textit{Energy accumulation during charging}}$$

$$\eta_2^B = \frac{\Delta E_1 + \Delta E_2}{\Delta E_1} = 1 - \frac{Q_{l,2}}{\Delta E_1} \tag{10.81}$$

$$\eta_2^D = \frac{\textit{Final energy in store}}{\textit{Initial energy in store}} = 1 - \frac{\textit{Energy loss during storing}}{\textit{Initial energy in store}}$$

$$\eta_2^D = \frac{E'_{f,2}}{E'_{i,2}} = 1 - \frac{Q_{l,2}}{E'_{i,2}} \tag{10.81a}$$

and the two corresponding exergy efficiencies are

$$\psi_2^B = \frac{\Delta\Xi_1 + \Delta\Xi_2}{\Delta\Xi_1} = 1 - \frac{X_{l,2} + I_2}{\Delta\Xi_1} \tag{10.82}$$

$$\psi_2^D = \frac{\Xi_{f,2}}{\Xi_{i,2}} = 1 - \frac{X_{l,2} + I_2}{\Xi_{i,2}} \tag{10.82a}$$

It can be shown that $\eta_2^B = \eta_2^D$ if $E'_{i,1} = 0$, and $\psi_2^B = \psi_2^D$ if $\Xi_{i,1} = 0$.

10.5.7 Discharging-period efficiencies

Four discharging-period energy efficiencies are considered:

$$\eta_3^A = \frac{Q_d}{\Delta E_1 + \Delta E_2} = 1 - \frac{\Delta E + Q_{l,3}}{\Delta E_1 + \Delta E_2} \tag{10.83}$$

where Q_d is energy recovered, ΔE_l and ΔE_2 are energy accumulation during charging and storing, ΔE is overall energy accumulation, and $Q_{l,3}$ is energy loss.

$$\eta_3^B = \frac{Q_d + \Delta E}{\Delta E_1 + \Delta E_2} = 1 - \frac{Q_{l,3}}{\Delta E_1 + \Delta E_2} \tag{10.83a}$$

where Q_d is energy recovered, ΔE_l and ΔE_2 are energy accumulation during charging and storing, ΔE is overall energy accumulation, and $Q_{l,3}$ is energy loss.

$$\eta_3^C = \frac{\textit{Energy recovered}}{\textit{Initial energy in store}} = 1 - \frac{\textit{Final energy in store} + \textit{Energy loss}}{\textit{Initial energy in store}}$$

$$\eta_3^C = \frac{Q_d}{E'_{i,3}} = 1 - \frac{E'_{f,3} + Q_{l,3}}{E'_{i,3}} \tag{10.83b}$$

$$\eta_3^D = \frac{\textit{Energy recovered} + \textit{Final energy in store}}{\textit{Initial energy in store}} = 1 - \frac{\textit{Energy loss}}{\textit{Initial energy in store}}$$

$$\eta_3^D = \frac{Q_d + E'_{f,3}}{E'_{i,3}} = 1 - \frac{Q_{l,3}}{E'_{i,3}} \tag{10.83c}$$

Table 10.6 Summary of possible efficiency definitions for closed TES systems.*

	Energy Efficiency, η			
Period	*Case A*	*Case B*	*Case C*	*Case D*
Overall	$\frac{Q_d}{Q_c}$	$\frac{Q_d + \Delta E}{Q_c}$	$\frac{Q_d}{Q_c + E'_i}$	$\frac{Q_d + E'_f}{Q_c + E'_i}$
Charging (1)	–	$\frac{\Delta E_1}{Q_c}$	–	$\frac{E'_{f,1}}{Q_c + E'_{i,1}}$
Storing (2)	–	$\frac{\Delta E_1 + \Delta E_2}{\Delta E_1}$	–	$\frac{E'_{f,2}}{E'_{i,2}}$
Discharging (3)	$\frac{Q_d}{\Delta E_1 + \Delta E_2}$	$\frac{Q_d + \Delta E}{\Delta E_1 + \Delta E_2}$	$\frac{Q_d}{E'_{i,3}}$	$\frac{Q_d + E'_{f,3}}{E'_{i,3}}$

	Exergy Efficiency, ψ			
Period	*Case A*	*Case B*	*Case C*	*Case D*
Overall	$\frac{X_d}{X_c}$	$\frac{X_d + \Delta\Xi}{X_c}$	$\frac{X_d}{X_c + \Xi i}$	$\frac{X_d + \Xi_f}{X_c + \Xi_i}$
Charging (1)	–	$\frac{\Delta\Xi_1}{X_c}$	–	$\frac{\Xi_{f,1}}{X_c + \Xi i_{,1}}$
Storing (2)	–	$\frac{\Delta\Xi_1 + \Delta\Xi_2}{\Delta\Xi_1}$	–	$\frac{\Xi_{f,2}}{\Xi_{i,2}}$
Discharging (3)	$\frac{X_d}{\Delta\Xi_1 + \Delta\Xi_2}$	$\frac{X_d + \Delta\Xi}{\Delta\Xi_1 + \Delta\Xi_2}$	$\frac{X_d}{\Xi_{i,3}}$	$\frac{X_d + \Xi_{f,3}}{\Xi_{i,3}}$

* Charging and storing efficiencies are not defined for Cases A and C. The cases listed across the top are denoted by superscripts, and the periods along the left side by subscripts (e.g. the lower left entry corresponds to η_3^A, and the upper right entry to ψ^D.

Four analogous exergy efficiencies are defined:

$$\psi_3^A = \frac{X_d}{\Delta\Xi_1 + \Delta\Xi_2} = 1 - \frac{\Delta\Xi + X_{l,3} + I_3}{\Delta\Xi_1 + \Delta\Xi_2} \tag{10.84}$$

$$\psi_3^B = \frac{X_d + \Delta\Xi}{\Delta\Xi_1 + \Delta\Xi_2} = 1 - \frac{X_{l,3} + I_3}{\Delta\Xi_1 + \Delta\Xi_2} \tag{10.84a}$$

$$\psi_3^C = \frac{X_d}{\Xi_{i,3}} = 1 - \frac{X_{l,3} + I_3 + \Xi_{f,3}}{\Xi_{i,3}} \tag{10.84b}$$

$$\psi_3^D = \frac{X_d + \Xi_{f,3}}{\Xi_{i,3}} = 1 - \frac{X_{l,3} + I_3}{\Xi_{i,3}} \tag{10.84c}$$

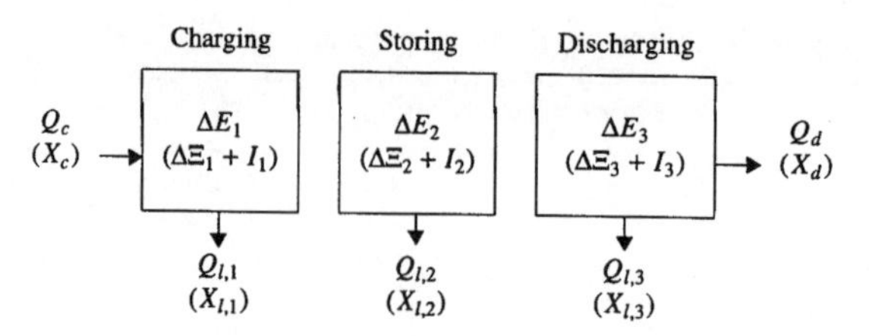

Figure 10.6 The three periods in the overall heat storage process (charging, storing and discharging), showing energy parameters (not in parentheses) and the exergy parameters (in parentheses).

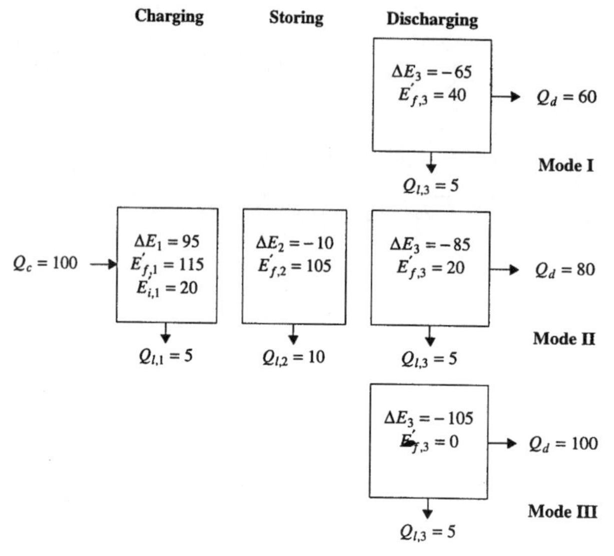

Figure 10.7 Energy values (in kJ) for the example, showing the three modes of operation (I, II and III), and the three storing periods (charging, storing and discharging).

10.5.8 Summary of efficiency definitions

The efficiency definitions in the four preceding subsections are summarized in Table 10.6. Although all the efficiencies are useful in specific applications, the choice of definition depends on the specific application and the aspect of performance considered critical. Many of the efficiencies discussed here are applied in practice and some are used in section 10.4, but with Q_c (or X_c) given as the difference between the total enthalpy (or exergy) of the fluid passing into and out of the store during charging, and Q_d (or X_d) as the corresponding difference during discharging.

For the assumptions considered in this paper, it can be shown for Cases *B* and *D* that the overall efficiency is the product of the three corresponding subprocess efficiencies, i.e.

$$\eta^B = \prod_{j=1}^{3} \eta_j^B \tag{10.85}$$

$$\eta^D = \prod_{j=1}^{3} \eta_j^D \tag{10.86}$$

$$\psi^B = \prod_{j=1}^{3} \psi_j^B \tag{10.87}$$

$$\psi^D = \prod_{j=1}^{3} \psi_j^D \tag{10.88}$$

By noting the similarities between Cases A and B and between Cases C and D, it can also be shown that

$$\eta^A = \eta_1^B \eta_2^B \eta_3^A \tag{10.89}$$

$$\eta^C = \eta_1^D \eta_2^D \eta_3^C \tag{10.90}$$

$$\psi^A = \psi_1^B \psi_2^B \psi_3^A \tag{10.91}$$

$$\psi^C = \psi_1^D \psi_2^D \psi_3^C \tag{10.92}$$

10.5.9 Illustrative example

Problem statement

A TES is considered which undergoes discrete charging, storing and discharging processes. Three modes of operation are considered, defined according to the sign of the energy and exergy accumulations during the overall storing process:

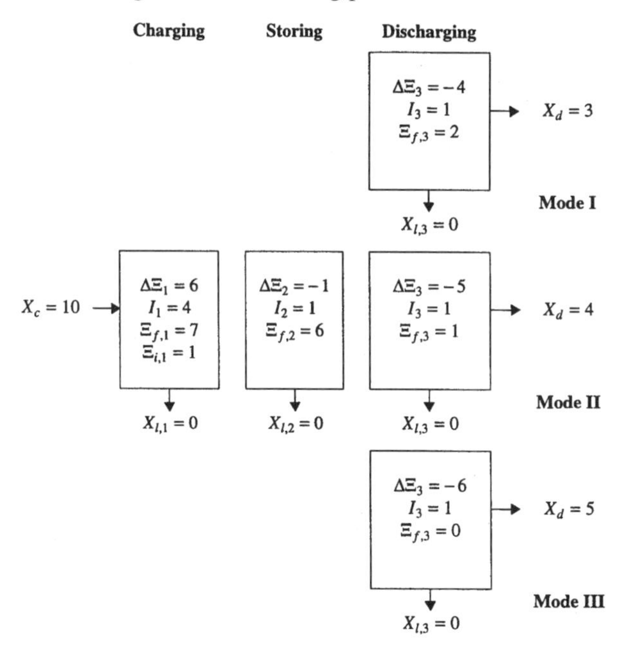

Figure 10.8 Exergy values (in kJ) for the example, showing the three modes of operation and the three storing periods.

Table 10.7 Efficiency values (in %) for the illustrative example.

	Energy efficiency, η				Exergy efficiency, ψ			
Period	A	B	C	D	A	B	C	D
Mode of operation I								
Overall	60	80	50	83	30	40	27	45
Charging (1)	–	95	–	96	–	60	–	64
Storing (2)	–	89	–	91	–	83	–	86
Discharging (3)	71	94	57	95	60	80	50	83
Mode of operation II								
Overall	80	80	67	83	40	40	36	45
Charging (1)	–	95	–	96	–	60	–	64
Storing (2)	–	89	–	91	–	83	–	86
Discharging (3)	94	94	76	95	80	80	67	83
Mode of operation III								
Overall	100	80	83	83	50	40	45	45
Charging (1)	–	95	–	96	–	60	–	64
Storing (2)	–	89	–	91	–	83	–	86
Discharging (3)	118	94	95	95	100	80	83	83

I. $\Delta E > 0$ and $\Delta\Xi > 0$;
II. $\Delta E = \Delta\Xi = 0$, i.e. a complete storage cycle; and
III. $\Delta E < 0$ and $\Delta\Xi < 0$.

For all modes of operation, and with values in kilojoules, $Q_c = 100$, $Q_{l,1} = Q_{l,3} = 5$, $Q_{l,2} = 10$, $E'_{i,1} = 20$, $X_{l,j} = 0$ for all j, $\Xi_{i,1} = 1$, $I_1 = 4$ and $I_2 = I_3 = 1$. For modes I, II and III, respectively, Q_d is 60, 80 and 100 kJ. Also, $X_c = 0.1 \times Q_c$ and $X_d = 0.05 \times Q_d$. Note that the specification of the exergy parameters involves assumptions regarding the temperatures associated with the heat transfers Q_c, Q_d, and $Q_{l,j}$ (e.g. the fact that $X_{l,j} = 0$ and $Q_{l,j} \neq 0$ for all j, implies that heat transfers $Q_{l,j}$ occur at the environmental temperature T_o). The specified parameter values are chosen so as to represent a realistic but simple case.

Results and discussion

Unknown parameter values are evaluated using the previously specified values and energy and exergy balances. Following the format of Figure 10.6, all relevant values are summarized illustratively in Figure 10.7 for energy parameters, and in Figure 10.8 for exergy parameters. Since the charging and storing processes are identical for all three modes of operation, they are illustrated only once, while the discharging process is shown for each operation mode. Values for ΔE and $\Delta\Xi$ are not shown in Figures 10.7 and 10.8, but by substituting subprocess values for ΔE_j and $\Delta\Xi_j$ from Figures 10.7 and 10.8 into Equations 10.70 and 10.71, can be shown to be (in kJ) as follows: $\Delta E = 20$ and $\Delta\Xi = 1$ for mode of operation I, $\Delta E = \Delta\Xi = 0$ for mode II, and $\Delta E = -20$ and $\Delta\Xi = -1$ for mode III. The exergy values associated with all heat transfers in Figure 10.8 are significantly less than the corresponding energy values in Figure 10.7, reflecting the fact that for most TES applications the temperatures associated with all heat transfers are relatively low (between T_o and $2T_o$). These values indicate that the usefulness of thermal energy transferred at a practical temperature is significantly less than an equal quantity of any work-equivalent energy form.

Efficiencies are evaluated according to the expressions in Table 10.6, and tabulated in the same format in the top, middle and bottom sections of Table 10.7 for operation modes I, II and III, respectively. The corresponding efficiencies for modes I to III are the same for charging and storing since these processes are independent of mode of operation. Several points are demonstrated in Table 10.7. First, all exergy efficiency values are less than the corresponding energy efficiency values, due to the degradation of temperature as heat is transferred and stored. Since this degradation is reflected in efficiencies based on exergy and not in those based on energy, exergy efficiencies are considered more meaningful. Secondly, for energy and exergy efficiencies for all periods, the values of the calculated efficiencies are different for Cases *A* to *D*. The achievement of consistent comparisons and evaluations of performance requires that care be exercised in selecting efficiency definitions. Thirdly, some efficiency values are not meaningful because they are not based on rational measures of performance for the operation mode considered. For example, consider, for Case *A* and operation mode III, the overall energy efficiency (η^A = 100%), the discharging energy efficiency (η_3^A = 118%), and the discharging exergy efficiency (ψ_3^A = 100%). These efficiency values consider the percentage of the input energy or exergy which is discharged, but are not rational measures because they do not account for the fact that part of the discharged energy and exergy is attributable to energy and exergy in the TES before the charging process began. Thus, the efficiency definitions presented for Cases *B* to *D* are more meaningful for mode of operation III.

Several generalizations can be made for each of the three modes of operation with respect to the validity and meaningfulness of the efficiency definitions for the different cases. For mode I, the definitions for Cases *B* and *D* are preferred. The definitions for Cases *A* and *C*, although valid, can be misleading because they neglect the fact that some of the charging energy or exergy increases the internal energy or exergy of the TES. For mode II (the complete storage cycle), the definitions for all cases are valid and meaningful. For mode III, the definitions for Cases *B* and *D* are preferred. As discussed in the previous paragraph, the definition for Case *A* is not rational for mode III, while that for Case *C*, although valid, can be misleading.

10.5.10 Closure

Several key points are discussed in this section. First, many valid and meaningful TES efficiency definitions exist. Differences in definitions normally depend on which quantities are considered to be products and inputs. Secondly, different efficiency definitions are appropriate in different situations and when different aspects of performance are being evaluated. Of course, efficiency comparisons for different TES systems are logical only if the efficiencies are based on common definitions. Thirdly, exergy efficiencies, because they measure how nearly a system approaches ideal performance and account for the loss of temperature in TESs, are generally more meaningful and illuminating than energy efficiencies, and may prove useful in establishing standards for TES evaluation and comparison.

Table 10.8 Relation between several temperature parameters for above-environment temperatures.*

T/T_o	T (K)	τ
1.00	283	0.00
1.25	354	0.20
1.50	425	0.33
2.00	566	0.50
3.00	849	0.67
5.00	1415	0.80
10.00	2830	0.90
100.00	28,300	0.99
$\equiv$	$\equiv$	1.00

* The reference-environment temperature is T_o = 10°C = 283 K.

Table 10.9 Values of the ratio ψ/η for a range of practical values for T_d and T_c.*

Discharging temperature, T_d (°C)	Charging temperature, T_c (°C)			
	40	70	100	130
40	1.00	0.55	0.40	0.32
70	–	1.00	0.72	0.59
100	–	–	1.00	0.81
130	–	–	–	1.00

* The reference-environment temperature is T_o = 10°C = 283 K.

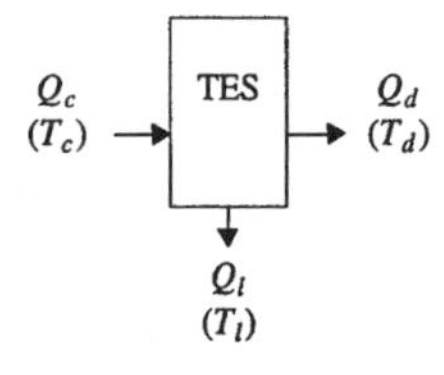

Figure 10.9 The overall heat storage process for a general TES system. Shown are heat flows, and associated temperatures at the TES boundary (terms in parentheses). The corresponding energy and exergy parameters are shown for this system in Figure 10.5.

10.6 Importance of Temperature in Performance Evaluations for Sensible TES Systems

Being energy-based, most existing TES evaluation measures disregard the temperatures associated with the heat injected into and recovered from a TES. Examining energy efficiencies alone can result in misleading conclusions because such efficiencies weight all thermal energy equally. Exergy efficiencies acknowledge that the usefulness of thermal energy depends on its quality, which is related to its temperature level, and are therefore more suitable for determining how advantageous is one TES relative to another.

This section discusses the importance of temperature in TES performance evaluations, following an earlier report (Rosen, 1991). The energy and exergy efficiencies for a simple sensible TES system are compared and the differences between them highlighted. It is demonstrated that exergy analysis weights the usefulness of thermal energy appropriately, while energy analysis tends to present overly optimistic views of TES performance by neglecting the temperature levels associated with thermal energy flows. The concepts are illustrated by examining several TES systems.

10.6.1 Energy, entropy and exergy balances for the TES system

Consider a process involving only heat interactions and occuring in a closed system for which the state is the same at the beginning and end of the process. Balances of energy and exergy, respectively, can be written for the system using Equations 10.3b and 10.5b as follows:

$$\sum_r Q_r = 0 \tag{10.93}$$

$$\sum_r Q_r - 1 = 0 \tag{10.94}$$

where I denotes the exergy consumption, and X_r denotes the exergy associated with Q_r, the heat transferred into the system across region r at temperature T_r. Note that the exergetic temperature factor τ is illustrated as a function of the temperature ratio T/T_o in Figure 10.1, and these parameters are compared with the temperature T in Table 10.8 for above-environmental temperatures (i.e. for $T \geq T_o$), the temperature range of interest for most heat storages.

10.6.2 TES system model considered

Consider the overall storage process for the general TES system shown in Figure 10.9. Heat Q_c is injected into the system at a constant temperature T_c during a charging period. After a storing period, heat Q_d is recovered at a constant temperature T_d during a discharging period. During all periods, heat Q_l leaks from the system at a constant temperature T_l and is lost to the surroundings.

For normal heating applications, the temperatures T_c, T_d and T_l exceed the environment temperature T_o, but the discharging temperature cannot exceed the charging temperature. Hence, the exergetic temperature factors for the charged and discharged heat are subject to the constraint $0 \leq \tau_d \leq \tau_c \leq 1$.

10.6.3 Analysis

The energy and exergy balances in Equations 10.93 and 10.94, respectively, can be written for the modeled system as

$$Q_c = Q_d + Q_l \tag{10.93a}$$

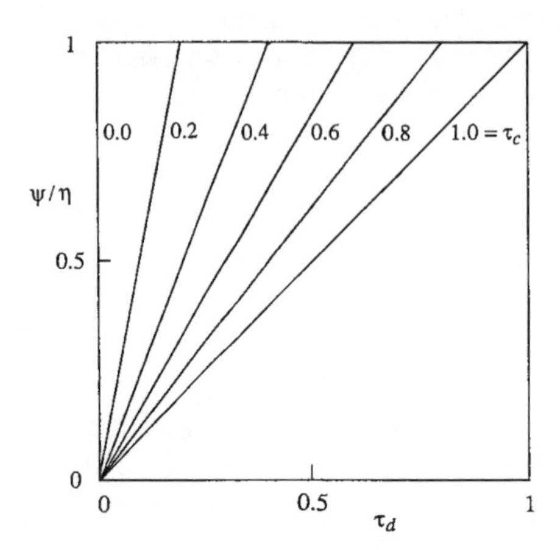

Figure 10.10 Energy-efficiency-to-exergy-efficiency ratio, ψ/η, as a function of the discharging exergetic temperature factor τ_d, for several values of the charging exergetic temperature factor τ_c.

and

$$X_c = X_d + X_l + I \tag{10.94a}$$

With Equation 10.25a, the exergy balance can be expressed as

$$Q_c\tau_c = Q_d\tau_d + Q_l\tau_l + I \tag{10.94b}$$

Following the general energy and exergy efficiency statements in Equations 10.32 and 10.33, the energy efficiency can be written for the modeled system as

$$\eta = \frac{Q_d}{Q_c} \tag{10.95}$$

and the exergy efficiency (with Equations 10.25a and 10.95) as

$$\psi = \frac{X_d}{X_c} = \frac{Q_d\tau_d}{Q_c\tau_c} = \frac{\tau_d}{\tau_c}\eta \tag{10.96}$$

10.6.4 Comparison of energy and exergy efficiencies

An illuminating parameter for comparing the efficiencies is the ratio ψ/η. For the general TES system above, Equation 10.96 implies that the energy-efficiency-to-exergy-efficiency ratio can be expressed as

$$\frac{\psi}{\eta} = \frac{\tau_d}{\tau_c} \tag{10.97}$$

With Equation 10.25a, Equation 10.97 can be alternatively expressed as

$$\frac{\psi}{\eta} = \frac{(T_d - T_o)T_c}{(T_c - T_o)T_d} \tag{10.97a}$$

The ratio ψ/η is plotted against τ_d for several values of τ_c in Figure 10.10. It is seen that ψ/η varies linearly with τ_d for a given value of τ_c. Also, if the product heat is delivered at the charging temperature (i.e. $\tau_d = \tau_c$), $\psi = \eta$, while if the product heat is delivered at the temperature of the environment (i.e. $\tau_d = 0$), $\psi = 0$ regardless of the charging temperature. In the first case, there is no loss of temperature during the entire storage process, while in the second there is a complete loss of temperature. The largest deviation between values of ψ and η occurs in the second case.

The deviation between ψ and η is significant for most present day TES systems. This can be seen by noting that (i) most TES systems operate between charging temperatures as high as $T_c = 130°C$ and discharging temperatures as low as $T_d = 40°C$, and (ii) a difference of about 30ºC between charging and discharging temperatures is utilized in most TES systems (i.e. $T_c - T_d = 30°C$). With Equation 10.26 and $T_o = 10°C$, the first condition can be shown to imply for most present day systems that $0.1 \leq \tau_d \leq \tau_c \leq 0.3$. Since it can be shown with Equation 10.26 that

$$\tau_c - \tau_d = \frac{(T_c - T_d)T_o}{T_c T_d} \tag{10.98}$$

the difference in exergetic temperature factor varies roughly between 0.06 and 0.08. Then the value of the exergy efficiency is nearly 50% to 80% of that of the energy efficiency.

10.6.5 Illustration

The ratio ψ/η is illustrated in Table 10.9 for a simple TES having charging and discharging temperatures ranging between 40ºC and 130ºC, and a reference-environment temperature of $T_o = 10ºC$. The energy and exergy efficiencies differ (with the exergy efficiency always being the lesser of the two) when $T_d < T_c$, and the difference becomes more significant as the difference between T_c and T_d increases. The efficiencies are equal only when the charging and discharging temperatures are equal (i.e. $T_d = T_c$), and no values of the ratio ψ/η are reported for the cases when $T_c < T_d$, since such situations are not physically possible. Unlike the exergy efficiencies, the energy efficiencies tend to appear overly optimistic, in that they only account for losses attributable to heat leakages but ignore temperature degradation.

10.6.6 Closure

This section demonstrates the importance of temperature in TES performance evaluations. Exergy efficiencies are seen to be more illuminating than energy efficiencies because they weight heat flows appropriately, being sensitive to both the fraction of the heat injected into a TES that is recovered and the temperature at which heat is recovered relative to the temperature at which it is injected. Energy efficiencies are only sensitive to the first of the above factors. TES energy efficiencies are good approximations to exergy efficiencies when there is little temperature degradation, as thermal energy quantities then have similar qualities. In most practical situations, however, thermal energy is injected and recovered at significantly different temperatures, making energy efficiencies both poor approximations to exergy efficiencies and prone to lead to erroneous interpretations and conclusions.

10.7 Exergy Analysis of Aquifer TES Systems

Underground aquifers are the storage type considered in this section (Jenne, 1992). The storage medium in many aquifer TES (ATES) systems remains in a single phase during the storing cycle, so that temperature changes are exhibited in the store as thermal energy is added or removed. To assess ATES systems and to integrate them into larger thermal systems properly, their characteristics must be accurately expressed. As for TES systems, standards have not been established for evaluating and comparing the performance of different ATES systems, and conventional energy-based performance measures are often misleading. Exergy methods are advantageous in this regard.

In this section, the application of exergy analysis to ATES systems is described, following a previous assessment (Rosen, 1999b). For an elementary ATES model, expressions are presented for the injected and recovered quantities of energy and exergy and for efficiencies. The impact is examined of introducing a threshold temperature below which residual heat remaining in the aquifer water is not considered worth recovering. ATES exergy efficiencies are demonstrated to be more useful and meaningful than energy efficiencies because the former account for the temperatures associated with thermal energy transfers and consequently assess how nearly ATES systems approach ideal thermodynamic performance. ATES energy efficiencies do not provide a measure of approach to ideal performance and, in fact, are often misleadingly high because some of the thermal energy can be recovered at temperatures too low for useful purposes. A case study using realistic ATES parameter values is presented.

10.7.1 ATES model

Charging of the ATES occurs over a finite time period t_c and after a holding interval discharging occurs over a period t_d. The working fluid is water, having a constant specific heat c, and assumed incompressible. The temperature of the aquifer and its surroundings prior to heat injection is T_o, the reference-environment temperature. Only heat stored at temperatures above T_o is considered, and pump work is neglected.

Charging and discharging

During charging, heated water at a constant temperature T_c is injected at a constant mass flow rate $\dot{m}_c$ into the ATES. Then, after a storing period, discharging occurs, during which water is extracted from the ATES at a constant mass flow rate $\dot{m}_d$. The fluid discharge temperature is taken to be a function of time, i.e. $T_d = T_d(t)$. The discharge temperature after an infinite time is taken to be the temperature of the reference-environment, i.e. $T_d(\infty) = T_o$, and the initial discharge temperature is taken to be between the charging and reference-environment temperatures, i.e. $T_o \leq T_d(0) \leq T_c$.

Many discharge temperature-time profiles are possible. The discharge temperature may be taken to decrease linearly with time from an initial value $T_d(0)$ to a final value T_o. The final temperature is reached at a time t_f and remains fixed at T_o for all subsequent times, i.e.

$$T_d(t) = \begin{cases} T_d(0) - (T_d(0) - T_o t / t_f, & 0 \leq t \leq t_f \\ T_0, & t_f \leq t \leq \infty \end{cases} \tag{10.99}$$

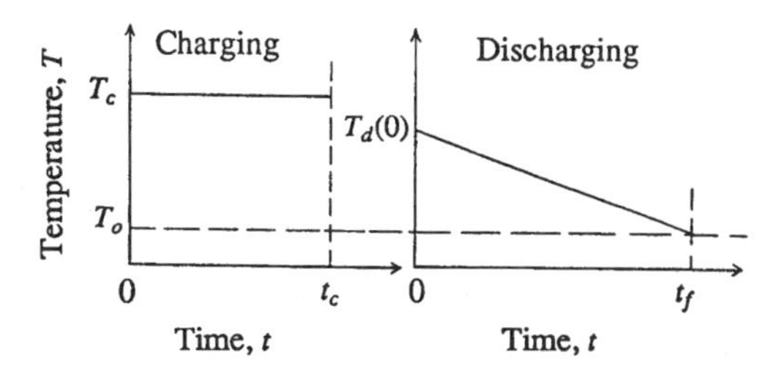

Figure 10.11 Temperature-time profiles assumed for the charging and discharging periods in the ATES model considered.

Alternatively, the discharge temperature can be expressed as

$$T_d(t) = T_o + (T_d(0) - T_o)e^{-\alpha t} \tag{10.99a}$$

Here, the discharge temperature decreases with time asymptotically towards the environment temperature T_o, at a rate controlled by the arbitrary constant α. Of course, these and other models of discharging temperature-time profiles are simply elementary representations of events that are often more complicated in actual devices.

The simple linear discharge temperature-time profile (Equation 10.99) is used here because, for the purposes of this section, most of the other possible discharge temperature-time profiles are inconveniently complex mathematically. The linear profile is sufficiently realistic and simple to illustrate the importance of using exergy analysis in ATES evaluation and comparison, without obscuring the central ideas of the section. The temperature-time profiles considered in the present model for the fluid flows during the charging and discharging periods are summarized in Figure 10.11.

Thermodynamic losses

The two main types of thermodynamic losses that occur in ATES systems are accounted for in the model:

- *Energy losses.* Energy injected into an ATES that is not recovered is considered lost. Thus, energy losses include energy remaining in the ATES at a point where it could still be recovered if pumping were continued, and energy injected into the ATES that is convected in a water flow or is transferred by conduction far enough from the discharge point that it is unrecoverable, regardless of how much or how long water is pumped out of the ATES. The effect of energy losses is that less than 100% of the injected energy is recoverable after storage.
- *Mixing losses.* As heated water is pumped into an ATES, it mixes with the water already present (which is usually cooler), resulting in the recovered water being at a lower temperature than the injected water. In the present model, this loss results in the discharge temperature T_d being at all times less than or equal to the charging temperature T_c, but not below the reference-environment temperature T_o (i.e. $T_o \leq T_d(t) \leq T_c$ for $0 \leq t \leq \infty$).

10.7.2 Energy and exergy analyses

To perform energy and exergy analyses of ATES systems, the quantities of energy and exergy injected during charging and recovered during discharging must be evaluated. The energy flow associated with a flow of liquid at a constant mass flow rate $\dot{m}$, for an arbitrary period of time with T a function of t, is

$$E = \int_t \dot{E}(t)dt \tag{10.100}$$

where the integration is performed over the time period, and the energy flow rate at time t is

$$\dot{E}(t) = \dot{m}c(T(t)-T_o) \tag{10.101}$$

Here c denotes the specific heat of the liquid. Combining Equations 10.100 and 10.101 for constant $\dot{m}, c$ and T_o,

$$E = \dot{m}c\int_t (T(t)-T_o)dt \tag{10.100a}$$

The corresponding exergy flow is

$$\in = \int_t \dot{\in}(t)dt \tag{10.102}$$

where the exergy flow rate at time t is

$$\dot{\in}(t) = \dot{m}c[(T(t)-T_o) - T_o\ln(T(t)/T_o)] \tag{10.103}$$

Combining Equations 10.102 and 10.103, and utilizing Equation 10.100a,

$$\in = \dot{m}c\int_t [(T(t)-T_o) - T_o\ln(T(t)/T_o)]dt = E - \dot{m}cT_o\int_t \ln(T(t)/T_o)dt \tag{10.102a}$$

Charging

The energy input to the ATES during charging, for a constant water injection rate $\dot{m}_c$ and over a time period beginning at zero and ending at t_c, is expressed by Equation 10.100a with $T(t) = T_c$. That is,

$$E_c = \dot{m}_c c \int_{t=0}^{t_c} (T_c - T_o)dt = \dot{m}_c c t_c (T_c - T_o) \tag{10.104}$$

The corresponding exergy input is expressed by Equation 10.102a, with the same conditions as for E_c. Thus, after integration,

$$\in_c = \dot{m}_c c t_c [(T_c - T_o) - T_o\ln(T_c/T_o)] = E_c - \dot{m}_c c t_c T_o \ln(T_c/T_o) \tag{10.105}$$

Discharging

The energy recovered from the ATES during discharging, for a constant water recovery rate $\dot{m}_d$ and for a time period starting at zero and ending at t_d, is expressed by Equation 10.100a with $T(t)$ as in Equation 10.99. Thus,

$$E_d = \dot{m}_d c \int_{t=0}^{t_d} (T_d(t) - T_o)dt = \dot{m}_d c[T_d(0) - T_o]\theta(2t_f - \theta)/(2t_f) \tag{10.106}$$

where

$$\theta = \begin{cases} t_d \ ; \ 0 \leq t_d \leq t_f \\ t_f \ ; \ t_f \leq t_d \leq \infty \end{cases} \tag{10.107}$$

The corresponding exergy recovered is expressed by Equation 10.102a, with the same conditions as for E_d. Thus,

$$\epsilon_d = \dot{m}_d c \int_{t=0}^{t_d} [(T_d(t) - T_o) - T_o \ln(T_d(t)/T_o)]dt = E_d - \dot{m}_d c T_o \int_{t=0}^{t_d} \ln(T_d(t)/T_o)dt \tag{10.108}$$

Here,

$$\int_{t=0}^{t_d} \ln(T_d(t)/T_o)dt = \int_{t=0}^{t_d} \ln(at+b)dt = [(a\theta+b)/a]\ln(a\theta+b) - \theta - (b/a)\ln b \tag{10.109}$$

where

$$a = [T_o - T_d(0)]/(T_o t_f) \tag{10.110}$$

$$b = T_d(0)/T_o \tag{10.111}$$

When $t_d \geq t_f$, the expression for the integral in Equation 10.109 reduces to

$$\int_{t=0}^{t_d} \ln(T_d(t)/T_o)dt = t_f \left[\frac{T_d(0)}{T_d(0) - T_o} \ln \frac{T_d(0)}{T_o} - 1 \right] \tag{10.109a}$$

Energy and exergy balances

An ATES energy balance taken over a complete charging-discharging cycle states that the energy injected is either recovered or lost. A corresponding exergy balance states that the exergy injected is either recovered or lost, where lost exergy is associated with both waste exergy emissions and internal exergy consumptions due to irreversibilities.

If f is defined as the fraction of injected energy E_c that can be recovered if the length of the discharge period approaches infinity (i.e. water is extracted until all recoverable energy has been recovered), then

$$E_d(t_d \to \infty) = fE_c \tag{10.112}$$

It follows from the energy balance that $(1 - f)E_c$ is the energy irreversibly lost from the ATES. Clearly, f varies between zero for a thermodynamically worthless ATES to unity for an ATES having no energy losses during an infinite discharge period. (Note that even if f = 1, the ATES can still have mixing losses that reduce the temperature of the recovered water and consequently cause exergy losses.) Since E_c is given by Equation 10.104 and $E_d(t_d \to \infty)$ by Equation 10.106 with $\theta = t_f$, Equation 10.112 may be rewritten as

$$\dot{m}_d c(T_d(0) - T_o)t_f / 2 = f \dot{m}_c c(T_c - T_o)t_c \tag{10.112a}$$

or, after rearranging,

$$f = \frac{t_f \dot{m}_d (T_d(0) - T_o)}{2t_c \dot{m}_c (T_c - T_o)} \tag{10.112b}$$

Since $T_d(0)$ can vary from T_o to T_c, the temperature-related term $(T_d(0) - T_o)/(T_c - T_o)$, like f, varies between zero and unity. The time ratio t_f/t_c and mass-flow-rate ratio $\dot{m}_d / \dot{m}_c$ can both take on any positive values, subject to the above equality.

Efficiencies and losses

For either energy or exergy, efficiency is defined as the fraction, taken over a complete cycle, of the quantity input during charging that is recovered during discharging, while loss is the difference between input and recovered amounts of the quantity. Hence, the energy loss as a function of the discharge time period is given by $[E_c - E_d(t_d)]$, while the corresponding exergy loss is given by $[\in_c - \in_d(t_d)]$. It is emphasized that energy losses do not reflect the temperature degradation associated with mixing, while exergy losses do.

The energy efficiency η for an ATES, as a function of the discharge time period, is

$$\eta(t_d) = \frac{E_d(t_d)}{E_c} = \frac{\dot{m}_d (T_d(0) - T_o)}{\dot{m}_c (T_c - T_o)} \frac{\theta(2t_f - \theta)}{2t_f t_c} \tag{10.113}$$

and the corresponding exergy efficiency ψ by

$$\psi(t_d) = \in_d (t_d) / \in_c \tag{10.114}$$

Note that the energy efficiency in Equation 10.113 simplifies when the discharge period t_d exceeds t_f, i.e. $\eta(t_d \geq t_f) = f$. Thus, for an ATES in which all injected energy is recoverable during an infinite discharge period, i.e. $f = 1$, the energy efficiency can reach 100% if the discharge period t_d is made long enough. The corresponding exergy efficiency, however, remains less than 100% because due to mixing losses, much of the heat is recovered at near-environmental temperatures. Only a thermodynamically reversible storage, which would never be an objective since such other factors as economics must be considered, would permit the achievement of an exergy efficiency of 100%.

10.7.3 Effect of a threshold temperature

In practice, it is not economically feasible to continue the discharge period until as much recoverable heat as possible is recovered. As the discharge period increases, water is recovered from an ATES at ever decreasing temperatures (ultimately approaching the reference-environment temperature T_o), and the energy in the recovered water is of decreasing usefulness. Exergy analysis reflects this phenomenon, as the magnitude of the recovered exergy decreases as the recovery temperature decreases. To determine the appropriate discharge period, a threshold temperature T_t is often introduced, below which the residual energy in the aquifer water is not considered worth recovering from an ATES. For the linear temperature-time relation used here (see Equation 10.99), it is clear that no thermal energy could be recovered over a cycle if the threshold temperature exceeds the initial discharge temperature, while the appropriate discharge period can be evaluated using Equation 10.99 with T_t replacing $T_d(t)$ for the case where $T_o \leq T_t \leq T_d(0)$. Thus,

$$t_d = \begin{cases} \dfrac{T_d(0) - T_t}{T_d(0) - T_o} t_f, & T_o \leq T_t \leq T_d(0) \\ 0, & T_d(0) \leq T \end{cases} \tag{10.115}$$

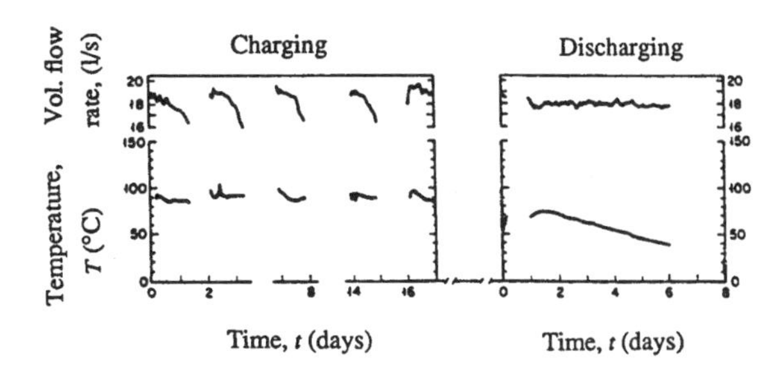

Figure 10.12 Observed values for the temperature and volumetric flow rate of water, as a function of time during the charging and discharging periods, for the experimental test cycles used in the ATES case study.

The effect of a threshold temperature in practice, therefore, is to place an upper limit on the allowable discharge time period. Utilizing a threshold temperature usually has the effect of decreasing the difference between the corresponding energy and exergy efficiencies.

10.7.4 Case study

Background

In this case study, experimental data are used from the first of four short-term ATES test cycles, using the Upper Cambrian Franconia-Ironton-Galesville confined aquifer. The test cycles were performed at the University of Minnesota's St. Paul campus from November 1982 to December 1983 (Hoyer *et al.,* 1985). During the test, water was pumped from the source well, heated in a heat exchanger and returned to the aquifer through the storage well. After storage, energy was recovered by pumping the stored water through a heat exchanger and returning it to the supply well. The storage and supply wells are located 255 meters apart.

For the test cycle considered here, the water temperature and volumetric flow rate vary with time during the injection and recovery processes as shown in Figure 10.12. The storage period duration (13 days) is also shown. Charging occurred during 5.24 days over a 17 day period. The water temperature and volumetric flow rate were approximately constant during charging, and had mean values of 89.4°C and 18.4 l/s, respectively. Discharging also occurred over 5.24 days, approximately with a constant volumetric flow rate of water and linearly decreasing temperature with time. The mean volumetric flow rate during discharging was 18.1 l/s, and the initial discharge temperature was 77°C, while the temperature after 5.24 days was 38°C. The ambient temperature was reported to be 11°C.

Assumptions and simplifications

In subsequent calculations, mean values for volumetric flow rates and charging temperature are used. Also, the specific heat and density of water are both taken to be fixed, at 4.2 kJ/kg K and 1000 kg/m^3, respectively. Since the volumetric flow rate (in litres/second) is equal to the mass flow rate (in kilograms/second) when the density is 1000 kg/m^3, $\dot{m}_c$ = 18.4 kg/s and $\dot{m}_d$ = 18.1 kg/s. Also, the reference-environment temperature is fixed at the ambient temperature, i.e. T_o = 11°C = 284 K.

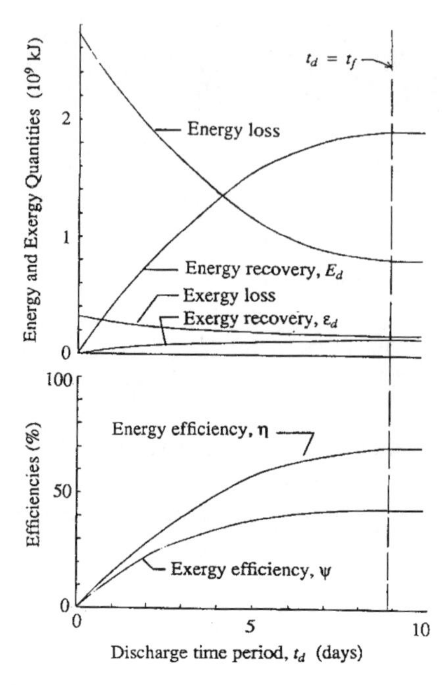

Figure 10.13 Variation of several calculated energy and exergy quantities and efficiencies, as a function of discharge time period, for the ATES case study.

Analysis and results

During charging, it can be shown using Equations 10.104 and 10.105, with t_c = 5.24 d = 453,000 s and T_c = 89.4°C = 362.4 K, that

$$E_c = (18.4\ kg/s)(4.2\ kJ/kg\ K)(453{,}000\,s)(89.4^oC - 11^oC) = 2.74 \times 10^9\ kJ$$

and

$$\begin{aligned}\epsilon_c &= 2.74 \times 10^9\ kJ - (18.4\ kg/s)(4.2\ kJ/kgK)(453{,}000\,s)(284K)\ln(362.4\ K/284K)\\ &= 0.32 \times 10^9\ kJ\end{aligned}$$

During discharging, the value of the time t_f is evaluated using the linear temperature-time relation of the present model and the observations that $T_d(t = 5.24\ d) = 38°C$ and $T_d(0) = 77°C = 350\ K$. Then, using Equation 10.99 with t = 5.24 d,

$$38^oC = 77^oC - (77^oC - 11^oC)(5.24\,d/t_f)$$

which can be solved to show that t_f = 8.87 d. Thus, with the present linear model, the discharge water temperature would reach T_o if the discharge period was lengthened to almost nine days. In reality, the rate of temperature decline would likely decrease, and the discharge temperature would asymptotically approach T_o.

The value of the fraction f can be evaluated with Equation 10.112b as

$$f = \frac{(8.87\,d)(18.1\,kg/s)(77^oC - 11^oC)}{2(5.24d)(18.4\ kg/s)(89.4^oC - 11^oC)} = 0.701$$

Thus, the maximum energy efficiency achievable is approximately 70%. With these values and Equations 10.110 and 10.111, it can be shown that

$$a = (11^{\circ}\mathrm{C} - 77^{\circ}\mathrm{C})/(284\,K \times 8.87\,d) = -0.0262\,d^{-1} \text{ and } b = (350\,\mathrm{K})/(284\,\mathrm{K}) = 1.232$$

Consequently, expressions dependent on discharge time period t_d can be written and plotted (see Figure 10.13) for E_d, $\in_d$, η and ψ using Equations 10.106 to 10.111, 10.113 and 10.114, and for the energy loss ($E_c - E_d$) and exergy loss ($\in_c - \in_d$).

Discussion

Both energy and exergy efficiencies in Figure 10.13 increase from zero to maximum values as t_d increases. Further, the difference between the two efficiencies increases with increasing t_d. This latter point demonstrates that the exergy efficiency gives less weight than the energy efficiency to the energy recovered at higher t_d values, since it is recovered at temperatures nearer to the reference-environment temperature.

Several other points in Figure 10.13 are worth noting. First, for the conditions specified, all parameters level off as t_d approaches t_f, and remain constant for $t_d \geq t_f$. Second, as t_d increases towards t_f, the energy recovered increases from zero to a maximum value, while the energy loss decreases from a maximum of all the input energy to a minimum (but non-zero) value. The exergy recovery and exergy loss functions behave similarly qualitatively, but exhibit much lower magnitudes.

- **Importance of temperature**

The difference between energy and exergy efficiencies is due to temperature differences between the charging and discharging fluid flows. As the discharging time increases, the deviation between these two efficiencies increases (Figure 10.13) because the temperature of recovered heat decreases (Figure 10.12). In this case, the energy efficiency reaches approximately 70% and the exergy efficiency 40% by the completion of the discharge period, even though the efficiencies are both 0% when discharging commences.

To further illustrate the importance of temperature, a hypothetical modification of the present case study is considered. In the modified case, all details are as in the original case except that the temperature of the injection flow during the charging period is increased from 89.4°C to 200°C (473 K), while the duration of the charging period is decreased from its initial value of 5.24 days (453,000 s) so that the energy injected does not change. By equating the energy injected during charging for the original and modified cases, the modified charging-period duration t'_c can be evaluated as a function of the new injection flow temperature T'_c as follows:

$$t'_c = t_c \frac{T_c - T_o}{T'_c - T_o} = (453{,}000\,s)\frac{(89.4^{\circ}\mathrm{C} - 11^{\circ}\mathrm{C})}{(200^{\circ}\mathrm{C} - 11^{\circ}C)} = 188{,}000\,s$$

The modified exergy input during charging can then be evaluated as

$$\begin{aligned}\in'_c &= 2.74 \times 10^9\,kJ - (18.4\,kg/s)(4.2\,kJ/kgK)(188{,}000\,s)(284\,K)\ln(473\,K/284K) \\ &= 0.64 \times 10^9\,kJ\end{aligned}$$

This value is double the exergy input during charging for the original case, so, since the discharging process remains unchanged in the modified case, the exergy efficiency (for any discharging time period) is half that for the original case. The altered value of exergy efficiency is entirely attributable to the new injection temperature, and occurs despite the fact that the energy efficiency remains unchanged.

- **Effect of threshold temperature**

If a threshold temperature is introduced and arbitrarily set at 38°C (the actual temperature at the end of the experimental discharge period of 5.24 d), then the data in Figure 10.13 for t_d = 5.24 d apply, and one can see that:

(i) the exergy recovered (0.127×10^9 kJ) is almost all (91%) of the exergy recoverable in infinite time (0.139×10^9 kJ), while the energy recovered (1.60×10^9 kJ) is not as great a portion (83%) of the ultimate energy recoverable (1.92×10^9 kJ);
(ii) the exergy loss (0.19×10^9 kJ) exceeds the exergy loss in infinite time (0.18×10^9 kJ) slightly (by 5.5%), while the energy loss (1.14×10^9 kJ) exceeds the energy loss in infinite time (0.82×10^9 kJ) substantially (by 39%); and
(iii) the exergy efficiency (40%) has almost attained the exergy efficiency attainable in infinite time (43.5%), while the energy efficiency (58%) is still substantially below the ultimate energy efficiency attainable (70%).

- **Verification of results**

To gain confidence in the model and the results, some of the quantities calculated using the linear model can be compared with the same quantities as reported in the experimental paper (Hoyer *et al.*, 1985):

(i) the previously calculated value for the energy injection during charging of 2.74×10^9 kJ is 1.1% less than the reported value of 2.77×10^9 kJ;
(ii) the energy recovered at the end of the experimental discharge period of t_d = 5.24 days can be evaluated with Equation 10.106 as

$$E_d(5.24\,\text{d}) = (18.1)(4.2)(77-11)[5.24\,(2\times 8.87-5.24)/(2\times 8.87)](86{,}400\,\text{s/d}) = 1.60\times 10^9\,kJ$$

which is 1.8% less than the reported value of 1.63×10^9 kJ; and
(i) the energy efficiency at t_d = 5.24 d can be evaluated with Equation 10.113 as

$$\eta(5.24\,\text{d}) = (1.60\times 10^9\,kJ)(2.74\times 10^9\,kJ) = 0.584$$

which is 1.0% less than the reported value of 0.59 (referred to as the 'energy recovery factor').

10.7.5 Closure

Although energy-based performance measures are normally used in ATES assessments, it can be seen using an elementary ATES model that ATES performance measures based on exergy are more useful and meaningful than those based on energy. Exergy efficiencies account for the temperatures associated with energy transfers to and from an ATES, as well as the quantities of energy transferred, and consequently provide a measure of how nearly ATES systems approach ideal performance. Energy efficiencies account only for quantities of energy transferred, and can often be misleadingly high, e.g. in cases where heat is recovered at temperatures too low to be useful. The use of an appropriate threshold recovery temperature can partially avoid the most misleading characteristics of ATES energy efficiencies. The analysis presented here for a simple ATES cycle can be extended to more complex systems, and is applicable to a wide range of ATES designs.

10.8 Exergy Analysis of Thermally Stratified Storages

Two key advantages of exergy analysis over energy analysis in TES applications are that exergy analysis recognizes differences in storage temperature, even for storages containing equivalent energy quantities, and evaluates quantitatively losses due to degradation of storage temperature towards the environment temperature (i.e. the cooling of heat storages and the heating of cold storages) and due to mixing of fluids at different temperatures. These advantages of the exergy method are particularly important for stratified storages due to the internal spatial temperature variations they exhibit. Since thermodynamic losses are incurred when storage fluids at different temperatures mix, the inhibition of mixing through appropriate temperature stratification is advantageous. Through carefully managing the injection, recovery and holding of heat (or cold) during a storage cycle so that temperature degradation is minimized, better storage-cycle performance can be achieved (as measured by better thermal energy recovery and temperature retention and accounted for explicitly through exergy efficiencies) (Hahne *et al.*, 1989; Krane and Krane, 1991).

This section focuses on the energy and exergy contents of stratified storages, which are usually evaluated numerically for arbitrary temperature distributions. Since numerical methods can be effort consuming and often do not provide practical insights into the physical systems, the temperature distributions may alternatively be modeled so that the corresponding TES energy and exergy values can be evaluated analytically. However, accurate mathematical expressions are usually too complex to permit closed-form solutions, and their solutions again normally necessitate the use of computational techniques. As analytical expressions are usually more convenient and provide greater physical insights, temperature-distribution models which achieve an optimal balance between the needs for accuracy, convenience and physical insight when evaluating storage energy and exergy contents are desirable. Such models can be particularly useful in economic design activities which are greatly simplified and enhanced if analytical expressions for storage energy and exergy contents are available.

In the first part of this section, which follows earlier reports (Rosen and Hooper, 1991b, 1992, 1994; Rosen *et al.*, 1991), several models are presented for the temperature distributions in vertically stratified thermal storages, which are sufficiently accurate, realistic and flexible for use in engineering design and analysis, yet simple enough to be convenient, and which provide useful physical insights. One-dimensional gravitational temperature stratification is considered, and temperature is expressed as a function of height for each model. To reduce effort, only distributions for which energy and exergy data can be obtained analytically are considered. Expressions are derived for TES energy and exergy contents in accordance with the models, which are discussed, compared and illustrated.

In the second part of this section, following previous studies (Rosen and Tang, 1997; Rosen and Dincer, 1999a), the increase in exergy storage capacity resulting from stratification is described.

10.8.1 General stratified TES energy and exergy expressions

The energy E and exergy Ξ in a TES can be found, following Equations 10.12 and 10.31, by integrating over the entire storage-fluid mass m within the TES as follows:

$$E = \int_m e\,dm \tag{10.116}$$

$$\Xi = \int_m \zeta\,dm \tag{10.117}$$

where e denotes specific energy, and ξ specific exergy. For an ideal liquid, the e and ξ are functions only of temperature T, and can be expressed as follows:

$$e(T) = c(T - T_o) \tag{10.118}$$

$$\zeta(T) = c[(T - T_o) - T_o \ln(T/T_o)] = e(T) - cT_o \ln(T/T_o) \tag{10.119}$$

Both the storage-fluid specific heat c and reference-environment temperature T_o are assumed constant here.

Consider now a TES of height H which has only one-dimensional stratification, i.e. temperature varies only with height h in the vertical direction. The horizontal cross-sectional area of the TES is assumed constant. A horizontal element of mass dm can then be adequately approximated as

$$dm = \frac{m}{H}dh \tag{10.120}$$

Since temperature is a function only of height (i.e. $T = T(h)$), the expressions for e and ξ in Equations 10.118 and 10.119, respectively, can be written as

$$e(h) = c(T(h) - T_o) \tag{10.118a}$$

$$\zeta(h) = e(h) - cT_o \ln(T(h)/T_o) \tag{10.119a}$$

With Equation 10.120, the expressions for E and Ξ in Equations 10.116 and 10.117, respectively, can be written as

$$E = \frac{m}{H}\int_0^H e(h)dh \tag{10.116a}$$

$$\Xi = \frac{m}{H}\int_0^H \zeta(h)dh \tag{10.117a}$$

With Equation 10.118a, the expression for E in Equation 10.116a can be written as

$$E = mc(T_m - T_o) \tag{10.116b}$$

where

$$T_m \equiv \frac{1}{H}\int_0^H T(h)dh \tag{10.121}$$

Physically, T_m represents the temperature of the TES fluid when it is fully mixed. This observation can be seen by noting that the energy of a fully mixed tank E_m at a uniform temperature T_m can be expressed, using Equation 10.118 with constant temperature and Equation 10.116, as

$$E_m = mc(T_m - T_o) \tag{10.122}$$

and that the energy of a fully mixed tank E_m is by the principle of conservation of energy the same as the energy of the stratified tank E:

$$E = E_m \tag{10.123}$$

Comparing Equations 10.116b, 10.122 and 10.123 confirms that T_m represents the temperature of the mixed TES fluid.

With Equation 10.119a, the expression for Ξ in Equation 10.117a can be written as

$$\Xi = E - mcT_o \ln(T_e / T_o) \tag{10.117b}$$

where

$$T_e \equiv \exp\left[\frac{1}{H}\int_0^H \ln T(h)dh\right] \tag{10.124}$$

Physically, T_e represents the equivalent temperature of a mixed TES that has the same exergy as the stratified TES. In general, $T_e \neq T_m$, since T_e is dependent on the degree of stratification present in the TES, while T_m is independent of the degree of stratification. In fact, $T_e = T_m$ is the limit condition reached when the TES is fully mixed. This can be seen by noting (with Equations 10.117, 10.117b, 10.122 and 10.123) that the exergy in the fully mixed TES, Ξ_m, is

$$\Xi_m = E_m - mcT_o \ln(T_m / T_o) \tag{10.125}$$

The difference in TES exergy between the stratified and fully mixed (i.e. at a constant temperature T_m) cases can be expressed with Equations 10.117b and 10.125 as

$$\Xi - \Xi_m = mcT_o \ln(T_m / T_e) \tag{10.126}$$

The change given in Equation 10.126 can be shown to be always negative. That is, the exergy consumption associated with mixing fluids at different temperatures, or the minimum work required for creating temperature differences, is always positive.

It is noted that when the temperature distribution is symmetric about the center of the TES such that

$$\frac{T(h) + T(H-h)}{2} = T(H/2) \tag{10.127}$$

the mixed temperature T_m is always equal to the mean of the temperatures at the top and bottom of the TES.

10.8.2 Temperature-distribution models and relevant expressions

Six stratified temperature-distribution models are considered: linear (denoted by a superscript L), stepped (S), continuous-linear (C), general-linear (G), basic three-zone (B)

and general three-zone (T). For each model, the temperature distribution as a function of height is given, and expressions for T_m and T_e are derived. The distributions considered are simple enough in form to permit energy and exergy values to be obtained analytically, but complex enough to be relatively realistic. Although other temperature distributions are, of course, possible, these are chosen because closed-form analytical solutions can readily be obtained for the integrals for T_m in Equation 10.121 and T_e in Equation 10.124.

Linear temperature-distribution model

The linear temperature-distribution model (see Figure 10.14) varies linearly with height h from T_b, the temperature at the bottom of the TES (i.e. at $h = 0$), to T_t, the temperature at the top (i.e. at $h = H$), and can be expressed as

$$T^L(h) = \frac{T_t - T_b}{H} h + T_b \tag{10.128}$$

By substituting Equation 10.128 into Equations 10.121 and 10.124, it can be shown that

$$T_m^L = \frac{T_t + T_b}{2} \tag{10.129}$$

which is the mean of the temperatures at the top and bottom of the TES, and that

$$T_e^L = \exp\left[\frac{T_t(\ln T_t - 1) - T_b(\ln T_b - 1)}{T_t - T_b}\right] \tag{10.130}$$

Stepped temperature-distribution model

The stepped temperature-distribution model (see Figure 10.15) consists of k horizontal zones, each of which is at a constant temperature, and can be expressed as

$$T^S(h) = \begin{cases} T_1; & h_0 \le h \le h_1 \\ T_2; & h_1 < h \le h_2 \\ \cdots & \\ T_k; & h_{k-1} < h \le h_k \end{cases} \tag{10.131}$$

where the heights are constrained as follows:

$$0 = h_0 \le h_1 \le h_2 \ldots \le h_k = H \tag{10.132}$$

It is convenient to introduce here x_j, the mass fraction for zone j:

$$x_j \equiv \frac{m_j}{m} \tag{10.133}$$

Since the TES-fluid density ρ and the horizontal TES cross-sectional area A are assumed constant here, but the vertical thickness of zone j, $h_j - h_{j-1}$, can vary from zone to zone,

$$m_j = \rho V_j = \rho A(h_j - h_{j-1}) \tag{10.134}$$

and

$$m = \rho V = \rho A H \tag{10.135}$$

where V_j and V denote the volumes of zone j and of the entire TES, respectively. Substitution of Equations 10.134 and 10.135 into Equation 10.133 yields

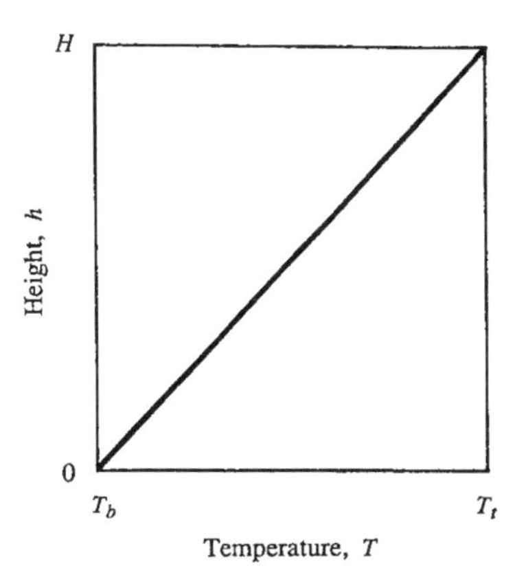

Figure 10.14 A vertically stratified storage having a linear temperature distribution.

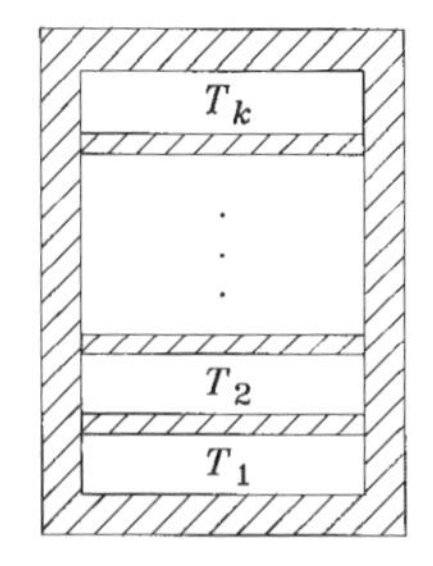

Figure 10.15 A vertically stratified storage having a stepped temperature distribution.

$$x_j = \frac{h_j - h_{j-1}}{H} \tag{10.133a}$$

With Equations 10.121, 10.124, 10.131 and 10.133a, it can be shown that

$$T_m^s = \sum_{j=1}^{k} x_j T_j \tag{10.136}$$

which is the weighted mean of the zone temperatures, where the weighting factor is the mass fraction of the zone, and that

$$T_e^S = \exp\left[\sum_{j=1}^{k} x_j \ln T_j\right] = \prod_{j=1}^{k} T_j^{x_j} \tag{10.137}$$

Continuous-linear temperature-distribution model

The continuous-linear temperature distribution consists of k horizontal zones, in each of which the temperature varies linearly from the bottom to the top, and can be expressed as

$$T^C(h) = \begin{cases} \phi_1^C(h), & h_0 \le h \le h_1 \\ \phi_2^C(h), & h_1 < h \le h_2 \\ \cdots\cdot & \\ \phi_k^C(h), & h_{k-1} < h \le h_k \end{cases} \tag{10.138}$$

where $\phi_j^C(h)$ represents the linear temperature distribution in zone j:

$$\phi_j^C(h) = \frac{T_j - T_{j-1}}{h_j - h_{j-1}} h + \frac{h_j T_{j-1} - h_{j-1} T_j}{h_j - h_{j-1}} \tag{10.139}$$

The zone height constraints in Equation 10.132 apply here. The temperature varies continuously between zones.

With Equations 10.121, 10.124, 10.133a, 10.138 and 10.139, it can be shown that

$$T_m^C = \sum_{j=1}^{k} x_j (T_m)_j \tag{10.140}$$

where $(T_m)_j$ is the mean temperature in zone j, i.e.

$$(T_m)_j = \frac{T_j + T_{j-1}}{2} \tag{10.141}$$

and that

$$T_e^C = \exp\left[\sum_{j=1}^{k} x_j \ln (T_e)_j\right] = \prod_{j=1}^{k} (T_e)_j^{x_j} \tag{10.142}$$

where $(T_e)_j$ is the equivalent temperature in zone j, i.e.

$$(T_e)_j = \begin{cases} \exp\left[\dfrac{T_j(\ln T_j - 1) - T_{j-1}(\ln T_{j-1} - 1)}{T_j - T_{j-1}}\right], & \text{if } T_j \neq T_{j-1} \\ T_j, & \text{if } T_j = T_{j-1} \end{cases} \tag{10.143}$$

General-linear temperature-distribution model

For the general-linear model, there are k horizontal zones, in each of which the temperature varies linearly. The temperature does not necessarily vary continuously between zones. That is,

$$T^G(h) = \begin{cases} \phi_1^G(h), & h_0 \leq h \leq h_1 \\ \phi_2^G(h), & h_1 < h \leq h_2 \\ \cdots \\ \phi_k^G(h), & h_{k-1} < h \leq h_k \end{cases} \tag{10.144}$$

where the zone height constraints in Equation 10.132 apply, and $\phi_j^G(h)$ represents the temperature distribution (linear) in zone j:

$$\phi_j^G(h) = \frac{(T_t)_j - (T_b)_j}{h_j - h_{j-1}} h + \frac{h_j (T_b)_j - h_{j-1}(T_t)_j}{h_j - h_{j-1}} \tag{10.145}$$

Here, $(T_t)_j$ and $(T_b)_j$ denote respectively the temperatures at the top and bottom of zone j, and h_j and h_{j-1} the heights at the top and bottom of zone j (relative to the bottom of the TES).

For the general-linear temperature-distribution model,

$$T_m^G = \sum_{j=1}^{k} x_j (T_m^G)_j \tag{10.146}$$

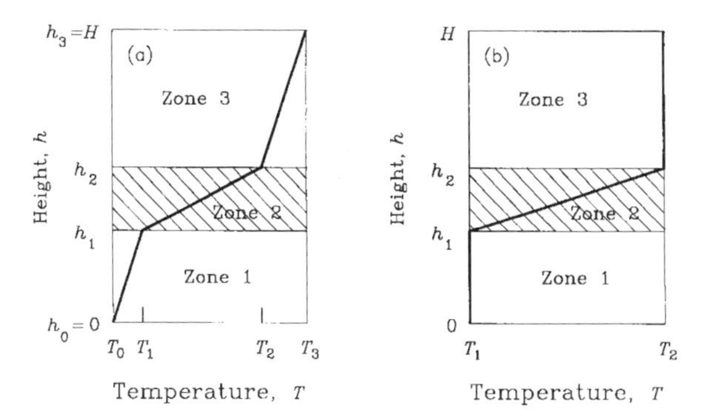

Figure 10.16 The three-zone temperature-distribution models. (a) General, (b) basic.

$$T_e^G = \exp\left[\sum_{j=1}^{k} x_j \ln(T_e^G)_j\right] = \prod_{j=1}^{k} (T_e^G)_j^{x_j} \tag{10.147}$$

where

$$(T_m^G)_j = \frac{(T_t)_j + (T_b)_j}{2} \tag{10.148}$$

$$\left(T_e^G\right)_j = \begin{cases} \exp\left[\dfrac{(T_t)_j(\ln(T_t)_j - 1) - (T_b)_j(\ln(T_b)_j - 1)}{(T_t)_j - (T_b)_j}\right] & \text{if } (T_t)_j \neq (T_b)_j \\ (T_t)_j, & \text{if } (T_t)_j = (T_b)_j \end{cases} \tag{10.149}$$

General three-zone temperature-distribution model

Two three-zone temperature-distribution models are considered in this and the next subsection: general and basic. Both of the three-zone models are subsets of the continuous-linear model in which there are only three horizontal zones (i.e. k = 3). The temperature varies linearly within each zone, and continuously across each zone.

The temperature distribution for the general three-zone model is illustrated in Figure 10.16a, and can be expressed as follows:

$$T^T(h) = \begin{cases} \phi_1^C(h), & h_0 \le h \le h_1 \\ \phi_2^C(h), & h_1 < h \le h_2 \\ \dots & \\ \phi_3^C(h), & h_2 < h \le h_k \end{cases} \tag{10.150}$$

where $\phi_j^C(h)$ represents the temperature distribution (linear) in zone j (see Equation 10.139), and where the heights are constrained as in Equation 10.132 with k = 3.

Expressions for the temperatures T_m and T_e can be obtained for the general three-one model with Equations 10.121, 10.124 and 10.150 (or from the expressions for T_m and T_e for the continuous-linear model with k = 3):

$$T_m^T = \sum_{j=1}^{3} x_j (T_m^C)_j \tag{10.151}$$

$$T_e^T = \exp\left[\sum_{j=1}^{3} x_j \ln(T_e^C)_j\right] = \prod_{j=1}^{3} (T_e^C)_j^{x_j} \tag{10.152}$$

where

$$(T_m^C)_j = \frac{T_j + T_{j-1}}{2} \tag{10.153}$$

$$(T_e^C)_j = \begin{cases} \exp\left[\dfrac{T_j(\ln T_j - 1) - T_{j-1}(\ln T_{j-1} - 1)}{T_j - T_{j-1}}\right], & \text{if } T_j \neq T_{j-1} \\ T_j, & \text{if } T_j = T_{j-1} \end{cases} \tag{10.154}$$

Basic three-zone temperature-distribution model

The basic three-zone temperature-distribution model (B) is a subset of the general three-zone model. In the top and bottom zones, the temperatures are constant at T_2 and T_1, respectively, and in the middle zone, the temperature varies linearly between T_2 and T_1. The temperature distribution for this model is illustrated in Figure 10.16b, and can be expressed as

$$T^B(h) = \begin{cases} T_1, & 0 \le h \le h_1 \\ \dfrac{T_2 - T_1}{h_2 - h_1} h + \dfrac{h_2 T_1 - h_1 T_2}{h_2 - h_1}, & h_1 \le h \le h_2 \\ T_2\,, & h_2 \le h \le H \end{cases} \tag{10.155}$$

where $0 \le h_1 \le h_2 \le H$.

By extension of the general three-zone model, it can be shown that

$$T_m^B = x_1 T_1 + x_2 \frac{T_1 + T_2}{2} + x_3 T_2 = f T_1 + (1 - f) T_2 \tag{10.156}$$

where f represents the mean height fraction in zone 2, expressible as

$$f \equiv \frac{h_1 + h_2}{2H} \tag{10.157}$$

It can also be shown that

$$T_e^B = \exp\left[x_1 \ln T_1 + x_2 \frac{T_2(\ln T_2 - 1) - T_1(\ln T_1 - 1)}{T_2 - T_1} + x_3 \ln T_2\right] \tag{10.158}$$

10.8.3 Discussion and comparison of models

Each of the six stratified temperature-distribution models has advantages and disadvantages, particularly with respect to ease of utilization and flexibility to approximate accurately different actual temperature distributions. The linear temperature-distribution model is simple to utilize but not flexible enough to fit the wide range of actual temperature distributions possible, while the stepped, continuous-linear and general-linear distribution models are flexible and, if the zones are made small enough, can accurately fit any actual temperature distribution. However, the latter models are relatively complex to utilize when the number of zones is large. Although the continuous and general-linear models involve the more complex equations, they usually require fewer zones than the stepped model to achieve similar accuracy in results.

For most normal situations, the three-zone temperature-distribution models may be optimal in terms of achieving an appropriate compromise between such factors as result accuracy, computational convenience, physical insight, etc. Both three-zone models are relatively easy to utilize, yet are flexible enough to simulate well the stratification distribution in most actual TES fluids, which possess lower and upper zones of slightly varying or approximately constant temperature, and a middle zone (the thermocline region) in which temperature varies substantially. The intermediate zone, which grows as thermal diffusion occurs in the tank being modeled, accounts for the irreversible effects of thermal mixing. The basic three-zone model is simpler while the general three-zone model is more accurate.

It is noted that simplified forms of the expressions for T_m and T_e can be written for the multizone temperature-distribution models when all zone vertical thicknesses are the same, since in this special case, the mass fractions for each of the k zones are the same (i.e. $x_j = 1/k$ for all j). Also, several relations exist between the model temperature distributions beyond those described earlier:

- The general-linear temperature distribution reduces to (i) the stepped distribution when $(T_t)_j = (T_b)_j$ for all j; (ii) the continuous-linear distribution when $(T_t)_j = (T_b)_{j+1}$ for $j = 1, 2, \ldots k-1$; and (iii) the linear distribution when $k = 1$.
- The continuous-linear model with $k = 1$ reduces to the linear model.

10.8.4 Illustrative example: the exergy-based advantage of stratification

This example illustrates the advantage of stratification by showing that the exergy of a stratified storage is greater than the exergy for the same tank when it is fully mixed, even though the energy content does not change. The case considered is a perfectly insulated, rigid tank of volume V, divided by an adiabatic partition into two equal-size compartments. The tank is filled with an incompressible fluid. The temperatures of the fluids in each compartment (T_1 and T_2) are initially different.

The partition is removed and a new equilibrium state is attained. No chemical reactions occur. Clearly, the initial state corresponds to a stratified tank with a stepped temperature distribution, and the final state to a fully mixed tank containing an identical quantity of energy. Thus, the TES energy and exergy contents can be expressed with Equations 10.116b and 10.117b for the initial state (denoted by subscript i) and Equations 10.122 and 10.125 for the final state (f), using the stepped temperature-distribution expressions for T_m^S and T_e^S (Equations 10.136 and 10.137) with $k = 2$ and $x_1 = x_2 = 0.5$.

Since neither work nor heat interactions occur, the total internal energy of the storage fluid does not change, i.e. $E_f - E_i = 0$. The exergy change of the fluid can be expressed with Equation 10.126 as

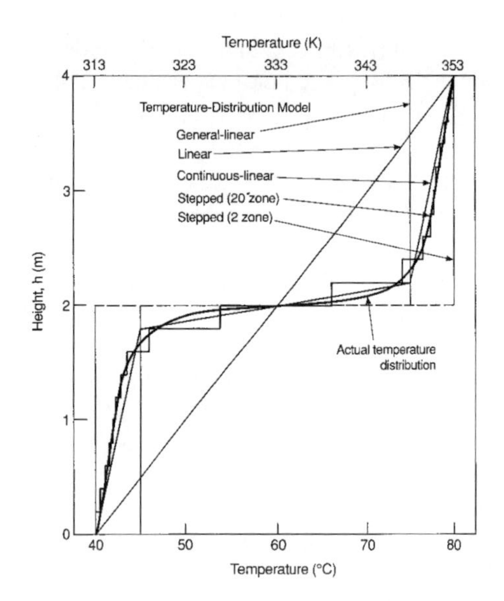

Figure 10.17 The realistic vertically stratified temperature distribution considered in the example, and some of the temperature-distribution models used to approximate it (linear, continuous-linear, general-linear, stepped with two zones, and stepped with 20 zones). The shown continuous-linear distribution is equivalent to a general three-zone distribution.

Table 10.10 Specified general data for the example.

• Temperatures (K)	
At TES top, $T(h = H)$	353
At TES bottom, $T(h = 0)$	313
Reference environment, T_o	283
• TES fluid parameters	
Height, H (m)	4
Mass, m (kg)	10,000
Specific heat, c (kJ/kg K)	4.18

$$\Xi_f - \Xi_i = -T_o mc \ln \frac{T_m}{T_e} \tag{10.159}$$

Equation 10.136 can be used to show that the final temperature of the store after adiabatic mixing is $T_m = (T_1 + T_2)/2$. Also, from Equation 10.137, $T_e = \sqrt{T_1 T_2}$ Equation 10.159 therefore simplifies for this case as follows:

$$\Xi_f - \Xi_i = -T_o mc_v \ln \frac{T_1 + T_2}{2\sqrt{T_1 T_2}} = -T_o mc_v \ln \frac{\sqrt{r} + \sqrt{1/r}}{2} \tag{10.159a}$$

where $r = T_1/T_2$, a positive quantity. The above expression can be further simplified by noting that, since the mixed temperature is given by the mean of the initial temperatures, the initial temperatures can be written as $T_1 = T_m + \Delta T$ and $T_2 = T_m - \Delta T$, where ΔT is an arbitrary temperature increment. Then,

$$\Xi_f - \Xi_i = T_o mc_v \ln \sqrt{1-\alpha^2} \tag{10.159b}$$

where $\alpha \equiv \Delta T/T_m$ and $0 \leq \alpha \leq 1$. Note that if $T_m = T_o$, the temperature of the store after mixing is T_o and the total exergy content of the store can be shown with Equation 10.116b to be zero. Noting that $\sqrt{r}+\sqrt{1/r} \geq 2$ and $\sqrt{1-a^2} \leq 1$, that the natural logarithm of a quantity greater than one is always positive and less than one is always negative, and that the other terms in the Equations 10.159a and 10.159b are positive, it can be seen that $\Xi_f - \Xi_i$ is zero for $r = 1$ and $\alpha = 1$, and negative for all other possible values of r and α.

Thus, while the total energy contained in the store (the sum of the energy in each of the two sections) before and after mixing is equal, the total exergy contained in the store (the sum of the exergy in each of the two sections) before mixing is greater than that contained in the store after mixing. A corollory to this point is that to reverse the mixing process, no net energy need be added to the tank, while some net exergy must be added. The exergy change is zero only if $T_1 = T_2$ (i.e. there are no exergy consumptions due to irreversibilities if fluids of the same temperature are mixed).

10.8.5 Illustrative example: evaluating stratified TES energy and exergy

Problem statement

Several energy and exergy quantities are determined for a realistic stratified temperature distribution, which was assembled by the authors based on many temperature observations for actual storages, using the linear, stepped, continuous-linear, general-linear, and general three-zone temperature-distribution models to approximate the actual distribution. For comparative purposes, the exact values for these quantities are determined by numerical integration of the integrals in Equations 10.121 and 10.124 for the actual temperature distribution. Three stepped distribution cases are considered, each with a different numbers of zones. The actual distribution is shown in Figure 10.17, with most of the model distributions.

The TES fluid is taken to be water. Specified general data are listed in Table 10.10, and additional data specific to the temperature-distribution models are shown in Figure 10.17 and summarized below:

- Continuous-linear: $k = 3$, $h_1 = 1.8$ m, $h_2 = 2.2$ m, $T_1 = 318$ K, $T_2 = 348$ K.
- General-linear: $k = 2$, $h_1 = 2.0$ m, $(T_t)_1 = 318$ K, $(T_b)_2 = 348$ K.
- Stepped: $k = 2$, $h_j - h_{j-1} = 2$ m for the first case; $k = 20$, $h_j - h_{j-1} = 0.2$ m for the second case; $k = 200$, $h_j - h_{j-1} = 0.02$ m for the third case. The temperatures for the stepped cases can be read from Figure 10.17.
- General three-zone: $h_1 = 1.8$ m, $h_2 = 2.2$ m, $T_1 = 318$ K, $T_2 = 348$ K. Note that this distribution is equivalent to the continuous-linear temperature distribution described above.

Table 10.11 Results for the stratification example.

	Temperature-distribution model						Results from numerical integration
			Stepped			Continuous-linear*	
	Linear	General-linear	$k = 200$	$k = 20$	$k = 2$		
Temperatures (K)							
T_m	333.000	333.000	333.000	333.000	333.000	3.000	333.000
T_c	332.800	332.540	332.550	332.560	332.400	2.570	332.550
Energy values (MJ)							
E	2090.000	2090.000	090.000	090.000	2090.000	90.000	2090.000
E_m	2090.000	2090.000	2090.000	2090.000	2090.000	0.000	2090.000
$E - E_m$	0.000	0.000	0.000	0.000	0.000	0.000	0.000
Exergy values (MJ)							
Ξ	172.500	181.800	181.400	181.000	186.700	80.700	181.400
Ξ_m	165.400	165.400	165.400	165.400	165.400	5.400	165.400
$\Xi - \Xi_m$	7.100	16.400	16.000	15.600	21.300	5.300	16.000
Percentage errors							
In values of T_e	+0.075	–0.030	0.000	+0.003	–0.045	+0.006	–
In values of Ξ	–4.900	+2.000	0.000	–0.200	+2.900	–0.400	–
In values of $\Xi - \Xi_m$	–55.600	+22.500	0.000	–2.500	+33.100	–4.400	–

* This case is also a general three-zone temperature-distribution model.

Results and discussion

The results (see Table 10.11) demonstrate the following points for all temperature-distribution models considered:

- The TES energy and exergy contents differ significantly, the exergy values being more than an order of magnitude less than the energy values. This observation is attributable to the fact that the stored thermal energy, although great in quantity, is at near-environmental temperatures, and therefore low in quality or usefulness.
- Values vary among the models for T_e, Ξ and $\Xi - \Xi_m$, while values do not vary for T_m, E, E_m, and Ξ_m. The latter point is expected since the symmetry condition of Equation 10.127 holds for all model temperature distributions considered. As expected from Equation 10.123, $E = E_m$.

The effort required to evaluate the results in Table 10.11 and the accuracy of the results are different for each temperature-distribution model, as explained below:

- Accuracy can be measured by comparing the results for the model distributions with the results obtained by numerical integration (see Table 10.11). Percentage errors are given for the quantities in Table 10.11 which vary from model to model (i.e. for T_e, Ξ and $\Xi - \Xi_m$), where

$$\%\,error = \frac{(Value) - (Numerical\,simulation\,value)}{(Numerical\,simulation\,value)} 100\% \tag{10.160}$$

  In Table 10.11, the percentage errors for T_e values are very small (ranging from –0.045% to +0.075%) and for Ξ values are fairly small (ranging from –4.9% to +2.9%), but for values of $\Xi - \Xi_m$ are in some cases large (ranging from –55.6% to +33.1%).
- Calculational effort expended increases as the number of zones required in multizone distributions for acceptable approximations to the actual temperature distribution increases, and as the complexity of the calculations involved increases. Less calculational effort is required to obtain the values in Table 10.11 for all the different temperature-distribution models considered, than for the values obtained by numerical integration.

In the present example, the general three-zone model has an acceptable accuracy and yet is simple to use. Although the stepped models (with 20 or 200 zones) are more precise, they are more complex to apply. The poor correlations between the actual and linear temperature distributions, and between the actual and stepped (with $k = 2$) distributions, cause these model distributions to provide results of significantly lower precision, although they are simpler to apply. Clearly, the selection of a particular distribution as a model involves a trade-off, and the three-zone temperature-distribution model appears to represent an appropriate compromise between high precision of results and an acceptable level of calculational effort.

10.8.6 Increasing TES exergy storage capacity using stratification

The increase in the exergy capacity of a thermal storage through stratification is described. A wide range of realistic storage-fluid temperature profiles is considered, and for each the relative increase in exergy content of the stratified storage compared to the same storage when it is fully mixed is evaluated. It is shown that, for all temperature profiles considered, the exergy storage capacity of a thermal storage increases as the degree of stratification, as represented through greater and sharper spatial temperature variations, increases. Furthermore, the percentage increase in exergy capacity is greatest for storages at temperatures near to the environment temperature, and decreases as the mean storage temperature diverges from the environment temperature (to either higher or lower temperatures).

Analysis

Thermal storages for heating and cooling capacity, having numerous temperature-distribution profiles, are considered. The general three-zone model, which as discussed previously is a suitable design-oriented temperature distribution model for vertically stratified thermal storages, is utilized to evaluate storage energy and exergy contents. For each case, the ratio is evaluated of the exergy of the stratified storage Ξ to the exergy of the same storage when fully mixed Ξ_m. Using Equations 10.116b, 10.117b and 10.125, this ratio can be expressed, after simplification, as

$$\frac{\Xi}{\Xi_m} = \frac{T_m / T_o - 1 - \ln(T_e / T_o)}{T_m / T_o - 1 - \ln(T_m / T_o)} \qquad (10.161)$$

This ratio increases, from as low as unity when the storage is not stratified, to a value greater than one as the degree of stratification present increases. The ratio in Equation 10.161 is independent of the mass m and specific heat c of the storage fluid. The ratio is also useful as an evaluation, analysis and design tool, as it permits the exergy of a stratified storage to be conveniently evaluated by multiplying the exergy of the equivalent mixed storage (a quantity straightforwardly evaluated) by the appropriate exergy ratio, where values for the exergy ratio can be determined separately (as is done here).

Several assumptions and approximations are utilized throughout this subsection:

- Storage horizontal cross-sectional area is taken to be fixed.
- The environmental temperature T_o is fixed at 20°C for all cases (whether they involve thermal storage for heating or cooling capacity).
- Only one-dimensional gravitational (i.e. vertical) temperature stratification is considered.
- Only temperature distributions which are rotationally symmetric about the center of the storage, according to Equation 10.127, are considered. This symmetry implies that zone 2 is centered about the central horizontal axis of the storage, and that zones 1 and 3 are of equal size, i.e. $x_1 = x_3 = (1 - x_2)/2$.

To model and then assess the numerous storage cases considered, two main relevant parameters are varied realistically:

- the principal temperatures (e.g. mean, maximum, minimum), and
- temperature-distribution profiles (including changes in zone thicknesses).

Specifically, the following characterizing parameters are varied to achieve the different temperature-distribution cases considered:

- The mixed-storage temperature T_m is varied for a range of temperatures characteristic of storages for heating and cooling capacity.
- The size of zone 2, which represents the thermocline region, is allowed to vary from as little as zero to as great as the size of the overall storage, i.e. $0 \leq x_2 \leq 1$. A wide range of temperature profiles can thereby be accommodated, and two extreme cases exist: a single-zone situation with a linear temperature distribution when $x_2 = 1$, and a two-zone distribution when $x_2 = 0$.
- The maximum and minimum temperatures in the storage, which occur at the top and bottom of the storage, respectively, are permitted to vary about the mixed-storage temperature T_m by up to 15°C.

Using the zone numbering system in Figure 10.16a, and the symmetry condition introduced earlier, the following expressions can be written for the temperatures at the top and bottom of the storage, respectively:

$$T_3 = T_m + \Delta T_{st} \quad \text{and} \quad T_0 = T_m - \Delta T_{st} \tag{10.162}$$

while the the following equations can be written for the temperatures at the top and bottom of zone 2, respectively:

$$T_2 = T_m + \Delta T_{th} \quad \text{and} \quad T_1 = T_m - \Delta T_{th} \tag{10.163}$$

where the subscripts 'th' and 'st' denote thermocline region (zone 2) and overall storage, respectively, and where

$$\Delta T \equiv |T - T_m| \tag{10.164}$$

According to the last bullet above, $0 \leq \Delta T_{th} \leq \Delta T_{st} = 15°C$. Also, ΔT_{th} is the magnitude of the difference, on either side of the thermocline region (zone 2), between the temperature at the outer edge of zone 2 and T_m, while ΔT_{st} is the magnitude of the difference, on either side of the overall storage, between the temperature at the outer edge of the storage and T_m. That is,

$$\Delta T_{th} = \Delta T_1 = \Delta T_2 \quad \text{and} \quad \Delta T_{st} = \Delta T_0 = \Delta T_3 \tag{10.165}$$

where the ΔT parameters in the above equations are defined using Equation 10.164 as follows:

$$\Delta T_j \equiv |T_j - T_m|, \quad \text{for } j = 0,1,2,3 \tag{10.166}$$

Effects of varying stratification parameters

- **Effect of varying T_m**

The variation of thermal-storage exergy with storage temperature for a mixed storage is illustrated in Figure 10.18. For a fixed storage total heat capacity (mc), storage exergy increases, from zero when the temperature T_m is equal to the environment temperature T_o, as the temperature increases or decreases from T_o. This general trend, which is illustrated here for a mixed storage, normally holds for stratified storages, since the effect on storage exergy of temperature is usually more significant than the effect of stratification.

- **Effect of varying minimum and maximum temperatures for a linear profile**

A linear temperature profile across the entire storage occurs with the three-zone model when $x_2 = 1$. Then, the upper and lower boundaries of zone 2 shift to the top and bottom of the storage, respectively, and correspondingly the temperature deviation ΔT_{th} occurs at those positions. For a linear temperature profile, the ratio Ξ/Ξ_m is illustrated in Figures 10.19a, 10.19b and 10.19c for three temperature regimes, respectively:

- high-temperature thermal storage for heating capacity, i.e. $T_m \geq 60°C$,
- low-temperature thermal storage for heating capacity, i.e. $20°C \leq T_m \leq 60°C$, and
- thermal storage for cooling capacity, i.e. $T_m \leq 20°C$.

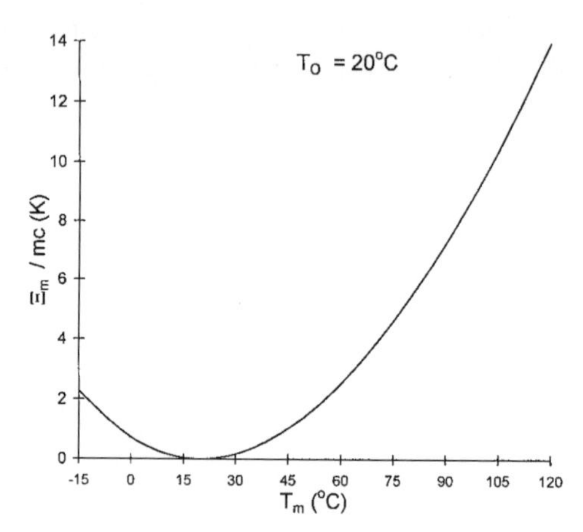

Figure 10.18 Variation with the mixed-storage temperature T_m of the modified exergy quantity Ξ_m/mc (where m and c are constant) for a mixed storage. When T_m equals the environment temperature T_o = 20°C, $\Xi_m = 0$.

The temperature range considered is above the environment temperature T_o = 20ºC for the first two cases, and below it for the third. Two key points are demonstrated in Figure 10.19. First, it is evident for all cases that, for a fixed mixed-storage temperature T_m, storage exergy content increases as level of stratification increases (i.e. as ΔT_{th} increases). Secondly, the percentage increase in storage exergy, relative to the mixed-storage exergy at the same T_m, is greatest when $T_m = T_o$, and decreases both as T_m increases from T_o (see Figures 10.19a and 10.19b) and decreases from T_o (see Figure 10.19c). The main reason for this second observation relates to the fact that the absolute magnitude of the mixed exergy for a thermal storage is small when T_m is near T_o, and larger when T_m deviates significantly from T_o (see Figure 10.18). In the limiting case where $T_m = T_o$, the ratio Ξ/Ξ_m takes on the value of unity when $\Delta T_{th} = 0$ and infinity for all other values of ΔT_{th}. Hence, the relative benefits of stratification as a tool to increase the exergy-storage capacity of a thermal storage are greatest at near-environment temperatures, and less for other cases.

- **Effect of varying thermocline-size parameter x_2**

The variation of the ratio Ξ/Ξ_m with the zone-2 size parameter and the temperature deviation at the zone-2 boundaries, ΔT_{th}, is illustrated in Figure 10.20 for a series of values of the mixed-storage temperature T_m. For a fixed value of ΔT_{th} at a fixed value of T_m, the ratio Ξ/Ξ_m increases as the zone-2 size parameter x_2 decreases. This observation occurs because the stratification becomes less smoothly varying and more sharp and pronounced as x_2 decreases.

- **Effect of varying temperature-distribution profile shape**

The temperature-distribution profile shape is varied, for a fixed value of T_m, primarily by varying values of the parameters x_2 and ΔT_{th} simultaneously. The behavior of Ξ/Ξ_m as x_2 and ΔT_{th} are varied for several T_m values is shown in Figure 10.20. For all cases considered, by varying these parameters at a fixed value of T_m (except for $T_m = T_o$), the ratio Ξ/Ξ_m increases, from a minimum value of unity at $x_2 = 1$ and $\Delta T_{th} = 0$, as x_2 decreases and ΔT_{th} increases. Physically, these observations imply that, for a fixed value of T_m, storage exergy increases as stratification becomes more pronounced, both through increasing the maximum temperature deviation from the mean storage temperature, and increasing the sharpness of temperature profile differences between storage zones.

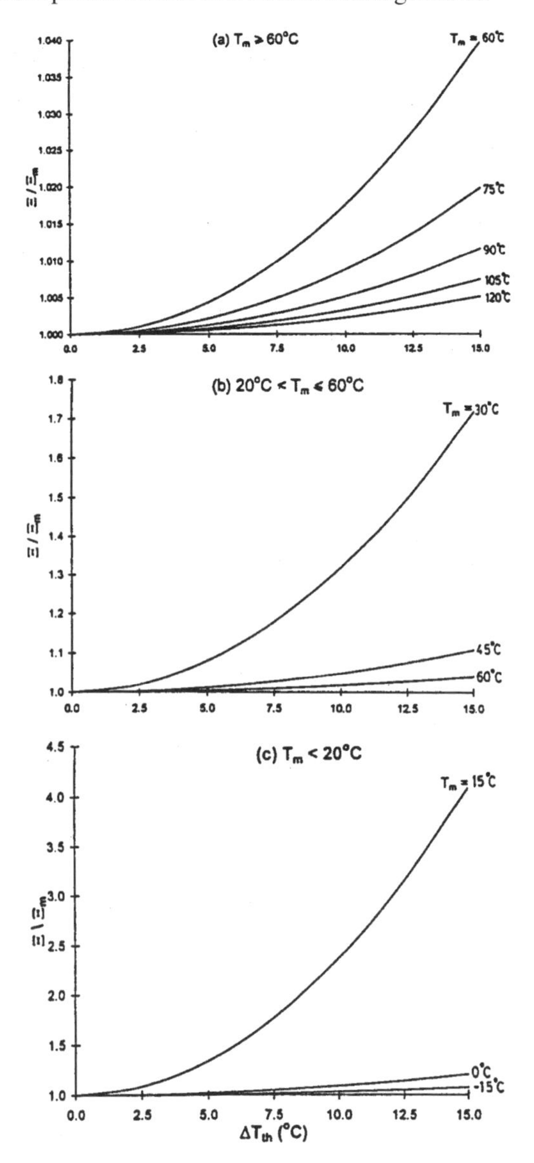

Figure 10.19 Illustration for three ranges of values of the mixed-storage temperature T_m (each corresponding to a different graph) of the variation of the ratio of the exergy values for stratified and fully mixed storages, Ξ/Ξ_m, with temperature deviation from T_m at the upper and lower boundaries of the thermocline zone (zone 2), ƛT_{th}. For all cases, a linear temperature profile is considered, i.e. the zone-2 mass fraction is fixed at $x_2 = 1$. (a) High-temperature thermal storage for heating capacity, i.e. $T_m \geq 60°C$; (b) low-temperature thermal storage for heating capacity, i.e. $20°C \leq T_m \leq 60°C$; (c) thermal storage for cooling capacity, i.e. $T_m \leq 20°C$.

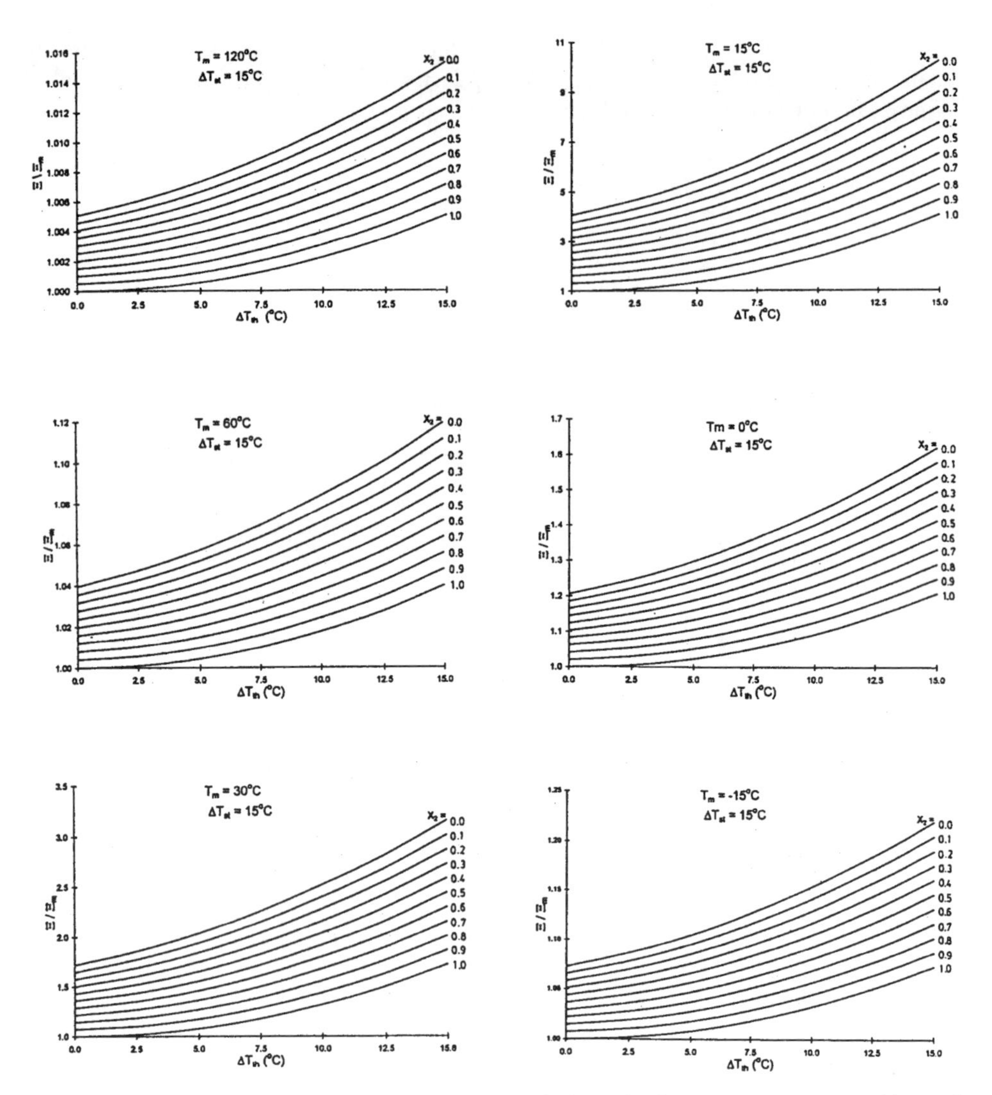

Figure 10.20 Illustration for a series of values of the mixed-storage temperature T_m (each corresponding to a different graph) of the variation of the ratio of the exergy values for stratified and fully mixed storages, Ξ/Ξ_m, with temperature deviation from T_m at the upper and lower boundaries of the thermocline zone (zone 2), ΔT_{th}, and with the zone-2 mass fraction x_2. The magnitude of the temperature deviation from T_m at the top and bottom of the storage, ΔT_{st}, is 15°C for all cases.

10.8.7 Illustrative example: increasing TES exergy with stratification

In this example, several energy and exergy quantities relevant to the previous subsection are determined using the general three-zone model, for a thermal storage using water as the storage fluid and having a realistic stratified temperature distribution. The use as a design tool of the data presented in this section is also illustrated.

The actual distribution is taken to be that from the illustrative example in Section 10.8.5 for the general three-zone case. The temperature distribution is shown in Figure 10.17, along with the general three-zone model distribution used to approximate the actual distribution. Specified general data are listed in Table 10.10.

The results of the example (see Table 10.11) demonstrate that, for the case considered, the ratio Ξ/Ξ_m = 180.7/165.4 = 1.09. This implies that the exergy of the stratified storage is about 9% greater than the exergy of the mixed storage. In effect, therefore, stratification increases the exergy storage capacity of the storage considered, relative to its mixed condition, by 9%.

Rather than determine values of the ratio Ξ/Ξ_m using the expressions in this paper, values can be read from figures such as those in Figure 10.20 (although the case here of T_m = 60°C, x_2 = 0.1 and ΔT_{th} = 20°C falls slightly outside of the range of values covered in Figure 10.20 for the case of T_m = 60°C). Then, such diagrams can serve as design tools from which one can obtain a ratio that can be applied to the value of the exergy of the mixed storage to obtain the exergy of the stratified storage.

10.8.8 Closure

The temperature-distribution models described here (linear, stepped, continuous-linear, general-linear, basic three-zone and general three-zone) facilitate the evaluation of energy and exergy contents of vertically stratified thermal storages. The selection of a particular distribution as a model involves a trade-off between result accuracy and calculational effort, and the three-zone models appear to provide the most reasonable compromise among these factors, and thus to be suitable as simple engineering aids for TES analysis, design and optimization.

TES exergy values, unlike energy values, change due to stratification, giving a quantitative measure of the advantage provided by stratification. The examples considered illustrate how (i) the quantities of energy and exergy contained in a stratified TES differ, and (ii) the exergy content (or capacity) of a TES increases as the degree of stratification increases, even if the energy remains fixed. The use of stratification can therefore aid in TES design as it increases the exergy storage capacity of a thermal storage.

10.9 Energy and Exergy Analyses of Cold TES Systems

An important application of TES is in facilitating the use of off-peak electricity to provide building heating and cooling. Recently, increasing attention has been paid in many countries to Cold Thermal Energy Storage (CTES), an economically viable technology that has become a key component of many successful thermal systems. In many CTES applications, inexpensive off-peak electricity is utilized during the night to produce with chillers a cold medium, which can be stored for use in meeting cooling needs during the day when electricity is more expensive. CTES is applied mainly in building cooling, particularly for large commercial buildings which often need year-round cooling due to the heat released by occupants, lighting, computers and other equipment. In most situations, CTES permits the mismatch between the supply and demand of cooling to be favorably altered (Dincer and Dost, 1996), leading to more efficient and environmentally benign energy use as well as reduced chiller sizes, capital and maintenance costs, and carbon dioxide and CFC emissions (Beggs, 1991).

To be economically justifiable the annual costs needed to cover the capital and operating expenses for a CTES (and related systems) should be less than the costs for primary generating equipment supplying the same service loads and periods (Dincer, 1999). A CTES can lead to lower initial and operating costs for associated chillers when cooling loads are intermittent and of short duration. Secondary system capital costs may also be lower with CTES, e.g. electrical service entrance sizes can sometimes be reduced because energy demand is lower.

Although CTES efficiency and performance evaluations are conventionally based on energy, energy analysis itself is inadequate for complete CTES evaluation because it does not account for such factors as the temperatures at which heat (or cold) is supplied and delivered. Exergy analysis overcomes some of these inadequacies in CTES assessments.

This section deals with the assessment using exergy and energy analyses of CTES systems, including sensible and/or latent storages, following earlier reports (Rosen *et al.*, 1999, 2000). Several CTES cases are considered, including storages which are homogeneous or stratified, and some which undergo phase changes. A full cycle of charging, storing and discharging is considered for each case. This section demonstrates that exergy analysis provides more realistic efficiency and performance assessments of CTES systems than energy analysis, and conceptually is more direct since it treats cold as a valuable commodity. An example and case study illustrate the usefulness of exergy analysis in addressing cold thermal storage problems.

10.9.1 Energy balances

Consider a cold storage consisting of a tank containing a fixed quantity of storage fluid and a heat-transfer coil through which a heat-transfer fluid is circulated. Kinetic and potential energies and pump work are considered negligible. An energy balance for an entire cycle of a CTES can be written following Equation 10.3 in terms of 'cold' as follows:

$$\textit{Cold input} - [\textit{Cold recovered} + \textit{Cold loss}] = \textit{Cold accumulation} \quad (10.167)$$

Here, 'cold input' is the heat removed from the storage fluid by the heat-transfer fluid during charging; 'cold recovered' is the heat removed from the heat transfer fluid by the storage fluid; 'cold loss' is the heat gain from the environment to the storage fluid during charging, storing and discharging; and 'cold accumulation' is the decrease in internal energy of the storage fluid during the entire cycle. Following Equation 10.3a, the overall energy balance for the simplified CTES system illustrated in Figure 10.21 becomes

$$(H_b - H_a) - [(H_c - H_d) + Q_l] = -\Delta E \quad (10.167a)$$

where H_a, H_b, H_c and H_d are the enthalpies of the flows at points a, b, c and d in Figure 10.21; Q_l is the total heat gain during the charging, storing, and discharging processes; and ΔE is the difference between the final and initial storage-fluid internal energies. The terms in square brackets in Equations 10.167 and 10.167a represent the net 'cold output' from the CTES, and ΔE = 0 if the CTES undergoes a complete cycle (i.e. the initial and final storage-fluid states are identical).

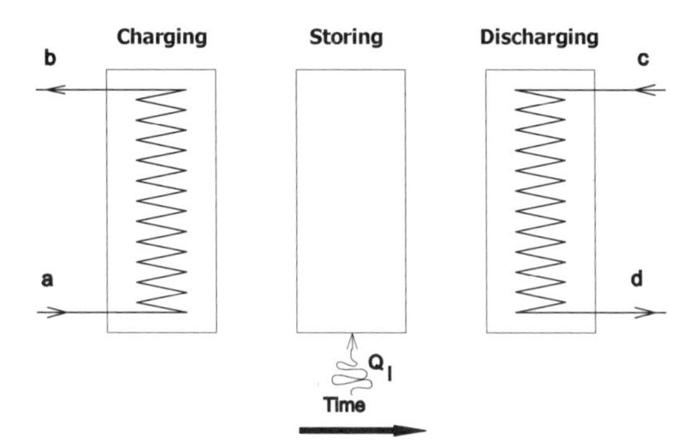

Figure 10.21 The three processes in a general CTES system: charging (left), storing (middle), and discharging (right). The heat leakage into the system Q_l is illustrated for the storing process, but can occur in all three processes.

The energy transfer associated with the charging fluid can be expressed as

$$H_b - H_a = m_a c_a (T_b - T_a) \tag{10.168}$$

where m_a is the mass flow of heat-transfer fluid at point a (and at point b), and c_a is the specific heat of the heat transfer fluid, which is assumed constant. A similar expression can be written for $H_c - H_b$. The energy content of a storage which is homogeneous (i.e. entirely in either the solid or the liquid phase) is

$$E = m(u - u_o) \tag{10.169}$$

which, for sensible heat interactions only, can be written as

$$E = mc(T - T_o) \tag{10.169a}$$

where, for the storage fluid, c denotes the specific heat (assumed constant), m the mass, u the specific internal energy and T the temperature. Also, u_o is u evaluated at the environmental conditions.

For a mixture of solid and liquid, the energy content of the solid and liquid portions can be evaluated separately and summed as follows:

$$E = m[(1 - F)(u_s - u_o) + F(u_t - u_o)] \tag{10.169b}$$

where u_s and u_t are the specific internal energies of the solid and liquid portions of the storage fluid, respectively, and F is the melted fraction (i.e. the fraction of the storage fluid mass in the liquid phase).

For a storage fluid which is thermally stratified with a linear temperature profile in the vertical direction, the energy content can be shown with Equations 10.116b and 10.129 to be

$$E = mc\left(\frac{T_t + T_b}{2} - T_o\right) \tag{10.169c}$$

where T_t and T_b are the storage fluid temperatures at the top and bottom of the linearly stratified storage tank, respectively.

The change in CTES energy content from the initial (*i*) to the final state (*f*) of a process can be expressed as in Equation 10.38.

10.9.2 Exergy balances

An exergy balance for a CTES undergoing a complete cycle of charging, storing and discharging can be written as in Equations 10.40 and 10.40a. The exergy content of a flow of heat transfer fluid at state k (where $k = a$, b, c, or d in Figure 10.21) can be expressed as in Equation 10.44. The exergy transfers associated with the charging and discharging of the storage by the heat-transfer fluid can be expressed by Equations 10.45 and 10.46, respectively.

The exergy loss associated with heat infiltration during the three storage periods can be expressed as in Equation 10.47a. The thermal exergy terms are negative for sub-environment temperatures, as is the case here for CTESs, indicating that the heat transfer and the accompanying exergy transfer are oppositely directed. That is, the losses associated with heat transfer are due to heat infiltration into the storage when expressed in energy terms, but due to a cold loss out of the storage when expressed in exergy terms.

The exergy content of a homogeneous storage can be shown with Equation 10.116 to be

$$\Xi = m[(u-u_o)-T_o(s-s_o)] \tag{10.170}$$

where s is the specific entropy of the storage fluid and s_o is s evaluated at the environmental conditions. If only sensible heat interactions occur, Equation 10.170 can be written with Equation 10.119 as

$$\Xi = mc[(T-T_o)-T_o \ln (T/T_o)] \tag{10.170a}$$

For a mixture of solid and liquid, the exergy content can be written as

$$\Xi = m\{(1-F)[(u_s- u_o)-T_o(s_s-s_o)]+F[(u_t-u_o)-T_o(s_t-s_o)]\} \tag{10.170b}$$

where s_s and s_t are the specific entropies of the solid and liquid portions of the storage fluid, respectively.

The exergy content of a storage which is linearly stratified can be shown with Equations 10.117b and 10.130 to be

$$\Xi = E-mcT_o\left[\frac{T_t(\ln T_t-1)-T_b(\ln T_b-1)}{T_t-T_b}-\ln T_o\right] \tag{10.170c}$$

The change in TES exergy content can be expressed as in Equation 10.43.

10.9.3 Energy and exergy efficiencies

For a general CTES undergoing a cyclic operation, the overall energy efficiency η can be evaluated as in Equation 10.32, with the word energy replaced by cold for understanding. Then, following Figure 10.21, the overall and charging-period energy efficiencies can be expressed as in Equations 10.48 and 10.52, respectively.

Energy efficiencies for the storing and discharging subprocesses can be written respectively as

Table 10.12 Specified temperature data for the cases in the CTES example.

Temperature (°C)	Case			
	I	II	III	IV
T_b	4.0	15	–1	–1
T_d	11.0	11	10	10
T_1	10.5	19/2*	0 (t)	8
T_2	5.0	17/–7*	0 (s)	–8
T_3	6.0	18/–6*	0 (t&s)	0 (t&s)

* When two values are given, the storage fluid is vertically linearly stratified and the first and second values are the temperatures at the top and bottom of the storage fluid, respectively.

$$\eta_2 = \frac{\Delta E_1 + Q_l}{\Delta E_1} \tag{10.171}$$

$$\eta_3 = \frac{H_c - H_d}{\Delta E_3} \tag{10.172}$$

where ΔE_1 and ΔE_3 are the changes in CTES energy contents during charging and discharging, respectively.

The exergy efficiency for the overall process can be expressed as in Equation 10.49, and for the charging, storing and discharging processes, respectively, as in Equations 10.55, 10.61 and 10.67.

10.9.4 Illustrative example

Cases considered and specified data

Four different CTES cases are considered. In each case, the CTES has identical initial and final states, so that the CTES operates in a cyclic manner, continuously charging, storing, and discharging. The main characteristics of the cold storage cases are as follows:

I. Sensible heat storage, with a fully mixed storage fluid.
II. Sensible heat storage, with a linearly stratified storage fluid.
III. Latent heat storage, with a fully mixed storage fluid.
IV. Combined latent and sensible heat storage, with a fully mixed storage fluid.

The following assumptions are made for each of the cases:

- Storage boundaries are nonadiabatic.
- Heat gain from the environment during charging and discharging is negligibly small relative to heat gain during the storing period.
- The external surface of the storage tank wall is at a temperature 2°C greater than the mean storage-fluid temperature.
- The mass flow rate of the heat transfer fluid is controlled so as to produce constant inlet and outlet temperatures.

- Work interactions, and changes in kinetic and potential energy terms, are negligibly small.

Specified data for the four cases are presented in Table 10.12 and relate to the diagram in Figure 10.21. In Table 10.12, T_b and T_d are the charging and discharging outlet temperatures of the heat transfer fluid, respectively. The subscripts 1, 2 and 3 indicate the temperature of the storage fluid at the beginning of charging, storing or discharging, respectively. Also, *t* indicates the liquid state and *s* indicates the solid state for the storage fluid at the phase-change temperature.

In addition, for all cases, the inlet temperatures are fixed for the charging-fluid flow at $T_a = -10$°C and for the discharging-fluid flow at $T_c = 20$°C. For cases involving latent heat changes (i.e. solidification), $F = 10\%$. The specific heat *c* is 4.18 kJ/kg K for both the storage and heat-transfer fluids. The phase-change temperature of the storage fluid is 0°C. The configuration of the storage tank is cylindrical with an internal diameter of 2 m and internal height of 5 m. Environmental conditions are 20°C and 1 atm.

Results and discussion

The results for the four cases are listed in Table 10.13, and include overall and subprocess efficiencies, input and recovered cold quantities, and energy and exergy losses. The overall and subprocess energy efficiencies are identical for Cases I and II, and for Cases III and IV. In all cases the energy efficiency values are high. The different and lower exergy efficiencies for all cases indicate that energy analysis does not account for the quality of the 'cold' energy, as related to temperature, and considers only the quantity of 'cold' energy recovered.

The input and recovered quantities in Table 10.13 indicate the quantity of 'cold' energy and exergy input to and recovered from the storage. The energy values are much greater than the exergy values because, although the energy quantities involved are large, the energy is transferred at temperatures only slightly below the reference-environment temperature, and therefore is of limited usefulness.

Table 10.13 Energy and exergy quantities for the cases in the CTES example.

	Energy quantities				Exergy quantities			
Period or quantity	I	II	III	IV	I	II	III	IV
Efficiencies (%)								
Charging (1)	100	100	100	100	51	98	76	77
Storing (2)	82	82	90	90	78	85	90	85
Discharging (3)	100	100	100	100	38	24	41	25
Overall	82	82	90	90	15	20	28	17
Input, recovered and lost quantities (MJ)								
Input	361.1	361.1	5,237.5	6,025.9	30.9	23.2	499.8	575.1
Recovered	295.5	295.5	4,713.8	5,423.3	4.6	4.6	142.3	94.7
Loss (external)	65.7	65.7	523.8	602.6	2.9	2.9	36.3	48.9
Loss (internal)	–	–	–	–	23.3	15.6	321.2	431.4

The cold losses during storage, on an energy basis, are entirely due to cold losses across the storage boundary (i.e. heat infiltration). The exergy-based cold losses during storage are due to both cold losses and internal exergy losses (i.e. exergy consumptions due to irreversibilities within the storage). For the present cases, in which the exterior surface of the storage tank is assumed to be 2°C warmer than the mean storage-fluid temperature, the exergy losses include both external and internal components. Alternatively, if the heat transfer temperature at the storage tank external surface is at the environment temperature, the external exergy losses would be zero and the total exergy losses would be entirely due to internal consumptions. If heat transfer occurs at the storage-fluid temperature, on the other hand, more of the exergy losses would be due to external losses. In all cases, the total exergy losses, which are the sum of the internal and external exergy losses, remain fixed.

The four cases demonstrate that energy and exergy analyses give different results for CTES systems. Both energy and exergy analyses account for the quantity of energy transferred in storage processes. Exergy analyses take into account the loss in quality of 'cold' energy, and thus more correctly reflect the actual value of the CTES.

In addition, exergy analysis is conceptually more direct when applied to CTES systems because cold is treated as a useful commodity. With energy analysis, flows of heat rather than cold are normally considered. Thus, energy analyses become convoluted and confusing as one must deal with heat flows, while accounting for the fact that cold is the useful input and product recovered for CTES systems. Exergy analysis inherently treats any quantity which is out of equilibrium with the environment (be it colder or hotter) as a valuable commodity, and thus avoids the intuitive conflict in the expressions associated with CTES energy analysis. The concept that cold is a valuable commodity is both logical and in line with one's intuition when applied to CTES systems.

10.9.5 Case study: thermodynamic performance of a commercial ice TES system

One type of CTES is ice storage, which has been increasingly utilized recently. In an Ice Thermal Energy Storage (ITES) system, off-peak electricity is utilized to create a large mass of ice, and during the day this ice store is melted by absorbing the heat from buildings needing cooling (De Lucia and Bejan, 1990; Beggs, 1991). ITES is a proven technology applicable to any plant that has a chilled water system.

Numerous economic studies of ITES systems and their applications have been undertaken (Althof, 1989; Beggs, 1991; Fields and Knebels, 1991). Chen and Sheen (1993) developed a model for a CTES system that evaulates the component size for a eutectic salt storage system. The related computer simulation determines energy consumption, and performs a cost/benefit analysis to estimate the system payback period. Chen and Sheen also determined the best system operating strategies for varying weather conditions and electricity rates, particularly in Taiwan. Wood *et al.* (1994) investigated TES technical and economic aspects, and presented a techno-economic feasibility evaluation method for new technologies in the commercial sector using a pseudo-data analysis approach. Some (Badar *et al.,* 1993; Domanski and Fellah, 1998) have applied to TES systems thermoeceonomic optimization techniques which minimize the combination of entropy generation cost and annualized capital cost of the thermal component.

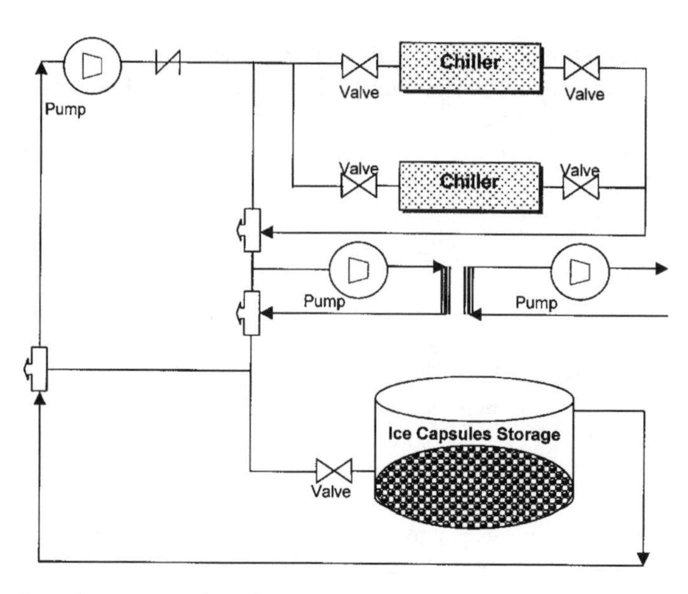

Figure 10.22 A schematic representation of an encapsulated ITES system.

In this case study, which is based on an earlier analysis (Rosen *et al.,* 2000), energy and exergy analyses are performed of a commercial encapsulated ITES. A full cycle, with charging, storing and discharging stages, is considered. The case study demonstrates the usefulness of exergy analysis in thermodynamic assessments of ITES systems, provides insights into their performances and efficiencies, and shows that energy analysis leads to misleadingly optimistic statements of ITES efficiency.

Types and operation of ITES systems

By making ice at night for use in the day when it is needed to provide cooling, ITES systems take advantage of low off-peak electricity rates and often permit building chillers to be shut down or operated at reduced levels during the day. Consequently, owners can realize lower utility bills, and electrical utilities attain a reduction in peak demand, which can defer or avoid the need for new power plants. Ice storage systems are capable of providing the same quantity and quality of cooling as conventional chillers.

There are two main cases of ice storage systems, differing in the levels of electrical load shifting achieved. With a 'full storage,' the entire cooling load is shifted, eliminating chiller operation during peak hours. In a 'partial storage,' part of the load is met from a downsized chiller, and part from the storage. In both cases, the sizes of the chiller and of the storage depend on the total amount of cooling needed to meet the building load over the entire day. Sizing is the main factor differentiating ice storage from conventional air conditioning, since in the latter case the chiller must be larger to meet the peak cooling load, which normally occurs during the hottest part of the day.

Four main types of ITES systems are commonly used in commercial and industrial applications (Beckman and Gilli, 1984; Beggs, 1991): ice-on-coil, static ice tank, ice harvesting and encapsulated ice store. The most common ITES type is the encapsulated ice store (Figure 10.22), which is considered in this case study. Encapsulated ITES systems consist of an insulated storage tank (normally constructed from either steel or concrete) which is filled with small plastic capsules containing deionized water and a nucleating agent (Carrier, 1990). Circulating around the capsules is a glycol/water brine solution. During charging, the brine solution is circulated through the tank at a sub-freezing temperature, causing the water in the capsules to freeze to form the ice store. To discharge the store when cooling is needed, the same brine solution is circulated through the tank but at a temperature above the freezing point of water. The ice in the capsules then melts, cooling the brine solution which is subsequently circulated to the air conditioning unit.

Description of ITES considered and its operation

The ITES system considered in this study (Figure 10.22) is designed to have the chiller operate continuously during the design day, and utilizes three operating modes (Carrier, 1990):

- *Charging*. The charging mode is the normal operating mode during no-load periods. The chiller cools the antifreeze solution to approximately –4°C. The antifreeze solution flows through the storage module, freezing the liquid water inside the encapsulated units. The circulating fluid increases in temperature (to a limit of 0°C), and returns to the chiller to be cooled again. During charging, the building loop is isolated so that full flow is achieved through the storage module.
- *Chilling*. The chilling mode is the same as for a non-storage (conventional) chiller system, with the entire building load being met directly by the chiller. The chiller operates at a warmer set point than for ice making, which results in an increased capacity and a higher coefficient of performance. In this mode there is no flow through the storage module, and ice in the store is kept in reserve for use later in the day.
- *Chilling and discharging*. Chilling and discharging is the normal operating mode during daytime hours. The chiller and storage module share the cooling load, often in a series configuration. Systems are normally designed with the chiller downstream of the storage. Then the storage module pre-cools the building return fluid before it is further cooled to the design supply temperature by the chiller. This sequence gives a higher effective storage capacity, since the exit temperature from the storage can be higher. The ice melting rate is controlled by modulating valves which cause some flow to bypass the storage module, usually to hold constant the blended fluid temperature downstream of the storage throughout discharging.

The design-day cooling load profile of a typical office building is considered here (Figure 10.23). The circles and squares in Figure 10.23 indicate the cooling load required for the building and the chiller loads, respectively. Data for the case, taken from Carrier (1990), is presented in Table 10.14 for a full 24-hour cycle. The ITES is designed for the chiller to operate continuously during the day (i.e. partial storage operation is used). The ITES module has nonadiabatic storage boundaries with a total thermal resistance of R_T = 1.98 m^2 K/W. The specific heat of the heat-transfer fluid (a glycol-based antifreeze solution) is 3.22 kJ/kg K at –6.6°C and 3.60 kJ/kg K at 15.5°C, and the specific gravity is 1.13. The storage fluid (deionized water) has a freezing point of 0°C, a mass of 144,022 kg and a density of 1000 kg/m^3. The storage module has a volume of 181.8 m^3, with 144.0 m^3 occupied by the storage fluid, and a surface area A of 241.6 m^2. The reference environment conditions are 20°C and 1 atm.

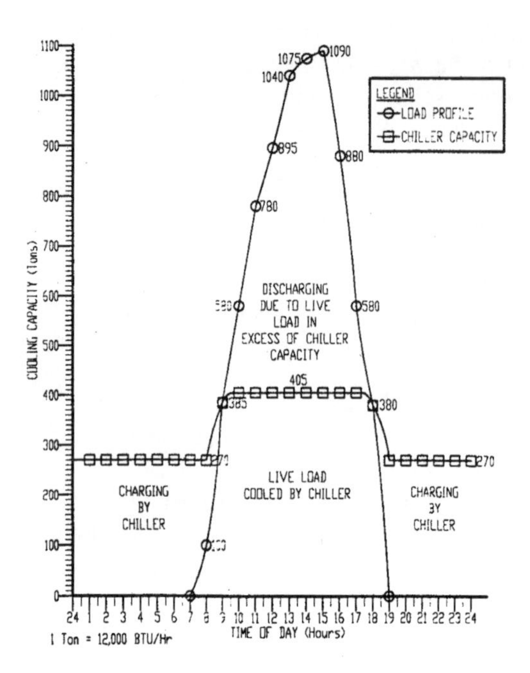

Figure 10.23 A load profile diagram for an encapsulated ITES system (see Rosen *et al.,* 2000).

Thermodynamic analysis

To reduce analysis complexity and to emphasize the significant factors in ice storage operation, it is assumed that the storage fluid remains isothermal at its melting point, fluids are frictionless, the pumping power is zero, and kinetic and potential energy terms are negligibly small. These assumptions neglect some of the irreversibilities in the encapsulated ITES system, notably those associated with the temperature gradients close to the freezing or thawing surfaces in the storage. The net effect of these assumptions is to increase the apparent exergy efficiencies, but probably not their relative values. Hence the results should be valid in ordering the performances of competitive systems in an optimization process. The analysis generally follows the discussions earlier in this section. Also, losses due to heat gain from the environment, Q_l, are determined here as follows:

$$Q_l = \frac{A\Delta T}{R_T} \tag{10.173}$$

where ΔT is the difference between the storage-fluid and environmental temperatures, A is the surface area through which the heat is transferred and R_T is the unit total thermal resistance of the storage module. Following Figure 10.21, the overall energy and efficiencies can be written with Equations 10.32 and 10.49 for the ITES module as it undergoes a cyclic operation over a 24-hour period, by summing over the 24 hours:

$$\eta = \frac{Product\,cold\,recovered}{Cold\,input} = \frac{\sum_{j=1}^{24}(H_c - H_d)_j}{\sum_{j=1}^{24}(H_b - H_a)_j} \tag{10.174}$$

$$\psi = \frac{Exergy\ recovered}{Exergy\ input} = \frac{\sum_{j=1}^{24}(\in_d - \in_c)_j}{\sum_{j=1}^{24}(\in_a - \in_b)_j} \tag{10.175}$$

Energy and exergy efficiencies for charging, storing and discharging are similarly evaluated, following section 10.9.3.

Results and discussion

The overall energy and exergy efficiencies are 99.5% and 50.9%, respectively, and the hourly energy and exergy efficiencies are listed in Table 10.14. The hourly exergy efficiencies range from 80–94% and average 86% for the overall charging period, range from 53–66% and average 60% for the overall discharging period, and range from 99–100% for the overall storing period. The hourly energy efficiencies exceed 99% for all periods.

Table 10.14 Specified and evaluated data for the ITES case study.

		Load (tons)			Melted fraction (%)	Efficiency (%)	
Hour	Process	Storage	Building	Chiller		Exergy	Energy
1	Charging	270	0	270	48.55	88.1	99.7
2	Charging	270	0	270	41.46	87.0	99.7
3	Charging	270	0	270	34.36	85.9	99.7
4	Charging	270	0	270	27.27	84.8	99.7
5	Charging	270	0	270	20.17	83.7	99.7
6	Charging	270	0	270	13.08	82.6	99.7
7	Charging	270	0	270	5.99	81.6	99.7
8	Charging	170	100	270	1.53	80.4	99.5
9	Storing	0	385	385	1.55	99.9	99.9
10	Discharging	175	580	405	6.12	66.0	99.7
11	Discharging	375	780	405	15.96	63.3	99.9
12	Discharging	490	895	405	28.83	59.9	99.9
13	Discharging	635	1040	405	45.53	58.5	99.9
14	Discharging	670	1075	405	63.15	57.1	99.9
15	Discharging	685	1090	405	81.16	55.7	99.9
16	Discharging	475	880	405	93.63	52.9	99.9
17	Discharging	175	580	405	98.21	63.9	99.7
18	Storing	0	380	380	98.22	99.9	99.9
19	Charging	270	0	270	91.13	93.5	99.7
20	Charging	270	0	270	84.03	92.6	99.7
21	Charging	270	0	270	76.94	91.8	99.7
22	Charging	270	0	270	69.84	90.9	99.7
23	Charging	270	0	270	62.74	90.1	99.7
24	Charging	270	0	270	55.65	89.3	99.7

The marked differences in the energy and exergy efficiencies for the overall process and the subprocesses merit emphasis and explanation. The energy efficiencies are high, since they only account for heat gains from the environment, which are small. The exergy efficiencies are much lower since they account for the 'usefulness' of the energy, which is related to the inlet and outlet temperatures and the mass flow rates of heat-transfer fluid. In the example, the charging fluid being at –4°C, a much lower temperature than that of the environment, is a high-quality cold flow. The cold flow recovered during discharging, however, is of much lower quality with a temperature much closer to that of the environment. Thus, the energy efficiencies, for each hour or for the entire cycle, are misleadingly high as they only account for energy recovery but neglect entirely the loss of quality of the flows. This quality loss is quantified with exergy analysis. Since the irreversibilities in an ITES process destroy some of the input exergy, ITES exergy efficiencies are always lower than the corresponding energy efficiencies.

Another interesting observation stems from the fact that exergy efficiencies provide a measure of how nearly a process approaches ideality, while energy efficiencies do not. The energy efficiencies being over 99% here for the overall process and all subprocesses erroneously implies that the ITES system is nearly ideal. The overall exergy efficiency of approximately 51%, as well as the subprocess exergy efficiencies, indicate that the ITES system is far from ideal, and has a significant margin for efficiency improvement. In this example, the same cooling capacity could be delivered from the ITES using about half of the input exergy if it were ideal. Thus, overall electrical use by the chillers could be greatly reduced while still maintaining the same cooling services. Such a reduction would reduce the necessary installed cooling power and electrical costs.

Further implications of the results follow:

- The fact that the exergy efficiencies are less than 100% implies that a mismatch exits between the quality of the thermal energy delivered by the ITES (and required by the cooling load) and the quality of the thermal energy input to the ITES. This mismatch, which is detectable through the temperature of the thermal energy flows across the ITES boundaries, is quantifiable with exergy analysis as the work potential lost during storage. The exergy loss, therefore, correlates directly with an additional use of electricity by the chillers than would occur without the exergy loss. When exergy efficiencies are 100%, there is no loss in temperature during storage.
- The non-ideal exergy efficiencies imply that excessively high-quality thermal energy is supplied to the ITES than is required given the cooling load. Thus, exergy analysis indicates that lower quality sources of thermal energy could be used to meet the cooling load. Although economic and other factors must be taken into account when selecting energy resources, the exergy-based results presented here can assist in identifying feasible energy sources that have other desired characteristics (e.g. environmentally benign or abundant).

Consequently, the results demonstrate that a more perceptive measure of comparison than energy efficiency is needed if the true usefulness of an ITES is to be assessed and a rational basis for the optimization of its economic value established. Energy efficiencies ignore the quality (exergy) of the energy flows, and so cannot provide a measure of ideal performance. Exergy efficiencies provide more comprehensive and useful efficiency measures for practical ITES systems, and facilitate more rational comparisons of different systems. In addition, exergy analysis can assist in the optimization of ITES systems, when combined with assessments of such other factors as resource-use reductions, environmental impact and emissions decreases, and economics.

10.9.6 Closure

Exergy analysis provides more meaningful and useful information than energy analysis about the efficiencies, losses and performance for CTES systems. A prime justification for this view is that the loss of low temperature is accounted for in exergy- but not in energy-based performance measures. Furthermore, the exergy-based information is presented in a more direct and logical manner, as exergy methods provide intuitive advantages when CTES systems are considered. Consequently, exergy analysis can likely assist in efforts to optimize the design of CTES systems and their components, and to identify appropriate applications and optimal configurations for CTES in general engineering systems. Several additional key points can be drawn from this section: (i) exergy analysis can assist in selecting alternative energy sources for CTES systems, so that the potential role can be properly considered for CTES in meeting society's preferences for more efficient, environmentally benign energy use in various sectors; and (ii) the application of exergy analysis to CTES systems permits mismatches in the quality of the thermal energy supply and demand to be quantified, and measures to reduce or eliminate reasonably avoidable mismatches to be identified and considered. The material presented in this section for cold TES parallels that presented in the rest of this chapter for heat storage systems, although the advantages of the exergy approach appear to be more significant when CTES systems are considered due to manner in which 'cold' is treated as a resource.

10.10 Exergy-based optimal discharge periods for closed TES systems

Since in many instances energy-based performance measures can be misleading while exergy-based measures provide more realistic TES evaluations, exergy results can be useful in design and optimization activities. Some researchers (Bejan, 1978; Rosen, 1990) have attempted to obtain optimum design and operating parameters for sensible TES systems using exergy techniques. Bejan (1978) considered the charging process of a TES system. Krane (1987), considering the entire charging-discharging cycle, found that high-efficiency systems had charging-discharging time ratios of 1:4 or higher. It may not be possible to incorporate such charging-discharging time ratios into many TES applications. Optimal efficiencies and temperatures for the phase change in latent TES have also been investigated (Saborio-Aceves *et al.*, 1994).

In this section, which follows an earlier report (Gunnewiek *et al.*, 1993), the usefulness of using exergy-based measures in optimization and design is demonstrated for TES discharge processes. Analytical expressions are developed using energy and exergy methods for the storage-fluid temperature during discharging and the TES discharge efficiencies. Although in much of this chapter pump work has been neglected with respect to thermal energy flows, one must account for work terms in optimization efforts. Thus pump work is accounted for in this section, and the optimum discharge period based on thermodynamic criteria is determined. An illustrative example is presented.

10.10.1 Analysis description and assumptions

A simple case is considered in which a sensible, closed, fully mixed TES undergoes a complete storage cycle where the final state of the TES is the same as the initial state. The TES system boundaries may be adiabatic or nonadiabatic, and the surroundings are at a constant temperature and pressure. Fluid flows are modeled as steady and one-dimensional. Kinetic and potential energy terms are considered negligible, as is the chemical component of exergy, because it does not contribute to the exergy transfers for a sensible TES system.

10.10.2 Evaluation of storage-fluid temperature during discharge

It is assumed that the recovery of thermal energy from the TES system is achieved with a heat exchanger (Figure 10.24). The heat recovery rate may be written as

$$\dot{Q} = \dot{m}_s c_s (T_{si} - T_{so}) \tag{10.176}$$

where $\dot{m}_s$ and c_s are the mass flow rate and specific heat of the storage fluid, and T_{si} and T_{so} are the heat exchanger inlet and outlet storage-fluid temperatures, respectively.

Knowing the effectiveness ε of the heat exchanger, which is dependant on the heat exchanger configuration and fluid flow conditions, and the minimum heat capacity rate $C_{\min}$ for the two fluids involved in the heat transfer process, the heat recovery rate may also be written as

$$\dot{Q} = C_{\min} \varepsilon (T_{si} - T_{wi}) \tag{10.177}$$

Assuming that the temperature of the storage fluid entering the heat exchanger T_{si} is the same as the storage-fluid temperature T_s, and that the working-fluid temperature entering the heat exchanger T_{wi} is the same as the reference-environment temperature T_o, then it can be shown with Equations 10.176 and 10.177 that

$$T_{so} = T_s - \frac{C_{\min}\varepsilon}{\dot{m}_s c_s}(T_s - T_o) \tag{10.178}$$

To evaluate the changing storage-fluid temperature during the discharge process, a mathematical model is introduced in which, for a finite time step of t^s, a mass of storage fluid at temperature T_s exits the TES while an equal mass of storage fluid at temperature T_{so} enters. The importance has recently been recognized of the time-step size used in the mathematical modeling and numerical simulation of TES systems (Fanny and Klein, 1988; Lightstone *et al.,* 1988). The accuracy of the results is partially dependant on the time-step size and flow rates. A decrease in the time-step size increases the accuracy of the solution to a certain point. In one particular study (Lightstone *et al.,* 1988), a decrease in a time step of ten seconds resulted in no significant difference in the results.

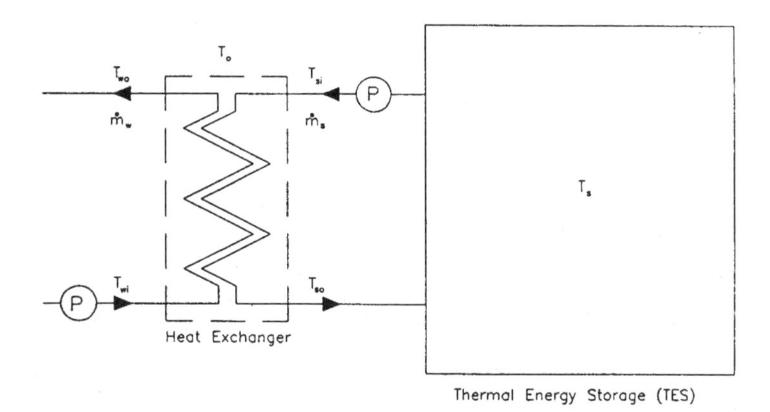

Figure 10.24 The closed TES system considered in evaluating optimal discharge periods.

Adiabatic TES case

After each time step of the discharging period, the new storage-fluid temperature T_{sa} for a fully mixed TES with adiabatic boundaries can be expressed as

$$T_{sa} = \frac{(m - \dot{m}_s t^s)T_s + (\dot{m}_s t^s)T_{so}}{m} \tag{10.179}$$

where m is the mass of the storage fluid and is constant for a closed TES system. Using Equation 10.178, Equation 10.179 becomes

$$T_{sa} = T_s - \frac{C_{\min} \varepsilon t^s}{mc_s}(T_s - T_o) \tag{10.179a}$$

Nonadiabatic TES case

For a nonadiabatic TES with an overall heat transfer coefficient U based on an outer-surface area A, the heat loss Q_l at any storage-fluid temperature T_s, for a finite time step t^s, is

$$Q_l = UAt^s(T_s - T_o) \tag{10.180}$$

and the TES energy content E_s is

$$E_s = mc_s(T_s - T_o) \tag{10.181}$$

Thus, after each finite time step during the discharging process, the new storage-fluid temperature T_{sn}, for a fully mixed TES with nonadiabatic boundaries, can be found with the following energy balance:

$$mc_s(T_{sn} - T_o) = mc_s(T_{sa} - T_o) - Q_l \tag{10.182}$$

After rearranging and substituting Equations 10.179a and 10.180, it can be shown that

$$T_{sn} = T_s - \frac{C_{\min} \varepsilon t^s}{mc_s}(T_s - T_o) - \frac{UAt^s}{mc_s}(T_s - T_o) \tag{10.183}$$

Note that Equation 10.183 expresses the new storage-fluid temperature after a finite time step for any fully mixed, closed TES with adiabatic or nonadiabatic boundaries.

10.10.3 Discharge efficiencies

Energy efficiency

The heat Q_3 recovered from a TES during the ith finite time step $t_s(i)$ can be written as

$$Q_3(i) = \dot{m}_s c_s t^s(i)(T_{sn}(i) - T_{so}(i)) \tag{10.184}$$

Equation 10.184 does not account for the work required to recover the heat. If pump work is considered, the net energy recovered from the TES system may be written as $Q_3(i) - Wt^s(i)$, where W is the shaft power used by the pump, and the discharge energy efficiency after n time steps is defined as

$$\eta = \sum_{i=1}^{n} \frac{Q_3(i) - \dot{W}t^s(i)}{E_s(i=0)} \tag{10.185}$$

Here $E_s(i = 0)$ is the initial energy content of the storage, when the storage fluid temperature is T_s. Note that the energy discharge efficiency is 100% when heat transfer is ideal and the TES is perfectly insulated.

Exergy efficiency

The thermal exergy X_3 recovered from the TES during time step i can be written as

$$X_3(i) = \dot{m}_s c_s t^s(i)\left[(T_{sn}(i) - T_{so}(i)) - T_o \ln \frac{T_{sn}(i)}{T_{so}(i)}\right] \tag{10.186}$$

Accounting for pump work, the net exergy recovered is $X_3(i) - Wt^s(i)$ and the discharge exergy efficiency can be written as

$$\psi = \sum_{i=1}^{n} \frac{X_3(i) - \dot{W}t^s(i)}{\Xi_s(i-0)} \tag{10.187}$$

where $\Xi_s(i = 0)$ is the initial exergy content of the TES, expressible as

$$\Xi_s = mc_s[(T_s - T_o) - T_o \ln(T_s / T_o)] \tag{10.188}$$

10.10.4 Exergy-based optimum discharge period

From a thermodynamic perspective, the optimum discharge period for a TES is that corresponding to the maximum discharge efficiency. The authors feel that the optimum discharge period is more meaningfully determined using exergy rather than energy efficiencies, because exergy analysis considers the quality or usefulness of storage-fluid energy, which is dependant on the fluid and ambient temperatures, and recognizes the difference in usefulness of pump work and recovered heat, whereas energy analysis treats these two energy forms as equal.

It is noted that the optimum discharge period here, which is constrained to being based solely on thermodynamic criteria, may not in general coincide with the optimum period when factors other than thermodynamics (e.g. economics, environmental impact) are taken into account.

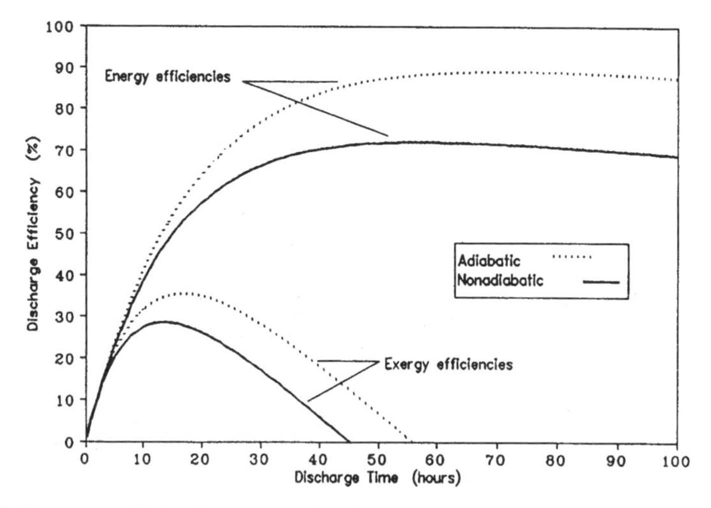

Figure 10.25 Discharge efficiencies, accounting for pump work, for adiabatic and nonadiabatic TES systems.

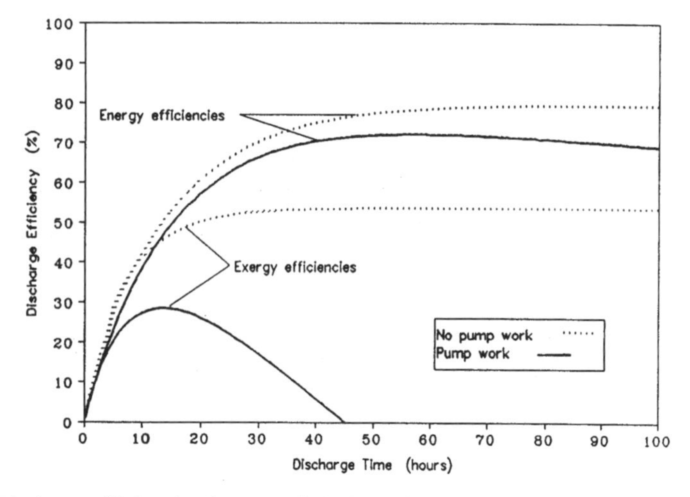

Figure 10.26 Discharge efficiencies for non-adiabatic TES systems with and without pump work.

10.10.5 Illustrative example

Consider an active solar space-heating system having a sensible, fully mixed TES which is charged during daylight hours and discharged at night. The storage medium, water, has a mass of m = 10,000 kg, a constant specific heat of c_s = 4.18 kJ/kg K, and a temperature at the beginning of the discharge period of T_s = 353 K. Heat transfer between the storage fluid and working fluid (air, with c_w = 1.007 kJ/kg K) occurs in a heat exchanger with a 'number of transfer units' of NTU = 2.5, an effectiveness of ε = 0.7, and mass flow rates of $\dot{m}_s = 0.22$ and $\dot{m}_w = 1.2$ kg/s. The temperature of the reference environment is T_o = 293 K. A time step of t^s = 600 seconds is used.

Energy and exergy discharge efficiencies are evaluated as a function of discharge period duration for several cases. Figure 10.25 compares storages having adiabatic and nonadiabatic boundary conditions, i.e. $UA = 0$ and 0.15 kW/K, when the pump shaft power is $W = 0.745$ kW. Figure 10.26 considers the influence of pump shaft power ($W = 0$ and 0.745 kW) for a nonadiabatic TES ($UA = 0.15$ kW/K).

In Figure 10.25, the top two curves show energy efficiencies for adiabatic and nonadiabatic storages, and the bottom two curves show the corresponding exergy efficiencies. The difference in discharge efficiencies for adiabatic and nonadiabatic TES systems is attributable to heat losses, which reduce the amount of heat that can be recovered in the heat transfer process. Two points of significance are noted in the differences between the energy and exergy efficiency curves. First, the maximum exergy and energy discharge efficiencies differ and occur at different times, e.g. for the nonadiabatic TES the maximum exergy efficiency (28.7%) occurs at 13.5 hours, while the maximum energy efficiency (72.1%) occurs at 57.3 hours. Secondly, the net exergy recovered from a TES becomes negative (and consequently, the exergy discharge efficiency becomes negative) before the maximum energy discharge efficiency is attained.

In Figure 10.26, it can be seen that if pump shaft power is considered negligible; maximum discharge efficiencies are higher for both energy and exergy analysis, relative to when the pump shaft power is considered non-zero. Also, with negligible pump shaft power, the maximum net energy recovery is not diminished by continued operation of the heat-exchanger pump. In addition, Figure 10.26 demonstrates that thermal energy and thermal exergy differ, depending on the temperatures involved, while pump shaft power is equivalent in energy and exergy terms. Hence in Figure 10.26, the difference between the exergy and energy efficiencies is much greater for the cases with pump work than without.

10.10.6 Closure

Using energy and exergy discharge efficiencies and a method for evaluating the changing storage-fluid temperature for the discharge process of a closed, fully mixed, sensible TES system, optimal discharge periods can be evaluated based on thermodynamic considerations. The results show that the difference between the results of energy and exergy analyses is significant, and that the impact of pump work on the optima can be important. The authors feel that, since exergy is a measure of the quality or usefulness of energy, exergy performance measures are more meaningful than energy performance measures, and should be considered in the evaluation of the optimum discharge period for TES systems and in related design activities.

10.11 Concluding Remarks

This chapter has demonstrated that the use of exergy analysis is very important in developing a good understanding of the thermodynamic behavior of TES systems, and for rationally assessing, comparing and improving their efficiencies.

Methods identified by exergy analysis as having high improvement potential for TES systems are only limited by the creativity and knowledge of engineers and designers, and can include

- reducing thermal losses (heat leakage from hot TESs and heat infiltration to cold TESs) by improving insulation levels and distributions;
- avoiding temperature degradation by using smaller heat-exchanger temperature differences, ensuring that heat flows of appropriate temperatures are used to heat cooler flows, and increasing heat-exchanger efficiencies;
- avoiding mixing losses by retaining and taking advantage of thermal stratification; and
- reducing pumping power by using more efficient pumps, reduced-friction heat-transfer fluids, and appropriate heat-recovery threshold temperatures.

The authors feel that the development is required of standard TES evaluation methodologies which take into account the thermodynamic considerations discussed in this chapter. By accounting for these considerations, meaningful methodologies can be developed for assessing the comparative value of alternative storages. In particular, the use of exergy analysis (and related concepts) is important because it clearly takes into account the loss of availability and temperature of heat in storage operations, and hence it more correctly reflects the thermodynamic and economic value of the storage operation.

Without methodologies capable of providing perceptive measures of technical, and ultimately of competitive economic, performance, the development of better technology will be unscientific and disorganized, and open to subjective assessments of accomplishment. The development of better assessment methodologies will ensure effective use of energy resources by providing the basis for identifying the more productive directions for development of TES technology, and identifying the better systems without the lengthy and inefficient process of waiting for them to be sorted out by competitive economic success in the marketplace.

Nomenclature

A	surface area
c	specific heat
c_p	specific heat at constant pressure
c_v	specific heat at constant volume
C	heat capacity rate
e	specific energy
E	energy
f	fraction; mean height fraction
F	fraction of storage fluid mass in liquid phase
h	specific enthalpy; specific base enthalpy; height (relative to TES bottom)
H	enthalpy; TES fluid height
i	time step increment
I	exergy consumption due to irreversibilities
ke	specific kinetic energy
m	mass
N	moles
NTU	number of transfer units
pe	specific potential energy

P absolute pressure
Q heat
R thermal resistance
s specific entropy
S entropy
t time
T temperature
u specific internal energy
U internal energy; overall heat transfer coefficient
v specific volume
V volume
W shaft work
x mass fraction
X thermal exergy (i.e. the exergy associated with heat Q)
y mole fraction

Greek Symbols

α constant parameter
ε specific flow exergy; heat exchanger effectiveness
$\in$ flow exergy
η energy efficiency
θ parameter
μ chemical potential
ς specific exergy
Ξ exergy
Π entropy production
ρ density
τ exergetic temperature factor
ϕ zone temperature distribution
ψ exergy efficiency

Subscripts

a inlet flow during charging; adiabatic; parameter
amb ambient
b outlet flow during charging; bottom; parameter
c injected during charging period; charging; inlet flow during discharging
d recovered during discharging period; discharging; outlet flow during discharging
e exit; equivalent
f final
i initial; i^{th} constituent; inlet
j zone j
k number of zones
min minimum

kin	kinetic component
l	loss
m	mixed
n	nonadiabatic
net	net
o	environmental state; chemical exergy; outlet
oo	dead state
p	product
ph	physical component
pot	potential component
r	region of heat interaction
s	solid state; storage fluid
st	storage (overall)
t	threshold; top; liquid state
T	total
th	thermocline zone (zone 2)
w	working fluid
1	charging period
2	storing period
3	discharging period

Superscripts

.	rate with respect to time
–	mean
′	modified case
A	case A
B	case B; basic three-zone model
C	case C; continuous-linear model
D	case D
G	general-linear model
L	linear model
s	step
S	stepped model
T	general three-zone model

Acronyms

ATES	aquifer thermal energy storage
CTES	cold thermal energy storage
ITES	ice thermal energy storage
TES	thermal energy storage

References

Adebiyi, G.A., Hodge, B.K., Steele, W.G., Jalalzedeh-Aza, A. and Nsofor, E.C. (1996). Computer simulations of a high temperature thermal energy storage system employing multiple families of phase-change storage materials, *Journal of Energy Resources Technology* 118, 102–111.

Ahern, J.E. (1980). *The Exergy Method of Energy System Analysis*, Wiley, New York.

Ahrendts, J. (1980). Reference states, *Energy–The International Journal* 5, 667–678.

Alefeld, G. (1990). What are thermodynamic losses and how to measure them? *A Future for Energy: Proc. of the Florence World Energy Research Symposium*, Firenze, Italy, pp. 271–279.

Althof, J. (1989). Economic feasibility of thermal storage. *Heating/Piping/Air Conditioning*, September, 159–163.

Badar, M.A., Zubair, S.M. and Al-Farayedhi, A.A. (1993). Second-law-based thermoeconomic optimization of a sensible heat thermal energy storage system, *Energy* 18, 641–649.

Baehr, H.D. and Schmidt, E.F. (1963). Definition und berechnung von brennstoffexergien (Definition and calculation of fuel exergy), *Brennst-Waerme-Kraft* 15, 375–381.

Barbaris, L.N., Hooper, F.C. and Rosen, M.A. (1988). The relationship between storage period duration and measures of the overall performance of sensible thermal energy storages, *Proc. Int. Conf. on Applied Geothermal Energy and Thermal Energy Storage*, France, pp. 723–727.

Bascetincelik, A., Ozturk, H.H., Paksoy, H.O. and Demirel, Y. (1998). Energetic and exergetic efficiency of latent heat storage system for greenhouse heating, *Renewable Energy* 16, 691–694.

Beckman, G. and Gilli, P.V. (1984). *Thermal Energy Storage*, Springer-Verlag, New York.

Beggs, C. (1991). The economics of ice thermal storage, *Building Research & Information* 19, 342–355.

Bejan, A. (1978). Two thermodynamic optima in the design and operation of thermal energy storage systems, *Journal of Heat Transfer* 100, 708–712.

Bejan, A. (1982). Thermal energy storage, Chapter 8 of *Entropy Generation through Heat and Fluid Flow*, Wiley, New York, pp. 158–172.

Bejan, A. (1995). *Entropy Generation Minimization: The Method of Thermodynamic Optimization of Finite-Size Systems and Finite-Time Processes,* CRC Press, Boca Raton, Florida.

Bjurstrom, H. and Carlsson, B. (1985). An exergy analysis of sensible and latent heat storage, *Heat Recovery Systems* 5, 233–250.

Bosnjakovic, F. (1963). Bezugszustand der exergie eines reagiernden systems (Reference states of the exergy in a reacting system), *Forsch. Ingenieurw.* 20, 151–152.

Brosseau, P. and Lacroix, M. (1998). Numerical simulation of a multi-layer latent heat thermal energy storage system, *International Journal of Energy Research* 22, 1–15.

Carrier (1990). *Encapsulated Ice Store*, USA.

Chase, M.W. *et al.*, Eds. (1985). *JANAF Thermochemistry Tables,* 3rd edition, American Chemical Society and American Inst. of Physics, Washington.

Chen, C.S. and Sheen, J.N. (1993). Cost benefit analysis of a cooling energy storage system, *IEEE Transactions on Power Systems* 8(4), 1504–1510.

Costa, M., Buddhi, D. and Oliva, A. (1998). Numerical simulation of a latent heat thermal energy storage system with enhanced heat conduction, *Energy Conversion and Management* 39, 319–330.

Crane, P., Scott, D.S. and Rosen, M.A. (1992). Comparison of exergy of emissions from two energy conversion technologies, considering potential for environmental impact, *International Journal of Hydrogen Energy* 17, 345–350.

De Lucia, M. and Bejan, A. (1990). Thermodynamics of energy storage by melting due to conduction or natural convection, *ASME J. Solar Energy Engineering* 112, 110–116.

Dincer, I. (1999). Evaluation and selection of energy storage systems for solar thermal applications, *International Journal of Energy Research* 23, 1017–1028.

Dincer, I. and Dost, S. (1996). A perspective on thermal energy storage systems for solar energy applications, *International Journal of Energy Research* 20, 547–557.

Dincer, I., Dost, S. and Li, X. (1997a). Performance analyses of sensible heat storage systems for thermal applications, *International Journal of Energy Research* 21, 1157–1171.

Dincer, I., Dost, S. and Li, X. (1997b). Thermal energy storage applications from an energy saving perspective, *International Journal of Global Energy Issues* 9, 351–364.

Domanski, R. and Fellah, G. (1998). Thermoeconomic analysis of sensible heat, thermal energy storage systems, *Applied Thermal Engineering* 18(8), 693–704.

Edgerton, R.H. (1982). *Available Energy and Environmental Economics*, D.C. Heath, Toronto.

Fanny, A.H. and Klein, S.A. (1988). Thermal performance comparisons for solar hot water systems subjected to various collector and heat exchanger flow rates, *Solar Energy* 40, 1–11.

Fields, W.M.G. and Knebel, D.E. (1991). Cost effective thermal energy storage, *Heating/Piping/Air Conditioning*, July, 59–72.

Gaggioli, R.A. (1983). Second law analysis to improve process and energy engineering, in *Efficiency and Costing: Second Law Analysis of Processes* (ed. R.A. Gaggioli), ACS Symposium Series 235, American Chemical Society, Washington, DC, pp. 3–50.

Gaggioli, R.A. and Petit, P.J. (1977). Use the second law first, *Chemtech* 7, 496–506.

Gretarsson, S.P., Pedersen, C.O. and Strand, R.K. (1994). Development of a fundamentally based stratified thermal storage tank model for energy analysis calculations, *ASHRAE Transactions* 100, 1213–1220.

Gunnewiek, L.H., Nguyen, S. and Rosen, M.A. (1993). Evaluation of the optimum discharge period for closed thermal energy storages using energy and exergy analyses, *Solar Energy* 51, 39–43.

Gunnewiek, L.H. and Rosen, M.A. (1998). Relation between the exergy of waste emissions and measures of environmental impact, *International Journal of Environment &Pollution* 10, 261–272.

Hafele, W., Ed. (1981). Energy, negentropy, and endowments, Chapter 21 of *Energy in a Finite World: A Global Systems Analysis*, Ballinger, Toronto, pp. 693–705.

Hahne, E. (1986). Thermal energy storage: some views on some problems, *Proc. 8th Int. Heat Transfer Conf.*, San Francisco, pp. 279–292.

Hahne, E., Kubler, R. and Kallewit, J. (1989). The evaluation of thermal stratification by exergy, in *Energy Storage Systems* (ed. B. Kilkis and S. Kakac), Kluwer, Dordecht, pp. 465–485.

Hevert, H.W. and Hervert, S.C. (1980). Second law analysis: an alternative indicator of system efficiency, *Energy-The International Journal* 5, 865–873.

Homan, K.O., Sohn, C.W. and Soo, S.L. (1996). Thermal performance of stratified chilled water storage tanks, *HVAC & R Research* 2(2), 158–170.

Hooper, F.C., Barbaris, L.N. and Rosen, M.A. (1988). An engineering approach to the evaluation of the performance of stratified thermal energy storages, *Proc. Int. Conf. on Applied Geothermal Energy and Thermal Energy Storage*, Versailles, France, pp. 155–160.

Hoyer, M.C., Walton, M., Kanivetsky, R. and Holm, T.R. (1985). Short-term aquifer thermal energy storage (ATES) test cycles, St. Paul, Minnesota, U.S.A., *Proc. 3rd Int. Conf. on Energy Storage for Building Heating and Cooling*, Toronto, Canada, pp. 75–79.

Ismail, K.A.R., Leal, J.F.B. and Zanardi, M.A. (1997). Models of liquid storage tanks, *Energy-The International Journal* 22, 805–815.

Ismail, K.A.R. and Stuginsky, R. (1999). A parametric study on possible fixed bed models for PCM and sensible heat storage, *Applied Thermal Engineering* 19, 757–788.

Jansen, J. and Sorensen B. (1984). *Fundamentals of Energy Storage*, Wiley, New York.

Jenne, E.A., Ed. (1992). *Aquifer Thermal Energy (Heat and Chill) Storage*, Pacific Northwest Lab, Richland, WA.

Keenan, J.H., Chao, J. and Kayle, J. (1992). *Gas Tables International Version. Properties of Air Products of Combustion and Component Gases Compressible Flow Functions*, Krieger, Florida.

Kestin, J. (1980). Availability: the concept and associated terminology, *Energy-The International Journal* 5, 679–692.

Kleinbach, E.M., Beckman, W.A. and Klein, S.A. (1993). Performance study of one-dimensional models for stratified thermal storage tanks, *Solar Energy* 50, 155–166.

Kotas, T.J. (1995). *The Exergy Method of Thermal Plant Analysis,* reprint edition, Krieger, Florida.

Kotas, T.J., Raichura, R.C. and Mayhew, Y.R. (1987). Nomenclature for exergy analysis, in *Second Law Analysis of Thermal Systems* (ed. M.J. Moran &E. Sciubba), ASME, New York, pp. 171–176.

Krane, R.J. (1985). A second law analysis of a thermal energy storage system with Joulean heating of the storage element, Paper 85-WA/HT-19, ASME Winter Annual Meeting, 17–21 Nov., Miami.

Krane, R.J. (1987). A second law analysis of the optimum design and operation of thermal energy storage systems, *International Journal of Heat and Mass Transfer* 30, 43–57.

Krane, R.J. and Krane, M.J.M. (1991). The optimum design of stratified thermal energy storage systems, *Proc. Int. Conf. on the Analysis of Thermal and Energy Systems*, Greece, pp. 197–218.

Laouadi, A. and Lacroix, M. (1999). Thermal performance of a latent heat energy storage ventilated panel for electric load management, *International Journal of Heat and Mass Transfer* 42, 275–286.

Lightstone, M., Hollands, K.G.T. and Hassani, A.V. (1988). Effect of plume entrainment in the storage tank on calculated solar energy performance, *Energy Solutions for Today: Proc. 14th Annual Conf. of Solar Energy Soc. of Canada*, pp. 236–241.

Lucca, G. (1990). The exergy analysis: role and didactic importance of a standard use of basic concepts, terms and symbols, *A Future For Energy: Proc. Florence World Energy Research Symposium*, Firenze, Italy, pp. 295–308.

Maloney, D.P. and Burton, J.R. (1980). Using second law analysis for energy conservation studies in the petrochemical industry, *Energy-The International Journal* 5, 925–930.

Mathiprakasam, B. and Beeson, J. (1983). Second law analysis of thermal energy storage devices, *Proc. AIChE Symp. Series*, National Heat Transfer Conference, Seattle, Washington.

Mavros, P., Belesiotis, V. and Haralambopoulos, D. (1994). Stratified energy storage vessels-characterization of performance and modeling of mixing behavior, *Solar Energy* 52, 327–336.

Moran, M.J. (1989). *Availability Analysis: A Guide to Efficient Energy Use*, ASME, New York.

Moran, M.J. (1990). Second law analysis. What is the state of the art? A Future For Energy: *Proceedings of the Florence World Energy Research Symposium*, Firenze, Italy, pp. 249–260.

Moran, M.J. and Keyhani, V. (1982). Second law analysis of thermal energy storage systems, *Proceedings of the 7th International Heat Transfer Conference*, Vol. 6, Munich, pp. 473–478.

Moran, M.J. and Shapiro, H.N. (2000). *Fundamentals of Engineering Thermodynamics*, 4th ed., Wiley, Toronto.

Morris, D.R., Steward, F.R., and Szargut, J. (1988). *Exergy Analysis of Thermal, Chemical and Metallurgical Processes*, Springer-Verlag, New York.

Nelson, J.E.B., Balakrishnan, A.R. and Murthy, S.S. (1999). Experiments on stratified chilled water tanks, *International Journal of Refrigeration* 22(3), 216–234.

Rodriguez, L.S.J. (1980). Calculation of available-energy quantities, in *Thermodynamics: Second Law Analysis* (ed. R.A. Gaggioli), American Chemical Society, Washington, DC, pp. 39–60.

Rosen, M.A. (1990). Evaluation of the heat loss from partially buried, bermed heat storage tanks, *International Journal of Solar Energy* 9, 147–162.

Rosen, M.A. (1991). On the importance of temperature in performance evaluations for sensible thermal energy storage systems, *Proc. Bienniel Congr. Int. Solar Energy Soc.* (ed. M.E. Arden, S.M.A. Burley and M. Coleman), Vol. 2, Part II, Pergamon, New York, pp. 1931–1936.

Rosen, M.A. (1992a). Evaluation of energy utilization efficiency in Canada using energy and exergy analyses, *Energy-The International Journal* 17, 339–350.

Rosen, M.A. (1992b). Appropriate thermodynamic performance measures for closed systems for thermal energy storage, *ASME Journal of Solar Energy Engineering* 114, 100–105.

Rosen, M.A. (1998a). A semi-empirical model for assessing the effects of berms on the heat loss from partially buried heat storage tanks, International Journal of Solar Energy 20, 57–77.

Rosen, M.A. (1998b). The use of berms in thermal energy storage systems: energy-economic analysis, *Solar Energy* 63(2), 69–78.

Rosen, M.A. (1999a). Second law analysis: approaches and implications, *International Journal of Energy Research* 23(5), 415–429.

Rosen, M.A. (1999b). Second-law analysis of aquifer thermal energy storage systems, *Energy-The International Journal* 24, 167–182.

Rosen, M.A. and Dincer, I. (1997a). On exergy and environmental impact, *International Journal of Energy Research* 21, 643–654.

Rosen, M.A. and Dincer, I. (1997b). Sectoral energy and exergy modeling of Turkey, *ASME J. Energy Resources Technology* 119, 200–204.

Rosen, M.A. and Dincer, I. (1999a). Thermal storage and exergy analysis: the impact of stratification, *Transactions of CSME* 23(1B), 173–186.

Rosen, M.A. and Dincer, I. (1999b). Exergy analysis of waste emissions, *International Journal of Energy Research* 23(13), 1153–1163.

Rosen, M.A., Dincer, I. and Pedinelli, N. (2000). Thermodynamic performance of ice thermal energy storage systems, *ASME Journal of Energy Resources Technology* 122(4), 205–211.

Rosen, M.A. and Hooper, F.C. (1991a). A general method for evaluating the energy and exergy contents of stratified thermal energy storages for linear-based storage fluid temperature distributions, *Proceedings of the 17th Annual Conference of the Solar Energy Society of Canada*, Toronto, pp. 182–187.

Rosen, M.A. and Hooper, F.C. (1991b). Evaluating the energy and exergy contents of stratified thermal energy storages for selected storage-fluid temperature distributions, *Proceedings of the Biennial Congress of International Solar Energy Society*, Denver, pp. 1961–1966.

Rosen, M.A. and Hooper, F.C. (1992). Modeling the temperature distribution in vertically stratified thermal energy storages to facilitate energy and exergy evaluation, in *Thermodynamics and the Design, Analysis and Improvement of Energy Systems*, AES-Vol 27/HTD-Vol. 228 (ed. R.F. Boehm), American Society of Mechanical Engineers, New York, pp. 247–252.

Rosen, M.A. and Hooper, F.C. (1994). Designer-oriented temperature-distribution models for vertically stratified thermal energy storages to facilitate energy and exergy evaluation, *Proc. 6th Int. Conf. on Thermal Energy Storage*, Espoo, Finland, pp. 263–270.

Rosen, M.A., Hooper, F.C. and Barbaris, L.N. (1988) Exergy analysis for the evaluation of the performance of closed thermal energy storage systems, *ASME Journal of Solar Energy Engineering* 110, 255–261.

Rosen, M.A. and Horazak, D.A. (1995). Energy and exergy analyses of PFBC power plants, Chapter 11 of *Pressurized Fluid Bed Combustion* (ed. M. Alvarez-Cuenca and E.J. Anthony), Chapman & Hall, London, pp. 419–448.

Rosen, M.A., Nguyen, S. and Hooper, F.C. (1991). Evaluating the energy and exergy contents of vertically stratified thermal energy storages, *Proc. 5th Int. Conf. on Thermal Energy Storage*, Scheveningen, The Netherlands, pp. 7.4.1–7.4.6.

Rosen, M.A., Pedinelli, N. and Dincer, I. (1999) Energy and exergy analyses of cold thermal storage systems, *International Journal of Energy Research* 23(12), 1029-1038.

Rosen, M.A. and Scott, D.S. (1987). On the sensitivities of energy and exergy analyses to variations in dead-state properties, in *Analysis and Design of Advanced Energy Systems* (ed. M.J. Moran *et al.*), AES-Vol. 3-1, ASME, New York, pp. 23–32.

Rosen, M.A. and Scott, D.S. (1998). Comparative efficiency assessments for a range of hydrogen production processes, *International Journal of Hydrogen* Energy 23, 653-659.

Rosen, M.A. and Tang, R. (1997). Increasing the exergy storage capacity of thermal storages using stratification, in *Proc. ASME Advanced Energy Systems Division*, AES-Vol. 37 (ed. M.L. Ramalingam, J.L. Lage, V.C. Mei and J.N. Chapman), ASME, New York, pp. 109–117.

Rosenblad, G. (1985). Quality loss from seasonal storage of heat in rock, magnitude and evaluation, *Proc. 3rd Int. Conf. on Energy Storage for Building Heating and Cooling*, Toronto, pp. 594–599.

Saborio-Aceves, S., Nakamura, H. and Reistad, G.M. (1994). Optimum efficiencies and phase change temperature in latent heat storage systems, *Journal of Energy Resources Technology* 116, 79–86.

Sussman, M.V. (1980). Steady-flow availability and the standard chemical availability, *Energy-The International Journal* 5, 793–804.

Sussman, M.V. (1981). Second law efficiencies and reference states for exergy analysis, *Proc. 2nd World Congress Chem. Eng.*, Canadian Society of Chemical Engineers, Montreal, pp. 420–421.

Szargut, J. (1967). Grenzen fuer die anwendungsmoeglichkeiten des exergiebegriffs (Limits of the applicability of the exergy concept), *Brennst.-Waerme-Kraft* 19, 309–313.

Taylor, M.J. (1986). Second Law Optimization of a Sensible Heat Thermal Energy Storage System with a Distributed Storage Element, MS Thesis, Department of Mechanical and Aerospace Engineering, Univ. of Tennessee, Knoxville.

Wepfer, W.J. and Gaggioli, R.A. (1980). Reference datums for available energy, in *Thermodynamics: Second Law Analysis* (ed. R.A. Gaggioli), ACS Symposium Series 122, American Chemical Society, Washington, DC, pp. 77–92.

Wood, L.L., Miedema, A.K. and Cates, S.C. (1994). Modeling the technical and economic potential of thermal energy storage systems using pseudo-data analysis, *Resource and Energy Economics* 16, 123–145.

Yoo, H., Kim, C-J. and Kim, C. (1998). Approximate analytical solutions for stratified thermal storage under variable inlet temperature, *Solar Energy* 66, 47–56.

Appendix: Glossary of Selected Exergy-Related Terminology

This glossary identifies exergy-related terminology from the literature that is of relevance to the TES discussions in this chapter. Most exergy terminology has only recently been adopted, and is still evolving. Often more than one name is assigned to the same quantity, and more than one quantity to the same name. Only exergy-related definitions are given for terms having multiple meanings. The glossary is based in part on previously developed broader glossaries (Kotas *et al.,* 1987; Kotas, 1995; Kestin, 1980).

- *Available energy.* See exergy
- *Available work.* See exergy
- *Availability.* See exergy
- *Base enthalpy.* The enthalpy of a compound (at T_o and P_o) evaluated relative to the stable components of the reference environment (i.e. relative to the dead state).
- *Chemical exergy.* The maximum work obtainable from a substance when it is brought from the environmental state to the dead state by means of processes involving interaction only with the environment.
- *Dead state.* The state of a system when it is in thermal, mechanical and chemical equilibrium with a conceptual reference environment (having intensive properties pressure P_o, temperature T_o, and chemical potential μ_{ioo} for each of the reference substances in their respective dead states).
- *Degradation of energy.* The loss of work potential of a system which occurs during an irreversible process.
- *Dissipation.* See exergy consumption.
- *Effectiveness.* See second-law efficiency.
- *Energy analysis.* A general name for any technique for analyzing processes based solely on the First Law of Thermodynamics. Also known as First-Law analysis.
- *Energy efficiency.* An efficiency determined using ratios of energy. Also known as thermal efficiency; First-Law efficiency.
- *Energy grade function.* The ratio of exergy to energy for a stream or system.
- *Entropy creation.* See entropy production.
- *Entropy generation.* See entropy production.
- *Entropy production.* A quantity equal to the entropy increase of an isolated system (associated with a process) consisting of all systems involved in the process. Also known as entropy creation; entropy generation.
- *Environment.* See reference environment.
- *Environmental state.* The state of a system when it is in thermal and mechanical equilibrium with the reference environment, i.e. at pressure P_o and temperature T_o of the reference environment.
- *Essergy.* See exergy. Derived from essence of energy.
- *Exergetic temperature factor.* A dimensionless function of the temperature T and environmental temperature T_o given by $(1 - T_o/T)$.
- *Exergy.* 1) A general term for the maximum work potential of a system, stream of matter or a heat interaction in relation to the reference environment as the datum state. Also known as available energy; availability; essergy; technical work capacity; usable energy; utilizable energy; work capability; work potential; xergy. 2) The unqualified term exergy or exergy flow is the maximum amount of shaft work obtainable when a steady stream of matter is brought from its initial state to the dead state by means of processes involving interactions only with the reference environment.
- *Exergy analysis.* An analysis technique in which process performance is assessed by examining exergy balances. A type of Second-Law analysis.
- *Exergy consumption.* The exergy consumed or destroyed during a process due to irreversibilities within the system boundaries. Also known as dissipation; irreversibility; lost work.
- *Exergy efficiency.* A second-law efficiency determined using ratios of exergy.
- *External irreversibility.* The portion of the total irreversibility for a system and its surroundings occurring outside the system boundary.
- *First-law analysis.* See energy analysis.
- *First-law efficiency.* See energy efficiency.

- *Ground state.* See reference state.
- *Internal irreversibility.* The portion of the total irreversibility for a system and its surroundings occurring within the system boundary.
- *Irreversibility.* 1) An effect, making a process non-ideal or irreversible. 2) See exergy consumption.
- *Negentropy.* A quantity defined such that the negentropy consumption during a process is equal to the negative of the entropy creation. Its value is not defined, but is a measure of order.
- *Non-flow exergy.* The exergy of a closed system, i.e. the maximum net usable work obtainable when the system under consideration is brought from its initial state to the dead state by means of processes involving interactions only with the environment.
- *Physical exergy.* The maximum amount of shaft work obtainable from a substance when it is brought from its initial state to the environmental state by means of physical processes involving interaction only with the environment. Also known as thermomechanical exergy.
- *Rational efficiency.* A measure of performance for a device given by the ratio of the exergy associated with all outputs to the exergy associated with all inputs.
- *Reference environment.* An idealization of the natural environment which is characterized by a perfect state of equilibrium, i.e. absence of any gradients or differences involving pressure, temperature, chemical potential, kinetic energy and potential energy. The environment constitutes a natural reference medium with respect to which the exergy of different systems is evaluated.
- *Reference state.* A state with respect to which values of exergy are evaluated. Several reference states are used, including environmental state, dead state, standard environmental state and standard dead state. Also known as ground state.
- *Reference substance.* A substance with reference to which the chemical exergy of a chemical element is calculated. Reference substances are often selected to be common, valueless environmental substances of low chemical potential.
- *Resource.* A material found in nature or created artificially in a state of disequilibrium with the environment.
- *Restricted equilibrium.* See thermomechanical equilibrium.
- *Second-law analysis.* A general name for any technique for analyzing process performance based solely or partly on the Second Law of Thermodynamics. Abbreviated SLA.
- *Second-law efficiency.* A general name for any efficiency based on a second-law analysis (e.g. exergy efficiency, effectiveness, utilization factor, rational efficiency, task efficiency). Often loosely applied to specific second-law efficiency definitions.
- *Task efficiency.* See second-law efficiency.
- *Technical work capacity.* See exergy.
- *Thermal efficiency.* See energy efficiency.
- *Thermal exergy.* The exergy associated with a heat interaction, i.e. the maximum amount of shaft work obtainable from a given heat interaction using the environment as a thermal energy reservoir.
- *Thermomechanical exergy.* See physical exergy.
- *Thermomechanical equilibrium.* Thermal and mechanical equilibrium.
- *Unrestricted equilibrium.* Complete (thermal, mechanical and chemical) equilibrium.
- *Usable and useful energy.* See exergy.
- *Utilizable energy.* See exergy.
- *Utilization factor.* See second-law efficiency.
- *Work capability.* See exergy.
- *Work potential.* See exergy.
- *Xergy.* See exergy.

11

Thermal Energy Storage Case Studies

I. Dincer and M.A. Rosen

11.1 Introduction

TES systems have been employed to cool buildings since Roman Imperial times, when snow was raced from the mountains to provide short-term cooling in sweltering villas. In the early part of the industrial era, natural ice was used in theaters to cool the stage and portions of the audience. Air was blown over a pit full of ice to remove heat. With the advent of commercial air conditioning equipment in the 1930s, some engineering practitioners used chilled-water storage to supplement cooling and downsize compressors. Advances in manufacturing and an abundance of inexpensive power often encouraged designers and owners to rely on direct cooling equipment. These systems were cost-effective but consumed energy at the peak period of the day (when it was typically generated with oil). The 1973 oil embargo radically altered energy prices and perceptions of energy availability in the world. Many policies were developed that encouraged the use of TES to shift demand to evening and other off-peak hours (when nuclear and coal were the primary fuels for electrical generation). Electric utilities were encouraged to promote TES systems through rebates, design assistance and demonstration. Now, TES finds use for heating and/or cooling purposes in a variety of applications.

Space heating using electric thermal storage has been used extensively in Europe and North America. The storage media include ceramic brick, crushed rock, water and building mass, and systems can be either room or centrally based. Many improvements have been introduced in the past few years, including new phase-change materials for latent heat storage which have recently become available commercially.

Cold storage using ice, water or eutectic salts as the storage media is widely applied where summer cooling requirements are high. It is also used in Europe, often in combination with heat recovery and hot water storage, and in Australia, Canada, Korea, Japan, Taiwan and South Africa.

TES can be installed in both residential and commercial buildings, and can be cost-effective. Results from many of the monitored projects demonstrate payback periods of less than three years. If time-of-use tariffs exist, electricity costs to the consumer can be reduced by shifting the main electrical loads to periods when electricity prices are lower. If demand charges are implemented, a shifting or spreading of the load can reduce these significantly. To be effective, each storage system must be sized and controlled to minimise electricity costs.

Benefits from TES use also accrue to electricity utilities. The shifting of loads to off-peak periods not only spreads the demand over the generating period, but may enable output from the more expensive generating stations to be reduced.

Worldwide there are many electricity utility programs promoting the use of storage technologies, many of them part of demand-side management programs. Such programs can greatly influence the economic feasibility of installing thermal storage through offers of financial rebates for equipment, information programs, or special electricity rates for consumers.

In this chapter, a wide range of case studies are presented to illustrate the benefits, as well as drawbacks, of TES. The cases consider applications from the commercial and institutional building sector, industry, and groups within the utility sector representing electricity generation and district heating and cooling. The different case studies illustrate the full context in which a given TES opportunity is viewed, rather than examining the TES in isolation. The types of TES represented through the case studies include both

- cold TES, using ice, chilled water and PCMs, and
- heat TES, using both sensible and latent storage techniques.

The material presented here, based on the actual applications, is drawn from various sources, e.g. company reports and catalogs. In this regard, for the most part the presentation and wording from the original sources is used to permit the views of the original writers to be provided.

11.2 Ice CTES Case Studies

11.2.1 Rohm and Haas, Spring House Research Facility, Pennsylvania, USA

In 1992, the manager of facilities and engineering, and the maintenance and utilities manager, Rohm and Haas, Spring House Research Facility, requested an in-depth energy review of the entire facility, considering both long- and short-term goals. The purpose of the study was to identify various ways to reduce electrical costs and optimize chilled water usage. The original chiller plant was built 32 years ago with 1200 ton-hour installed capacity and served two buildings. By 1987, the plant had grown to 4000 ton-hour installed capacity and served 13 buildings. The last research building was added in 1987. By 1990, the available cooling capacity at peak ambient conditions could barely meet the peak demand. Setting the supply chilled water temperature to 7.2°C from 10°C and hydraulically rebalancing the entire system permitted Rohm and Haas to meet cooling demands. Thus, more time was gained, permitting operation for the following two years; even 1991, which had unusually high ambient peak temperatures, presented only minor problems. This time allowed for a total re-evaluation and review of the chilled water plant, which uses almost 50% of the total electric demand. Therefore, the energy utilization study was done in 1992.

Scope of the project

Rohm and Haas was concerned with the environmental impact of the refrigerant that was to be used for cooling, since additional refrigeration capacity was required. The existing four chillers' refrigerant is either CFC-11 or R-114, which mandated replacement. Another concern was the ever-increasing electric summer peak demands, largely derived from the chillers and their auxiliaries, with the corresponding increase in operating cost.

Many options were evaluated, including absorption refrigeration, high-efficiency centrifugal chillers, and cogeneration. After investigation and evaluation, a TES system was chosen, which produces ice during off-peak hours and, if elected, chilled water during the peak hours. The selection was reached based on the capability to shave in excess of 50% or more of the electric peak demand resulting from operating the chillers and auxiliaries.

Peco Energy, the electric company serving Spring House, has a CTES rider, which reduces the peak demand hours from 12 to 10 hours Monday through Thursday and 6 hours on Friday. In addition, the peak demand for each peak month is averaged, and this results in a lower annual peak billing demand. This provision is valid provided that the total cooling demand is reduced by 50% or more (preliminary results obtained during June and July 1995 indicate a reduction of about 70%).

In February 1994, Frank V. Radomski & Sons, Inc., a general contracting firm, was selected, and began preliminary engineering and cost estimate activities. By September 1994, all engineering services were selected and all major equipment vendors were chosen. In September 1994, the project was put on a fast track with strict scheduling and cost control, and construction commenced. The plant went into successful operation at the beginning of June 1995, exceeding all the peak demand shaving requirements during June and July.

The goal was to reduce peak electrical demand by a minimum of 1.6 MW. The payback calculations were based on 2.0 MW peak demand shaving and 10-month operation. An average peak demand shaving of about 2.3 MW was achieved, a 15% increase over the 2.0 MW design value. All project aspects, including project management, architectural, electrical and mechanical design and engineering, and plant operation, were coordinated. All final decisions were open to review and solved cooperatively. This approach led to a well-designed, architecturally attractive addition to the utility plant, and the research facility began operation on time with a minimum of field changes and operating problems.

Description of the system

The ice CTES system selected for the facility was an ice harvester-type system. A weekly load-shift strategy was incorporated in the system design to shift electric cooling load from the peak hours to the less expensive evening and weekend off-peak hours. The system consisted of four Mueller 250 ton-hour evaporators mounted on top of a rectangular poured-in-place concrete ice water storage tank. Completely assembled units were shipped to the plant.

The system was chosen over other thermal energy storage technologies for the following reasons:

- The flexibility to optimize the system's efficiency under various load conditions.
- The ability to operate as a chiller as well as an ice maker.

- The ability to maintain consistent low water temperature from the ice storage, thus providing lower water temperatures to the existing air handling units. This allows an increase in heat transfer and compensates for increased loads for some of the units and thus saves in replacement costs.

American Industrial Refrigeration (AIR) supplied the assembled the high-side refrigeration package system, including a four-cell evaporator condenser furnished by Evapco, two screw compressor packages, and a PC-based control system was supplied by FES. The control system includes the integration and programming of the ice system's controls, compressor package controls, refrigeration system controls, and the ice and chilled water system controls.

This system is enclosed in an insulated double-wall enclosure, 12.8 m long by 8.5 m wide. The refrigeration high-side package, including all electrical switchgear, wiring, and pipe insulation, was completely fabricated at AIR's shop facilities. The work also included the installation of the compressor packages. The system was trucked from Minnesota to the Rohm and Haas plant site. The package was designed and built to be split into two halves and reassembled at the job site. The structure was designed to support the evaporative condensers, which were shipped directly in two packages. The condensers were installed and piped at the job site, using a preassembled structural steel catwalk assembly and piping. The on-site construction time for the assembly of these packages and evaporative condenser was four weeks. The refrigeration system is an HCFC-22 liquid recirculation system with a capacity of 1280 tons of refrigeration during the ice making mode and 1720 tons of refrigeration during the water chilling mode.

The goal of the project to reduce energy consumption economically was considered throughout the project. For example, the compressor packages were provided with oversized oil separators and suction valve assemblies to reduce pressure losses and allow operation at higher suction pressures and lower discharge pressures. Thermosyphon oil cooling was selected for further economy of operation. The system utilizes the economizer cycle available on screw compressors. A flash-type economizer vessel is used to subcool the HCFC-22. The flash gas goes to a sideport connection on the screw compressor. This further increases the overall efficiency of the system. The evaporative condenser was oversized to allow overall operation at lower condensing temperatures. All equipment was selected with zero negative performance allowance. All electric motors were selected for high efficiency. The evaporator fan motors and one of the chilled water pumps have variable speed drives.

All operating functions are automatically controlled by the PC-based control system, which controls the ice harvesters and the water-side system. Compressors, condenser fans, condenser water pumps, refrigerant pumps, and control valves can all be operated on local control. In addition to its general features, such as dual pressure relief valves and high- and low-level alarms, the refrigeration high-side package also includes refrigerant detectors and an oxygen detector. All of the alarm signals are sent to the control system for operator display and acknowledgment.

The ice water storage tank includes a spray distribution system at the top of the tank to provide for an even melting of the ice. This produces low temperatures at the suction header. The suction channel is a 0.457 by 0.457 m formed channel in the bottom of the tank. This channel is covered by a 0.0127 m galvanized plate that has perforations to draw water evenly across the bottom of the tank. The mechanical equipment room is located between the ice storage tank and the existing utility building and is 7.62 by 24.38 m and 7 m high. Located in the mechanical equipment room are three ice water pumps nominally rated at 0.157 m^3/s each, circulating ice water through a single plate-type heat exchanger rated for 0.466 m^3/s and 9.5°C or 4625 ton-hour back to the ice harvesters. There are also three chilled water pumps rated at 0.157 m^3/s. One of the chilled water pumps has a variable speed drive. The electrical switchgear is located on the second floor.

The ice water storage tank is an above-ground concrete tank, poured in place to hold more than 45,000 ton-hour of latent cooling. The tank's nominal internal dimensions are 27.5 m long by 18.3 m with a usable height of 6.7 m. The tank floor is 0.3 m thick and the walls are 0.46 m. Galvanized structural steel, fully welded to wall channels, is anchored to the 0.46 m concrete tank walls. The tank top is designed to support the four 250 ton/hour harvester-type evaporators and two future evaporators. The tank interior is coated with a commercial industrial membrane, which is liquid applied urethane coating. All masonry exposed walls are insulated with 0.076 m of polyisocyanurate insulation and 0.1 m of split face block exterior. The roof insulation has 0.076 m of polyisocyanurate insulation with 0.46 by 0.46 m concrete pavers for ballast.

Further information on this project may be obtained from Kent (1996).

11.2.2 A Cogeneration Facility, California, USA

Combustion gas turbines are constant-volume engines for which shaft horsepower is proportional to the combustion air mass flow. Engine output improves if the air temperature is depressed at the compressor inlet to increase the air density. When a combustion turbine generator is used in a power plant, increased engine output increases the electrical generating capacity. That is the concept presented in this case study of an inlet air-chilling system installed in a cogeneration plant in California. The plant also uses a TES system with the inlet air chiller to optimize the plant's economic performance.

The facility considered here is a 36 MW gas turbine topping cycle cogeneration plant that began commercial operation in November 1991, producing electricity for sale to a regulated utility and generating steam for sale to an enhanced oil recovery operation in a local oil field.

The plant operates all year at base load. The summer season is when power sales are most valuable. The plant makes almost 80% of its electrical revenues between May 1 and October 31, yet the plant power output is substantially reduced by the high ambient temperature. Inlet air chilling with TES was installed at the facility to increase output during critical peak hours in the summer when maximum unit performance is required.

The chiller/TES system consists of a mechanical vapor-compression refrigeration cycle driving an ice harvester that is operated in the evening hours to stockpile ice in a thermal energy storage tank. Chilled water from the tank is circulated through cooling coils at the gas turbine air inlet during the heat of the day to increase the plant's electrical output.

Plant description

The plant's prime mover is a single-combustion gas turbine. It is an industrial-frame unit of single-shaft design driving a synchronous generator through a load gear. The compressor is a 17-staged, axial-flow type with variable-inlet guide vanes. The turbine is three-staged and is designed for an 1104.5°C firing temperature. It has ten combustion chambers arranged in can-annular design and, in this application, is fired on natural gas with NOx combustors for emission control. A Heat Recovery Steam Generator (HRSG) captures the 542.7°C waste heat from the turbine exhaust to generate 61.6 MW of 75% quality steam for enhanced oil recovery. In new and clean conditions, the unit is rated for 36.24 MW gross output at the generator terminals with ambient air at ISO conditions (dry-bulb temperature 15°C; relative humidity 60%). Approximately 1 MW is used for the plant's station light and power requirements. At the rated output, 132.4 m/s of air is consumed by the engine.

Inlet chilling concept

Electric power is sold to the utility under the terms of a purchase agreement. This agreement recognizes peak periods when high consumer demand places a premium value on generating capacity. Peak period occurs weekdays from noon to six p.m. from May 1 through October 31 each year. During these periods, central California temperatures exceed 37.77°C and air conditioning units create the highest demand for power.

The power purchase agreement is structured so that the utility pays an energy payment for every kWh delivered, a capacity payment for delivering power at no less than 85% of a dedicated firm capacity level, and a bonus payment based on how well the plant meets the remaining 15% of the dedicated firm capacity during peak hours. No bonus is earned on kWh delivered above the dedicated firm capacity. The dedicated firm capacity level was set by a test of plant output at the time commercial operation began. That was, as it happened, in winter months when cool temperatures and the new and clean condition of the unit allowed a 35.5 MW dedicated firm capacity level to be set.

The gas turbine air-inlet system was originally equipped with an evaporative cooler to reduce inlet dry-bulb temperature. With the evaporative cooler operating at 85% effectiveness on a typical 35°C day with 20% relative humidity, the net output of the plant is at 'best' 34 MW. Further reduction of output occurs due to unit degradation, e.g. the effect of fouling, erosion, corrosion, and foreign object damage that inevitably degrades performance by reducing compressor airflow. Typically, such degradation will advance very rapidly during the first two or three years of operation to as much as 6% of output capability. Hence, performance of the plant would not meet the 35.5 MW dedicated capacity level, and was subject to the loss of a large share of potential bonus revenue.

To compensate for temperature and degradation effects, inlet air chilling was installed at a design temperature 5.5°C. In addition to increased output, chilled inlet air improves the gas turbine heat rate. The net plant heat rate is lowered when additional station power is used to generate ice at night, but the effect is almost completely mitigated by the heat rate improvement when chilling. Overall, net plant output with inlet air chilled to 5.5°C during peak hours now satisfies the dedicated capacity level. Operating points are indicated for typical summer peak ambient conditions (35°C, 20% RH) with and without evaporative cooling and with chilling.

Design considerations

For the turbine generator, inlet air chilling is limited to 5.5°C. Inlet air temperatures that were any lower at a nearly saturated condition could cause icing at the compressor inlet, resulting in damage to the engine. As air enters the bell mouth of the axial compressor, the velocity increases. Air enthalpy is transformed to kinetic energy in an adiabatic process as the velocity increases. Air at 5.5°C accelerated to 106.7 m/s results in an approximate −12.2°C drop in temperature, to 0°C. With this limitation in mind, three types of inlet air chilling were considered: direct mechanical refrigeration, absorption refrigeration, and TES.

The overall effect on net power produced for each scheme is shown in Table 11.1. Direct mechanical refrigeration consists of a vapor-compression refrigeration system to chill inlet air to 5.55°C during the peak hours, without TES. This system has the added benefit of chilling capability during all hours of the day. Furthermore, there is a 1300 kW penalty associated with running the refrigeration compressor while chilling. The compressor load penalty would lower the net plant power output during peak hours. The plant would not meet 35.5 MW with 5.5°C inlet air if an additional 1300 kW station load were subtracted from the net output. Furthermore, compared to TES, the installed refrigeration capacity required for direct refrigeration would be three times larger. Direct refrigeration would have to be sized to deliver the full instantaneous chilling duty for a turbine generator. With TES, the chilling duty for 6 hours is accumulated and stored over 18 hours, thus reducing the refrigeration required by a factor of 6/18, or one-third the size required for direct refrigeration.

Absorption refrigeration (e.g. a lithium bromide system) was considered. It would have required modifying the existing HRSG to provide low-pressure steam to drive the absorption system, thereby avoiding the power penalty associated with a direct mechanical refrigeration system. A drawback, however, is that there is no cooling water available at the site. A closed-loop cooling system would have been required, with the heat rejected to the air with an aerial cooler. As a result, during peak hours when the ambient temperature is above 32.22°C, it would be difficult to chill the inlet air below 10.5°C. This system would not maximize power output during peak hours.

TES was selected because it allows power output to be maximized during peak hours. The TES configuration allows operation of the refrigeration system when the value for power is lowest. In turn, the refrigeration system is turned off during peak hours to minimize station load, and inlet chilling is accomplished with the stored energy. The size of the refrigeration equipment is optimized with TES, since it is allowed 18 hours of operation to store only 6 hours of chilling capacity. The method of TES with ice was evaluated against cold water and cold brine storage. The latent heat of fusion for ice, 333.8 kJ/kg, substantially reduces the mass required to store energy, providing a more compact and economical system. After ice storage design considerations, such as air space and circulating water space, are accounted for, the volume required to store a given amount of energy with ice is about one-fifth the volume required to store an equivalent amount of energy with liquid. With ice storage, chilled water can be consistently supplied to the inlet air coil at 0°C to 1.2°C, whereas if storing chilled liquid, the temperature of the liquid gradually increases during the six-hours-per-day use cycle, creating a transient heat transfer problem at the air inlet.

System design basis

Weather data was analyzed to design the system. The inlet-air chiller coil is designed to cool the inlet air for an average peak temperature condition of 35°C and 20% RH. This corresponds to a duty of 4.73 MW. Under these conditions, the refrigeration system will not totally replace the ice melted during a six-hour peak period overnight; however, the storage tank is sized for a five-day weekly chilling cycle. The ice will be nearly depleted by the end of the week, but by operating over the weekend, the tank can again be filled with ice for the following week's cycle. The system was not designed to fully accommodate extreme weather conditions of more than 37.7°C with greater than 15% relative humidity, which only periodically occur at the site. Under these conditions, the system may be operated to chill inlet air to a point above 5.5°C in order to maintain enough ice storage to last the week. The extra ice storage capacity, or refrigeration tonnage, necessary to accommodate extreme conditions was deemed too costly for optimum economic return on investment.

Another option, reducing the size of the refrigeration equipment to a capacity of 300–350 tons, was considered. With this scenario, an ice storage tank 1.7 times larger than that selected would be required. The savings in the size of the refrigeration equipment did not offset the cost of the increased storage. For this reason, as well as fear of ice/water distribution problems in a tank this size, this option was not selected.

Table 11.1 Plant performance data with inlet air cooling options.

Ambient conditions (35°C, 20% RH)	Evaporator cooler (85% efficiency)	Direct mechanical refrigeration	Absorption refrigeration	TES
Inlet air (°C)	21.1	5.5	10.5	5.5
Gross power (MW)	34.7	38.8	37.3	38.8
Station power (MW)	1.0	2.3	1.0	1.0
Net power (MW)	33.7	36.5	36.3	37.8
Heat rate (Btu/kWh)	11,190	10,900	10,980	10,900

Source: Hall *et al.* (1994).

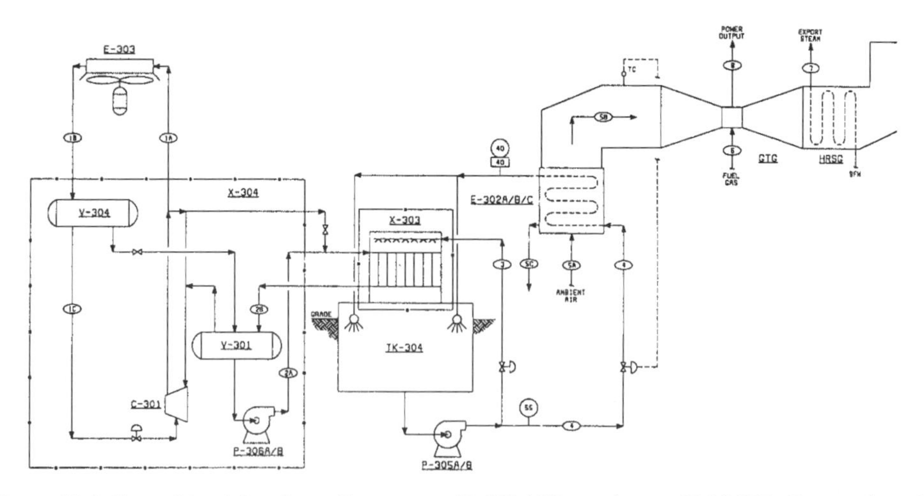

Figure 11.1 Gas turbine inlet air cooling system (E-303 NH_3 condenser, V-304HP pilot receiver, P-306A/b refrigerant pumps, X-303 evaporator package, TK-304 ice water storage tank, C-301 NH_3 compressor, X-304 refrigeration package, V-301 LP receiver, P-305A/B chilled water pumps, E-302A/B/C inlet air cooling coils) (Reprinted from Hall *et al.* (1994) by permission of ASHRAE).

Inlet air chiller coil design

The inlet air coils (Figure 11.1) were required to fit up to the existing inlet air-filter house for the gas turbine. This required a configuration that would add no new structural load, which could compromise the integrity of the existing structure. The design of the coils was also required to minimize the airflow pressure drop.

The resultant design in three sections of horizontal tube bundles, with each bundle having a tube length of 8.12 m. The bundles are aluminum finned tubes with a diameter of 3.175 cm; the fin diameter is 6.35 cm with five fins per cm. Tubes are arranged four deep in a triangular pitch with 176 tubes per bundle. Each bundle is set in a galvanized steel framework that mounts on a concrete ring wall built around the outside perimeter of the inlet housing. The coils are supported by the ring wall on the front and sides of the filter house. Aluminum sheet metal was installed along the back of the filter house and between the coil housings and the filter house to prevent airflow from bypassing the coils. The coil arrangement allowed for ease of installation and a minimum reconfiguration of the existing plant. Pressure loss across the coils is only 1.27 cm H_2O due to the low air velocity. There was concern about placing the coils upstream of the inlet air filters when chilling below the dew point, since this would expose the filters to possible carry-over of condensed water droplets from the coils. Carry-over has not proved to be a problem; condensate forms on the coils and falls down into the basin formed inside the ring wall and is drained away.

TES design

The ice harvester selected for this project utilizes flat inflated stainless steel plates for the evaporator surface. Cold ammonia is circulated through the annular space inside the plates, and water is circulated out of the tank and over the outside surface of each plate. Ice is formed on both sides of the plate. When the ice reaches 1 cm thickness, hot gas is injected directly from the discharge of the compressor into the plate. The hot gas breaks the bond between the ice and the plate and the ice falls off (is harvested) into the tank. As ice falls into the tank, it forms a mound similar to sand in an hourglass. The angle of the mound, called the angle of repose, for ice built on the ice harvester is 15°. The location and size of the ice opening relative to the tank-top area is critical. The evaporator plates for this application were positioned to minimize tank void volumes and optimize tank utilization. The size of the ice storage tank was determined based on the ice density of 916.22 kg/m^3 and the 560,926.08 kg of maximum ice storage required for the design condition. Assuming a 50% water/ice ratio below the water level and a 50% air/ice ratio above, the tank minimum volume requirement (*V*) was determined to be

$$V = 2\times(560{,}926.08/916.22) = \underline{1224.43\ \text{m}^3}$$

Additional space allowance was added for the ice mounding at the top of the tank. The final internal dimensions of the tank are 14 m long by 12 m wide by 8 m high. The ice storage tank is cast-in-place concrete. The tank is installed partially in the ground on a hillside. One end of the tank is exposed from the hillside; this is where circulating pumps are installed, avoiding the need for a pump vault below ground. The tank is initially filled with water to the 60% level. As ice is dropped into the tank it floats, with 91.7% of the ice below the water level and the remainder above. The water level remains constant in the initial stages of the charge cycle. When the ice level meets the bottom of the tank, the water level starts to drop. When the high ice level has been reached, the water is at the 20% level.

During the charging (ice-making) cycle, 0.126 m^3/s of water is pumped over the evaporator. During the discharge (melting) cycle, 0.126–0.252 m^3/s is pumped through the system as needed to meet the chilling demand. During the discharge cycle, when the chilled water feeds the turbine inlet air coil, the water bypasses the evaporator and flows directly into the tank. The 'warm' return water is distributed in the tank via a spray header. The spray header is mounted at the top of the tank and evenly distributes the water over the ice pile. This even distribution of return water over the ice is necessary to maintain a constant supply water temperature to the coil through the entire discharge cycle.

Refrigeration design

The refrigeration system is a pumped liquid overfeed system. Liquid refrigerant is pumped from the low-pressure receiver into the evaporator plates. The plates are overfed with refrigerant at a rate of 3:1. Both liquid and gas refrigerant is then returned from the plates back to the low-pressure receiver. Gas is drawn out of the top of the low-pressure receiver and into the compressor. The gas is compressed and discharged into an air-cooled condenser. The condensed liquid refrigerant then flows through a high-side float and back into the low-pressure receiver. A diagram of the system is given in Figure 11.1. This type of refrigeration system was chosen due to its simplicity, high reliability and low operating costs, as well as the fact that it avoids slugging in the compressor.

Due to the large size of the system, a single screw compressor was chosen over multiple reciprocating compressors. While in the ice-making mode, the compressor operates with a coefficient of performance (COP) of 2.8. Ammonia was chosen as the refrigerant due to the operators' familiarity with it (ammonia is stored on-site for a selective catalytic reduction system), and because it has environmentally benign qualities, including zero ozone depletion and greenhouse effect factors.

System operation

During the first three months of operation (June, July and August 1993), the chiller/TES system maintained the plant's net electrical output above the dedicated firm capacity level of 35.5 MW for 100% of peak hours, fully satisfying its intended purpose of capturing full bonus revenues for the plant. A sister plant, a cogeneration plant situated two miles away, is identical to this facility in every respect except that it has evaporative cooling instead of an inlet chilling/TES system. The sister plant experiences the same ambient air conditions, and thus provides an ideal yardstick by which to measure the benefit of inlet chilling at the original facility. During June, July, and August, the sister plant generated an average net output of 34.26 MW compared to the facility's average net output of 36.57 MW during peak hours. The inlet air temperature averaged 19.5°C at the sister plant with the evaporative cooler. The inlet air temperature averaged 10.5°C at this facility with the chiller/TES system. In future summer periods, the output of the sister plant will continue to degrade without the ability to make up the loss with chilled inlet air. The inlet temperature at this facility will be depressed further with the chiller/TES system, thus overcoming degradation in order to keep output above 35.5 MW.

Closing remarks

Inlet air chilling is a viable means to enhance turbine generator performance, provided revenues associated with the incremental output are cost-effective. This is particularly the case when hot summer weather conditions cause a peak power demand that raises generating capacity value to a premium, and at the same time inlet chilling can produce a significant performance improvement. TES provides a means to maximize chilled inlet air performance gains needed during peak hours by deferring the refrigeration parasitic load to night-time hours. TES allows optimized refrigeration equipment sizing, since it spreads chilling duty for weekday peak hours over night-times and weekends.

Further information is available elsewhere (Hall *et al.*, 1994).

11.2.3 A power generation plant, Gaseem, Saudi Arabia

This case study provides some insight into a relatively new application of industrial refrigeration technology. Significant benefits can be realized by electricity generating organizations employing gas turbine generators, particularly when located in regions of high ambient temperature, using the methods discussed here. Several systems are considered including some that use CTES.

Some motivations for this project follow:

- The Middle East countries (e.g. Saudi Arabia) have a rapidly developing economy, with ever increasing power use. There is presently a shortage of generating capacity to meet the peak demands of the summer period. Investment in new plants is expensive and the timescale to implement new plant is typically 18–24 months.
- Gas turbines are used for the majority of the power generation capacity throughout the Middle East and a large proportion of the world, particularly where gas or liquid fuels are readily available. These drivers are fixed-speed, constant inlet-air-volume machines, designed generally for steady-state operation. Whether operating on simple or combined cycles, their output power is directly linked to the rate of fuel that they can consume efficiently. This in turn is a function of air mass flow, which is a function of air density. It follows that if the air density can be increased, the turbine is likely to be able to burn more fuel and therefore produce more power.
- Due to the high electricity load for air conditioning in this region, the demand peaks between 12:00 and 17:00 hours. Outside these hours, particularly at night, demand drops considerably.
- At the time of the peak demand, the available capacity from the turbines is at its lowest, because of the high ambient temperatures (and corresponding low air density).

The profile of the electricity demand, particularly in the Middle East, introduces opportunities for economically attractive, innovative plant designs. Thermal Energy Storage Turbine-Inlet-Air Cooling (TESTIAC) offers such alternate design.

Cooling the combustion air to the turbines is beneficial because it increases the density and therefore increases the mass flow of air. This enables more power to be generated. Figures 11.2a and 11.2b illustrate the original process flow diagrams, and indicate the effect of turbine inlet air cooling on a typical gas turbine generating set, respectively.

Summer ambient design temperatures at the site under consideration are 50°C dry bulb, 10% RH. The site is at an altitude (relative to sea level) of 650 m, where the air density is approximately 1k g/m^3. If this air is cooled to 10°C, the density is increased to 1.15kg/m^3.

This increases the mass flow by 15%, and enables an increase in turbine output power of about 33%.

The improvement in output capacity has to be balanced against additional costs of the enhancements to the system. Normally, the costs of implementing TESTIAC to increase electricity production capacity are meritted if they are less than the costs of new turbine equipment. The development of capacity enhancement of gas turbine power generation by TESTIAC has been progressing for about 10 years. In North America, there are now approximately 10 plants in operation, of various sizes and designs, operating with varying degrees of success. The development process has stimulated many differing configurations, some of which are illustrated in Figures 11.3–11.5.

The simplest example (Figure 11.3a) is where the refrigeration system is matched to the instantaneous cooling requirement. This involves no TES, but requires the greatest refrigeration capacity. This arrangement may be appropriate when the requirement for capacity enhancement is relatively small and the peak load duration is a large proportion of the available hours. Such a system will ordinarily employ centrifugal compressor-chiller packages. These are more efficient, due to the small temperature differential across which they operate, and less expensive, per unit of cooling capacity. Because of the smaller water temperature differential available from this system, water flow rates and pumping costs are higher than with other systems.

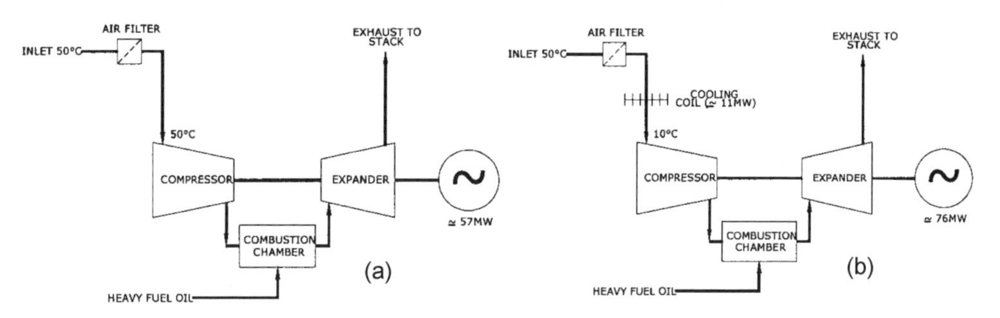

Figure 11.2 Gas turbine generator system (a) without precooling and (b) with precooling (*Courtesy of WS Atkins Consultants Limited*).

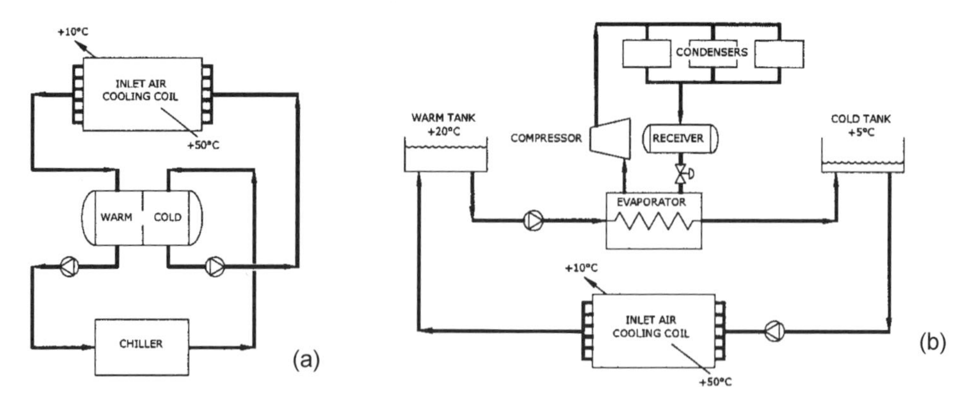

Figure 11.3 A simple on-line inlet air cooling system (a) without TES (b) with TES using chilled water tanks (*Courtesy of WS Atkins Consultants Limited*).

The next example (Figure 11.3b) utilizes chilled water storage. This method uses only the sensible heat in the storage fluid. The method can be economical for small to medium sized loads, but the storage volumes involved can increase the cost dramatically for larger applications. The disadvantages of the previous example also apply if packaged chillers are used.

The third example as shown in Figure 11.4 uses an ice CTES system. This system requires a more expensive refrigeration plant, due to the lower evaporating temperatures and specialized evaporator design configurations. The significant advantage is that the latent thermal storage capacity of the ice reduces the storage volume considerably, although this design requires extensive evaporator surface area to handle the necessary ice storage volume. An additional advantage with some ice storage systems is that very peak loads can be readily managed. This is usually only possible with flake/slurry/sheet ice systems that offer high surface area-to-volume ratios within the ice storage system. In practical applications, external or internal melt ice-on-coil systems cannot normally meet this requirement as effectively.

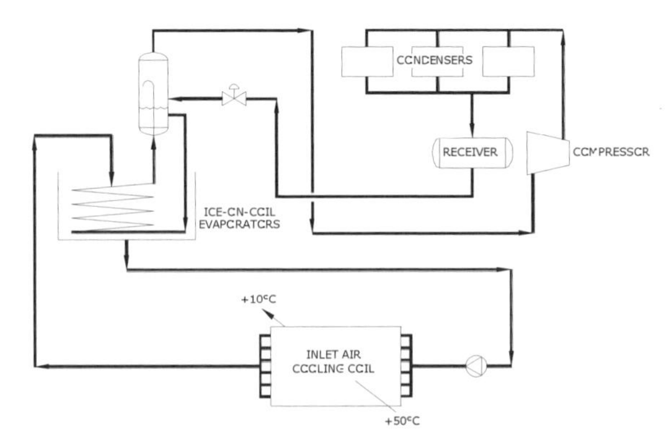

Figure 11.4 A simple on-line inlet air cooling system with TES using ice-on-coil system (*Courtesy of WS Atkins Consultants Limited*).

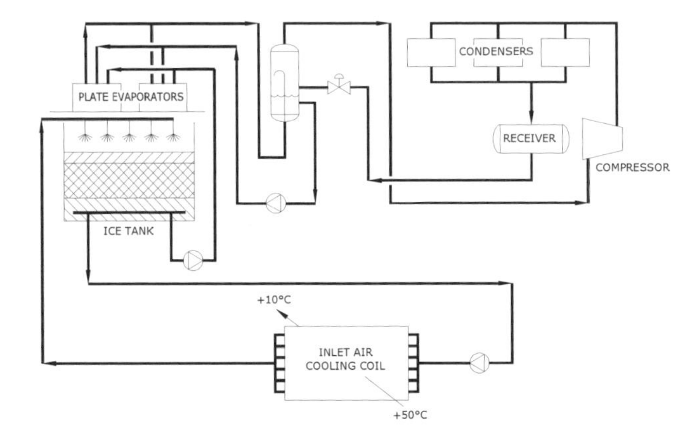

Figure 11.5 A simple on-line inlet air cooling system with TES using flake or slurry ice in tank (*Courtesy of WS Atkins Consultants Limited*).

Table 11.2 Comparison of three main types of TES systems for the project.

System	Storage volume (m^3)	Capital estimate (M£)	Daily energy use (MWh)
Chilled water	20,000	20	250
Ice-on-coil	10,000	18	350
Ice in tank	8,000	14	314

Source: Palmer (2000).

Figure 11.6 Gaseem TESTIAC system (*Courtesy of MARCO*).

In this project, many systems were evaluated. These included ice-on-coil, panel (sheet), slurry ice and various configurations of vacuum ice generators. In this particular evaluation, ice-on-coil systems were discounted as expensive, inflexible and difficult to monitor in terms of available/residual storage capacity. Slurry ice systems had many advantages, but due to their relatively short development and application time, were discounted by the client. They also normally require the use of a eutectic solution, which adds cost. Ice slurry systems generally also offer advantages of two-phase pumping to the process load, which significantly lowers the parasitic pumping load. Vacuum ice generators offer the potential for large, future installations, pending the development of suitable compressors. On balance, the client's preferred option was the sheet-ice evaporator system, as indicated in Figure 11.5. This system represents a reasonable compromise between the competing factors of efficiency, flexibility, cost and reliability. This design does, however, involve complex water and ice management, and also numerous solenoid valve actuations for the frequent defrosting of the plate packs. Another method, already operational in India, is to use the heat in the turbine exhaust gas to generate cooling via a lithium bromide (LiBr) absorption refrigeration plant. This approach would require that the system either be on-line, as Figure 11.3a, or use water storage, as Figure 11.3b, because the chilled temperature is limited to around 5°C. Such systems may also require an intermediate steam generator, because of control difficulties experienced to-date when applying the exhaust gases directly to the absorption plant generator.

A simple comparison of the main system types, relevant to this application, is indicated in Table 11.2.

The overall client specification was of a functional nature. Six new turbines were to be installed, each having a nominal capacity of 342 MW at the summer design ambient condition. The cooling plant had to produce an air temperature of 10°C off the inlet-air cooling coils, for the full 5-hour peak period, across all six turbines. This precooling would enable a full load capacity of 455 MW with an increase of 113 MW.

Basis of design

In order to enable the additional output power generation as specified by the gas turbine manufacturer, the inlet air-cooling system was designed on the following basis:

- Ambient dry bulb temperature: 50°C
- Ambient air humidity: 10% RH
- Air mass flow rate per turbine (excluding air cooler): 275 kg/s
- Number of turbines: 6
- Required turbine inlet air dry bulb temperature: 10°C
- Peak load period: 5 hours
- Ice re-generation period: 19 hours

The plant configuration, as shown in Figure 11.6, is an integrated ammonia refrigeration plant, with four screw compressors, multiple air-cooled condensers, four air-cooled oil coolers and eight plate-pack sheet-ice evaporators arranged for pumped circulation from a single suction accumulator and a flash economizer vessel. The high ambient temperatures and the shortage of a water source precluded the use of conventional thermosyphon oil cooling and evaporative condensers. Instead, remote air-cooled oil coolers and air-cooled condensers were used.

The evaporators are multiple vertical stainless steel plate evaporators, located on the roof of an ice storage tank. Water is passed over the plates by a sparge system that ensures an even film of water is applied to each side of each plate. During the ice-build period, some of this flow is frozen to the plate, which is maintained at approximately –8°C by the re-circulation of liquid ammonia from the suction accumulator. After a specified period of time, or on ice thickness measurement, a hot-gas defrost cycle is initiated. This injects a flow of ammonia discharge vapor into a section of the plate pack, melting a small proportion of the ice and causing it to fall directly into the storage tank below. The cycle repeats for each of the evaporators units until the ice inventory is replenished.

The tank is a reinforced concrete structure, measuring 34m×34m×13m high externally and initially containing over 8000 m^3 of water. The walls are 1200mm thick at the base. There are sleeved perforations through the wall base, to allow the pump suction connections to join the inner ring header. The tank is lined internally with a plastic membrane to protect the concrete and to reduce the risk of leakage.

Two sets of water circulation pumps are provided. Each of them draws from the specially designed header within the tank. This header regulates the flow of water within the tank, to ensure even distribution through the ice inventory. The water flow diagram shows that during the accumulation period, water is collected via the inner ring pipework and circulated over the evaporators. The disposition of the ice build-up within the tank is an important factor. If the ice store becomes unevenly distributed, water bypassing can result. This will prevent sufficient cooling contact between warm water and ice and cause over-temperature water to be circulated to the cooling coils. For this reason, during the accumulation period, circulation of water to the evaporators is controlled by motorized valves and sensors, ensuring that the ice within the tank is evenly distributed. This generation process continues over the 19-hour build period, or until the desired ice storage volume has been achieved. The refrigeration plant is then turned off. During the five-hour demand period, the second set of water pumps are run, also drawing from the inner ring and then circulating to the inlet air cooling coils. Flow to each turbine is controlled locally, to maintain the desired air temperature. Flow to the coils in general is regulated by control of the pumps and a bypass control valve. Return water is routed back to the ice tank and distributed through a matrix of nozzles, controlled by motorized valves to ensure even distribution over the ice within the tank. The warm return water passes through the broken sheet ice, melting it in the process of being cooled. The cooled water is then collected in the internal header and re-circulated.

The pipework system was comprehensively analyzed for stress during the design period, to determine the range of forces to which the foundations, support structures, pipework and fittings would be subjected, over the range of operational temperatures. The complete water and refrigerant pipework systems and all the necessary structural supports, anchor points and spring hangers, were designed on the basis of these analyses and site surveys.

The ammonia refrigerant is to be circulated by hermetic pumps from the suction accumulator. These operate constantly during the build cycle. Evaporated vapor is drawn off by the four screw compressor packages. These operate through a centralized controller to maintain a constant evaporating temperature by slide valve control and off cycling, as required by the system demand. The discharge vapor from these is condensed in the air-cooled condensers. Condensate is collected in a control level receiver, which modulates a control valve and feeds the liquid into the combined flash economizer/liquid storage vessel. Vapor from this is taken to the compressor economizer port and the liquid is fed to the suction accumulator to maintain a working level.

The refrigeration plant has been designed to work at the maximum average ambient condition when required, but also to take advantage of the diurnal swing in ambient temperature. Due to these temperature swings, more capacity is available at night than during the day. The parasitic power consumption is also reduced. Over the 19 hours of ice-build time, the maximum requirement accumulation of ice can be achieved.

The operating economics are less significant than the capital cost of such a plant. The cost of fuel (and hence electricity) is low, during off-peak periods. The benefit gained by increasing the net peak capacity is of great significance, whereas the cost of the energy expended in enabling it during the off-peak period is low. In this example, the energy cost of operating the plant to accumulate the ice capacity is about 304 MWh (16 MW for 19 hours) per day (during the peak ambient conditions). The additional energy cost of circulating the chilled water to the cooling coils during the peak 5-hour period (a direct parasitic cost), is about 10 MWh. The generating benefit gained during this period is about 565 MWh (113 MW for 5 hours). This illustrates that the net benefit of this system, during peak, is an increase in generating capacity of 555 MWh per day, equivalent to two additional turbines. Even over the whole day, the net benefit is 251 MWh. The plant began operation in June 2000.

Since this project was initiated, there have been continuing developments in the technologies that contribute to the TESTIAC concept. In particular, slurry-ice production and handling systems have been developed. Suitable alternative eutectic materials have also been developed that lower costs. Test rigs have been constructed from which the pumping characteristics of various ice concentrations of these materials can be determined. Investigations are continuing and, as a result, the probability of slurry ice displacing sheet ice systems is increasingly likely.

Further details can be obtained from Palmer (2000) and Abusaa (2000).

11.2.4 Channel Island power station, Darwin, Australia

A dramatic population increase in the city of Darwin, Australia's power demand since the early 1990s has been met by the introduction of an ammonia ice TES system at the city's Channel Island power station.

Motivation

The power station was built in 1985, and is operated by a Northern Territory Government utility. The electricity consumption of Darwin has been steadily monitored over the years by the Power and Water Authority (PAWA). An assessment in 1995 showed that demand was growing at a faster rate than earlier projections had identified, and that action was needed to avoid electricity shortages before 2000. The decision made to utilize an ice TES has not only eased the pressure, but also resulted in considerable savings for PAWA and the government. The TES system cost is one-third the installation cost of an extra turbine. In order to fully appreciate the actions taken, it is essential to study the background of the power station's previous capability.

Background

Channel Island has five gas-fired General Electric 'Frame 6' combustion turbines. Each turbine has a nominal capability of 40 MW when operating at 15°C air-inlet temperature. The gas is piped a distance of almost 1500 km from the city of Alice Springs, Australia. Two of the five gas turbines recover heat from the exhaust gas, passing it across heat exchangers to generate steam under pressure. The steam drives a steam-turbine generator, providing additional electrical power through this combined cycle operation. Darwin's power demand is dominated by air conditioning usage, which accounts for 70% of the peak load, a higher ratio than in cooler climates. In part, this high demand ratio can be attributed to population growth, increasing affluence, and more widespread use of air conditioning. Peak load customarily occurs around 2:00 p.m., when the ambient temperature (summer average 35°C is warmest, and the air-conditioning plant is greatest.

Power production challenges and options

A challenge to be addressed in the overall system is that gas turbine output is affected significantly by changes in climatic conditions. An increase in inlet air temperature results in an output drop of the turbine. This loss can be as high as 8 MW per turbine when electricity demand is peaking. Considerable studies by PAWA to solve these problems and meet adjusted anticipated demands identified to three major options. One option was for consumers to employ TES systems, allowing ice to build overnight. The daytime ice melt would provide the necessary cooling, and would avoid the need for a larger refrigeration plant. As a result, the Northern Territory University retrofitted some facilities to include TES. For PAWA, the second option was to purchase extra turbines to increase output. This cost was estimated at $40 million Australian dollars. The third option, ultimately chosen and successful in other hot climatic conditions, was to cool the inlet air to the turbines. This had the dual advantages of cost-effectiveness and quick installation and operation.

Description of the selected system

The final system chosen by PAWA employs an ammonia ice TES system plant (Figure 11.7), supplying ice water during peak periods to wet air coolers on the inlets of three turbines. The first stage of the project includes the following major equipment:

- Five 500 kW Sullair screw compressors, of 1500 kW (426 tons) capacity each.
- Five BAC Model CXV 435 evaporative condensers.
- 96 BAC ice-builder coils, capacity of building 1400 tons of ice that will provide 130,000 kWh (36,932 ton-hours) of storage.
- Three B.A.C. wet air cooler modules, cooling inlet air from 37°C to 9°C.
- Two concrete tanks 15 m×15 m×7 m high.

The system has simple operation. During the night and morning off-peak periods, the refrigeration plant builds ice on the ice builder coils. At peak demands (customarily from noon to 4:00 p.m.), ice water is pumped from the storage tanks and across a direct-contact heat exchanger medium to cool air before it enters the turbine. The ultimate cost of the installation was ~$12 million Australian dollars. The system was expected to be running by 1998, and to yield considerable financial savings by PAWA and a gain of 20 MW extra power production at peak periods. The ice-build period is off-peak, and has no effect on the daily maximum demand. Also, the control of ice build can be co-ordinated with spinning reserve needs. The wet air cooler provides extra benefits in washing the air, removing dust, insects and smoke. The dirt is collected in a sump that is easily accessible for filtration or cleaning. There are no finned coils to clog. The use of ammonia leads to an environmentally benign system that does not contribute to the greenhouse effect and has very little effect on the ozone layer. The use of natural gas system rather than a coal-fired station also reduced greenhouse gas production. A final benefit is that the ice TES system can be activated in minutes, allowing operating flexibility.

Figure 11.7 Evaporative condensers, plant room and ice TES system (*Courtesy of Baltimore Aircoil International N.V.*).

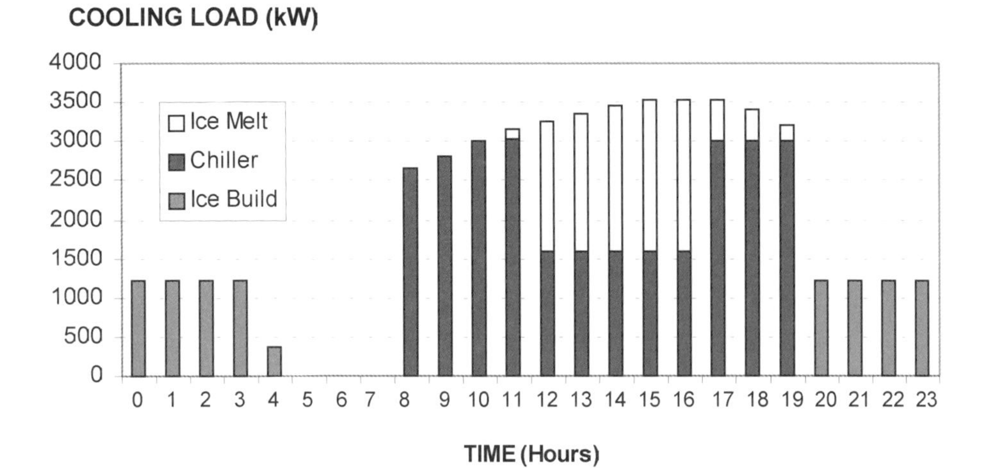

Figure 11.8 Daily building cooling requirements (*Courtesy of Baltimore Aircoil International N.V.*).

Further information is available elsewhere (BAC, 1999a).

11.2.5 The Abraj Atta'awuneya ice CTES project, Riyadh, Saudi Arabia

A state-of-the-art ice CTES system which has been designed of perform under all circumstances has been installed at the Abraj Atta'awuneya, which is an impressive example of contemporary architectural design. It comprises twin 17-storey triangular shaped towers joined together by two bridges. The building, located on a 7440 m^2 site, is the first high-rise building in Saudi Arabia. The architectural marvel provides 44,500 m^2 of office space, a 5200 m^2 business center and 600 car spaces.

TES was required because of the Saudi Consolidated Electric Company's (SCECO) new regulations regarding power supply for large retail areas and office buildings, limiting the power supply to the building to only 50% of the total connection capacity between 1:00 and 5:00 p.m. Therefore, building owners had to limit the power supply for chillers to 50% and have a system which is capable of handling the remaining 50% of the cooling load by other means. A well-accepted technique is the use of TES. With a TES system, it is possible to produce and store cooling capacity during periods of low cooling demands and when sufficient electrical power to operate the chillers is available during the night. The accumulated energy is then used during the next cooling cycle the following day.

Figure 11.8 shows the daily cooling requirements for the Abraj Atta'awuneya. Also shown on the diagram is the cooling delivered by the chillers during the day and the cooling to be delivered by the TES system, which should deliver daily 26 MWh. During night-time, the chillers operate and accumulate this cooling capacity. Various TES techniques were available on the market, and it was the task given by the owner to the designers to integrate a TES system which could meet the following design criteria:

- minimum occupied space,
- low initial cost,

- possiblility of using in high-rise buildings,
- maximum use of standard components,
- easy maintenance of the system and low operating cost,
- user friendly and full integration in the building management system, and
- reliable operation and thermal performance.

Detailed comparisons between different types of TES systems to fully assess their overall impact on the initial costs of the complete installation and future operating were made with the help of value engineering techniques.

- *Minimum occupied space*: two different TES techniques for cooling capacity are available: chilled water CTES and ice CTES. Chilled water CTES systems make use of the sensible cooling capacity of water. For normal air conditioning applications, the cool storage capacity of chilled water is limited to 8 kWh per m^3 of storage volume. Ice CTES systems make use of the latent storage capacity of water/ice. Therefore, ice CTES systems can accumulate about 48 kWh per m^3 of storage volume. Depending on the type of CTES system selected for the Abraj Atta'awuneya, the occupied space could vary from 600 m^3 for an ice CTES system, that is, a room of 20 by 10 m and 3 m high to 4000 m^3 for a system using chilled water storage, that is, a room of 20 m by 10 m and 20 m high. As space is costly in this high-quality type of building, the owner's preference went rather quickly to an ice CTES concept.
- *Low first cost*: the selected CTES system made full use of the full concrete basement which was used as the tanks for the thermal storage units. By doing this, factory assembled tanks could be eliminated and strong heavy-gauge hot dip galvanized steel coils were used. The low glycol temperature coming from ice CTES units allowed the designer to design the glycol loop with a maximum temperature difference-reducing glycol flows, pipe sizes and pumps. The chillers will first cool the warm glycol from the heat exchanger, which allows them to operate in the most favorable conditions.
- *Possible use in high-rise buildings*: all the mechanical HVAC equipment located in the building basement had to be able to withstand the pressures caused by the total static height in the system. Since the building is 21 floors high and the mechanical equipment was installed in the basement at minus 3 level, all equipment including the TES heat exchangers and TES units required a design pressure of 16 bar. This had to be considered when selecting the equipment for the thermal storage system.
- *Maximum use of standard components*: an important criterion was the maximum use of materials that are commonly used in air conditioning systems. This was done to keep the initial cost reasonable, but also to assure controllable maintenance and replacement costs in the future. The final selection included pumps, standard valves and control equipment of a standard range made of materials suitable for glycol/water mixtures. The chillers used in the CTES system are standard air-cooled packaged chillers.
- *Easy maintenance and low operating cost*: the aim was to have a system not requiring any additional special skills of the operators or maintenance staff. Ice CTES systems, making use of standard products can be maintained as any other piece of mechanical equipment. Chilled water storage systems, with their large water volumes, would have required expensive and complicated maintenance of the stored water and storage tanks.

The system design was to be executed in a manner that operating costs were at a lowest possible level.

- *User-friendly and full integration in the building management system*: to keep the complete TES system manageable a full automated control system for the TES system was needed. To assure a reliable communication between the building management system and the TES it was decided to fully integrate the TES system controls within the overall building management system. Uniform communication language and single source responsibility for the system control was obtained.
- *Reliable operation and thermal performance*: in a building of such a high standard, it was an absolute must that the air conditioning system should perform under all circumstances. Detailed analyses of the hourly cooling loads were made and the performance of every piece of equipment including the ice storage (Figure 11.9) was scrutinized.

The design had to be in such a way that all possible operating modes, including ice building at night and cooling, were possible. Also any potential risk for freezing the heat exchangers must be eliminated. The performance of the TES system on an hourly basis was fully guaranteed by the manufacturer. The owner made the manufacturer of the TES units, Baltimore Aircoil full system responsible for design, selection of all TES components and commissioning. This was to ensure that the performance for the complete TES system was guaranteed by a reputable manufacturer.

Detailed information can be found in BAC (1999b).

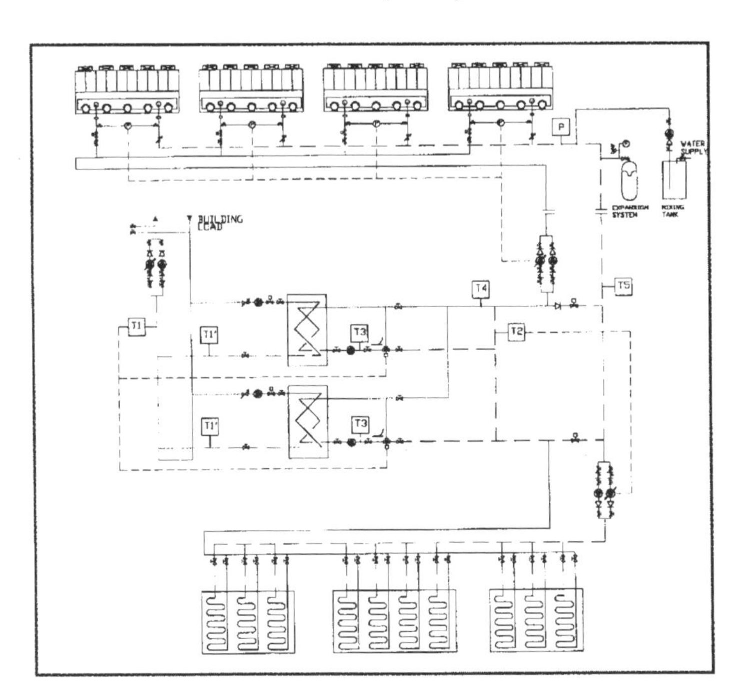

Figure 11.9 A schematic of ice CTES system (*Courtesy of Baltimore Aircoil International N.V.*).

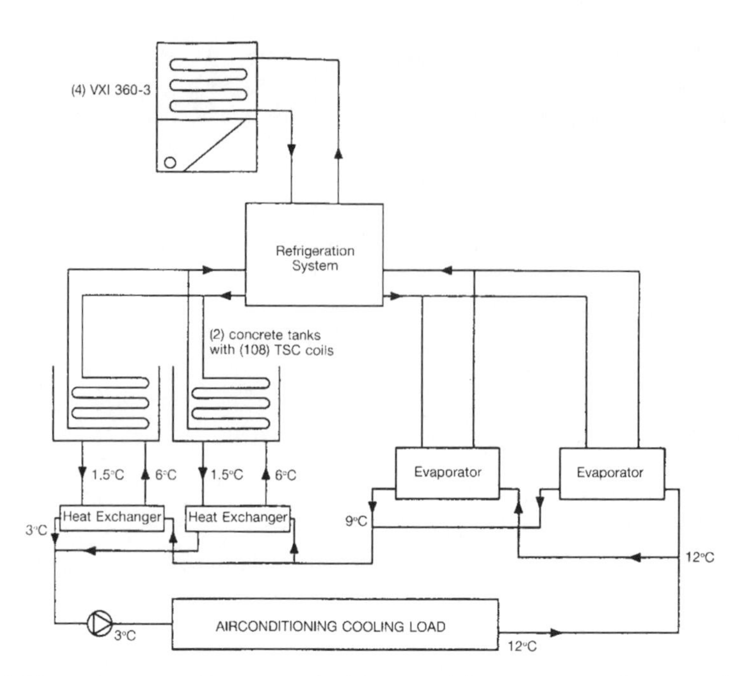

Figure 11.10 Thermal storage coils integrated with air conditioning system (*Courtesy of Baltimore Aircoil International N.V.*).

11.2.6 Alitalia's Headquarters Building, Rome, Italy

In June 1991, Alitalia, Italy's largest airline company, officially opened its new headquarters building in the south-west area of Rome. With a total plan area of 130,000 m^2, this building contains 55,000 m^2 of office space, 72,000 m^2 of service and technical rooms and 26 conference rooms, surrounded by a large parking area for more than 3000 cars. For the air conditioning and computer cooling needs of the entire complex, ice chiller thermal storage coils (Figure 11.10) were selected to meet a total storage capacity of 65,000 kWh. This makes Alitalia the largest ice storage installation in Europe, and one of the largest in the world. Due to the magnitude of this project, the designers opted to install thermal storage coils in two concrete tanks each 34 m long by 12 m wide. During the night, the central refrigeration system builds a complete charge of ice on the coils over a 12 hour off-peak period. During the daytime, the ice is melted to provide the building's base cooling requirements, thereby minimizing the peak demand. The refrigeration system is used during the day to provide any extra required cooling capacity upstream of the ice thermal storage coils. This results in higher refrigeration system efficiency, and lower operating costs. Unlike other types of thermal storage units, the ice chiller with consistent low temperature supply can be located downstream from the chillers for highest overall system efficiency. In addition to the thermal storage coils, four large industrial fluid coolers and six heat exchangers on this largest European thermal storage project were also provided.

Detailed information can be found in BAC (1999c).

11.3 Ice-slurry CTES Case Studies

11.3.1 The Stuart C. Siegel Center at Virginia Commonwealth University, Richmond, USA

During the summer/fall of 1998, a 380-ton ice slurry generating system was installed to cool the Stuart C. Siegel Center, a 190,000-square-foot basketball arena and athletic complex at Virginia Commonwealth University in Richmond, Virginia. The arena has a seating capacity of 7500 people and the total complex peak design cooling load is 1290 tons. Engineering and economic evaluations were undertaken and led to the decision to install the slurry CTES system. Ice slurry storage has two main characteristics which provide the potential for cost savings in HVAC projects. These characteristics are:

- *Ability to provide chilled water at temperatures as low as –1.1°C.* This characteristic reduces the size; and therefore, the first cost, of chilled-water distribution piping, air handlers, heat exchangers, and other components.
- *Ability to produce and store cooling when the cooling load is small and then use the stored cooling when the cooling load peaks.* This characteristic can reduce the peak electrical demand charges and can reduce first cost for many projects depending on the ratio of peak cooling load to average cooling load over a day or week.

These two characteristics played a part in the decision for ice thermal storage in the project discussed here. The chilled-water supply temperature is 1.1°C and the return temperature 11.6°C. The low temperature supply water resulted in downsizing of the air-side equipment and offered additional dehumidification via the variable-volume air-supply system. The design peak cooling load is 1290 tons and the peak 24-hour design load is 6776 ton-hours. An ice-slurry generator system of 380-ton capacity was selected to meet this peak load.

System schematic

Figure 11.11 illustrates the operation of an ice slurry generator and the application to this project. The slurry solution is 7% glycol and has a freeze point of –2.22°C. The solution is pumped from the bottom of the tank and delivered to the top of the slurry generator. The solution moves through the evaporator with an average temperature of –8.33°C which causes crystals to form in the solution as it falls into the tank. The slurry floats and therefore accumulates in the tank as this process continues. The slurry generators and storage tank are located about 30 m away from the building. The ice-slurry storage provides a 1.1°C solution to the building heat exchanger which, in turn, provides 2.22°C supply water to the building air handlers. The outside air units are equipped with variable-speed drives which operate to maintain the proper fresh air in the 7500 seat arena. Control of these outside-air units is based on levels of occupancy. The outside dampers are set at maximum volumetric flow rate for 'event' operation only. During normal occupancy, supply fans are ramped down to 30% speed and outside air dampers are reduced to 10% of full open capacity. During unoccupied periods and in the evenings, the air systems are shut down.

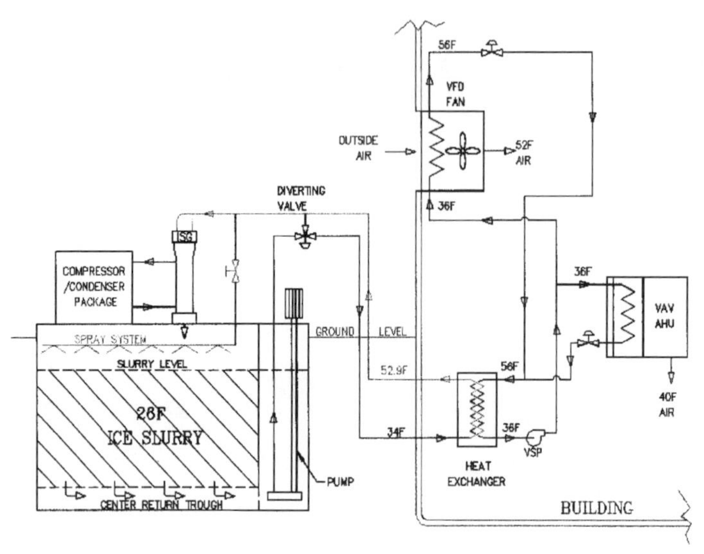

Figure 11.11 Ice-slurry system for the project (*Courtesy of Paul Mueller Company from Proceedings at the IDEA College-University Conference, New Orleans, February 1999*).

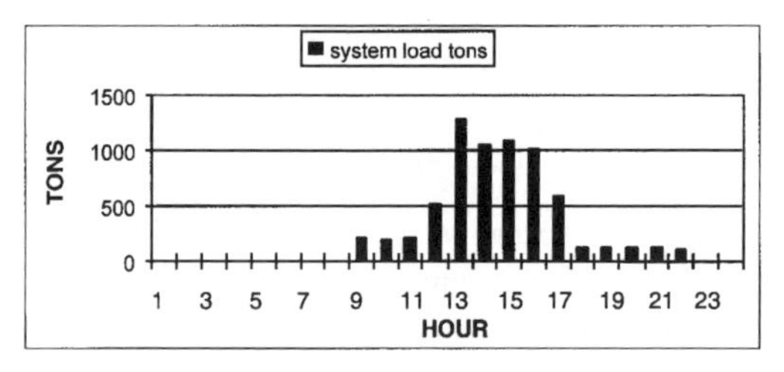

Figure 11.12 System loads (*Courtesy of Paul Mueller Company from Proceedings at the IDEA College-University Conference, New Orleans, February 1999*).

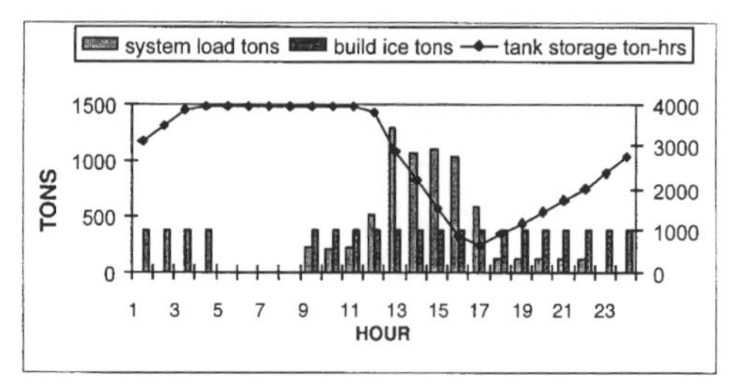

Figure 11.13 System loads during ice-slurry CTES operation (*Courtesy of Paul Mueller Company from Proceedings at the IDEA College-University Conference, New Orleans, February 1999*).

System operation

Figure 11.12 illustrates the design load which peaks at 1290 tons at hour 13. Most of this cooling load is due to the need for fresh air for 7500 occupants during a function in the arena. The timing of a major event is arbitrary. Since the total 24-hour load is 6776 ton-hours, it was evident that the load could be spread out over 24 hours with a CTES system. This system incorporates with a 380-ton slurry generator.

Figure 11.13 illustrates the operation of the slurry system. The tank has 3927 ton-hours of cooling capacity stored at 4:00 a.m., and holds that amount until noon when the cooling load exceeds the capacity of the slurry generator. Over the next five hours, slurry is melted to meet the cooling load, reducing the storage to about 640 ton-hours of cooling capacity at 17:00. The load is less than 380 tons at 18:00, and slurry begins to accumulate in the tank, which attains about 2700 ton-hours of cooling capacity by midnight. The ice generator shuts off at 4:00 a.m. when the tank again is full of ice slurry.

Economics

The ice system provides annual operating-cost savings of approximately $75,000 through reduced demand charges. Electrical demand is reduced due to:

- a 380-ton ice machine versus a 1290-ton chiller,
- reduced water pump sizes, and
- smaller fan motors.

The first cost of this ice slurry system was less than a conventional system because the 1.1°C supply water permitted the following:

- reduced duct/pipe and insulation sizes,
- reduced refrigeration capacity (and smaller air coils, water pumps, etc.),
- reduced motor sizes, and
- reduced electrical service size.

Compared to a conventional 6.66°C supply water, significant savings in first cost are realized. For this application, the ice-slurry TES system proved to be an attractive alternative providing both first- and operating-cost savings.

Further information on this project can be found elsewhere (Nelson *et al.*, 1999).

11.3.2 A slurry-ice rapid cooling system, Boston, UK

Slurry-ice is a crystallized water-based ice solution which can be pumped, and offers a secondary cooling medium for TES while remaining fluid enough to pump. It flows like conventional chilled water whilst providing 5–6 times the cooling capacity.

System description

The installed system consists of an 88 kW remote condensing slurry-ice machine and an associated 10 m^3 ice storage tank to satisfy a peak load of 180 kW (Figure 11.14). Ice machine is designed to operate until the tank is full or during rapid cooling periods in order to introduce ice into the cooling circuit. The stored energy over off-peak/low load periods is later utilized to satisfy the short and sharp peak loads. Hence, the installed refrigeration machinery is one-third of the equivalent capacity of a conventional direct cooling system. Harvested fresh vegetables are subject to rapid cooling within the cold storage facility whereby the energy stored during off-peak periods is recovered by circulating solution within the air spray heat exchanger in order to provide 0–1°C air off temperatures during the rapid cooling periods.

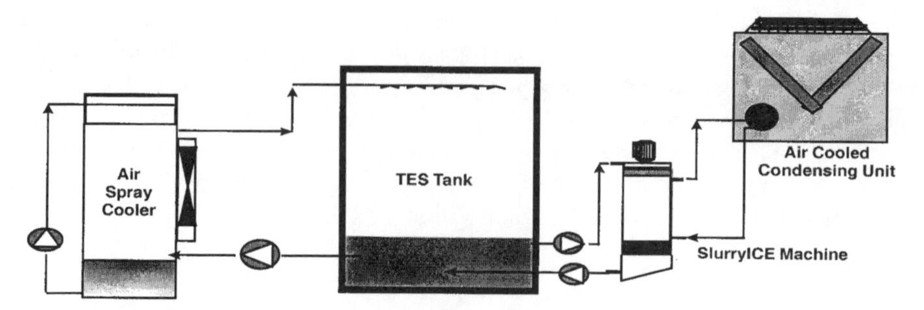

Figure 11.14 The slurry-ice system (*Courtesy of Environmental Process Systems Ltd*).

Technical benefits

- *Cost effective installation.* Smaller pipework, flexible storage tank coupled with smaller pump sizes result in lower initial installation cost.
- *Reduced running cost.* Reduced refrigeration machinery results in reduced maximum demand and availability charges coupled with night-time low ambient and off-peak electricity prices offers unmatched overall running cost savings.
- *Quick response.* Fine ice crystals offers unmatched thermal efficiency of the system. Hence, large peak loads can be handled without effecting the system leaving temperatures.
- *Flexible system.* Any future capacity or change of operational patterns can be easily handled without the need for additional refrigeration machinery.
- *Full stand-by capability.* The stored energy can be used to operate the process in case of breakdown or regular maintenance shutdowns.
- *Green solution.* Reduced refrigeration machinery leads to reduced refrigerant volume.

Further information can be obtained from EPS (2000).

11.4 Chilled Water CTES Case Studies

11.4.1 The Central Chilled-Water System at the University of North Carolina, Chapel Hill, USA

The Central Chilled-Water System (CCWS) at the University of North Carolina at Chapel Hill (UNC-CH) is responsible for chilling water and pumping it throughout the main campus, where it is used as a heat sink in air conditioning. Both electric and steam-driven chillers are used to chill the water. Thus electric and steam consumption represents a significant cost. Electricity is purchased from the Duke Power Company on its Hourly Pricing (HP) rate. Under this plan, the electric rate may change at the top of each hour, but remains constant through the hour. The rates for each day from Tuesday through Saturday are known at 4:00 p.m. on the previous day. Sunday and Monday rates are known on Friday. Steam is produced on campus at the University Cogeneration Facility at a constant rate based on fuel prices. The use of steam allows UNC-CH to generate its own electricity, reducing the amount purchased from Duke Power.

To reduce utility costs, a TES unit is to be installed. With this system, chilled water can be stored for later use, allowing the CCWS to take advantage of lower nightly HP rates and cooling loads. A study was undertaken to assess when and how much chilled water to store in the TES, and when the TES should be discharged in order to take advantage of the hourly HP rates while satisfying university demands for chilled water. The Department of Operations Research at UNC-CH carried out a study which is described here.

Purpose of the study

This study examines the use of a TES by developing a computer-based tool capable of producing optimal cooling strategies. An optimal cooling strategy is an hourly schedule indicating how the TES should be utilized (charged and discharged) in conjunction with existing equipment to satisfy cooling load at minimum overall cost to the university.

Determining such a strategy involves comparing the relative cooling costs with a TES and with chillers. The total cost of providing one ton of cooling via the TES during a particular hour, is the sum of the charging and discharging costs, where the former depends on the utility rates during the hour the TES is charged. The cost of supplying the same cooling ton through electric and absorption chillers depends only on the utility rates during that hour. An optimal strategy ensures the most cost-effective use of the TES.

The optimization tool may be used on a daily basis. In determining the strategies, the program accounts for all current system parameters and attributes, including:

- Each chiller has a maximum cooling capacity (given in tons). These capacities, along with pumping constraints, limit the rate at which the TES may be charged.
- Each absorption (steam-driven) chiller must be 'ramped' up to capacity. That is, the rate at which steam levels are changed cannot exceed an upper bound.
- Absorption chillers which have been off-line for an extended period must be primed before being turned back on. This leads to the practice of base loading. Absorbers are kept on-line, regardless of their cost effectiveness, to ensure the satisfaction of anticipated high demands.
- The cogeneration facility produces steam at a pressure higher than is needed by the absorption chillers. This excess pressure is used to generate electricity, reducing the amount purchased from Duke Power and altering the overall costs to the university.
- Parameters for the TES suggest that it should be fully charged (discharged) before being discharged (charged).

The tool is designed to aid CCWS plant operators in determining 32- and 110-hour cooling schedules (for weekdays and weekends, respectively) corresponding to periods of known HP rate schedules. However, it may be used over a different horizon provided that the HP rates and cooling loads for each hour are known (or predicted). Furthermore, the tool may be easily modified, allowing for system changes such as the addition or removal of chillers, parameter changes to the TES, and changes in cost calculations.

Methods and model development

The development of the optimization tool was divided into two steps. The first step involved finding a method of predicting cooling loads. This was done through linear regression on historical data provided by the CCWS. It was determined that using wet-bulb temperature predictions allows accurate predictions of cooling demand.

Table 11.3 Comparison of four different options for base loading.

Base loading options	1. Current system (No TES)	2. Industry operating procedure	3. Optimal cooling schedule	4. Two electric chillers
a. Current	1,017,766	999,978	910,710	971,185
b. Current less one	918,245	900,297	821,074	868,318
c. Zero	711,812	702,812	612,506	650,213

Source: UNC-CH (2000).

Given estimated demand and known HP rates, a model was developed of the system, which provides a quantitative means of determining optimal chilling strategies. The model uses an operations research technique called dynamic programming.

Analyses

The study examined and compared four options regarding a TES:

1. Operation under the current sytem (without a TES).
2. Operation with a 20,000 ton-hour capacity TES running under a standard industry operating procedure, as suggested by the client. Under this policy, the TES is charged and discharged at constant rates, starting at midnight and noon, respectively.
3. Operation with a 20,000 ton-hour capacity TES running according to the policies generated by the optimization tool.
4. As an alternative to a TES, the final option examines the effect of adding two high-efficiency electric chillers.

In addition, for each of the four scenarios above, the benefits were examined of reducing the number of base loaded absorbers. Three cases were studied:

a. Current base loading operations.
b. Reducing the number of base-loaded absorbers by the previous step.
c. Eliminating the need to base-load absorption chillers.

Findings

Based on 152 days of demand and corresponding HP data from the summer of 1996, the study found that the optimized operating procedure (TES Option 3) results in significant benefits over Options 1, 2 and 4. Furthermore, it was determined that by reducing the number of base loaded chillers, the CCWS can reduce its costs by up to 30%. Table 11.3 gives the total costs of operating under the different options.

Several results are worth noting:

- For each base loading option, the optimized operating procedure (TES Option 3) results in the lowest cost to the university.
- Given base loading Option 1, the optimized operating procedure represents a $107,056 saving over the current operation with no TES. For a 7% interest rate, the present value of the savings over five years is $469,677. (Note that this represents savings during the summer months only.)

- Given base loading Option 2, the optimized operating procedure represents a $97,171 savings over the current system. The present value of the savings over five years is $426,309.
- Given base loading Option 3, the optimized operating procedure represents a $98,992 savings over the current system. The present value of the savings over five years is $434,299.
- Combining TES Option 3 (optimization of the TES) with base loading option 3 (elimination of the need to base-load absorbers) results in a $1,017,766–$612,506 = $404,260 (or 39.8%) saving when compared to the current system and base-loading practices. Over five years, the savings are $1,777,961.

Closing remarks

Given the advantages of the optimized operating procedure (TES Scenario 3), it was recommended that the CCWS Plant Operators utilize the optimization tool on a daily basis. Following the schedules generated by the tool will result in substantial savings to the university. It was also recommended that the CCWS examine ways of minimizing the number of base loaded absorption chillers.

The model was developed under two primary restrictions that limit the scope of this study. First, the method of predicting demands was based only on summer data. Secondly, the TES was assumed to have no 'dedicated' chillers. That is, the load placed on the current system to charge the TES is divided equally among all operating chillers. A dedicated system involves separating the TES piping system from the main system. Thermal energy is transferred to the main piping system through a heat exchanger. This allows the TES to store colder water, though some energy is lost through heat transfer. Further study is needed to determine an optimal operating policy for the TES over non-summer months and with dedicated chillers.

Further information on this case study is available from UNC-CH (2000).

11.4.2 Chilled-Water CTES in a Trigeneration Project for a World Fair (EXPO'98), Lisbon, Portugal

This case study considers a trigeneration (simultaneous generation of electricity, heat and cold) project, which incorporates CTES, and which was constructed for the World Fair EXPO'98 in Lisbon, Portugal. The project was completed on a fast-tack basis in good time, the first distribution of chilled water being achieved nearly nine months before the opening of EXPO'98. The client was Parque EXPO'98 and contractors were Climaespço (Elyo, Climespaçe, Gaz de France & RAR Ambiante) and EIG (Entrepose, Ingérop and GTMH).

Technical details

The plant consists of a gas turbine, waste heat boiler, two absorption chillers, two compression chillers, chilled water storage and auxiliary equipment (Figure 11.15a and 11.15b). The exhaust gases from the gas turbine are reheated by direct combustion in the exhaust duct and then sent to the waste heat boiler to produce steam. An additional boiler produces the extra steam needed during the winter, and also acts as a stand-by unit.

Steam is supplied for following purposes:

- for use in a shell-tube exchanger in which water is heated from 65°C to 100°C, and
- for use in absorption chillers, providing the first stage of refrigeration from 12°C down to 8°C.

Compression chillers then reduce the chilled-water temperature from 8°C to 4°C. A closed circuit treated water system cools the chiller condensers and the machinery, and the treated water is cooled by river water through a heat exchanger.

Electricity produced by the gas turbine generator is partly utilized within the plant, with the excess being sold to the EDP, the Portuguese national electricity company.

Chilled water thermal storage is provided for peak demands. Pumps for chilled-water service (both chiller loop, and chilled-water distribution) have variable frequency drives.

The chilled and hot water are distributed through two distinct networks to over 40 users throughout the exhibition area, some as far distant as 5 km. Each user has its own plate exchanger and closed circuit system.

Chilled water CTES

The chilled water storage is undoubtedly the most significant and innovative concept of the Trigeneration plant. Table 11.4a describes the capacity and Table 11.4b the main equipment of the plant. Apart from its feature as a unique application in trigeneration, it is the largest stratified chilled water thermal storage to-date in Europe, having a thermal capacity of approximately 140 MWh (39,807 ton-hours).

With a diameter of 35 m and a height of 17 m, the bottom 6 m of which is below ground, it consists of a cylindrical reinforced concrete tank, cast in situ, having a wall thickness of 45 cm. It is coated internally with a sealant and painted externally. The steel roof is thermally insulated (Figure 11.15b).

The tank is equipped with instrumentation to monitor the system performance and current status. The actual vertical profile of the water temperature is measured by thermocouples. The 26 temperature sensors, at 60 cm intervals, allow the operator to monitor the thermal capacity and to confirm that stratification is maintained. The temperature profiles recorded during the plant operation, both for reduced capacity and for design capacity, have shown that the gravitational separation between the lower zone and warm upper zone appears to be as good as can be expected (for details, see Figures 11.16a and 11.16b).

Designed for both charge and discharge rates of 18 MWr (5118 tons), the storage depends on effective stratification. The chilled water is introduced and withdrawn near the tank bottom, and warm water is withdrawn and introduced from the just under the water surface by two piping diffusers. Each diffuser is composed of three octagonal rings of piping with calibrated, equally spaced holes along the top of the straight sections for the upper diffuser and along the bottom for the lower diffuser. During the design, a techno-economic comparison was also made between single-pipe and double-pipe diffusers (Table 11.5). Although double-pipe diffusers provided certain advantages, such as higher charge and discharge rates, single-pipe diffusers were selected because of their lower material and labor costs. The inlet Froude number is not well defined for single-pipe diffusers, whereas the Reynolds number does not depend upon single or double pipe, but rather on total length and flow rate. The single-pipe diffusers are designed to keep the inlet Reynolds number below 1200.

Table 11.4a Present and planned future capacity of the Trigeneration plant.

	Present capacity	Planned capacity
Electricity	5 MWe	5 MWe
Refrigeration (including TES)	40 MWr (11,373 tons)	60 MWr (17,060 tons)
Heat	23 MWth	44 MWth

Source: Dharmadhikari (2000), Dharmadhikari *et al.* (1999; 2000).

Table 11.4b Main equipment and their capacities of the Trigeneration plant.

Equipment	Main characteristics
Gas turbine	Solar Taurus 60 (5.2 MWe)
Waste heat boiler	12 MWth (41 MM BTU/h), 1000 kPa (145 psi) steam
Auxillary boiler	15.3 MWth (52.2 MM BTU/h), 1000 kPa (145 psi) steam
Absorption chillers	2×5.1 MWr (1450 tons) double effect (lithium bromide)
Compression chillers	2×5.8 MWr (1649 tons) ammonia screw compressors
Chilled-water storage	Stratified storage, 15,000 m^3 (4 million gallons) concrete cylindrical tank
	Inner diameter 35 m×17 m high (bottom 6 m below ground)
Chilled-water circuit	Capacity: 21.9 MWr (6230 tons)
	Planned extension: 42 MWr (11,945 tons)
	Chiller feed pumps: (2 + 1) 1190 m^3/h (5240 gpm), 270 kPa (39 psi)
	Variable frequency motors: 132 kWe
	Distribution pumps: (2 + 1) 2050 m^3/h (9027 gpm), 460 kPa (67 psi)
	Variable frequency motors: 355 kWe
Hot water circuit	Shell-tube heat exchanger: 11.5 MWth (39.2 MM BTU/h)
	Distribution pumps: (1 + 1) 555 m^3/h (2444 gpm), 600 kPa (87 psi)
	Variable-frequency motors: 160 kWe
Cooling-water circuit	4 plate exchangers
	Cooling water pumps: (2 + 1) 1750 m^3/h (7706 gpm), 250 kPa (36 psi)
	Fixed speed motors: 160 kWe
	River pumps: (2 + 1) 1830 m^3/h (8058 gpm), 380 kPa (55 psi)
	Fixed-speed motors: 260 kWe

Source: Dharmadhikari (2000), Dharmadhikari *et al.* (1999; 2000).

Table 11.5 Techno-economic comparison of diffusers.

Expo'98 Thermal Storage - Single-Pipe vs. Double-Pipe Diffusers							
		Single-Pipe			Double-Pipe		
Charge/discharge rate	MWr (tons)	20 (5687)			25 (7108)		
	m^3/h (gpm)	2156 (9494)			2725 (12,000)		
Octagon		Inner	Middle	Outer	Inner	Middle	Outer
Reynolds number	Re	1200	690	980	1510	870	1230
(flow per unit length over diffuser)							
Relative material and labor costs		100			120		

Source: Dharmadhikari (2000), Dharmadhikari *et al.* (1999; 2000).

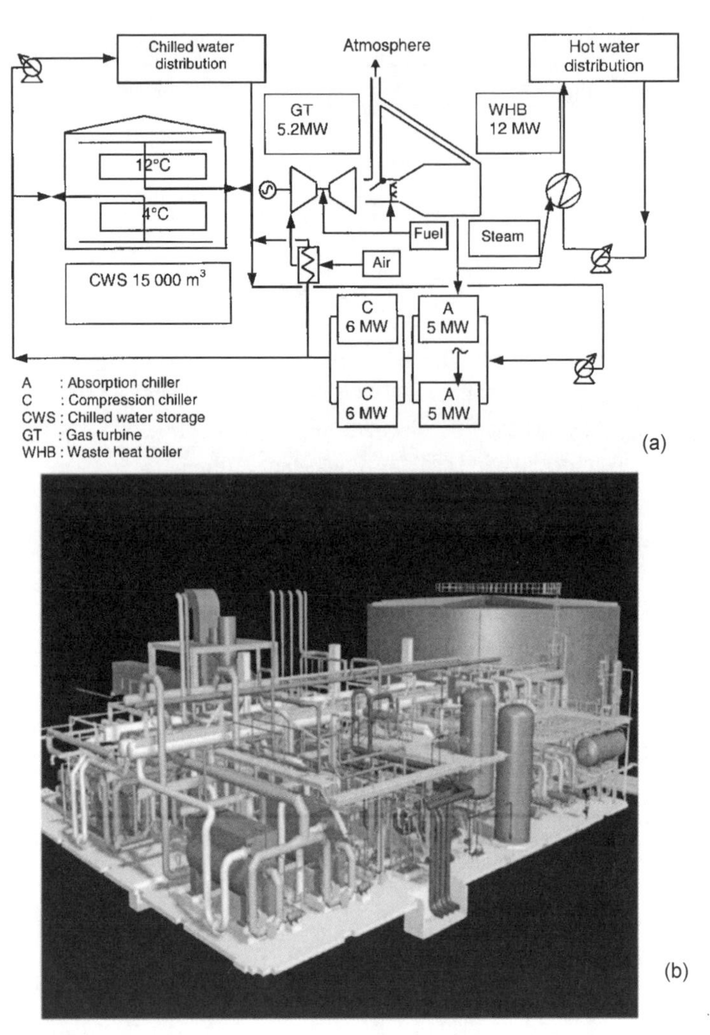

Figure 11.15 The EXPO'98 Trigeneration plant (a) its schematic and (b) its computerized view (*Courtesy of Paragon-Litwin*).

Cost saving achieved with TES

The cost of the initial phase of the entire trigeneration plant was approximately 40 million US$. The thermal storage considerably reduced the chiller plant capacity, which would otherwise have been required, giving an estimated capital cost saving of 2.5 million US$ (Table 11.6). Decreasing the chiller plant capacity also considerably reduced utility (electricity and steam) consumption and, consequently, loads on the power and steam generation. It resulted in savings in fuel consumption and also a reduction of pollutant emissions (CO_2 by 17,500 tons per year and NOx by 45 tons per year). Considering only the saving in fuel, the estimated reduction in annual operating cost is over 1.6 million US$. In these calculations, no credit is taken for chiller operation during the night in order to charge the thermal storage, which copes with the daytime peak demand (Figure 11.17).

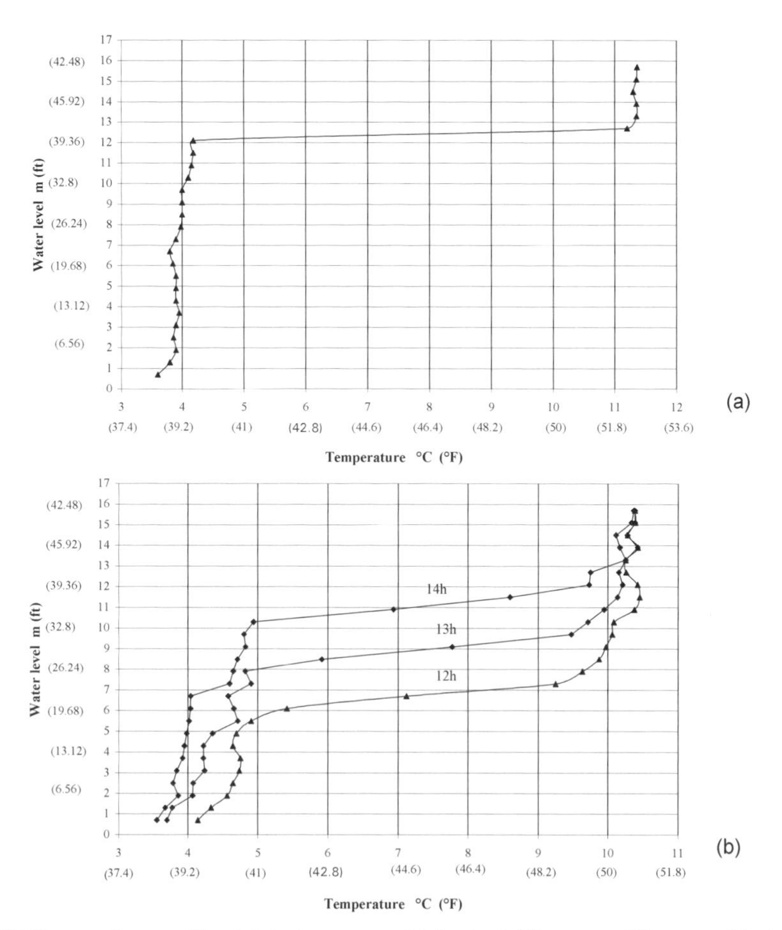

Figure 11.16 Temperature profiles (a) design case and (b) actual (*Courtesy of Paragon-Litwin*).

Other related innovations

Throughout the design, a refrigerant was sought that would not adversely effect the environment. Besides its inherent characteristics, including zero ozone depletion, ammonia has the advantages of high thermal capacity and high latent heat of vaporization. The ammonia charge was reduced by minimizing the size of the plate heat exchanger for both the condenser and evaporator, resulting in lighter and more compact chiller packages.

A trigeneration plant, by definition, includes a refrigeration product. Thus, a cold source for cooling the turbine inlet air is readily available. For the project, during hot weather, less than 1% of the chilled water is used for the gas-turbine inlet-air cooling, but this measures increases the power output by nearly 17% and reduces the specific fuel consumption by nearly 7%.

The chilled water requirement is met by four chillers, two absorption units and two compression units. All units are mounted in a series-parallel arrangement, i.e. two absorption chillers in parallel, connected in series with two compression chillers in parallel. A by-pass is also provided for each group to enhance operational flexibility. This arrangement not only reduced the investment cost, but increased the plant overall efficiency, and also flexibility by modulating its operating mode of the plant to energy demands.

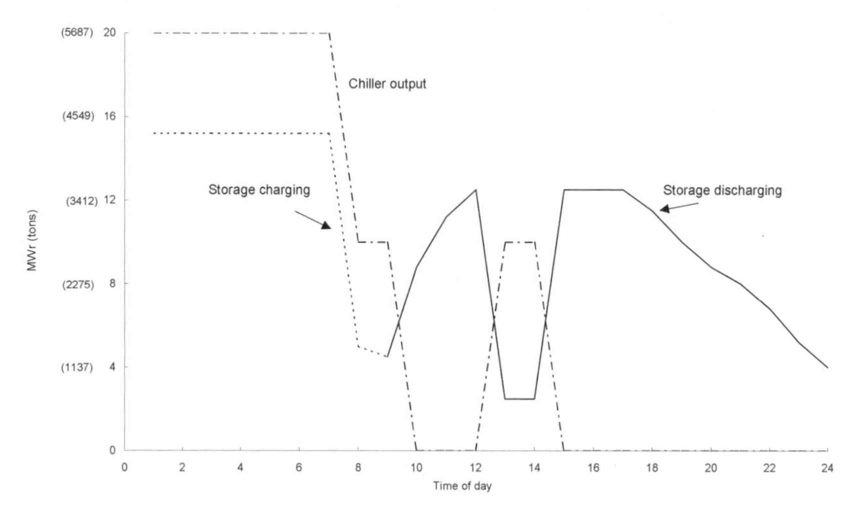

Figure 11.17 TES charging and discharging (*Courtesy of Paragon-Litwin*).

Table 11. 6 Equipment and capital costs with and without TES.

		With storage	Without storage	Equip. cost reduction (Thousand US$)
- Chillers	MWr (tons)	20 (5687)	40 (11,373)	1870
. Compression		2 x 5 MWr (1422)	4 x 5 MWr (1422)	
. Absorption		2 x 5 MWr (1422)	4 x 5 MWr (1422)	
- Chilled water storage	m^3 (million gallons)	15,000 (4)	-	(-920)
	MWh (ton-hour)	140 (39,807)	-	
- Gas turbine	MWe	5	8	970
- Waste heat boiler	MWth (MM Btu/h)	12 (41)	20 (68.3)	105
- Pumps				185
. Chiller feed pumps	m^3/h (gpm)	(2+1) x 1100 (4844)	(4+1) x 1100 (4844)	
. Cooling water (closed circuit)	m^3/h (gpm)	(2+1) x 1800 (7926)	(4+1) x 1800 (7926)	
. Cooling water (open circuit)	m^3/h (gpm)	(2+1) x 1800 (7926)	(4+1) x 1800 (7926)	
. Boiler feed water pumps	m^3/h (gpm)	(1+1) x 15 (66)	(1+1) x 30 (132)	
- Cooling water exchangers	MWth (MM Btu/h)	8.1 (27.6)	16.2 (55.2)	290
Equipment cost	million US $	5.6	8.1	
Operating cost (fuel gas)	million US $/year	1.84	3	
Cost reduction using storage				
- Investment	million US $	2.5	-	
- Operating cost (fuel gas)	million US $/year	1.16	-	
Reduction in CO_2 emission	tonnes/year	17,500	-	
Reduction in NOx emission	tonnes/year	45	-	

Source: Dharmadhikari (2000), Dharmadhikari *et al.* (1999; 2000).

Benefits from centralized system

A centralized trigeneration plant not only helped to reduce investment cost, but also contributed to a significant reduction in operating expenses. It achieved an energy saving of 45%, compared to a conventional system in which the chiller plant is installed in each building, resulting in an estimated annual saving of around 6000 TOE (ton oil equivalent). Also, it completely eliminated use of CFC/HCFC refrigerants in individual buildings, and resulted in a reduction of environment pollutants (including CO_2) by 20,000 tons per year, corresponding to a reduction of more than 54%. Furthermore, there are additional substantial savings in costs related to civil engineering, machine maintenance, etc.

All connected buildings also receive numerous other benefits, including suppression of noise nuisance, fewer maintenance staff, more space available in the building for rental or sale purposes, and total absence of exhaust gas ducting and chimneys.

Detailed information on this case study can be obtained from Dharmadhikari (2000), Dharmadhikari *et al.* (1999; 2000).

11.4.3 TES at a Federal Facility, Dallas, USA

The Dallas, Texas Veterans Administration (VA) Medical Center was the first VA medical facility in the USA to use TES technology to reduce operating costs. A partnership with Texas Utilities Electric Company (TU Electric) made implementing this technology possible. The center now shifts a significant portion of its energy demand away from the peak cost period to a lower cost period. Supporting this technology benefits the utility because it relieves pressure to construct new, increasingly more costly power plants. Shifting a large energy demand away from the peak period enables the utility to optimize the use of its generating plants. Thus, TES saves money for both the client and the utility.

Background

Like most large commercial utility customers, the VA Medical Center's annual electrical costs were based on a rate that is 80% of the highest demand charged during the year. If the facility could find a way to lower the demand, it could reduce its overall annual expense for electrical service. The system selected to reduce demand was a stratified chilled-water storage tank (Figure 11.18). The tank provides 24,628 ton-hours of thermal storage. In the summer of 1997 thermal storage reduced peak demand by 2934 kW, cutting annual electricity costs by $223,650. Installing the system also allowed the utility to reduce its costs by getting more use out of its existing generating plants. The need to build new generating capacity lessened considerably. The hospital can meet cooling requirements by producing chilled water during the period of lowest electric cost, then storing it for use during periods of high electric cost.

The project

On February 6, 1996, the center contracted with TU Electric for financial support of the project. It was agreed that TU Electric would build the tank, finance the cost, and let the center pay for the thermal storage on its electric bill. The utilities provided $500,000 of the total cost of $2.2 million required for design and installation. Savings resulting from installation of the thermal storage technology will allow the VA to recoup its investment within seven years.

Figure 11.18 The VA Medical Center's 12,491 m^3 stratified water storage tank (DOE, 1999).

Initially, a feasibility study was conducted on how to add thermal storage to the existing central plant. This study offered six alternate solutions, each with estimated construction costs, maintenance costs, utility savings, and lifecycle cost analysis. The partners selected a stratified water thermal storage system with a 12,491 m^3 tank at its core, which has 3000 tons of chiller capacity for eight hours. (The tank is considerably larger than an olympic-size swimming pool.) The system became operational in the fall of 1996.

The storage tank is filled with chilled water every night. Then every day, from noon until 8:00 p.m., the hospital draws its chilled water from the tank to meet its cooling load, instead of requiring continuous electric service to operate on-site chillers.

As an added benefit, the TES system also doubles the capacity of the hospital's central cooling plant. If the hospital expands, the VA could use the thermal storage system and its existing cooling system at the same time, avoiding additional capital investment.

Benefits of utility contracting

Although TU Electric was involved in about 200 thermal storage systems before this project, this was its first project at a VA facility. It was also TU Electric's first partnership arrangement in which the utility agreed to design, build, and help find third-party financing for a thermal storage system. The authority for this partnership was existing Federal legislation that allows government to enter into sole-source arrangements with utility companies for services generally available to other customers.

Some important lessons were learned through the project:

- *Participant education.* The Department of Energy's Federal Energy Management Program (FEMP) worked closely with both parties to facilitate this project. A knowledgeable FEMP liaison staff member provided necessary training and support, greatly assisting this project.
- *Quality subcontractor selection.* To improve the project, the utility chose to serve only as the project's prime contractor. TU Electric retained a professional architectural and engineering firm, Carter & Burgess, Inc., of Ft. Worth, Texas, to provide design, architecture, and construction management services.

Detailed information on this case and its benefits can be obtained from DOE (1999).

11.5 PCM Used CTES Case Studies

11.5.1 Minato Mirai 21, Yokohama

Minato Mirai 21 (MM 21) is a new urban area in Yokohama. Mitsubishi Petrochemical Engineering Co. chose a District Heating and Cooling (DHC) system to provide heating and cooling to the whole district. Use of DHC is a new Japanese policy for protecting the environment. MM 21 DHC includes a Cristopia Energy Systems (STL latent CTES) manufactured by Mitsubishi under Cristopia licence. This STL (one of the biggest latent TESs in the world), as shown in Figure 11.25, is the basic component of the system and permits efficient energy management.

Technical data

- Cooling energy stored: 120,000 kWh
- Storage volume: 2200 m^3
- Storage temperature: 0°C
- Nodule type: S-00
- Number of tanks: 2 (vertical)
- Tank height: 28 m
- Tank diameter: 7.3 m
- Tank volume: 1100 m^3

Characteristics

The thermal storage is a Cristopia STL-00-2200, which is composed of two 1100 m^3 vertical tanks filled with S-00 nodules. The nodules are filled with PCM allowing for thermal phase transition at 0°C. 120,000 kWh of cooling energy are stored daily in two 1100 m^3 vertical tanks. The cooling energy is provided by two centrifugal chillers. The height of the tanks is equivalent to a nine storey building.

Technical advantages

The use of CTES in the DHC system yields several technical benefits:

- Smaller chiller capacity.
- Smaller heat rejection from the plant.
- Reduced maintenance.
- Increased system efficiency.
- Increased system reliability.
- Reduced electrical installation.

Financial advantages

The use of CTES also provides significant economic benefits such as lower initial investment and high savings on electricity costs.

Further information on this project can be obtained from Cristopia (2001a).

Figure 11.19 CTES system located on the roof of the technical building in Minato Mirai 21 (*Courtesy of Cristopia Energy Systems*).

11.5.2 Harp Brewery, Dundalk, Ireland

Harp Brewery in Dundalk, Ireland (Figure 11.20a), produces 1,000,000 hectoliters of lager per year and consumes 12,000,000 kWh of electricity (4,000,000 kWh for the refrigerant plant). In 1992 Harp decided to modernize its refrigeration plant.

The existing system consisted of a wide range of process cooling loads satisfied by two circuits, a high-temperature circuit at –8°C and a low-temperature circuit at –12°C. During the design stage it was decided to install a Cristopia TES System (STL) on the high-temperature circuit. The STL reduces the instantaneous peak refrigeration demand by as much as 45% by shifting daily 10,000 kWh from day to night. The STL has reduced the maximum demand by 1 MW and the cooling consumption by 16%.

Technical data

- Daily cooling energy consumption: 44,000 kWh
- Maximum cooling demand: 4000 kW
- Cooling energy stored: 10,000 kWh
- Storage temperature: –10.4°C
- Store type: STL - N10 - 200 (Figure 11.20b)
- Nodules type: SN.10
- Number of tanks: 2
- Tank diameter: 3 m
- Tank height: 14 m

Characteristics

To improve the system efficiency and to reduce electrical cost, it was proposed to install a Cristopia STL-N10-200 TES system to shift electrical consumption to off-peak periods and to reduce the peak period electrical demand. To achieve this, a STL-N10-200 was specified storing 10,000 kWh at –10°C during the night-time off peak tariff period. The STL also reduces day-time chiller power by 1000 kWe.

Figure 11.20 (a) Two STL-N10-100 systems, and (b) Harp Brewery in Dundalk (*Courtesy of Cristopia Energy Systems*).

The STL-N10-200 is composed of two 100 m³ vertical tanks filled with SN.10 (77 mm diameter) nodules. The nodules are filled with a PCM allowing thermal storage at −10°C.

Technical advantages

- Smaller chiller capacity.
- Stable flow temperature.
- Electrical supply reduced by 1000 kWe.
- Refrigerant efficiency improved.
- Servicing simplified.
- Back-up at disposal.
- Smaller cooling towers.

Financial advantages

- The system with STL costs less than a traditional system with chillers.
- Saving on maintenance costs.
- Electrical supply reduced by 1000 kWe.
- Saving on demand charge and energy cost.

Environmental advantages

- Refrigerant charge reduced (leading to ozone layer protection).
- Emission of CO_2, SO_2 and N_2O reduced (avoiding contributions to the greenhouse effect and other environmental impacts).

Further information on this project can be obtained from Cristopia (2001b).

11.5.3 Korean Development Bank, Seoul

An office building, occupied by the Korean Development Bank (KDB), is located at Yoido island in the business center of Seoul, South Korea. It is the computer center for Korean and overseas activities. For this building, the KDB chose to use advanced technologies having high performance. The KDB decided to install a Cristopia TES System using STL to reduce nearly 50% of the cooling demand and to make system operation more reliable. In addition to the technical advantages (better back-up, reduced maintenance, higher efficiency and reliability) the KDB anticipated significant operating and maintenance cost savings.

Technical data

- Daily cooling energy consumption: 18,132 kWh
- Maximum cooling demand: 1948 kW
- Energy stored: 10,100 kWh
- Store type: STL - C -184
- Storage temperature: 0°C
- Nodule type: C - 00
- Number of tanks: 4
- Tank diameter: 3.6 m
- Tank length: 5 m

Characteristics

The STL system is composed of four 46 m^3 horizontal tanks for a total storage of 10,100 kWh. This capacity represents 55% of the daily cooling consumption, and allows for financial incentives from KEPCO (Korean Electricity Power Co.). Two 566 kW chillers with reciprocating compressors are used in the direct cold production and in the charge mode of the TES system. The control and energy management are achieved using a sophisticated BMS (Building Management System) to ensure the highest operating cost savings.

Technical advantages

- Smaller chiller capacity (reduced by 47%).
- Smaller cooling towers.
- Simplified servicing.
- Back-up at disposal.
- Reduced electrical installation.

Financial advantages

- Lower initial investment.
- KEPCO incentive for load shifting.
- Large savings on electricity costs (energy cost and demand charge).

Environmental advantages

- Use of refrigerant reduced (leading to ozone layer protection).
- Emissions of CO_2, SO_2 and N_2O reduced due to off peak consumption (mitigating greenhouse-effect and other pollution contributions).

Further information on this project can be obtained from Cristopia (2001c).

11.5.4 Museum of Sciences and Industry, La Villette, France

At the site of the old Paris slaughterhouse in La Villette park, France a modern leisure center was built in 1983 including (i) the sciences & industry museum, (ii) an audio-visual center, and (iii) the GEODE 3-D cinema. The center (Figure 11.21a) is one of the largest science and technology centres in the world, and one of the most innovative (150,000 m² of displays, shows, exhibitions). A cooling and heating plant supplies air conditioning to a distribution network. This plant includes an STL-00-550 (storing 31 MWh) and three 2700 kW chillers.

Technical data

• Main AC system type:	Air handing unit
• Building volume:	1,350,000 m³
• Daily cooling energy consumption:	163,000 kWh
• Maximum Cooling Demand	12,500 kW
• Chiller Capacity:	
- Direct mode:	8100 kW (6/12°C)
- Charge mode:	4400 kW (-6/-2°C)
• Energy stored	31,750 kWh
• Volume	549 m³
• Store type	STL - 00 - 550 (11.21b)
• Number of tanks	3
• Tank diameter	4 m
• Tank height	18 m

Characteristics

The STL system is used to reduce 50% of the peak cooling demand. The storage is charged between 22:00 and 7:00 when during electricity rates are low. Since 1985, significant operating cost savings have been achieved.

(a)

(b)

Figure 11.21 (a) The center, and (b) the TES tanks (*Courtesy of Cristopia Energy Systems*).

Figure 11.22 The STL system (*Courtesy of Cristopia Energy Systems*).

Technical advantages

- Smaller chiller capacity.
- Smaller heat rejection plant.
- Reduced maintenance.
- System efficiency and reliability increased.
- Increased plant life-time expectancy.

Environmental advantages

- Refrigerant charge reduced.
- Emission of CO_2, SO_2 and N_2O reduced.

Further information on this project can be obtained from Cristopia (2001d).

11.5.5 Rueil Malmaison Central Kitchen, France

The Rueil Malmaison Central Kitchen is a new modern kitchen using the latest technologies for producing meals and for managing the production process. It supplies 12,000 meals daily to schools in Rueil Malmaison (Parisian suburbs). To meet food hygienic standards the meals are rapidly cooled leading to high cooling peaks during a relatively short time. During the design stage, it appeared that the use of the STL latent storage had the potential to reduce the system capital cost compared to a classical installation without storage. An STL (Figure 11.22) was used and sized to reduce more than 80% of the peak electrical demand. In addition, the STL permits efficient energy management with a very high reliability and yields savings on operating costs.

Technical data

- Daily cooling energy consumption: 440 kWh
- Maximum cooling demand: 139 kW at 1/7°C
- Energy stored: 440 kWh
- Storage temperature: –10.4°C

- Store type: STL-N10-10
- Nodules type: SN.10
- Number of tanks: 1
- Tank diameter: 1.6 m
- Tank length: 5.2 m

Characteristics
The cooking is done in autoclaves, cooking pots or streaming water systems at temperatures between 65°C and 95°C. The meals are packed in small baskets under vacuum. An initial cooling down to 20°C is achieved with waste water, then a second cooling down to 2°C is achieved with the STL. The meals are then stored in cold rooms. The rapid cooling of food for hygienic reasons introduces high load peaks into the cooling profile. The use of latent CTES helps reduce these peaks.

Technical advantages
- Smaller chiller capacity.
- Smaller heat rejection plant.
- Flexible system available for efficient energy management.
- Servicing simplified.
- Back-up at disposal.

Financial advantages
The STL system costs less than a more conventional system with chillers, but not TES. Also, the use of CTES leads to lower operating costs (because of reduced demand charges and off-peak electricity consumption) and financial savings associated with the plant's efficiency, life of plant, and maintenance.

Environmental advantages
- Use of refrigerant reduced.
- Emission of CO_2, SO_2 and N_2O reduced due to off peak consumption.
- Primary fuel consumption of generator power plant reduced.

Further information on this project can be obtained from Cristopia (2001e).

11.5.6 The Bangsar District Cooling Plant, Malaysia

The district cooling system supplies chilled water to a district that comprises: the Cygal Hotel and the Cygal Towers A&B, the Atlas Towers A to F, Menara Telekom and Wisma Telekom, and Tenaga Head Quarters (TNB) (Figure 11.23a).

Objective
To take advantage of the lower electricity tariff during the night, the Cristopia TES System (STL) (Figure 11.23b) is used for storing cold during the night for use during the day. It has enabled the client to greatly reduce the installed cooling capacity and to increase the plant efficiency.

Figure 11.23 (a) The Bangsar district cooling plant, and (b) the STL system (*Courtesy of Cristopia Energy Systems*).

Technical data

- Daily cooling energy consumption: 450,000 kWh
- Maximum cooling demand: 40,000 kW
- Cooling energy stored: 110,000 kWh
- STL storage volume: 1900 m^3
- Number of tanks: 5

Characteristics

The plant consists of five centrifugal chillers (3500 kW each) working in conjunction with five cylindrical STL steel tanks of 380 m^3 (3.80 m diameter, 35 m long). Two conventional water chillers are used for the base load. Each brine chiller operates with one STL and one heat exchanger to provide brine at 3.3°C at the primary side of the heat exchanger. Each of the five loops operates independently of the others. The chillers and the STL's can be operated singularly and separately or in any combination to meet the demand, and the decision for their operating status during the day is based on the objective of minimizing the use of the chillers and depleting the energy stored.

Technical advantages

- Smaller chiller capacity.
- Smaller heat rejection plant.
- Reduced maintenance.
- Efficient and reliable system.
- Increase of the plant lifetime.
- Flexible system available for efficient energy management.

Financial advantages

- Saving on operating costs (24%), maintenance, demand charge and off-peak consumption.
- Lower initial investment.

Further information on this project can be obtained from Cristopia (2001f).

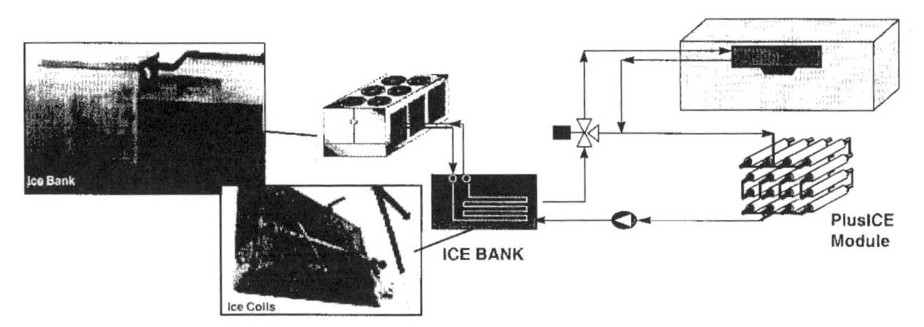

Figure 11.24 TES system using PlusIce modules (*Courtesy of Environmental Process Systems Ltd.*).

11.5.7 Dairy TES application using eutectic solutions, Dorset, UK

The original system (Figure 11.24) consisted of an ice bank and room cooler circulation system. Following a yoghurt incubation process around 45°C, the warm return water penetrated an ice block, and therefore ice bank response was very poor. Hence, it took between six and eight hours to cool the product. A combination of 7°C and 10°C PlusIce modules were applied in a two-phase program on the return leg of the system as a thermal buffer being charged by the ice bank over night. During the rapid cooling process the stored energy in the PlusIce beams is released back to the return leg, reducing the return temperature back to the ice bank considerably. As a result of this buffer action, the product cooling period is reduced by half, and therefore the client doubled their production capacity with an estimated annual saving of 30%.

Benefits

- *Maintenance free installation.* PlusIce has no moving parts, and therefore it is a virtually maintenance free concept.
- *Reduced running cost.* Reduced refrigeration machinery results in reduced maximum and availability charges, coupled with night-time low ambient and off-peak electricity rates, offers unmatched overall running cost savings.
- *Quick response.* Large peak loads can be handled without affecting the system leaving temperatures.
- *Stand-by capability.* The stored energy can be used to operate the process in case of breakdown or regular maintenance shut downs.
- *Flexible system.* Any future capacity or change of operational patterns can be easily handled by simply adding more beams.
- *Green solution.* PlusIce using salts has no environmental impact.

Detailed information is available in EPS (2000).

11.6 PCM Used Latent TES for Heating Case Studies

11.6.1 Solar Power Tower in Sandia National Laboratories, Albuquerque, USA

Solar power towers use solar radiation to heat the working fluid of an electricity generation cycle to high-temperatures (see Figure 11.25).

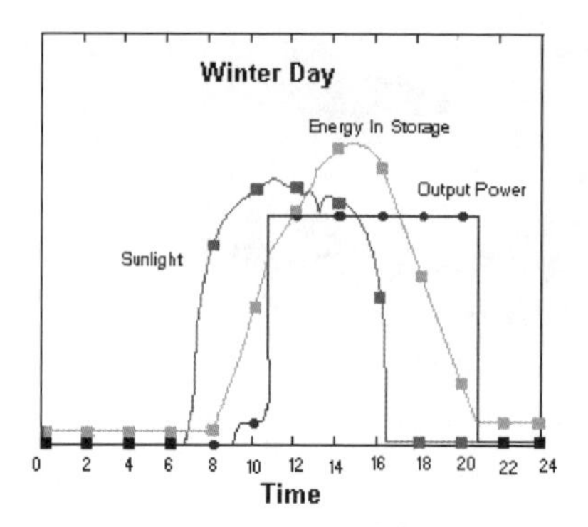

Figure 11.25 Capacity profiles of the solar power tower (Mahoney, 2000).

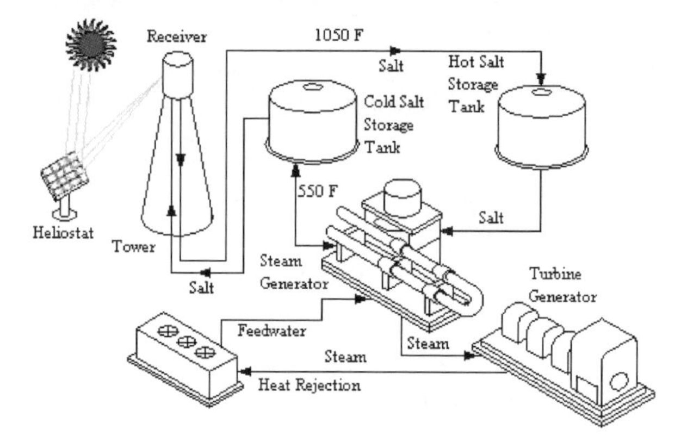

Figure 11.26 Schematic of the solar power tower (Kolb, 2000).

Desirable features of power towers for utilities

Because of their practical energy storage, solar power towers (Figure 11.26) have two features that are particularly desirable for utilities: flexible capacity factors and a high degree of dispatchability. Power towers can be designed with annual capacity factors up to 60%, and as high as 80% in summer when the days are longer. This means a power tower can operate at capacity for up to 60% of the year without using fossil fuel as a back-up, thus being able to deliver power during most peak demands.

Without energy storage, the annual capacity factor of any solar technology is generally limited to about 25%. A solar power tower's high capacity factors are achieved by building the solar portion of the plant with extra heliostats so that during daylight, sufficient energy is collected to power the turbine, while extra energy can be put into the TES system. At night or during extended cloudy periods, the turbine is powered with stored thermal energy. The dispatchability of a solar power tower (its ability to deliver electricity on demand) is illustrated in Figure 11.33, where three different parameters are plotted against the time of day: the intensity of sunlight (insolation), the amount of energy stored in the hot-salt tank, and the output power from the turbine generator. In this example, sunrise on a winter's day is around 7:00 a.m., and the intensity of sunlight rises quickly to reach its maximum at noon and drops off at sunset around 5:00 p.m.

The solar plant begins collecting energy shortly after sunrise and stores it in the hot-salt tank (the level of energy in storage increases during daylight hours). The turbine is brought on-line not at sunrise, but when the power is needed, in this example at 11:00 a.m. The output power of the plant is constant throughout the day, even though there are fluctuations in the intensity of sunlight. After sunset, the turbine continues to operate on energy from the storage tank; note the level of energy in storage declines after sunset. The turbine operates continuously until 9:00 p.m. using the thermal energy in storage. In the summer when the days are longer, the turbine would be able to operate for a larger fraction of each day.

In designing a power tower, the size of the turbine, the fraction of the day it is in operation, and the period when it is operated are flexible. The plant's TES system provides dispatchability, and by adjusting the size of the solar field and the size of the storage tanks, the capacity factor can be tailored to meet the specific needs of a utility.

Advantages of using molten salt as a heat transport and storage medium

A variety of fluids was tested to transport the sun's heat, including water, air, oil and sodium, before molten salt was selected. Molten salt is used in solar power tower systems because it is liquid at atmosphere pressure, it provides an efficient, low-cost medium in which to store thermal energy, its operating temperatures are compatible with today's high-pressure and high-temperature steam turbines, and it is non-flammable and nontoxic. In addition, since molten salt is used in the chemical and metals industries as a heat-transport fluid, experience with molten-salt systems exists for non-solar applications.

The molten salt is a mixture of 60% sodium nitrate and 40% potassium-nitrate, commonly called saltpeter. The salt melts at 221.1°C and is kept liquid at 287.7°C in an insulated cold storage tank. The salt is them pumped to the top of the tower, where concentrated sunlight heats it in a receiver to 565.5°C. The receiver is a series of thin-walled stainless steel tubes. The heated salts then flow back down to a second insulated hot storage tank. The size of this tank depends on the requirements of the utility; tanks can be designed with enough capacity to power a turbine from two to twelve hours. When electricity is needed from the plant, the hot salt is pumped to a conventional steam-generating system to produce superheated steam for a turbine/generator.

The uniqueness of this solar system is in de-coupling the collection of solar energy from producing power. Electricity can be generated in periods of inclement weather, or even at night using the stored thermal energy in the hot salt tank. The tanks are well insulated and can store energy for up to a week. As an example of their size, tanks that provide enough thermal storage to power a 100 MW turbine for four hours would be about 9.1 m tall and 24.4 m in diameter. Studies show that the two-tank storage system could have an annual efficiency of over 90%.

11.7 Sensible TES Case Studies

11.7.1 New TES in Kumamuto, Kyushu

Wide daily and seasonal fluctuations in power consumption, which have long troubled electricity providers, are a problem. Kumamoto University and Kyushu Electric Power Co.

have jointly developed a system for storing electricity obtained during cheaper late night hours as thermal energy, making it available for climate control during the day, when power grids are often overtaxed. The storage medium for the thermal energy is common dirt, keeping installation costs as well as daytime power consumption down. Electricity providers need sufficient generation capacity to handle peak demand; increased night-time use, by spreading consumption more evenly throughout the day, lowers the number of generators needed, and therefore leads to reduced emissions of gases like carbon dioxide.

System description

The new thermal-storage climate-control system is designed around the use of common soil as the heat storage medium (Figure 11.27). The first such system was installed in spring 1997 in the city of Kumamoto, Kyushu, where it is being used for the climate control of an indoor exercise ground's lounge area. Four layers of flexible 25 mm plastic water pipe, with a total length of 4800 m, are buried up to a meter beneath the earth's surface. Using electricity during late-night hours, when rates are much lower, the system adjusts the temperature of the water running through the pipe, cooling the surrounding earth to 10°C in the summer and heating it to 45°C in the winter. This thermal energy is used to cool or heat the lounge during the daytime hours of high power usage.

The system actually uses 20% to 30% more electricity than a standard climate-control system. But by switching its hours of power use from day (8:00 a.m. to 10:00 p.m.) to night (10:00 p.m. to 8:00 a.m.) under a discounted pricing scheme, the exercise ground pays only one-third to one-fourth the normal rate for the electricity it converts into thermal energy.

The TES system occupies about 200 m^2 (the same area as the lounge it is used to heat and cool). A wall of thermal insulation surrounds the earth to a depth of 1.8 m; deeper than this, temperatures remain quite stable, negating the need for an insulating layer at the bottom of the heat storage area. The system boasts the merit of low installation cost, as its relatively simple structure requires no laying of concrete or other elaborate construction techniques. Moreover, the system occupies no more area than the space to be heated or cooled (it can be installed directly below the building, solving problems of where to put the climate-control unit).

Levelling peak electrical loads

Once generated, electricity cannot be stored economically. Power companies must therefore install enough generating capacity to cover the highest peaks of consumption, which come at midday on the hottest days of summer. Recent years have seen rises in both household and industrial power consumption, and demand has accordingly become more extreme. This has also led to the problems of excess capacity and inefficiency, however, as some generators are only called into service for those few hours of peak usage during the summer, and lie idle for much of the year. Still, the need for the capacity is forcing power providers to invest heavily in new generating plants, which can be costly. Nuclear power plants can be built almost anywhere, and have low greenhouse gas emissions, but anti-nuclear movements often make the planning of new reactors difficult. So power providers are seeking alternative means to reduce peak demand.

The soil-based thermal storage system, by transferring some demand to night-time hours, is seen as a way to flatten the daytime peaks in electricity consumption and bring about more constant usage levels. This in turn improves the efficiency of existing power plants and reduces the need for the construction of any new capacity. Customers, too, should find the system attractive due to its lower power costs. Systems using inexpensive electricity to chill water at night and use it to cool a building during the day were first put in use in 1952. Similar systems that freeze the water and then use the transfer of heat as the ice melts for daytime climate control have been on the market since 1995. This latter system proved to be a great step forward, since ice absorbs 10 times the thermal energy of liquid water for its weight, and its use made possible the miniaturization of such cooling systems.

Although these units may have grown smaller, they often require space on a building's roof for installation. They also cost between 20% and 30% more to install than conventional air conditioning systems. Moreover, they become a less attractive option for small- to mid-sized buildings where economies of scale cannot be brought to bear. These factors make a new thermal storage climate-control system, that uses dirt as its medium, advantageous. With its low construction costs and minimal external space requirements, it is seen as likely to contribute to greatly expanded off-peak power consumption by smaller buildings, and to more efficient energy consumption in Japan as a result.

Further information can be found in JIN (2000).

11.7.2 The World's first passive annual heat storage home, Montana, USA

Passive Annual Heat Storage (PAHS) is a method of collecting heat in the summer, by passively cooling a home, storing the heat in the earth passively, then returning that heat to the home in the winter. PAHS includes extensive use of natural heat flow methods, and the arrangement of building materials so as to direct heat from where its is produced to where it is needed without using machinery.

Use of PAHS has resulted in the creation of homes and other structures that are able to self-maintain an internal temperature that varies only a few degrees up and down from a comfortable average of about 20°C. Such homes have actually been able to collect and save more energy than they need for operation. Such homes have been built and operated successfully in the USA (e.g. Montana and Alaska), New Zealand and Europe.

The goal of PAHS is to provide a method of building-materials placement and construction organization such that continuously comfortable environments are produced, and which are able to extract all of their energy needs from the natural environment without using any commercial energy sources or mechanical devices and without causing disruption to global ecosystems.

Figure 11.27 The network of water pipes with TES for heating and cooling buildings (JIN, 2000).

Figure 11.28 The PAHS residence (RMRC, 2000).

To-date, a number of homes around the world have actually achieved full annual heat storage. That is, they collect the heating and cool that the homes need for the winters and summers. They thereby reduce energy consumption and sometimes provide a surplus of energy that is used to provide partial domestic water heating, and power to run heat recovery systems for fresh air.

The world's first PAHS home (Figure 11.28) was built in the winter of 1980–1981 in Montana. Through its first summer, the specially insulated earth around it extracted the summer heat from the home keeping it a cool 18.5°C on the lower floor and 23.5°C on the second floor ceiling at the dome's zenith. The following winter the home obtained its heating needs by conduction back into the home. Although the ground gradually cooled off, by April the home had not dropped below 19°C. This cold climate building met all of its cooling and heating needs throughout the entire year without the aid of either a furnace or air conditioner. The layout of building materials controls temperature naturally on an annual basis. The dome's shape was chosen because of its strength, which is capable of supporting a load of earth over 1.22 m thick. The PAHS method, however, does not require the use of a dome, but can be used with any earth-sheltered shape.

Further information on this project and its applications is available in RMRC (2000).

11.8 Concluding Remarks

A wide range of case studies drawn from reports in the literature and elsewhere are presented here. Included are CTES systems using chilled water, ice and PCMs, as well as sensible and latent TES systems for heating capacity. The case studies demonstrate how TES is normally not addressed in an isolated manner, but as a part of the overall energy infrastructure for a facility. The applications represented by these case studies also vary widely, and include institutional facilities, such as government and educational buildings, commercial facilities, industrial uses and such broader uses as part of a district heating and cooling network.

Each TES system and installation has its own advantages and disadvantages. The energy sources represented also vary widely, ranging from conventional fuels to solar energy. The case studies demonstrate in many different ways how TES is an efficient and effective way of storing thermal energy.

References

Abusaa, G. (2000). *Combustion Turbine Inlet Air Cooling*, Lecture Notes, presented at: Mechanical Engineering Seminar, KFUPM, Dhahran, 31 October.

BAC (1999a). *Channel Island Power Station Opts for Ice Thermal Storage System*, Project Report, No. PRJ44/99, Baltimore Aircoil Company.

BAC (1999b). Thermal storage is state of the art, *Gulf Construction*, February, Baltimore Aircoil Company.

BAC (1999c). *Ice Keeps Alitalia Cool*, Application Leaflet, No. MN-92-10, Baltimore Aircoil Company.

Cristopia (2001a). *MM21 in Yokohama (Japan)*, Cristopia Energy Systems, France, http://www.cristopia.com/english/project/yokohama.html.

Cristopia (2001b). *Harp Guinness Subsidiary (Ireland)*, Cristopia Energy Systems, France, http://www.cristopia.com/english/project/harpguinness.html.

Cristopia (2001c). *Korean Development Bank (Korea)*, Cristopia Energy Systems, France, http://www.cristopia.com/english/project/koreanbank.html.

Cristopia (2001d). *Museum of Sciences and Industry: 'La Villette' (France)*, Cristopia Energy Systems, France, http://www.cristopia.com/english/project/lavillette.html.

Cristopia (2001e). *Rueil Malmaison Central Kitchen (France)*, Cristopia Energy Systems, France, http://www.cristopia.com/english/project/rueil.html.

Cristopia (2001f). *Bangsar District Cooling Plant (Malaysia)*, Cristopia Energy Systems, France, http://www.cristopia.com/english/project/bangsar.html.

Dharmadhikari, S. (2000). Une installation de trigénération exemplaire (in French). *Gaz d'aujourd'hui*, N° 6, pp. 7–12.

Dharmadhikari, S., Pons, D. and Principaud, F. (1999). Trigeneration for the World Fair Expo'98, Lisbon, Presented at: *the 20th International Congress of Refrigeration*, IIR/IIF, Sydney.

Dharmadhikari, S., Pons, D. and Principaud, F. (2000). Contribution of stratified thermal storage to cost-effective trigeneration project, *ASHRAE Transactions* 106(2).

DOE (1999). *Thermal Energy Storage at a Federal Facility*, the U.S. Department of Energy (DOE) Federal Energy Management Program, DOE/GO-10098-439.

EPS (2000). *Thermal Energy Storage Systems*, Environmental Process Systems Limited, Berkshire, http://www.epsltd.co.uk.

Hall, A.D., Stover, J.C. and Breisch, R. (1994). *Gas turbine inlet-air chilling at a cogeneration facility*, *ASHRAE Transactions* Vol. 100, Part 1, 595–600.

JIN (2000). Going Underground, Japan information Network, http://jin.jcic.or.jp/trends98/honbun/ntj980217.html.

Kent, H.S. (1996). *Thermal Energy Storage Increases Research Center's Cooling Capacity*, Technical Report, Levittown, Pennsylvania.

Kolb, G. (2000). *Desirable Features of Power Tower for Utilities*, Sandia National Laboratories, Albuquerque, http://www.sandia.gov/Renewable_Energy/solarthermal/feature.html.

Mahoney, A.R. (2000). *Advantages of Using Molten Salt*, Sandia National Laboratories, Albuquerque, http://www.sandia.gov/Renewable_Energy/solarthermal/salt.html.

Nelson, K.P., Pippin, J. and Dunlap, J. (1999). *University Ice Slurry Systems*, Applications Catalog, Paul Mueller Company, Springfield.

Palmer, M. (2000). *Case Study in Saudi Arabia, Project Report*, WS Atkins Consultants Limited, Surrey, UK.

RMRC (2000). The World's First Passive Annual Heat Storage Home, Rocky Mountain Research Center (RMRC), Montana, http://www.rmrc.org/dome1.htm.

UNC-CH (2000). A Recent Operations Research Public Service Project, University of North Carolina at Chapel Hill, http: //www.unc.edu/depts/or/dp/or350/bradford.htm.

APPENDIX A

Conversion Factors

Table A.1 Conversion factors for commonly used quantities.

Quantity	SI to English	English to SI
Area	$1\ m^2 = 10.764\ ft^2$ $= 1550.0\ in^2$	$1\ ft^2 = 0.00929\ m^2$ $1\ in^2 = 6.452 \times 10^{-4}\ m^2$
Density	$1\ kg/m^3 = 0.06243\ lb_m/ft^3$	$1\ lb_m/ft^3 = 16.018\ kg/m^3$ $1\ slug/ft^3 = 515.379\ kg/m^3$
Energy	$1\ J = 9.4787 \times 10^{-4}\ Btu$	1 Btu = 1055.056 J 1 cal = 4.1868 J $1\ lb_f \cdot ft = 1.3558\ J$ $1\ hp \cdot h = 2.685 \times 10^6\ J$
Energy per unit mass	$1\ J/kg = 4.2995 \times 10^{-4}\ Btu/lb_m$	$1\ Btu/lb_m = 2326\ J/kg$
Force	$1\ N = 0.22481\ lb_f$	$1\ lb_f = 4.448\ N$ 1 pdl = 0.1382 N
Gravitation	$g = 9.80665\ m/s^2$	$g = 32.17405\ ft/s^2$
Heat flux	$1\ W/m^2 = 0.3171\ Btu/h \cdot ft^2$	$1\ Btu/h \cdot ft^2 = 3.1525\ W/m^2$ $1\ kcal/h \cdot m^2 = 1.163\ W/m^2$ $1\ cal/s \cdot cm^2 = 41\ 870.0\ W/m^2$
Heat generation (volum.)	$1\ W/m^3 = 0.09665\ Btu/h \cdot ft^3$	$1\ Btu/h \cdot ft^3 = 10.343\ W/m^3$
Heat transfer coefficient	$1\ W/m^2 \cdot K = 0.1761\ Btu/h \cdot ft^2 \cdot °F$	$1\ Btu/h \cdot ft^2 \cdot °F = 5.678\ W/m^2 \cdot K$ $1\ kcal/h \cdot m^2 \cdot °C = 1.163\ W/m^2 \cdot K$ $1\ cal/s \cdot m^2 \cdot °C = 41870.0 W/m^2 \cdot K$
Heat transfer rate	1 W = 3.4123 Btu/h	1 Btu/h = 0.2931 W
Length	1 m = 3.2808 ft = 39.370 in 1 km = 0.621 371 mi	1 ft = 0.3048 m 1 in = 2.54 cm = 0.0254 m 1 mi = 1.609344 km 1 yd = 0.9144 m
Mass	$1\ kg = 2.2046\ lb_m$ 1 ton (metric) = 1000 kg $1\ grain = 6.47989 \times 10^{-5}\ kg$	$1\ lb_m = 0.4536\ kg$ 1 slug = 14.594 kg
Mass flow rate	$1\ kg/s = 7936.6\ lb_m/h$ $= 2.2046\ lb_m/s$	$1\ lb_m/h = 0.000126\ kg/s$ $1\ lb_m/s = 0.4536\ kg/s$
Power	1 W = 1 J/s = 3.4123 Btu/h $= 0.737\ 562\ lb_f \cdot ft/s$ 1 hp (metric) = 0.735 499 kW 1 ton of refrig. = 3.516 85 kW	1 Btu/h = 0.2931 W 1 Btu/s = 1055.1 W $1\ lb_f \cdot ft/s = 1.3558\ W$ $1\ hp^{UK} = 745.7\ W$

(Continued)

Table A.1 (Continued) Conversion factors for commonly used quantities.

Quantity	SI to English	English to SI
Pressure and stress (Pa = N/m^2)	1 Pa = 0.020886 lb_f/ft^2 = 1.4504 × 10^{-4} lb_f/in^2 = 4.015 × 10^{-3} in water = 2.953 × 10^{-4} in Hg	1 lb_f/ft^2 =47.88 Pa 1 lb_f/in^2 = 1 psi = 6894.8 Pa 1 stand. atm. = 1.0133 × 10^5 Pa 1 bar = 1 × 10^5 Pa
Specific heat	1 J/kg·K=2.3886×10^{-4} Btu/lb_m·°F	1 Btu/lb_m·°F = 4187 J/kg·K
Surface tension	1 N/m = 0.06852 lb_f/ft	1 lb_f/ft = 14.594 N/m 1 dyn/cm = 1 × 10^{-3} N/m
Temperature	T(K) = T(°C) + 273.15 = T(°R)/1.8 = [T(°F) + 459.67]/1.8 T(°C) = [T(°F) -32]/1.8	T(°R) = 1.8T(K) = T(°F) + 459.67 = 1.8T(°C) + 32 = 1.8[T(K) - 273.15] + 32
Temperature difference	1 K = 1°C = 1.8°R = 1.8°F	1°R = 1°F = 1 K/1.8 = 1°C/1.8
Thermal conductivity	1 W/m·K = 0.57782 Btu/h·ft·°F	1 Btu/h·ft·°F = 1.731 W/m·K 1 kcal/h·m·°C = 1.163 W/m·K 1 cal/s·cm·°C = 418.7 W/m·K
Thermal diffusivity	1 m^2/s = 10.7639 ft^2/s	1 ft^2/s = 0.0929 m^2/s 1 ft^2/h = 2.581 × 10^{-5} m^2/s
Thermal resistance	1 K/W = 0.52750 °F·h/Btu	1 °F·h/Btu = 1.8958 K/W
Velocity	1 m/s = 3.2808 ft/s 1 km/s = 0.62137 mi/h	1 ft/s = 0.3048 m/s 1 ft/min = 5.08 × 10^{-3} m/s
Viscosity (dynamic) (kg/m·s = N·s/m^2)	1 kg/m·s = 0.672 lb_m/ft·s = 2419.1 lb_m/fh·h	1 lb_m/ft·s = 1.4881 kg/m·s 1 lb_m/ft·h = 4.133 × 10^{-4} kg/m·s 1 centipoise (cP) = 10^{-2} poise = 1 × 10^{-3} kg/m·s
Viscosity (kinematic)	1 m^2/s = 10.7639 ft^2/s = 1 × 10^4 stokes	1 ft^2/s = 0.0929 m^2/s 1 ft^2/h = 2.581 × 10^{-5} m^2/s 1 stoke = 1 cm^2/s
Volume	1 m^3 = 35.3134 ft^3 1 L = 1 dm^3 = 0.001 m^3	1 ft^3 = 0.02832 m^3 1 in^3 = 1.6387 × 10^{-5} m^3 1 gal^{US} = 0.003785 m^3 1 gal^{UK} = 0.004546 m^3
Volumetric flow rate	1 m^3/s = 35.3134 ft^3/s = 1.2713 × 10^5 ft^3/h	1 ft^3/s = 2.8317 × 10^{-2} m^3/s 1 ft^3/min = 4.72 × 10^{-4} m^3/s 1 ft^3/h = 7.8658 × 10^{-6} m^3/s 1 gal^{US}/min = 6.309 × 10^{-5} m^3/s

APPENDIX B

Thermophysical Properties

Table B.1 Thermophysical properties of pure water at atmospheric pressure.

T (°C)	ρ (kg/m³)	μ×10³ (kg/m·s)	ν×10⁶ (m²/s)	k (W/m·K)	β×10⁵ (1/K)	c_p (J/kg·K)	Pr
0	999.84	1.7531	1.7533	0.5687	-6.8140	4209.3	12.976
5	999.96	1.5012	1.5013	0.5780	1.5980	4201.0	10.911
10	999.70	1.2995	1.2999	0.5869	8.7900	4194.1	9.2860
15	999.10	1.1360	1.1370	0.5953	15.073	4188.5	7.9910
20	998.20	1.0017	1.0035	0.6034	20.661	4184.1	6.9460
25	997.07	0.8904	0.8930	0.6110	20.570	4180.9	6.0930
30	995.65	0.7972	0.8007	0.6182	30.314	4178.8	5.3880
35	994.30	0.7185	0.7228	0.6251	34.571	4177.7	4.8020
40	992.21	0.6517	0.6565	0.6351	38.530	4177.6	4.3090
45	990.22	0.5939	0.5997	0.6376	42.260	4178.3	3.8920
50	988.04	0.5442	0.5507	0.6432	45.780	4179.7	3.5350
60	983.19	0.4631	0.4710	0.6535	52.330	4184.8	2.9650
70	977.76	0.4004	0.4095	0.6623	58.400	4192.0	2.5340
80	971.79	0.3509	0.3611	0.6698	64.130	4200.1	2.2010
90	965.31	0.3113	0.3225	0.6759	69.620	4210.7	1.9390
100	958.35	0.2789	0.2911	0.6807	75.000	4221.0	1.7290

Source: D.J. Kukulka, *Thermodynamic and Transport Properties of Pure and Saline Water*, MSc Thesis, State University of New York at Buffalo (1981).

Table B.2 Thermophysical properties of air at atmospheric pressure.

T (K)	ρ (kg/m³)	c_p (J/kg·K)	μ×10⁷ (kg/m·s)	ν×10⁶ (m²/s)	k×10³ (W/m·K)	a×10⁶ (m²/s)	Pr
200	1.7458	1.007	132.5	7.59	18.10	10.30	0.737
250	1.3947	1.006	159.6	11.44	22.30	15.90	0.720
300	1.1614	1.007	184.6	15.89	26.30	22.50	0.707
350	0.9950	1.009	208.2	20.92	30.00	29.90	0.700
400	0.8711	1.014	230.1	26.41	33.80	38.30	0.690
450	0.7740	1.021	250.7	32.39	37.30	47.20	0.686
500	0.6964	1.030	270.1	38.79	40.70	56.70	0.684
550	0.6329	1.040	288.4	45.57	43.90	66.70	0.683
600	0.5804	1.051	305.8	52.69	46.90	76.90	0.685
650	0.5356	1.063	322.5	60.21	49.70	87.30	0.690
700	0.4975	1.075	338.8	68.10	52.40	98.00	0.695
750	0.4643	1.087	354.6	76.37	54.90	109.00	0.702
800	0.4354	1.099	369.8	84.93	57.30	120.00	0.709
850	0.4097	1.110	384.3	93.80	59.60	131.00	0.716
900	0.3868	1.121	398.1	102.90	62.00	143.00	0.720
950	0.3666	1.131	411.3	112.20	64.30	155.00	0.723

Source: I. Dincer, *Heat Transfer in Food Cooling Applications*, Taylor & Francis, Washington, DC. (1997); and C. Borgnakke and R.E. Sonntag, *Thermodynamic and Transport Properties*, Wiley, New York (1997).

Table B.3 Thermophysical properties of ammonia (NH_3) gas at atmospheric pressure.

T (K)	ρ (kg/m³)	c_p (J/kg·K)	μ×10⁷ (kg/m·s)	ν×10⁶ (m²/s)	k×10³ (W/m·K)	a×10⁶ (m²/s)	Pr
300	0.6994	2.158	101.5	14.70	24.70	16.66	0.887
320	0.6468	2.170	109.0	16.90	27.20	19.40	0.870
340	0.6059	2.192	116.5	19.20	29.30	22.10	0.872
360	0.5716	2.221	124.0	21.70	31.60	24.90	0.870
380	0.5410	2.254	131.0	24.20	34.00	27.90	0.869
400	0.5136	2.287	138.0	26.90	37.00	31.50	0.853
420	0.4888	2.322	145.0	29.70	40.40	35.60	0.833
440	0.4664	2.357	152.5	32.70	43.50	39.60	0.826
460	0.4460	2.393	159.0	35.70	46.30	43.40	0.822
480	0.4273	2.430	166.5	39.00	49.20	47.40	0.822
500	0.4101	2.467	173.0	42.20	52.50	51.90	0.813
520	0.3942	2.504	180.0	45.70	54.50	55.20	0.827
540	0.3795	2.540	186.5	49.10	57.50	59.70	0.824
560	0.3708	2.577	193.5	52.00	60.60	63.40	0.827
580	0.3533	2.613	199.5	56.50	63.68	69.10	0.817

Source: I. Dincer, *Heat Transfer in Food Cooling Applications*, Taylor & Francis, Washington, DC. (1997); and C. Borgnakke and R.E. Sonntag, *Thermodynamic and Transport Properties*, Wiley, New York (1997).

Table B.4 Thermophysical properties of carbon dioxide (CO_2) gas at atmospheric pressure.

T (K)	ρ (kg/m³)	c_p (J/kg·K)	μ×10⁷ (kg/m·s)	ν×10⁶ (m²/s)	k×10³ (W/m·K)	a×10⁶ (m²/s)	Pr
280	1.9022	0.830	140.0	7.36	15.20	9.63	0.765
300	1.7730	0.851	149.0	8.40	16.55	11.00	0.766
320	1.6609	0.872	156.0	9.39	18.05	12.50	0.754
340	1.5618	0.891	165.0	10.60	19.70	14.20	0.746
360	1.4743	0.908	173.0	11.70	21.20	15.80	0.741
380	1.3961	0.926	181.0	13.00	22.75	17.60	0.737
400	1.3257	0.942	190.0	14.30	24.30	19.50	0.737
450	1.1782	0.981	210.0	17.80	28.20	24.50	0.728
500	1.0594	1.020	231.0	21.80	32.50	30.10	0.725
550	0.9625	1.050	251.0	26.10	36.60	36.20	0.721
600	0.8826	1.080	270.0	30.60	40.70	42.70	0.717
650	0.8143	1.100	288.0	35.40	44.50	49.70	0.712
700	0.7564	1.130	305.0	40.30	48.10	56.30	0.717
750	0.7057	1.150	321.0	45.50	51.70	63.70	0.714
800	0.6614	1.170	337.0	51.00	55.10	71.20	0.716

Source: I. Dincer, *Heat Transfer in Food Cooling Applications*, Taylor & Francis, Washington, DC. (1997); and C. Borgnakke and R.E. Sonntag, *Thermodynamic and Transport Properties*, Wiley, New York (1997).

Table B.5 Thermophysical properties of hydrogen (H_2) gas at atmospheric pressure.

T (K)	ρ (kg/m^3)	c_p (J/kg·K)	μ×10^7 (kg/m·s)	ν×10^6 (m^2/s)	k×10^3 (W/m·K)	a×10^6 (m^2/s)	Pr
100	0.2425	11.23	42.1	17.40	67.00	24.60	0.707
150	0.1615	12.60	56.0	34.70	101.00	49.60	0.699
200	0.1211	13.54	68.1	56.20	131.00	79.90	0.704
250	0.0969	14.06	78.9	81.40	157.00	115.00	0.707
300	0.0808	14.31	89.6	111.00	183.00	158.00	0.701
350	0.0692	14.43	98.8	143.00	204.00	204.00	0.700
400	0.0606	14.48	108.2	179.00	226.00	258.00	0.695
450	0.0538	14.50	117.2	218.00	247.00	316.00	0.689
500	0.0485	14.52	126.4	261.00	266.00	378.00	0.691
550	0.0440	14.53	134.3	305.00	285.00	445.00	0.685
600	0.0404	14.55	142.4	352.00	305.00	519.00	0.678
700	0.0346	14.61	157.8	456.00	342.00	676.00	0.675
800	0.0303	14.70	172.4	569.00	378.00	849.00	0.670
900	0.0269	14.83	186.5	692.00	412.00	1030.00	0.671

Source: I. Dincer, *Heat Transfer in Food Cooling Applications*, Taylor & Francis, Washington, DC. (1997); and C. Borgnakke and R.E. Sonntag, *Thermodynamic and Transport Properties*, Wiley, New York (1997).

Table B.6 Thermophysical properties of oxygen (O_2) gas at atmospheric pressure.

T (K)	ρ (kg/m^3)	c_p (J/kg·K)	μ×10^7 (kg/m·s)	ν×10^6 (m^2/s)	k×10^3 (W/m·K)	a×10^6 (m^2/s)	Pr
100	3.9450	0.962	76.4	1.94	9.25	2.44	0.796
150	2.5850	0.921	114.8	4.44	13.80	5.80	0.766
200	1.9300	0.915	147.5	7.64	18.30	10.40	0.737
250	1.5420	0.915	178.6	11.58	22.60	16.00	0.723
300	1.2840	0.920	207.2	16.14	26.80	22.70	0.711
350	1.1000	0.929	233.5	21.23	29.60	29.00	0.733
400	0.9620	0.942	258.2	26.84	33.00	36.40	0.737
450	0.8554	0.956	281.4	32.90	36.30	44.40	0.741
500	0.7698	0.972	303.3	39.40	41.20	55.10	0.716
550	0.6998	0.988	324.0	46.30	44.10	63.80	0.726
600	0.6414	1.003	343.7	53.59	47.30	73.50	0.729
700	0.5498	1.031	380.8	69.26	52.80	93.10	0.744
800	0.4810	1.054	415.2	86.32	58.90	116.00	0.743
900	0.4275	1.074	447.2	104.60	64.90	141.00	0.740

Source: I. Dincer, *Heat Transfer in Food Cooling Applications*, Taylor & Francis, Washington, DC. (1997); and C. Borgnakke and R.E. Sonntag, *Thermodynamic and Transport Properties*, Wiley, New York (1997).

Table B.7 Thermophysical properties of water vapor (steam) gas at atmospheric pressure.

T (K)	ρ (kg/m³)	c_p (J/kg·K)	μ×10⁷ (kg/m·s)	ν×10⁶ (m²/s)	k×10³ (W/m·K)	a×10⁶ (m²/s)	Pr
380	0.5863	2.060	127.1	21.68	24.60	20.40	1.060
400	0.5542	2.014	134.4	24.25	26.10	23.40	1.040
450	0.4902	1.980	152.5	31.11	29.90	30.80	1.010
500	0.4405	1.985	170.4	38.68	33.90	38.80	0.998
550	0.4005	1.997	188.4	47.04	37.90	47.40	0.993
600	0.3652	2.026	206.7	56.60	42.20	57.00	0.993
650	0.3380	2.056	224.7	66.48	46.40	66.80	0.996
700	0.3140	2.085	242.6	77.26	50.50	77.10	1.000
750	0.2931	2.119	260.4	88.84	54.90	88.40	1.000
800	0.2739	2.152	278.6	101.70	59.20	100.00	1.010
850	0.2579	2.186	296.9	115.10	63.70	113.00	1.020

Source: I. Dincer, *Heat Transfer in Food Cooling Applications*, Taylor & Francis, Washington, DC. (1997); and C. Borgnakke and R.E. Sonntag, *Thermodynamic and Transport Properties*, Wiley, New York (1997).

Table B.8 Thermophysical properties of some solid materials.

Composition	T (K)	ρ (kg/m³)	k (W/m·K)	c_p (J/kg·K)
Aluminum	273–673	2720	204–250	895
Asphalt	300	2115	0.0662	920
Bakelite	300	1300	1.4	1465
Brass (70% Cu + 30% Zn)	373–573	8520	104–147	380
Carborundum	872	—	18.5	—
Chrome brick	473	3010	2.3	835
	823	—	2.5	—
Diatomaceous silica, fired	478	—	0.25	—
Fire clay brick	478	2645	1.0	960
	922	—	1.5	—
Bronze (75% Cu + 25% Sn)	273–373	8670	26.0	340
Clay	300	1460	1.3	880
Coal (anthracite)	300	1350	0.26	1260
Concrete (stone mix)	300	2300	1.4	880
Constantan (60% Cu + 40% Ni)	273–373	8920	22–26	420
Copper	273–873	8950	385–350	380
Cotton	300	80	0.06	1300
Glass				
Plate (soda lime)	300	2500	1.4	750
Pyrex	300	2225	1.4	835
Ice	253	—	2.03	1945
	273	920	1.88	2040
Iron (C ≈ 4% cast)	273–1273	7260	52–35	420

(Continued)

Table B.8 (Continued) Thermophysical properties of some solid materials.

Composition	T (K)	ρ (kg/m^3)	k (W/m·K)	c_p (J/kg·K)
Iron (C ≈ 0.5% wrought)	273–1273	7850	59–35	460
Lead	273–573	—	—	—
Leather (sole)	300	998	0.159	—
Magnesium	273–573	1750	171–157	1010
Mercury	273–573	13400	8–10	125
Molybdenum	273–1273	10220	125–99	251
Nickel	273–673	8900	93–59	450
Paper	300	930	0.18	1340
Paraffin	300	900	0.24	2890
Platinum	273–1273	21400	70–75	240
Rock				
Granite, Barre	300	2630	2.79	775
Limestone, Salem	300	2320	2.15	810
Marble, Halston	300	2680	2.80	830
Rubber, vulcanized				
Soft	300	1100	0.13	2010
Sandstone, Berea	300	2150	2.90	745
Hard	300	1190	0.16	—
Sand	300	1515	0.27	800
Silver	273–673	10520	410–360	230
Soil	300	2050	0.52	1840
Steel (C ≈ 1%)	273–1273	7800	43–28	470
Steel (Cr ≈ 1%)	273–1273	7860	62–33	460
Steel (18% Cr + 8% Ni)	273–1273	7810	16–26	460
Snow	273	110	0.049	—
Teflon	300	2200	0.35	—
Tin	273–473	7300	65–57	230
Tissue, human				
Skin	300	—	0.37	—
Fat layer (adipose)	300	—	0.2	—
Muscle	300	—	0.41	—
Tungsten	273–1273	19350	166–76	130
Wood, cross grain				
Fir	300	415	0.11	2720
Oak	300	545	0.17	2385
Yellow pine	300	640	0.15	2805
White pine	300	435	0.11	—
Wood, radial				
Fir	300	420	0.14	2720
Oak	300	545	0.19	2385
Zinc	273–673	7140	112–93	380

Source: I. Dincer, *Heat Transfer in Food Cooling Applications*, Taylor & Francis, Washington, DC. (1997); and F.P. Incropera and D.P. DeWitt, *Fundamentals of Heat and Mass Transfer*, Wiley, New York (1998).

APPENDIX C

Glossary

All terms in this section follow standard industry definitions which were adapted from ARI Standard 900 (1998). These definitions are of importance in practice and are intended to provide clarity.

Ambient Air. The air in the space surrounding a thermal energy storage device, or outside air.

Ambient Heat Load. The thermal load (typically expressed in tons [kW]) imposed on a storage device due to heat gain.

Build Period. The operating period of a thermal storage generator during which ice is produced.

Charge Fluid. The heat transfer fluid used to remove heat from a thermal storage device or generator during the charge or build period, or to add heat to a heat storage.

Charge Period. The period of time during which energy (heat) is removed from a cold storage or added to a heat storage.

Charge Rate. The rate (typically expressed in tons [kW]) at which energy (heat) is removed from or added to a storage device during the charge period.

Discharge Fluid. The heat transfer fluid used to add heat to the thermal storage device.

Discharge Period/Cycle. The period of time when energy (heat) is added to the storage device.

Discharge Rate. The rate (typically expressed in tons [kW]) at which energy (heat) is added to the storage device during the discharge period.

Fouling Factor. A thermal resistance included in heat transfer calculations to account for the fouling expected over time on a heat transfer surface.

Hermetic (Sealed Unit) Compressor. A motor-compressor assembly contained within a gas-tight casing through which no shaft extends, which uses the refrigerant as the motor coolant.

Interval. A time span, such as the period between individual test readings.

Latent Heat of Fusion. The change in energy accompanying the conversion of a unit mass of a solid to a liquid at its melting point, at constant pressure and temperature.

Load Intensity. The ratio of the instantaneous load imposed upon the storage device to the net usable storage capacity of the unit, typically expressed in tons/ton·hour [kW/kW·h].

Melt Period. That period of a test during which the ice produced by a thermal storage generator is melted to determine its quality.

Net Ice-Making Capacity. The net ice-producing capability of a thermal storage generator operating in a charge mode, typically expressed in tons [kW].

Net Usable Storage Capacity. The actual amount of stored cooling that can be supplied from the storage device at or below the specified cooling supply temperature for a given charge and discharge cycle, typically expressed in ton·h [kW·h].

Nominal Storage Capacity. A theoretical capacity of the storage device as defined by the storage device manufacturer (which in many cases is greater than the net usable storage capacity).

Open-type Compressor. A refrigerant compressor with a shaft or other moving part extending through its casing to be driven by an outside source of power, thus requiring a shaft seal or equivalent rubbing contact between fixed and moving parts.

Period. A time duration during which a storage process occurs, often a period such as the total duty cycle of a thermal storage system is divided for the purpose of analysis and evaluation into one-hour time segments.

Phase Change Material (PCM). A substance that undergoes changes of state while absorbing or rejecting thermal energy, normally at a constant temperature.

Published Ratings. Published ratings for thermal storage equipment are the data and methodologies which are used to develop supplier specified data for a specified duty cycle. They may take the form of tables, graphs or computer programs, as elected by the manufacturer, and are intended to apply to all units of like nominal size and type (identification) produced by the same manufacturer.

Standard Rating. A rating based on tests performed at standard rating conditions.

Mapped Rating(s). Mapped ratings are ratings falling within certain specified limits that are provided for products which do not have a standard rating condition. They are based upon tests performed across a range of operating conditions as defined by the product manufacturer.

Application Rating. A rating, based on tests, at application rating conditions (other than standard or mapped rating conditions).

Saturated Evaporator Temperature. The dew point temperature of the refrigerant at the pressure at the outlet connection of the evaporator.

Secondary Coolant. Any liquid cooled by a refrigerant and used for heat transmission without a change in its state. Sometimes referred to as simply 'coolant'.

Thermal Storage Device. Equipment which stores cooling capacity using sensible and/or latent heat. May consist solely of a storage means or be packaged with one or more components of a mechanical refrigeration package.

Thermal Storage Equipment. Any one of, or a combination of, thermal storage devices and/or generators, that may include various other components of a mechanical refrigeration package.

Thermal Storage Generator. An assemblage of components packaged by the manufacturer (but not necessarily shipped as one piece) to provide refrigeration to a thermal storage system which includes an evaporator, and may include compressor(s), controls, heat rejection devices, etc. whose overall performance as a thermal storage generator is rated by the manufacturer.

Ton-Hour. A quantity of thermal energy typically used to describe the capacity of a thermal storage device, in tons, absorbed or rejected in one hour [3.517 kW].

Source: ARI Standard 900. *Standard for Thermal Storage Equipment Used for Cooling, Air-Conditioning & Refrigeration Institute*, Arlington, Virginia (1998).

Subject Index